Methods in Enzymology

Volume 371
RNA Polymerases and Associated Factors
Part D

METHODS IN ENZYMOLOGY

EDITORS-IN-CHIEF

John N. Abelson   Melvin I. Simon

DIVISION OF BIOLOGY
CALIFORNIA INSTITUTE OF TECHNOLOGY
PASADENA, CALIFORNIA

FOUNDING EDITORS

Sidney P. Colowick and Nathan O. Kaplan

Methods in Enzymology

Volume 371

RNA Polymerases and Associated Factors

Part D

EDITED BY

Sankar L.Adhya

Susan Garges

NATIONAL INSTITUES OF HEALTH
NATIONAL CANCER INSTITUTE
LABORATORY OF MOLECULAR BIOLOGY
BETHESDA, MARYLAND

ELSEVIER
ACADEMIC
PRESS

AMSTERDAM • BOSTON • HEIDELBERG • LONDON
NEW YORK • OXFORD • PARIS • SAN DIEGO
SAN FRANCISCO • SINGAPORE • SYDNEY • TOKYO
Academic Press is an imprint of Elsevier

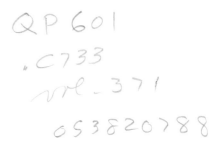
This book is printed on acid-free paper. ∞

Elsevier Academic Press
525 B Street, Suite 1900, San Diego, California 92101-4495, USA
84 Theobald's Road, London WC1X 8RR, UK
http://www.academicpress.com

International Standard Book Number: 0-12-182274-5

PRINTED IN THE UNITED STATES OF AMERICA
03 04 05 06 07 08   9 8 7 6 5 4 3 2 1

Table of Contents

Section I. Transcription Elongation

Section II.   Transcription Termination

Section III.   Chromatin

Contributors to Volume 371

Article numbers are in parentheses and following the names of contributors.
Affiliations listed are current.

TODD E. ADAMSON (19), *Department of Biochemistry, University of Iowa, Iowa City, Iowa 52242*

MICHAEL ANIKIN (9), *Morse Institute of Molecular Genetics, Department of Microbiology and Immunology, SUNY Downstate Medical Center, 450 Clarkson Avenue, Brooklyn, New York 11203-2098*

EKATERINA AVETISSOVA (13), *Skirball Institute, New York University Medical Center, New York, New York 10016*

PAUL BABITZKE (30), *Department of Biochemistry and Molecular Biology, The Pennsylvania State University, University Park, Pennsylvania 16802*

GIL BAR-NAHUM (11), *Department of Biochemistry, New York University Medical Center, New York, New York 10016*

IRINA BASS (14), *Institute of Molecular Genetics, Russian Academy of Sciences, Moscow, 123182, Russia*

JODI BECKER (17), *NCI Center for Cancer Research, National Cancer Institute— Frederick Cancer Research and Development Center, Frederick, Maryland 21702*

OXANA BERESHCHENKO (14), *Public Health Research Institute, 455 First Avenue, New York, New York 10016*

PHILIP C. BEVILACQUA (30), *Department of Chemistry, The Pennsylvania State University, University Park, Pennsylvania 16802*

KISHOR K. BHAKAT (22), *Sealy Center for Molecular Science and the Department of Human Biological Chemistry and Genetics, University of Texas Medical Branch, Galveston, Texas 77555*

YEHUDIT BIRGER (39), *Protein Section, Laboratory of Metabolism, Center for Cancer Research, National Cancer Institute, National Institutes of Health, Bethesda, Maryland 20892-4255*

SERGEI BORUKOV (16), *Department of Microbiology and Immunology, SUNY Health Science Center at Brooklyn, 450 Clarkson Avenue Brooklyn, New York 11203-2098*

ZACHARY F BURTON (18), *Department of Biochemistry, Michigan State University, East Lansing, Michigan 48824*

MICHAEL BUSTIN (39), *Protein Section, Laboratory of Metabolism, Center National Cancer Institute, National Institutes of Health, Bethesda, Maryland 20892-4255*

MARIAN CARLSON (45), *Department of Genetics & Development and Institute of Cancer Research, Columbia University, New York, New York 10032*

MICHAEL J. CARROZZA (40), *Howard Hughes Medical Institute, Department of Biochemistry and Molecular Biology, The Pennsylvania State University, University Park, Pennsylvania 16802-4500*

MICHAEL CASHEL (44), *Laboratory of Molecular Genetics, NICHD, NIH Building 6, Room 3B314 Bethesda, Maryland 20892-2785*

HYON CHOY (38), *Genome Center for Enteropathogenic Bacteria, Department of Microbiology, Chonnam National University Medical College, Dong-Gu, Hak-1-Kwang-Ju 501-746, South Korea*

ADRIENNE CLEMENTS (41), *The Wistar Institute, Philadelphia, Pennsylvania 19104*

RONALD CONAWAY (20), *Stowers Institute for Medical Research, Kansas City, Missouri 64110; Department of Biochemistry and Molecular Biology, Kansas University Medical Center, Kansas City, Kansas 66160*

JOAN WELIKY CONAWAY (20), *Stowers Institute for Medical Research, Kansas City, Missouri 64110; Department of Biochemistry and Molecular Biology, Kansas University Medical Center, Kansas City, Kansas 66160*

CIARÁN CONDON (35), *CNRS UPR9073, Institute de Biologie Physico-Chimigue, 75005 Paris, France*

ASIS DAS (33), *University of Connecticut Health Center, Farmington, Connecticut 06030-3205*

SAUL A. DATWYLER (6), *Abbott Laboratories, 100 Abbott Park Rd, Abbott Park, Illinois 60064*

RICHARD H. EBRIGHT (10), *Waksman Institute, Howard Hughes Medical Institute, Rutgers University, 190 Frelinghuysen Road, Piscataway, New Jersey 08854*

YON W. EBRIGHT (10), *Waksman Institute, Howard Hughes Medical Institute, Rutgers University, 190 Frelinghuysen Road, Piscataway, New Jersey 08854*

VITALY EPSHTEIN (14), *Institute of Molecular Genetics, Russian Academy of Sciences, Moscow, 123182, Russia*

DOROTHY A. ERIE (5), *Department of Chemistry, CB 3290, Venable and Kenan Laboratories, University of North Carolina at Chapel Hill, Chapel Hill, North Carolina 27599-3290*

PEGGY J. FARNHAM (43), *McArdle Laboratory for Cancer Research, University of Wisconsin Medical School 1400 University Avenue, Madison, Wisconsin 53706-1599*

JANE FELLOWS (36), *Cancer Research UK, London Research Institute, Clare Hall Laboratories, South Mimms, Hertfordshire, EN6 3LD United Kingdom*

J. ESTELLE FOSTER (5), *Department of Chemistry, CB 3290, Venable and Kenan Laboratories, University of North Carolina at ChapelHill, ChapelHill, North Carolina 27599-3290*

DAVID I. FRIEDMAN (32), *Department of Cell and Structure Biology, University of Colorado School of Medicine, 4200 E. Ninth Avenue, UCHSC Campus, Box B-111 Denver, Colorado 80262*

DMITRY V. FYODOROV (37), *Section of Molecular Biology, University of California, San Diego, La Jolla, California 92093-0347*

MICHELL E. GARBER (24), *Regulatory Biology Laboratory, Salk Institute for Biological Studies, 10010 North Torrey Pines Road La Jolla, California 92037-1099*

ALEX GOLDFARB (14), *Public Health Research Institute, 455 First Avenue, New York, New York 10016*

PAUL GOLLNICK (31), *Department of Biological Sciences, Hochstetter Hall 613, State University of New York Buffalo, New York 14260*

NATHAN GOMES (24), *Regulatory Biology Laboratory Salk Institute for Biological Studies, 10010 North Torrey Pines Road, La Jolla, California 92037-1099*

XUE Q. GONG (18), *Department of Biochemistry and Molecular Biology, Michigan State University, East Lansing, Michigan 48824*

FENG GONG (29), *Department of Biological Sciences, Stanford University, Stanford, California 94305-5020*

MAX E. GOTTESMAN (26), *Institute of Cancer Research, Department of Biochemistry and Molecular Biophysics, Columbia University, New York, New York 10032-2798*

MIKHAIL GRACHEV (14), *Limnological Institute, Russian Academy of Sciences, Ulanbatorskaya 3 664033, Russia*

IVAN GUSAROV (11, 28), *Department of Biochemistry, New York University Medical Center, New York, New York 10016*

MARTIN GUTHOLD (3), *Department of Physics, Wake Forest University Winston-Salem, North Carolina 27109-7507*

HIROSHI HANDA (18), *Frontier Collaborative Research Center, Tokyo Institute of Technology, Yokohama 226-8503, Japan*

AHMED H. HASSAN (40), *Howard Hughes Medical Institute, Department of Biochemistry and Molecular Biology, The Pennsylvania State University, University Park, Pennsylvania*

SHANNON F. HOLMES (4), *Department of Chemistry, CB 3290, Venable and Kenan Laboratories, University of North Carolina at Chapel Hill, Chapel Hill, North Carolina 27599-3290*

MASAYORI INOUYE (34), *Department of Biochemistry, Robert Wood Johnson Medical School, Piscataway, New Jersey 08854-5635*

AKIRA ISHIHAMA (6), *Nippon Institute for Biological Science, Shin-machi, 9-221 Ome, Tokyo 198-09924, Japan*

NANDAN JANA (33), *University of Connecticut Health Center, Farmington, Connecticut 06030-3205*

MANLI JIANG (9), *Morse Institute of Molecular Genetics, Department of Microbiology and Immunology, SUNY Downstate Medical Center 450 Clarkson Avenue, Brooklyn, New York 11203-2098*

KATHERINE A. JONES (24), *Regulatory Biology Laboratory, Salk Institute for Biological Studies, 10010 North Torrey Pines Road, La Jolla, California 92037-1099*

JAMES KADONAGA (37), *Section of Molecular Biology, University of California, San Diego, La Jolla, California 92093-0347*

CHANGWON KANG (12), *Department of Biological Sciences, Korea Advanced Institute of Science and Technology, 373-1 Guseong-dong, Yuseong-gu, Daejeon 701, Republic of Korea*

ACHILLEFS N. KAPANIDIS (10), *Department of Chemistry and Biochemistry, UCLA, Los Angeles, California 90095*

VALERI N. KARAMYCHEV (7), *Department of Nuclear Medicine, Warren G. Magnuson Clinical Center, National Institutes of Health, Bethesda, Maryland 20892*

MIKHAIL KASHLEV (7, 14, 17, 42), *NCI Center for Cancer Research, National Cancer Institute—Frederick Cancer Research and Development Center, Frederick, Maryland 21702*

HYEONG C. KIM (26), *Department of Biochemistry and Molecular Biophysics Columbia University New York, New York 10032-2798*

JAE-HONG KIM (38), *Aging and Apoptosis, Research Center, Dept. of Biochemistry and Molecular Biology, College of Medicine, Seoul National University, 28 Yongon-Dong, Chongno-GU, Seoul 110-799, South Korea*

RODNEY A. KING (15), *Laboratory of Molecular Genetics, NICHD, NIH, Building 6B, Room 3B-308, Bethesda, Maryland 20892*

MARIA L. KIREEVA (17, 42), *NCI Center for Cancer Research, National Cancer Institute—Frederick Cancer Research Development Center, Frederick, Maryland 21702*

NATALIA KOMISSAROVA (7, 17), *NCI Center for Cancer Research National Cancer Institute—Frederick Cancer Research and Development Center, Frederick, Maryland 21702*

STEPHANIE KONG (20), *Stowers Institute Medical Research, 1000 E. 50th Street, Kansas City, Missouri 64110*

EKATERINE KORTKHONJIA (10), *Waksman Institute, Howard Hughes Medical Institute, Rutgers University, 190 Frelinghuysen Road, Piscataway, New Jersey 08854*

NATALIYA KORZHEVA (13, 14), *Public Health Research Institute, 225 Warren Street, Newark, New Jersey 07103-3535*

MAXIM KOZLOV (14), *Public Health Research Institute, 455 First Avenue, New York, New York 10016*

SERGEI KUCHIN (45) *Institute of Cancer Research, Columbia University, New York, New York 10032*

OLEG LAPTENKO (16), *Morse Institute of Molecular Genetics, Department of Microbiology and Immunology, SUNY Health Science Center at Brooklyn, 450 Clarkson Avenue, Brooklyn, New York 11203-2098*

DAVID LAZINSKI (33), *University of Connecticut Health Center, Farmington, Connecticut 06030-3205*

LASSE LINDAHL (27), *Department of Biological Sciences, University of Maryland, Baltimore Country, 1000 Hilltop Circle, Baltimore, Maryland 21250*

CUIHUA LIU (2), *Wyeth Research, 87 Cambridgepark Drive, Cambridge, Massachusetts 02140*

EUGENY LUKHTANOV (14), *Epoch Pharmaceuticals, Inc., Bothell, Washington 98021*

KAIYU MA (9), *Morse Institute of Molecular Genetics, Department of Microbiology and Immunology, SUNY Downstate Medical Center, Brooklyn, New York 11203-2098*

VADIM MARKOVTSOV (14), *Public Health Research Institute, 455 First Avenue, New York, New York 10016*

RONEN MARMORSTEIN (41), *The Wistar Institute, Philadelphia, Pennsylvania 19104*

CRAIG T. MARTIN (2), *Department of Chemistry, University of Massachusetts, Amherst, Massachusetts 01003-9336*

TATYANA MAXIMOVA (14), *Limnological Institute, Russian Academy of Sciences, Ulanbatorskaya 3 664033, Russia*

WILLIAM T. MCALLISTER (9), *Morse Institute of Molecular Genetics, Department of Microbiology and Immunology, SUNY Downstate Medical Center 450 Clarkson Avenue Brooklyn, New York 11203-2098*

CLAUDE F. MEARES (6), *Department of Chemistry, University of California, Davis, One Shields Avenue, Davis, California 95616-5295*

VLADIMIR MEKLER (10), *Waksman Institute, Howard Hughes Medical Institute, Rutgers University, 190 Frelinghuysen Road, Piscataway, New Jersey 08854*

JAIME GARCIA MENA (33), *University of Connecticut Health Center, Farmington, Connecticut 06030-3205*

GREGORY MICHAUD (33), *University of Connecticut Health Center, Farmington, Connecticut 06030-3205*

SANKAR MITRA (22), *Sealy Center for Molecular Science, University of Texas, Medical Branch, 6.136 Medical Research Building, Galveston, Texas 77555*

MARK MORTIN (46), *Laboratory of Molecular Genetics NICHD, NIH, Building 6B, Room 3B-331, Bethesda, Maryland 20892-4255*

JAYANTA MUKHOPADHYAY (10), *Waksman Institute, Howard Hughes Medical Institute, Rutgers University, 190 Frelinghuysen, Road, Piscataway, New Jersey 08854*

HELEN MURPHY (44), *Laboratory of Molecular Genetics, NICHD, NIH Building 6, Room 3B-314, Bethesda, Maryland 20892-2785*

ARKADY MUSTAEV (13, 14), *Public Health Research Institute, 455 First Avenue, New York, New York 10016*

YURI A. NEDIALKOV (18), *Department of Biochemistry, Michigan State University, East Lansing, Michigan 48824*

MELODY N. NEELEY (32), *Department of Microbiology and Immunology, The University of Michigan Medical School, Ann Arbor, Michigan 48109-0620*

RONALD D. NEUMANN (7), *Department of Nuclear Medicine, Warren G. Magnuson Clinical Center, National Institutes of Health, Bethesda, Maryland 20892*

VADIM NIKIFOROV (14), *Institute of Molecular Genetics, Russian Academy of Sciences, Moscow, 123182, Russia*

EVGENY NUDLER (11, 13, 28), *Department of Biochemistry, New York University Medical Center, New York, New York 10016*

MATTHEW J. OBERLEY (43), *Department of Oncology, Room 417A McArdle Laboratory of Cancer Research, University of Wisconsin-Madison, 1400 University Avenue, Madison, Wisconsin 53706-1599*

JEFFREY OWENS (6), *IDEC Pharmaceuticals, 3031 Science Park Road, San Diego, CA 92191*

IGOR G. PANTYUTIN (7), *Department of Nuclear Medicine, Warren G. Magnuson Clinical Center, National Institutes of Health, Bethesda, Maryland 20892*

SANG CHUL PARK (38), *Aging and Apoptosis Research Center, Department of Biochemistry and Molecular Biology, College of Medicine, Seoul National University, 28 Yongon-Dong Chongno-GU, Seoul 110-799, South Korea*

THODORIS G. PETRAKIS (36), *Cancer Research UK, London Research Institute, Clare Hall Laboratories, South Mimms, Hertfordshire EN6 3LD, United Kingdom*

SANGITA PHADTARE (34), *Department of Biochemistry, Robert Wood Johnson Medical School, 675 Hoes Lane Piscataway, New Jersey 08854-5635*

YURI V. POSTNIKOV (39), *Protein Section, Laboratory of Metabolism, Center for Cancer Research, National Cancer Institute, National Institutes of Health, Bethesda, Maryland 20892-4255*

DAVID H. PRICE (19), *Department of Biochemistry, University of Iowa, Iowa City, Iowa 52242*

DANIEL REINES (21), *Department of Biochemistry, Emory University School of Medicine, 4023 Rollins Research Center, Atlanta, Georgia 30322*

CLAUDIO RIVETTI (3), *Dipartimento di Biochimica e Biologia, Molecolare Universita degli Studi, di Parma Parco Area, delle Scienze 23/A 43100, Parma, Italy*

JEFFREY W. ROBERTS (8), *Department of Molecular Biology and Genetics, Cornell University, Ithaca, New York 14853*

AZIZ SANCAR (23), *Department of Biochemistry and Biophysics, University of North Carolina, School of Medicine, Chapel Hill, North Carolina 27599-7260*

THOMAS J. SANTANGELO (8), *Department of Molecular Biology and Genetics, Cornell University, Ithaca, New York 14853*

JANELL SCHAAK (30), *Department of Biochemistry and Molecular Biology and Department of Chemistry, The Pennsylvania State University, University Park, Pennsylvania 16802*

BRIAN D. SCHMIDT (6), *Onyx Pharmaceuticals, 3031 Research Drive, Richmond, California 94806*

C. P. SELBY (23), *Department of Biochemistry and Biophysics, University of North Carolina, School of Medicine, Chapel Hill, North Carolina 27599-7260*

RANJAN SEN (15), *Laboratory of Molecular Genetics, NICHD, NIH, Bldg. 6B, Rm. 3B-308, Bethesda, Maryland 20892*

SIBANI SENGUPTA (33), *University of Connecticut Health Center, 263 Farmington Avenue, Farmington, Connecticut 06030-3205*

HYUK-KYU SEOH (35), *Department of Molecular Biology and Microbiology, Tufts University School of Medicine, Boston, Massachusetts 02111-1800*

KONSTANTIN SEVERINOV (14, 34), *Waksman Institute of Microbiology, Rutgers, The State University of New Jersey, 190 Frelinghuysen Road, Piscataway, New Jersey 08854*

HAIHONG SHEN (12), *Department of Biological Sciences, Korea Advanced Institute of Science and Technology, 373-1 Guesong-dong, Yuseong-gu, Daejeon 305-701 Korea*

ALI SHILATIFARD (20), *Edward A. Doisey, Department of Biochemistry, St. Louis University School of Medicine, St. Louis, Missouri 63104*

NUBUO SHIMAMOTO (4), *National Institute of Genetics, School of Life Science, The Graduate University for Advanced Studies, Mishima, 411-8540 Japan*

SARAH M. SHORE (19), *Department of Biochemistry, University of Iowa, Iowa City, Iowa 52242*

IGOR SIDORENKOV (17), *NCI Center for Cancer Research, National Cancer Institute—Frederick, Cancer Research Development Center, Frederick, Maryland 21702*

YOUNGHEE SOHN (12), *Department of Biological Sciences, Korea Advanced Institute of Science and Technology, 373-1 Guseong-dong Yuseong-gu, Daejeon 305-701, Republic of Korea*

RUI SOUSA (1), *Department of Biochemistry, University of Texas Health Science Center, 7703 Floyd Court Drive, San Antonio, Texas 78284-7760*

CATHERINE L. SQUIRES (35), *Department of Molecular Biology and Microbiology, Tufts University, School of Medicine, Boston, Massachusetts 02111-1800*

VASILY M. STUDITSKY (42), *Department of Biochemistry and Molecular Biology, Wayne State University School of Medicine, Detroit, Michigan 48201*

JESPER Q. SVEJSTRUP (36), *Cancer Research UK, London Research Institute, Clare Hall Laboratories, South Mimms, Hertfordshire EN6 3LD, United Kingdom*

ALEXI TATUSOV (7), *Department of Nuclear Medicine, Warren G. Magnuson Clinical Center, National Institutes of Health, Bethesda, Maryland 20892*

VLADIMIR TCHERNAJENKO (566), *Department of Biochemistry and Molecular Biology, Wayne State University School of Medicine, Detroit, Michigan 48201*

DMITRI TEMIAKOV (9), *Morse Institute of Molecular Genetics, Department of Microbiology and Immunology, SUNY Downstate 450 Clarkson Avenue, Brooklyn, New York 11203-2098*

IGOR TSAREV (14), *Limnological Institute, Russian Academy of Sciences, Ulanbatorskaya 3 664033, Russia*

ANDREA ÚJVÁRI (2), *Department of Molecular Biology, Lerner Research Institute, Cleveland Clinic Foundation, Cleveland, Ohio 44195*

SUSAN UPTAIN (25), *Department of Molecular Genetics and Cell Biology, University of Chicago, 5841 South Maryland Avenue, Chicago, Illinois 60637*

WENDY WALTER (7), *Department of Biochemistry and Molecular Biology, Wayne State University School of Medicine, Detroit, Michigan 48201*

ROBERT A. WEISBERG (15), *Laboratory of Molecular Genetics, NICHD, NIH Building 6B, Room 3B-308, Bethesda, Maryland 20892*

KATHERINE L. WEST (39), *Protein Section, Laboratory of Metabolism, Center for Cancer Research, National Cancer Institute, National Institutes of Health, Bethesda, Maryland 20892-4255*

JERRY L. WORKMAN (40), *Howard Hughes Medical Institute, Department of Biochemistry and Molecular Biology, The Pennsylvania State University, University Park, Pennsylvania 16802-4500*

ALEXANDER V. YAKHNIN (30), *Department of Biochemistry and Molecular Biology, The Pennsylvania State University, University Park, Pennsylvania 16802*

YUKI YAMAGUCHI (18), *Graduate School of Bioscience and Biotechnology, Tokyo Institute of Technology, Yokohama 226-8503, Japan*

SUK HOON YANG (22), *Sealy Center for Molecular Science and the Department of Human Biological Chemistry and Genetics, University of Texas Medical Branch, Galveston, Texas 77555*

CHARLES YANOFSKY (29), *Herzstein Professor of Biology, Department of Biological Sciences, Stanford University, Stanford, California 94305-5020*

EUGENY ZAYCHIKOV (14), *Max-Planck-Institute for Biochemie 8033, Martinsreid bei, Munchen, Germany*

JANICE M. ZENGEL (27), *Department of Biological Sciences, University of Maryland Baltimore Country, Baltimore, Maryland 21250*

ZUO ZHANG (33), *University of Connecticut Health Center, Farmington, Connecticut, 06030-3205*

VICTOR ZHURKIN (7), *Laboratory of Computational and Experimental Biology, National Cancer Institute, National Institutes of Health, Bethesda, Maryland 20892*

METHODS IN ENZYMOLOGY

Section I

Transcription Elongation

[1] On Models and Methods for Studying Polymerase Translocation

By RUI SOUSA

When trying to understand how macromolecules move, it is tempting to extrapolate from our everyday experience, and to imagine that they must have active mechanisms to push or pull themselves along. However, it is important to keep in mind that at the microscopic level, and at temperatures at which water is a liquid, molecular movement is rapid and unceasing.[1] What *is* required are mechanisms to give direction to this random, Brownian motion. Our studies of RNA polymerase translocation have therefore focused on two questions: (1) How can the random collisions and jostling that characterize the microscopic world be harnessed to drive an RNA polymerase in a directed fashion along a template? (2) What are the consequences of such mechanisms of movement for the enzymology of transcription and for regulation of polymerase activity?

Model for Polymerase Translocation

In developing a model for RNA polymerase translocation we sought to obey the dictum that mechanism should be as simple as possible (but no simpler). We imagined a mechanism in which the polymerase would be bound to the DNA in a manner that allows it to be pushed back and forth by intermolecular collisions.[2] At the same time, the requirement for processivity in synthesis meant that the polymerase might be able to slide on the template, but that once engaged in transcript elongation, it shouldn't disengage from the DNA until it encounters a terminator. Structural studies of polymerases provide a basis for believing in such a binding mode since they reveal that these enzyme have deep template binding clefts surrounded by flexible elements which might wholly or partially wrap around the DNA so as to allow polymerase sliding while preventing polymerase dissociation.[3–6] Most dramatic in this respect are the DNA

[1] A. Einstein, *Annalen Der Physik* **19,** 371–381 (1906).

[2] R. Guajardo and R. Sousa, *J. Mol. Biol.* **265,** 8–19 (1997).

[3] R. Sousa, J. Rose, and B. C. Wang, *J. Mol Biol.* **244,** 6–12 (1994).

[4] D. Jeruzalmi and T. A. Steitz, *EMBO J.* **17,** 4101–4113 (1998).

[5] G. Zhang, E. A. Campbell, L. Minakhin, C. Richter, K. Severinov, S. A. Darst, *Cell* **98,** 811–824 (1999).

[6] P. Cramer, D. A. Bushnell, R. D. Kornberg, *Science* **292,** 1863–1876 (2001).

polymerase processivity factors which form rings with DNA threading the central hole.[7] In addition to structures which keep the polymerase from releasing the DNA, we also considered that interactions with DNA and RNA might limit the range of sliding. For example, base pairing between DNA and RNA and interactions between RNA and polymerase would keep the polymerase from sliding away from the 3′-end of the RNA in the downstream direction, while extensive upstream sliding could be restricted if the template binding cleft immediately downstream of the active site (which normally binds single-stranded template) is unable to accommodate an RNA:DNA hybrid. We therefore end up with the model shown in Fig. 1A, where the polymerase can slide on the template over a range that restricts it to primarily occupy two positions: the pre-translocated position, in which the 3′-nt of the RNA occupies the site, which is bound by the substrate NTP during catalysis, and the post-translocated position, in which the RNA has cleared the NTP binding site. This

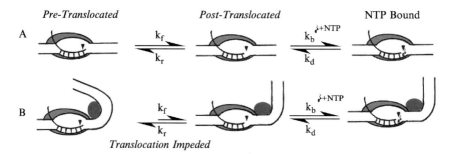

FIG. 1. (A) A model for polymerase translocation.[9] The RNA polymerase is assumed to be able to slide passively back and forth on the DNA template, driven by intermolecular collisions and fluctuations. The *pre-translocated position* corresponds to the position of the polymerase immediately after completion of bond formation: the 3′-nucleotide of the RNA occupies the NTP binding site (indicated by the triangle). Forward sliding clears the RNA 3′-nt out of the NTP binding site (*post-translocated position*), allowing NTP binding to occur. Because the NTP can only bind to the post-translocated polymerase, the apparent K_{NTP} is equal to: $(k_d/k_b)(1+k_r/k_f)$, where k_d and k_b are, respectively, the rates of NTP dissociation and NTP binding and k_f is the rate of sliding from pre- to post-translocated position, while k_r is the rate of sliding from post- to pre-translocated position. Sliding upstream of the pre-translocated position, or downstream of the post-translocated position, can occur and may lead to transcription arrest or disruption of the elongation complex but is disfavored because it would cause loss of RNA:RNAP interaction or RNA:template base-pairing, or lead to unfavorable interactions between RNAP and the RNA:DNA hybrid.[2] (B) A DNA bound protein can impede polymerase translocation leading, in turn, to an increase in apparent K_{NTP}.

[7] J. Kuriyan and M. O'Donnell, *J. Mol. Biol.* **234**, 915–925 (1993).

model has been tested both by developing and evaluating its kinetic implications, and by monitoring the position of the polymerase during transcript elongation with exonucleases.

Kinetic Consequences and Tests of the Model

The NTP can only bind to the transcription complex in the post-translocated position, i.e., after the RNA has cleared the NTP binding site (Fig. 1A). If it is assumed further that movement between pre- and post-translocated positions is faster than the rate of bond formation (a reasonable assumption for macromolecular Brownian motions over distances of 1–10 Å), then the effect on the kinetics of the reaction is to increase the apparent K_{NTP} by the factor: $(1+k_r/k_f)$, where k_f is the rate of movement from pre- to post-translocated positions, and k_r is the rate of the reverse movement (Fig. 1A).[2] The $(1+k_r/k_f)$ term should seem familiar: it is analogous to the $(1+[I](k_b/k_d))$ term that increases the apparent K_m of an enzyme in classic competitive inhibition, where $[I]$ is the inhibitor concentration, k_b is the rate of inhibitor binding to enzyme and k_d is the rate of inhibitor: enzyme dissociation. In the mechanism shown in Fig. 1A the competitive inhibitor may be considered to be the 3'-nt of the RNA (which is at a concentration of unity), while k_r and k_f are analogous to k_b and k_d, the rates of the steps that either fill or clear the substrate binding site of inhibitor.

The kinetic consequence of this mechanism is that anything which impedes the forward movement of the transcription complex (slows k_f) will increase the local apparent K_{NTP} of the NTP which is incorporated following translocation. The simplest and probably most physiologically relevant test of this is to measure K_{NTP} on collision of the transcription complex with a DNA bound protein (Fig. 1B), since such encounters must occur frequently in vivo. We did this by using a T7 promoter containing lac repressor binding sites centered at either +13 or +15.[8] Lac repressor blocked transcript extension beyond +4 or +6 when bound to the +13 or +15 centered sites, respectively. However, transcription extension from +3 to +4, or from +5 to +6, was impeded, but not blocked, by repressor bound at the +13 or +15 centered sites, respectively.[9] This impedance was measured as an increase in the % abortion (termination) of transcription occurring at RNA lengths of 3 and 5 nt, relative to what was seen with the same templates in the absence of bound repressor (Fig. 2). Consistent with the kinetic consequences of the translocation mechanism outlined in Fig. 1, this impedance could be overcome by increasing the concentration

---

[8] P. J. Lopez, J. Guillerez, R. Sousa, and M. Dreyfus, *J. Mol. Biol.* **269,** 49–51 (1997).
[9] R. Guajardo, P. J. Lopez, M. Dreyfus, and R. Sousa, *J. Mol. Biol.* **281,** 777–792 (1998).

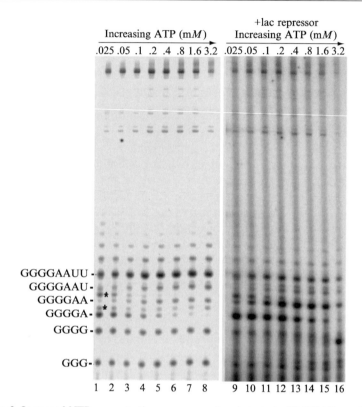

Fig. 2. Increased NTP concentrations overcome the obstacle to T7 RNAP translocation created by bound lac repressor.[9] Lanes 1-8: Transcript patterns from a linearized template containing a T7 promoter. The sequences of the transcripts generated during abortive transcription (3–8mers) are given. GTP, CTP, UTP are at .5 mM (transcript were labeled by including α^{32}P-GTP in the reaction). ATP concentrations were varied as indicated on the top of each gel lane. Lanes 9-16: As in lanes 1-8 except that lac repressor is present at 0.3 μM (30-fold excess over template). The template contains a lac repressor binding site centered at +15. Bound repressor impedes transcript extension from 5 to 6 nt, as evidenced by an increase in the amount of 5mer product at low ATP concentrations. However, increases in ATP concentrations overcome the impedance due to lac repressor, as evidenced by the decrease in 5mer at high ATP concentrations.

of the NTP utilized in the extension step (i.e., the NTP used to extend the RNA from 3 to 4 nt, or from 5 to 6 nt). Similar effects were observed when a barrier to translocation was formed by using NTP limitation to halt a T7RNAP elongation complex at +13.[9]

In these experiments barriers were arranged to impede transcript extension while the polymerase was engaged in "abortive" transcription, during which short RNAs are released from the transcription complex as a significant rate. This was done so as to allow impedance to be detected as an increase in the amount of termination at a particular transcript length (see Fig. 2). With appropriately constructed templates such methods could be used to measure impedance to translocation as a function of [NTP] for any RNA polymerase and DNA binding protein barrier (the situation in which pol II encounters a histone is of obvious interest), so long as the barrier interferes with translocation during the abortive phase of transcription. Without a quench- or stop-flow apparatus capable of millisecond time resolution, it may be difficult to measure apparent K_{NTP} effects during transcript elongation. Consider that transcript extension by *E. coli* or T7 RNAP occurs at rates of 10-100 nt/sec or $\sim$200 nt/sec, respectively, and that transcription is highly processive during elongation.[10,11] In a manual experiment, it would not be possible to detect impedance that, for example, slows a single nucleotide extension step by a factor of 10. However, the feasibility of characterizing *E. coli* RNAP kinetics with millisecond time resolution using quench-flow methods has been recently demonstrated.[12] Using such methods, together with appropriate templates that would place a DNA bound protein or other obstacle in front of an elongation complex, it should be possible to determine whether an obstacle that impedes—but does not completely block—translocation causes the type of local increases in apparent K_{NTP} predicted from the mechanism in Fig. 1.

The kinetic consequences of this mechanism are also relevant to single-molecule studies in which RNAPs transcribe against an applied force. Such studies have generally found that tugging on the RNAP does not slow it down, unless one tugs hard enough to disrupt and arrest the complex.[13,14] This is consistent with the translocation mechanism shown in Fig. 1, because in this mechanism, translocation is a rapid, diffusive step (i.e., it is not rate-limiting for extension), so slowing translocation down (by tugging on the polymerase) does not slow the overall rate of extension. However, the mechanism in Fig. 1 predicts that the effects of such tugging will be felt as increase in apparent K_{NTP}. Therefore, single-molecule studies of transcription against an applied force should look not at whether transcription is slowed at high NTP concentrations, but at whether the application of

[10] O. V. Makarova, E. M. Makarov, R. Sousa, M. Dreyfus, *Proc. Natl. Acad. Sci. USA* **92,** 12250–12254 (1995).

[11] J. Huang, J. Villemain, R. Padilla, and R. Sousa, *J. Mol. Biol.* **293,** 457–475 (1999).

[12] J. E. Foster, *et al.*, *Cell* **106,** 242–252 (2001).

[13] M. D. Wang, M. J. Schnitzer, H. Yin, R. Landick, J. Gelles, and S. M. Block, *Science* **282,** 902–906 (1998).

[14] A. D. Mehta, M. Rief, J. A. Spudich, D. A. Smith, and R. M. Simmons, *Science* **283,** 1689 (1999).

force slows transcription at low NTP concentrations up to the range of the normal K_{NTP} of the enzyme.

Using Exonucleases to Study Translocation

Kinetic effects rarely allow for unambiguous interpretation of mechanism, so it is important to use a more direct probe of polymerase position to test models for translocation mechanism. Exonucleases may be more useful than area footprinting reagents in this respect because a single nucleotide shift in the position of a diffuse footprint that extends over 20 or 30 nucleotides can be difficult to observe, while a similar shift in the position at which digestion by an exonuclease is blocked is readily detected (Fig. 3). We used exonuclease III (which digests in the 3' to 5' direction) to detect the upstream and downstream boundaries of T7RNAP elongation complexes on 5'-end labeled template or nontemplate DNA strands, respectively. Lambda exonuclease (which digests in the 5' to 3' direction)

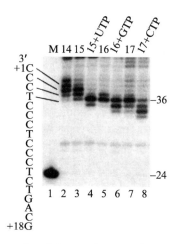

FIG. 3. Exonuclease III mapping of the template strand upstream boundary of T7 RNAP elongation complexes as a function transcript length and NTP binding[15]: Elongation complexes with 3'-dNMP terminated transcripts were formed on templates in which the 5'-end of the template strand was α^{32}P-labeled. NTPs were removed by ultrafiltration. Transcripts ranged from 14 to 17 nts in length, as indicated over each gel lane. NTPs complementary to the template base immediately downstream of the RNA 3'-end were added back to the complexes as indicated in lanes 4, 6, and 8. Complexes were treated with 1 u/μl of Exo III for 15 min at r.t. The template strand sequence from +1 to +18 is given on the left. The bands between −1 and +6 in lanes 2-8 show where exo III digestion halts due to collision with the upstream boundary of the elongation complex. Lane 1 contains a marker derived from restriction digestion. DNA lengths are given on the right.

was used to detect upstream and downstream boundaries on 3′-end labeled non-template or template DNA strands, respectively.[15] To evaluate the effects of NTP binding on translocation we prepared elongation complexes in which the transcript was 3′-terminated by incorporation of a 3′-dNMP (3′-dNTPs are available from Trilink Biotechnologies (T7RNAP readily incorporates 3′-dNMPs, but it is not known how readily *E. coli* RNAP or pol II will incorporate these chain terminators). Excess NTPs were then washed away by ultrafiltration, using Amicon microcon ultrafiltration units with 100 kD cutoff membranes (other methods such as using His-tagged RNAPs or biotinylated DNAs immobilized on solid supports would probably work as well). The position of the elongation complex, as monitored by exonuclease digestion, was then examined in the absence or presence of saturating concentrations of the NTP complementary to the template base immediately downstream of the RNA 3′-end (addition of a non-complementary NTP provides a control for non-specific effects). As seen in Fig. 3, downstream movement of the elongation complex is seen, not upon completion of the transcript extension step, but upon NTP binding (i.e., the complex with the 15mer + UTP in lane 4 is at nearly the same position as the complex with 16mer in lane 5). This is consistent with the translocation mechanism in Fig. 1, and with the additional conclusion that, in the rapid equilibrium between pre- and post-translocated positions, the pre-translocated position is favored until the NTP binds to and stabilizes the post-translocated complex.[2,15]

Because the elongation complexes can slide they do not present an impenetrable barrier to the processive action of exonucleases. It is therefore essential that a range of exonuclease concentrations and/or digestion times be evaluated so as to identify appropriate digestion conditions. It is also critical that comparisons only be made between elongation complexes in parallel reactions with identical exonuclease concentrations and digestion times. Exonuclease III is better for mapping the actual boundaries of the elongation complex since it is able to digest almost to the edge of the complex, while lambda exonuclease stops 8 nt away from the complex.[15] This probably reflects the position of the exonuclease active site relative to its leading edge. In addition to effects on the position of the complex, NTP binding increases the resistance of the elongation complex to sliding. This is especially apparent at the downstream boundary of the elongation complex, where extended digestion with lambda exonuclease can push the boundary of the complex lacking NTP upstream (Fig. 4). However, when the NTP binds to the complex it becomes much more resistant to the processive action of the exonuclease (see Fig. 4).

[15] J. Huang and R. Sousa, *J. Mol. Biol.* **303**, 347–358 (2000).

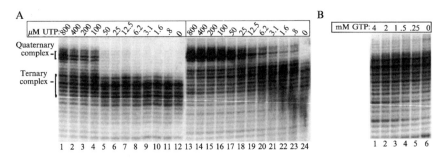

FIG. 4. Lambda exonuclease mapping of the template strand downstream boundary of a T7 RNAP elongation complex with a 3'-dNMP terminated 15 nt RNA.[15] A: The NTP complementary to the template base immediately downstream of the RNA 3'-end (UTP) was added to the reactions at concentrations indicated over each gel lane. Complexes were digested with either 0.4 (lanes 1-12) or 0.08 (lanes 13-24) u/μl of lambda exonuclease for 60 min at r.t. The downstream boundary of the ternary complex (DNA+RNAP+RNA) is diffuse and is pushed upstream at higher exonuclease concentrations. The quaternary complex (DNA+RNAP+RNA+NTP) is more resistant to the processive exonuclease, as evidenced by a sharper boundary, which is not pushed back at the higher exonuclease concentration. B: Control for nonspecific NTP effects. Digestion done as in lanes 13-24 detects only the ternary complex boundary if non-complementary NTP (GTP) is added.

The resistance to sliding seen upon binding NTP may be directly due to template:NTP:RNAP interactions, or to an NTP-induced isomerization that causes the polymerase to close in around the template. Johnson outlined a translocation mechanism for DNAPs in which the polymerase, in the absence of bound NTP, assumes an "open" conformation, which can slide passively on the DNA.[16] NTP binding induces isomerization to a "closed" conformation, which is resistant to sliding. These NTP binding driven transition—initially inferred from kinetic studies—were subsequently observed directly in crystal structures of DNAPs with and without bound NTPs.[17,18] T7 RNAP is highly homologous to the DNAP I family[19] and probably follows a similar mechanism.

Exonuclease digestion and other footprinting reagents have also been used to obtain evidence for sliding by *E. coli* RNAP and pol II.[20–22] Though

[16] K. A. Johnson, *Annu. Rev. Biochem.* **62**, 685–713 (1993).
[17] S. Doublié, S. Tabor, A. M. Long, C. C. Richardson, and T. Ellenberger, *Nature* **391**, 251–258 (1998).
[18] Y. Li, S. Korolev, and G. Waksman, *EMBO J* **17**, 7514–7525 (1998).
[19] R. Sousa, *Trends in Biochem. Sci.* **21**, 186–190 (1996).
[20] R. Landick, *Cell* **88**, 741–744 (1997).
[21] N. Komissarova and M. Kashlev, *J. Biol. Chem.* **272**, 15329–15338 (1997).
[22] N. Komissarova and M. Kashlev, *Proc. Natl. Acad. Sci. USA* **25**, 14699–14704 (1998).

strict quantitative comparisons have not been done, the multisubunit RNAPs seem more prone than T7 RNAP to engage in long-range upstream sliding in which the transcription bubble and RNA:DNA hybrid track the position of the polymerase and the 3′-end of the RNA is displaced from the active site.[20] Such sliding can result in formation of arrested complexes that require RNA cleavage to resume elongation.[21,22] While these observations are supportive of the general idea that RNAPs can slide on their templates, it must be kept in mind that these longer-range sliding events that lead to formation paused or arrested elongation complexes are unlikely to be in rapid equilibrium with the pre- and post-translocated states illustrated in Fig. 1 and therefore, they will not increase local $K_{NTP,app}$. Instead, escape from the paused or arrested state becomes locally rate-limiting.[23]

Studying Translocation with RNAses

Translocation of the polymerase can also be monitored with RNAse, because downstream movement of the polymerase relative to the RNA exposes more RNA to digestion. Experiments are done as described for the exonuclease digestions: ECs are prepared with 3′-dNMP terminated RNAs, NTPs are washed away, and then digestion is monitored as a function of the presence of the NTP complementary to the template base immediately downstream of the RNA 3′-end (Fig. 5).[15] The RNA can be labeled at either the 5′-end using a γ-labeled initiating NTP, or near the 3′-end by using an appropriately designed template (we used templates where C and U were incorporated only at +15 or +16,[24] so that transcripts terminated at +16 or +17 could be 3′-end labeled by inclusion of α-^{32}P CTP or UTP in the transcription reaction). As with the exonuclease digestions, different RNAse concentrations and digestion times must be tested, and any comparisons must be made between ECs treated in parallel under identical conditions. Translocation induced by NTP binding is detected as an increase in the RNAse sensitivity of the RNA at the point where it emerges from the transcription complex (see Fig. 5).

Future Studies

The methods described here, which have been used to study T7 RNAP translocation, can be readily adapted to study translocation by other polymerases. An important question will be whether cells take advantage

[23] P. Urali and R. Landick, *J. Mol. Biol.* **311,** 265–282 (2001).
[24] P. E. Mentesanas, S. T. Chin-Bow, R. Sousa, and W. T. McAllister, *J. Mol. Biol.* **302,** 1049–1062 (2000).

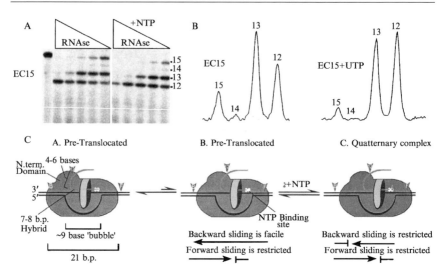

FIG. 5. A: RNAse T1 sensitivity of a T7 RNAP elongation complex with a 3′-dNMP terminated 15 nt RNA.[15] Transcript sequence is complementary to the +1 to +15 template strand sequence given in Fig. 3. The transcript is labeled by inclusion of α^{32}P-GTP in the reaction. Transcript lengths are given on the right of the gel. Elongation complexes were treated with 0 (lane 1), 0.5 (lanes 2, 7), 0.25 (lanes 3, 8), 0.15 (lanes 4, 9), 0.075 (lanes 5, 10), or 0.03 (lanes 6, 11) u/μl RNAse T1 for 10 min. at r.t. with either 0 (lanes 1-6) or 0.5 mM (lanes 7-11) of the complementary NTP (UTP) present. B: Scans of lanes 6 (left) and 11 (right) of the gel shown in A. Increased digestion 12 nt away from the RNA 3′-end in the presence of UTP indicates forward translocation of the polymerase and consequent increased exposure of RNA 12 nt away from the 3′-end upon NTP binding to the elongation complex. C: Interpretation of the exonuclease and RNAse digestions patterns. The polymerase slides between pre- and post-translocated positions—spending most of its time in the pre-translocated position—unless NTP binds to the post-translocated complex. In the post-translocated complex, DNA (which is protected in the pre-translocated complex) at the back end ("B") becomes acessible, while DNA at the front end ("F") becomes covered. In addition, RNA 12 nt away from the 3′-end ("12") is protected in the pre-translocated complex but acessible in the post-translocated complex. NTP binding therefore leads to a downstream shift in exo III (Fig. 3) and lambda exo (Fig. 4) digestion, and increased RNAse accessibility of the RNA 12 nt from the 3′-end (Fig. 5).

of the kinetics of these processes to create mechanisms for regulation. It is sometimes said that steps in a reaction which are not rate-limiting are "kinetically invisible." This seems to suggest that they are irrelevant for control or regulation. Nothing could be less true. Polymerase translocation does not appear to be rate-limiting during either RNA[2,13,14] or DNA[16] synthesis, but as emphasized here it can affect the reaction by its effect on apparent K_{NTP} (for that matter, the substrate binding and dissociation steps in any rapid equilibrium enzyme reaction are "kinetically invisible" because they are not rate-limiting; however, these steps are hardly kinetically inconsequential because their rates determine K_S). The equilibrium between pre- and post-translocated states could theoretically be affected by sequence so as to cause the local apparent K_{NTP} to vary over an indefinitely large range. Such sequence-dependent variations in apparent K_{NTP} have been reported for *E. coli* RNAP.[25] Specific regulation at such "high K_{NTP}" sequences could then be achieved through variation in NTP concentrations, or by factors which alter the affinity of the polymerase for NTPs. Specific regulation of *E. coli* ribosomal RNA promoters appears to operate through a mechanism[26] of the former type (though in this case a high K_{NTP} is a consequence of a rapid equilibrium between open and closed complexes, rather than between pre- and post-translocated states, but the principle is equivalent). Regulation of T7RNAP activity provides an example of the latter mechanism. In this case, regulation is achieved by the binding of T7 lysozyme to the polymerase. This decreases its affinity for NTPs and therefore represses high K_{NTP} (class II) promoters more than low K_{NTP} (class III) promoters.[27,11]

Acknowledgments

Work in the author's laboratory is supported by NIH grant GM52522 (to Rui Sousa) and funds from the Welch Foundation.

[25] J. R. Levin and M. J. Chamberlin, *J. Mol. Biol.* **196,** 61–84 (1987).
[26] T. Gaal, M. Bartlett, W. Ross, C. L., Jr. Turnbough, and R. Gourse, *Science* **278,** 2092–2097 (1997).
[27] J. Villemain and R. Sousa, *J. Mol. Biol.* **281,** 793–802 (1998).

[2] Evaluation of Fluorescence Spectroscopy Methods for Mapping Melted Regions of DNA Along the Transcription Pathway

By Craig T. Martin, Andrea Újvári, and Cuihua Liu

RNA polymerases follow a complex series of events as they progress from an initially bound protein-DNA complex, through the opening of the DNA strands, the *de novo* initiation of RNA synthesis, and progression to a stably cycling elongation complex. It is clear that the "initially transcribing complex" is fundamentally different from the elongation complex clear of the promoter, and the structural transitions that occur along the pathway from one to the other are poorly understood. Finally, elongation complexes are likely to vary under specific conditions. A complete understanding of the mechanisms involved requires an understanding of the structural transitions occurring throughout the various stages of transcription.

Fluorescence has been a staple of the biophysical chemist for years. It is a highly sensitive technique that is able to report on relatively local interactions. The four normal bases of DNA are not fluorescent, which allows specific probes to be engineered to provide information on structural changes local to the probe. The fluorescent base analog of adenine, 2-aminopurine (2AP), has been used for some time to report on local DNA melting, both in transcription, and in a variety of enzymes involved in nucleic acid metabolism.[1–10] As illustrated in Fig. 1, while the fluorescence quantum yield of the free nucleoside is quite high, formation of a polymer of DNA (or RNA) reduces the quantum yield substantially, through both static and dynamic quenching.[11] Finally, the quantum yield is reduced still further on formation of a duplex structure, where neighbor bases are more substantially ordered. It is this latter effect that is exploited in probing local melting within DNA.[12]

More recently, other fluorescent base analogs have been shown to have melting-dependent fluorescence behavior similar to that of 2-aminopurine.[15–17] This opens up substantially the range of sequences that can be probed with this general approach.

In the application of this method to following the progress of transcription, we make use of classic "walking" approaches, in which DNA constructs are prepared lacking a specific encoded nucleotide through to a desired stall position. Thus if the first occurrence of U in a transcript

[1] C. A. Dunlap and M. D. Tsai, *Biochemistry* **41,** 11226 (2002).
[2] R. Hochstrasser, T. Carver, L. Sowers, and D. Millar, *Biochemistry* **33,** 11971 (1994).
[3] K. Raney, L. Sowers, D. Millar, and S. Benkovic, *Proc. Natl. Acad. Sci. USA* **91,** 6644 (1994).
[4] K. S. Dunkak, M. R. Otto, and J. M. Beechem, *Anal. Biochem.* **243,** 234 (1996).
[5] Y. Jia, A. Kumar, and S. Patel, *J. Biol. Chem.* **271,** 30451 (1996).
[6] S. Sastry and B. Ross, *Biochemistry* **35,** 15715 (1996).
[7] A. Újvári and C. T. Martin, *Biochemistry* **35,** 14574 (1996).
[8] A. M. Fedoriw, H. Liu, V. E. Anderson, and P. L. deHaseth, *Biochemistry* **37,** 11971 (1998).
[9] J. J. Sullivan, K. P. Bjornson, L. C. Sowers, and P. L. deHaseth, *Biochemistry* **36,** 8005 (1997).
[10] T. M. Nordlund, S. Andersson, L. Nilsson, R. Rigler, A. Graslund, and L. W. McLaughlin, *Biochemistry* **28,** 9095 (1989).
[11] E. L. Rachofsky, R. Osman, and J. B. Ross, *Biochemistry* **40,** 946 (2001).
[12] D. Xu, K. Evans, and T. Nordlund, *Biochemistry* **33,** 9592 (1994).

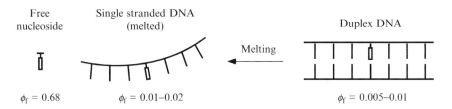

FIG. 1. Illustration of the principle. Fluorescence of the base analog is largest as the free nucleoside, is reduced substantially on incorporation into DNA, and is reduced further on formation of a duplex. The latter distinction, duplex vs. single-stranded, is exploited to monitor regions of the DNA melted open by RNA polymerase. Quantum yields reported are representative values for 2-aminopurine.[13,14]

occurs at position +16 of the transcript, transcription in the presence of ATP, CTP, and GTP will drive transcription to stall at position +15. As will be detailed later, this is expected to work better at some positions than at others and investigators must take care to demonstrate that the complex is homogeneous.

Fluorescent Base Analogs

While 2-aminopurine is the most widely used fluorescent base analog, other potentially useful probes are becoming available.[15–17] Table I presents a listing of fluorescent base analogs that have potential utility in probing regions of melted DNA. These probes expand the range of sequences that can be studied and provide the advantage of having excitation and emission maxima farther from those of Trp in the protein. This results in lower background fluorescence from the protein and simplifies data analysis.

Benefits of Fluorescence over Other Approaches

More traditional assays of local melting include $KMnO_4$, endonuclease, and related footprinting assays. A fundamental difference between fluorescence and footprinting is the time scale of the measurement. Regions of the

[13] E. L. Rachofsky, E. Seibert, J. T. Stivers, R. Osman, and J. B. Ross, *Biochemistry* **40,** 957 (2001).

[14] D. C. Ward, E. Reich, and L. Stryer, *J. Biol. Chem.* **244,** 1228 (1969).

[15] M. E. Hawkins, W. Pfleiderer, O. Jungmann, and F. M. Balis, *Anal. Biochem.* **298,** 231 (2001).

[16] M. E. Hawkins, *Cell Biochem. Biophys.* **34,** 257 (2001).

[17] M. E. Hawkins, W. Pfleiderer, F. M. Balis, D. Porter, and J. R. Knutson, *Anal. Biochem.* **244,** 86 (1997).

TABLE I
FLUORESCENT BASE ANALOGS POTENTIALLY USEFUL FOR THE STUDY OF LOCAL HELIX MELTING*

Base	Analog	Structure	$\lambda_{max}^{excitation}$ (nm)	$\lambda_{max}^{emission}$ (nm)	Source
A	2AP	2-aminopurine	310	370	Glen Research, IDT DNA
A	6-MAP	4-amino-6-methyl-pteridone	330	435	TriLink Biotechnologies, Toronto Research Chemicals
G	6-MI	6-methyl-isoxanthopterin	340	430	TriLink Biotechnologies, Toronto Research Chemicals
G	3-MI	3-methyl-isoxanthopterin	348	431	TriLink Biotechnologies, Toronto Research Chemicals
C	py-C	pyrrolo-cytosine	350	460	Glen Research
—	Trp	—	280	348	—

*The fluorescence properties of the amino acid Trp are shown for comparison. Sources for phosphoramidites and/or synthetic oligonucleotides containing the analog are shown (TriLink: http://www.trilinkbiotech.com; Glen Research: http://www.glenres.com /; Toronto Research Chemicals: http://www.trc-canada.com; IDT DNA: http://www.idtdna.com).

DNA may be open only transiently, so that relatively slow footprinting reactions might show no signal changes. Conversely, for a situation in which
the complexes interconvert more slowly between an open and closed state,
footprinting reagents may over time have access to all complexes and
therefore show a disproportionate "signal." Similarly, movement of the
complex along the DNA during the course of the measurement might yield
a footprint larger than the actual size of the melted region at any instant in
time. In contrast, the time scale of the fluorescence measurement is the lifetime of the excited state: typically 1–10 nsec, much faster than the motions
described previously. Thus fluorescence provides a much more accurate
"snap shot" of the average behavior of the system.

A more detailed understanding of the structures represented by changes
in steady state fluorescence can potentially be obtained by analyses of
fluorescence lifetime distributions.[11–13,18] Similarly, the use of external
quenchers, such as potassium iodide, can provide information on the accessibility of a melted base to solvent.[13] These approaches have the potential
to provide expanded insight into structural details of probed complexes.

Beyond the mapping of statically melted regions within various transcription complexes, fluorescence is also ideally suited for stopped-flow
kinetic measurements of dynamic changes in those complexes.[7,19–23] Although fluorescence from these probes is not sufficiently sensitive for use
in single molecule kinetic measurements, there is at least the potential
for their future use in this rapidly expanding area as well.

Preparation of Samples

T7 RNA polymerase is readily prepared from overproducing strains, in
the completely native form,[24] and with a His-tag for rapid purification.[25] In
either case, yields from a 4 or 0.5 L culture, respectively, are sufficient
for many fluorescence measurements. It is convenient to maintain stock
solutions at concentrations of 20 μM or higher, in a buffer of 30–40 mM
HEPES, pH 7.8, 150–200 mM NaCl, 0.05% Tween–20, and 1 mM EDTA.

[18] P. Wu, H. Li, T. M. Nordlund, and R. Rigler, *Proc. SPIE-Int. Soc. Opt. Eng.* **1204,**
262 (1990).
[19] R. P. Bandwar, Y. Jia, N. M. Stano, and S. S. Patel, *Biochemistry* **41,** 3586 (2002).
[20] R. P. Bandwar and S. S. Patel, *J. Biol. Chem.* **276,** 14075 (2001).
[21] N. M. Stano, M. K. Levin, and S. S. Patel, *J. Biol. Chem.* **277,** 37292 (2002).
[22] Y. Jia and S. S. Patel, *Biochemistry* **36,** 4223 (1997).
[23] Y. Jia and S. S. Patel, *J. Biol. Chem.* **272,** 30147 (1997).
[24] P. Davanloo, A. H. Rosenberg, J. J. Dunn, and F. W. Studier, *Proc. Natl. Acad. Sci. USA* **81,**
2035 (1984).
[25] B. He, M. Rong, D. Lyakhov, H. Gartenstein, G. Diaz, R. Castagna, W. T. McAllister, and
R. K. Durbin, *Protein Expr. Purif.* **9,** 142 (1997).

DNA containing fluorescent base analogs can be synthesized via the traditional phosphoramidite approach, with no modifications in the synthesis protocols. The phosphoramidite derivatives of 2-aminopurine (2AP) and pyrrolo-dC (py-dC) are available from Glen Research (Sterling, VA). Oligonucleotides containing 2-aminopurine, 6-MI, 3-MI, or 6-MAP can be purchased directly from TriLink Biotechnologies (San Diego, CA). Single strands can be synthesized trityl-on and then purified by use of an Amberchrom CG-16C reverse phase resin[26] or can be purified by electrophoresis on a 15% to 20% polyacrylamide gel containing *Tris*-borate buffer and 7 M urea. In the latter case, recovery of the purified DNA using an Elu-Trap device provides a high yield (Schleicher & Schuell, Keene, NH).

In our early use of py-dC, we incorporated the phosphoramidite derivative of furano-dT (Glen Research, Sterling, VA) during oligonucleotide synthesis. During the subsequent deprotection (standard conditions, 30% ammonium hydroxide at 55 °C for 6 hours), furano-dT is quantitatively converted to py-dC. The direct phosphoramidite for the latter is now available from Glen Research.

Double-stranded DNA is prepared by mixing equal molar amounts of template and nontemplate strands, at reasonably high stock concentration (20 μM or more), in TE buffer (10 mM Tris, pH 7.8, 1 mM EDTA). To ensure correct annealing, samples are typically heated briefly to 75 °C, followed by slow cooling to room temperature. Stocks may be stored frozen at -20 °C.

It is important that the concentrations of complementary strands be equal, as an excess of labeled single strand will yield anomalously high fluorescence from otherwise "duplex" solutions. Concentrations of each strand are initially estimated from an extinction coefficient derived from the sum of the extinction coefficients of the individual bases, or by more complex calculations that recognize nearest-neighbor effects. Prior to preparing the double-stranded construct, unlabeled DNA is titrated onto the labeled strand, following changes in fluorescence to confirm the equivalence point.

Conversely, it is important that the single-stranded controls not have stable secondary structure in the region surrounding the fluorescent base analogs because this could lead to anomalously low fluorescence for the single-stranded control. The T7 ϕ10 promoter sequence out to about position $+26$ shows substantial secondary structure in its single strands (this is not a problem for DNA constructs truncated at less than about position $+15$).[27]

[26] C. Schick and C. T. Martin, *Biochemistry* **32**, 4275 (1993).
[27] C. Liu and C. T. Martin, *J. Mol. Biol.* **308**, 465 (2001).

Fluorescence Measurements

Fluorescence measurements can be carried out in any research grade L- or T-format fluorescence spectrometer. Experiments here used a dual monochrometer Photon Technologies fluorimeter with a 75 watt arc lamp for excitation. The use of small volume cells is convenient for minimization of sample usage; a 75 μL (light path 3 × 3 mm) ultramicro cell is available from Hellma (Müllheim, Germany). Table I provides guidance for placement of the excitation and emission monochrometers. For fluorescent probes with excitation and emission maxima far from the protein, slits can be set fairly large (>5 nm). For measurements with 2-aminopurine, care should be taken to minimize fluorescence from protein tryptophan groups (see later). This can be achieved by setting excitation and emission monochrometers 5–10 nm to the red of the maxima shown in Table I (but with some loss in signal intensity). In this case, fluorescence background from the protein should be subtracted, using parallel experimental runs containing analogous oligonucleotides lacking 2-aminopurine. Such corrections are not necessary for the other probes (although background fluorescence should always be verified experimentally).

Fluorescence measurements are carried out in a final buffer of 30 mM HEPES, pH 7.8, 15 mM magnesium acetate, 25 mM potassium acetate, 0.25 mM EDTA, and 0.05% (v/v) Tween—20 (Calbiochem, 10% protein grade). Temperature is a variable which should be considered carefully. Measurements are most readily carried out at 25 °C, where the enzyme is stable for extended periods. Measurements at 37 °C, where the enzyme is more active can be carried out, but stability of the protein should be verified.

Final DNA concentrations of 0.5–2 μM provide reasonable signal levels. Typically, equimolar amounts of enzyme and DNA are employed, which given the K_d = 10 nM for complex formation[28] should yield about 90% complex formation. As noted later, addition of GTP should drive this to close to 100% (T7 RNA polymerase preparations are typically very close to 100% active, though this should be verified). An excess of enzyme can be used to drive more complete complex formation, but nonspecific binding to DNA outside of the promoter may perturb the complex.

Considerations

While the approach outlined here has provided excellent agreement with other experimental measurements, a variety of issues should be considered when using fluorescent base analogs as monitors of local helix melting.

---

[28] A. Újvári and C. T. Martin, *J. Mol. Biol.* **273,** 775 (1997).

Base Analogs Are Not Identical to the Bases They Replace

By definition, a base analog is not an exact replacement for the physiologically relevant base that it replaces. The extent to which this presents a problem in interpretation must be weighed based on other data. For example, a substitution that alters an energetically important protein-DNA contact is likely to lead to a perturbed binary complex (both in binding energetics and potentially in structure). This is expected to be most important in the upstream recognition element within the promoter: from positions -17 to -5.[29–31] In contrast, during elongation, where the complex is presumably evolved to reasonably accommodate all four base pairs, subtle changes to base structure should have very little effect on the system.

Of the probes described here, 2-aminopurine perturbs the chemical nature of both the minor and major groove faces of the duplex, while py-dC introduces added steric bulk in the major groove. To a first approximation, the pteridine derivatives do not alter major or minor groove faces, but they have enlarged ring systems, which will certainly alter their packing into the duplex. Satisfyingly, most of these probes have only a minor effect on duplex stability (with the exception of 3-MI, which places a methyl group at the Watson-Crick interface, leading to duplex destabilization equivalent to the introduction of a mismatch).[32,33]

While one must consider that base analogs might perturb the system, footprinting assays can similarly perturb the system in ways which could lead to false representations of the system. Reagents that bind to double-stranded DNA (e.g., nucleases or the minor groove binding methydium propyl EDTA) can, in principle, competitively alter the equilibrium of the system.

Fluorescent Probes Report Only Indirectly on Melting

It is important to remember that the fluorescence changes observed for these probes arise from net changes in quenching of fluorescence and report only indirectly on melting. Quenching of fluorescence is thought to arise from base stacking and collisions with neighboring bases, and these interactions decrease only partially in the transition from double- to single-stranded DNA.[11,34] Hence, melting-induced increases in fluorescence are often on

[29] K. E. Joho, L. B. Gross, N. J. McGraw, C. Raskin, and W. T. McAllister, *J. Mol. Biol.* **215,** 31 (1990).
[30] T. Li, H. H. Ho, M. Maslak, C. Schick, and C. T. Martin, *Biochemistry* **35,** 3722 (1996).
[31] D. Imburgio, M. Rong, K. Ma, and W. T. McAllister, *Biochemistry* **39,** 10419 (2000).
[32] K. Wojtuszewski, M. E. Hawkins, J. L. Cole, and I. Mukerji, *Biochemistry* **40,** 2588 (2001).
[33] S. M. Law, R. Eritja, M. F. Goodman, and K. J. Breslauer, *Biochemistry* **35,** 12329 (1996).
[34] P. G. Wu, T. M. Nordlund, B. Gildea, and L. W. McLaughlin, *Biochemistry* **29,** 6508 (1990).

the order of two-fold (but can vary). Other structural or dynamic changes that lead to decreased quenching will also lead to an increase in fluorescence.[32,35] In fact, in mapping of the initially bound binary complex, protein binding to promoter DNA leads to an increase in fluorescence for 2-aminopurine placed at position −10, a region known not to melt in the binary complex.[7] Analysis of the crystal structure[36] shows a slight kink in the DNA at the base step involving the −10 adenine, confirming that the fluorescence change arises from a slight kinking of the DNA on polymerase binding.

That the analyses can be less than straightforward is further illustrated by the fact that the absolute quantum yield (and precise mechanism of quenching) of fluorescent base analogs is sequence dependent.[13,37–39] This illustrates the fact that the approach is monitoring potentially subtle changes in base-base interactions. This also presents a potential problem in that some substitutions yield significantly lower overall fluorescence or show smaller fractional increases in fluorescence associated with melting. To aid in comparison of melting at different positions, fluorescence values for a specific substitution are usually normalized to the fluorescence of double-stranded DNA containing that substitution.

In spite of the above caveats, recent applications of the approach have yielded remarkably good agreement with other experimental approaches, including crystallography. With the exception of position −10, mentioned above, the initial binary complex bubble defined by fluorescence is in complete agreement with the subsequently reported crystal structure.[36] Interestingly, the fluorescence changes of 2-aminopurine placed at position −4 are anomalous in that fluorescence increases substantially *more* than the two-fold amount typically observed.[7,20] Analysis of the crystal structure shows that this base is removed from its neighbor bases by complex formation, predicting less base neighbor-mediated quenching and the larger increase in fluorescence observed. In general, for analogs flipped out of the duplex, the potential for protein-mediated quenching or energy transfer exists, but has not been demonstrated conclusively.[40] Shifts in the excitation and emission maxima have been used to argue for the placement of a probe in a more or less hydrophobic environment.[11]

[35] E. Seibert, J. B. Ross, and R. Osman, *Biochemistry* **41,** 10976 (2002).

[36] G. M. Cheetham, D. Jeruzalmi, and T. A. Steitz, *Nature* **399,** 80 (1999).

[37] J. T. Stivers, *Nucleic Acids Res.* **26,** 3837 (1998).

[38] S. L. Driscoll, M. E. Hawkins, F. M. Balis, W. Pfleiderer, and W. R. Laws, *Biophys. J.* **73,** 3277 (1997).

[39] M. Kawai, M. J. Lee, K. O. Evans, and T. M. Nordlund, *J. Fluorescence* **11,** 23 (2001).

[40] S. S. Szegedi, N. O. Reich, and R. I. Gumport, *Nucleic Acids Res.* **28,** 3962 (2000).

Estimating Extent of Melting

In some cases, the complex being probed may not be homogeneous. For example, one might have a mixture of open and closed complexes. In this case, the measured fluorescence, relative to appropriate controls, can provide an estimate of the percent melting. For the controls, fluorescence from labeled duplex DNA presents the lower limit for fluorescence intensity, while to a first approximation, fluorescence from single-stranded DNA under similar conditions can be used to define the upper limit. The extent of melting in a complex can then be estimated as the fractional increase in fluorescence, relative to this range. In some cases, such as at the initial binding stage (see later), other controls might be more appropriate.

This estimation becomes more complex when one considers incomplete binding. For example, in static complexes of T7 RNA polymerase and its promoter, one must consider that the K_d for complex dissociation is about 10 nM,[28] so that at 1 μM polymerase and 1 μM promoter only 90% of the DNA is expected to be bound to enzyme. If the measured fluorescence for the complex is 90% of the control range, then the (bound) complex is estimated to be 100% melted. This analysis is potentially complicated by the fact that incorporation of the probe can alter the dissociation constant. For more accurate estimations of melting, an independent measure of the dissociation constant should be performed for DNA containing the base analog.

One can drive complex formation by addition of excess RNA polymerase, but then background fluorescence from the protein becomes a larger contribution to the measured fluorescence (as discussed earlier, for 2-aminopurine, protein background fluorescence must always be subtracted from measured fluorescence values; for the other probes listed in Table I, this is less of an issue).

In any case, multiple estimations of fractional melting from complexes containing probes at various positions in the melted region should increase the certainty of such estimations.

Mapping Melted Bubbles in Transcription

Mapping the Bubble in the Initial Promoter Complex

Initial studies mapping the melted region of the T7 RNA polymerase:-promoter complex, summarized in Fig. 2, used only 2-aminopurine.[7] To illustrate the approach, fluorescence from a construct containing 2-aminopurine at position −6 shows no significant change on addition of enzyme, consistent with models in which the upstream bases from positions −17 through −5 form a duplex recognition element.[30,41] In contrast,

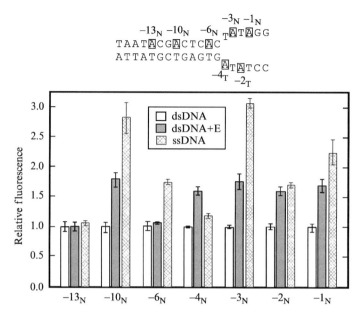

FIG. 2. Mapping of the initial transcription bubble. In separate experiments, 2-aminopurine is placed at the positions indicated (N: nontemplate strand; T: template strand; numbering relative to the transcription start site, +1). Concentrations of DNA and enzyme (E) were 0.50 and 0.64 μM, respectively. Adapted with permission from A. Újvári and C. T. Martin, *Biochemistry* **35,** 14574 (1996).

fluorescence from probes at positions -4, -3, -2, and -1 (on both template and nontemplate strands) increases on addition of RNA polymerase, consistent with models in which this region is melted. The subsequent crystal structure of the binary complex has confirmed the conclusions from the fluorescence study.[36] More recently, we have used 2-aminopurine, 6-MI, and py-dC to map the downstream edge of the bubble, a region not present in the crystal structure (Liu, Hawkins, and Martin, unpublished). These studies confirm the earlier picture and show that melting extends to about position $+3$ or $+4$ downstream, information not available from current crystal structures, which lack downstream DNA.

These results illustrate some considerations in the use of fluorescent base analogs. The data at position -13 demonstrate the sequence dependence of the fluorescence behavior. In this context, there is very little difference in fluorescence between single- and double-stranded control DNA

[41] K. A. Chapman and R. R. Burgess, *Nucleic Acids Res.* **15,** 5413 (1987).

(suggesting that the double stranded DNA in this sequence context might have an unusual conformation). Indeed, the differences are less than the errors in the measurement, making this position not useful in mapping melted regions. In contrast, the increase in fluorescence from a probe at position −4 is *larger* than that of the single-stranded DNA (note that a recent study reports an even larger relative increase at this position[20]), suggesting either secondary structure in the single-stranded control (not likely here) or an unusual structure in the binary complex. In fact, an unusual structure is observed in the crystal structure, in that the A at position −4 of the template strand is completely unstacked from its neighbors, consistent with the large increase in fluorescence from 2-aminopurine at that position.[36] Finally, as noted above, although the bases near position −10 do not melt on complex formation, there is a slight kink in the base step at that position, yielding the otherwise unexpected increase in fluorescence at this position.

For the initially melted complex, a more appropriate control may be derived by the addition of the initial substrate GTP, which drives opening by a mass action effect.[21] However, forward motion of the polymerase might alter the environment of the probe, particularly if located on the template strand (this strand must move relative to the protein during translocation). For probes placed on the template strand, an alternate approach is to use partially single-stranded DNA constructs (lacking completely the nontemplate strand from position −4 to the downstream 3′ end).[42]

Finally, the data in Fig. 2 illustrate cautions regarding estimating the extent of open complex formed from the fluorescence intensities. Using the data from positions −3, −2, and −1, the fractional change in fluorescence in the complex (relative to the double- and single-stranded controls) is 37%, 85%, and 55%, respectively. Remembering that under these conditions we expect that only 90% of the DNA should be bound (see previous), the data from position −2 predict almost complete melting, while the other data predict substantially smaller levels of melting.

Mapping the Bubble in a Stably Stalled Elongation Complex

A similar approach has been used to map the melted region within a stalled elongation complex, clear of the promoter.[27] In this case, RNA polymerase can be "walked" to various positions along the DNA by synthesizing DNA encoding RNA which is "U-less" through to the desired position. In principle, transcription in the presence of GTP, ATP, and CTP (only) leads to stalling at the position immediately prior to the first

[42] M. Maslak and C. T. Martin, *Biochemistry* **32**, 4281 (1993).

encoded U. In practice, if dissociation of the complex at that position is non-zero, one achieves a steady state distribution of species, and it is important to demonstrate that one does not have, for example, a substantial population of complexes sitting back at the promoter. In the cited study, fluorescence from 2-aminopurine at position −2 returned to duplex levels, indicating that, in fact, all complexes had cleared the promoter (and as detailed below) had in fact translocated beyond position +8.

As shown in Fig. 3, a variety of constructs are prepared, all directing the stall to the same position (+15, relative to the start site), but incorporating fluorescent base analogs (one per construct) to different positions in the DNA. The figure shows the variety of controls that one can perform in these measurements. Of course, the double-stranded form of the DNA,

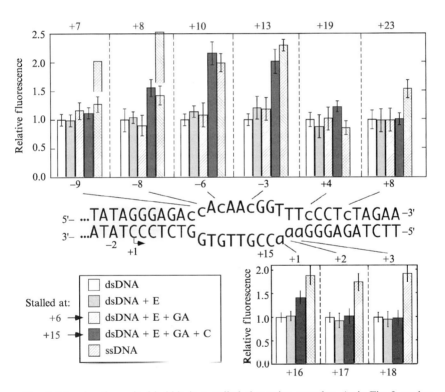

FIG. 3. Mapping the melted bubble in a stalled elongation complex. As in Fig. 2, probes were placed as indicated above (either (a) 2-aminopurine or (c) py-dC). Complexes were then prepared and walked sequentially to positions +6 (by addition of only ATP and GTP) or +15 (by subsequent addition of CTP). Positional numbering in black is relative to the last base incorporated; numbering in grey is relative to the promoter start site. Adapted with permission from C. Liu and C. T. Martin, *J. Mol. Biol.* **308,** 465 (2001).

in the absence of polymerase, forms the control for fully duplex DNA. Addition of enzyme is not expected to result in changes in fluorescence for the probes indicated, as the enzyme remains fixed at the promoter—this is observed. Addition of GTP and ATP should allow transcription to position +6, and again, melting in the probed region is not expected and is not observed. Addition of CTP to the mixture containing GTP and ATP next generates the desired stalled complex. Finally, addition of UTP to the mix should drive transcription to the end of the template (+28 in this case) and the bubble in this region is observed to collapse, as expected (results not shown).

The fluorescence pattern for the complex stalled at position +15 shows a simple, clear pattern. Fluorescence from probes at positions −8, −6, and −3 relative to the last incorporated base (−1) at position +15 increases relative to double-stranded levels, consistent with melting at these positions. In contrast, fluorescence from probes at positions −9, +2, +3, +4, and +8 does not change, consistent with maintenance of duplex at these positions. There is some evidence of melting at position +1, one base past the last incorporated base. Thus the elongation bubble is estimated to extend from about 8 bases upstream of the stall site and at most one base downstream. This interpretation has recently been confirmed by crystal structures of artificially constructed elongation complexes.[43,44]

It is important to note that in this application, at least some of the probes placed in the template strand upstream of the stall site are expected to be in a heteroduplex with the nascent RNA and ought to yield low levels of fluorescence (unpublished results). Indeed, this approach (but also see later) should allow direct mapping of the size of the hybrid.

Note in the above that in the $\phi10$ promoter sequence used, positions −9 and −8 relative to the stall site are expected to be in duplex secondary structure in the single-stranded control, as there is substantial predicted secondary structure in the DNA. Thus, the simple controls are not appropriate. In this case, we synthesized DNA constructs containing the same analog in the same local sequence context (5 base pairs on either side should be more than enough), but engineered not to contain secondary structure. Those fluorescence values (for the same molar concentration of single-stranded DNA) are shown in Fig. 3 as bars above the initial controls.

One ought to consider whether the size of the bubble so measured is a fundamental property of the elongation complex or is in some way

[43] Y. W. Yin and T. A. Steitz, *Science* **298,** 1387 (2002).
[44] T. H. Tahirov, D. Temiakov, M. Anikin, V. Patlan, W. T. McAllister, D. G. Vassylyev, and S. Yokoyama, *Nature* **420,** 43 (2002).

dependent on the sequence chosen. This question is easily addressed by altering the sequence of the DNA in this region. The simplest approach to determining if the upstream edge of the bubble is defined by the sequence is to add or delete a base close to the stall site (far from the upstream edge). For T7 RNA polymerase, an 8-base bubble was maintained through this simple substitution, suggesting that the observed bubble is a fundamental property of the system. Very recent crystal structures of artificially created transcription bubbles are fully consistent with the fluorescence results.[43,44]

Finally, the fluorescence from a probe in the nontemplate strand at position −8 relative to the stall site is intermediate between that of the double- and single-stranded controls. This observation is conserved in a parallel construct with a different sequence context −8 relative to the stall site. There are two interpretations for this result. Intermediate fluorescence might arise from heterogeneity in the precise positioning of the stalled complex. Alternatively, the base at the edge of the bubble may be only slightly displaced from its normal stacking interactions (partially melted), or be dynamically interconverting between open and closed (but on a time scale still slow by fluorescence), such that only intermediate quenching is observed.

Monitoring Bubble Collapse on Promoter Clearance

Of fundamental interest to the process of transcription is how the complex makes the transition from an initially transcribing, promoter-bound complex, to a stably elongating complex free of the promoter (transcription complexes translocated to less than about position +10 are notably less stable than complexes stalled farther out).[45] This is particularly interesting given the large structural changes recently observed between initially transcribing and elongating complexes.[43] Fluorescent base analogs present a very simple way to assess promoter clearance.[46] A probe (or probes) placed in the initially melted bubble (see above) can report on the collapse of that bubble which presumably accompanies promoter clearance.

The results presented in Fig. 4A show that on translocation to positions +7 and +8, the fluorescence from 2-aminopurine placed back at the initial bubble (position +2) remains at the level of single-stranded DNA—the initially melted bubble remains fully open. On translocation to position +9 the fluorescence decreases to an intermediate level, and on translocation to position +10, the fluorescence is near that of the double-stranded

[45] G. A. Diaz, M. Rong, W. T. McAllister, and R. K. Durbin, *Biochemistry* **35,** 10837 (1996).
[46] C. Liu and C. T. Martin, *J. Biol. Chem.* **277,** 2725 (2002).

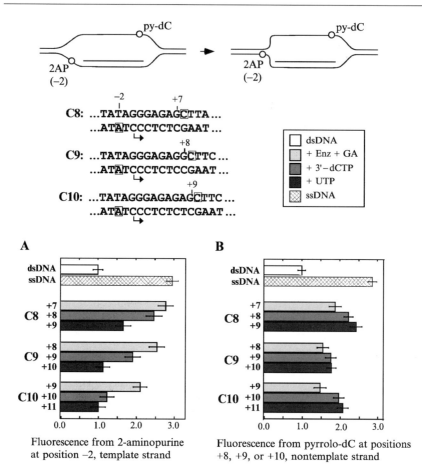

FIG. 4. Simultaneous monitoring of two ends of the bubble in initially transcribing complexes. (A) Observation of the collapse of the initial bubble by following the fluorescence quenching of 2-aminopurine placed near the start site, at position −2 of the template strand. Concentrations of enzyme and DNA are 1 μM. Controls representing fully duplex (dsDNA) and fully single-stranded (ssDNA) are shown. The sequences of the DNA templates are such that transcription in the presence of GTP and ATP will walk the complex out to the position shown in the top number of each pair along the y-axis. Subsequent addition of 3′–dCTP allows the complex to walk one base pair further, as indicated by the middle number of each group. Finally, addition of UTP should drive translocation still one base farther. High fluorescence, as for complexes stalled at positions +7 and +8, indicates a melted bubble. Low fluorescence, as for complexes stalled at positions +10 and beyond, indicates collapse to a duplex. (B) Simultaneous monitoring of the downstream bubble near the stall site. In the same DNA samples, changing excitation and emission wavelengths allows monitoring of downstream melting, near the stall site. Adapted with permission from C. Liu and C. T. Martin, *J. Biol. Chem.* **277**, 2725 (2002).

control. This indicates that the initially melted bubble has collapsed back to a duplex on translocation from position +8 through to position +10. The intermediate decrease of the fluorescence again suggests either heterogeneity in the translocation of the stalled complex or collapse to a structure that is not yet fully duplex at position −2.

Fig. 4A demonstrates that careful construction of the DNA allows walking to desired stall positions in steps. In particular, the sequence of the encoded RNA is

$$5'\text{-}[G, A \ only]_n CU \ldots \text{-}3'$$

Transcription in the presence of GTP and ATP only should allow walking to position +n. Subsequent addition of 3'–dCTP then allows walking forward to position n + 1. Finally, subsequent addition of UTP allows occupancy of the elongating substrate site, driving translocation to position n + 2. In this way, with a minimal number of DNA constructs, positionally redundant data are obtained, enhancing the strength of the conclusions.

The 2-aminopurine data above clearly demonstrate collapse of the upstream bubble (and strongly suggest loss of promoter contacts) on translocation beyond position +8. Simultaneous placement of py-dC downstream allows confirmation that the steady state distribution of complexes is indeed translocated as desired, as shown in Fig. 4B. In all three cases, translocation to position +n leads to a not-quite maximal increase in fluorescence from a probe placed at position n + 1, consistent with other data that suggest that the bubble is melted at most one base beyond the stall site.

While one cannot be certain that free single-stranded DNA represents the ideal control for a fully melted complex, the data are consistent with the conclusion that under the steady state conditions described, most of the complexes are stalled as predicted. Independent, parallel assays of transcription show rapid synthesis of the expected RNA product with very little turnover, also consistent with the preparation of a relatively homogeneous population of stalled complexes.

Monitoring Initial Dissociation of the 5' End of the RNA

The approaches described utilize the fact that fluorescence is quenched more in double-stranded B-form DNA than in single-stranded DNA. It is also true that fluorescence is quenched more in double-stranded A-form nucleic acid than in the single-stranded form. In principle, then, one should be able to use this approach to monitor resolution of the RNA-DNA hybrid. In other words, one should be able to watch the RNA peel away from the hybrid by placing a fluorescent base analog in the templating DNA strand. However, if the complementary nontemplate strand

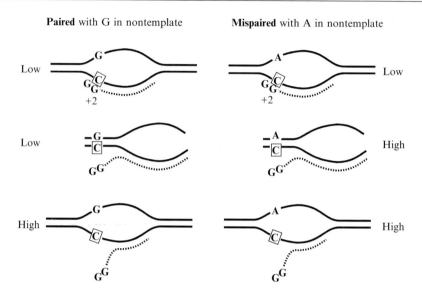

FIG. 5. Use of parallel matched and mismatched constructs allows determination of RNA hybrid presence. The fluorescent base analog (py-dC in this case) is indicated by a box. "Low" and "high" refer to fluorescence values similar to that from double- and single-stranded controls, respectively. Analysis of a complex in parallel constructs allows the three possible structures to be distinguished.

reanneals to the template strand coincident with loss of the RNA, then the fluorescence will remain low in both cases. One is replacing an A-form duplex for a B-form duplex.

This problem can be overcome by the fact that a fluorescent base analog mispaired within otherwise duplex DNA shows higher fluorescence than in the duplex, presumably because stacking interactions are disrupted by the constraints of the mismatch.[46,47] As shown in Fig. 5, construction of parallel constructs in which the fluorescent base analog in the template strand is either paired or mispaired with its partner in the nontemplate strand allows the two possibilities to be distinguished. If fluorescence is low in both constructs, the fluorescent base is paired with the properly encoded RNA. If the fluorescence is high in both constructs, the analog is paired with neither the DNA nor the RNA. Finally, if the fluorescence is low for the correctly paired construct but high in the mismatch construct, then the DNA duplex has reformed, and therefore, the RNA has peeled away.

[47] L. C. Sowers, Y. Boulard, and G. V. Fazakerley, *Biochemistry* **39**, 7613 (2000).

Placement of py-dC at position +2 on the template strand provides a signal for the separation of the RNA from the initially formed hybrid. Measuring fluorescence as the complex is walked along the template (as previously stated) provides valuable new insight into structural changes occurring during the early stages of transcription.

Results for the construct in which C in the template strand is correctly paired with G in the nontemplate strand are presented in Figure 6A. In complexes stalled at positions +10, +11, and +12, the fluorescence from py-dC at position +2 is comparable to that of the control duplex, indicating that the structure occurs as either of the top two possibilities in Fig. 5. Fluorescence from parallel experiments in which the nontemplate strand across from the probe is a mismatched A is shown in Figure 6B. In this case, fluorescence is at duplex levels for complexes translocated to positions +10 and +11, but rises to near single-stranded levels on translocation from position +11 to +12. These results argue that the hybrid at position +2 remains as the complex translocates forward to position +11. On translocation to position +12, RNA dissociates from the hybrid and the template strand returns to a DNA:DNA duplex; the bubble has collapsed.

As before, this result was confirmed in two different constructs. In the first, the DNA sequence allows initial walking to position +10, with subsequent step-wise addition of 3′–dCTP, and then UTP. In the second construct the initial walking is to position +11. That the results agree for both constructions supports the generality of the conclusions.

Similar results (not shown) demonstrate that the RNA one back, at position +1, similarly dissociates on translocation from position +10 to +11. Together these results suggest a smooth, step-by-step peeling off of the RNA, as the complex translocates first to position +11, and then to position +12. They also show that at this stage of transcription, the hybrid reaches a maximal length of 10 base pairs, slightly larger than the hybrid during elongation.

In this class of experiment, an artificial mismatch is constructed and one must consider the possibility that the mismatch might alter the nature of the associated bubble edge. Nevertheless, consistent observations with probes placed at positions +1 and +2 and in different walking constructs support the interpretations presented above.

Concluding Remarks

Fluorescent base analogs provide a relatively non-perturbing probe of very local structural changes within the DNA and/or RNA. They can be used to map melted bubbles in various static states of transcription, either the initially bound promoter complexes, or through walking experiments,

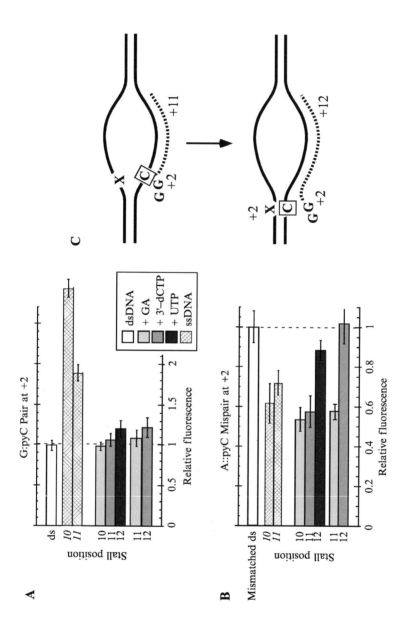

in stalled complexes approximating dynamic movement along the DNA. Furthermore, with appropriately prepared constructs, one can also map the RNA:DNA hybrid in any of these states. The probes, however, are not completely non-perturbing and the approach measures melting only indirectly by its effects on stacking-mediated quenching, so redundant measurements with differently prepared complexes provide added strength to the interpretations derived.

The advent of a variety of base analogs, all with similar duplex-dependent quenching, opens up the range of sequences that can be probed and even allows simultaneous measurements at different positions along the DNA.[15,16,46] Fluorescence, of course, is also an ideal tool to use in time-dependent kinetic assays, so the dynamics of transcription bubbles (and of RNA resolution) can be monitored in stopped-flow kinetic assays.[19,22]

The application of these approaches is limited only by sensitivity. Experiments described here typically used 1 μM concentrations of enzyme and DNA, but one may be able to get reliable data at somewhat lower concentrations. Site-specific incorporation of fluorescent probes requires chemical synthesis, but if substantially longer DNA templates are required, these might be introduced by clever applications of the polymerase chain reaction, in which the probe is introduced in one of the primers. Thus, this approach should increasingly become useful in analyses of the multi-subunit RNA polymerases.[9]

Acknowledgments

Financial support from the National Institutes of Health (1R01GM55002) and the National Science Foundation (MCB-9308670) is gratefully acknowledged. We thank Mary E. Hawkins (NCI) and John Randolph (Glen Research) for many very helpful discussions.

FIG. 6. Monitoring dissociation of the RNA from the hybrid. (A) Fluorescence from fully double stranded. (B) Parallel measurements on DNA constructs in which py-dC is intentionally mismatched opposite an A in the nontemplate strand. Double- and single-stranded controls, in the absence of protein, are shown. In the first set of experimental bars in each panel, addition of GTP and ATP allows walking to a stalled complex at position +10. Subsequent addition of 3'–dCTP walks the complex to position +11. Finally, addition of UTP drives translocation one base farther (without bond formation) to position +12. For the DNA constructs portrayed in the second set of bars in each panel, initial walking with GTP and ATP is to position +11, with 3'–dCTP providing translocation to +12. (C) The conclusion: RNA at position +2 dissociates from the hybrid on translocation from position +11 to +12. Adapted with permission from C. Liu and C. T. Martin, *J. Biol. Chem.* **277,** 2725 (2002).

[3] Single DNA Molecule Analysis of Transcription Complexes

By Claudio Rivetti and Martin Guthold

Recently, the high-resolution crystallographic structures of the bacterial RNAP core and holo enzymes,[1–3] together with the structure of a holo-enzyme-DNA complex[4] have been determined. Also, detailed structural models, based on DNA footprinting, protein-DNA, and protein-RNA cross-linking data, have been developed to predict the conformation of the open promoter and elongation complexes.[5–8] The structure of the yeast RNAP II elongation complex at 3.3 Å resolution has also been determined.[9] Both bacterial and eukaryotic RNA polymerases have a shape resembling a crab claw with the two pincers defining a central cleft where the active site is located. In elongation complexes, the transcription bubble and the surrounding DNA lie within the cleft with a consequently highly bent conformation of the DNA backbone. Although the nucleic acid structure is well defined within the active-center cleft of the Pol II enzyme, the electron density map of the upstream and downstream DNA is weak, and detailed DNA-Pol II interactions made in these regions are not resolved. The short DNA template that was used and possible crystal packing forces contribute to the lack of structural information about the DNA conformation in transcription complexes.[9]

New insights into the structure of the open promoter and elongation complexes also come from high-resolution microscopy studies. Both the electron microscope (EM) and the atomic force microscope (AFM) make it possible to look at the global conformation of protein-nucleic acid complexes and can provide information about the spatial relationships between

[1] G. Zhang, E. A. Campbell, L. Minakhin, C. Richter, K. Severinov, and S. A. Darst, *Cell* **98**, 811 (1999).

[2] D. G. Vassylyev, S. I. Sekine, O. Laptenko, J. Lee, M. N. Vassylyeva, S. Borukhov, and S. Yokoyama, *Nature* **417**, 712 (2002).

[3] K. S. Murakami, S. Masuda, and S. A. Darst, *Science* **296**, 1280 (2002).

[4] K. S. Murakami, S. Masuda, E. A. Campbell, O. Muzzin, and S. A. Darst, *Science* **296**, 1285 (2002).

[5] E. Nudler, *J. Mol. Biol.* **288**, 1 (1999).

[6] N. Korzheva, A. Mustaev, M. Kozlov, A. Malhotra, V. Nikiforov, A. Goldfarb, and S. A. Darst, *Science* **289**, 619 (2000).

[7] S. A. Darst, *Curr. Opin. Struct. Biol.* **11**, 155 (2001).

[8] N. Naryshkin, A. Revyakin, Y. Kim, V. Mekler, and R. H. Ebright, *Cell* **101**, 601 (2000).

[9] A. L. Gnatt, P. Cramer, J. Fu, D. A. Bushnell, and R. D. Kornberg, *Science* **292**, 1876 (2001).

protein, DNA, and RNA during transcription initiation and elongation.[10–13] In particular, the AFM offers a number of advantages over other types of microscopes. First, sample deposition in the AFM can be controlled and carried out in relatively mild conditions. Second, samples can be imaged without the aid of contrast agents at high or low humidity and in a variety of salt conditions. Third, DNA molecules of any arbitrary size can be used. Finally, as with other single molecule visualization methods, AFM studies yield not only the mean of a molecular parameter but also the overall distribution.[13–17]

This chapter describes procedures developed to assemble, image, and analyze initiation and elongation transcription complexes of both bacterial and eukaryotic RNA polymerases. In particular, we will describe methods to deposit transcription complexes onto a substrate, methods to measure the bend angle induced by the RNAP, methods to measure the contour length of DNA molecules and transcription complexes, and methods to identify single RNAP subunits within the complexes.

Methods

Surface Equilibration of DNA Molecules

One prerequisite of the conformational analysis of transcription complexes by AFM is the ability to transfer the complexes from solution to the two-dimensional surface without altering the spatial relationship between the RNAP and the DNA, thus allowing meaningful quantitative characterization of the images. Using DNA as a probe, it has been shown that the mean square end-to-end distance ($<R^2>$) of DNA molecules deposited onto a mica surface in a low salt buffer containing Mg^{+2} (e.g., 10 mM NaCl, 2 mM $MgCl_2$, 4 mM Hepes, pH 7.4) (Fig. 1A), is in very good agreement with that of worm-like chain polymers at equilibrium in two dimensions (Fig. 1B).[18] Consequently, molecules deposited under

[10] B. ten Heggeler-Bordier, W. Wahli, M. Adrian, A. Stasiak, and J. Dubochet, *EMBO J.* **11,** 667 (1992).

[11] W. A. Rees, R. W. Keller, J. P. Vesenka, G. Yang, and C. Bustamante, *Science* **260,** 1646 (1993).

[12] J. Bednar, V. M. Studitsky, S. A. Grigoryev, G. Felsenfeld, and C. L. Woodcock, *Mol. Cell* **4,** 377 (1999).

[13] C. Rivetti, M. Guthold, and C. Bustamante, *EMBO J.* **18,** 4464 (1999).

[14] C. Bustamante and C. Rivetti, *Ann. Rev. Biophys. Biomol. Struc.* **25,** 395 (1996).

[15] C. Bustamante, C. Rivetti, and D. J. Keller, *Current Opinion in Structural Biology* **7,** 709 (1997).

[16] C. Bustamante, D. Keller, and G. Yang, *Current Opinion in Structural Biology* **3,** 363 (1993).

[17] K. Rippe, M. Guthold, P. H. von Hippel, and C. Bustamante, *J. Mol. Biol.* **270,** 125 (1997).

[18] C. Rivetti, M. Guthold, and C. Bustamante, *J. Mol. Biol.* **264,** 919 (1996).

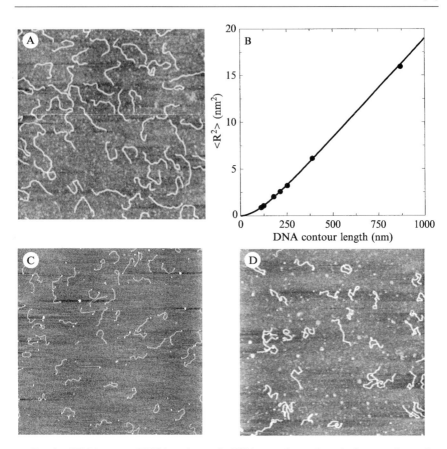

FIG. 1. AFM images of DNA and protein-DNA complexes deposited onto mica under different conditions. A: DNA molecules 1258 bp long deposited in deposition buffer as described in the text. Under these conditions, DNA equilibrates on the surface before adsorption as shown by the mean square end-to-end distance which is in good agreement with that predicted by the worm-like chain model. B: Plot of the $<R^2>$ values as a function of the DNA contour length. The symbols represent experimental data with DNA molecules of 350, 400, 565, 681, 825, 1258, and 2712 bp. The line represents the $<R^2>$ of worm-like chain polymers with a persistence length of 53 nm at equilibrium in two dimensions, as given by equation 9 in Rivetti et al.[18] C: DNA fragments 1048 bp long end-labeled with a Streptavidin-horseradish peroxidase fusion protein. These molecules have been used as a model for sample depositions involving protein-surface interactions. Analysis of their $<R^2>$ revealed that the end-bound proteins do not interfere with equilibration of the molecules on the surface. Conversely, harsh deposition procedures like glow-discharged mica (D) impede the DNA to freely move and equilibrate onto the surface. The molecules attain trapped configurations for which quantitative interpretations are difficult.

these conditions can equilibrate onto the substrate as in a two-dimensional solution. Similarly, DNA molecules with an end-bound protein can still equilibrate on the surface without interference by the protein (Fig. 1C). On the other hand, DNA deposited onto glow-discharged mica is trapped in non-equilibrium conformations by the strong DNA-surface interaction resulting from this mica treatment (Fig. 1D). Thus, glow-discharged mica substrates should be avoided for a reliable analysis of DNA-protein complexes. From this study, it has also emerged that during the deposition process, molecules are transferred from the solution to the surface solely by diffusion, and they cannot go back into solution once they are adsorbed on the surface. Thus, the number of molecule adsorbed onto the surface grows with the square root of time as shown by equation (1) given in Rivetti *et al.*[18]

Sample Preparation

For bacterial promoter complexes, good images can be obtained with DNA fragments of about 1000 bp and having the promoter located near the center of the molecule (at about 2/5 from one end). This arrangement will make it possible to distinguish between the upstream and downstream DNA in the images. To assemble and image RNAP elongation complexes, it is necessary to halt the RNAP in a particular position along the DNA template, and this can be obtained either by roadblock techniques[19] or nucleotide omission.[20] As in the case of promoter complexes, it is useful to position the stalling site near the center of the DNA fragment.

Initiation complexes of eukaryotic RNAP are more difficult to image by AFM because they require the concomitant assembly of several protein factors at the promoter region in order to favor RNAP binding. Conversely, stalled elongation complexes of eukaryotic RNAPs can be easily assembled by allowing the RNAP to initiate transcription from one DNA end with a 3' overhang[21] or a dC tail structure.[22] Clean DNA suitable for complex formation and AFM imaging can be prepared either by restriction digestion or PCR, followed by gel purification, electroelution, phenol extraction, and ethanol precipitation.

Open promoter complexes of bacterial RNAP are obtained by mixing 200 fmoles of promoter DNA and 200 fmoles of RNAP in 10 μl of transcription buffer A (20 mM Tris-HCl, pH 7.9, 50 mM KCl, 5 mM MgCl$_2$, 1 mM Dithiothreitol). The reaction is incubated for 15 minutes at 37°C.

[19] P. A. Pavco and D. A. Steege, *Nucleic Acids Res.* **19,** 4639 (1991).

[20] J. R. Levin, B. Krummel, and M. J. Chamberlin, *J. Mol. Biol.* **196,** 85 (1987).

[21] C. Bardeleben, G. A. Kassavetis, and E. P. Geiduschek, *J. Mol. Biol.* **235,** 1193 (1994).

[22] T. R. Kadesch and M. J. Chamberlin, *J. Biol. Chem.* **257,** 5286 (1982).

2 μl of the reaction are diluted into 18 μl of deposition buffer (4 mM Hepes, pH 7.4, 10 mM NaCl, 2 mM MgCl$_2$) and immediately deposited onto a freshly-cleaved ruby mica surface (Mica New York, NY). After about 2 minutes incubation, the mica is rinsed with 18-MΩ water and dried with a weak flux of nitrogen.

The sample must be imaged within a few hours from its preparation, since long storage can alter the DNA structure with consequent modification of its contour length. Good images can be obtained with a scan size of one or two microns and scan rates of 2-5 scans/s using an AFM operating with the tapping mode "in air."

Using the method of nucleotide omission, stalled elongation complexes of bacterial RNAP are prepared by mixing 200 fmoles of promoter DNA harboring an X-less cassette and 200 fmoles of RNAP in 10 μl of transcription buffer A containing 20 U ribonuclease inhibitor (RNasin–Promega). After a 15-minute incubation at 37 °C to facilitate open promoter complex formation, a mixture of three nucleotides, depending on the DNA sequence to be transcribed, is added to the reaction to a final concentration of 100 μM each. Transcription is carried out at room temperature for 20 minutes and the reaction is immediately used for AFM imaging as described previously.

Using a DNA template with the 3' overhang followed by a G-less cassette,[23] stalled elongation complexes of yeast RNAP III can be formed in a reaction mixture containing: 200 fmoles of DNA, 350 fmoles of RNAP III enzyme, 400 μM CpU dinucleotide, 200 μM ATP, 200 μM UTP, 200 μM CTP, 20 U ribonuclease inhibitor in transcription buffer B (40 mM Tris-HCl, pH 8, 100 mM KCl, 6 mM MgCl$_2$, 1 mM DDT). Transcription is carried out at 30 °C for 45 minutes, and the reaction is immediately used for AFM imaging as described previously.

DNA Contour Length Measurements

Determination of the DNA contour length from AFM images, in which molecules are described by a subset of pixels in a two-dimensional grid, is a nontrivial task. In fact, during the digitization process, the exact contour of the original molecule is lost, and only an approximation of it remains. Therefore, the contour length of DNA molecules can only be estimated rather than exactly determined.[24] The accuracy of such an estimate will depend on both image resolution and the method employed to compute the contour length from the string of pixels representing the molecule. While the resolution is an inherent property of the microscope or of the

[23] C. Rivetti, S. Codeluppi, G. Dieci, and C. Bustamante, *J. Mol. Biol.* **326,** 1413 (2003).
[24] L. Dorst and A. W. M. Smeulders, *Comput. Graphics Image Process.* **40,** 311 (1987).

microscope-sample system,[25] two operational steps of image processing can significantly affect the reliability of DNA contour length determination from AFM images. These are the correct identification of the subset of pixels that best describe the backbone of the DNA molecule in the image and the use of an algorithm capable of yielding the most accurate estimate of the DNA contour length.

Identification of a DNA Molecule. Several methods can be used to identify the subset of pixels that compose a DNA molecule. These methods can be completely automated, relying only on image segmentation procedures[26] or can involve user action for the digitization process.[27] Fully automated procedures are very efficient and can detect thousands of DNA molecules in minutes. They perform best when dealing with images of DNA molecules alone and with little noise. When imaging transcription complexes, the scenario is complicated by the presence in the image of several types of molecules: (i) specific complexes in which the RNAP is bound at the expected position along the DNA template; (ii) complexes in which the RNAP is non-specifically bound to the DNA; (iii) complexes with more then one RNAP bound to the same DNA template; and (iv) unbound DNA and RNAP molecules. These images are difficult to interpret by blind image processing routines. Procedures in which the molecules to be measured are selected by user action are usually preferred. Once a molecule has been identified, software can be used to detect the subset of pixels describing the DNA backbone.

Relying on user action, the trace of a DNA molecule can be digitized as follows: using dedicated or commercially available software, the two ends and several points along the DNA are selected with the mouse, and a first DNA trace is obtained by interpolating the selected points with steps of one pixel (Fig. 2A). The position of each point is then adjusted to that of the pixel with the highest intensity (in the case of AFM images, this corresponds to the DNA backbone) within a window five-pixels wide. Then, each digitized DNA molecule is skeletonized to an eight connected chaincode (Fig. 2B) in which the pixels of a molecule are contacted by no more than two neighbor pixels. The chaincode, connecting the center of each pixel, is defined based on the scheme given in Fig. 2B, inset. In the case of transcription complexes where the DNA contacting the RNAP is hidden by the protein, the DNA trace is drawn passing through the center of the protein. The position of the protein along the DNA template is defined by the user who

[25] C. Bustamante and D. Keller, *Physics Today* **48**, 32 (1995).

[26] A. Sanchez-Sevilla, J. Thimonier, M. Marilley, J. Rocca-Serra, and J. Barbet, *Ultramicroscopy* **92**, 151 (2002).

[27] C. Rivetti and S. Codeluppi, *Ultramicroscopy* **87**, 55 (2001).

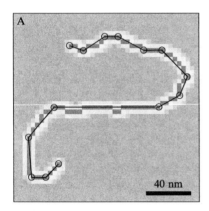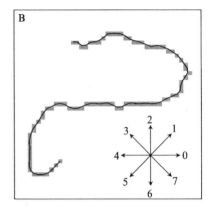

Fig. 2. Schematic representation of the digitization procedures of DNA molecules from AFM images. A: A DNA molecule is first localized by selecting several points along its contour from one end to the other (thin line with open circles). The approximated trace is then used to seek the DNA backbone represented by the pixels with the highest intensity within a window five pixels wide. The DNA backbone is that skeletonized to a eight-connected string of pixels shown in B, and the chaincode is obtained using the scheme shown in the inset. The DNA contour length can then be determined using the different estimators described in the text and with details in Rivetti and Codeluppi.[27] The thin line in B represents the smoothed DNA trace obtained by polynomial fitting of the pixel coordinates.

manually selects the center of the protein. The position of the RNAP, with respect to both ends of the DNA trace, is used for distinguishing specific and nonspecific complexes, as well as for the bend-angle measurements.

Computation of the DNA Contour Length. The subset of pixels and the corresponding chaincode are used to compute the contour length of the "free" DNA molecules and those that are in complex with a RNAP. The most frequently used algorithm to compute the contour length of objects from digital images is the so-called Freeman estimator, by which the contour length L is given by: $L = ne + 1.414no$, where ne and no are the number of even and odd codes respectively.[28]

Several other algorithms exist to determine the contour length of objects from digital images, and we have recently tested six of them for the determination of the DNA contour length from AFM images.[27] The results indicate that the Freeman estimator overestimates the contour length of a DNA molecule by about 4% and that better estimates, with errors of ~1%, are obtained with the Kulpa estimator ($L = 0.948ne + 1.343no$). The Kulpa estimator has an additional advantage that it can be easily

[28] H. Freeman, *in* "Picture Processing and Psychopictorics" (B. S. Lipkin and A. Rosenfeld, ed.), p. 241. Academic Press, New York, 1970.

implemented.[29] More accurate estimates, with errors of less than 0.4%, are obtained with the corner count estimator ($L = 0.980ne + 1.406no - 0.091nc$, where nc is the number of consecutive odd-even or even-odd sequences in the chaincode)[30] or with a routine that smooths the coordinates of the string of pixels by polynomial fitting of degree 3 over a moving window of 5 points.[27] In Fig. 2B, the skeletonized DNA trace is shown together with the smoothed trace represented by the thin line. The contour length is then calculated from the sum of the Euclidean distance between consecutive points in the smoothed trace.

DNA Bend-Angle Determination

The DNA bend angle of transcription complexes is defined as the deviation from linearity of the DNA helix induced by the RNAP. We describe here two methods to determine the DNA bend angle of transcription complexes by AFM (Fig. 3). With the first method, the bend angle is measured by drawing tangents along the two DNA arms in proximity of the protein. This method has the advantage to be simple, it can be applied to any fibrous molecule (even those that cannot be described by the wormlike chain model), and it gives the mean and the standard deviation of the bend-angle distribution.

A second method for bend-angle determination utilizes the mean-square, end-to-end distance, $<R^2>$, of an ensemble of complexes. It has been demonstrated that the $<R^2>$ of bent wormlike chain polymers depends, among other parameters, also on the DNA bend angle.[31] Accordingly, for bent DNA molecules in two dimensions, the DNA bend angle β is given by:

$$\cos\beta = \frac{\langle R^2\ \beta \rangle - 4PL + 8P^2(1 - e^{-\ell/2P}) + 8P^2(1 - e^{-(L-\ell)/2P})}{8P^2(1 - e^{-\ell/2P})(1 - e^{-(L - \ell)/2P})} \tag{1}$$

where L is the contour length of the polymer, ℓ and L-ℓ are the distances of the bend from the two DNA ends, and P is the DNA persistence length ($\sim$50 nm).

To achieve higher accuracy and to limit user bias, it is good practice to use both methods for bend-angle determination. In addition, AFM methods could be complemented by standard biochemical methods such as circular permutation, phasing, and cyclization analysis.[32]

[29] Z. Kulpa, *Comput. Graphics Image Process.* **6,** 434 (1977).
[30] A. M. Vossepoel and A. W. M. Smeulders, *Comput. Graphics Image Process.* **20,** 347 (1982).
[31] C. Rivetti, C. Walker, and C. Bustamante, *J. Mol. Biol.* **280,** 41 (1998).
[32] J. D. Kahn, E. Yun, and D. M. Crothers, *Nature* **368,** 163 (1994).

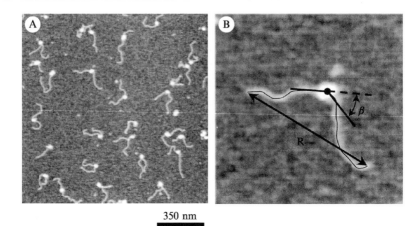

350 nm

FIG. 3. Image analysis of transcription complexes. A: Representative AFM image of open promoter complexes of *E. coli* RNAP. The DNA template 1054 bp long harbors a λ_{PR} promoter near the center of the fragment. From the image, it appears that several molecules are sharply bent at the protein site. B: A complex close-up showing the measurements that are performed on the images: β is the DNA bend angle measured with tangents. The thin line represents the smoothed DNA trace obtained as described in the text and in Fig. 2. R is the end-to-end distance corresponding to the Euclidean distance between the first and the last points of the smoothed DNA trace. The black dot represents the position of the RNAP along the DNA trace and is used to select specific complexes from those in which the RNAP is bound at positions other than the cognate site.

As mentioned, one of the advantages of bend-angle determination by high-resolution microscopy compared with other techniques is the possibility to obtain the whole distribution of bend angles. This is particularly important in those cases where more than one population of bent molecules are present in the sample under investigation. This was the case of promoter complexes between $\sigma^{54}-$ RNAP and the *glnA* promoter region where a two-modal distribution with peaks at $50 \pm 24°$ and $114 \pm 18°$ was determined by AFM.[17] The two-modal distribution was attributed to the concomitant presence of closed and open promoter complexes in the sample preparation. Bulk measurement methods would have failed to detect the two populations of complexes, and only an average bend angle would have been determined. Nonetheless, care needs to be taken when determining bend-angle distribution by high-resolution microscopies. Meaningful bend-angle measurements can only be obtained if the deposition process does not interfere with the conformation of the complex. In fact, kinetic trapping of the complexes[18] can distort the bend angle, thus causing altered means and standard deviations.

In cases where the induced bend angle is coplanar and the complexes are allowed to equilibrate on the surface without being trapped, the surface effect on the bend-angle distribution can be theoretically analyzed.[14] Assuming a bending energy of the complex in solution given by: $E = 1/2\,K < \beta >^2$ with fluctuations of $\Delta E = 1/2K(\beta - <\beta>)^2$ around this energy value, the distribution of bend angles in solution can be written as:

$$P(\beta) = \frac{K}{2\pi k_B T} e^{-K(\beta - <\beta>)^2 / 2k_B T} \tag{2}$$

where K is the constant of bending rigidity of the complex. The standard deviation of this distribution is given by:

$$\sigma_c = \frac{k_B T}{K} \tag{3}$$

On transferring these complexes from solution onto a surface, they will lose one degree of freedom. This can be accounted for by multiplying K by a factor of 2. Moreover, the complexes are now also interacting with the surface and thus another energy term, E^{surf} has to be added to the energy E of the bent complex. If E^{surf} is independent of the angle β (most likely a situation given the nonspecific, electrostatic nature of the complex-surface interaction), its inclusion amounts only to a multiplicative factor in the bending distribution. In this case, the surface energy will not affect the angular distribution.

In AFM images, additional broadening of the distribution is introduced by the fact that the DNA very close to the protein is hidden by the tip effect. This additional term is given by: $\sigma_b = \ell/P$, where ℓ is the length of hidden DNA and P the DNA persistence length. For transcription complexes, where about 20 nm of DNA are hidden by the RNAP and with $P = 53$ nm, $\sigma_b \approx 22°$. Thus, the total standard deviation is given by:

$$\sigma = \sigma_c + \sigma_b = \frac{k_B T}{K} + \frac{\ell}{P} \tag{4}$$

Using equation 4, it is possible to estimate the rigidity of the protein-DNA complex from the standard deviation of the bend-angle distribution.

We have used the tangent method and the $<R^2>$ method to determine the DNA bend angle of open promoter complexes at the λ_{PR} promoter.[13] When using the tangents method, bend-angle distributions were obtained that displayed a mean of about $55° \pm 55°$ for DNA fragments harboring one promoter (Fig. 4). When using the $<R^2>$ method, an angle of about $70°$ was obtained, while an angle of about $74°$ was determined from gel mobility experiments. These results are largely consistent with other data.

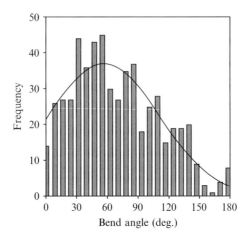

Fig. 4. DNA bend-angle distributions of *E. coli* RNAP open promoter complexes determined with the tangents method. The line represent the Gaussian fitting of the distribution.

Using neutron scattering[33] and quantitative electro-optics,[34] a DNA bend angle of about 45° was determined for open promoter complexes of the T7 promoter.[35] Two other AFM studies determined bend angles of 54° for λ_{PL} open promoter complexes[11] and 114° for the *glnA* promoter, which requires σ^{54} – RNAP.[17]

DNA Wrapping in Transcription Complexes

When imaging transcription complexes by AFM, the large size of the RNAP and the tip broadening effect hides the DNA in contact with or in the vicinity of the protein. Nevertheless, it is possible to estimate the amount of DNA interacting with the RNAP by measuring its contour length and comparing the value with that of unbound DNA molecules. Extended interactions of the DNA over the protein surface produce a compaction of the DNA and consequently, the overall contour length of the complex will be less than the contour length of unbound DNA molecules (Fig. 5). To reduce experimental variability, several hundred molecules must be measured. Ideally, complexes and free DNA molecules are measured from the same set of images. Furthermore, because the DNA contour length

[33] H. Heumann, M. Ricchetti, and W. Werel, *EMBO J.* **7,** 4379 (1988).
[34] F. J. Meyer-Alme, H. Heumann, and D. Porschke, *J. Mol. Biol.* **236,** 1 (1994).
[35] R. C. Williams and M. J. Chamberlin, *Proc. Natl. Acad. Sci. USA* **74,** 3740 (1977).

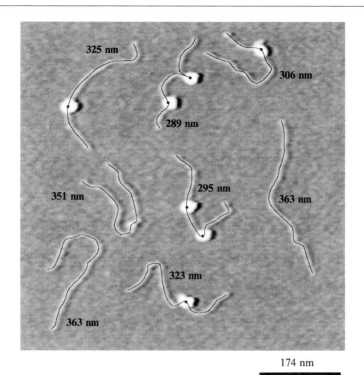

174 nm

FIG. 5. AFM image of *E. coli* RNAP open promoter complexes. The DNA template used to assemble the complexes is 1150 bp long and harbors two λ_{PR} promoters separated by 298 bp. In the image are present complexes with both promoters occupied by one RNAP, complexes in which only one promoter is bound by a RNAP and unbound DNA molecules. The thin line represents the smoothed DNA trace obtained as described in the text and in Fig. 2. The black dots represent the position of the RNAP along the DNA trace. The DNA contour length, written near each molecule, shows that the DNA is compacted by the binding of one or two RNAPs.

may be altered during sample deposition, conditions that facilitate equilibration of the molecules onto the surface must be used. Fig. 6 shows the results of a series of experiments with transcription complexes in which the DNA compaction produced by RNAP binding has been analyzed. Open promoter complexes of bacterial RNAP produce a DNA compaction of about 30 nm (Fig. 6A and B) and when two OPC are assembled on the same DNA template, the DNA compaction doubles (Fig. 6B).[13] These results have been confirmed by protein-DNA photocrosslinking experiments showing an extended interaction between the upstream DNA region (up to position − 90) and the αCTD of the *E. coli* RNAP.[8]

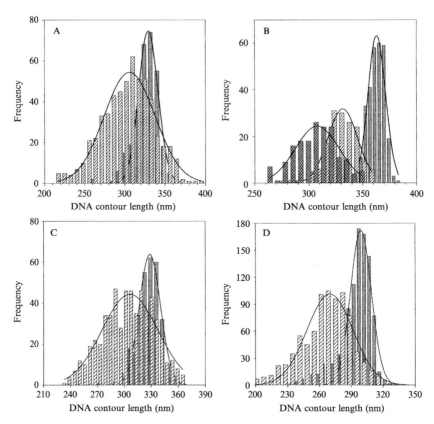

FIG. 6. Comparison between the contour length of free DNA molecules (filled gray bars) and that of transcription complexes (hatched bars). A: Open promoter complexes of *E. coli* RNAP assembled with a 1054 bp long DNA fragment. The DNA compaction observed in this case is of 28 nm. B: Open promoter complexes of *E. coli* RNAP assembled with a DNA fragment 1150 bp long containing two promoters. The DNA compaction is of 31 nm for complexes with one RPo (hatched bars) and 55 nm for complexes with two RPos (crossed-hatched bars). C: Stalled elongation complexes of *E. coli* RNAP on a 1054 bp fragment. In this case, the DNA compaction is of 22 nm less compared to that observed with the corresponding RPo. D: Stalled elongation complexes of yeast RNAP III showing a DNA compaction of 31 nm.

DNA compaction is also observed in the case of elongation complexes of bacterial RNAP (Fig. 6C) and yeast RNAP III (Fig. 6D)[23]. Based on the dimensions of the RNAP, on the different complexes we have analyzed, and on the variability of the measurements, caution must be taken in drawing conclusions when the DNA contour length reduction is less than 10 nm.

Tagging a specific RNA polymerase subunit

For many years, microscopists have developed methods to label target proteins with antibodies or functionalized beads for easy identification under the microscope. However, even though antibodies are available for the α^{36} and β'^{37} subunits, this method has never been applied to identify the location of the RNAP subunits. Although this technique could readily be applied to AFM imaging of transcription complexes and protein-DNA complexes in general, the large size of the label, relative to the size of the protein target, often masks the region of interest. Additionally, when the target is a particular protein subunit, the globular shape of the label prevents precise localization of the subunit within the complex. In these cases, small sharp tags that can be clearly visible under the microscope should be preferred.

Here we describe an approach to label RNAP subunits with a small DNA tag, which is easily discernible in the AFM images. The tag is made of a 200 bp DNA fragment with a Streptavidin molecule (MW 47 kDa) attached to one biotinylated DNA end. Because Streptavidin is a tetramer that can bind up to four biotin molecules, the DNA-Streptavidin tag can be used to label biotinylated protein subunits (Fig. 7A). DNA tags are small and their presence has little or no effect on the overall conformation of the RNAP alone, or in complexes with DNA. In addition, the linear shape of the DNA tag acts as an arrow pointing in the direction of the labeled subunit. Consequently, the interpretation and the quantitative data analysis of the images is straightforward. Fig. 7B-D shows a close-up of an open promoter complex of *E. coli* RNA, in which the biotinylated β' subunit was labeled with a 196 bp DNA-Streptavidin tag (Guthold *et al.*, in preparation). Quantitative analysis of the images to localize the labeled subunit within the complexes can be obtained by measuring the angle formed by the DNA tag and the upstream or downstream arm of the DNA template.

Preparation of RNAP with a Biotinylated β' Subunit. To increase the specificity of the target subunit, an *in vivo* biotinylation method was used to attach a biotin at the CTD of the RNAP β' subunit (the protein was obtained from R. Landick, University of Wisconsin, Madison, Wisconsin). The 85 amino acid biotinylation signal was inserted at the Carboxyl-terminal of the β' gene (plasmid pRL625, R. Landick, University of Wisconsin, Madison, Wisconsin). On expression in *E. coli*, (strain JM109) residue 51 (Lys) of the biotinylation signal was post-translationally biotinylated by

[36] F. Riftina, E. DeFalco, and J. S. Krakow, *Biochemistry* **28**, 3299 (1989).
[37] J. Luo and J. S. Krakow, *J. Biol. Chem.* **25**, 18175 (1992).

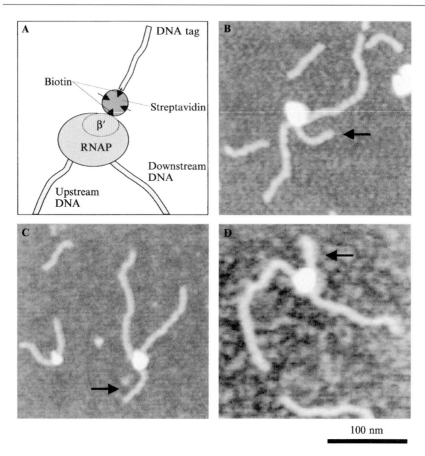

FIG. 7. Labeling the β' RNAP subunit with Streptavidin-DNA tags. A: Schematic representation of a tagged transcription complex. The Streptavidin tetramer joins the biotinylated DNA tag to the biotinylated β' subunit. B to D: Close-up of labeled open promoter complexes. The white globular feature is the RNAP from which emerge the DNA template and the DNA tags. The shorter DNA fragment pointed by the arrow is the tag; the intermediate and longer DNA fragments are the upstream and downstream DNA arms, respectively. Also visible in the images are free DNA tags.

the bacterium apparatus.[38] The holoenzyme was then purified following standard procedures. The labeling efficiency was about 40% as estimated from the images. Commercially available biotinylation kits are also available.

[38] J. E. Cronan, *J. Biol. Chem.* **265,** 10327 (1990).

Preparation of Biotinylated DNA Tag. To label RNAP subunits within a transcription complex, the DNA tag must be long enough to be seen under the microscope (more then 100 bp) and significantly different from the length of either upstream or downstream arms to avoid misinterpretation of the complexes. For the open promoter complexes we are reporting here, with a 439 bp upstream DNA and 615 bp downstream DNA, a 196 bp DNA tag was chosen.

The biotinylated DNA tag is obtained by PCR reaction from a random sequence DNA using one 5'-biotinylated oligonucleotide. The amplified DNA is passed through a Microcon 50 filter several times to clear the solution from free primers and nucleotides. This step is crucial since the presence of free biotin or biotinylated primers will compete with the biotinylated DNA in the binding of Streptavidin. The filtered DNA is phenol/chloroform extracted, ethanol-precipitated, and the pellet is resuspended in Tris-EDTA buffer.

Streptavidin-DNA fusion reaction is carried out with an excess of protein to avoid multiple DNA binding to a single Streptavidin molecule. In a 100 μl volume, the reaction is as follows: 800 nM Streptavidin (Gibco, Grand Island, NY), 33 nM DNA, 150 mM NaCl, 10 mM sodium phosphate, pH 7.2. The reaction is incubated for 1 hour at room temperature. Free Streptavidin is separated from Streptavidin-DNA fusions by several filtrations with a Microcon-100 which retains the latter. Also, this second filtration is crucial, since free Streptavidin will compete with the tags in binding the biotinylated RNAP subunit. The efficiency of the reaction can be measured by a band-shift assay in agarose gel electrophoresis. Streptavidin-DNA tags can be stored for several days at 4°C.

Labeling RNAP Open Promoter Complexes with DNA Tags. Open promoter complexes are obtained by mixing 200 fmoles of DNA template harboring a λ_{PR} promoter with 200 fmoles of β'-biotinylated RNAP from *E. coli* in 10 μl of transcription buffer A. The reaction is incubated for 15 minutes at 37°C to facilitate open complex formation. Then, 600 fmoles of Streptavidin-DNA tag are added and the reaction is incubated for 40 minutes at room temperature. The reaction is immediately used for AFM imaging as described above.

Concluding Remarks

This chapter describes basic methods for the conformational analysis of transcription complexes by means of AFM. Buffer conditions for the equilibration of DNA and protein-DNA complexes onto mica surface are first presented. Equilibration, as opposed to trapping onto the surface, is an essential prerequisite for a meaningful analysis, since only the structure of equilibrated molecules can be reliably related to their three-dimensional solution structure. Next are presented methods to analyze the contour length of DNA molecules. These are non-trivial image processing procedures that can significantly affect AFM measurements. As an example, it has been shown how a careful analysis of the DNA contour length of transcription complexes can give information on the extent of RNAP-DNA interaction. The determination of protein-induced DNA bend angles, using either tangents or the mean-square end-to-end distance, is then described. Finally, a novel DNA-Streptavidin tag method for the labeling of individual RNAP subunits is presented. Even though the focus is on transcription, the methodology described here is of general practical use for imaging protein-DNA complexes by high-resolution microscopy.

Acknowledgments

We are grateful to Carlos Bustamante for his wonderful support, advice, and friendship. Financial support from the Italian Ministry of University and of Scientific and Technological Research (Cofin 2001) is gratefully acknowledged. Martin Guthold was supported by an American Cancer Society grant (IRG-93-035-6), by an award from Research Corporation, Wake Forest University Start-up funds, and a Science Research Fund. We would like to thank R. Landick for providing us with the β'-biotinated RNA polymerase.

[4] Assay for Movements of RNA Polymerase along DNA

By Nobuo Shimamoto

As an inevitable consequence of successive reading of the DNA sequence, all DNA-dependent RNA polymerases must move along DNA during transcription. The movement is named *translocation* and accompanied by the synthesis of RNA. In addition to this mode of movement, the enzyme could move along DNA as thermal motions without RNA synthesis[1] as several other DNA-binding proteins.[2] By the latter movement, such a DNA-binding protein reaches its specific site, a promoter in the case of RNA polymerase. The movement is generically called *one-dimensional diffusion*, or simply *sliding*.[3] After binding to a promoter, RNA polymerase starts synthesizing and releasing short RNA typically shorter than 12 base,

---

[1] H. Kabata, O. Kurosawa, I. Arai, M. Washizu, S. A. Margarson, R. E. Glass, and N. Shimamoto, *Science* **262,** 1561–1563 (1993).

[2] P. H. von Hippel and O. G. Berg, *J. Biol. Chem.* **264,** 675–678 (1989).

[3] N. Shimamoto, *J. Biol. Chem.* **274,** 15293–15296 (1999).

namely abortive products,[4] without much movement of the front end of the enzyme along DNA.[5] The entire complex in this period is conventionally supposed to become elongation complex[6] after a step named *promoter clearance*.[7] However, at least at several promoters including λp_R, *lac*UV5, and *mal*T promoters, the complex is composed of two fractions: one quickly becomes elongation complex, while the other only directs abortive synthesis and the inactivated with an arrest at the promoter.[8,9] The fractions and reversibility between them depend on promoters and could be regulated by trans-acting factors such as GreA/B[10] and CRP.[11] This new notion predicts that some fraction of RNA polymerase could be arrested at the promoter, and some transcription factors could work as activator or inhibitor of initiation by changing the fraction.

There is a more restricted definition of sliding: the movement along DNA continuously keeping contacts with DNA (Fig. 1A). The one-dimensional diffusion is composed of the sliding in this restricted sense and the inter-segment transfer,[2] which is defined as iterative contacts with DNA through two or more interacting surfaces of a protein molecule (Fig. 1B). The inter-segment transfer has been proved only for the protein that has two separated domains for recognizing specific sequences such as

A. Sliding in restricted sense

semi-bound state

B. Inter-segment transfer

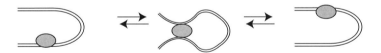

FIG. 1. Translocation of DNA binding protein by one-dimensional diffusion along DNA.

[4] D. E. Johnston and W. R. McClure, *in* "RNA polymerase," pp. 413–428. Cold Spring Harbor Laboratory Press, Cold Spring Harbor, 1976.

[5] W. Metzger, P. Schickor, and H. Heumann, *Embo J.* **8,** 2745–2754 (1989).

[6] A. J. Carpousis and J. D. Gralla, *Biochemistry* **19,** 3245–3253 (1980).

[7] W. Kammerer, U. Deuschle, R. Gentz, and H. Bujard, *Embo J.* **5,** 2995–3000 (1986).

[8] T. Kubori and N. Shimamoto, *J. Mol. Biol.* **256,** 449–457 (1996).

[9] R. Sen, H. Nagai, and N. Shimamoto, *J. Biol. Chem.* **275,** 10899–10904 (2000).

[10] R. Sen, H. Nagai, and N. Shimamoto, *Gene. Cells.* **6,** 389–402 (2001).

[11] H. Tagami and H. Aiba, *Embo J.* **17,** 1759–1767 (1998).

LacI but not for the protein with a single domain for DNA binding. For this reason, the term *sliding* is used here in the broader sense, one-dimensional diffusion. One of the roles of sliding is to accelerate the search of the specific site of a protein, a promoter in the case of RNA polymerase, by compressing the space to diffuse.[2,3] Therefore, a sliding protein molecule should move smoothly without large friction with DNA; otherwise the trial-and-error search for a specific site by repeated association and dissociation would slow down the searching process. In fact, a sliding RNA polymerase molecule is dragged by a bulk flow with a velocity close to the flow.[1]

Assay for Sliding by Single-Molecule Dynamics

The most direct assay of a movement of a protein is carried out by visualization of individual molecules under an optical microscope, which is called *single-molecule dynamics*. When a protein molecule is injected at an angle to DNA that is extended and fixed in parallel at an enough concentration, sliding of the molecule is detected as its trace parallel to DNA, which is distinct from drift by bulk flow (Fig. 2A). Individual molecules in aqueous solution are visualized by fluorescent labeling. A single-molecule experiment does not necessarily mean a single fluorophore experiment, although a fusion protein with the green fluorescent protein (GFP) is commonly used as a single-fluorophore labeling. Labeling with

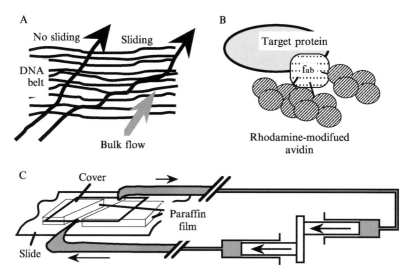

FIG. 2. Detection of sliding motion. (A) Distinction of sliding. (B) Fluorescent labeling of single molecules. (C) The observation chamber and liquid delivery system.

single fluorophore molecule is very sensitive to photobleaching and the lifetime of a single fluorophore is only several seconds. Therefore, labeling of single molecules with several more than ten fluorophores is required for observation for a time longer than several seconds. For this purpose, avidin or streptavidin conjugated with several amine-modifying fluorescent residues or a fluorescent bead is generally used. Theoretically, up to 40 fluorophores can be introduced to a tetrameric avidin molecule. RNA polymerase is conjugated with the labeled avidin by biotinylated Fab fragment of IgG that was raised against a peptide of the C-terminus of β' subunit[1](Fig. 2B). As an alternative to fluorescent labeling, single molecules are directly observed by atomic force microscopy (AFM), if they are adsorbed on a surface. *Escherichia coli* RNA polymerase and DNA molecules adsorbed on a surface carry out a slow-motion version of the movements in solution, including sliding,[12] but this technique is not within the scope of this chapter.

Multiple Introduction of Tetramethylrhodamine Into Avidin

The following method for introduction is a combination of the original[13] with a highly efficient method for refolding denatured protein,[14] and can be applied to other amine-directed fluorescent reagents.

1. Dissolve 1 mg avidin (or streptavidin) in 600 μl of B buffer (10 mM Bicine-HCl [pH 8] and 10 mM NaCl).

2. Dialyze the solution against more than 50 volumes of B buffer at 4° for more than 2 h to remove trace of primary amines such as AS (ammonium sulfate) included in the lot of avidin. Change the buffer at least once.

3. Adjust the concentration of avidin to 1 mg/ml by adding the buffer. The concentration is determined by the $E_{1\%}$ of 15.4.[15] Transfer an exact volume (700–900 μl) of the solution into a 1.5-ml sample tube. Keep the tube at room temperature for Step 6.

4. Take 5 mg of tetramethylrhodamine isothiocyanate isomer R (Sigma) in a 0.5-ml sample tube and add 200 μl of acetone. Close the lid and vibrate the tube for 10 mm to make a saturated solution at room temperature.

[12] M. Guthold, X. Zhu, C. Rivetti, G. Yang, N. H. Thomson, S. Kasas, H. G. Hansma, B. Smith, P. K. Hansma, and C. Bustamante, *Biophys. J.* **77,** 2284–2294 (1999).

[13] N. Shimamoto, H. Kabata, O. Kurosawa, and M. Washizu, *in* "Structural tools for the analysis of protein-nucleic acid complexes" (D. Lilley, H. Heumann, and D. Suck, eds.), pp. 241–253. (Birkhäuser Verlag AG, Basel, 1992).

[14] N. Shimamoto, T. Kasciukovich, H. Nagai, and R. S. Hayward, *Tech. Tips Online.* t01576 (1998).

[15] N. M. Green, *Biochem. J.* **89,** 585–591 (1963).

5. Centrifuge the tube at more than 10,000 rpm for 1 min and add one tenth of the volume of the saturated solution to the avidin solution at room temperature. Vortex the mixture gently for 1 h at room temperature. To take an exact volume of the acetone solution, you have to repeat aspirating and blowing three times to fill the cavity of the pipetting tool with acetone vapor.

6. Add the same volume of the saturated solution to the mixture, and gently vortex the mixture overnight at room temperature.

7. Precipitate the modified avidin by adding acetone to 50%, and centrifuge it. Wash the precipitate with acetone and repeat the washing until no color is recognized when the supernatant is diluted with water.

8. Dry the precipitate by opening the lid for 10 min at 37°. Suspend the precipitate in 0.5 ml of 6M guanidium hydrochloride. Add 0.63 g (0.5 ml) of glycerol.

9. Put the suspension with 0.3 ml of Toyoperl Phenyl-650S (Tosoh) or Phenyl Sepharose (Amersham Biosciences) resin that has been saturated with 75G buffer, which is 3:1 mixture of glycerol and TE (10 mM Tris-HCl pH 7.5 and 1 mM EDTA) into a dialysis bag. Dialyze the mixture against 75G buffer for overnight at 4°. This dialysis against the 75% glycerol solution can be carried out by rotating a 50-ml disposable tube containing the dialysis bag. Make sure that the content in the dialysis bag is mixed by the rotation.

10. Dialyze the bag against 75G buffer saturated with ammonium sulfate for 1 h. At every 1 h, change the solution to 50G, 25G buffer (1:1 and 1:3 mixture of glycerol and TE, respectively) saturated ammonium sulfate, and finally to 2M ammonium sulfate in TGE buffer (10 mM Tris-HCl pH 7.5, 1 mM EDTA, and 5% glycerol).

11. Take out the resin and pack it in a small empty column or an empty spin column. The column is either installed in a liquid chromatography system or eluted stepwise manner by centrifugal force at different AS concentrations. Unmodified avidin passes through the column at 2M ammonium sulfate, but modified protein is eluted at less than 1M ammonium sulfate according to the degree of modification. The heavily labeled protein should be eluted in the first two peaks; a linear gradient of ammonium sulfate is applied.

12. The eluate is centrifuged at more than 10,000 × g for 1 h at 4° for removing aggregates. The modified protein can be stored for years in S buffer (50% glycerol, 50 mM Tris-HCl [pH7.8, and 0.1 M KCl]) at −20°.

13. The protein content is determined by SDS PAGE and silver staining, and its rhodamine content by spectroscopy. If the staining of the protein and the maximum absorption of the chromophore at 500–560 nm were not changed by the modification, the average number of fluorophores was calculated to be 5–10 per avidin monomer for fractions eluted at

lowest salt concentrations.[13] At least half of this fraction bound to the biotin-agarose, but we did not determine whether the modified avidin keeps the intact tetrameric conformation or not.

Preparation of Fluorescently Labeled RNA Polymerase

The casein used in the following sections is partially hydrolyzed in 1 N sodium hydroxide at 55° for 12 h to maximize the anti-adsorption activity and to remove RNase activity. The hydrolysate is neutralized and dialyzed against water containing 0.05% diethylpyrocarbonate, followed by removing precipitate by centrifugation.

1. Mix 12.5 μg anti-β' IgG with 630 ng papain (Sigma) in 50 μl PBS containing 20 mM cysteine and 20 mM EDTA at 37° typically for 1–5 h.[16] The reaction time must be determined by a pilot experiment of 1/25 scale followed by SDS PAGE with quick reverse staining.[17] Digestion is stopped by adding 50 mM iodoacetoamide (Sigma).

2. The digested mixture is injected into a Superdex 75 column of SMART System and fractionated with B buffer. Fractions are analyzed by SDS PAGE with reverse staining.

3. The fractions containing fab are combined ($\sim$0.2 ml) and 2 μg NHS-biotin (Pierce) is added on ice and incubated for 2 h.[18] The reaction is stopped by adding one-tenth volume of 1M Tris-HCl (pH 7.5) buffer. A maximum of two biotin residues are introduced per a Fab molecule.

4. The unreacted biotin is removed either by the Superdex column or dialysis, and then concentrated to ca. 50 μl by a Speed Vac.

5. Dialyze RNA polymerase against TGE Buffer containing 0.1 M KCl to remove DTT.

6. Titrate the Fab fragment with the RNA polymerase by using ELISA to determine the stoichiometry.

7. The Fab is incubated with a stoichiometric amount of RNA polymerase at 4° for 30 min and then 20-fold excess of the labeled avidin for further 30 min. This two-step addition is essential to prevent RNA polymerase from oligomerization through avidin and biotinylated fab.

8. Prepare a 2-ml tube for ultracentrifugation and 10%, 30%, and 60% glycerol in TGE buffer containing 50 mM KCl and 0.1 mg/ml of the casein.

---

[16] J. E. Coligan, A. M. Kruisbeek, D. H. Marglies, E. M. Shevach, and W. Strober, *in* "Current protocols in immunology," pp. 2.0.3–2.8.10. Greene Pub. Associates and Wiley-Interscience, New York, 1991.

[17] M. L. Ortiz, M. Calero, C. Fernandez Patron, C. F. Patron, L. Castellanos, and E. Mendez, *FEBS Lett.* **296,** 300–304 (1992).

[18] M. D. Savage, G. Mattson, S. Desai, G. W. Nielander, S. Morgensen, and E. J. Conklin, "Avidin-biotin chemistry: a handbook." Pierce, Rockford, 1992.

Put 0.1 ml of the 60% solution in a tube and then overlay a gradient of 30%–10% in total 1.8 ml by using a mixing device. Load the mixture onto the gradient and centrifuge for 4 h at 4° at 150,000 × g.

9. Collect zones from the bottom and detect RNA polymerase by SDS PAGE with silver staining, Fab by ELIZA with anti-rabbit antibody, and the labeled avidin by fluorometry. The unbound-Fab remains at the top, and the unbound and labeled avidin is collected at 14%; the unbound holoenzyme at 23%; and the ternary complex of RNA polymerase, Fab, and avidin sediments at the boundary of 30% and 60%. This fraction is stored at −20° in S buffer containing 0.01 mM DTT, but dissociation of Fab resulting from reduction of cystine by DTT has not been examined.

Fixing of Extended DNA in Parallel

DNA can be extended with a flow or electric field. Since the assay of sliding includes the binary reaction of binding, the fixed DNA should be sufficiently concentrated. If there is only a single molecule of λDNA in a typical vision of a microscope (0.1 mm × 0.1 mm × 10 μm), a protein molecule has to wait for a day to bind to it. The only available technique to extend and fix DNA at a concentration of the order of 100 μg/ml is dielectrophoresis,[19] which is electrophoresis of dielectric, of a set of DNA, and the accompanying counter ions. A long DNA such as lambda is stretched and positioned in parallel at an electrode in the alternating electric field of 4 × 10^6 V/m at 1 MHz, forming a pair of DNA belts parallel to the electrodes. The DNA is fixed on the surface of a glass slide by adsorption of avidin molecules attached to the biotinylated ends of DNA. The fixing of DNA at a local concentration of more than 300 μg/ml required a pretreatment of the glass slide such as coating with a silane.[20,21] The fixing of DNA lower than 50 μg/ml or the level of single molecules does not require any pretreatment of both glass slide and DNA, and an unknown reaction between the end of DNA and alminum electrode fixes the DNA.[22]

The electrodes on a glass slide (Fig. 3A) are formed either by vapor deposition of aluminum with a metal mask of the electrode pattern or photo-lithography of an aluminum-coated slide. The former provides the maximum amount of fixed DNA putatively resulting from sharper edges of the electrodes, and the latter offers the ability to try different shapes

[19] M. Washizu, O. Kurosawa, I. Arai, S. Suzuki, and N. Shimamoto, *IEEE Trans. on Indst. Appl.* **31,** 447–456 (1995).
[20] E. M. Southern, U. Maskos, and K. Elder, *Genomics* **13,** 1008–1017 (1992).
[21] U. Maskos and E. M. Southern, *Nucleic Acid Res.* **20,** 1679–1684 (1992).
[22] T. Yamamoto, O. Kurosawa, H. Kabata, N. Shimamoto, and M. Washizu, *IEEE Trans. on Indst. Appl.* **36,** 1010–1017 (2000).

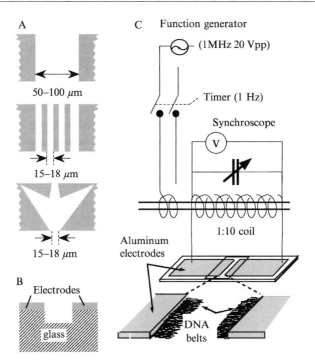

Fig. 3. Preparation of DNA belts by dielectrophoresis. (A) Various shapes of electrodes for T7 DNA (15 μm) or λDNA (18 μm). A simple gap type (*the top*), a floating potential type (*the middle*), and a variable gap type (*the bottom*). (B) A vertical section of etched electrode to prevent adsorption of DNA on the surface. (C) Connection of the instruments used.

of electrodes. For vapor deposition, the most accessible instrument in a biological department is an evaporator used for electron microscopy (JEOL JEE-400, for example). Various patterns of electrodes other than a simple gap of 50–100 μm are available. Floating potential type is convenient for detection of single DNA molecule, and variable gap one is good for the first trial to fix DNA (Fig. 3A). For prevention of post-fixing adsorption of DNA onto the surface at a magnesium concentration higher than 1 mM, the glass surface is etched by hydrofluoric acid (Fig. 3B).

Because a strong electric field is applied to the DNA solution, the solution should be free from salt. In the absence of salt, not much power is required for dielectrophoresis and a function generator combined with a pull-up coil is satisfactory (Fig. 3C). The application of the field is pulsed at every second to distinguish the fixed and unfixed DNA. The pulsed application of the field facilitates the diffusion of DNA into the unfixed area and enables the real-time monitoring of fixing under fluorescent microscope. For real-time monitoring, DNA is stained with ethidium bromide that can be removed by washing after fixing.

In the assay of sliding, protein molecules are driven into the DNA belt by bulk flow in an observation chamber with spacers of Parafilm (American National Can). For mechanical control of the flow, a pair of injecting and ejecting pumps is required for the flow rate of 5 μl/h. Two 5-μl microsyringes can be fixed to a dual syringe pump (such as Harvard) with minor modifications. Each needle is connected to polyethylene tubing that has a sharp tip. Both tips must contact the buffer between the slide and the cover from opposite sides (Fig. 2C). Manual injection by a pipetting tool is also available, but the speed and the direction of flow are hard to control. Furthermore, manual mixing eventually makes a flow parallel to DNA, which hampers distinction of sliding from simple drift.

The fluorescently labeled molecules are observed with a 100 × objective lens with a vision size of 50 μm × 50 μm. By using a silicon-intensified target video camera (Hamamatsu Photonics C2400-08) or an equally sensitive CCD camera, their traces can be recorded on videotape or DVD.

For a protein molecule to collide with DNA, its residence time in the DNA belt should be long. In the case of a protein that has the association rate constant of 10^5 M^{-1} sec^{-1}, such as RNA polymerase, the average time cost in binding is 0.6 sec in the DNA belt of 100 μg/ml, which requires a flow rate slower than 30 μm/sec. At a flow rate less than 0.5 μm/sec, the sliding movement cannot be distinguished from Brownian motion of free molecules. Therefore, the flow rate of 1–10 μm/sec is used for the assay.

Preparation of a Slide with Electrodes

1. Prepare clean glass slides or silane-treated glass slides,[20,21] and clip masks with electrode patterns with them in case of direct patterning. Put 4–8 pairs of them at an equal distance from the tungsten boat in a bell jar of an evaporator. Put 1–3 cm of aluminum wire of 1 mmD on the boat. The wire length determines the thickness.

2. Run the vacuum pump and heat the boat when the vacuum reach 10^{-4} Pa. The best thickness of aluminum layer is 3–5 μm.

3. For photo-lithography of an electrode pattern, a glass slide is set on a spin coater (Mikasa 1H-D3) and covered by posiresist (Shipley S1400-25). The glass is spun at 500 rpm for 5 sec and at 4000 rpm for 40 sec, and dried at 110° for 50 sec.

4. The slide is exposed to a pattern for 15 sec with a light source unit for fluorescent microscopy with a 50-W high-pressured mercury lamp, and then dipped in the developer (Shipley MF319). The slide is washed with

distilled water, heated at $140°$ for 5 min, etched for 20 min in 6:1 mixture of 40% NH_4F and 49% HF, and then washed with water and acetone.[22]

Preparation of DNA Belts by Dielectrophoresis

1. Dialyze DNA against water to remove trace amounts of salt, and make up a solution at 2 μg/ml. Centrifuge the solution for more than 1 h at 13,000 $\times$ g at $4°$ to remove aggregated DNA.

2. Put a drop of the solution at the center of electrodes. If real-time observation is required, mix it *in situ* with an equal volume of 5 μg/ml ethidium bromide solution, which has been centrifuged or filtered to remove aggregates, and put a coverslip on the mixture. If real-time monitoring is not necessary, there is no need for a coverslip.

3. Apply a sinusoidal electric field and monitor the output level. Adjust the variable capacitor and the attenuator of the function generator to get enough extension. Typically apply 200 Vp-p for 100 μm total gap and increase the voltage 20%–50% when the electrodes gap is inhomogeneous.

4. In real-time monitoring, the DNA synchronizing with the pulse is not yet fixed but the DNA belt with no change of fluorescence is completed. The fixing will be completed in a minute when a good electrode is formed on a clean or pretreated glass.

5. Turn off the pulse generator and remove the slide. Wash the electrode 10 times with the buffer used in transcription to remove unbound DNA and ethidium bromide. Put a coverslip on the slide and keep the slide wet in a sealed box with water until the assay of sliding.

Assay and Analysis of Sliding Motion

1. The fluorescently labeled protein molecules are injected mechanically or manually so that the injection causes a flow at an angle to DNA. In the case of manual injection, inject 5 μl with the tip touching the solution. The cover may not be fixed, because fixing causes less chance for the flow to have a speed suitable for the observation.

2. Record all the traces on focus. Usually the thickness of the DNA belt is close to the focusing depth of 2 μm, so that only the traces of molecules on focus are the target of the analysis.

3. An obtained trace in a DNA belt is classified into the following categories in this order:

> This is simply an example of the criteria of motions.
> Sliding: translocation longer than 3 μm in parallel with DNA within $\pm10°$.
> Vertical jumping: vertical movement from on focus to off focus or the opposite.

Trapping: translocation shorter than 2 μm for more than 10 sec.

Simple drift: translocation longer than 3 μm in parallel with bulk flow within $\pm 30°$.

Unidentified: others including translocation in the case where bulk flow is parallel with DNA.

4. Collect more than 100 samples and make a histogram.

Detection of Rotational Motion and its Application for Groove Tracking During Sliding[23]

Because of the helical structure of double-stranded DNA, a translational motion of a protein molecule along DNA could accompany its rotational motion relative to DNA. In the case of elongation of transcription this rotational motion is inevitable. During sliding, however, it is not inevitable and there could be two major models for sliding of a protein molecule along DNA. In one model, the sliding complex keeps most the interaction between the protein and DNA that is found in a specific complex. Therefore, the relative distance between DNA and the protein surface is equal to or a little larger than that of the specific complex, and thus the protein molecule tracks a groove of DNA. Therefore, sliding is a helical motion along DNA. In an alternative model, the charges of the phosphates and the counter ions make a tunnel with an atmosphere that traps a protein by electrostatic interaction and by breaking the structure of the surrounding water clusters. Sliding in this model is a Brownian motion confined in the tunnel, and thus the protein molecule will move along the axis of helix without tracking a groove. The first model may have a higher efficiency of recognition of specific site than the latter, but it is difficult to determine the efficiency because no absolute standard of the efficiency of recognition is available.

Detection of a rotational motion of the protein relative to DNA can be made in two ways: a labeled protein with a fixed DNA or a labeled DNA with a fixed protein. The former has several technical problems. The fixing of DNA should limit the freedom of rotation of every segment of DNA, and thus the entire region of DNA has to be fixed. Such a fixing could sacrifice its intact interaction with the protein. Moreover, the detection of rotational motion of DNA is best performed by introduction of single fluorophore moiety, but this accompanies a short lifetime as well as a requirement of refined detection system.[24] Therefore, the easiest way to assay is by detecting the rotational motion of DNA with a fixed protein. The

---

[23] K. Sakata-Sogawa and N. Shimamoto, unpublished results.

[24] H. Noji, R. Yasuda, M. Yoshida, K. Kinosita, Jr., *Nature* **386,** 299–302 (1997).

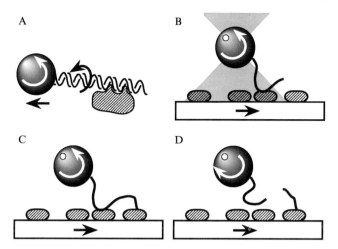

FIG. 4. The assay of tracking a groove. The tracking can be detected as a rotation of a bead attached at an end of DNA (A). The bead with a tiny fluorescent bead (the small circle) is fixed in space by laser tweezers with free rotation (B). By fixing another end on the surface, the sliding complex is stabilized (C), but rewinding of DNA occurs upon dissociation (D).

sliding motion of a protein along DNA generates a torque to DNA, if a groove is tracked (Fig. 4A). In the assay of rotational movement, a bead is used as rotational indicator with its center fixed in space by laser tweezers (Fig. 4B). The putative torque is transmitted to the large bead that is attached at the end of DNA, and the rotational motion of the bead is detected by the movement of a small fluorescent bead attached to the large bead. If several rotational indicators need to be observed simultaneously, magnetic beads with a magnetic field are used to hold many beads.[25]

Preparation of a Rotational Indicator

The fixing of a bead at the end of DNA is usually made by the binding of a biotin moiety, which has been introduced at the end, to streptavidin fixed on the bead. Since all biotin-labeling reagents link the biotin moiety and the phosphodiester backbone with a rotational-free aliphatic chain, several biotin moieties have to be introduced to transmit torque to the bead. The introduction can be made by PCR with a primer that has more than four biotin residues.[24] Alternatively, a DNA fragment of several hundred base pair can be chemically labeled with Carbobiotin (Nisshinbo,

[25] Y. Harada, O. Ohara, A. Takatsuki, H. Itoh, N. Shimamoto, and K. Kinoshita, Jr., *Nature* **409,** 113–115 (2000).

Tokyo) or a photoactivatable biotin reagent. This modified fragment is ligated to an end of a 2-20–kbp unmodified DNA, and the designed DNA is purified on an agarose gel. The length of the modified fragment as well as the ends to be ligated must be designed so that you can isolate the expected DNA from other possible side products.

A rotational indicator is composed of a large bead of about 1 μm, a tiny bead with an emission bright enough to be detected, and streptavidin coating the large bead as well as a biotinylated DNA fixed on the surface of the large bead. For the starting material for beads we selected calboxyl beads of 1-μm diameter (Polysciences) and of 93-nm diameter (Molecular Probes) that contains Texas Red, for example. A smaller single bead such as 20 nm is hardly visible unless it forms aggregates. In the preparation of a rotational indicator, it is essential to establish the best conditions by yourself on the basis of trial-and-error, because the nature of surfaces of the commercially available carboxyl beads is variable. Therefore, the detailed conditions are replaced by general technical hints in this article. A typical size of a trial is 0.1 mg–1 mg solid of the beads. The most frequent troubles in the preparation is aggregation of beads during preparation and storage. The aggregated beads can be dispersed by sonication with a micro-tip (e.g., Tomy UR-20P), but covalently conjugated pairs of large and small beads are also dissociated by such sonication. Furthermore, all the treatment of the fluorescent beads must be done in a dark place, although complete darkness is not required.

The conjugation of the large beads with the small fluorescent beads is carried out in two steps. At first, the small beads are converted into amino beads by coupling with 1,2-diaminomethane. This coupling is carried out by activating a carboxyl residue by 1-ethyl-3-(3-dimethylaminopropyl)-carbodiimide (EDC) in the presence of large excess of the diamine. The unreacted diamine has to be completely removed either by washing with centrifugation followed by sonication or by using a spin column that allows the beads to pass through but retains the diamine. In the second step, the large carboxyl beads are thoroughly washed and suspended in a buffer. Since the formation of aggregates depends on the pH of the buffer, several pH between 5 and 9 should be tried in this procedure by using Na-phosphate, MES, and Na-Bicin. Note that once the small beads are conjugated with the large beads, no more sonication is allowed. The large beads are treated with EDC, and then EDC is removed by washing. Various amounts of the amino-derived fluorescent beads are then added and incubated for a short time. While the surface carboxyl is still activated, dialyzed streptavidin is added and incubated for 2 h in the dark. The coupling reaction is quenched by adding 50 mM glycine, and the beads are washed twice with 0.5STE buffer (0.5 M NaCl, 10 mM Tris-HCl [pH 7.5] and 10 mM

EDTA). Inspect the samples under a fluorescent microscope and select the sample that has the maximum fraction of 1:1 conjugation of the large and small beads, which can be used as the rotational indicator. Incubate the best aliquot of the conjugated beads with an equal number of DNA molecules with biotin moieties at one end overnight. An anti-adsorbent such as BSA or partially hydrolyzed casein should be added.

Laser tweezers are conventionally composed of a laser beam from an Nd-YAG laser emitting at 1064 nm, and are introduced to the microscope through its epifluorescence port to be focused with a $40\times - 100\times$ oil-immersion objective lens. In the case of the rotational indicator, the largest magnification is required because the center of the bright spot must be determined with an accuracy on the order of 10 nm, which is higher than the optical resolution. A power of 10–50 mW, which is measured at the aperture of the objective, is required for easy manipulation of a 1-μm bead. Increase in the local temperature with this power is negligible.

There could be many artifacts specific to single-molecule–based measurements, especially including the use of beads, because many uncontrollable phenomena such as adsorption of protein are accompanied. One should realize that the design of an experiment such as that shown in Figure 4B is merely an image of an optimistic scientist in principle. Therefore, one has to confirm that the object being observed has the same structure as the design. It is noted that a relatively large size of bead is required for the detection of rotational motion, but the torque transmitted from DNA is too weak to reflect exactly the rotational motion where the torque is generated. Therefore, this method should be used qualitatively rather than quantitatively when the rotational motion is rapid, typically more than 0.1 revolution/sec for DNA of several kbp long.

Assay of Groove Tracking During Sliding

A simple introduction of a rotational indicator was not successful, because a generated torque, if any, disappears while RNA polymerase is dissociated from nonspecific sites on DNA. To enhance the nonspecific binding, we fixed the end of the DNA on the surface by introducing a sequence of ultra-high affinity for *E. coli* RNA polymerase[26] at the end (Fig. 4C). As a drawback of this improvement, the groove tracking no longer rotates DNA in one direction and the opposite rewinding of DNA while a protein molecule dissociates from DNA (Fig. 4D). The tracking is thus detected as a rotational motion in both directions more coherent than the Brownian motion.

---

[26] T. Gaal, W. Ross, S. T. Estrem, L. H. Nguyen, R. R. Burgess, and R. L. Gourse, *Mol. Microbiol.* **42,** 939–954 (2001).

An observation chamber similar to that used in the sliding assay is prepared by sandwiching Parafilm as a spacer. The bottom coverslip is treated with collodion, and RNA polymerase is fixed on the surface by incubating for 20 min before the assembly of the chamber. For reliability of the assay, the density of the functional protein fixed has to be measured. In the case of RNA polymerase that maintained the DNA binding activity, active molecules can be counted by using a linearized plasmid DNA containing the ultra-high affinity promoter[26] with post-stain YOYO-1 (Molecular Probe).

The DNA with the rotational indicator is then introduced in a fresh chamber, and unbound DNA is washed away. The confirmation of the structure of the targets for observation is carried out by examining the maximum moving distance of the bead when it is moved by laser tweezers. Select only the rotational indicator giving the distance equal to twice of the DNA length. After the confirmation, hold the indicator 1 μm above the surface and then move the stage at a velocity of 5–10 nm/sec by a piezo-flexure nanopositioner (Physik Instrumente, P-720). The rotational motion of bead is observed in a similar instrument used in the assay of sliding.

Assay of Arrest at a Promoter (Formation of Moribund Complex)

Abortive synthesis is observed for all DNA-dependent RNA polymerases ever tested. This period in the lifetime of a transcription complex has long been assumed to be the precursor of the elongation complex. However, a homogeneous preparation of E. coli RNA polymerase makes two kinds of complex at several promoters.[8–11] One synthesizes long RNA whereas the other synthesizes only abortive products and is named *moribund complex*. The moribund complex is the major source of abortive products, although the absence of abortive synthesis by the former has not been proved. At least *in vitro*, the moribund complex is converted into an inactive complex while retaining its short transcript, dead-end complex,[8,9] forming a branched pathway of arrest at a promoter. The method to detect the moribund complex, and to determine its fraction, uses the movement of the enzyme.

Kinetic Assay for Moribund Complex[8–10,27]

This assay detects the inconsistency in the precursor-product relationship of the moribund complex and the elongation complex. If the former remains longer than the latter, the former is not the precursor of the latter.

[27] M. Susa, R. Sen, and N. Shimamoto, *J Biol. Chem.* **277**, 15407–15412 (2002).

Therefore, this assay examines a sufficient condition for the existence of moribund complex, and thus a negative result does not deny the existence. This is the most sensitive assay for the existence of moribund complex and can detect the complex formed at the T7A1 promoter, which otherwise shows no evidence in another standard condition.[27]

The best labeling nucleotide for this assay is γ-^{32}P–labeled initiation nucleotide, because it eliminates the danger of misunderstanding cleaved products as abortive ones. If the promoter to examine is too weak to give clear bands for the full-length product in the 5 min–reaction in the condition described below, [α-^{32}P]UTP can be used. In both cases, at the concentration of the labeling nucleotide, usually 5 μM, the absence of anomalous transcripts must be confirmed by using gel electrophoresis. This assay is a kind of single-round transcription assay and requires inhibitor of turnover, such as heparin, as a competitor for DNA binding.[27] In the case of γ-^{32}P–labeled reaction, one can instead use a DNA fragment harboring a strong promoter with the other initiation nucleotide, the A-starting T7A1 promoter in the examination of G-starting promoter, for example. The amount of the inhibitor has to be determined by the complete loss of transcription activity if it is premixed with the template DNA used. This pilot experiment is critical, although it was sometimes neglected in previous literature. In the case of heparin, 40–100 μg/ml is usually sufficient but some promoter requires more than 1 mg/ml.

1. Prepare a 20% polyacrylamide sequencing gel containing 7M urea in TBE buffer. Fill the slots with 1 mM methionine to quench the residual radicals, and pre-run the gel.

2. When unfixed DNA is used as a template, 0.4–0.8 pmole of DNA and the stoichiometric amount of RNA polymerase (in terms of activity) are preincubated for 10 min at 37°. When immobilized DNA is used, 50% excess of RNA polymerase is preincubated for 10 min at 37°, and unbound enzyme is then removed by centrifugation at 10,000 rpm for 5 sec (prewash). All the necessary DNA-enzyme solution is made up in one tube, and 5.0 μl is dispensed to every 0.5 ml sample tube by using a fresh dispenser tip.

3. Start transcription by adding unlabeled substrate solution to the preincubated mixture of the enzyme and DNA. In the standard condition, the reactions are carried out in 7 μl of T buffer (50 mM Tris-HCl pH 7.9, 100 mM KCl, 10 mM MgCl$_2$, 1 mM DTT, 0.1 mg/ml partially hydrolyzed casein) at 37° with all cold substrates at 0.1–0.4 mM 3NTP as well as usually 5 μM NTP for labeling in the next step.

4. At various time points from 5 sec to 20 min, 0.2 μCi of the labeling nucleotide of high specific radioactivity is added in 1 μl, and incubated for a further 5 min. For the zero time reaction, the labeled nucleotide is

premixed with cold substrates added in Step 3 and then added to the enzyme-DNA mixture.

5. Stop the reaction by adding 50 μl water-saturated phenol/chloroform/isoamyl alcohol mixture and then vortex vigorously.

6. After a series of reactions, all the tubes are centrifuged at 13,000 $\times$ g for 5 min. The aqueous phase forms a drop in the tube. Collect the whole drop as much as possible, because this recovery determines the reproducibility unless a recovery marker is used. Load the collected aqueous phase, as well as a small amount of phenol eventually collected, into the slot of the gel. Run the electrophoresis.

7. Dry the gel and apply it to autoradiography by using a phosphoimager (Fuji BAS2500). Quantify the bands of the full-length transcript and representative abortive transcripts. Be careful because the apparent density is not linear to the radioactivity.

8. Calculate the ratio of the amount of the full-length transcripts to that of abortive transcripts at every time point. An increasing ratio in time indicates the presence of moribund complex. However, a decreasing or constant ratio does not prove its absence.

Isolation and Examination of Moribund and Dead-End Complex[9,10,27]

This assay is performed on the template DNA attached to a bead. The template must have a restriction site at more than 30 base downstream of the promoter. For isolation of moribund and dead-end complexes, the template DNA has to be fixed at the downstream end. For isolation of productive complex, the upstream end of DNA is fixed and the template should have the early transcribed sequence lacking one nucleotide, a nucleotideless cassette. The sites for NotI (GC/GGCCGC) and SspI (AAT/ATT) that work in T buffer may be useful for this purpose.

1. Prepare the template DNA by PCR with a primer biotinylated at its 5′-end. Purify the amplified DNA by polyacrylamide gel electrophoresis and by recovering from the crushed gel slice.

2. Incubate the beads with the partially hydrolyzed casein at 37° for overnight, and then wash at least twice with 0.5STE buffer (0.5 M NaCl, 10 mM Tris-HCl [pH 7.5] and 10 mM EDTA).

3. DNA must be fixed with a minimum amount of the beads in 0.5STE buffer containing 0.1 mg/ml of the casein. Add an aliquot of the bead suspension to DNA solution and incubate the mixture for 30 min. Centrifuge the mixture and estimate the amount of unbound DNA semiquantitatively by 1.5% agarose gel electrophoresis for only 5–10 min so that the present band can be compared with the previous ones. If no more decrease of free DNA is observed, the mixture is incubated overnight and

then washed twice with the 0.5STE buffer. This immobilized DNA can be stored for at least a year at 4°.

4. For an assay, 0.4–0.8 pmole of DNA and four times the stoichiometric amount of RNA polymerase are preincubated for 10 min at 37°, and unbound enzyme is then removed by centrifugation at 10,000 rpm for 5 sec. A recovery marker of a labeled 20–30 bp DNA for RNA assay or a protein, such as soybean trypsin inhibitor, can be added to improve the accuracy. Suspend the beads with T buffer. The volume at this stage is limited so that the entire amount of a suspension can be loaded onto an RNA or protein gel.

5. The reaction is started by adding the substrates. For RNA assay, the concentration of the nucleotide to be labeled should be kept low (see Step 2 of the previous section). If the DNA is fixed at its downstream end, or if RNA polymerase is stalled by a nucleotide-less cassette, there is no need to add heparin, because the enzyme is not released.

6. When the single-round synthesis of the full-length products levels off, usually 10–20 min, 1U of the restriction enzyme is added for 1 min. The time of the leveling off has to be determined beforehand by a conventional single-round transcription assay.

7. The reaction mixture is centrifuged and 70–80% of supernatant is collected with caution not to collect beads. When the beads need to be collected, remove the rest of the supernatant as much as possible, add 0.5 ml of preincubated T buffer, voltex the mixture, and centrifuge again. This essential washing process reduces the yield of the beads, but the amount may be corrected by quantifying the amount of streptavidin either by PAGE or immunological detection.

Single-Molecule Based Techniques Used in Elongation

The most conventional elongation assay is electrophoresis in a 6–20% polyacrylamide gel in the presence of 7M urea and can detect a single step of elongation that corresponds to the translocation along DNA of 1 bp, namely 3 angstrom. By immobilization of RNA polymerase or template DNA, one can rapidly change substrate composition for elongation, enabling stepwise elongation, namely walking. The technique called *fluorescence resonance energy transfer (FRET)* can sense a displacement of 1–10 nm, a sufficient resolution to detect the single step. For the evaluation of a displacement, the relative orientation between a donor fluorophore and an acceptor must be maintained during the displacement. This problem on orientation as well as the doubled side effect by introducing two chemical modifications limits its application to a specific problem in transarscription.[28] By the same reason and the limited lifetime of single fluorophores, the single-molecule FRET becomes less useful for analyzing phenomena including both translational and rotational displacement such as elongation. An alternative single-molecule assay of elongation is the detection of the rotational motion accompanied with elongation. A trial along this line was carried out by using a similar system shown in Figure 4.[24] The limitations of the accuracy are the imperfect transmission of the torque to the bead and that rapid elongation could not be measured. Therefore, the detection of the single step of single molecules of elongation complex is still a technical challenge.

At present, the best single-molecule method to measure elongation is still the oldest one.[29,30] Elongation is assayed by measuring the distance of translocation by attaching a 0.2–1 μm bead to the end of the template DNA. The observed fluctuation of the center of the bead along an axis (e.g., x-axis) are recorded. The mean square displacement, $<x^2(t)>$, is calculated at every 10–30 seconds. Since this displacement is parallel to the amount of translocation at the time (t) as a result of random fluctuation, the elongation can be calculated by using a suitable standard. In fact, the value of $<x^2(t)>$ also depends on the time-response of the detection system used; the requirement of standard is absolute, although a simulation using the persistence length of DNA gave a similar result obtained by a system with rapid time-response.[24] The standard could be the bead tethered by a known length of DNA, or the value obtained for a known pause site. The resolution of the length is typically several hundreds base.[30]

There are several problems that can be addressed only by single-molecule–based assays. Studies on the behavior of single molecules can decide whether there are heterogeneous fractions in a group of molecules or all the molecules have the common nature that has been found as a nature of a mass. For example, the existence of such a heterogeneity in elongation complex was suggested by single-molecule–based technique.[31,32] It is interesting that two-state models are proposed for three stages of transcriptions by *E. coli*, RNA polymerase, initiation, elongation, and termination. The branched pathway of initiation is a two-state mode.[8–11,27] Two states during elongation[33] are also observed, and only a fraction of elongation complex is

[28] J. Mukhopadhyay, A. N. Kapanidis, V. Mekler, E. Kortkhonjia, Y. W. Ebright, and R. H. Ebright, *Cell* **106**, 453–463 (2001).

[29] D. A. Schafer, J. Gellas, M. P. Sheetz, and R. Landick, *Nature* **352**, 444–448 (1991).

[30] H. Yin, R. Landick, and J. Gelles, *Biophys. J.* **67**, 2468–2478 (1994).

[31] R. J. Davenport, G. J. Wuite, R. Landick, and C. Bustamante, *Science* **287**, 2497–2500 (2000).

[32] N. R. Forde, D. Izhaky, G. R. Woodcock, G. J. Wuite, and C. Bustamante, *Proc. Natl. Acad. Sci. USA* **99**, 11682–11687 (2002).

[33] D. A. Erie, O. Hajiseyedjavadi, M. C. Young, and P. H. von Hippel, *Science* **262**, 867–873 (1993).

considered to terminate.[34] These two states found in these stages may have a common cause. Interestingly, the arrested states in initiation and elongation share a common feature that the relative relationship between polymerase and DNA is different from the productive fraction[9,27,35] and some of them are commonly relieved by the same factors, GreA and GreB.[10,36,37] These two states could be an example that a complex composed of homogeneous fractions of RNA polymerase and DNA has a molecular memory as multiple conformations and uses the memory in regulation.

One of the main applications of single-molecule technique to elongation is getting a force-elongation diagram, which indicates the mechanism of elongation and termination.[31,32,38] This combination of enzyme catalysis with mechanics is a pioneering field, which may contribute a lot in the elucidation of molecular mechanism of energy-driven catalytic reactions. Since a typical force in this field is pN rather than nN, laser tweezers rather than AFM are preferred.

Cautions of Application of Single-Molecule Techniques

Single molecule techniques are accepted as a tool to open new fields of biology. It could be the extreme of high-sensitive method, but it may be a wrong use of the method. High sensitivities are usually sought after quantitativity has been satisfied. Although the new methods are quite refined, most of single-molecule techniques actually sacrifice quantitativeness. The largest problem is the inclusion of uncharacterized process in the experimental set-up. For example, at least one component has to be fixed on the surface. This fixing is carried out simply by adsorption or more carefully by chemical reaction with a linker, but uncharacterized interaction of the fixed component always happens irrespective of the fixing method as long as the fixed molecules exist near a surface. The active fraction of protein molecules adsorbed on a clean glass surface is usually less than 10%, as *E. coli* RNA polymerase, although much more seems to be active on nitrocellulose. In the case of detection of heterogeneity, the residual activity is of the second importance, because homogeneity of the active molecules fixed on a surface is most critical. However, examination of this homogeneity is neglected in more than half of the literature using single-molecule

[34] R. Landick, *Science* **284**, 598–599 (1999).

[35] N. Komissarova and M. Kashlev, *Proc. Natl. Acad. Sci. USA* **94**, 1755–1760 (1997).

[36] S. Borukhov, A. Polyakov, V. Nikiforov, A. Goldfarb, *Proc. Natl. Acad. Sci. USA* **89**, 8899–902 (1992).

[37] S. Borukhov, V. Sagitov, and A. Goldfarb, *Cell* **72**, 459–466 (1993).

[38] M. D. Wang, M. J. Schnitzer, H. Yin, R. Landick, J. Gelles, and S. M. Block, *Science* **282**, 902–907 (1998).

technique, and the homogeneity is merely assumed. Even in some publications, the measurements giving data that fit to a macroscopic measurement are only selected and presented as rationalization of the single-molecule method.

The measurement of fluorescence may also introduce misinterpretation. Changes in the level of fluorescence are caused by many factors, but some of them are neglected in literature. Significant fractions of most versions of GFP molecules are non-fluorescent because of immature fluorophore, but this effect is rarely considered in quantitative discussions. Even in the case where a protein fused with GFP is little cleaved, a trace amount of the cleaved product could be the major active species in cells, causing a discrepancy between fluorescence and activity. In some FRET studies, the effect of orientation between donor and acceptor molecules is neglected. The disappearance of fluorescence from single molecules is not fully understood and thus is not always interpreted as disappearance or appearance of the molecular interaction in concern. In addition to the dangerous alteration of chemical nature introduced by the fluorescent labeling, these artifacts must be removed before conclusion.

Caution is required on the mechanical parameters obtained by the single-molecule technique. In the macroscopic world, a mechanical parameter, such as Young's modulus, is a constant determined by a substance. However, this is not true in the world of single molecules in solution. The molecules in solution are always thermally fluctuated, and thus any large distortion could happen as a rare large fluctuation. Therefore, distortion of single molecules not only depends on a substance, but also on the time region of observation, or speed of application of a perturbation. This dependence on time should be incorporated when the microscopic mechanical parameters are compared.

During a pioneer period of the methods, some inconsistency could be allowed, although it is not recommended, but subjective selection of data or conclusion should not be introduced at present. Therefore, three ways of introduction of single-molecule technique are preferential. The first is the introduction of objective statistics on data based on clear criteria. The second is the combination of single-molecule technique and conventional solution methods. The third is the introduction of various control experiments as biological science. The comparison with macroscopic results obtained by conventional methods in solution is important, but the agreement with them should not be used solely as rationale for the single-molecule method, because a contradiction may indicate a hidden truth. Along this track, new techniques may open up not only new aspects of measurements, but also new sciences.

[5]   Kinetics of Multisubunit RNA Polymerases: Experimental Methods and Data Analysis

By SHANNON F. HOLMES, J. ESTELLE FOSTER, and DOROTHY A. ERIE

Several single-molecule and kinetic studies have shown that RNA polymerase (RNAP) can catalyze synthesis in different conformational states.[1–5a] Several regulatory events, such as pausing and termination, require RNAP to occupy different conformational states. To understand the kinetic mechanism of such a conformationally diverse enzyme, it is essential to identify the rate limiting step(s), which are the most likely targets of regulation. Only transient-state kinetic methods can identify individual rate-limiting steps.

This chapter summarizes the transient-state kinetic methods and subsequent data analysis that can be used to determine the kinetic mechanisms of transcription elongation. We have used these methods to previously determine the mechanism of single nucleotide addition using *Escherichia coli* RNAP.[5] Therefore, these methods and subsequent data analysis will be discussed as they apply to this particular system; however, the same analysis can be applied to similar systems. These methods could also be used to study eukaryotic systems, provided the appropriate transcription factors and conditions are present. (See Chapter 18).

The order of the following sections follows the order in which data collection and kinetic analysis should occur.

Formation of Stalled Elongation Complexes

To observe single or multiple nucleotide incorporation during the elongation phase of transcription, it is essential to form stalled complexes. Such complexes permit the synchronization of RNAP on the DNA template. Stalled elongation complexes can be formed using nucleotide starvation.[6,7] For example, a biotinylated DNA template in which the first 30 nucleotides of the transcript are

---

[1] R. J. Davenport, G. J. Wuite, R. Landick, and C. Bustamante, *Science* **287**, 2497 (2000).

[2] H. Yin, I. Artsimovitch, R. Landick, and J. Gelles, *Proc. Natl. Acad. Sci. USA* **96**, 13124 (1999).

[3] H. Matsuzaki, G. A. Kassavetis, and E. P. Geiduschek, *J. Mol. Biol.* **235**, 1173 (1994).

[4] D. A. Erie, O. Hajiseyedjavadi, M. C. Young, and P. H. von Hippel, *Science* **262**, 867 (1993).

[5] J. E. Foster, S. F. Holmes, and D. A. Erie, *Cell* **106**, 243 (2001).

[5a] S. F. Halmes and D. A. Erie, *J. Biol. Chem.* **278**, 3559 (2003).

$$+25$$
pppAUGUAGUAAG GAGGUUGUAU GGAACAACGC

has been used.[4] Stable-stalled elongation complexes at +24 can be formed by carrying out transcription in the absence of CTP.[4] The ensuing complexes are stable with the ternary complex (RNAP, DNA template, and nascent RNA) remaining intact and in a homogenous population. Templates with a stall site prior to the incorporation of an A or C facilitate the formation of complexes stalled at a single site. Readthrough at stall sites prior to G or U incorporation is often observed because ATP and CTP stocks are usually contaminated with their deamination products, ITP and UTP, respectively.[6]

Open promoter complexes can be formed by incubating RNAP with the DNA template at 37° prior to the addition of NTPs. Stalled complexes should be formed by the addition of low concentrations of NTPs (3-20 μM). During the formation of the stalled complexes, aliquots should be removed and extended to a full-length transcript by the addition of 1 mM of all four NTPs (chase). This control monitors the percentage of arrested complexes throughout the reaction.

The stalled elongation complexes must be purified before single nucleotide incorporation experiments can be conducted. By using a biotinylated DNA template bound to streptavidin-coated magnetic beads, the complexes can be purified by placing the reaction tube next to a magnetic Eppendorf tube rack and washing away the NTPs. The magnet holds the complexes to the side of the tube and allows for gentle but thorough washing. Other methods, such as using an Ni^{+2}-chelating agarose resin to bind to a his-tag on the RNAP, require using centrifugation to purify the complexes[8] and are not as gentle of a method and may perturb the conformational state of the elongation complexes.[4] Another advantage of using the magnetic beads is that they are small enough to be used in a rapid quench apparatus (see the next section). If necessary, because the beads are attached to the complexes via the DNA template, they can also be removed from the complexes by restriction enzyme digestion.

After washing, the complexes can be used in kinetic experiments to observe single or multiple nucleotide addition. Alternatively, the RNAP

[6] J. R. Levin, B. Krummel, and M. J. Chamberlin, *J. Mol. Biol.* **196,** 85 (1987).
[7] M. Guthold, X. Zhu, C. Rivetti, G. Yang, N. H. Thomson, S. Kasas, H. G. Hansma, B. Smith, P. K. Hansma, and C. Bustamante, *Biophys. J.* **77,** 2284 (1999).
[8] M. Kashlev, E. Martin, A. Polyakov, K. Severinov, V. Nikiforov, and A. Goldfarb, *Gene* **130,** 9 (1993).

can be "walked" down the template to the next position or subsequent positions to start at a different stall site. This procedure simply involves incubating the complexes with the next NTP(s) and purifying the complexes again.[5,6]

Rapid Quench Kinetic Experiments

Performing transient-state kinetic studies on RNAP, which exhibits an average rate of RNA synthesis of ∼35 nucleotides per second[9], requires the use of a rapid quench flow apparatus. A KinTek Rapid Quench Flow apparatus was designed to perform these types of experiments[5,10,11] (see Fig. 1). For a detailed description of this instrument, see references 10 and 11. The complexes should be kept on ice when not being used in the rapid quench to prevent the formation of arrested complexes. For

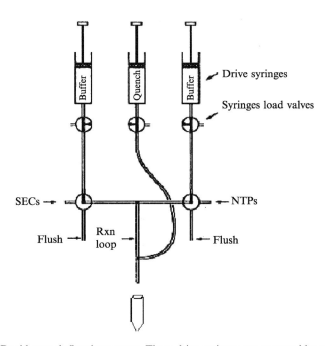

FIG. 1. Rapid quench flow instrument. Three drive syringes are operated by a step motor and can be precisely controlled. There are seven different length-reaction loops that allow the instrument to produce a wide range of time points. The stalled elongation complexes and the NTPs are loaded into the apparatus via two separate loops shown and allowed to mix in the reaction loop for a specified time. The reaction reaches the quench and is pushed out the exit line. The samples are collected in microfuge tubes and analyzed via gel electrophoresis. Each time point represents a separate experiment.

each time point, the purified elongation complexes are injected into one reactant loop and the designated NTP(s) into the other reactant loop (see Fig. 1). For the template shown in the previous section, because the complexes are stalled at position +24, CTP is added to the reaction loop to observe the single nucleotide addition of CMP at position +25. Each time point is quenched with 0.5 M EDTA, which chelates the Mg^{+2} required for synthesis at the active site. To ensure that the results are not dependent on the length of time the complexes remained on ice, time points should be taken in different orders. In addition, at designated times during the reactions, a portion of the purified elongation complexes should be chased (as described previously) to establish that the complexes are still active. After the quenched reactions are collected in tubes, centrifugation is used to pellet the magnetic beads, and a magnet is used to retain the complexes while the solution is removed. The complexes are then resuspended in 80% formamide, heated to 95°, and run on 8 M urea, 20% polyacrylamide sequencing gels (0.75 mm thick and 40 cm long).

Nucleoside Triphosphate Concentration Dependence

A primary measurement in any kinetic analysis is the examination of the change in the rate of the reaction as a function of substrate concentration. The concentration-dependent analysis allows one to observe the dissociation constants (K_ds) of productive nucleotide binding for each intermediate in the pathway. For the experiments performed with the template shown previously, the rate of CMP incorporation at position +25 using various [CTP] was measured. The experiments should be conducted at least three times for each concentration of NTP used to ensure reproducibility. Some concentrations may be more reproducible than others.

Data Analysis

Quantification and Normalization of Rate Data

The percentage of complexes at each position on the template is measured by dividing the amount of radioactivity in the indicated band by the total amount of radioactivity in all the bands the length of the stall site

---

[9] M. J. Chamberlin, W. C. Nierman, J. Wiggs, and N. Neff, *J. Biol. Chem.* **254,** 10061 (1979).
[10] K. A. Johnson, *Methods Enzymol.* **134,** 677 (1986).
[11] K. A. Johnson, *Methods Enzymol* **249,** 38 (1995).

and longer. In most experiments, some complexes may misincorporate during stalled elongation complex formation. There may also be some complexes that do not elongate over the time course of the experiment but do elongate in the chase reactions. Such complexes have been observed previously.[5,12] Because these complexes catalyze synthesis at such slow rates (minutes), they do not contribute to the observed kinetics and must be removed from the population. To compare data from different experiments, it is necessary to normalize the data such that at 0 time, there is 0% incorporation, and upon completion, there is 100% incorporation. Accordingly, the percentage of complexes that misincorporate prior to the addition of the correct NTP should be subtracted from each time point. Subsequently, the data should be normalized to 100% by dividing each time point by the highest percentage of complexes that incorporate the correct NTP.

Single and Double Exponential Fits

After obtaining normalized data for each experiment, the percentage of complexes that incorporated the correct NTP is plotted as a function of time (see Fig. 2 for an example). If possible, each data curve is fit to a single exponential $[y = A \exp(-k_{app}t) + C]$ to determine the pseudo-first order rate constant, k_{app}. To examine the substrate interaction with RNAP, the pseudo-first order rate constants, k_{app}, are plotted against [NTP]. To further examine the data, it is useful to plot $[NTP]/k_{app}$ versus [NTP] (called a *Hanes plot*). For a classic Michaelis-Menten mechanism, in which $E + S \rightleftharpoons ES \rightarrow E + P$ describes the reaction, the former plot is hyperbolic and the latter plot is linear with a positive slope. For a steady state approach, this analysis allows one to determine the apparent K_m for nucleotide binding. With a pre-steady state approach, such as described here, this analysis yields the apparent dissociation constants (K_dS) of productive nucleotide binding. For cases in which more than once molecule of substrate binds to the enzyme, the substrate saturation plot will not be hyperbolic and the Hanes plot will be nonlinear. As an example, Figure 3 shows the differences among plots that characterize reactions described by classical Michaelis-Menten kinetics, substrate inhibition, and positive cooperativity resulting in sigmoidal kinetics. For a detailed explanation of each plot, refer to Chapter 12 of reference 13.

For the data described here, the plot of k_{app} versus [CTP] exhibits sigmoidal behavior, and the $[CTP]/k_{app}$ versus [CTP] plot is nonlinear with an

[12] N. Komissarova and M. Kashlev, *J. Biol. Chem.* **272,** 15329 (1997).

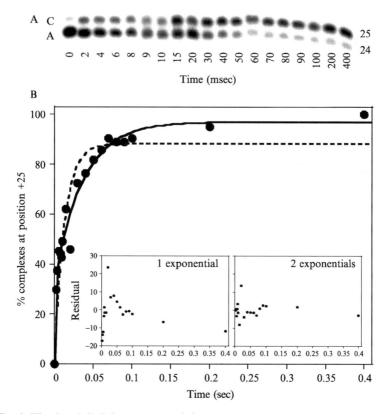

FIG. 2. Kinetics of CMP incorporation. (A) Distribution of the transcription products on a 20% acrylamide, 8M urea gel as a function of time after the addition of 20 μM CTP to stalled elongation complexes. The lengths of the transcripts are indicated on the right, the identities of the nucleotides incorporated are on the left, and the times at which the reactions were quenched, in milliseconds, are below each lane. (B) Plot of the percentage of complexes (normalized) that incorporated a CMP at position +25 for the gel shown in (A). The dashed and undashed curves are the best single- and double-exponential fits, respectively. The plots in the insert show the residuals for the single- and double-exponential fits. The quality of the double-exponential fit is clearly better than that of the single-exponential fit [χ^2 (1 exp.) = 1550 and χ^2 (2 exp.) = 332; F = 32]; however, it is useful to use both the single- and double-exponential fits to examine the data.

initial negative slope (see Fig. 3C and reference 5). The initial negative slope indicates a sigmoidal dependence of k_{app} on [CTP]. The Hanes plot is extremely helpful in this analysis because it is the most sensitive indicator of sigmoidal kinetics.[13] The sigmoidal substrate-saturation curve indicates that the rate of CMP incorporation has a quadratic dependence on [CTP]

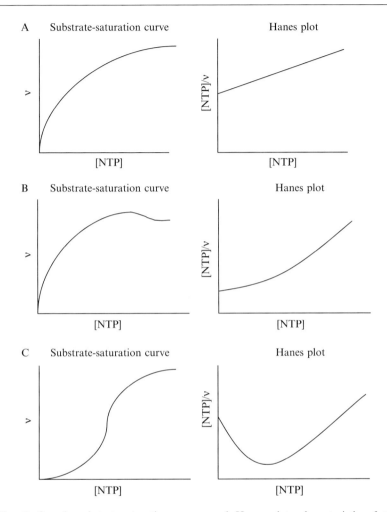

Fig. 3. Sample substrate-saturation curves and Hanes plots characteristic of three different types of enzyme-catalyzed reactions. (A) Plots characteristic of typical Michealis-Menton kinetics (i.e., $E + S \rightleftharpoons ES \rightleftharpoons E + P$). (B) Plots characteristic of substrate inhibition. (C) Plots characteristic of sigmoidal kinetics.

and that there is positive cooperativity. Accordingly, RNAP must contain two binding sites, presumably the catalytic site and an allosteric site, to which CTP can bind, with binding to the allosteric site causing an increase

---

[13] A. R. Schulz, *Enzyme Kinetics: From Diastase to Multi-enzyme Systems*, Cambridge: Cambridge University Press, New York, 1994.

in the rate of CMP incorporation.[5,13,14] In other words, CTP acts as both a substrate and an allosteric effector, and the kinetic mechanism must reflect this observation.[5,5a]

In some cases, the data cannot be fit to a single-exponential and instead fit better to a double-exponential $[y = A_1\exp(-k_1t) + A_2\exp(-k_2t) + C]$ (see Fig. 2B). Fitting to a double-exponential can be done with conventional graphing programs. Alternatively, kinetics fitting programs such as Dynafit[15] and simulation programs such as KinSim[16] can be used to determine k_1 and k_2 for each rate curve. Because the double-exponential fit has two additional parameters relative to the single-exponential fit, the quality of the fit will necessarily be better for the double-exponential fit. Consequently, it is necessary to use F statistics to determine if a double-exponential fit is warranted. For the comparison of a four-parameter versus a two-parameter fit, the definition of F is

$$F = 1/2[(N - 2)\chi_1^2/\chi_2^2 - (N - 4)]$$

where N is the number of points in the data set and χ_1^2 and χ_2^2 are the chi-squared goodness-of-fit statistics for the single- and double-exponential fits, respectively. As an example, for data sets containing 15 or more points, an F value >3.7 indicates that a double exponential fits the data better than a single exponential with a confidence of greater than 95%, and an F value >6.4 indicates $>99\%$ confidence.[17] Data that require a double-exponential fit indicate that there are two states that catalyze synthesis at fast and slow rates and that the two states are not in rapid equilibrium with one another. These biphasic kinetics could result from two sequential steps ($E \rightarrow E' \rightarrow P$) or from two forms of the enzyme catalyzing the reaction at different rates ($E \rightarrow P$ and $E' \rightarrow P$).

The data shown here could be fit to single-exponentials; however, for all concentrations between 5 and 100 μM CTP, a double-exponential fit was significantly better (with greater than 99% confidence) as judged by the F-test (see Fig. 2B).[5,17,18] These data are biphasic, and in the context of

[14] I. H. Segel, *Enzyme Kinetics: Behavior and Analysis of Rapid Equilibrium and Steady-State Enzyme Systems*, John Wiley and Sons, Inc., New York, 1993.

[15] P. Kuzmic, *Anal. Biochem.* **237**, 260 (1996).

[16] K. S. Anderson, J. A. Sikorski, and K. A. Johnson, *Biochemistry* **27**, 7395 (1988).

[17] D. P. G. Shoemaker, C. W., and Nibler, J. W., *Experiments in Physical Chemistry*, 4th edition, McGraw Hill, 1981.

[18] D. S. Moore, *The Basic Practice of Statistics*, W. H. Freeman and Co., New York, 1995.

allosteric regulation, this result is consistent with a nonessential activation mechanism in which the fast (or activated state) (RNAP bound with substrate in the allosteric site) catalyzes synthesis more rapidly than the slow (or unactivated state) with a slow interconversion between states after substrate binding in the catalytic site.[5]

The data from the double-exponential fits of the individual rate curves can be used to obtain initial estimates for the dissociation and the rate constants. The pseudo-first order rate constants, k_1 and k_2, can be plotted as a function of [NTP] and fit to a hyperbolic function to yield an estimate of the dissociation and rate constants for the two paths. For example, k_1 and k_2, obtained from the double-exponential fits to the CTP data, can be plotted as a function of [CTP] and fit to a hyperbolic function to yield the dissociation constants for the catalytic site and the polymerization rate constants for the slow and fast states, respectively. Similarly, the amplitude of the burst phase (i.e., percentage of complexes in the fast state [E_{fast}]) can be plotted against [CTP] and fit to a hyperbolic function to yield the dissociation constant for binding to the allosteric site. These parameters can be used as starting points for fitting the data to a mechanism.

Determination of the Final Kinetic Mechanism

The program Dynafit[15] can be used to determine the rate and binding constants that fit the data at all [NTP]. When using a kinetics fitting program such as Dynafit,[15] all [NTP] should be fit simultaneously; that is, the fits should be globally optimized. In these fits, most of the rate and binding constants should be allowed to vary simultaneously. The parameters determined from the analysis of each curve (see previous discussion) should be used as starting points for fitting the data to a mechanism. In addition, the starting parameters should be varied over several orders of magnitude. It should be noted that neither KinFit nor DynaFit[15,16] can fit data to a mechanism with branched pathways in which one of the equilibrium constants is not independent. In this case, the data should be fit manually as discussed below.

KinSim[16] can be used to fit the data "manually"; that is, the data can be simulated using many different combinations of rate and binding constants until the best fits are obtained. Fitting the data manually can be very time consuming; however, it is the only way to ensure that the kinetic curves produced can accurately fit the data. This program can also be used to double-check the fits obtained from DynaFit by simulating curves generated with the parameters from DynaFit and overlaying these curves with the data (see Fig. 4).

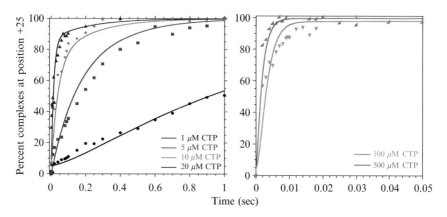

FIG. 4. Plots of the percentage of complexes that incorporated CMP at position +25. Data for reactions at six CTP concentrations (1, 5, 10, 20, 100, and 500 μM) are shown. The curves were generated using the program KinSim[16,19,20] and plotted on the same graph as the data points to evaluate the quality of the fits. (See color insert.)

These programs can be used to attempt to fit the data to all possible mechanisms that can accurately describe the system. For example, biphasic kinetics could result from two sequential steps (E $\rightarrow$ E' $\rightarrow$ P) or from two forms of the enzyme catalyzing the reaction at different rates (E $\rightarrow$ P and E' $\rightarrow$ P). The two simplest mechanisms that can reflect such observations are essential and nonessential activation mechanisms[14], with the substrate acting as the activator (see Fig. 4A in reference 5). (For a full discussion on the analysis, see the texts by Segel and Schulz.) These two mechanisms differ only in that in nonessential activation, the reaction can occur in the absence of the activator, whereas in essential activation, the activator must be bound for the reaction to occur. Fitting programs such as DynaFit and KinFitSim can be used to fit the data to the essential activation mechanism. Consequently, this mechanism should be tested exhaustively before choosing the non-essential activation mechanism. For the example shown here (Figure 4A in reference 5), a single set of constants could not be found that produced reasonable fits to data for all [CTP] using the essential activation mechanism, suggesting two pathways for synthesis.[5] Several important points should be made about fitting data:

[19] B. A. Barshop, R. F. Wrenn, and C. Frieden, *Anal. Biochem.* **130,** 134 (1983).
[20] C. T. Zimmerle and C. Frieden, *Biochem. J.* **258,** 381 (1989).

1. Although the reported mechanism should be the simplest mechanism that can fit the data, it is generally underdetermined. Specifically, there are usually more than one set of rate and binding constants that can fit the data. Therefore, the constants that seem to be the most reasonable should be reported.[5] If a good fit to the concentration dependence of the apparent rate and binding constants to a double exponential were obtained, the "most reasonable" values would be the ones that are the most similar to these data. It should always be noted, however, that there is usually not a unique set of parameters that will fit the data. For example, it was discovered that several sets of parameters can fit the data as long as the ratios of constants between the activated and unactivated states remain the same (personal observation).

2. When fitting biphasic data, changing parameters such as equilibrium binding steps can dramatically change the shapes of the curves and how well they fit the biphasic nature of the data. A program such as KinSim illustrates this point by simulating the different curves generated by changing the on and off rates of the equilibrium-binding constants.

3. In experiments with *E. coli* RNAP, a small percentage of complexes ($\sim$5%) always seem to be predisposed toward the fast state of synthesis. This percentage of complexes only shows up in the data from the lower [NTP], because at higher [NTP], they are embedded in the burst phase. If this percentage of complexes is accounted for in the program KinSim, the resulting simulated curves fit the data (especially at the lower [NTP]) significantly better.

To fit multiple nucleotide incorporation data, all of these methods can be used. The data for the first nucleotide incorporation should be fit first to a mechanism. Then, holding these fits constant, the data for the second nucleotide incorporation can be fit, again using similar methods. The fits for the second nucleotide can then be held constant while the third nucleotide is fit, etc. By using these methods, the mechanism for single or multiple nucleotide addition can be established. This mechanism, when combined with structural information, can give us vital details about transcription elongation and function.

[6] Principles and Methods of Affinity Cleavage in Studying Transcription

By CLAUDE F. MEARES, SAUL A. DATWYLER, BRIAN D. SCHMIDT, JEFFREY OWENS, and AKIRA ISHIHAMA

The development of chemical reagents capable of cleaving protein and/or nucleic acid backbones has provided the means to probe the structures of complex biomolecular systems in solution.[1–4] One useful technique includes iron-EDTA (ethylenediaminetetraacetate) cleavage, which takes place in the presence of an oxidizing agent such as hydrogen peroxide (H_2O_2) and a reducing agent such as ascorbate (Vitamin C). The products of redox chemistry at the iron center can induce the cleavage of nearby protein, DNA, and RNA backbones. This cleavage activity of iron-EDTA has proved useful for the identification of protein-binding sites on DNA (DNA footprinting)[5] and, more recently, the binding surfaces of proteins (protein footprinting).[6]

Conjugation of the cutting reagent to a site on a biological molecule provides a more powerful way to map proximity relations within or between macromolecules (explained later in this chapter). Bifunctional chelating agents contain (1) a strong metal-binding moiety and (2) a chemically reactive electrophile that reacts with nucleophilic sites on a biological molecule. Rana and Meares[7,8] first described the use of a bifunctional chelating agent, iron (S)-1-(p-bromoacetamidobenzyl)EDTA (FeBABE), tethered at a cysteine residue, to cleave polypeptide chains at sites proximal to the chelate, apparently independent of the amino acid residue. The tethered metal ion serves as a localized source for generation of reactive species to cleave the peptide backbone. Conditions are available for which cleavage of the polypeptide backbone predominates over oxidation of amino acid side chains, under physiological conditions of pH, temperature, and ionic strength. The polypeptide chain is cleaved at

[1] D. S. Sigman, D. R. Graham, V. D'Aurora, and A. M. Stern, *J. Biol. Chem.* **254,** 12269 (1979).

[2] M. W. Van Dyke, R. P. Hertzberg, and P. B. Dervan, *Proc. Natl. Acad. Sci. USA* **79,** 5470 (1982).

[3] P. G. Schultz and P. B. Dervan, *Proc. Natl. Acad. Sci. USA* **80,** 6834 (1983).

[4] T. D. Tullius and B. A. Dombroski, *Proc. Natl. Acad. Sci. USA* **83,** 5469 (1986).

[5] G. B. Dreyer and P. B. Dervan, *Proc. Natl. Acad. Sci. USA* **82,** 968 (1985).

[6] E. Heyduk and T. Heyduk, *Biochemistry* **33,** 9643 (1994).

[7] T. M. Rana and C. F. Meares, *J. Amer. Chem. Soc.* **112,** 2457 (1990).

[8] T. M. Rana and C. F. Meares, *J. Amer. Chem. Soc.* **113,** 1859 (1991).

sites that are nearby in the three-dimensional structure, though possibly quite distant in the primary sequence.[8,9] Interestingly, the hydroxyl radical is thought to be responsible for the scission of nucleic acids,[10] whereas peptide backbones can be cleaved by either hydrolytic[11] or oxidative mechanisms. Some anecdotal evidence exists that oxidation predominates when only ascorbate is added, whereas inclusion of peroxide contributes to the hydrolytic mechanism. Other groups have developed related EDTA-based reagents capable of cleaving proteins.[12–14]

Like other biological macromolecules, proteins are linear polymers, folded into unique structures. Affinity cleavage with tethered iron-EDTA derivatives, such as FeBABE and iron (S)-1-(p-isothiocyanatobenzyl) EDTA (FeCITC), can be used to identify the segments of a protein or nucleic acid that are within close proximity to another molecule.[8,15–19] FeBABE can be tethered to a specific site, such as the cysteine side chain of a protein containing a single cysteine, or distributed among lysine side chains via a 2-iminothiolane (2IT) linker. FeCITC can be attached directly to the ε-amine of lysine side chains. Other strategies are also possible; for example, FeBABE can be tethered to DNA by attachment to phosphorothioate residues.[20]

Obtaining Bifunctional Chelating Agents

Recent advances in the use of protein-cutting reagents for mapping protein–protein and protein–nucleic acid interactions have made their use more universal and has created a need for an easy, large-scale preparation of these reagents. The preparation of two protein-cutting reagents, (S)-1-[p-(bromoacetamido)benzyl]EDTA (BABE) and (S)-1-[p-(isothiocyanato)benzyl]EDTA (CITC),[21] is briefly referenced here. The synthesis of

[9] D. P. Greiner, R. Miyake, J. K. Moran, A. D. Jones, T. Negishi, A. Ishihama, and C. F. Meares, *Bioconjugate Chem.* **8,** 44 (1997).

[10] W. K. Pogozelski, T. J. McNeese, and T. D. Tullius, *J. Am. Chem. Soc.* **117,** 6428 (1995).

[11] T. M. Rana and C. F. Meares, *Proc. Natl. Acad. Sci. USA* **88,** 10578 (1991).

[12] D. Hoyer, H. Cho, and P. G. Schultz, *J. Am. Chem. Soc.* **112,** 3249 (1990).

[13] M. R. Ermácora, J. M. Delfino, B. Cuenoud, A. Schepartz, and R. O. Fox, *Proc. Natl. Acad. Sci. USA* **89,** 6383 (1992).

[14] Y. W. Ebright, Y. Chen, P. S. Pendergrast, and R. H. Ebright, *Biochemistry* **31,** 10664 (1992).

[15] G. M. Heilek and H. F. Noller, *Science* **272,** 1659 (1996).

[16] K. Murakami, J. T. Owens, T. A. Belyaeva, C. F. Meares, S. J. W. Busby, and A. Ishihama, *Proc. Natl. Acad. Sci. USA* **94,** 11274 (1997).

[17] R. Miyake, K. Murakami, J. T. Owens, D. P. Greiner, O. N. Ozoline, A. Ishihama, and C. F. Meares, *Biochemistry* **37,** 1344 (1998).

[18] J. T. Owens, R. Miyake, K. Murakami, A. J. Chmura, N. Fujita, A. Ishihama, and C. F. Meares, *Proc. Natl. Acad. Sci. USA* **95,** 6021 (1998).

[19] S. L. Traviglia, S. A. Datwyler, and C. F. Meares, *Biochemistry* **38,** 4259 (1999).

[20] B. D. Schmidt and C. F. Meares. *Biochemistry* **41,** 4186 (2002).

BABE was first described by DeRiemer et al.[22]; this synthesis has since been modified by others, including Hayward et al.[23] and Greiner et al.[9] More recently, Chmura et al.[24] have updated the procedure. Iron is added to BABE and CITC as described in references 9 and 19 to yield the desired protein cleavage agents, FeBABE and FeCITC. An inert gallium BABE chelate can be prepared to measure the steric consequences of conjugation, without complications from oxidative damage by iron (gallium(III) is the same size as iron[III]).[19] At the present time, reagents such as BABE, CITC, and FeBABE are commercially available from various sources.

Principles and Practical Aspects

When comparing strategies such as chemical cross-linking, untethered iron-EDTA footprinting, and tethered iron-EDTA mapping, the appropriate choice of chemical probe is determined by the extent to which the system is understood and by the information that is sought from the experiment. If the identities of all interacting components within the system of interest are not known, chemical cross-linking experiments may be ideal to identify previously unknown players. However, it is often difficult to resolve exactly where the cross-link occurs between two macromolecules. If the goal is to identify a binding site on a DNA molecule, DNA footprinting with untethered iron-EDTA can be used; for technical reasons, the analogous approach to identifying a binding site on a protein is generally not practical for large protein molecules.[25] To obtain a picture of the point-to-point interactions between protein domains, and for protein–nucleic acid orientations, tethered iron-EDTA may be the most useful approach.

BABE has an EDTA metal–binding moiety and a bromoacetamido-functional group to target cysteines (Fig. 1). The most reactive nucleophile that a protein is likely to contain is the sulfur of a cysteine side chain, and the bromoacetamido group has proven to be an effective electrophile.[26] The EDTA moiety of BABE has a high affinity for the Fe(III) ion, not allowing the metal to dissociate from the chelate at an appreciable rate under physiological conditions (Fe[III] is the stable oxidation state of iron

[21] C. F. Meares, M. J. McCall, D. T. Reardan, D. A. Goodwin, C. A. Diamanti, and M. McTigue, Anal. Biochem. 142, 68 (1984).

[22] L. H. DeRiemer, C. F. Meares, D. A. Goodwin, and C. I. Diamanti, J. Labelled Compd. Radiopharm. 18, 1517 (1981).

[23] M. M. Hayward, J. C. Adrian, and A. Schepartz, J. Org. Chem. 60, 3924 (1995).

[24] A. J. Chmura, B. D. Schmidt, D. T. Corson, S. T. Traviglia, C. F. Meares, J. Controlled Release 78, 249 (2002).

[25] S. A. Datwyler and C. F. Meares, Trends Biochem. Sci. 25, 408 (2000).

[26] P. Scheltè, C. Boeckler, B. Frisch, and F. Schuber, Bioconjug. Chem. 11, 118 (2000).

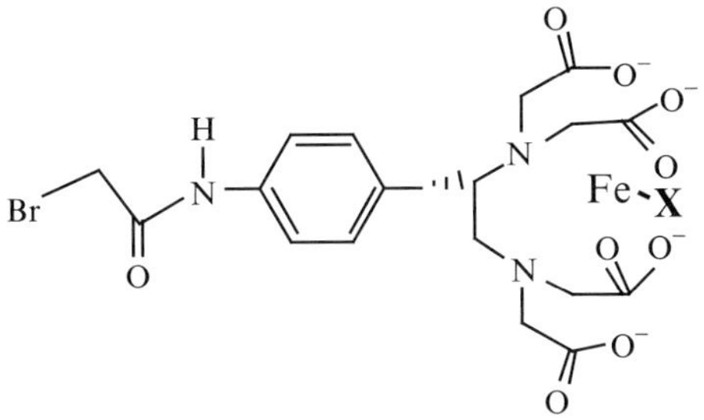

FIG. 1. Structural drawing of iron (S)-1-(p-bromoacetamidobenzyl)EDTA (FeBABE), showing the site (X) at which reactive species such as peroxide are thought to bind to the iron.

in this case). The reactive electrophilic end can be changed to react with different nucleophilic groups of a protein, such as an isothiocyanate to react with N-terminal amino groups or lysine ε-amino groups.

The success of cleavage experiments depends in part on the ability: (1) to obtain the desired molecular complex in relatively high yield from pure components, (2) to develop a sensitive detection system specific for the target proteins or nucleic acids, and (3) to reliably determine the specific site(s) of fragmentation (e.g., to generate internally consistent molecular markers from the biomolecules under study). DNA end-labeling and sequencing reactions[27] have been fully developed and can be applied to DNA footprinting experiments. For protein footprinting experiments, Western blotting and immunostaining using N- or C-terminus specific antibodies can provide a sensitive end-label detection method for multi-subunit proteins and their fragments (Fig. 2). To identify sites of protein cleavage, molecular weight markers of the same protein(s) are typically generated by genetic truncation, enzymatic digestion, or amino acid–specific chemical digestions to match fragments in spite of sequence-dependent effects on the migration rate of proteins in sodium dodecyl sulfate-polyacrylamide gel electrophoresis (SDS-PAGE).[17,18] For example, in reference 18, the assignment of cleavage sites on the β and β' subunits of RNA polymerase was carried out by comparison with sequence markers generated from the same protein. As reference markers, β and β' were cleaved at either Cys or Met residues by chemical digestions of RNAP to produce known marker fragments.[28,29]

[27] A. M. Maxam and W. Gilbert, *Methods Enzymol.* **65,** 499 (1980).

[28] M. A. Grachev, E. A. Lukhtanov, A. A. Mustaev, E. F. Zaychikov, M. N. Abdukayumov, I. V. Rabinov, V. I. Richter, Y. S. Skoblov, and P. G. Chistyakov, *Eur. J. Biochem.* **180,** 577 (1989).

[29] A. Mustaev, M. Kozlov, V. Markovtsov, E. Zaychikov, L. Denissova, and A. Goldfarb, *Proc. Natl. Acad. Sci. USA* **94,** 6641 (1997).

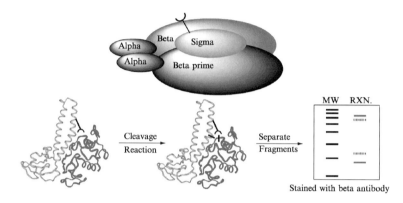

FIG. 2. Determining protein-protein interactions via protein cleavage reactions. The ⊃— symbol represents a protein cleavage reagent that is conjugated to the sigma subunit. For simplicity, the beta subunit is shown being cleaved; beta-prime is also cleaved prominently by conjugated sigma. The cleavage reaction products are subsequently separated via SDS-PAGE and transferred to a membrane, where they are stained with a specific antibody (directed against the N terminus of the beta chain in this example).[18,47] (See color insert.)

Conjugation of FeBABE to a Single Site

Conjugation to a single site (e.g., cysteine) is diagrammed in Figure 3. This requires a protein with either a naturally occurring or engineered free cysteine available for conjugation. The first demonstration from our lab was by Rana and Meares,[7] where the single free cysteine (Cys34) of the 66 kDa protein bovine serum albumin (BSA) was modified with FeBABE. In the presence of ascorbate and hydrogen peroxide for 10 s at pH 7, cleavage was observed at two sites in the BSA polypeptide backbone. When the products were resolved by SDS-PAGE, fragments of mass 45, 17, and 5 kDa were observed. By performing both N- and C-terminal sequencing of the fragments using standard methods, the sites of cleavage were identified to be at the peptide bonds between Ala150-Pro151 and Ser190-Ser191. Unlike DNA cleavage with iron chelates, there was no obvious loss of a residue to oxidation.[4] Thus, standard protein analytical procedures gave results consistent with hydrolysis rather than oxidation. These observations also suggested that cleavage does not depend on the chemical reactivity of the amino acid being cleaved and indicated that the cleavage reaction is catalytic, because more than one cleavage event occurred per protein molecule. Cleavage was not observed under these conditions when the protein was unfolded or when the chelate was untethered.

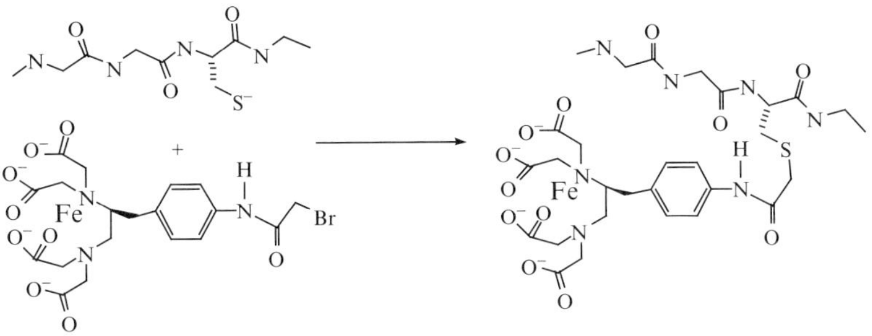

FIG. 3. Conjugation of FeBABE with a cysteine side chain,[18] generally carried out around pH 8, near the pK$_a$ of the side chain. The conjugation reaction works best when the macromolecule is somewhat concentrated. We have gotten best results when the protein concentration is around 100 μM, but it has also worked for concentrations as low as 5 μM. The final concentration of FeBABE in the conjugation reaction should be $\approx$ 1 mM. To avoid protein precipitation or denaturation, we choose to add a volume of a DMSO-FeBABE stock solution that is $\leq$ 10% of the total reaction volume. The conjugation reaction is usually performed at room temperature or 37° for 1 h. The reaction can be stopped by (1) removing the excess FeBABE using a size exclusion spin column, or (2) quenching with an equal volume of 1 M Tris buffer, pH 8, and exchanging the sample buffer by dialysis. The conjugation yield is determined by comparing the amount of free cysteine before and after conjugation (we use a CPM[9,30] test).

Conjugation to Multiple Sites

Preparing a large enough set of single-Cys mutants should allow mapping the entire periphery of a protein-protein interaction site. However, we have devised an abbreviated procedure that approaches this goal without molecular cloning.[19,37] This involves attaching cutting reagents (in low yield) randomly to lysine residues on a protein surface, and then using this lysine-labeled protein to cleave the polypeptide backbone of another protein at exposed residues adjacent to its binding site. The reagent 2-iminothiolane (2IT, Traut's reagent) has been used to tether radioactive chelates and nucleases by way of lysine side chains.[31–34] The basic scheme for preparing and using a 2IT-FeBABE conjugate library is shown in Figure 4. Modification with 2IT moves the existing charge of the lysine two atoms away from the original site, preserving the overall charge of the side chain. Studies with peptides show that labeling of lysines with

[30] R. Parvari, I. Pecht, and H. Soreq, *Anal. Biochem.* **133,** 450 (1983).
[31] J. M. Lambert, R. Jue, and R. R. Traut, *Biochemistry* **17,** 5406 (1978).
[32] M. J. McCall, H. Diril, and C. F. Meares, *Bioconjug. Chem.* **1,** 222 (1990).
[33] C. H. Chen and D. S. Sigman, *Science* **237,** 1197 (1987).
[34] B. G. Lavoie and G. Chaconas, *Genes Dev.* **7,** 2510 (1993).

Fɪɢ. 4. Conjugation of FeBABE with a lysine side chain, via 2IT. Usually carried out around pH 8 (near the pK_a of the thiol intermediate). In a typical example, the protein was transferred from storage buffer into conjugation buffer (10 mM MOPS, pH 8, 100 mM NaCl, 0.1 mM EDTA, 5% glycerol) using a gel filtration spin column.[36] FeBABE ($\approx$20 mM) in DMSO was added to a final concentration of 1.4 mM, immediately followed by freshly prepared 2IT (final concentration 0.7 mM). The reaction was incubated at 37° for 1 h. Excess reagents were removed by gel filtration spin column equilibrated in 2× cleavage buffer (1×: 10 mM MOPS, pH 8, 120 mM NaCl, 10 mM MgCl$_2$, 1 mM EDTA). An equal volume of glycerol was added, and the conjugate library was stored at −70°.

2IT is effective even though there is a considerable difference in pK_a between the amino terminus of the protein and ε-amino groups of lysines.[35]

Lysine side chains are widely distributed on protein surfaces. The side chain ε-amino groups of lysine residues tend to have similar chemical reactivity, presumably because of their relatively large separation from the polypeptide backbone (four –CH$_2$–groups), which facilitates their location on the surface of proteins. Examining the crystal structures of several proteins, we find that the average distance from a lysine ε-amino group to its nearest neighbor lysine ε-amino group is 12 ± 4 Å. If one conjugates 2IT-FeBABE to lysine side chains that are distributed on average 12 Å apart, the 12 Å reach of FeBABE plus the 6 Å extension caused by 2IT means that, on average, the cutter will reach past the next nearest lysine side chain. Thus a collection of macromolecules, each bearing one to two lysine-tethered cutters, is likely to have sufficient members to cut almost any nearby site on a target macromolecule.

If we use a reagent such as 2IT that conjugates to a residue type such as lysine, and if all such residues on the macromolecule are equally and independently reactive, we can describe the mixture of conjugation products using the binomial formula.[37] A few simple examples of the relation

[35] R. Wetzel, R. Halualani, J. T. Stults, and C. Quan, *Bioconjug. Chem.* **1**, 114 (1990).
[36] H. S. Penefsky, *Methods Enzymol.* **56**, 527 (1979).
[37] S. L. Traviglia, S. A. Datwyler, and C. F. Meares, *Biochemistry* **38**, 4259 (1999).

between the *average* number of FeBABE molecules per protein and the percentage of protein molecules with 0, 1, or > 1 FeBABE per protein are given in Table I. Results are easiest to interpret when the macromolecule carries no more than one label, because this minimizes perturbation of the system. Unlabeled macromolecules generally do not interfere with the experiments, except to reduce the yield of cleavage products.

Changing any of the following can vary the conjugation yield of a 2IT-FeBABE lysine–labeled protein: the pH of the conjugation buffer, the conjugation reaction time, or the concentrations of 2IT and FeBABE. A titration of increasing amounts of 2IT and FeBABE is used to demonstrate the minimal conjugation yield needed to produce clearly visible product bands. The concentration of 2IT should be less than the concentration of FeBABE (this helps prevent disulfide formation and encourages rapid coupling of FeBABE to the newly generated thiol group resulting from the 2IT ring opening reaction). The conjugation yield should be high enough for easy detection of all possible cuts, but low enough to preserve biological activity. If some molecules can no longer bind after conjugation, they should not interfere with the experiment because the conjugate must bind in order to cleave. The 2IT and the bifunctional chelating agent should be in a large enough molar excess over the protein concentration that the conjugation reaction follows pseudo–first order kinetics. In this circumstance, the rate of protein conjugation is controlled by the concentration of the 2IT.

One can also use the reagent FeCITC to label lysines, leading to a shorter tether between protein and chelate (Fig. 5).[19]

TABLE I
THE NATURE OF THE PRODUCTS OF A TYPICAL RANDOM PROTEIN-LABELING REACTION.[a]

Average number of labels/macromolecule	Macromolecules unlabeled (%)	Macromolecules with one label (%)	Macromolecules with > one label (%)
0.1	90	9	<1
0.3	74	22	4
0.5	60	30	9
1	36	37	26

[a] For a macromolecule with 60 reactive groups, when the reactive groups are equivalent and independent (e.g., lysine). (Note: Because of round-off error, numbers do not always total 100.) Binomial formulas are given in reference 37. A similar model is expected to apply to random DNA labeling of phosphorothioate residues.

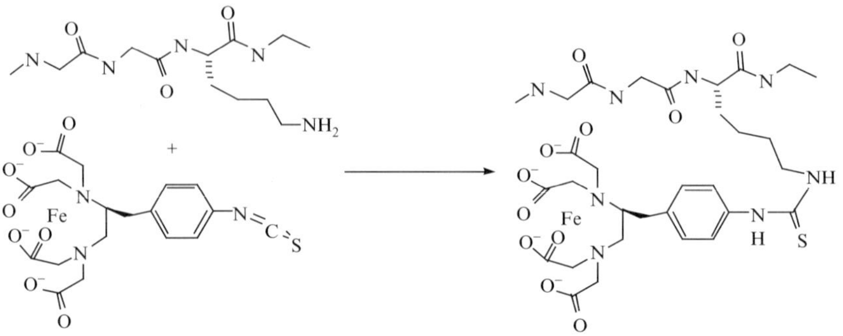

FIG. 5. Conjugation scheme of FeCITC with lysine. For a typical FeCITC conjugation, the protein is transferred into 50 mM tetramethylammonium phosphate, pH 10, 100 mM NaCl, 1 mM EDTA, and 5% glycerol, to a final protein concentration of ≈20 μM. FeCITC is added to a final concentration of ≈67 μM, and the reaction is allowed to incubate at room temperature for 1 h.

Conjugation to DNA

Comparing the common appearance of lysine side chains on the surface of proteins to the appearance of phosphodiester groups on the surface of DNA, we reasoned that mapping DNA binding sites on proteins could be achieved by placing FeBABE on the phosphodiester groups, thus transforming DNA into a chemical protease.[20] Substituting a non-bridging oxygen on the phosphodiester DNA backbone with a sulfur results in a phosphorothioate oligodeoxynucleotide (Fig. 6). We have used PCR to create a set of four DNA fragments,[38] each containing a 5'-thiophosphoryl nucleotide at the A, T, G, or C positions. Under appropriate conditions,[39] a small percentage of the nucleophilic sulfur atoms on the phosphodiester backbone react with FeBABE to produce a collection of DNA molecules conjugated at various positions (Fig. 7). Each of these fragments contains a limited number of FeBABE molecules scattered across the surface at different locations. If these conjugated DNA fragments are taken as a set, almost any part of the DNA surface has the possibility of containing a FeBABE molecule. Thus, upon forming a protein-DNA complex, every region of the protein surface near the DNA should be susceptible to polypeptide backbone cleavage (Fig. 7). Because of the extension of FeBABE from the surface of the DNA, initiation of the cleavage reaction is expected to produce cut sites at the periphery of the protein-DNA binding site. Because of the

[38] D. J. King, D. A. Ventura, A. R. Brasier, and D. G. Gorenstein, *Biochemistry* **37**, 16489 (1998).
[39] J. A. Fidanza and L. W. McLaughlin, *J. Am. Chem. Soc.* **111**, 9117 (1989).

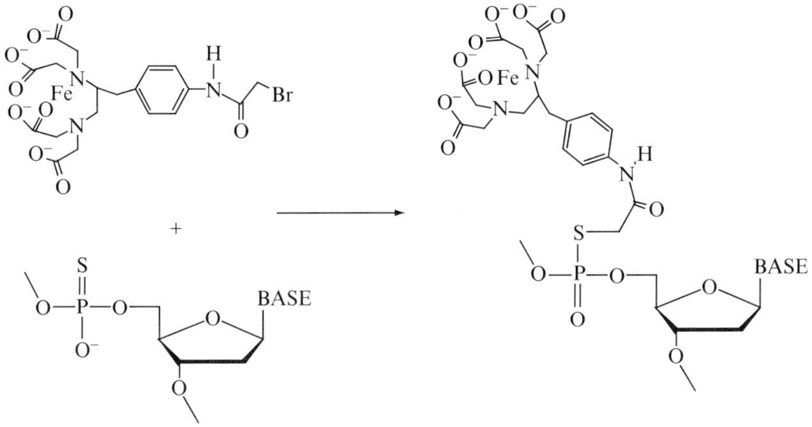

FIG. 6. Alkylation of DNA phosphorothioate groups with FeBABE. Typical conjugation reactions were performed in 20 mM MOPS, 2 mM EDTA, pH 7, containing 6 mM FeBABE and ≈0.2 mM phosphorothioate residues in DNA. After 3 h of incubation at 50°, excess FeBABE was removed by a spin column equilibrated with 15% glycerol in cleavage buffer (10 mM MOPS, 120 mM NaCl, 10 mM MgCl$_2$, 1 mM EDTA, pH 7.9). Conjugates were stored at −70° until use. For control experiments, non-phosphorothioate DNA was treated in the same manner.

reach of FeBABE, regardless of which dNTPαS is used, cleavage reagents are expected to reach similarly across the DNA surface.[20]

Conjugation Yield

To determine the approximate number of FeBABE molecules that were conjugated to each double-stranded *lac*UV5 fragment [a 228-bp fragment of pLAC12 (−157 to +71) containing the *lac*UV5 promoter], a competitive ELISA was performed as described below. Fifty μL of 1.5 μM biotin-FeCITC conjugate[40] was added to each well of a neutravidin coated 96-well microtiter plate and incubated at 37° for 1 h. After washing the plate three times with PBST,[41] serial dilutions of Fe-(S)-*p*-nitrobenzyl-EDTA and the DNA conjugate were performed in separate lanes, over a range from 90 μM to 0.3 nM Fe-(S)-*p*-nitrobenzyl-EDTA and 0.8 μM to 3 pM DNA conjugate. Twenty-five μL of each dilution, followed by

[40] O. Renn, D. A. Goodwin, M. Studer, J. K. Moran, V. Jacques, and C. F. Meares, *J. Control. Release* **39**, 239 (1996).

[41] E. Harlow and D. Lane, "Antibodies: A Laboratory Manual," Cold Spring Harbor Laboratory, New York, 1988.

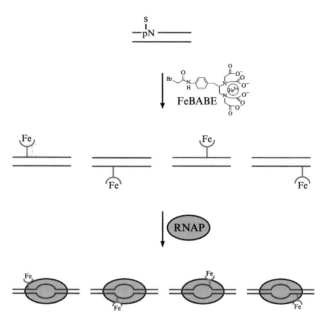

Fig. 7. DNA residue N (representing either A, G, C, or T) is lightly conjugated to produce a random labeling of FeBABE on the DNA backbone. A small percentage of the total number of residues are labeled. After formation of the RNAP-DNA complex, cleavage of the protein occurs where it is proximal to a FeBABE group.[20] [Reprinted in part with permission from B. D. Schmidt and C. F. Meares, *Biochemistry* 41, 4186 (2002). Copyright 2002 American Chemical Society.] (See color insert.)

25 μL of 10 nM anti-chelate antibody CHA255,[42] was added to each well, and the plate was shaken for 2 h at room temperature. After washing three times with PBST, 50 μL of a 1/2000 dilution of anti-mouse λ–alkaline phosphatase was added to each well and incubated for 2 h at room temperature. The plate was washed three times with PBST and rinsed with 10 mM diethanolamine, 0.5 mM MgCl$_2$, pH 9.5. Then 100 μL of *p*-nitrophenyl phosphate dissolved in diethanolamine was added to each well. The plate was centrifuged for 2 min at 1000 × g, and kinetic measurements were taken at 405 nm with a microplate reader over 1 h. All assays were done in triplicate. An example is shown in Figure 8.

[42] D. T. Reardan, C. F. Meares, D. A. Goodwin, M. McTigue, G. S. David, M. R. Stone, J. P. Leung, R. M. Bartholomew, and J. M. Frincke, *Nature* **316**, 265 (1985).

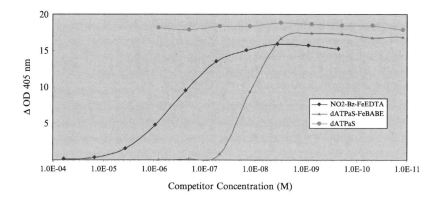

FIG. 8. Competitive ELISA comparing conjugated phosphorothioate DNA (dATPαS-FeBABE) to non-conjugated phosphorothioate DNA (dATPαS). The non-conjugated phosphorothioate DNA does not compete for the anti-chelate antibody. The number of chelates per DNA was derived by dividing the concentration of the Fe-(S)-p-nitrobenzyl-EDTA standard at 50% saturation by the concentration of DNA in the DNA-FeBABE conjugate at 50% saturation. Adapted in part with permission from B. D. Schmidt and C. F. Meares, Biochem. **41**, 4186 (2002). Copyright 2002 American Chemical Society. (See color insert.)

Transcription Assay Using Phosphorothioate DNA

Single-round runoff transcription assays were performed on both conjugated and unconjugated *lac*UV5 phosphorothioate DNA.[43–45] Briefly, an open complex was formed by the addition of phosphorothioate template DNA to holoenzyme and incubating at 37° for 15 min. Transcription was initiated by the addition of heparin and a mixture of NTPs, including $[\alpha\text{-}^{32}\text{P}]$UTP, and allowed to proceed at 37° for 5 min. Reactions were quenched, precipitated, and resolved by PAGE. The 71 nt RNA transcript was visualized on a phosphorimager. Transcriptional activities relative to non-phosphorothioate DNA were measured by densitometry and averaged for triplicate experiments. Conjugating FeBABE to the DNA had a minimal effect on the ability of RNAP to transcribe from the promoter DNA.

Cleavage of RNAP Holoenzyme by lacUV5-FeBABE Conjugate

Site-specific cleavage by *lac*UV5-FeBABE produced a great number of cleavage fragments of β and β', followed by σ^{70} and α. An example Western blot of an SDS-PAGE gel showing cleavage of β is shown in Figure 9.

[43] K. Igarashi and A. Ishihama, *Cell* **65,** 1015 (1991).
[44] M. Kajitani and A. Ishihama, *Nucleic Acids Res.* **11,** 3873 (1983).
[45] M. Kajitani and A. Ishihama, *Nucleic Acids Res.* **11,** 671 (1983).

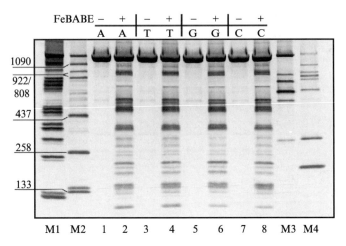

FeBABE − + | − + | − + | − +

A A | T T | G G | C C

1090

922/
808

437

258

133

M1 M2 1 2 3 4 5 6 7 8 M3 M4

Fig. 9. Protein fragments generated from FeBABE conjugated phosphorothioate *lac*UV5 cleaving RNAP holoenzyme. Immunostained Western blots of SDS-PAGE detecting β using an antibody against the *N*-terminus of β. Lanes 1, 3, 5, and 7 correspond to unconjugated *lac*UV5 phosphorothioate DNA modified at the A, T, G, and C bases, respectively. Lanes 2, 4, 6, and 8 correspond to FeBABE conjugated *lac*UV5 DNA phosphorothioate modified at the A, T, G, and C bases, respectively. Lanes M1, M2, M3, and M4 correspond to respective CNBr, NH_2OH, NTCB, and NCS chemical digests of the respective subunit.[20] Reprinted in part with permission from B. D. Schmidt and C. F. Meares, *Biochemistry 41*, 4186 (2002). Copyright 2002 American Chemical Society.

All four sets of *lac*UV5-FeBABE conjugates, corresponding to conjugation either at the A, T, G, or C nucleotides, displayed the same cleavage fragments within experimental error. This is consistent with the relatively uniform coverage of the DNA surface expected for all four phosphorothioate templates. The RNAP/DNA open complex was challenged with heparin to eliminate RNAP/DNA complexes not resulting from promoter-specific complexes. It is possible that some non-specific RNAP/DNA complexes (involving binding of RNAP at the DNA ends) may be resistant to heparin.[46] To identify cleavage fragments that are a result of end-bound complexes, FeBABE was conjugated to DNA containing phosphorothioate nucleotides located at only the ends of the DNA (prepared using phosphorothioate pcr primers and normal NTP substrates in the per reaction). The cleavage sites resulting from reactions with this DNA and the RNAP holo complex were a subset of the fragments normally seen when only the middle of the DNA is labeled. Omitting these cut sites still

[46] N. Naryshkin, A. Revyakin, Y. Kim, V. Mekler, and R. H. Ebright, *Cell* 101, 601 (2000).

produces a complete, consistent description of where the DNA interacts with the holoenzyme.

Methods

The Reagent FeBABE

Chemical Form (Typical). 26 mM FeBABE in DMSO (dimethylsulfoxide).

Storage. Store at $-70°$ in the dark. Just prior to use, the chelate solution should be thawed in a small beaker filled with room temperature water. After use, the FeBABE should be aliquoted into single-use quantities and then frozen in liquid nitrogen and stored at $-70°$. The reagent can undergo several freeze/thaw cycles with only a small loss of alkylating capacity.

Preparing the Protein

Demetalation. As shown in Scheme I, unless the macromolecule is stored in an EDTA buffer, the procedure begins by first demetalating the protein by incubating in a 0.1 M NaP$_i$, pH 7.0, 20 mM EDTA buffer. Previous work has shown that unchelated iron binds extensively to proteins such as RNA polymerase, but iron-EDTA does not.[47] Addition of EDTA for a brief period removes only a part of the bound iron. Because the side chains of amino acids such as Cys, His, Glu, Asp, etc. enable proteins to bind metals, the binding of metal ions can change the conformation of the metal-binding groups and the net electric charge of the protein; therefore metal binding introduces uncertainty into the results. In our hands, removing EDTA from buffer solutions in order to study protein cleavage by unchelated iron led to varying amounts of cleavage of the large subunits of RNA polymerase without the addition of iron, ascorbate, or peroxide. This is almost certainly due to the traces of transition metals present as contaminants in the purest biochemicals and cleanest containers.

Buffer Exchange into Conjugation Buffer. Following the demetalation, the macromolecule is buffer exchanged (via size exclusion spin columns or dialysis) into a conjugation buffer. Since the conjugation reaction proceeds by nucleophilic attack of a free thiol on FeBABE, the conjugation buffer should not contain other nucleophiles.

---

[47] D. P. Greiner, K. A. Hughes, A. H. Gunasekera, C. F. Meares, *Proc. Natl. Acad. Sci. USA* **93,** 71 (1996).

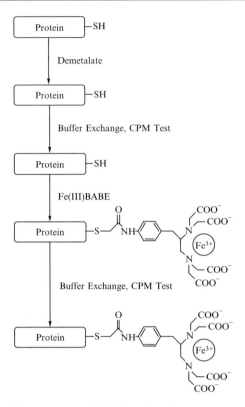

SCHEME. I. Flow chart for FeBABE conjugation to a cysteine side chain.

COMPATIBLE WITH CONJUGATION. Fifty mM MOPS or HEPES, 6 M urea, 20% glycerol, 1 mM EDTA, and 0.1 M simple salts such as NaCl, KCl, MgCl$_2$, NaOAc, etc.

NOT COMPATIBLE WITH CONJUGATION. Tris(hydroxymethyl)amino-methane (Tris), dithiothreitol (DTT), β-mercaptoethanol (βME), glu-tathione, or any other nucleophilic reagents such as those containing thiols or primary amines.

The buffer should be at pH 8, and the presence of $\geq$0.1 mM EDTA is necessary to scavenge any unbound iron. An example of a typical conjuga-tion buffer is 20 mM MOPS, pH 8.0, 0.1 M NaCl, 5% glycerol, 0.1 mM EDTA.

DEGREE OF MODIFICATION. This can be measured by the CPM {N-[4-(7-diethylamino-4-methylcoumarin-3-yl)phenyl]maleimide} assay[30] as modified by Greiner *et al.*,[9] or Ellman's[48] reagent for cysteine modification. In principle, lysine modification can be measured by the TNBS (2,4,6-trinitrobenzenesulfonic acid) assay[49] or the fluorescamine assay,[50] but this is not very useful at low degrees of conjugation.

Cleavage Reaction: Initiation

The cleavage reaction is initiated by the rapid sequential addition of ascorbate and hydrogen peroxide. Ascorbic acid (e.g., Fluka, microselect grade) is prepared fresh as a $10\times$ stock solution titrated to pH 7.0 with dilute NaOH. EDTA is added to 10 mM to scavenge adventitious metals. Hydrogen peroxide (H_2O_2) (e.g., J. T. Baker, ultrex grade) is prepared fresh as a $10\times$ stock solution containing 10 mM EDTA. Final concentrations of 5 mM ascorbate and 5 mM peroxide generally give good results. The cleavage reaction is typically complete in less than 1 min, but can be allowed to proceed longer.

For cutting caused by diffusible species such as hydroxyl radicals, the reach of each tethered reagent will be greater than the length of the tethered FeBABE. In our experiments, the distance a diffusible hydroxyl radical can travel can be moderated by the presence of a hydroxyl radical scavenger such as 10% glycerol. Because multiple cuts on a single molecule may be difficult to interpret, conditions should be chosen so that multiply cleaved molecules are rare. The same binomial distribution used to calculate the conjugation numbers in Table I can be used to predict that when 90% of all the molecules remain uncut, 9% of the molecules have been cut once and $<1\%$ of all the molecules have been cut more than once. Thus, there should be a large amount of uncut protein observed on the gel. The same mathematics described here can be applied to DNA footprinting.[51]

Quenching

Care must be taken to prevent additional cleavage of large proteins during sample workup. For protein analysis by SDS-PAGE, the reaction can be stopped by adding sample loading buffer and flash freezing in liquid nitrogen until the sample is ready to be analyzed.[19] The SDS, β-mercaptoethanol, and glycerol in sample application buffer will both quench the reaction and unfold the proteins. Avoiding the usual 95°

[48] G. L. Ellman, *Arch. Biochem. Biophys.* **82,** 70 (1959).
[49] A. F. Habeeb, *Anal. Biochem.* **14,** 328 (1966).
[50] S. J. Stocks, A. J. M. Jones, C. W. Ramey, and D. E. Brooks, *Anal. Biochem.* **154,** 232 (1986).
[51] M. Brenowitz, D. Senear, M. A. Shea, and G. K. Ackers, *Methods Enzymol.* **130,** 133 (1986).

heating step often gives better results. For nucleic acid cleavage reactions, the reaction is typically stopped by the addition of thiourea; then protein is extracted with phenol/chloroform and the sample is prepared for electrophoresis by standard methods.[52]

Analysis of Products

Gel Electrophoresis of Proteins

In contrast to nucleic acids, gel electrophoresis of proteins has yet to achieve resolution to the single-residue level. Protein fragments are separated by SDS-PAGE. The quenched cleavage reaction is loaded directly onto a Tris-glycine gel. The percentage of the gel can be varied depending on the desired range of resolution.

Western Blots

After electrophoresis, the gel is equilibrated for 10 min in 10 mM CAPS, pH 11, 10% methanol, electroblotting buffer. The proteins are electroblotted onto polyvinylidene difluoride (PVDF) membrane for 2 h at 50 V in electroblotting buffer. The membrane is then blocked in a solution of 5% nonfat dry milk in TBS for at least 1 h. The membrane is washed $4 \times$ (5 min each) with TBST before immunostaining.

Markers

For accuracy, cut sites should be assigned by comparison to the migration of known fragments created by chemical or enzymatic digestion, rather than by commercial markers. This helps eliminate potential problems associated with the anomalous migration of protein fragments. It is best to try a variety of proteases and chemicals and choose those that will give the best standard curve to assign the cleaved fragments. Reagents that we have found useful include NTCB to cut at the amino terminal side of cysteines,[53] cyanogen bromide to cut at the carboxyl terminal side of methionines,[54] N-chlorosuccinimide to cleave at tryptophan residues[55] and hydroxylamine to cut between asparagine and glycine.[56] The details

[52] J. Sambrook, E. F. Fritsch, and T. Maniatis, "*Molecular Cloning: A Laboratory Manual,*" 2nd ed., Cold Spring Harbor Laboratory Press, New York, 1989.
[53] G. R. Jacobson, M. H. Schaffer, G. R. Stark, and T. C. Vanaman, *J. Biol. Chem.* **248,** 6583 (1973).
[54] E. Gross, *Methods Enzymol.* **11,** 238 (1967).
[55] M. A. Lischwe and M. T. Sung, *J. Biol. Chem.* **252,** 4976 (1977).
[56] P. Bornstein and G. Balian, *Methods Enzymol.* **47,** 132 (1977).

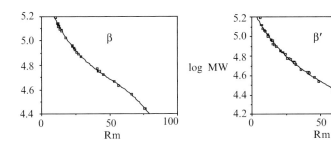

FIG. 10. Examples of polynomial fits of the electrophoretic migration of defined fragments of *E. coli* RNAP subunits, used to assign the molecular weights of cleavage fragments (see text). Adapted from J. T. Owens *et al., Proc. Natl. Acad. Sci. USA* **95,** 6021 (1998). Copyright 1998 by the National Academy of Sciences of the United States of America.

of assigning cleavage sites and determining uncertainty from molecular weight standards have been discussed.[17,18] In one example, the cleavage sites for each fragment were assigned by using a third-order polynomial fit of log molecular weight versus migration distance (Rm) on SDS-PAGE for a set of known marker fragments (Fig. 10). The errors of these assignments were estimated by refitting after removal of each point in turn and reanalyzing the data; β errors ranged from ±2 to ±9 residues (average ±6); β' errors ranged from ±7 to ±16 residues (average ±11).

Peptide-Terminus Specific Antibodies

When assigning cut sites using gel electrophoresis, it is useful to have an end-labeling scheme to visualize all the fragments that possess the original N- or C-terminal polypeptide sequence. This is conceptually equivalent to labeling one end of a nucleic acid chain with ^{32}P. We have prepared antibodies that are specific for either the amino or carboxyl terminal end of the polypeptide being cleaved. Alternatively, a commercially available antibody against an epitope tag that has been inserted into the protein sequence can be used.

We use the multiple antigenic peptide (MAP) method[57] to produce highly sensitive rabbit polyclonal antibodies against the native polypeptide terminus. The peptides attached to the eight-branched core matrix are either the first 10 to 15 residues of the protein for an amino terminal antibody or the last 10 to 15 residues for a carboxyl terminal antibody. The MAP is then injected into rabbits and serum is collected by standard methods.[41] The antibodies must be affinity-purified from the antiserum before use.[47]

---

[57] J. P. Tam, *Proc. Natl. Acad. Sci. USA* **85,** 5409 (1988).

Comparison with Structural Data

Recent high-resolution X-ray crystal structures of *Thermus aquaticus* RNA polymerase holoenzyme[58] and the *Taq* holoenzyme-fork junction promoter complex,[59] as well as the *Thermus thermophilus* holoenzyme,[60] provide the opportunity to compare structural information gained using FeBABE with an X-ray crystal structure of a related transcription complex. We will consider *Taq* in detail; the *T. thermophilus* structure appears to be in substantial agreement.

Because the FeBABE experiments were performed on *E. coli* RNAP in solution while the crystal structure was determined on *Taq* RNAP, care is needed in comparing the two. For our working assumptions, we use homology matches of *E. coli* to *Taq* sequences generously supplied by Seth Darst and Katsuhiko Murakami, or obtained using ClustalW software. We also must assume that the important parts (e.g., conserved regions) of *E. coli* RNAP in solution and *Taq* RNAP in the crystal have matching 3D structures, even though the overall sizes of the subunits are significantly different (Table II). In many cases the cut sites we observed on *E. coli* subunits were in regions that lacked homology with *Taq* because of long insertions or deletions in one sequence relative to the other, so they were omitted from consideration. Other potential sources of error include the effects of uncertainty in the assignment of FeBABE cut sites from the migration of gel bands, any chemical consequences (e.g., steric blockage, radical scavenging, or lack thereof) resulting from inserted or deleted peptide segments in *E. coli* RNAP relative to *Taq*, the possible consequences of multiple cuts on the target, and steric effects resulting from the presence

TABLE II

COMPARISON OF THE LENGTHS OF THE POLYPEPTIDE CHAINS OF THE SUBUNITS IN *E. COLI* AND *TAQ* RNAP.

Subunit	*E. coli*	*Taq*
α	329aa	314aa
β	1342	1118
β'	1407	1524
σ	613	438

[58] K. S. Murakami, S. Masuda, and S. A. Darst, *Science* **296,** 1280 (2002).

[59] K. S. Murakami, S. Masuda, E. A. Campbell, O. Muzzin, and S. A. Darst, *Science* **296,** 1285 (2002).

[60] D. G. Vassylyev, S.-I. Sekine, O. Laptenko, J. Lee, M. N. Vassylyeva, S. Borukhov and S. Yokoyama, *Nature* **417,** 712 (2002).

of attached FeBABE. The results in Table III should be interpreted with these limitations in mind.

Sigma/Core Interactions Using Single-Cys Mutants of Sigma

Here we compare protein interactions between conserved regions of σ and the core enzyme subunits identified using FeBABE-conjugated single-Cys mutants of *E. coli* σ^{70}, with those identified in the recent crystal structure of the *T. aquaticus* holoenzyme containing σ^{A}.[58] Despite the differences between the two methods, and the potential structural differences between thermophilic and non-thermophilic enzymes themselves, agreement is relatively high. For example, both structural elucidation methods identified conserved regions 2.1, 2.2, and 3.1 of both sigmas as having interactions with conserved region B, the coiled-coil region, and region C on the β' subunit of each polymerase. In addition, both systems identified conserved region 3.1 of σ as interacting with the amino-terminal end of conserved region D on the β subunit. Further, both FeBABE mapping and the crystal structure indicate conserved region 4.2 of σ interacts with the *C*-terminal end of conserved region G on β. Finally, the regions in or near conserved region 4.2 of both σ subunits were identified as interacting with the *C*-terminal end of β' conserved region C. The *absence* of substantial α-σ interactions in both model systems is also worth noting.

However, FeBABE mapping suggested that conserved regions 3.1 and 3.2 of σ^{70} interact with region G of β,[18] whereas the crystal structure of *Taq* σ^{A} did not identify these as primary contacts. In addition, the FeBABE probe positioned on mutant residue 496C of *E. coli* σ^{70} did not map to the amino-terminal end of conserved region H of the β subunit, as expected from the *Taq* crystal structure. Instead, FeBABE positioned downstream at mutant residue 517C identified this contact. It should be noted that the Western-blot detection method using *N*- or *C*-terminal–specific antibodies becomes problematic for identifying cut sites near either terminus. For example, identifying cut sites near the *C*-terminus using an *N*-terminal antibody is difficult, because the large peptide fragment may not be resolved from the dominant uncut band. On the other hand, cut sites near the amino terminus would produce a small peptide containing the *N*-terminal epitope, which could run off the SDS-PAGE or could blot poorly, and a large *C*-terminal fragment lacking the *N*-terminal epitope, which would go undetected. Each of these two scenarios might be exemplified by contacts between region 3.2 of σ and β conserved region I, and between σ region 3.1 and β region B, that were identified in the crystal structure but not using FeBABE mapping.

These discrepancies may also simply reflect differences in the two σs and their respective polymerases, but it is instructive to consider the special properties and limitations of affinity cleavage that may affect its results even when detection schemes other than Western blotting are used. A FeBABE probe may be situated near an important contact site but have an improper geometric orientation to achieve peptide backbone cleavage. This leads to the non-detection of an interaction or proximity relationship. Other chemical probe techniques such as cross-linking share this property; the occurrence of a chemical event provides good evidence for proximity, whereas the absence of a chemical event is weaker evidence for remoteness.

Further insight is provided by considering how well experimentally assigned FeBABE cut sites on *E. coli* RNAP map to inferred sites on the *Taq* crystal structure. Using only the strong cuts assigned by Owens *et al.*[18] and the computed homologies between the two enzymes, we find that 18 out of 19 strong cuts assigned by purely biochemical means have plausible sites on the structure (Table III). We arbitrarily chose plausible sites as being within 25 Å of the homologous σ residue on the crystal structure (allowing for cleavage by diffusible hydroxyl radical), or being at least 50% exposed to solvent and within several residues of the inferred cut site in β or β' (allowing for error in assignment of the cut site), or both. The cut for which there is no match may reflect a difference between the two enzymes, or it may be a second cut on the target (the molecule may change its shape after the first cut).

Sigma/DNA Interactions Using Single-Cys Mutants of Sigma

As an extension of the comparison of protein/protein interactions identified by FeBABE-σ^{70} and those determined in the *Taq* σ^A holo enzyme crystal structure, here we compare σ/DNA interactions identified in both systems. In general, *Taq* σ^A/DNA contacts in the crystal structure agree fairly well with those identified using FeBABE conjugated *E. coli* σ^{70}. In agreement with previous genetic studies,[61] both systems position σ-conserved region 4 along the -35 recognition element, including regions further upstream corresponding to potential DNA UP-element sites. Both models also identify the single-stranded, nontemplate DNA of the -10 element as positioned to interact with exposed aromatic residues of σ region 2.3. A DNA bend just upstream of the transcription bubble near position -16 is also consistent with both models. In agreement with previous FeBABE mapping studies which located region 2.5 of *E. coli* σ^{70}

[61] M. Lonetto, M. Gribskov, and C. A. Gross, *J. Bacteriol.* **174,** 3843 (1992).

TABLE III

COMPARISON OF STRONG CUT SITES ON *E. COLI* WITH HOMOLOGOUS SITES ON THE *TAQ* CRYSTAL STRUCTURE. ENTRIES (E.G., 396/219) GIVE THE *E. COLI* SEQUENCE NUMBER FOLLOWED BY THE HOMOLOGOUS *TAQ* SEQUENCE NUMBER. STRONG CUTS ON β AND β' WERE OBSERVED WITH FOUR SINGLE-CYS *E. COLI* σ[70] MUTANTS (396, 496, 517, 581).

σ coli/Taq	β' coli/Taq Strong cuts			β coli/Taq Strong cuts			
396/219	344/633	307/582		490/370	467/347		
Taq xtal str	606[b], 637[a]	582[b], 596[a]		365[a,b]	353[a,b]		
496/321	330/606	298/573	278/553	900/772	854/726	489/369	
Taq xtal str	596[a], 601[b]	no match	549[b], 556[a]	772[b]	716[a,b], 726[a]	365[a], 366[b]	
517/342(346)	328/604	294/569	276/553	913/785	858/730	550/430	493/373
Taq xtal str	596[a], 604[b]	556[a], 575[b]	550[b], 556[a]	781[a], 786[b]	716[a,b], 720[b]	420[b]	365[a], 380[b]
581/406	328/604			913/785			
Taq xtal str	596[a]			777[b], 781[a]			

[a] 50% exposed in crystal structure.
[b] Within 25 Å in crystal structure.

as interacting with the extended −10 promoter sequence,[62] the crystal structure also identified *Taq* σ^A residues His278 and Glu281 (corresponding to His455 and Glu458 of *E. coli*) as interacting with the extended −10 element.

However, a range of *Taq* σ^A/fork-junction DNA contact sites identified in the crystal structure were not observed using the relatively small number of cysteine conjugated FeBABE mutants of *E. coli* σ^{70} on the *lac*UV5 promoter.[63] Many of these unidentified contact sites likely arose from the practical constraints of using the single-Cys tethered FeBABE mapping strategy itself. When choosing amino acid residues within σ^{70} for site-directed mutagenesis and FeBABE conjugation, attempts were made to minimize structural and functional perturbations resulting from mutagenesis and probe conjugation. For this reason, sites that flanked conserved regions were chosen, and invariant residues within the σ family were avoided. In addition, the structure of the *E. coli* σ^{70} open complex positioned on the lacUV5 promoter, which includes both upstream and downstream DNA sequences, may be different from the crystallized *T. Aquaticus* σ^A/fork-junction complex.

Sigma/Core Interactions Using Lysine Conjugates of Sigma

The results of *E. coli* σ^{70} conjugates at lysine using 2IT-FeBABE are similar to the sum of the results with single-Cys mutants discussed previously.[37] The value of the lysine-conjugated σ should be to give an outline of the binding interface in a single experiment. A simple illustration of the power and the limitations of these results is provided by the comparison of the inferred *Taq* cut sites on β and β' with the location of *Taq* σ (Fig. 11). The binding surface of *Taq* σ on core is crudely outlined by the inferred cut sites from the *E. coli* enzyme. Further analysis is in progress.

DNA/Holoenzyme Interactions Using Proteolytic DNA

The results of protein cutting with *lac*UV5 phosphorothioate conjugates using FeBABE provide an overall map of those regions of the enzyme subunits that are at the periphery of the DNA-binding site,[20] giving an outline of the binding interface in a single experiment. It is notable that no cuts were observed in *E. coli* σ regions 4.1 or 4.2 (σ_4), which interact with the −35 region of the promoter[20]; this is probably due to the small

[62] J. A. Bown, J. T. Owens, C. F. Meares, N. Fujita, A. Ishihama, S. J. W. Busby, and S. D. Minchin, *J. Biol. Chem.* **274,** 2263 (1999).

[63] J. T. Owens, A. J. Chmura, K. Murakami, N. Fujita, A. Ishihama, and C. F. Meares, *Biochemistry* **37,** 7670 (1998).

FIG. 11. *Taq* RNAP structure showing cuts inferred using lysine-conjugated *E. coli* 2IT-FeBABE. The panels show the α-carbon trace with cuts rendered as 20aa runs of α-carbon spheres on β (green) and β' (red), first without σ (*left*) and then with σ (lavender) (*right*).[64] The two α subunits at the top of the structure are colored yellow and blue, and the omega subunit is gray. (See color insert.)

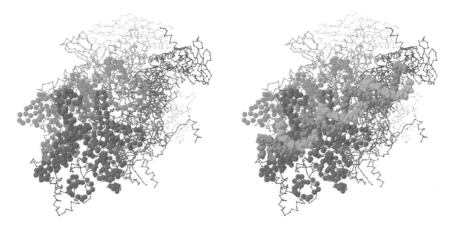

FIG. 12. The cut sites observed by proteolytic *lac*UV5 DNA on *E. coli* RNAP,[20] mapped onto the homologous regions of the structure of the *Taq* RNAP-DNA fork-junction complex,[59] which contains the upstream promoter region but not the downstream template. The panels show the α-carbon trace with cuts rendered as 20aa runs of α-carbon spheres on β (green), β' (red), and σ (lavender), first without DNA (*left*) and then with DNA (*right*). DNA strands brown (*upper*) and light blue (*lower*) run from upper right (*upstream*) to lower left (*fork*). The σ_4 region lies under the upstream end of the DNA (see text). Other colors match those in Figure 11. (See color insert.)

[64] N. Guex and M. C. Peitsch, *Electrophoresis* **18,** 2714 (1997).
http://www.expasy.org/spdbv/

fragments ($\sim$50aa) produced when σ is cut in this region and analyzed with a C-terminal antibody. The comparison of the inferred Taq cut sites on σ, β, and β' with the location of fork-junction DNA in the structure[59]—which should show the upstream interactions—is shown in Fig. 12. The binding surface of the upstream DNA on the Taq subunits is crudely outlined by the inferred cut sites from the $E.\ coli$ enzyme, which also include cuts caused by downstream DNA that is not included in the structure.

[7] Iodine-125 Radioprobing of *E. coli* RNA Polymerase Transcription Elongation Complexes

By VALERI N. KARAMYCHEV, ALEXEI TATUSOV, NATALIA KOMISSAROVA, MIKHAIL KASHLEV, RONALD D. NEUMANN, VICTOR B. ZHURKIN, and IGOR G. PANYUTIN

Radioprobing is based on an analysis of DNA or RNA strand breaks produced by decay of Iodine-125 (^{125}I) incorporated into one of the strands. ^{125}I belongs to a class of radioisotopes called "Auger electron emitters" that decay by electron capture or internal conversion.[1-3] Both of these processes create a vacancy in an inner atomic shell that results in a cascade of electron transitions accompanied by the emission of a number of orbital electrons, named Auger electrons after Pierre Auger, who first discovered them in the 1920s. Decay of ^{125}I produces, on average, 21 such electrons. The energy of the majority of these electrons is below 1 kev, which is considerably less than the energy of nuclear-generated beta particles. Auger electrons produce breaks in DNA or RNA strands either directly or by generation of OH radicals in "bound" water. It is assumed that the strand breaks are due to the energy deposition or radical attack in the sugar-phosphate moiety because damage to the DNA bases generally does not lead to strand scission.[4] Due to the extremely low energy of Auger electrons, most of the breaks (90%) are located within approximately one helical turn from the ^{125}I incorporation site. In addition, the bases located next to that site are affected by the positively charged (+21 on average for ^{125}I) daughter nucleus, which, by stripping electrons from the neighboring bonds,

[1] L. E. Feinendegen, *Radiat. Environ. Biophys.* **12**, 85–99 (1975).
[2] K. S. Sastry, *Med. Phys.* **19**, 1361–70 (1992).
[3] K. G. Hofer, *Acta. Oncol.* **35**, 789–96 (1996).
[4] H. Nikjoo, C. A. Laughton, M. Terrissol, I. G. Panyutin, and D. T. Goodhead, *Int. J. Radiat. Biol.* **76**, 1607–15 (2000).

can ultimately result in breakage of DNA strands. At distances approximately 7 nucleotides from the decay site, the damage produced by diffusible OH radicals generated in "free" water becomes comparable with that of Auger electrons.[5] The frequency of breaks produced by Auger electrons depends primarily on the distance from ^{125}I to the target atom (or the group of atoms) because nucleic acids and proteins are as "transparent" for the low-energy Auger electrons as water. Therefore, radioprobing, like "low resolution" NMR, allows one to obtain information on the interatomic distances, and, in principle, to reconstruct the 3-dimensional (3-D) structure of nucleic acids in complexes with proteins.

Due to the complex nature of DNA breaks and multiple mechanisms involved, there is no simple analytical relationship between the frequency of breaks at a given base and the distance to the sugar attached to this base. By analyzing breaks in the DNA strands in a DNA-protein complex with known 3-D structure,[6] we established an empirical relationship between the frequency of breaks and the distance from the site of ^{125}I decay.[7] Thus, by measuring the intensities of breaks in two strands of the DNA duplex in the RecA protein-mediated synaptic complex, it was possible to estimate the distances from the ^{125}I position in the invading strand to the strands of the duplex.[8] In addition, the profile of the distribution of breaks along a DNA strand depends on the conformation of this strand. By comparing the distributions of breaks in free DNA and in the complex with cyclic AMP receptor protein (CRP), we were able to detect the kink in DNA induced by the protein.[6] Therefore, two types of information can be obtained from a radioprobing experiment: the distance from the ^{125}I-labeled strand and the target strand, and the conformation of the target strand. Here we present sample results on the interstrand distances and conformation of the RNA/DNA heteroduplex in *E. coli* RNA polymerase (RNAP) transcription elongation complex.

Experimental Scheme

Transcription elongation complexes (EC) were either assembled on short DNA and RNA oligonucleotides or obtained on a promoter-containing template as reported earlier[9] and described in detail in Chapter 17 of this

[5] P. N. Lobachevsky and R. F. Martin, *Radiat. Res.* **153**, 271–8 (2000).

[6] V. N. Karamychev, V. B. Zhurkin, S. Garges, R. D. Neumann, and I. G. Panyutin, *Nat. Struct. Biol.* **6**, 747–50 (1999).

[7] V. N. Karamychev, I. G. Panyutin, R. D. Neumann, and V. B. Zhurkin, *J. Biomolec. Struct. and Dynamics* **11**, 156–67 (2000).

[8] V. A. Malkov, I. G. Panyutin, R. D. Neumann, V. B. Zhurkin, and R. D. Camerini-Otero, *J. Mol. Biol.* **299**, 629–40 (2000).

volume.[10] In the first case, a 55-nt long transcribed strand DNA oligonucleotide (T-strand) was annealed to a 8-nt long RNA oligonucleotide, followed by addition of RNAP and a 55-nt long nontranscribed strand DNA oligonucleotide (N-strand). EC was attached to Ni^{2+}-NTA agarose beads through RNAP histidine tag. Transcription was initiated by the addition of $[^{125}I]ICTP$ to the beads followed by an extensive wash with NTP-free transcription buffer. An aliquot of the resulting EC1 (here and below, the index indicates a distance in nucleotides between the $[^{125}I]IC$ and the 3′ end of the RNA) containing 9-nt long RNA transcript with $[^{125}I]IC$ residue in the active center of RNAP was removed at this point for the subsequent breaks analysis. EC3 to EC10 were obtained similarly by adding the appropriate NTPs, washing and removing aliquots as outlined in Fig. 1A. In all

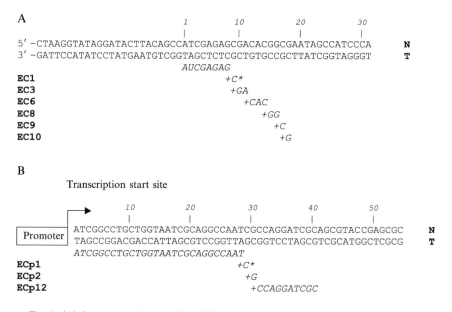

FIG. 1. (A) Sequences of transcribed (T) and nontranscribed (N) strands of DNA duplex and RNA primer (italic). (B) Sequence of the transcribed part of the promoter-containing DNA fragment and the RNA transcript (italic). "+" show the nucleotides added at the consecutive steps of RNAP walking to obtain the corresponding ECs and ECps (listed on the left). C^* denote the $[^{125}I]IC$ residues.

[9] I. Sidorenkov, N. Komissarova, and M. Kashlev, *Mol. Cell.* **2,** 55–64 (1998).

[10] N. Komissarova, M. I. Kireeva, J. Becker, I. Sidorenkov, and M. Kashlev, *Methods Enzymol.* **371,** 233–251 (2003).

the complexes, ^{125}I was in the same position relative to the transcribed strand but moved farther away from the active site as the RNAP was "walked" along the template as shown in Figs. 2A, B. DNA strands in the EC were labeled with ^{32}P either at the 5′ or 3′ end for detection of breaks. The EC samples were stored at −80°C for 14 days for accumulation of DNA breaks (^{125}I half-life is 60 days). The breaks in DNA strands were analyzed in 12% denaturing polyacrylamide gels (see Fig. 3). The intensity of the bands in the gel reflects the frequency of breaks at the corresponding nucleotides. These intensities were measured by fitting the intensity profile of each lane with a series of the Lorentzian peaks representing the individual bands. An example of such a fit for EC1 (see Fig. 3, lane "1") is shown

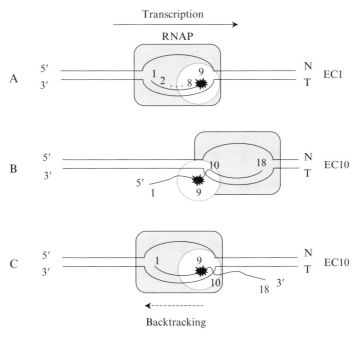

Fig. 2. Experimental scheme. RNAP EC was assembled with two complementary DNA strands and 8-mer RNA primer. (A) At the first step, the ^{125}I]ICTP (shown as an "explosion") was added and incorporated into the active center of RNAP, resulting in EC1. The ribonucleotides are numbered in the 5′-3′ direction. (B) Walking of RNAP along the DNA template results in elongation of the RNA transcript and movement of the ^{125}I]IC residue relative to the EC. (C) Backtracking of the EC leads to exposure of the 3′-end of the RNA transcript and reverse movement of the ^{125}I]IC residue relative to the EC. The circles around the ^{125}I]IC residues show the approximate range of ^{125}I decay-induced breaks (i.e., the range of measurable RNA-DNA distances).

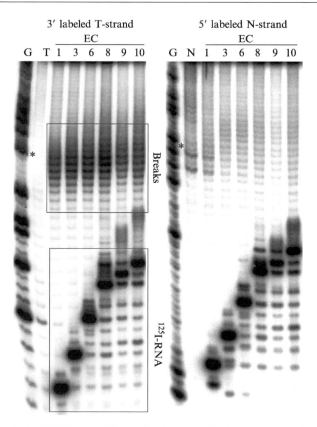

FIG. 3. Analysis of DNA strand breaks in the transcribed and nontranscribed strands in 12% denaturing PAGE. The numbers on the top of the lanes correspond to the analyzed EC. "G"—Maxam-Gilbert G sequencing ladder. "T" and "N" are the respective strands from the assembled ECs before addition of [^{125}I]ICTP. Bands corresponding to ^{125}I-produced breaks and ^{125}I-labeled RNA transcripts are outlined with boxes on the panel of the transcribed strand. Stars mark the positions of the guanine in the T-strand complementary to [^{125}I]IC, and the corresponding cytosine in the N-strand. The T-strand was ^{32}P-labeled at the 3' end and the N-strand was labeled with ^{32}P at the 5' end.

in Fig. 4. The resulting values were normalized and shown as graphs of break distributions for EC1 to EC10 (see Figs. 5 and 6). Based on these distributions, we analyzed DNA and RNA fold in the RNAP EC.

Alternatively, ECs were formed on a 124-bp long DNA fragment containing A1 promoter of bacteriophage T7. Transcription was initiated from the promoter and 28-nt long RNA transcript was synthesized by RNAP walking as described above. At this point, [^{125}I]ICTP was incorporated into

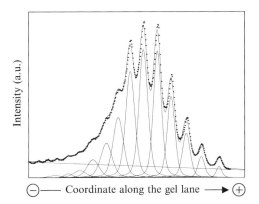

FIG. 4. Deconvolution of the experimental (dots) intensity profile (lane EC1 for the T-strand, Fig. 3) with a series of Lorentz-type peaks corresponding to individual bands. The resulting best fit of the experimental data (the sum of all peaks) is also shown. X axis—distance along the gel; Y axis—intensity in arbitrary units (a.u.).

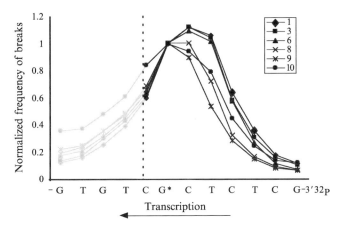

FIG. 5. Relative frequencies of breaks at the individual bases in the T-strand labeled at the 3′ end normalized on the intensity at the G* opposite [^{125}I]IC. Symbols are assigned as follows: "diamond"—EC1, "square"—EC3, "rectangle"—EC6, "cross"—EC8, "star"—EC9, "circle"—EC10. Data points on the left are grayed out (see text for details). Note that the direction of transcription in this orientation of the T-strand is from right to left.

the transcript and two aliquots of ECp1 ("p" stands for promoter-initiated, to distinguish from the previously described EC1) were removed for the subsequent RNA length and DNA breaks analysis. The remaining ECp was walked farther along DNA template. DNA breaks in the T-strand of ECp1, ECp2, and ECp12 were analyzed in denaturing PAGE.

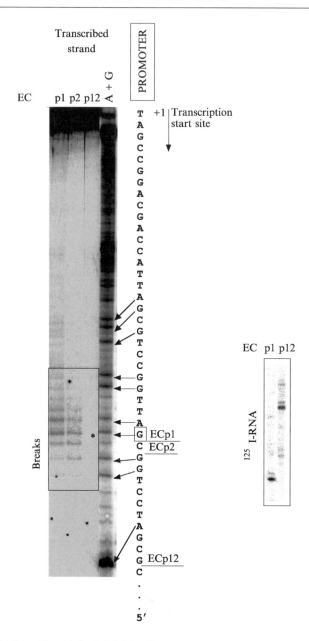

FIG. 6. Radioprobing of the EC initiated from the T7 A1 promoter (ECp). Left panel: 5′-labeled template DNA strand breaks in the ECp. The numbers above the lanes of the 20% denaturing PAGE images indicate the analyzed ECp. "A + G"—Maxam-Gilbert AG

Materials and Equipment

Oligonucleotides were synthesized on an ABI-394 DNA synthesizer (Applied Biosystems, Foster City, California) and purified by polyacrylamide gel electrophoresis (PAGE). T4 polynucleotide kinase and bovine serum albumine were purchased from New England Biolabs (Beverly, MA). Hexahistidin-tagged *E. coli* RNAP was purified and immobilized on Ni^{2+}-nitrilotriacetate (NTA)-agarose as described.[11] The concentration of the protein was determined by BSA* Protein Assay (Pierce, Rockford, IL). Terminal deoxynucleotidyl transferase (TdT) was purchased from Fermentas (Hanover, MD). RNase A was from Sigma (St. Louis, MO). All chemicals were at least of analytical grade and were purchased from Sigma. Radiochemicals were purchased from PerkinElmer Life Sciences (Boston, MA), except where indicated otherwise. HPLC was performed on model 1050 chromatography system (Hewlett Packard, Palo Alto, CA) equipped with UV and γ-RAM (IN/US Systems, Fairfield, NJ) detectors using PRP-1 (150 × 4.1 mm, 3 μm) C18 column (Hamilton, Reno, NV).

Synthesis of [^{125}I]ICTP

Solution of $Tl(NO_3)_3$ in HNO_3 was prepared by addition of 20 mg Tl_2O_3 to 500 μl of concentrated HNO_3 in a quartz tube. The mixture was heated on the top of a magnetic stirrer until $Tl(NO_3)_3$ was completely dissolved. The solution is ready to use after cooling down to room temperature and can be stored at room temperature for two weeks. One microliter of $Tl(NO_3)_3$ in HNO_3 was mixed with 30 μl of 0.5 M acetate buffer (pH 4.5), 1 μl (10 nmol) of 10 m*M* CTP, and 3 μl of Na^{125}I (0.5 mCi) in 0.1 M NaOH (0.2 nmol, 2200 Ci/mmol). After 15 min at 75 °C, the reaction was quenched with 10 μl of sodium bisulfite (10 mg/ml). The [^{125}I]ICTP was purified from unreacted CTP and other components of the reaction by reverse phase HPLC. Separation was performed on a 150 × 4.6 mm C18 column using 2% to 10% gradient of acetonitrile in 50 m*M* triethylammonium

---

[11] M. Kashlev, E. Nudler, K. Severinov, S. Borukhov, N. Komissarova, and A. Goldfarb, *Methods Enzymol.* **274**, 326–34 (1996).

---

sequencing ladder. The transcribed sequence of the promoter template is shown on the right of the gel. The position of the guanine in the T-strand complementary to [^{125}I]ICMP, and the corresponding cytosine in the N-strand are boxed and starred in the sequence. The 3′ ends of the RNA in ECp1, ECp2, and ECp12 are also indicated on the sequence. The [^{125}I-labeled RNA is not visible on the gel because the samples were treated with RNase A before the electrophoresis. The RNA was analyzed on a separate gel (right panel).

bicarbonate (pH 7.5). The peak corresponding to the product (retention time 12 min) was collected and lyophilized in a vacuum concentrator. The samples were dissolved in 3 μl H_2O and kept at $-80\,^\circ$C. Radio-HPLC and TLC analysis showed 96% purity. The stock solution can be stored for 2 days; after that, degradation of [^{125}I]ICTP reduces the yield of incorporation of [^{125}I]ICTP into RNA. Note that the presence of unreacted CTP in the [^{125}I]ICTP preparation also reduces the incorporation of [^{125}I]ICTP into RNA, leading to low frequency of breaks in DNA targets. The complete removal of CTP (retention time 4 min) from [^{125}I]ICTP (retention time 12 min) was achieved by shallow gradient of acetonitrile (from 2% to 10% for 20 min).

Labeling of DNA Oligonucleotides

The template strand (see Fig. 1B), was labeled at the 3' end with [$3'$-α-^{32}P]dATP and TdT or at the 5' end with [$5'$-γ-^{32}P]ATP and T4 polynucleotide kinase. The nontemplate strand was labeled at the 5' end with [$5'$-γ-^{32}P]ATP and T4 polynucleotide kinase. The maximum specific activity of labeled oligonucleotides was achieved by using 2 to 3 times molar excess of fresh [$5'$-γ-^{32}P]ATP and [$3'$-α-^{32}P]dATP relative to oligonucleotide. The ^{32}P-labeled oligonucleotides were purified on a MicroSpin G-50 column (Amersham Biosciences, Piscataway, NJ), evaporated to desired concentration, and equilibrated with transcription buffer (TB; 20 mM Tris-HCl (pH 7.9), 40 mM KCl, 5 mM $MgCl_2$, 1 mM $\beta\beta$-mercaptoethanol). The incorporation of ^{32}P into oligonucleotides was measured in a liquid scintillation counter (Packard model 2200CA).

Assembly of Transcription Elongation Complex and RNAP "Walking"

The T-strand (30 pmol), ^{32}P-labeled at the 5' or 3' end, was incubated with RNA-oligonucleotide (70 pmol) in 100 μl reaction mixture in 1 $\times$ TB buffer for 5 min at 45 $^\circ$C. After the samples were cooled down to room temperature, 6 μl (26 pmol) of RNAP were added, and the samples were mixed and kept for 10 min at room temperature, followed by the addition of 3 μl (120 pmol) of N-strand and incubation at 37 $^\circ$C for 10 min. Similarly, ECs were assembled with ^{32}P-labeled N-strand. In this case the T-strand was not ^{32}P-labeled. The concentration of all reagents in the experiments with ^{32}P-labeled N-strand was the same as in the experiments with ^{32}P-labeled T-strand. The assembled ECs were transferred to a 1.5 ml plastic tube containing 50 μl Ni^{2+}-NTA agarose prewashed with TB, mixed immediately, and agitated for 5 min to ensure uniform binding of EC to the solid phase. The Ni^{2+}-NTA agarose pellet was then washed 4 times with 1 ml of ice-cold TB, and the final reaction volume was adjusted to 30 μl with TB. The elongation was started by adding 0.4 mCi [^{125}I]ICTP in 3 μl H$_2$O. After 5 min, the agarose pellet was washed 4 times with 1 ml of ice-cold TB and two 1 μl aliquots were removed. One of the aliquots was diluted with 2.5 μl of RNA loading dye (9.8 M urea, 50 mM EDTA, 0.1% xylencyanol) and analyzed by 20% denaturing PAGE for ^{125}I incorporation. Another 1 μl aliquot was immediately frozen at $-80\,^{\circ}$C. The remaining assembled complex was "walked" in the presence of 5 μM of appropriate NTP subsets for 5 min at room temperature, then washed with TB 4 times, and 2 1-μl aliquots were analyzed/frozen as in the previous step. The procedure was repeated to obtain the RNA transcripts of various lengths. The sequence of the addition of NTPs is shown in Fig. 1A. The analysis of viability of the EC complexes after storage at $-80\,^{\circ}$C for 14 days showed that 80–90% of the complexes were able to produce runoff transcript after thawing and addition of all 4 NTPs (data not shown).

Analysis of Breaks in DNA

After 14 days of storage at $-80\,^{\circ}$C the samples were thawed, immediately mixed with 1 μl of formamide-based Stop solution (USB, Cleveland, OH), and analyzed by 12% denaturing PAGE for fragmentation of the ^{32}P-labeled DNA strands. Maxam-Gilbert sequencing reactions were performed as described.[12] The gel was fixed in 10% acetic acid, dried on Whatman 3-mm Chr paper with DrygelSr (Hofer Scientific, San Francisco, CA) and digitized with a BAS 1500 BioImaging Analyzer (FUJI Medical Systems, Stanford, CT). To measure the intensity of the individual bands, the intensity profile of each lane was generated from the digitized gel image using Fuji Lab Image Gauge software (FUJI Medical Systems, Stanford, CT). The profile was deconvoluted on a series of the Lorentz-type peaks corresponding to individual bands as described in detail in.[13] The best-fit curves were produced with the PeakFit software package for PC (SPSS Inc., Richmond, CA). In the ECs assembled on 5'-^{32}P-labeled T-strand, ^{125}I-labeled RNA transcripts comigrated in the gel with the DNA fragments produced by ^{125}I (data not shown). To degrade the RNA and to clean up the gel for a more accurate measurement of break distribution, we treated these samples with RNAse A before loading on the gel.

[12] A. M. Maxam and W. Gilbert, *Methods Enzymol.* **65**, 499–560 (1980).
[13] S. E. Shadle, D. F. Allen, H. Guo, W. K. Pogozelski, J. S. Bashkin, and T. D. Tullius, *Nucleic. Acids Res.* **25**, 850–60 (1997).

Radioprobing of Promoter-Initiated Elongation Complexes

A stable EC containing 14-nt long RNA was obtained by preincubating 25 pmoles of the template containing T7 A1 promoter (124-nt PCR-generated, agarose-purified fragment; see Fig. 1B for the sequence of the transcribed region) with 10 pmoles of *E. coli* RNAP in 120 μl TB for 5 min at 37 °C and subsequently adding 100 μM of trinucleotide RNA primer ApUpC and 10 μM of UTP, CTP, and GTP for 5 min at 37 °C. The EC was next immobilized on 45 μl of Ni^{2+}-NTA agarose beads suspension (50% v/v) prewashed with TB. After a 5-min incubation at 25 °C, the NTPs were washed off with TB, the reaction volume was brought to 50 μl, and the DNA was labeled by a 10-min incubation of the immobilized complex with 50 units of T4 polynucleotide kinase (New England Biolabs) and 1 mCi of [γ-^{32}P] ATP (7000 Ci/mmol; ICN Biomedicals, Inc.) at 25 °C. The complex was washed with TB, incubated for 5 min with TB containing 1M KCl to remove nonspecifically bound DNA, and washed again with regular TB. The complex was walked in the presence of 5 μM of appropriate NTP subsets until a 28-nt long RNA was synthesized. Then [^{125}I]ICTP was incorporated as described above. The resulting complex, ECp1, was walked to form ECp2 and, subsequently, ECp12. The DNA breaks were analyzed by PAGE as described previously.

Breaks in the T-strand and N-strand of DNA

Analysis of DNA strand breaks produced by decay of ^{125}I incorporated in RNA strand of RNAP ECs is shown in Fig. 3. The left panel represents the experiments with the 3′-end-labeled T-strand and the right panel with the 5′-end-labeled N-strand. Bands corresponding to the ^{125}I-labeled RNA transcripts were located in the lower part of the gels (^{125}I-RNA box). The positions of the G complementary to the [^{125}I]IC in the T-strand and the corresponding C in the N-strand are marked with asterisks (G* and C*). Bands corresponding to the ^{125}I-induced breaks in the T-strand are outlined with a box (left panel). The bands appear as ladders with a sharp bell-shaped distribution of intensities. For the N-strand, the intensity of the bands corresponding to the breaks is considerably lower and they are spread along the lanes without defined maxima.

We observed a similar pattern of breaks previously in the displaced identical strand of the RecA synaptic complex.[8] Therefore, radioprobing data indicate that in the RNAP EC, the newly transcribed RNA is in a close contact with the T-strand for at least 10 nt upstream from the RNA 3′ end, while the corresponding fragment of the N-strand is apparently displaced and does not interact with RNA.

Distribution of Breaks in the T-strand

The frequencies of breaks at individual nucleotides in EC1 through EC10 (calculated as outlined in Fig. 4) are presented in Fig. 5. For comparison of the break distribution in different ECs, the intensities of individual Lorentzian peaks were normalized, such that the intensity for G^*, the nucleotide complementary to [^{125}I]IC, equals 1 for all the profiles (Fig. 5). Note that the intensity of a band does not directly correspond to the frequency of breaks at a given base because one decay can produce multiple breaks in the same DNA strand. Strictly speaking, the intensity of a band corresponds to the frequency of breaks at that base under the condition that there are no other breaks between that base and the labeled 3' end. Therefore, we consider the data points close to the 5' end unreliable, and, thus, they are grayed out in Fig. 5.

The break distributions in the T-strand for EC1, EC3, and EC6 are almost identical and resemble the break distributions we observed earlier for the DNA/RNA hybrid in the T7 RNAP elongation complex.[7] Such a distribution of breaks, where the maximum is shifted from the G^* position towards the 3' end, is characteristic for A-DNA.[7,14] This result is consistent with formation of a relatively stable and uniform DNA/RNA duplex with a conformation close to the canonical A-form. The break distributions in EC8, EC9, and EC10 deviate from this pattern; in particular, the cleavage intensities in the 3'-end shoulder gradually go down, thereby indicating increase in the distance between RNA and the DNA T-strand[7] (Fig. 5). In EC8 and EC9, the [^{125}I]IC is located at the edge of the 8-9-bp long DNA/RNA duplex,[15,16] therefore, the observed effect (increase in the RNA-DNA distance) can be explained by separation of the 2 strands.

The cleavage pattern for EC10 deserves special discussion, however. In EC10, the [^{125}I]IC should be located beyond the 8-9-bp long DNA/RNA duplex (see "explosion" in Fig. 2B). Nevertheless, the total amount of breaks in the T-strand of EC10 is still comparable to that of EC1 (Fig. 3). This apparent contradiction can be explained by RNAP backtracking along the template with simultaneous translocation of the transcription bubble and the RNA/DNA hybrid.[17] The ECs assembled on short templates are known to be particularly susceptible to backtracking.[18] For example, in

[14] E. K. Gaidamakova, R. D. Neumann, and I. G. Panyutin, *Nucl. Acids Res.* **30,** 4960–65 (2002).
[15] A. L. Gnatt, P. Cramer, J. Fu, D. A. Bushnell, and R. D. Kornberg, *Science* **292,** 1876–82 (2001).
[16] N. Korzheva and A. Mustaev, *Curr. Opin. Microbiol.* **4,** 119–25 (2001).
[17] N. Komissarova and M. Kashlev, *Pros. Natl. Acad. Sci. USA.* **94,** 1755 (1997).
[18] M. L. Kireeva, N. Komissarova, D. S. Waugh, and M. Kashlev, *J. Biol. Chem.* **275,** 6530 (2000).

EC10, the backtracking would relocate the [^{125}I]IC to the downstream edge of the hybrid, where the "explosion" is positioned in Fig. 2C. As a result, after backtracking the [^{125}I]IC-DNA distance would be as in the duplex. On average, the intensity of breaks in the EC10 would be intermediate between those for the positions in Figs. 2B, and 2C and thus close to the intensity for EC1.

Notice that backtracking would not affect the breaks distribution in EC1, EC3, and EC6 because, in all possible positions of RNAP, the [^{125}I]IC always stays within the DNA/RNA hybrid. In the "forward" position the [^{125}I]IC is located between the ribonucleotides 10 and 18 (Fig. 2B), and in the "backward" position the [^{125}I]IC is located as shown in Fig. 2C.

Breaks in the T-strand of the Promoter-Initiated EC

To prevent backtracking, we obtained ECs of the *E. coli* RNAP on a 124-nt long double-stranded linear template. Transcription was initiated not by assembly but rather in a conventional way, from the strong A1 promoter of bacteriophage T7 inserted in the template. First, the 28-nt long RNA capable of forming a secondary structure was synthesized by walking, then [^{125}I]IC was incorporated into the 3' end. The ^{125}I decay induced breaks in the complementary T-strand of the DNA strand (see ECp1 in Fig. 6). Importantly, no breaks were detected when ECp1 was walked to form ECp12, which did not backtrack because a strong secondary structure formed in the nascent RNA behind RNAP. Therefore, in the absence of backtracking, the [^{125}I]IC located 12 nt from the RNA 3' end was no longer involved in the complementary interactions with the template DNA strand (Fig. 6, ECp12).

DNA Conformation in RNAP Active Center

The breaks distribution in ECp1, where [^{125}I]IC is located in the RNAP active center, is different from that of ECp2 (Fig. 6). Indeed, the frequency of breaks at the nucleotides located to the 5' end from the G* (C and G) seems lower in ECp1 than in ECp2, implying that these nucleotides were at greater distance from the [^{125}I]IC.

To further investigate this effect, we obtained the break distributions for assembled ECs with the T-strand labeled at the 5' end (Fig. 7). These data complement the results for the 3'-end-labeled T-stand (Fig. 5) by revealing the break distributions between G* and the 5' end of the T-strand. The break distribution for EC1 is radically different from the other ECs; in particular, the frequency of breaks in the nucleotides shifted to the 5'-end from the RNAP active center is abruptly reduced. This reduction in the

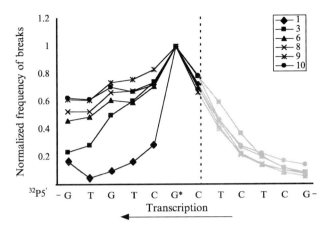

FIG. 7. Relative frequencies of breaks at the individual bases in the T-strand labeled at the 5' end normalized on the intensity at the G^* opposite [^{125}I]IC. Symbols are assigned as in Fig. 5. Data points on the right are grayed out.

DNA cleavage is consistent with the sharp turn in the T-strand of DNA similar to that observed in the yeast RNA polymerase II (Pol II) elongation complex[15] and proposed for the bacterial RNAPs as well.[16]

Concluding Remarks

Radioprobing of the *E. coli* RNAP elongation complex demonstrate that approximately 9 bp upstream from the active site of RNAP, the DNA/RNA hybrid undergoes a drastic conformational change, and the two strands become separated. This result is in agreement with the crystal structure of the Pol II elongation complex,[15] where the heteroduplex is 9 bp long, and the crystal/crosslinking-based model for the bacterial elongation complex,[16] where the length of the heteroduplex is assumed to be 8-9 bp. In addition, our data allow direct comparison between two RNA polymerases, the single subunit T7 RNAP, and the multi-subunit *E. coli* RNAP. Indeed, the breaks distribution presented here for the EC9 and EC10 (Fig. 5) is similar to that observed by the same method for the EC7 in the case of T7 RNAP.[7] This non-A-like distribution of the breaks reflects the local separation of the RNA and the DNA T-strand.[7] Therefore, we conclude that the DNA/RNA hybrid formed inside the *E. coli* RNAP is 1-2 bp longer than the hybrid inside the T7 RNAP. Again, this is consistent with the crystal structure of the T7 elongation complex,[19] where the

[19] Y. W. Yin and T. A. Steitz, *Science* **298,** 1387 (2002).

DNA/RNA hybrid is shown to be 7 bp (that is, less than in the multisubunit Pol II elongation complex[15]).

In summary, the ^{125}I radioprobing is a unique approach useful for revealing the DNA and RNA trajectories in nucleoprotein complexes, which are too large and flexible for the conventional high-resolution methods such as x-ray crystallography and NMR in solution. The advantage of radioprobing over other footprinting and crosslinking methods is the ability of the DNA/RNA breaking agent—Auger electron—to freely penetrate proteins and nucleic acids.

Acknowledgments

We are grateful to Jodi Becker for her expert technical support.

[8] Formation of Long DNA Templates Containing Site-Specific Alkane–Disulfide DNA Interstrand Cross-Links for Use in Transcription Reactions

By THOMAS J. SANTANGELO and JEFFREY W. ROBERTS

The ability to "walk" transcription complexes to discrete positions by performing repetitive rounds of nucleotide starvation has proven extremely useful.[1] However, generating a uniform population of elongation complexes is complicated on some natural templates, because some sequences require far too many rounds of walking to produce the desired complex. Other templates have positions that undergo significant backtracking or arrest, complicating the simple walking procedure. We present a technique to generate a uniform population of elongation complexes at any template position on any template in a single step by employing interstrand DNA cross-links.

Advances in oligonucleotide synthesis allow incorporation of a wide range of chemically modified bases. Verdine and colleagues introduced the use of "convertible nucleosides" during standard oligonucleotide synthesis at internal positions.[2–4] Conversion of these modified bases allows production of a wide range of alkylamine derivatized nucleotides;

---

[1] B. Krummel and M. J. Chamberlin, *J. Mol. Biol.* **225,** 221 (1992).
[2] A. E. Ferentz and G. L. Verdine, *J. Am. Chem. Soc.* **113,** 4000 (1991).
[3] A. M. MacMillan and G. L. Verdine, *J. Org. Chem.* **55,** 5931 (1990).
[4] A. M. MacMillan and G. L. Verdine, *Tetrahedron* **47,** 2603 (1991).

importantly, the conversion reaction does not yield unwanted modification of the standard bases comprising the rest of the oligonucleotide. The use of thiol-bearing alkylamines allows formation of interstrand or intrastrand disulfide linkages, as well as disulfide linkages between nucleic acids and proteins. Interstrand nucleic acid cross-links are most commonly formed between adjacent sites in opposite strands. Intrastrand cross-links can connect almost any two bases, given the flexibility of single-stranded DNA; when introduced to double-stranded DNA, intrastrand cross-links can be used to bend DNA.[5] The chemistry required for the conversion is relatively simple, and now that the phosphoramidite of each convertible nucleoside is commercially available, this technique is available to a broad range of investigators.

To date, the use of convertible nucleosides has been limited to synthesis of relatively small DNAs, usually designed for crystallographic or nuclear magnetic resonance (NMR) studies of nucleic acids, for use in protein purification, or for the analysis of protein–nucleic acid complexes.[5–16] In this article we review convertible nucleoside chemistry (a brief summary of many papers on this subject, primarily by Verdine and colleagues) and then demonstrate a simple technique to generate long, >300 bp, fully cross-linked DNA templates suitable for *in vitro* transcription reactions. We use these DNAs to show that *Escherichia coli* RNA polymerase is capable of synthesis up to and including the more upstream base involved in the cross-link. DNAs prepared in this manner should be applicable to the study of not only RNA polymerase, but also any macromolecule or complex that separates the DNA duplex.

Convertible Nucleosides

Convertible nucleosides contain good leaving groups that can be substituted by reaction with an appropriate primary amine, resulting in the linkage of the amine to the base (Fig. 1). Convertible nucleoside analogs for

[5] S. A. Wolfe, A. E. Ferentz, V. Grantcharova, M. E. Churchill, and G. L. Verdine, *Chem. Biol.* **2**, 213 (1995).
[6] A. E. Ferentz, T. A. Keating, and G. L. Verdine, *J. Am. Chem. Soc.* **115**, 9006 (1993).
[7] L. K. Dow, S. A. Wolfe, A. E. Ferentz, V. Grantcharova, M. E. Churchill, and G. L. Verdine, *Biochemistry* **39**, 9725 (2000).
[8] A. E. Ferentz and G. L. Verdine, *in* "Nucleic Acids and Molecular Biology" D. M. J. Eckstein, ed. p. 14. Springer-Verlag, Berlin, 1994.
[9] C. J. Larson and G. L. Verdine, *Nucleic Acids Res.* **20**, 3525 (1992).
[10] H. Huang, R. Chopra, G. L. Verdine, and S. C. Harrison, *Science* **282**, 1669 (1998).
[11] H. Huang, S. C. Harrison, and G. L. Verdine, *Chem. Biol.* **7**, 355 (2000).
[12] A. M. MacMillan, R. J. Lee, and G. L. Verdine, *J. Am. Chem. Soc.* **115**, 4921 (1993).
[13] D. M. Noll, A. M. Noronha, and P. S. Miller, *J. Am. Chem. Soc.* **123**, 3405 (2001).
[14] A. M. Noronha, C. J. Wilds, and P. S. Miller, *Biochemistry* **41**, 8605 (2002).
[15] A. M. Noronha, C. J. Wilds, and P. S. Miller, *Biochemistry* **41**, 760 (2002).
[16] D. M. van Aalten, D. A. Erlanson, G. L. Verdine, and L. Joshua-Tor, *Proc. Natl. Acad. Sci. USA* **96**, 11809 (1999).

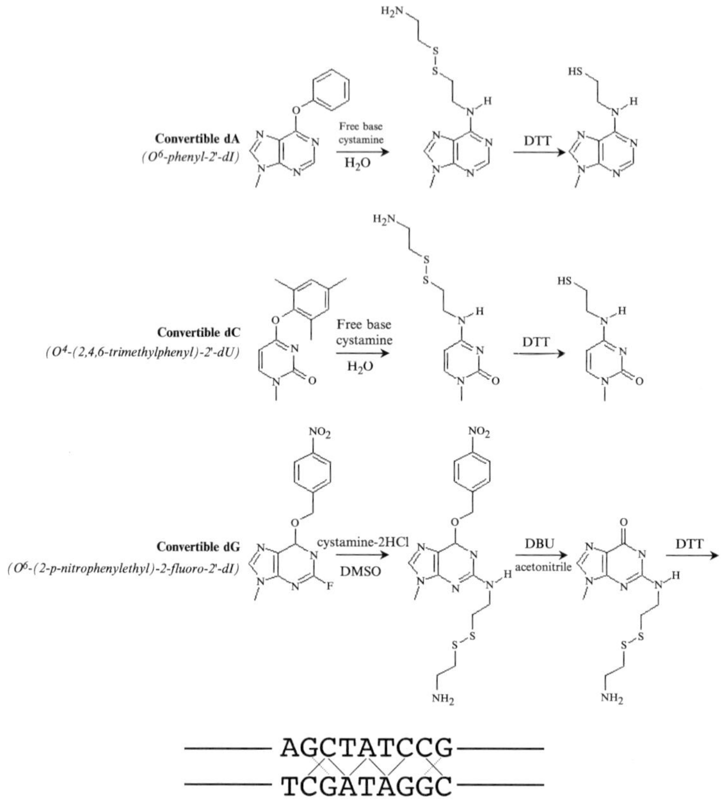

FIG. 1. Convertible nucleoside chemistry. Reaction of each convertible nucleoside with cystamine; only the nucleoside is shown for simplicity. Lower panel shows each available pairing of adjacent bases in opposite strands using convertible nucleotides. Major groove cross-links are shown as a solid line; minor groove cross-links are shown as a dashed line.

dA, dC, and dG are commercially available as standard phosphoramidites; dT does not contain any primary amines and therefore is not suitable for this technique. The phosphoramidite of each convertible nucleoside can be incorporated at any position (except the 3′ terminus) during standard oligonucleotide synthesis ("oligonucleotide" is used here to mean 2′-deoxyoligonucleotide, although the convertible nucleoside approach is also applicable to solid-state RNA synthesis[17]). Importantly, synthesis of the remaining bases is completely unaffected and can be of any sequence.

Conversion reactions can be carried out with virtually any primary amine, although this article only concerns the use of thiolalkylamines. Symmetric

[17] C. R. Allerson and G. L. Verdine, *Chem. Biol.* **2**, 667 (1995).

disulfides are used when introducing thiol groups, because free thiols lead to unwanted side reactions.[8] The substitution reaction is followed by a reducing step to yield the free thiol (Fig. 1). By converting nucleosides in two complementary oligonucleotides, one can form a disulfide cross-link through either the major or minor groove. Both convertible dA $(O^6$-phenyl-2'-deoxy-inosine) and convertible dC $(O^4$-[2,4,6-trimethlphenyl]-2'-deoxyuridine) are derivatized on the major groove side, dA at7 the N^6 position and dC at the N^4 position. In contrast, convertible dG $(O^6$-[2-p-nitrophenylethyl]-2-fluoro-2'-deoxyinosine) is substituted at the N^2 position, resulting in a converted base where the alkylamine substitution extends into the minor groove. Almost any sequence context can be cross-linked using a combination of the convertible bases (Fig. 1). By using different thiolalkylamines, one can vary the tether length and therefore vary the bases that can be cross-linked; thus cross-links between bases separated by one, two, or more nucleotides on opposite strands are possible,[8] although this study addresses only the use of cystamine (aminoethanethiol; $NH_2CH_2CH_2SSCH_2$ CH_2NH_2), which cross-links bases in adjacent positions.

The conversion reactions for convertible dA and convertible dC are carried out after complete synthesis, deprotection, and cleavage of the oligonucleotide from the solid support; they require the strongly nucleophilic free-base form of cystamine. Because of unwanted side reactions, specific and efficient derivation of convertible dG must be done while the oligonucleotide is still protected and bound to the column. This is partially advantageous because one can use cystamine-2HCl in dimethylsulfoxide (DMSO) (Glen Research Technical Report for 2-F-dI, 2002); the free-base form of cystamine must be made from the dihydrochloride salt, a process that takes several days.[18] After substitution at the N^2 position of convertible dG to yield the mixed disulfide, treatment with DBU (1,8-diazabicyclo-[5.4.0]-undec-7-ene) restores the carbonyl group at position O^6; standard deprotection and cleavage of the oligonucleotide from the resin yield the final product. Substitution of convertible dG requires that the researcher have access to the oligonucleotide while it is still on the column, although the chemistry of conversion is no harder than for convertible dA or convertible dC.

Construction of Templates with Site-Specific Interstrand
  DNA Cross-Links

Interstrand DNA–DNA cross-links are formed by mixing an equal molar ratio of complementary thiolalkylamine-substituted oligonucleotides in solution, reducing the derivatized bases to yield free thiols, and then

---

[18] J. A. Fidanza and L. W. McLaughlin, *J. Org. Chem.* **57,** 2340 (1992).

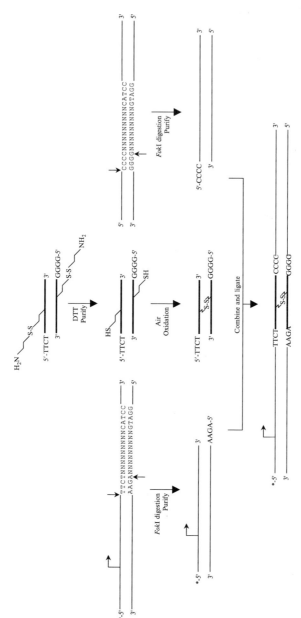

FIG. 2. Construction of a long DNA template containing a site-specific internal alkane-disulfide interstrand DNA cross-link. Complementary DNA oligonucleotides with nonpalindromic overhangs are combined, reduced, purified, and air oxidized to generate small cross-linked templates. Larger DNA fragments (generated by *Fok*I digestion of PCR fragments) are purified and ligated to the cross-linked oligonucleotides to generate the final template. (*) 5' biotin conjugate used to attach the final templates to streptavidin-coated beads.

purifying the oligonucleotides away from the noncovalently attached amine (Fig. 2). Subsequent removal of the reducing agent by dialysis, or purification of the oligonucleotides through commercial spin columns, followed by air oxidation of the purified oligonucleotides, leads to the formation of the disulfide linkage. By mixing together derivatized oligonucleotides that leave nonpalindromic overhangs, one can ligate a series of larger DNA fragments onto the cross-linked oligonucleotides to produce very long DNA templates. Purification of fully ligated, fully cross-linked material is achieved by isolation of the template from a denaturing gel, where both strands of the cross-linked template co-migrate more slowly than the separated strands (Fig. 3B).

Use of nonpalindromic overhangs greatly facilitates ligation specificity. Type II restriction enzymes that specifically recognize one sequence and then cleave the DNA at precise distances from the recognition sequence are a very convenient way to generate nonpalindromic overhangs on the larger DNA pieces. The studies presented here use FokI (New England Biolabs, Beverly, MA), which recognizes 5'-GGATG and then cleaves the DNA 9 nt downstream on the 5' strand and 13 nt downstream on the 3' strand, leaving a 4-nt sticky end. Other type II restriction enzymes that will generate a similar, potentially nonpalindromic $\geq$3-nt overhang include BbsI, BbvI, BfuAI, BsaI, BsmAI, BsmBI, BsmFI, BspMI, EarI, HgaI, SapI, and SfaNI. Different nonpalindromic sequences on each overhang allow specific ligation to two different DNAs; therefore the cross-linked sequence can be placed internally within a large DNA template (Fig. 2).

Preparation of Disulfide Cross-linked Templates

The oligonucleotides used to create templates are: TJS121, 5'-[P]TTGTGTA-TATATAGATA; TJS120, 5'-[P]GGGGTATCTATATATAC; TJS117CCCC, 5'-[P]CCCCTGTTGACATACCATCCATA; and TJS118, 5'-TATG-GATGGTATGTCAACGA (Fig. 3A). The derivatized oligonucleotides are synthesized with the phosphoramidite of O^6-phenyl-2'-dI in place of dA at the point of cross-linking and are otherwise identical in sequence to either TJS120 or TJS121. 0.1–1 nmoles of each oligonucleotide containing O^6-phenyl-2'-dI is converted to the mixed disulfide by reaction with 20 μl 1 M free-base cystamine for 18 h at 65° under mineral oil.[6] Reactions are neutralized by addition of 5% acetic acid (typically 1–3 μl), brought to 200 μl with 25 mM triethylammonium bicarbonate (TEAB), pH 8.0, and then dialyzed twice at 4° against 200 ml 25 mM TEAB, pH 8.0, in 1000 dalton molecular weight cut-off Tube-O-Dialyzer (Geno Technology, Inc., St. Louis, MO) to remove excess amine. Some samples are alternatively purified with Qiagen (Chatsworth, CA) Nucleotide Purification

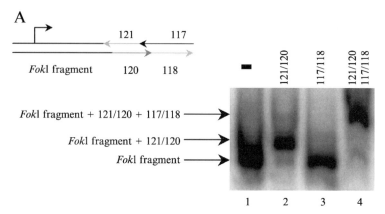

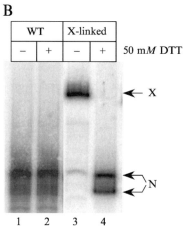

Fig. 3. Ligation and purification of templates. (A) Ligation products are resolved through a 4% native polyacrylamide gel to demonstrate the specificity and efficiency of the ligation reactions. Each lane contains a radiolabeled *Fok*I fragment ligated to the DNA fragments listed above each lane. Lane 1, none; lane 2, DNA fragments 121 and 120 are added to the ligation reaction (note the nearly quantitative shift caused by the ligation of these fragments to the *Fok*I fragment); lane 3, DNA fragments 117 and 118 are added to the ligation reaction (note that these fragments are not ligated to the *Fok*I fragment because of a lack of complementarity between the overhangs); lane 4, DNA fragments 121, 120, 117, and 118 are added to the ligation reaction (note the substantial and nearly quantitative shift resulting from ligation of the DNA fragments to the *Fok*I fragment). (B) Products from the complete ligation reaction in lane 4 of (A) are resolved through a 4% denaturing gel to demonstrate co-migration of the cross-linked strands. Non-cross-linked templates show no response to DTT addition and run as two separate strands (N). Cross-linked templates show a marked decrease in migration because of the co-migration of both strands (X). Treatment of the cross-linked templates with DTT yields two strands (N) indistinguishable from the non-cross-linked template. Fully cross-linked, fully ligated material is purified by excising the slowly migrating material (X) in lane 3 and eluting the DNA by a crush and soak procedure.

Spin Columns according to manufacturer's instructions. Mixed disulfide samples are precipitated with 10 mM MgCl$_2$ and 5 volumes 100% EtOH. Complementary derivatized pairs of oligonucleotides are mixed in equal molar amounts, DTT is added to a final concentration of 100 mM, and samples are incubated for 20 min at RT to form N^6-thioethyl-2'-dA. Samples are heated to 65° for 10 min, cooled slowly to 4°, and dialyzed as before to remove reducing agent and allow formation of the alkane disulfide interstrand DNA cross-link; alternatively, some samples are purified with Qiagen Nucleotide Purification Spin Columns according to manufacturer's instructions. Samples are alcohol precipitated and added to ligation reactions as described. Derivation in this manner gives >95% conversion to N^6-thioethyl-2'-dA (data not shown). Preparation of free-base cystamine (*aminoethanethiol*) is as described.[18]

Ligation Reactions

Large DNA fragments are generated by polymerase chain reactions (PCRs) using an upstream 5'-biotinylated primer so that the final products can be attached to streptavidin-coated paramagnetic beads (see Fig. 2). Standard gel purification of the large DNA fragments by commercially available kits, low melt agarose, or polyethylene glycol (PEG)-precipitation yields suitably clean material for the ligation reactions. *Fok*I digested gel-purified PCR products are ligated to a complementary series of oligonucleotides by combining each DNA fragment—a ten-fold molar excess of each oligonucleotide is used with respect to the *Fok*I digested PCR product—and heating the mixture to 65° for 10 min. The mixture is cooled to RT, and T4 DNA ligase buffer (New England Biolabs) and T4 DNA ligase (New England Biolabs) are added. A typical reaction contains 40–100 pmoles of the purified *Fok*I digested DNA fragment and 400–1000 units of T4 DNA ligase. Reactions are incubated at 37° for 60 min, and then at 16° overnight. Reactions are extracted with an equal volume phenol/chloroform/isoamyl alcohol (25:24:1), and DNA is precipitated from the aqueous phase with 10 mM MgCl$_2$ and 5 volumes 100% EtOH. Ligation in this manner is specific and generated >95% fully ligated product (Fig. 3A). Templates without interstrand DNA cross-links are used for *in vitro* transcription reactions without further purification. Templates containing interstrand DNA cross-links are further purified (see the following).

Purification of Pure Cross-linked Template

Ligation reactions involving derivatized oligonucleotides are extracted with an equal volume of phenol/chloroform/isoamyl alcohol (25:24:1), alcohol precipitated, resuspended in formamide loading buffer, boiled, and loaded onto 4% denaturing polyacrylamide gels to separate cross-linked from non-cross-linked DNA. The slower migrating band (X in Fig. 3B), containing both strands of the cross-linked template, is excised from the gel, and the DNA is eluted by a crush and soak procedure. The disulfide linkage is reversible; upon reduction with DTT the sample yields two strands indistinguishable from wild type (WT)-DNA (N in Fig. 3B), although each contains a single modified adenosine. Recovered samples are alcohol precipitated as before and used without further purification. Treatment in this manner yields templates that are >95% cross-linked (data not shown).

In Vitro Transcription Conditions

Reactions are carried out essentially as in Yarnell and Roberts.[19] Streptavidin-coated paramagnetic beads (Promega, Madison, WI) are washed three times with buffer #1 (B1 = 10 mM Tris–HCl, pH 8.0, 150 mM NaCl, 100 μg/ml BSA) and combined with biotinylated templates for 10 min at RT. Unbound template is removed by two additional washes with B1 and one wash with B2 (20 mM Tris–HCl, pH 8.0, 0.1 mM EDTA, 250 mM KCl, 5% glycerol, 100 μg/ml BSA).

Transcription Conditions for Fig. 4A

Open complex is formed in the presence of 75 μM ApC, 25 μM ATP, 100 μM GTP, 25 μM UTP, ^{32}P-α-UTP ($\sim$ 5 μC), 10 nM bound template, 40 nM RNAP, 20 mM Tris–HCl, pH 8.0, 0.1 mM EDTA, 50 mM KCl, and 100 μg/ml BSA for 10 min at 37°. Complexes are advanced to +25 by addition of MgCl$_2$ (4 mM final), incubated 8' at 37°, washed three times with B3 (20 mM Tris–HCl, pH 8.0, 0.1 mM EDTA, 250 mM KCl, 5% glycerol, 50 μg/ml BSA, 4 mM MgCl$_2$), challenged on ice for 5 min with B4 (as B3, with 200 μg/ml poly dA/dT and 350 mM NaCl) to remove any remaining open complex, and washed two times with B5 (as B3, with 20 μg/ml poly dA/dT). Complexes are advanced from +25 to +97 by 3-min incubation at 37° in B3 supplemented with 25 μM ATP, CTP, GTP, followed by three washes with B3. Complexes are advanced to +105 by 3' incubation in B3 with 100 μM CTP, GTP, UTP at 37°. Chase reactions (5') used +97 complexes as starting material and contained 100 μM NTPs in B3.

[19] W. S. Yarnell and J. W. Roberts, *Science* **284,** 611 (1999).

Transcription Conditions for Fig. 4B

Open complex is formed in the presence of all four NTPs (200 μM ATP, CTP, GTP, 50 μM UTP), ^{32}P-α-UTP ($\sim$ 5 μC), 1 to 2 nM bound template, 20 nM RNAP, 20 mM Tris–HCl, pH 8.0, 0.1 mM EDTA, 50 mM KCl, and 100 μg/ml BSA for 10 min at 37°. Transcription (5′) is initiated by the simultaneous addition of MgCl$_2$ (4 mM final) and rifampicin (10 μg/ml final) to limit transcription to a single round. When present, DTT is added during formation of open complex.

All transcription reactions are placed at 37° until 30 seconds before the indicated sample time, when they are removed to a magnetic stand to separate RNA associated with paramagnetic streptavidin-coated beads (pellet) from RNA released to solution (supernatant). Each transcription pellet sample is supplemented with 120 μl of 0.5 M Tris–HCl, pH 8.0, 10 mM EDTA, and 67 μg/ml tRNA to stop the reaction. Supernatant samples are supplemented with 100 μl 0.6 M Tris–HCl, pH 8.0, 12 mM EDTA, and 80 μg/ml tRNA to stop the reaction. All samples are extracted with an equal volume of phenol/chloroform/isoamyl alcohol (25:24:1), and the nucleic acids are precipitated from the aqueous phase by addition of 2.6 volumes 100% EtOH. Pellets are resuspended in formamide loading buffer, loaded into denaturing polyacrylamide gels, exposed, and quantitated using ImageQuant software (storm model 840; Molecular Dynamics, Inc., Piscataway, NJ).

Results

We first demonstrate that blocking separation of the transcription bubble via alkane disulfide interstrand DNA cross-linking inhibits translocation and transcription of elongation complexes beyond the site of the cross-link. All templates used in Fig. 4 contain a 5′-biotin on the nontemplate strand, which, in conjunction with streptavidin-coated paramagnetic beads, allowed us to separate transcripts associated with the ternary elongation complex from those released to solution. The pellet fraction (P) contains all of the DNA and the fraction of transcripts associated with intact elongation complexes; the supernatant fraction (S) contains only those transcripts released from the elongation complex.

The templates in Fig. 4A differ only by an interstrand cross-link between +106 of the nontemplate strand and +105 of the template strand. We used a three-stage traditional "walk" to generate discrete complexes on each template; +105 complexes, which are also the predicted stop position on the cross-linked template, were also made on each template. The walking serves both to demonstrate that the cross-linked template allows

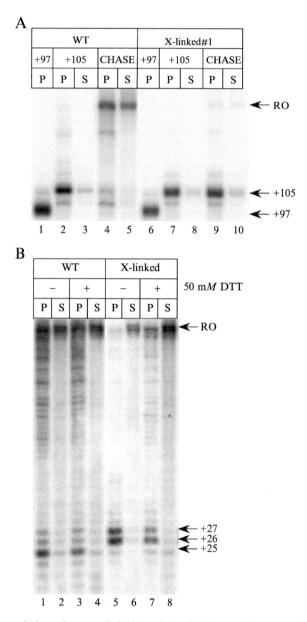

FIG. 4. Transcription using cross-linked templates. (A) Transcripts generated from a series of nucleotide starvation walks are shown for each template. The WT and cross-linked templates differ only by an interstrand cross-link between +106 nontemplate strand and +105 template strand on the cross-linked template. The position to which each elongation complex

normal transcription up to the cross-link and to provide markers of known length. When +97 complexes on the non-cross-linked template are supplemented with all four NTPs ("chased"), they produce full-length run-off (RO) products. In contrast, +97 complexes chased on the cross-linked template do not elongate beyond +105 (Fig. 4A, lanes 9 and 10). An elongation complex that has reached an interstrand DNA cross-link is stable and retains the nascent transcript (Fig. 4A, lane 9); all of the +105 transcript in the chase reaction on the cross-linked template remains in the pellet. A minor fraction of complexes transcribe past the cross-link ($\leq 2\%$), which we ascribe to a minor amount of reduced template in the reactions. Care should be taken to remove as much reducing agent from the reaction buffer as possible; addition of 1 mM DTT to the transcription reactions reduces $\sim 30\%$ of the DNA template and allows read-through of the cross-linked position under these conditions (data not shown).

To determine exactly how close to the cross-link *E. coli* RNAP can transcribe, we introduced a cross-link in a template that allows single nucleotide resolution of the resulting transcripts (Fig. 4B). We used a template that contains a strong pause at position +25[19] to mark the exact length of transcripts made on an equivalent template containing a cross-link between +27 of the template strand and +28 of the nontemplate strand. RNAP makes transcripts from the cross-linked DNA that are +26 and +27 nucleotides long; therefore RNAP must be able to transcribe up to and including the upstream base involved in the disulfide cross-link. RNAP was also shown to elongate up to the bases involved in a psoralen interstrand DNA cross-link.[20,21] The addition of high concentrations of DTT to transcription reactions reduces the disulfide linkage joining the two DNA strands (Fig. 3B, lanes 3 and 4) and allows RNAP to transcribe beyond the bases previously involved in the cross-link (Fig. 4B, lanes 7

---

was walked is shown above each lane. Chase lanes used +97 complexes as starting material. Transcript length is noted to the right of the gel. Note that elongation complexes cannot transcribe beyond the cross-link (lanes 9 and 10) when supplemented with all 4 NTPs. P, pellet; RO, run-off; S, supernatant. (B) *In vitro* transcription demonstrating that *E. coli* RNAP can transcribe up to and including the more upstream base involved in an interstrand alkane-disulfide cross-link. The WT and cross-linked templates differ only by an interstrand cross-link between +27 nontemplate strand and +26 template strand on the cross-linked template. Reduction of the cross-link (lanes 7 and 8) allows the majority ($\sim 80\%$) of RNAP to transcribe beyond the bases previously involved in the cross-link to produce run-off transcripts. Transcript length is shown to the right of the gel. +25 is a naturally occurring pause site on this template[19] and serves as a marker. RO, run-off transcript.

[20] Y. B. Shi, H. Gamper, and J. E. Hearst, *Nucleic Acids Res.* **15,** 6843 (1987).
[21] Y. B. Shi, H. Gamper, B. Van Houten, and J. E. Hearst, *J. Mol. Biol.* **199,** 277 (1988).

and 8). Transcription reactions are unaffected by the addition of DTT to concentrations as high as 0.5 M (data not shown). Elongation past the bases previously involved in the cross-link demonstrates that RNAP can efficiently transcribe N^6-thioethyl-dA, which is still capable of forming all W–C hydrogen bonds when the cross-linking arm occupies the *syn* position.[8] The halt to elongation is therefore due to the inability to open the downstream DNA rather than an inability to template the modified base. Introduction of the +27/+28 cross-link alters the efficiency of the σ^{70}-dependent pause at +25 for unknown reasons (Fig. 4B).

Importantly, once the elongation complex has elongated to the site of the cross-link, the cross-link becomes very resistant to reduction (data not shown). Thus use of these cross-links is an efficient way to generate complexes, but does not easily allow for continued transcription.

Concluding Remarks

Formation of long templates with site-specific alkane disulfide interstrand DNA cross-links provides a simple technique for one-step positioning of RNA polymerase at any downstream site. Templates containing interstrand DNA cross-links are currently being used to study transcription termination/antitermination reactions; in addition, these templates should be useful for study of any factor or complex that normally separates the duplex. Templates containing interstrand alkane disulfide cross-links may prove useful for study of transcription machinery, DNA replication machinery, DNA recombination factors, and DNA repair proteins. The use of convertible nucleotides has several advantages over other techniques that have been used to inhibit separation of the DNA duplex: absolute specificity of conversion/attachment; conservation of standard W–C base pairing; and conservation of the B-form canonical nature of the DNA duplex.

Acknowledgment

This work was supported by a National Institutes of Health grant (GM21941) to J.W.R.

[9] Probing the Organization of Transcription Complexes Using Photoreactive 4-Thio-Substituted Analogs of Uracil and Thymidine

By Dmitri Temiakov, Michael Anikin, Kaiyu Ma,
Manli Jiang, and William T. McAllister

Photoreactive nucleotide analogs that may be selectively incorporated into RNA or DNA at specific positions are increasingly being used to probe the structure and organization of transcription complexes. A number of analogs have been employed in such studies. These include 4-thiouridine (or thymidine), 6-thioguanidine, 5-iodouridine, 5-bromouridine, azido derivatives such as 2- or 8-azidoadenine, and 5-azidouridine (Fig. 1).

For studies of protein:nucleic acid interactions, 4-thio-substituted uridine or thymidine analogs have a number of advantages: (1) they interact with the target protein at a short range (i.e., within van der Waal's or chemical bond distances); (2) compared with other analogs, they provide a relatively high efficiency of cross-linking; (3) they are commercially available and relatively inexpensive; (4) in general, they exhibit less specifically for particular amino acids than other reagents (c.f., iodo- and bromo-substituted uridine analogs, which have a tendency to react with aromatic amino acids); (5) they may be efficiently incorporated into RNA or DNA by most commonly used RNA or DNA polymerases (in fact, 4-thio-U is a naturally occurring analog of uridine); this is in contrast to some nucleotide analogs (such as azido adenosine triphosphate [ATP]), which act as chain terminators and can therefore only be incorporated at the 3' end of the nucleic acid; (6) the cross-link formed upon ultraviolet (UV) irradiation is stable from pH 2–10 and is therefore resistant to conditions that are commonly used for protein cleavage and mapping of cross-linking sites (e.g., NTCB, CNBr); this is in contrast to the cross-links formed by other analogs such as azido derivatives, which are sensitive in this range.

There are, however, some limitations to the use of thio-substituted analogs. First, although the short radius of action of these agents affords a higher precision, it limits the range of interactions that may be visualized. Thus if the thio-substitution of the analog is not within chemical bonding distance of a target residue in the protein, the cross-linking efficiency will be low. A cross-linking reagent with a longer radius of action might be more revealing in such circumstances, although less precise.

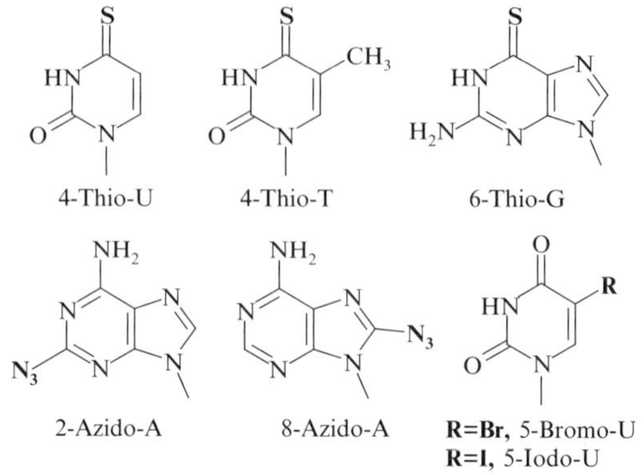

FIG. 1. Photoreactive nucleotide analogs commonly used to probe protein:nucleic acid contacts. For review, see A. Favre. *In* H. Morrison, *Bioorganic Photochemistry*, New York, John Wiley (1990).

Materials

RNA Polymerase, Nucleotide Analogs, and Template Assembly

Histidine-tagged T7 RNA polymerase (RNAP) is overexpressed and purified as previously described.[1,2] Synthetic DNA oligonucleotides are obtained from Macromolecular Resources (Fort Collins, CO) or IDT (Coralville, IA). RNA oligonucleotides may be purchased from Dharmacon Research, Inc. (Lafayette, CO). The photoreactive analogs 4-thio-UTP and 4-thio-dTTP are available from Trilink Biotechnologies (San Diego, CA).

To assemble DNA templates, oligonucleotides are taken up in water, mixed together at a concentration of 10 μm each, heated to 70° for 5 min, and cooled slowly to room temperature; after assembly the templates may be stored at −20°. Templates containing 4-thio-dTMP are constructed by annealing together template (T) or non-template (NT) strand oligomers and extension by DNA polymerase I (Klenow fragment; New England Biolabs, Beverly, MA) under standard conditions[3] as described in Fig. 3. After extension, the duplex templates should be purified using a QIAquick Nucleotide Removal Kit (Qiagen, Chatsworth, CA).

[1] B. He *et al.*, *Protein Expr. Purif.* **9**, 142 (1997).
[2] D. Temiakov, P. E. Karasavas, and W. T. McAllister, *in* "Protein: DNA Interactions: A Practical Approach" (A. A. Travers and M. Buckle, eds.), Chap. 25. Oxford University Press, Oxford, UK, 2000.
[3] J. Sambrook, E. F. Fritsch, and T. Maniatis, "Molecular Cloning: A Laboratory Manual," ed. 2, Cold Spring Harbor Laboratory Press, Cold Spring Harbor, New York, 1989.

Methods to advance His$_6$-tagged T7 RNAP along the template in a controlled manner have been described.[2,4] To prepare halted elongation complexes for RNA cross-linking studies, 20 pmol of T7 RNAP is incubated with an equimolar concentration of template, 0.3 mM GTP, 0.1 mM ATP, and 50 μm UTP analog (see the following) in 20 μl transcription buffer (40 mM Tris–acetate, pH 7.9; 8 mM magnesium acetate; 5 mM β-mercaptoethanol; 0.1 mM EDTA) for 5 min at 37°. The startup complexes are immobilized on Ni^{2+}-agarose beads (Qiagen) and extended in the presence of limiting mixtures of nucleoside triphosphates at a final concentration of 10 μm each. Samples are chilled on ice, and cross-linking is activated by exposure to a UV lamp (Cole-Palmer, Vernon Hills, IL, 6 watts) at 365 nm for 10 min. Cross-linked samples should be analyzed by electrophoresis in 6% polyacrylamide gels in the presence of 0.1% SDS–PAGE and visualized by autoradiography before peptide mapping to verify their quality. For DNA cross-linking experiments, samples may be treated with DNase [50 U DNase 1 and 25 U micrococcal nuclease (Roche Molecular Biochemicals, Indianapolis, IN, USB, Cleveland, OH) per 35 μl reaction; 2 h at 37°] before protein digestion if the labeling nucleotide has been introduced adjacent to the cross-link in a nuclease-resistant manner (see Fig. 3).

Experimental Design

Preparation of Transcription Complexes with Specifically Positioned Photoreactive Analogs

To map specific protein:nucleic acid interactions, the photoreactive analog must be placed at a particular position in the RNA product or the DNA template within the transcription complex. In general, there are three approaches to accomplish this. The first involves "walking" of immobilized transcription complexes along the template by the sequential addition of limiting mixtures of substrate triphosphates. This method was first developed to analyze elongation complexes formed by the multisubunit *Escherichia coli* RNA polymerase[5] and has since been adapted to studies of single subunit RNA polymerases such as T7 RNAP.[6,7] Immobilization of transcription complexes on a solid matrix can be achieved either through the use of an appropriately modified enzyme (e.g., a histidine-tagged enzyme that may be immobilized on Ni^{++}-agarose beads[6]) or by

[4] D. Temiakov *et al., Proc. Natl. Acad. Sci. U.S.A.* **97,** 14109 (2000).
[5] M. Kashlev *et al., Methods Enzymol.* **274,** 326 (1996).
[6] H. Shen and C. Kang, *J. Biol. Chem.* **276,** 4080 (2001).
[7] B. He *et al., Protein Expression Purification* **9,** 142 (1997).

the use of a biotin-tagged DNA template that may be immobilized on streptavidin beads.[7] In either case, successive cycles of walking of the polymerase allow the enzyme to be moved along the template in a precisely controlled manner (Fig. 2).

Although the walking approach provides great flexibility, it is often more convenient to design a template that allows the analog to be placed at a specific position in a halted elongation complex (EC) in one step, without immobilization. This may be accomplished by designing a promoter template that allows elongation to proceed to a defined position in the presence of a limited mixture of substrates, or that directs the incorporation of a chain-terminating nucleotide such as 3' dNTP at a particular position (see, for example, the template shown in Fig. 3). Use of a chain-terminating nucleotide may be advantageous, because it allows an examination of the organization of the complex in the presence or absence of the next incoming nucleotide (which may enhance the stability of the complex and/or favor the formation of a posttranslocated state[8,9]).

Lastly, one can design and assemble nucleic acids scaffolds that reflect the organization of the transcription complex by annealing together appropriate synthetic oligomers of RNA and DNA.[10,11] Nucleotide analogs may be incorporated into the DNA oligomers or into the RNA primer during chemical synthesis before assembly of the scaffold, or into the RNA product by extension of the primer after assembly of the transcription complex.[10,12]

Template Design and Construction

The design of the template is dictated by the purpose of the experiment and the position in the template or product that is being analyzed. In the case of RNA cross-linking, the template must direct the incorporation of 4-thio-UMP at a specific position in the transcript relative to the active site. An example of this is given in Fig. 2, where 4-thio-U is incorporated at various positions upstream from the 3' end of the growing transcript by the "walking" approach.

In the case of DNA templates, the 4-thio-substituted analog may be incorporated into the template (T) or nontemplate (NT) strands of the DNA

---

[8] V. Gopal, L. G. Brieba, R. Guajardo, W. T. McAllister, and R. Sousa, *J. Mol. Biol.* **290,** 411 (1999).

[9] P. E. Karasavas, thesis, State University of New York (1998).

[10] K. Ma, D. Temiakov, M. Jiang, M. Anikin, and W. T. McAllister, *J. Biol. Chem.* **277,** 43206 (2002).

[11] I. Sidorenkov, N. Komissarova, and M. Kashlev, *Molecular Cell* **2,** 55 (1998).

[12] D. Temiakov, M. Anikin, and W. T. McAllister, *J. Biol. Chem.* **277,** 47035 (2002).

```
                                          +1
DT6 : CTTAT CATCG ATCTG CAGTA ATACG ACTCA CTATA GGGAG AGAAG AAGTC GATCA GACTA ATGCA GGTCC
DT7 : GAATA GTAGC TAGAC GTCAT TATGC TGAGT GATAT CCCTC TCTTC TTCAG CTAGT CTGAT TACGT CCAGG
```

Step	Substrate(s)	Transcript	RNA length	Position of Analog
1	G,A,*U	GGGAGAGAAGAAG *U	14	−1
2	C,G,A	GGGAGAGAAGAAG *U CGA	17	−4
3	U,C,A	GGGAGAGAAGAAG *U CGAUCA	20	−7
4	G,A	GGGAGAGAAGAAG *U CGAUCAGA	22	−9
5	C,U	GGGAGAGAAGAAG *U CGAUCAGACU	24	−11
6	A	GGGAGAGAAGAAG *U CGAUCAGACUAA	26	−13
7	U,G	GGGAGAGAAGAAG *U CGAUCAGACUAAUG	28	−15
8	C	GGGAGAGAAGAAG *U CGAUCAGACUAAUGC	29	−16

FIG. 2. Site-specific incorporation of 4-thio-UMP by "walking" of T7 RNAP immobilized on Ni^{++}-agarose beads. A template that contains a consensus T7 promoter (shaded box) was constructed by annealing together synthetic oligomers DT6 and DT7; the start site of transcription is at +1. Startup complexes in which the RNAP is extended 14 bp downstream from the start site were formed by incubation with His_6-tagged T7 RNAP in the presence of GTP, α-^{32}P-ATP, and 4-thio-UTP (*U) (step 1). This places the *U analog at the 3′ end of the growing transcript (position −1). The complexes were immobilized on Ni^{++}-agarose beads, washed with transcription buffer, and advanced 3 nt by the addition of CTP, GTP, and ATP, placing the U analog at position −4 (step 2). Repeated washing and extension steps allows the placement of *U at other positions as indicated (steps 3–8). Samples were removed after each extension step, and the RNA products were analyzed by gel electrophoresis (bottom panel).

by primer extension as shown in Fig. 3. This method was first developed by Bartholomew et al. and was used for characterizing protein:DNA cross-links in halted yeast pol III RNA polymerase transcription complexes.[13] Here we illustrate the method for T7 RNAP by placing the analog in the T strand 17 nt downstream from the start site in a T7 promoter. The incorporation of a labeled nucleotide (α-^{32}P-CTP) adjacent to the analog allows the cross-linked DNA to be "trimmed" by subsequent incubation

---

[13] B. Bartholomew, B. R. Braun, G. A. Kassavetis, and E. P. Geiduschek, *J. Biol. Chem.* **269**, 18090 (1994).

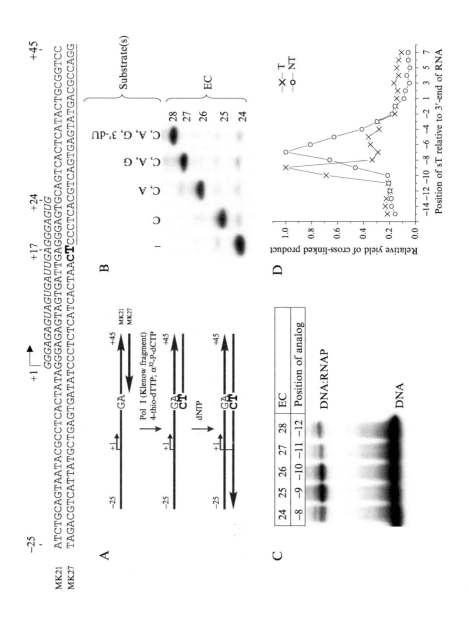

with DNase, which may be helpful in later peptide mapping strategies (see the following). Extension of the modified primer by DNA polymerase in the presence of a large excess (100-fold) of unmodified dTTP and unlabeled dCTP completes the synthesis of the T strand while minimizing the incorporation of the label or analog at other positions. In the presence of GTP, ATP, and UTP the template shown allows the direct formation of an EC halted at +24 (EC24) in which the photoreactive analog is positioned 9 nt upstream from the 3′ end of the RNA in the T strand (position T-9). Alternatively, the complex may be advanced along the template to examine cross-linking at other positions by "walking" immobilized complexes (Fig. 3B,C; EC25–26).

Modified DNA oligonucleotides that have 4-thio-T at specific positions may also be synthesized directly (e.g., Dharmacon) and used to assemble templates or RNA–DNA scaffolds.[12]

Activation of the Cross-Link and Analysis of the Sample

Before commencing mapping studies, it is important to demonstrate that the cross-link is specific and to verify the homogeneity of the transcription complexes. Appropriate controls should include a sample without UV

FIG. 3. Cross-linking of DNA template to T7 RNAP in halted elongation complexes. (A) Construction of modified DNA template. A template strand primer (MK27; underlined portion only) was annealed to a complete nontemplate (NT) strand (MK21) and extended 2 nt by DNA polymerase I in the presence of the photoreactive nucleotide analog 4-thio-dTTP (10 μm; bold, large font) and [α-^{32}P]dCTP (0.5 μm, bold); note that this places the ^{32}P label adjacent to the photoreactive analog. Synthesis of the modified T strand was completed by the addition of dGTP, dATP, and dCTP (100 μm each) and dTTP (1 mM). (B) Positioning of halted elongation complexes by walking of T7 RNAP. Elongation complexes halted 24 nt downstream from the start site (EC24) were formed by incubation of the modified template with His-T7 RNAP, GTP, α-^{32}P-ATP, and UTP. The complexes were immobilized on Ni^{++}-agarose beads, washed to remove unincorporated substrates, and advanced to positions 25–28 by the addition of substrates as indicated. Gel electrophoresis of the labeled RNA products demonstrates the stepwise advancement of the complex as predicted by the sequence of the template. (C) Cross-linking of template to RNA polymerase. EC halted at the positions indicated was irradiated with UV while still immobilized on the Ni^{++}-agarose beads, eluted with 0.1% SDS, and analyzed by gel electrophroresis. The position of the photoreactive analog relative to the 3′ end of the RNA in each elongation complex is indicated. (D) Relative cross-linking efficiency versus position of analog. Experiments similar to those described in panels A–C were carried to position the photoreactive analog at other locations in both the T and NT strands (not shown). The efficiency of cross-linking at each position [the percent of labeled DNA that is incorporated into the cross-linked complex relative to the maximum observed for each strand (14% for the T strand, 60% for the NT strand)] is presented as a function of the distance from the 3′ end of the RNA. Adapted from K. Ma, D. Temiakov, M. Jiang, M. Anikin, W. T. McAllister, *J. Biol. Chem.* **277,** 43206–43215 (2002). With permission.

irradiation, a sample in which the analog is replaced by UTP or dTTP, and a sample in which transcription is prevented (in the absence of other substrates). Verification of the homogeneity of the transcription complexes can be accomplished by incorporating a radioactive label (e.g., α-^{32}P-ATP) into the RNA during the formation of the startup complex (see Fig. 2 and Fig. 3C) or alternatively, during subsequent elongation steps. The latter approach has the advantage that abortive products are not labeled and are therefore not visible in subsequent analytical steps.

The optimum wavelength for excitation of 4-thio-substituted analogs is 312 nm, which is readily provided by a hand-held UV lamp. In our studies we have used a 6W UV lamp from Cole-Palmer with a filter cutoff of 512 nm. The efficiency of cross-linking may be examined by running the products in a gel electrophoresis system that resolves the protein from nucleic acid components (Fig. 3C). The efficiency of cross-linking may vary with the position of the analog in the complex, which presumably reflects changes in the environment of the RNA or DNA during its passage through the EC. For example, in the case of 4-thio-T in the T strand in a T7 EC, we observed maximum reactivity when the analog was positioned at −9 (Fig. 3C, D). Cross-linking efficiencies obtained with 4-thio analogs may be very high (we have observed efficiencies as high as 40%). The ease and reliability of subsequent mapping studies decreases with efficiency; in general, we have found it difficult to use samples with cross-linking efficiencies <10%.

Resolution and Purification of Cross-Link Products

For most purposes, initial cleavage of the complex and mapping of the cross-linked sample may require merely separating the labeled nucleic acid components from the cross-linked complex. This may be done by precipitation with ammonium sulfate (for DNA cross-linked complexes) or by acetone precipitation for RNA samples.[2,4] Precipitation not only removes most of non-cross-linked DNA or RNA, but it is also useful because it greatly concentrates the sample (which is essential when the yield of cross-linked products is low) and allows the sample to be resuspended in appropriate buffers for subsequent chemical or enzymatic cleavage. After purification, the sample should be run again in an analytical gel to ensure its homogeneity and purity, and to ensure that contaminating bands caused by degradation of the RNA or DNA, or the protein, are not present.

For some purposes it may be necessary to further purify the cross-linked protein (or cleavage fragments) for subsequent analysis. This step may also be required if the cross-linking reaction targets more than one subunit in multisubunit polymerase complexes, or more than one site in a single subunit enzyme. Such purification can be accomplished by preparative PAGE, followed by electrophoretic elution or by extraction of the protein–nucleic acid cross-links from a gel slice.[2]

Cleavage and Mapping

The exact strategy for mapping the site of the cross-link depends on the suitability of various chemical or enzymatic cleavage reagents with the particular protein being examined. Commonly used chemical cleavage agents include cyanogen bromide (CNBr; cleaves after methionine), 2-nitro-5-thiocyano-benzoic acid (NTCB; cleaves at cysteine), *N*-chlorosuccinomide (NCS) or BNPS skatole (cleaves at tryptophan), and hydroxylamine (cleaves between asparagine [N] and glycine [G] residues in NG pairs). Highly specific proteases such as factor XA and the outer membrane protease ompT from *E. coli*[10,14] may also prove useful. However, we have found that some proteases may not recognize the predicted cleavage site because of folding of the protein, or may utilize minor cleavage sites. Proteases that cleave more frequently, such as trypsin or subtilisin, may also be used, and the products may be analyzed by either gel electrophoresis or mass spectrometry. Specific methods to map proteins using such cleavage agents are beyond the scope of this article and are described elsewhere.[2,4,12]

In interpreting the electrophoretic pattern of the cleaved protein, a number of considerations must be kept in mind. First, the RNA or DNA oligomer that is attached to the cleaved peptide may alter its mobility in different ways in different gel systems. Thus the expected change in apparent mass of a peptide caused by the presence of a cross-linked oligomer in a typical Laemmli protein gel may not be observed. The use of other gel systems, such as the NuPAGE system and other buffer systems (MES; Invitrogen, Carlsbad, CA), may afford better resolution in certain ranges or more readily interpretable cleavage patterns. An additional advantage of the MES buffer system is that it is carried out at pH 6.5; other systems that utilize a higher pH may result in degradation of the nucleic acid component and consequent heterogeneity. This effect can be minimized by trimming the unlabeled RNA or DNA "tail" with appropriate nucleases before mapping. Placing the nucleotide that introduces the radioactive label close to the cross-linking analog in such a way that it is not cleaved by the nuclease is an important consideration in the design of the template (see Fig. 3).

In addition to mapping by exhaustive chemical cleavage, single hit digestion may allow the resolution of a nested set of partially cleaved cross-linked products that arise from the N or C terminus of the

[14] J. Grodberg and J. J. Dunn, *J. Bacteriol.* **170**, 1245 (1988).

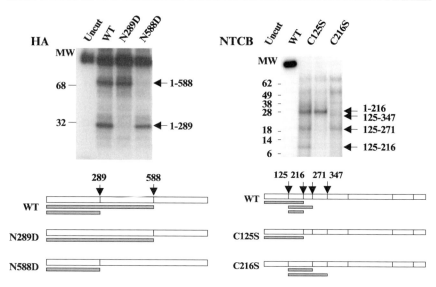

FIG. 4. Mapping of DNA:protein cross-links. Halted elongation complexes in which 4-thio-dTMP was positioned at -9 in the template strand (EC25) were formed as described in Fig. 3, and cross-linking was initiated by UV irradiation. The samples were treated with DNase to reduce the length of the cross-linked DNA (the ^{32}P label in the T strand is not removed by this process), subjected to proteolysis, and the labeled products were resolved by gel electrophoresis. Positions of molecular weight markers visible in the stained gels are indicated (MW; in kDa). In each panel, the predicted cleavage sites in WT and mutant RNAPs are indicated as vertical lines within an open box. The filled bars below represent the fit of the labeled peptides observed to the predicted cleavage pattern. Adapted from K. Ma, D. Temiakov, M. Jiang, M. Anikin, W. T. McAllister, *J. Biol. Chem.* **277**, 43206–43215 (2002). With permission. *Left*: Cleavage with hydroxylamine (HA). Cross-linked products were digested with HA as described,[10] and the products were resolved by electrophoresis in Nu-PAGE gels (4–12%) in MOPS buffer and identified by comparison to stained molecular weight markers (not shown). Results are shown for the wild type (WT) enzyme, as well as for two mutants in which the HA cleavage sites (NG pairs) at position 289 or 588 have been eliminated. *Right*: Cleavage with 2-nitro-5-thiocyano-benzoic acid (NTCB). As before, except gels were run in MES buffer. Results are shown for the WT enzyme, as well as for two mutants (C125S and C216S) in which the cleavage sites at 125 and 216, respectively, were eliminated. Conditions for NTCB cleavage have been previously described. Adapted from K. Ma, D. Temiakov, M. Jiang, M. Anikin, W. T. McAllister, *J. Biol. Chem.* **277**, 43206–43215 (2002).

protein.[15–17] Affinity tagged proteins, such as proteins that have been modified to include a histidine or GST tag at the N or C terminus, may facilitate

[15] N. Korzheva *et al.*, *Science* **289**, 619 (2000).
[16] V. Markovtsov, A. Mustaev, and A. Goldfarb, *Proc. Natl. Acad. Sci. USA* **93**, 3221 (1996).
[17] E. Nudler, E. Avetissova, V. Markovtsov, and A. Goldfarb, *Science* **273**, 211 (1996).

the retention and characterization of terminal fragments. It may also be useful to construct a modified form of the protein in which a protein kinase site has been incorporated at the N or C terminus to provide suitable molecular weight markers.

Complications

For some complexes, there may be multiple sites in the protein to which the analog may be cross-linked (for example, the structure of the protein may be flexible, or the trajectory of the nucleic acid may not be rigidly constrained). This can give rise to cleavage patterns that are difficult to interpret. For these reasons, it is usually necessary to combine two or even three cleavage strategies to reliably map a particular cross-link. Confirmation of the presumed site of cross-linking, and in some cases finer resolution, may be obtained by mutagenesis of the target protein to introduce or eliminate certain cleavage sites. This approach has been profitably used in our laboratory to characterize cross-links in T7 RNAP (Fig. 4).[4,10]

Concluding Remarks

The use of 4-thio-substituted UTP or dTTP analogs provides a powerful method to probe the organization of transcription complexes. These analogs may be placed at defined positions in the transcript or the DNA, and on UV activation form cross-links to nearby regions of the protein with high efficiency. Subsequent mapping of the site of the cross-link by peptide mapping provides important information concerning the organization of the complex and the trajectory of the nucleic acid components over the surface of the protein.

Acknowledgments

This work was supported by NIH grant GM38147 to W.T.M. We wish to thank Dr. Sergei Borukhov for stimulating discussions and for assistance in developing this approach in the T7 system.

[10] Fluorescence Resonance Energy Transfer (FRET) in Analysis of Transcription-Complex Structure and Function

By Jayanta Mukhopadhyay, Vladimir Mekler, Ekaterine Kortkhonjia, Achillefs N. Kapanidis, Yon W. Ebright, and Richard H. Ebright

In recent work we have developed methods that use fluorescence resonance energy transfer (FRET) to monitor movement of RNA polymerase (RNAP) relative to DNA during transcription,[1] and to define three-dimensional structures of transcription complexes in solution.[1,2]

FRET is a physical phenomenon that permits measurement of distances.[3–5] FRET occurs in a system in which a fluorescent probe serves as a donor and a second fluorescent probe serves as an acceptor, where the emission spectrum of the donor overlaps the excitation spectrum of the acceptor. In such a system, upon excitation of the donor with light of the donor excitation wavelength, energy can be transferred from the donor to the acceptor, resulting in excitation of the acceptor and emission at the acceptor emission wavelength. FRET readily can be detected, and its efficiency readily can be quantified by exciting the donor excitation wavelength with light and monitoring emission of the donor, emission of the acceptor, or both. The efficiency of energy transfer, E, is a function of the Förster parameter, R_o, and of the distance between the donor and the acceptor, R:

$$E = [1 + (R/R_o)^6]^{-1} \qquad (1)$$

Thus if one performs experiments under conditions where R_o is constant, measured changes in E permit detection of changes in R, and if one performs experiments under conditions where R_o is constant *and known*, the measured absolute magnitude of E permits determination of the absolute magnitude of R. With commonly used fluorescent probes, FRET permits accurate determination of distances in the range of $\sim$20 to $\sim$100 Å. For reference, the diameter of a transcription initiation or elongation

[1] J. Mukhopadhyay, A. Kapanidis, V. Mekler, E. Kortkhonjia, Y. Ebright, and R. Ebright, *Cell* **106,** 453 (2001).

[2] V. Mekler, E. Kortkhonjia, J. Mukhopadhyay, J. Knight, A. Revyakin, A. Kapanidis, W. Niu, Y. Ebright, R. Levy, and R. Ebright, *Cell* **108,** 599 (2002).

[3] T. Förster, *Ann. Phys.* **2,** 55 (1948).

[4] D. Lilley and T. Wilson, *Curr. Opin. Chem. Biol.* **4,** 507 (2000).

[5] P. Selvin, *Nature Structl. Biol.* **7,** 730 (2000).

METHODS IN ENZYMOLOGY, VOL. 371

complex is $\sim$150 Å.[1,2,6–11] Thus FRET permits accurate determination of distances up to more than one-half the diameter of a transcription complex.

Here we present protocols for FRET experiments that permit measurement of distances between positions on upstream DNA and positions within RNAP ("trailing-edge FRET"), distances between positions on downstream DNA and positions within RNAP ("leading-edge FRET"), and distances between positions on RNAP core and positions within σ^{70} ("core-σ^{70} FRET").

Preparation of DNA Fragments

DNA fragments corresponding to positions −40 to +25 of the *lacUV5–11* promoter (Fig. 1A)[1] are constructed as follows.

Oligodeoxyribonucleotides corresponding to positions −40 to +25 of the top strand (for use as polymerase chain reaction [PCR] template), positions −40 to −19 of the top strand (for use as PCR top-strand primer), and positions +6 to +25 of the bottom strand (for use as PCR bottom-strand primer) are prepared by automated β-cyanoethyl-phosphoramidite synthesis. Cy5-labeled oligodeoxyribonucleotides are prepared using Cy5-CE phosphoramidite (Glen Research, Inc., Sterling, VA) per the procedure of the manufacturer (i.e., with dG^{dmf} phosphoramidite, which can be deprotected under relatively mild conditions, in place of the standard dG^{ibu} phosphoramidite, and with deprotection by incubation in concentrated ammonium hydroxide for 12 h at 22°). Oligodeoxyribonucleotides (100 nmol aliquots in 200 μl water) are purified by reversed-phase high-performance liquid chromatography (HPLC) on a C18, 5 μm, 300 Å column (Rainin, Inc., Woburn, MA), with solvent A = 50 mM triethylamine acetate (pH 7.0) and 5% acetonitrile, solvent B = 100% acetonitrile, gradient = 0–30% solvent B in solvent A in 45 min, and flow rate = 1 ml/min. Unlabeled oligodeoxyribonucleotides elute at $\sim$17% of solvent B in solvent A; Cy5-labeled oligodeoxyribonucleotides elute at $\sim$23% of solvent B in solvent A. Eluted fractions are lyophilized, redissolved in 200 μl of water, and stored at −20°.

[6] G. Zhang, E. Campbell, L. Minakhin, C. Richter, K. Severinov, and S. Darst, *Cell* **98**, 811 (1999).
[7] P. Cramer, D. Bushnell, and R. Kornberg, *Science* **292**, 1863 (2001).
[8] A. Gnatt, P. Cramer, J. Fu, D. Bushnell, and R. Kornberg, *Science* **292**, 1876 (2001).
[9] K. Murakami, S. Masuda, and S. Darst, *Science* **296**, 1280 (2002).
[10] K. Murakami, S. Masuda, E. Campbell, O. Muzzin, and S. Darst, *Science* **296**, 1285 (2002).
[11] D. Vassylyev, S. Sekine, O. Laptenko, J. Lee, M. Vassylyeva, S. Borukhov, and S. Yokoyama, *Nature* **417**, 712 (2002).

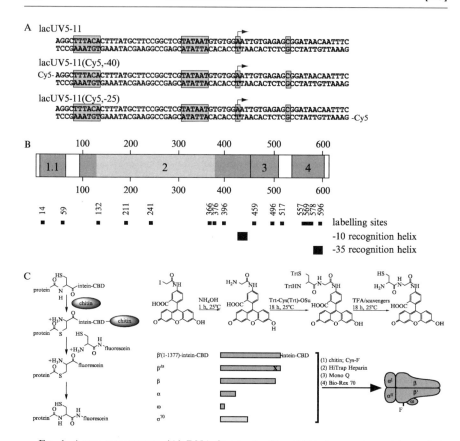

FIG. 1. Assay components. (A) DNA fragments. *Top*: DNA fragments used in analysis of RD$_{e,11}$ (*lacUV5* derivatives having no guanine residues on the template strand from +1 to +11 and having no fluorescent probe, the fluorescent probe Cy5 at −40, or the fluorescent probe Cy5 at +25).[1] *Shaded boxes*, transcription start site (*with arrow*), promoter −10 element, promoter −35 element, and halt site. (B) σ^{70} derivatives. Map of *E. coli* σ^{70} showing conserved regions 1.1 through 4 (*shaded boxes*; nonconserved insertion within *E. coli* σ^{70} region 2 shown in lighter shading),[34] determinants for interaction with promoter −10 and −35 elements (*solid bars*),[34] and sites at which the fluorescent probe TMR is incorporated (*filled squares*) (σ^{70} conserved regions defined as in Ref. 9). (C) Labeling of *E. coli* RNAP core (subunit composition $\beta'\beta\alpha^{I}\alpha^{II}\omega$). *Left*: Intein-mediated C-terminal labeling with Cys–fluorescein (see Preparation of RNAP Core).[1,2] *Top right*: Synthesis of Cys–fluorescein. *Bottom right*: Strategy for site-specific incorporation of fluorescein within intact, fully assembled RNAP core (at residue 1377 of β'). CBD, chitin-binding domain; Cys–F, Cys–fluorescein.

[34] C. Gross, C. Chan, A. Dombroski, T. Gruber, M. Sharp, J. Tupy, and B. Young, *Cold Spring Harbor Symp. Quant. Biol.* **63**, 141 (1998).

DNA fragments are prepared by PCR [reaction volume $= 1200\ \mu l$; reaction composition $= 1$ nM template, 500 nM top-strand primer, 500 nM bottom-strand primer, 60 units AmpliTaq DNA polymerase (Roche Molecular Biochemicals, Inc., Indianapolis, IN), 500 μM each dNTP, 10 mM Tris–HCl (pH 8.3), 50 mM KCl, and 1.5 mM MgCl$_2$; initial incubation $= 2$ min at 94°; number of cycles $= 40$; cycle profile $= 30$ s at 94°, 30 s at 55°, 30 s at 72°; final extension $= 5$ min 72°]. PCR products are precipitated by addition of one-tenth volume of 3 M sodium acetate (pH 5.2) and one volume of isopropanol, centrifuged (20,000 g; 5 min at 22°), redissolved in 20 μl of 5% glycerol and electrophoresed on 10% polyacrylamide slab gels (30:1 acrylamide/bisacrylamide; $6 \times 9 \times 0.1$ cm) in TBE [90 mM Tris–borate (pH 8.0), 0.2 mM EDTA] (20 V/cm; 30 min). Gel regions containing unlabeled DNA fragments are identified using UV shadowing[12]; gel regions containing labeled DNA fragments are identified using an x/y fluorescence scanner (Storm 860; Molecular Dynamics, Inc., Piscataway, NJ). Gel regions containing DNA fragments are excised and equilibrated in 400 μl 0.5 M ammonium acetate, 1 mM EDTA for 12 h at 4° with gentle rocking. Samples are centrifuged (5000 g; 1 min at 22°), supernatants are collected, and DNA fragments are precipitated by addition of one volume of isopropanol and centrifuged (20,000 g; 5 min at 22°). DNA fragments are redissolved in 20 μl water and stored at $-20°$. Concentrations and labeling efficiencies of DNA fragments are determined from UV/Vis absorption spectra as follows:

$$[\text{DNA}] = [\text{OD}_{260}]/\varepsilon_{\text{DNA},260} \tag{2}$$

$$\text{eff}_{\text{Cy5}} = (\text{OD}_{650}/\varepsilon_{\text{Cy5},650})/[\text{DNA}] \tag{3}$$

where $\varepsilon_{\text{DNA},260}$ is the molar extinction coefficient for the DNA fragment at 260 nm (850,000 M^{-1} cm^{-1}; calculated as in Ref. 13) and $\varepsilon_{\text{Cy5},650}$ is the molar extinction coefficient for Cy5 at 650 nM (250,000 M^{-1} cm^{-1}).[14] Concentrations typically are $\sim 5\ \mu M$. Labeling efficiencies typically are 80–90%.

Preparation of σ^{70}

Transformants of strain BL21(DE3)[15] with pGEMD(-Cys) or a pGEMD(-Cys) derivative [which encode, respectively, a σ^{70} derivative with no Cys residue, and a σ^{70} derivative with a single Cys residue at position 14,

[12] F. M. Orson, *Bio Techniques* **16,** 592 (1994).
[13] G. Felsenfeld and S. Hirschman, *J. Mol. Biol.* **13,** 407 (1965).
[14] R. Mujumdar, L. Ernst, S. Mujumdar, C. Lewis, and A. Waggoner, *Bioconj. Chem.* **4,** 105 (1993).
[15] F. Studier, A. Rosenberg, J. Dunn, and J. Dubendorff, *Methods Enzymol.* **185,** 60 (1990).

59, 132, 211, 241, 366, 376, 396, 459, 496, 517, 557, 569, 578, or 596 (Fig. 1B)][1,2,16–18] are shaken at 37° in 1 L LB[19] containing 200 μg/ml ampicillin until $OD_{600} = 0.6$, supplemented with IPTG to 1 mM, and further shaken at 37° until $OD_{600} = 1.5$. Cells are harvested by centrifugation (5000 g; 20 min at 4°), resuspended in 50 ml lysis buffer [40 mM Tris–HCl (pH 7.9), 300 mM NaCl, 1 mM EDTA, 1 mM phenylmethysulfonyl fluoride (PMSF), and 0.2% deoxycholate], and lysed by emulsification (Emulsiflex-C5; Avestin, Inc., Ottawa, Canada). Inclusion bodies containing σ^{70} derivatives are isolated by centrifugation (10,000 g; 20 min at 4°), washed with 20 ml lysis buffer containing 0.2 mg/ml lysozyme and 0.5% Triton X-100, and washed with 20 ml lysis buffer containing 0.5% Triton X-100 and 1 mM DTT [with each wash step involving sonication 2 × 1 min at 4° in wash buffer, incubation 10 min at 4° in wash buffer, and centrifugation (10,000 g; 20 min at 4°)]. Washed inclusion bodies containing σ^{70} derivatives are solubilized in 40 ml 6 M guanidine–HCl, 50 mM Tris–HCl (pH 7.9), 10 mM MgCl$_2$, 10 μM ZnCl$_2$, 1 mM EDTA, 10 mM DTT, and 10% glycerol, and dialyzed against 2 L TGED [20 mM Tris–HCl (pH 7.9), 0.1 mM EDTA, 0.1 mM DTT, and 5% glycerol] containing 0.2 M NaCl (20 h at 4°; two changes of buffer). The sample is centrifuged (10,000 g; 20 min at 4°) to remove particulates and applied to a Mono-Q HR 10/10 column (Amersham-Pharmacia Biotech, Piscataway, NJ) preequilibrated in the same buffer. The column is washed with 16 ml of the preequilibration buffer and eluted in 2-ml fractions of a 160-ml linear gradient of 200–600 mM NaCl in TGED (with σ^{70} derivatives typically eluting at ~360 mM NaCl in TGED). Fractions containing the σ^{70} derivative are identified by SDS–PAGE and Coomassie staining and are pooled. The σ^{70} derivative is precipitated by addition of 0.25 g/ml ammonium sulfate, followed by centrifugation (10,000 g; 20 min at 4°), and is stored as an ammonium-sulfate pellet at 4°. Yields typically are 50–60 mg. Purities typically are >98%.

Labeled σ^{70} derivatives are prepared using Cys-specific chemical modification. Reaction mixtures for labeling of σ^{70} contain (1 ml): 20 μM single-Cys σ^{70} derivative [from ammonium-sulfate pellet; washed with 0.5 ml ice-cold saturated ammonium sulfate, redissolved in 1 ml 100 mM sodium phosphate (pH 8.0), 1 mM EDTA, and subjected to solid-phase reduction on Reduce-Imm (Pierce, Inc., Rockford, IL) per the manufacturer's

[16] J. T. Owens, R. Miyake, K. Murakami, A. J. Chmura, N. Fujita, A. Ishihama, and C. Meares, *Proc. Natl. Acad. Sci. USA* **95**, 6021 (1998).
[17] S. Callaci, E. Heyduk, and T. Heyduk, *J. Biol. Chem.* **273**, 32995 (1998).
[18] J. Bown, J. Owens, C. Meares, N. Fujita, A. Ishihama, S. Busby, and S. Minchin, *J. Biol. Chem.* **274**, 2263 (1999).
[19] J. H. Miller, "Experiments in Molecular Genetics." Cold Spring Harbor Laboratory Press, Cold Spring Harbor, New York, 1972.

instructions, immediately before use], 200 μM tetramethylrhodamine-5-maleimide (TMR-5-maleimide; Molecular Probes, Inc., Eugene, OR), 100 mM sodium phosphate (pH 8.0), and 1 mM EDTA. After 1 h on ice, products are applied to a Bio-Gel P6DG column (Bio-Rad, Hercules, CA) preequilibrated in TGED containing 0.2 M NaCl and are eluted in the 15 × 0.5-ml fractions of the same buffer. Fractions containing labeled σ^{70} derivative are identified by UV/Vis absorbance, pooled, and stored in 20 mM Tris–HCl (pH 7.9), 100 mM NaCl, 0.1 mM EDTA, 0.1 mM DTT, and 50% glycerol in aliquots at −20°. The efficiency of labeling is determined from UV/V is absorption spectra as follows:

$$\text{eff}_{\text{TMR}} = (\text{OD}_{555}/\varepsilon_{\text{TMR,555}})/[\sigma^{70}] \tag{4}$$

where $\varepsilon_{\text{TMR,555}}$ is the molar extinction coefficient for TMR at 555 nm $(75,000\ M^{-1}\ \text{cm}^{-1})$[20] and where $[\sigma^{70}]$ is the concentration of σ^{70} determined by Bradford assay.[21] The site-specificity of labeling is determined by comparison to products of a parallel control reaction with a Cys-free σ^{70} derivative. Efficiencies of labeling typically are ~90%. Site-specificities of labeling typically are ~90%.

Preparation of RNAP Core

Escherichia coli strain XE54[22] transformed with plasmid pREII-NHα (which encodes an *N*-terminally hexahistidine-tagged derivative of RNAP α subunit)[23] is shaken at 37° in 6 L 4 × LB[19] containing 170 mM NaCl and 200 μg/ml ampicillin until OD$_{600}$ = 0.6, supplemented with IPTG to 1 mM, and further shaken at 37° until OD$_{600}$ = 1.5. Cells are harvested by centrifugation (5000 g; 20 min at 4°), resuspended in 150 ml lysis buffer [50 mM Tris–HCl (pH 7.9), 200 mM NaCl, 2 mM EDTA, 5% glycerol, 0.1 mM DTT, 1 mM PMSF, and 0.25% deoxycholate], and lysed by emulsification (Emulsiflex-C5; Avestin, Inc.). The lysate is cleared by centrifugation (8000 g; 20 min at 4°), and protein is precipitated by addition of 7 ml 10% Polymin P (pH 7.9) [incubation 10 min at 4° with stirring, followed by centrifugation (6000 g; 20 min at 4°)].[24] The precipitate is washed with 100 ml TGD [20 mM Tris–HCl (pH 7.9), 0.1 mM DTT, and 5% glycerol] containing 0.5 M NaCl [incubation 40 min at 4°, followed by centrifugation

[20] G. Kwon, A. Remmers, S. Datta, and R. Neubig, *Biochemistry* **32,** 2401 (1993).
[21] M. Bradford, *Anal. Biochem.* **72,** 248 (1976).
[22] H. Tang, K. Severinov, A. Goldfarb, D. Fenyo, B. Chait, and R. Ebright, *Genes Dev.* **8,** 3058 (1994).
[23] W. Niu, Y. Kim, G. Tau, T. Heyduk, and R. Ebright, *Cell* **87,** 1123 (1996).
[24] R. Burgess and J. Jendrisak, *Biochemistry* **14,** 4634 (1975).

(6,000 g; 20 min at 4 °)], and extracted with 100 ml of TGD containing 1 M NaCl [incubation 40 min at 4 °, followed by centrifugation (8,000 g; 20 min at 4 °)]. Protein in the extract is precipitated by addition of 30 g ammonium sulfate [incubation 30 min at 4 ° with stirring, followed by centrifugation (10,000 g; 20 min at 4 °)], and is redissolved in 30 ml buffer C [20 mM Tris–HCl (pH 7.9), 500 mM NaCl, and 5% glycerol], applied to a 6-ml column of Ni^{++}-NTA agarose (Qiagen, Chatswon CA) preequilibrated with buffer C. The column is washed with 30 ml buffer C and eluted with 5 × 6 ml each of buffer C containing 5 mM, 10 mM, 20 mM, 40 mM, and 100 mM imidazole (with RNAP core typically eluting at ∼40 mM imidazole). Fractions containing RNAP core are identified by SDS–PAGE and Coomassie staining, pooled, and dialyzed against 2 L TGED buffer containing 240 mM NaCl (12 h at 4 °). The sample (∼10 ml) is cleared by centrifugation (12,000 g; 20 min at 4 °) and applied to a Mono-Q HR 10/10 column (Amersham-Pharmacia Biotech.) preequilibrated with TGED containing 300 mM NaCl. The column is washed with 24 ml of the preequilibration buffer and eluted in 1-ml fractions of a 160-ml linear gradient of 300–500 mM NaCl in TGED (with RNAP core typically eluting at ∼350 mM NaCl). Fractions containing RNAP core are identified by SDS–PAGE and Coomassie staining, pooled, dialyzed against 200 ml 40 mM Tris–HCl (pH 7.9), 200 mM NaCl, 0.1 mM EDTA, 0.1 mM DTT, and 50% glycerol (8 h at 4 °), and stored in aliquots at −20 °. Yields typically are ∼12 mg.

Labeled RNAP core (with fluorescein incorporated at residue 1377 of β') is prepared by intein-mediated C-terminal labeling (Fig. 1C).[1,2] The procedure involves (1) co-expression of genes encoding β'(1-1377)-intein-CBD, β'^{ts397c} (temperature-sensitive, assembly-defective β' derivative),[25,26] β, α, ω, and σ^{70}; (2) affinity-capture of RNAP holoenzyme containing β'(1-1377)-intein-CBD on chitin; (3) cleavage, elution, and concurrent labeling with Cys–fluorescein; and (4) removal of σ^{70}. Cultures of strain 397c [$rpoC397$ $argG$ thi lac (λcI$_{857}$h$_{80}$S$_{t68}$ dlac^+)][25] transformed with plasmid pVMβ'1377-IC (which encodes β'(1-1377)-intein-CBD[1]) are shaken at 37 ° in 1 L 4 × LB[19] containing 170 mM NaCl, 3 mM IPTG, and 200 μg/ ml ampicillin, until OD$_{600}$ = 1.5, and are harvested by centrifugation (5000 g; 15 min at 4 °). Cell lysis, Polymin P precipitation, and ammonium sulfate precipitation are performed as described for preparation of

[25] G. Christie, S. Cale, L. Isaksson, D. Jin, M. Xu, B. Sauer, and R. Calendar, *J. Bacteriol.* **178**, 6991 (1996).
[26] L. Minakhin, S. Bhagat, A. Brunning, E. Campbell, S. Darst, R. Ebright, and K. Severinov, *Proc. Natl. Acad. Sci. USA* **98**, 892 (2001).

unlabeled RNAP, except that 5 ml protease-inhibitor mixture P8465 (Sigma, Inc., St. Louis, MO) is included in the lysis buffer, and 1 ml 0.8 mM σ^{70} is added immediately after lysis. The samples are suspended in 25 ml buffer D [20 mM Tris–HCl (pH 7.9), 0.1 mM EDTA, and 5% glycerol] containing 100 mM NaCl, and applied to a 10-ml column of chitin (New England Biolabs, Beverly, MA) in buffer D containing 100 mM NaCl. The column is washed with 50 ml buffer D containing 500 mM NaCl, washed with 25 ml buffer E [20 mM sodium phosphate (pH 7.3), 200 mM NaCl, and 0.5 mM tris(2-carboxyethyl)phosphine (TCEP; Pierce, Inc.)], and equilibrated with 10 ml buffer E containing 150 μM Cys–fluorescein (summary of synthesis in Fig. 1C),[2] 0.5% (v/v; saturating) thiophenol, and 0.1 mM PMSF. After 8 h at 4°, 10–20 ml buffer D containing 200 mM NaCl and 0.1 mM DTT is applied to the column, and 1-ml fractions are collected. Fractions containing labeled RNAP holoenzyme are pooled, centrifuged to remove fine-particulate chitin (15,000 g; 15 min at 4°), and filtered through 50 ml Sephadex LH-20 (Amersham-Pharmacia Biotech) in buffer D containing 200 mM NaCl and 0.1 mM DTT to remove excess Cys–fluorescein. The sample (~20 ml) is loaded onto a 1-ml HiTrap Heparin column (Amersham-Pharmacia Biotech) preequilibrated in TGED containing 100 mM NaCl. The column is washed with 5 ml of the preequilibration buffer and eluted in 1-ml fractions of a 30-ml linear gradient of 100–500 mM NaCl in TGED (with labeled RNAP typically eluting at ~400 mM NaCl). Fractions containing labeled RNAP are identified by SDS–PAGE and Coomassie staining, pooled, and, optionally, further chromatographed on Mono-Q HR 10/10 (methods as described previously for unlabeled RNAP core). The sample is diluted with TGED to yield [NaCl] = 100 mM, and applied to a 1-ml column of Bio-Rex 70 (Bio-Rad) preequilibrated in TGED buffer containing 100 mM NaCl. The column is washed with 5 ml of the preequilibration buffer and eluted in 0.5-ml fractions of a 20-ml linear gradient of 100–500 mM NaCl in TGED (with labeled RNAP core eluting at ~260 mM NaCl). Fractions containing labeled RNAP core are identified by SDS–PAGE and Coomassie staining, pooled, dialyzed against 250 ml 40 mM Tris–HCl (pH 7.9), 200 mM NaCl, 0.1 mM EDTA, 0.5 mM TCEP, and 50% glycerol (8 h at 4°), and stored in aliquots at −20°. Concentrations and efficiencies of labeling are determined from UV/Vis absorption spectra as follows:

$$[\text{RNAP, core}] = [\text{OD}_{280} - \varepsilon_{F,280}(\text{OD}_{494}/\varepsilon_{F,494})]/\varepsilon_{\text{RNAP},280} \qquad (5)$$

$$\text{eff}_F = (\text{OD}_{494}/\varepsilon_{F,494})/[\text{RNAP, core}] \qquad (6)$$

where $\varepsilon_{RNAP,280}$ is the molar extinction coefficient for RNAP core at 280 nm (199,000 M^{-1} cm^{-1}; calculated as in Ref. 27), and $\varepsilon_{F,280}$ and $\varepsilon_{F,494}$ are the molar extinction coefficients for fluorescein at 280 nm and 494 nm (14,000 M^{-1} cm^{-1} and 70,800 M^{-1} cm^{-1}).[28] Site-specificities of labeling are determined by comparison to the products of a control reaction with an RNAP derivative not containing intein-CBD. Yields typically are ~0.1 mg. Efficiencies of labeling typically are ≥95%. Site-specificities of labeling typically are ≥95%.

Preparation of RNAP Holoenzyme

Holoenzyme derivatives are prepared by incubation of 4 pmol unlabeled core or labeled core derivative with 8 pmol unlabeled σ^{70} or labeled σ^{70} derivative in 20 μl transcription buffer [TB: 50 mM Tris–HCl (pH 8.0), 100 mM KCl, 10 mM MgCl$_2$, 1 mM DTT, 10 μg/ml bovine serum albumin (BSA), and 5% glycerol] for 20 min at 25°. Transcriptional activities of RNAP holoenzyme derivatives are determined in fluorescence-detected abortive initiation assays[29] using the *lac(ICAP)UV5* promoter.[30] Transcriptional activities of labeled holoenzyme derivatives typically are indistinguishable from those of unlabeled holoenzyme ($K_B \sim 3 \times 10^8 \ M^{-1}$; $k_f \sim 0.01$ s^{-1}).

Preparation of RNA Polymerase-Promoter Open Complex (RP$_o$)

Reaction mixtures contain (20 μl): 200 nm RNAP holoenzyme derivative and 50 nm DNA fragment in transcription buffer (TB: 50 mM Tris–HCl (pH 8.0), 100 mM KCl, 10 mM MgCl$_2$, 1 mM DTT, 10 μg/ml BSA, and 5% glycerol). After 15 min at 37°, 0.5 μl 1 mg/ml heparin is added (to disrupt nonspecific complexes),[31] and, after a further 5 min at 37°, reaction mixtures are applied to 5% polyacrylamide slab gels (30:1 acrylamide/bisacrylamide; 6 × 9 × 0.1 cm) and electrophoresed in TBE (20 V/cm; 30 min at 37°). Gel regions containing RP$_o$ are identified using an x/y fluorescence scanner [Storm 860 (for Cy5) or FluorImager 595 (for fluorescein or TMR), Molecular Dynamics, Inc.] and excised.

[27] S. Gill, S. Weitzel, and P. von Hippel, *J. Mol. Biol.* **220,** 307 (1991).
[28] R. Cerione, R. McCarty, and G. Hammes, *Biochemistry* **22,** 769 (1983).
[29] W. Suh, S. Leirmo, and M. Record, Jr., *Biochemistry* **31,** 7815 (1992).
[30] N. Naryshkin, A. Revyakin, Y. Kim, V. Mekler, and R. Ebright, *Cell* **101,** 601 (2000).
[31] C. Cech and W. McClure, *Biochemistry* **19,** 2440 (1980).

DNA–RNAP FRET (Trailing-Edge/Leading-Edge FRET)

Gel slices containing RP_o prepared from labeled RNAP holoenzyme and labeled DNA (donor-acceptor experiment, DA), from labeled RNAP holoenzyme and unlabeled DNA (donor-only control, D), and from unlabeled RNAP holoenzyme and labeled DNA (acceptor-only control, A), are mounted in submicro fluorometer cuvettes (Starna, Inc., Atascadero, CA) containing 92 μl TB. For each gel slice, fluorescence emission intensities [excitation wavelengths = 530 nm and 620 nm; excitation and emission slit widths = 5 nm; QuantaMaster QM1 spectrofluorometer (PTI, Inc., South Brunswick, NJ)] are measured after incubation for 5 min at 37° (data labeled "RP_o" in Fig. 2A, B) and are again measured after addition of 8 μl 6 mM ApA, 1.25 mM ATP, 1.25 mM GTP, and 1.25 mM UTP, and incubation for 10 min at 37° (data labeled "RP_o + NTPs" in Fig. 2A, B). Fluorescence emission intensities (F) are corrected for background by subtraction of fluorescence emission intensities for equivalent-size gel slices not containing RP_o and are corrected for wavelength-dependence of lamp output. Fluorescence emission intensities attributable to FRET ($F_{665,530}^{FRET}$), efficiencies of FRET (E), Förster parameters (R_o; typically 61.4 Å in this article),[1,2] and donor-acceptor distances (R) are calculated as follows[1,2,32]:

$$F_{665,530}^{FRET} = F_{665,530}^{DA} - \frac{F_{580,530}^{DA} \cdot F_{665,530}^{D}}{F_{580,530}^{D}} - \frac{F_{665,620}^{DA} \cdot F_{665,530}^{A}}{F_{665,620}^{A}} \tag{7}$$

$$E = \frac{F_{665,530}^{FRET} \cdot \varepsilon_{620}^{A}}{F_{665,620}^{DA} \cdot \varepsilon_{530}^{D} \cdot d} \tag{8}$$

$$R_o = 9780(n^{-4}\kappa^2 Q_D J)^{1/6} \tag{9}$$

$$R = R_o[(1/E) - 1]^{1/6} \tag{10}$$

where d is the efficiency of donor labeling (eff$_{TMR}$ in Eq. (5); typically 0.8–1),[1] ε_{530}^{D} is the extinction coefficient of the donor at 530 nm (42,600 M^{-1} cm^{-1}),[1] ε_{620}^{A} is the extinction coefficient of the acceptor at 620 nm (110,000 M^{-1} cm^{-1}),[1] n is the refractive index of the medium (1.4),[32] κ^2 is the orientation factor relating the donor emission dipole and acceptor emission dipole (approximated as 2/3, justified by fluorescence-anisotropy measurements indicating that the donor reorients on the time scale of the excited state lifetime),[2,32] Q_D is the quantum yield of the donor

[32] R. Clegg, *Methods Enzymol.* **211**, 353 (1992).

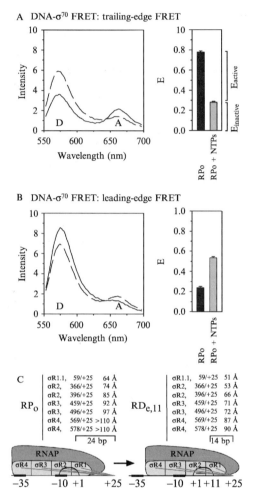

A DNA-σ^{70} FRET: trailing-edge FRET

B DNA-σ^{70} FRET: leading-edge FRET

FIG. 2. DNA–RNAP FRET (leading-edge/trailing-edge FRET). (A) Representative data for trailing-edge-FRET experiments assessing RP_o and RD_e [data for experiments with RNAP holoenzyme derivative labeled at σ^{70} position 569 and DNA fragment *lacUV5-11* (Cy5 −40)]. *Left*: Solid line, fluorescence emission spectrum of RP_o; dashed line, fluorescence emission spectrum 10 min after addition of A_pA, ATP, GTP, and UTP to RP_o. *Right*: Black bar, FRET in RP_o; gray bar, FRET 10 min after addition of A_pA, ATP, GTP, and UTP to RP_o. (B) Representative data for leading-edge-FRET experiments assessing RP_o and RD_e [data for experiments with RNAP holoenzyme derivative labeled at σ^{70} position 366 and DNA fragment *lacUV5-11*(Cy5, +25)]. *Left*: Solid line, fluorescence emission spectrum of RP_o; dashed lines, fluorescence emission spectrum 10 min after addition of A_pA, ATP, GTP, and UTP to RP_o. *Right*: Black bar, FRET in RP_o; gray bar, FRET 10 min after addition of A_pA, ATP, GTP, and UTP to RP_o. (C) Probe-probe distances in RP_o and $RD_{e,11}$, calculated from leading-edge FRET-[see Eqs. (10)(12)]. *Left*: Probe-probe distances in RP_o with Cy5 at +25 (Cy5 24 bp from RNAP active center). *Right*: Probe-probe distances in $RD_{e,11}$ with Cy5 at +25 (Cy5 14 bp from RNAP active center).

in absence of the acceptor (0.32; determined using fluorescein in 0.1 M NaOH as the reference[33]), and J is the spectral overlap integral of the donor emission spectrum and the acceptor excitation spectrum (1.1×10^{-12} M^{-1} cm^3; determined using corrected spectra for donor-only and acceptor-only controls[32]).

Efficiencies of FRET within RD_e (E_{RDe}), and donor-acceptor distances within RD_e (R_{RDe}), are calculated as:

$$E_{RDe} = [E_{RPo+NTPs} - (1-f)E_{RPo}]/f \tag{11}$$

$$R_{RDe} = R_o[(\theta_{\sigma70}/E_{RDe}) - 1]^{1/6} \tag{12}$$

where $\theta_{\sigma70}$ is the fractional occupancy of σ^{70} in $RD_{e,11}$ (1.0 for $RD_{e,11}$),[1] and where f is the fraction of molecules competent to undergo the transition to elongation, calculated as:

$$f = \frac{E_{active}}{E_{active} + E_{inactive}} \tag{13}$$

where E_{active} and $E_{inactive}$ are determined from trailing-edge-FRET experiments and correspond to, respectively, the FRET efficiency attributable to the subpopulation of complexes competent to undergo the transition to elongation (trailing-edge E_{RPo} minus trailing-edge $E_{RPo+NTPs}$), and the FRET efficiency attributable to the subpopulation of complexes not competent to undergo the transition to elongation (trailing-edge $E_{RPo+NTPs}$) (see Fig. 2A, far right)[1]; or as:

$$f = (\text{mols RNA}_n)/(\text{mols DNA}) \tag{14}$$

where mols RNA$_n$ and mols DNA are determined in parallel biochemical experiments and correspond to, respectively, the number of moles of RNA of length n synthesized, and the number of moles of DNA template.[1]

Core-σ^{70} FRET

Core-σ^{70} FRET experiments are performed analogously to DNA–RNAP FRET experiments. Excitation wavelengths are 470 nm and 540 nm; emission is monitored at 520 nm and 580 nm. The fluorescence intensity attributable to FRET ($F_{580,470}^{FRET}$), the efficiency of FRET (E), the Förster parameter (R_o; typically 57.9 Å in this article), and the donor-acceptor distance (R), are calculated using Eqs. (7)–(14) [substituting 530, 620, 580, and 665 by 470, 540, 520, and 580, respectively; and using $d = \text{eff}_F$ (Eq. (6); typically 0.95–1.0), $\varepsilon_{470}^D = 26,600$ M^{-1} cm^{-1}, and $\varepsilon_{540}^A = 52,000$ M^{-1} cm^{-1}].[1,2]

[33] J. Lakowicz, "Principles of Fluorescence Spectroscopy." Kluwer, New York, 1999.

Results

DNA–RNAP FRET (Trailing-Edge/Leading-Edge FRET)

DNA–RNAP FRET permits monitoring of movement of RNAP relative to DNA during promoter escape and elongation (differences in distances in left and right panels of Fig. 2C) and provides information about three-dimensional structures of RP_o and RD_e (absolute magnitudes of distances in left and right panels of Fig. 2C).[1,2]

The approach involves two complementary sets of experiments: (1) trailing-edge-FRET experiments, which monitor the distance between a fluorescent probe in RNAP and a fluorescent probe on DNA upstream of RNAP; and (2) leading-edge-FRET experiments, which monitor the distance between a fluorescent probe in RNAP and a fluorescent probe on DNA downstream of RNAP. The approach detects movement as changes in distances between fluorescent probes and concomitant changes in efficiency of FRET (which is proportional to the inverse sixth power of distance between fluorescent probes).[3–5]

To permit preparation of RP_o and of RD_e containing 11 nt of RNA ($RD_{e,11}$, for which $\theta_{\sigma70} = 1$),[1] we use derivatives of the *lacUV5* promoter that have the first template-strand guanine residue in the transcribed region at position +12 (Fig. 1A).[1] With these DNA templates, upon formation of RP_o and addition of A_pA, ATP, GTP, and UTP, RNAP initiates transcription, proceeds to position +11 and halts (because of absence of CTP, the next required NTP). The resulting halted complexes are *bona fide* elongation complexes; they are stable, they retain RNA, and they can be restarted on addition of CTP.[1] For trailing-edge-FRET experiments, we incorporate Cy5, serving as acceptor, upstream of the core promoter, at position −40; for leading-edge-FRET experiments, we incorporate Cy5, serving as acceptor, downstream of the core promoter, at position +25.

To incorporate fluorescent probe at single sites within RNAP holoenzyme, we prepare σ^{70} derivatives that have single Cys residues at positions 14, 59, 132, 211, 241, 366, 376, 396, 459, 496, 517, 557, 569, 578, and 596; we react each single-Cys σ^{70} derivative with TMR-5-maleimide under conditions that result in highly selective, highly efficient derivatization of Cys, and we reconstitute RNAP holoenzyme from each labeled σ^{70} derivative and RNAP core.[1,2] This yields a set of 15 fully functional labeled RNAP holoenzyme derivatives, each having TMR incorporated at a single site within σ^{70}, with the sites spanning the length of σ^{70}, including σ^{70} conserved regions 1.1, 2, 3, and 4 (Fig. 1B).

Representative data from trailing-edge-FRET experiments with RP_o and $RD_{e,11}$ are presented in Fig. 2A [data for RNAP holoenzyme derivative labeled at σ^{70} residue 569, within σ^{70} region 4, and DNA fragment *lacUV5-11* (Cy5, −40)]. Upon addition of the NTP subset that permits formation of an RNA product of 11 nt and incubation for 10 min, there is a 64% decrease in trailing-edge FRET, from 0.78 to 0.28 (Fig. 2A, right panel).[1] The residual trailing-edge FRET of 0.28 remains constant, even after 30 min or longer.[1] Extensive control experiments, involving both analysis of fluorescence and direct extraction and analysis of RNA products, establish that the residual trailing-edge FRET of 0.28 is attributable to the subpopulation of complexes not competent to undergo the transition to elongation.[1] Equivalent results are obtained with all σ^{70} derivatives labeled at sites sufficiently close to DNA position −40 to yield significant trailing-edge FRET in RP_o.[1] Equivalent results also are obtained with DNA fragments and/or NTP subsets permitting synthesis of longer RNA products.[1] We conclude that under our conditions, in 64% of complexes, RNAP holoenzyme leaves the promoter.

Representative data from leading-edge-FRET experiments with RP_o and $RD_{e,11}$ are presented in Fig. 2B [data for RNAP holoenzyme derivative labeled at σ^{70} residue 366, within σ^{70} region 2, and DNA fragment *lacUV5-11* (Cy5, +25)]. Upon addition of the NTP subset that permits formation of $RD_{e,11}$ and incubation for 10 min, there is a 2.3-fold increase in leading-edge FRET, from 0.24 to 0.54 (Fig. 2B, right panel).[1] Thus molecules of RNAP holoenzyme that leave the promoter, as documented in the trailing-edge-FRET experiments (Fig. 2A, right panel), arrive downstream (Fig. 2B, right panel).

From the measured efficiencies of leading-edge FRET in RP_o, it is possible to calculate the distances in RP_o between each probe in RNAP holoenzyme and the probe on DNA at position +25, 24 bp downstream of the RNAP active center [distances in Fig. 2C, left panel; Eq. (10)]. Analogously, from the measured efficiencies of leading-edge FRET in $RD_{e,11}$, it is possible to calculate the distance in $RD_{e,11}$ between each probe in RNAP holoenzyme and the probe on DNA at position +25, 14 bp downstream of the RNAP active center [distances in Fig. 2C, right panel; Eq. (12)]. For each probe in RNAP holoenzyme, the distance in $RD_{e,11}$ is significantly shorter than the distance in RP_o, reflecting the movement of RNAP holoenzyme relative to DNA in formation of $RD_{e,11}$ from RP_o (Fig. 2C). For each probe in RNAP holoenzyme, the distance in $RD_{e,11}$ is identical, or nearly identical, to the distance in a "reference-construct" RP_o derivative with the probe on DNA at position +15, 14 bp downstream of the RNAP active center.[1,2] We conclude that the positions of the tested σ^{70} regions relative to the remainder of the transcription complex in $RD_{e,11}$ are identical to those in RP_o[1], with σ^{70} region 1.1 located near the downstream end of the transcription bubble, σ^{70} region 2 located near the center and upstream end of the transcription bubble, and σ^{70} regions 3 and 4 are located upstream of the transcription bubble.[1,2,10]

Core-σ^{70} FRET

Core-σ^{70} FRET provides information about three-dimensional structures of RP_o and RD_e (absolute magnitudes of distances in Fig. 3B).[1,2] For core-σ^{70}-FRET experiments, we prepare RNAP holoenzyme derivatives having fluorescein (serving as donor) incorporated at single sites within RNAP core and having TMR (serving as acceptor) incorporated at single sites within σ^{70}. We then measure efficiencies of core-σ^{70} FRET in RP_o and RD_e.

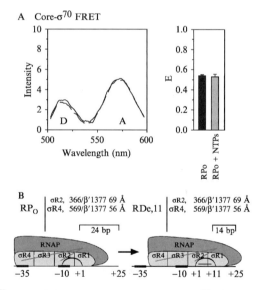

FIG. 3. Core-σ^{70} FRET. (A) Representative data for core-σ^{70}-FRET experiments assessing RP_o and $RD_{e,11}$ (data for experiments with σ^{70} derivative labeled at position 569, RNAP core labeled at the residue 1377 of β', and DNA fragment *lacUV5-11*). *Left*: Solid line, fluorescence emission spectrum of RP_o; dashed lines, fluorescence emission spectrum 10 min after addition of A_pA, ATP, GTP, and UTP to RP_o. *Right*: Black bar, FRET in RP_o; gray bar, FRET 10 min after addition of A_pA, ATP, GTP, and UTP to RP_o. (B) Probe-probe distances in RP_o and $RD_{e,11}$, calculated from core-σ^{70}-FRET [see Eqs. (10)(12)]. *Left*: Probe-probe distances in RP_o. *Right*: Probe-probe distances in $RD_{e,11}$.

To incorporate fluorescein at single sites within RNAP core, we use a novel intein-mediated C-terminal labeling strategy.[1,2] To incorporate TMR at single sites within σ^{70}, we use the same Cys-specific-chemical-modification strategy as in the preceding section.

Representative data from core-σ^{70}-FRET experiments with RP_o and $RD_{e,11}$ are presented in Fig. 3A (data for RNAP holoenzyme derivative labeled at β' residue 1377 and σ^{70} residue 569; DNA fragment *lacUV5-11*). Upon addition of the NTP subset that permits formation of $RD_{e,11}$ and incubation for 10 min, there is no change in core-σ^{70} FRET (Fig. 3A). From the efficiencies of core-σ^{70} FRET in RP_o and $RD_{e,11}$, it is possible to calculate the distances in RP_o and $RD_{e,11}$ between the probe in RNAP core and the probes in σ^{70} [distances in Fig. 3B; Eqs. (10),(12)]. In each case, the distances in RP_o and $RD_{e,11}$ are identical, consistent with the proposal that the orientations of the tested σ^{70} regions relative to the remainder of the transcription complex RNAP in RP_o and $RD_{e,11}$ are identical (see preceding section; see also Ref. 1).

Prospect

Straightforward extensions of the approaches described here permit systematic determination of large numbers of distances—tens to hundreds of distances—in transcription complexes and, in conjunction with auto-mated distance-constrained docking, can yield accurate three-dimensional structural models of transcription complexes.[2,9–11] Further straightforward extensions of the approaches described here permit analysis of kinetics of structural changes in transcription complexes (by use of stopped-flow spectroscopy; J. Mukhopadhyay and R. H. Ebright, unpublished data) and single-molecule analysis of kinetics of structural changes in transcription complexes (by use of confocal optical microscopy; A. Kapanidis, R. H. Ebright, and S. Weiss, unpublished data). In current work, we are using these approaches to carry out structural and mechanistic analysis of each step in transcription initiation, elongation, termination, and regulation.

Acknowledgments

This work was supported by a National Institutes of Health (NIH) grant (GM41376), a Howard Hughes Medical Institute Investigatorship to R.H.E., and by NIH grant GM64375 to R. Levy and R.H.E.

[11] Methods of Walking with the RNA Polymerase

By Evgeny Nudler, Ivan Gusarov, and Gil Bar-Nahum

Transcription elongation, the actual process of copying DNA into RNA by RNA polymerase (RNAP), is also a crucial point of gene regulation in all organisms. Transcription elongation is not a monotonic process and can be interrupted by transient pauses, arrests, or termination signals. Originally described as an essential step in controlling gene expression in bacteria and phages, elongation was later recognized as a critical regulatory target in eukaryotic cells as well. Despite the ever-growing interest in the mechanisms of transcription elongation, the process remains far less well understood than earlier steps of transcription cycle, promoter binding, and initiation. Dozens of factors that modulate elongation have been identified in bacteria and higher organisms, but the mechanisms by which those molecules transmit their signals to RNAP, changing its properties, are largely unknown. One of the major obstacles that restricted our understanding of elongational mechanisms is isolation of stable homogeneous intermediates, the stalled elongation complexes (ECs), in quantities suitable for biochemical and structural analysis.

The principal structural determinants of RNAP basic activity are highly conserved in evolution,[1–3] suggesting that the basic mechanisms of elongation and its control in bacteria and eukaryotes are the same. Much of the recent progress in understanding those mechanisms stems from the studies on a prototypic RNAP from *Escherichia coli* utilizing stepwise transcription in solid phase ("walking"). The walking system provides an opportunity to study ECs stalled virtually at any desired position along the DNA. The walking procedure was initially based on a fully functional 6-histidine-tagged *E. coli* RNAP immobilized on Ni^{++}-chelating agarose beads.[4,5] This technique has been used successfully in elucidating the mechanisms of pausing, arrests, intrinsic termination, and antitermination.[6–9]

[1] A. L. Gnatt, P. Fu, J. Cramer, D. A. Bushnell, and R. D. Kornberg, *Science* **292,** 1876 (2001).

[2] S. A. Darst, *Curr. Opin. Struct. Biol.* **11,** 155 (2001).

[3] R. H. Ebright, *J. Mol. Biol.* **304,** 687 (2000).

[4] M. Kashlev, E. Martin, A. Polyakov, K. Severinov, V. Nikiforov, and A. Goldfarb, *Gene* **130,** 9 (1993).

[5] M. Kashlev, E. Nudler, K. Severinov, S. Borukhov, N. Komissarova, and A. Goldfarb, *Methods Enzymol.* **274,** 326 (1996).

[6] E. Nudler, A. Mustaev, E. Lukhtanov, and A. Goldfarb, *Cell* **89,** 33 (1997).

[7] N. Komissarova and M. Kashlev, *Proc. Natl. Acad. Sci. USA* **94,** 1755 (1997).

[8] I. Gusarov and E. Nudler, *Mol. Cell* **3,** 495 (1999).

[9] I. Gusarov and E. Nudler, *Cell* **107,** 437 (2001).

Since the originally reported procedure,[4] the method underwent continuous refinement and optimization. The purpose of this article is to complement the previously published procedure by summarizing the most significant recent improvements in the walking technology and to demonstrate its flexibility and availability for many biochemical and protein chemical tools.

General Considerations

The principle behind solid-phase walking is that the initial EC immobilized onto a solid support undergoes rounds of washing (to remove the unincorporated nucleotide triphosphate [NTP] substrates) followed by addition of the incomplete set of NTPs (three or less) that allow transcription to proceed to the next DNA position corresponding to the first missing NTP. In theory, this procedure can be repeated numerous times without loss of material, and the only limitation in obtaining the desired EC would be a stretch of the same bases in DNA to be transcribed in one step. The efficiency of walking, however, depends on many factors. Here we describe the walking procedure for the ternary EC containing pure RNAP from *E. coli*, the polymerase chain reaction (PCR) generated DNA template, and RNA transcript. Similar rules are applicable for walking with other bacterial and eukaryotic RNAPs.[10,11]

Each component of EC (RNAP, DNA, or RNA) can be tagged for walking. Because binding to beads may interfere with certain properties of EC (e.g., sterically inhibit binding of an elongation factor), it is important to consider alternative tags. 6-histidine tag (His-tag) that supports walking can be located at the COOH terminus of the β' subunit,[4] NH_2 terminus of β, COOH terminus of α, (E. Nudler, unpublished observations), or NH_2 terminus of the σ^{70} subunit.[12] Biotin-tag can be put at the 5' terminus of the DNA strand (see the following). For the His-tag, two types of metal-chelating beads are available: Ni^{2+}-NTA-agarose (Qiagen Chatsworth, Palo Alto, CA) and TALON Co^{2+}-sepharose beads (BD Biosciences Clontech, Palo Alto, CA). The latter resin has less nonspecific binding and requires less time and concentration of imidazole to elute the EC off the beads, if necessary. For the biotin-tag, the best resin for walking is NuatrAvidine UltraLink from Pierce (Rockford, IL). Unlike other biotin-binding solid support (e.g., streptavidin-coated magnetic beads), NuatrAvidine

[10] M. Palangat, T. I. Meier, R. G. Keene, and R. Landick, *Mol. Cell* **1,** 1033 (1998).
[11] A. S. Mironov, I. Gusarov, R. Rafikov, L. Errais Lopez, K. Shatalin, R. A. Kreneva, D. A. Perumov, and E. Nudler, *Cell* **111,** 747 (2002).
[12] G. Bar-Nahum and E. Nudler, *Cell* **106,** 443 (2001).

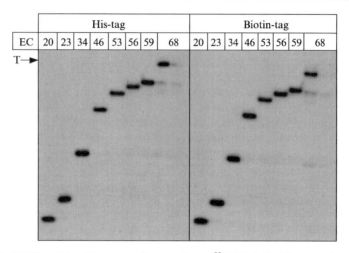

FIG. 1. RNAP walking. The autoradiogram shows [^{32}P]-labeled RNA transcripts isolated from the indicated ECs. The left and right panels show walking using His$_6$-tagged RNAP and biotinylated DNA template, respectively (see text for details). EC68 corresponds to the termination point of the λ tR2 terminator. The first and second lanes of EC68 show RNA recovery before and after washing the beads with TB, respectively.

beads have very low nonspecific binding and support walking, as well as the His-tag-binding resin (Fig. 1). A representative walking protocol, which is described here for the β' His-tag, can be utilized with minimal changes for other tags.

Preparation of RNAP

Although it is tempting to take advantage of the His-tag for RNAP purification, we recommend isolating the enzyme using a nonaffinity method. The affinity chromatography using Me^{++}-chelating beads, even after preliminary partial purification with Polymin P and heparin, produces RNAP that is still contaminated with nucleases and other proteins that interfere with walking. For better results, His-tagged-based purification requires further chromatography steps. Thus it does not save time and only limits the yield of the enzyme. We recommend omitting the affinity column step for a large-scale preparation of RNAP and following the procedure below. One-step His-tagged-based purification is useful for multiple small-scale purifications (e.g., for rapid assessment of mutant RNAP activities).[13]

---

[13] G. Bar-Nahum, V. Epshtein, and E. Nudler, manuscript in preparation.

All chemicals used are from Sigma (St. Louis, MO) unless specifically indicated.

1. GROWING CELLS. *E. coli* RL 721 strain (F-l-*thr*-1 *leu*-6 *pro*-A2 *his*-4 *thi*-1 *argE*3 *lacY*1 *galK*2 *ara*-14 *syl*-5 *mtl*-1 *tsx*-33 *rpsL*31 *supE*37 *recB*21 *recC*22 *sbcB*15 *sbcC*201 *rpoC*3531 [His6] *zja::kan*) carries a chromosomal copy of the *rpoC* gene encoding β' subunit with the His-tag at its COOH terminus. RL 721 is a Km^R derivative of JC7623 strain[14] that was made and kindly provided by R. Landick and co-workers (University of Wisconsin, Madison, Wisconsin). The use of this strain guarantees that all purified RNAP would be His-tagged. An overnight culture (50 ml) of RL 721 is diluted in 4 L LB broth containing 50 μg/ml kanamycine and grown at 37° with aeration to OD$_{600}$ $\sim$1.0 (usually 3–4 h). Cells are harvested by centrifugation in the Sorvall JA-10 (Du Pont, IN) rotor at 6000 rpm for 7 min. The cell pellet can be stored at $-70°$ for later purification.

2. CELL LYSIS. Cells are resuspended in 100 ml of grinding buffer (300 mM NaCl, 50 mM Tris–HCl, pH 7.9, 5% glycerol, 10 mM EDTA, 0.1% PMSF, 10 mM mercaptoethanol, 133 μg/ml lysozyme) and incubated for 20 min on ice, followed by addition of sodium deoxycholate (10%) to 0.2%. 20-ml aliquots of cells are lysed during 5–7 rounds of sonication with the Sonic Dismembrator 60 (Fisher Scientific, Chicago, IL) in 30-s pulses on the maximum output, followed by 30 s rest. The completion of the lysis is monitored by disappearance of viscosity. The lysate is then clarified in the Sorvall JA-20 rotor at 12,000 rpm for 30 min.

3. POLYMIN P EXTRACTION. Polymin P (10% solution) is slowly added to the supernatant with stirring to the final concentration 0.35%. Stirring continues for another 10 min, followed by centrifugation in the Sorvall JA-20 rotor at 6000 rpm for 5 min. The pellet is thoroughly resuspended by a glass homogenizer in 100 ml TGED buffer (10 mM Tris–HCl, pH 7.9, 0.5 mM EDTA, 5% glycerol, 0.1 mM DTT) plus 0.3 M NaCl. The suspension is centrifuged as before, and the supernatant is discarded. The pellet is then resuspended in 100 ml TGED buffer plus 0.5 M NaCl, homogenized, and centrifuged as before. Supernatant is discarded. To extract RNAP from polymine P, the pellet is resuspended in 100 ml of TGED plus 1 M NaCl for 1 h. The suspension is clarified at 6000 rpm for 5 min. Ground NH$_2$SO$_4$ is slowly added to the supernatant to 65% w/v (39.8 g/100 ml) and stirred for 30 m. NH$_2$SO$_4$ precipitate can be left overnight without loss of RNAP activity. The suspension is centrifuged in the Sorvall JA-10 rotor at 14,000 rpm for 45 min. The pellet is dissolved

---

[14] S. R. Kushner, H. Nagaishi, A. Templin, and A. J. Clark. *Proc. Natl. Acad. Sci. USA* **68,** 824 (1971).

in 50 ml TGED and centrifuged again to remove the insoluble material. If the protein concentration is too high, sedimentation may be compromised. In this case, the solution should be diluted with water to become colorless. The final protein concentration should not exceed 5 mg/ml.

4. HEPARIN COLUMN. The protein solution is loaded onto 10 ml Heparin–sepharose column (Amersham-Pharmacia Biotech, Piscataway, NJ) preequilibrated with TGED plus 50 mM NaCl. Loading proceeds at 0.5 ml/min using 50 ml superloop (Amersham-Pharmacia Biotech) in 15–20 ml portions. Next, the column is washed with 20 ml of the same buffer. The flow rate is changed to 1 ml/min, and the column is washed by 20 ml TGED plus 0.3 M NaCl. RNAP is eluted from the column with 20 ml TGED plus 0.6 M NaCl. The eluate from several protein-containing 1-ml fractions is concentrated with a Centricon-50 filter device (Millipore Corp., Bedford, MA) at 5,000 rpm so that the protein concentration will not exceed 10 mg/ml (1–0.5 ml).

5. GEL FILTRATION COLUMN. RNAP containing fractions from the heparin column is loaded onto a Superose-6 10/30 HR column (Amersham-Pharmacia Biotech) in 0.5-ml portions (usually twice) at 0.3 ml/min in TGED plus 0.2 M NaCl. Elution is carried out at the same rate for 90 min in 1-ml fractions. RNAP-containing fractions make a separate peak after ~35 min. Those fractions are combined, concentrated to 2 ml with a Centricon-50 filter device, and diluted to 10 ml by TGED plus 50 mM NaCl.

6. ION-EXCHANGE COLUMN. RNAP containing fractions is loaded onto a Mono Q 5/5 HR column (Amersham-Pharmacia Biotech) in TGED plus 50 mM NaCl at 0.2 ml/min using the 50-ml superloop. After washing the column with 5 ml of the same buffer, the 0.05–1 M NaCl gradient is applied as follows: 0–20% −2 ml, 21–80% −60 ml, 81–100% −2 ml. 1-ml fractions are collected. The RNAP core enzyme and holoenzyme are eluted as separate peaks between 30% and 45% of salt. The fractions are concentrated with a Centricon-50 filter device, mixed with equal volume of glycerol, and stored at −20°.

This method yields ~1 mg of RNAP-σ^{70} holoenzyme and ~1 mg of RNAP core per 10 g of cells. The enzymatic activity of the holoenzyme is ~2 fold higher than that of a commercially available prep of *E. coli* RNAP (Amersham-Pharmacia Biotech).

Preparation of NTP Substrates

None of the commercially available NTPs are pure enough to support walking on every DNA sequence. K_m for NTPs at different DNA positions varies by several orders of magnitude. Therefore a read-through of certain positions due to their unusually low K_m and small contamination in the

NTP stocks is common during walking. To avoid this problem, it is strongly recommended to purify original NTP stocks. The purest commercial NTPs are those from Amersham-Pharmacia Biotech.

1. COLUMN PREPARATION. 10 ml of DOWX 200–400 mesh resin (Aldrich, St. Louis, MO) is washed consequently with 0.5 M NaOH (5 min), water, 0.5 M HCl (5 min), and water. A 10-ml plastic pipette (Nalgene, Rochester, NY) with its tip clogged with a filter is packed with swollen resin and connected to a peristaltic pump, gradient mixer, and UV detector (Amersham-Pharmacia Biotech). The column is washed with 30 ml of 1 M LiCl plus 3 mM HCl (pH 2.5) solution. The flow rate is adjusted to 1 ml/ min and kept the same throughout the procedure. Next, the column is washed with 30 ml of 3 mM HCl. A separate column should be prepared for each NTP to be purified.

2. SEPARATION. 100–150 μl of 0.1 M NTP solution is loaded onto the column. Elution of NTPs is carried out in 3 mM HCl (400 ml) with specific LiCl gradient for each NTP: 0–1.0 M GTP; 0–0.8 M ATP; 0–0.5 M UTP; 0–0.5 M CTP. NTPs are eluted from the column at the following LiCl concentration: 0.7 M for GTP, 0.5 M for ATP, 0.3 M for UTP, and 0.2 M for CTP. In each case, only the middle part of the main peak is collected.

3. CONCENTRATION. NTP solutions are first concentrated by shaking with an equal volume of butanol, followed by centrifugation at 2000 rpm in the Sorvall JA-20 rotor for 5 min. The supernatant is discarded, and the procedure is repeated until the volume of NTP solution decreases to ~3 ml. Equal volume of acetone containing 2% NaI is added to the NTP solution, followed by centrifugation as before. Supernatant is discarded. The pellet is dried completely. NTPs are diluted in water to the final concentration up to 10 mM.

Preparation of the DNA Template

Promoter containing supercoiled plasmids or linear DNA templates can be used for walking. We prefer PCR-generated DNA fragments of 100–800 bp. They can be easily produced and purified in relatively large concentrations (e.g., 1 pmol/μl). At least twofold molar excess of DNA over RNAP is recommended to achieve the maximum yield of the initial EC. A typical amplification reaction (50 μl) includes 41.5 μl H$_2$O, 5 μl 10 × *Pwo* reaction buffer (Roche Molecular Biochemicals, Indianapolis, IN), 1 μl dNTP (1 mM stock, Amersham-Pharmacia Biotech), 1 μl of each oligo from 50 pmol/μl stock, and 0.5 μl *Pwo* DNA polymerase (Roche Molecular Biochemicals). The PCR reaction is programmed as follows: 95° 45 s, 55° 1 min, 72° 45 s (29 cycles). Bulk mixture is usually prepared for

20 PCR reactions and then distributed in 50-μl aliquots in 250 μl thin-walled plastic tubes (Perkin Elmer, Shelton, CT). The amplified product is resolved in a 1.2% (w/v) agarose gel (SeaKem FMC, Rockland, ME), followed by electroelution, extraction with phenol-chloroform, and precipitation with ethanol. The DNA pellet is diluted in 50 μl TE (10 mM Tris–HCl, pH 7.9, and 1 mM EDTA). The final concentration of DNA is expected to be $\sim$2 pmol/μl as determined by spectrometry. Phenol and high temperature ($>$70°) treatments affect the quality of DNA templates. To obtain DNA for biotin-tagged-based walking, the polyacrylamide gel electrophoresis (PAGE) purified upstream DNA oligo carrying the 5' biotin is used for PCR.

Walking

The following procedures exemplify the walking technique using the phage T7 A1 promoter template with the initial transcribed sequence as shown below: ATCGAGAGGG$_{10}$ACACGGCGAA$_{20}$TAGCCATCCC$_{30}$ AATCGAACAG$_{40}$GCCAGCTGGT$_{50}$AATCGCTGGC$_{60}$CTTTTTATT T$_{70}$GGATCCCCGG$_{80}$GTA

All reactions are performed in siliconized Eppendorf plastic tubes (USA Scientific Plastics, Ocalar, FL). The standard round of washing (i.e., pelleting and resuspending the beads) includes 3–5 s centrifugation in a table-top microcentrifuge (Fisher Scientific, Chicago, IL), removal of supernatant leaving $\sim$50 μl above the pellet, and resuspension in 1.5 ml of the appropriate transcription buffer.

His-Tagged-Based Walking

1. PREPARATION OF THE INITIAL EC. In a standard reaction (for 10–15 samples), 1 pmol ($\sim$0.5 μl) of His-tagged RNAP, 2 pmol ($\sim$1 μl) of T7A1 promoter DNA fragment ($\sim$200 bp), and 8 μl of transcription buffer (TB) (10 mM Tris–HCl, pH 7.9, 10 mM MgCl$_2$, and 100 mM KCl) are incubated for 5 min at 37°, followed by an addition of 1.5 μl of the 10 $\times$ starting mix (100 μm ApUpC RNA primer (Oligos, Wilsonville, OR), 250 μm ATP and GTP) and 5 μl of TB-preequilibrated Ni^{++}–NTA (Qiagen, Valencia, CA) or TALON resin (Clontech, Shelton, CA). After 8 min incubation at 37° with slight agitation, the beads are washed four times with 1.5 ml TB and resuspended in 10 μl TB containing 0.3 μm [α-^{32}P]CTP (3000 Ci/mmol; NEN Life Science) with or without ATP and GTP (25 μm). The reaction proceeds for 8 min at room temperature, followed by two rounds of washing with "high salt" TB (1 mM KCl) and two rounds of standard TB washing. Without ATP and GTP, the initial [α-^{32}P]EC is stalled at position

+11 (EC12); with ATP and GTP, the initial EC is stalled at position +20 (EC20). It is expected that the washed EC11 and EC20 count 1000–2000 and 3000–5000 cpm, respectively, if the bottom of the tube is placed next to the membrane of the Geiger counter (RPI Corp, Mt. Prospect, IL).

2. WALKING THE EC. The appropriate composition of NTP (1 μl) is added to the washed initial EC ($\sim$20 μl) for 3–5 min, followed by washing with TB as described previously. This procedure can be repeated as many times as needed to move the EC to a desired position. For example, to obtain the EC stalled at position +59 using EC20 as a starting point, the following walking steps are performed: AUG(EC23)→AUC(EC34) →AGC(EC46)→AUG(EC53)→GC(56)→AG(EC59) (Fig. 1). All walking steps are usually done at room temperature. The concentration of NTPs for each step has to be selected empirically to minimize the read-through, as well as incomplete chase. If purified NTPs are used, 20 μm of each NTP is generally recommended. If nonpurified NTPs are used, 3–5 μm NTP and longer elongation time ($\sim$5–8 min) are more appropriate. At the last walking step, TB can be changed for any other buffer that suits the purpose of the experiment (e.g., Cl^- can be substituted with acetate or glutamate to study the effects of elongation/termination factors).

Biotin-Tagged-Based Walking

In this case, the DNA template carries a biotin-tag. Biotin is usually attached at one or the 5$'$ terminus using the 5$'$-biotinylated DNA oligo for PCR. Preparation of the initial EC and walking procedures using biotinylated DNA are essentially the same as described previously for the His-tagged-based walking, except that 5 μl of TB-equilibrated NuatrAvidin beads (Pierce) are used instead of Me^{++}-chelating beads (Fig. 1B).

The EC can be also immobilized and walked via RNA using the 3$'$-end biotinylated DNA oligo complementary to 5$'$-terminal 12–15 bases of the nascent transcript. This procedure is useful for separating the active EC from all contaminants. For instance, it can be utilized for a rapid isolation of the EC from crude cell extracts.[12,13] Tethering the EC via RNA is performed as follows: The initial EC32 is prepared in solution as described previously for EC20 except that no beads were added. The modified T7A1 template is used in this case, which has T21A substitution to allow the EC32 formation in one step.[12] 10 pmol of the polyacrylamide gels (PAGE) purified biotinilated DNA oligo (5$'$-GTCCCTCTCGATGCTCT CTCAAGCTTTCTCTC-biotin-3$'$) complementary to the initial transcribed region is added for 2$'$ at 37$°$ for annealing to the nascent RNA. Next, 2 μl of TB equilibrated NeutrAvidin beads are added to the reaction mix for 5 min, followed by multiple washing with TB700 and TB. EC32 is now ready for walking. The recovery of the EC32 with this method is 60–70% that of the His-tagged-based recovery. To release the EC from the beads, ribonuclease A (Sigma) is added to 120 ng/μl for 10 min at 37° in TB. More than 50% of the EC is released under these conditions.

Roadblocking

The use of regular walking for obtaining the EC halted far from the promoter (e.g., after position +100) is laborious and time consuming. More importantly, the yield of the EC decreases with distance because of an inevitable loss of material (beads) during multiple washing steps. Furthermore, fractions of the complex get stuck along the way at the arrest prone positions, thus further decreasing the yield of the final product. To avoid these complications, a transcriptional roadblock can be used. The roadblock is a site-specific DNA binding protein that is able to stop the EC completely without termination at any distance from the promoter. Two proteins have been described as removable roadblocks, the *lac* repressor[15,16] and the mutant form of *Eco*RI restriction endonuclease, *Eco*RQ111.[17–19] Both proteins can be removed from DNA without interfering with the EC. *lac* repressor is removed by adding IPTG to 50 μm and *Eco*RQ111 by the high-salt buffers (e.g., 500 mM KCl). The use of *Eco*RQ111 has an advantage over the *lac* repressor in that it requires smaller changes in DNA for generating its binding site. Also, an ability of *Eco*RQ111 to halt the EC depends less on surrounding DNA sequences than that of the *lac* repressor (E. Nudler, unpublished observations). *Eco*RI site can be engineered by PCR just a few bp upstream of the regulatory signal of interest.

1. *ECO*RQ111 PROTEIN. Strains M5248 (λ *bio275 c1857* $\Delta H1$) and JC4588 (F^- *endoI$^-$ recA56 gal his322 thi*) and plasmid p*Eco*RQ111,[20] which is a derivative of pSCC2 (Ap^R), are kindly provided by Dr. Paul Modrich (Duke University). JC4588 is used to prepare p*Eco*RQ111 stock. p*Eco*RQ111 is introduced into M5248 expressing strain just before preparation of the enzyme. The protocol for *Eco*RQ111 isolation is described in detail by Cheng *et al.*[20] The protein ($\sim$1 mg/ml) is stored at −20° in 20 mM KPO$_4$ (pH 7.4), 0.4 M KCl, 1 mM EDTA, 0.2 mM DTT,

[15] U. Deuschle, R. A. Hipskind, and H. Bujard, *Science* **248,** 480 (1990).
[16] D. Reines and J. Jr. Mote, *Proc. Natl. Acad. Sci. USA* **90,** 1917 (1993).
[17] D. J. Wright, K. King, and P. Modrich, *J. Biol. Chem.* **264,** 11816 (1989).
[18] P. A. Pavco and D. A. Steege, *J. Biol. Chem.* **265,** 9960 (1990).
[19] E. Nudler, M. Kashlev, V. Nikiforov, and A. Goldfarb, *Cell* **81,** 357 (1995).
[20] S. C., Cheng, R. Kim, K. King, S. H. Kim, and P. Modrich, *J. Biol. Chem.* **259,** 11571 (1984).

and 50% glycerol. The working concentration of EcoRQ111 is about 8 pmol per 1 pmol of the EcoRI site.

2. PREPARATION OF THE ROADBLOCKED EC. The initial EC is prepared on the template containing the EcoRI site, using either the His-tagged- or biotin-tagged-based protocol as described previously. The EC is washed with TB (100 mM KCl, 10 mM MgCl$_2$, 5 mM Tris–HCl, pH 7.5) at 25°. The supernatant is discarded leaving ~10 μl of TB covering the beads. An aliquot of EcoRQ111 is diluted in TB to prepare 10 × stock (~0.02 mg/ml), which can be stored at +4° for a few days. 1 μl from 10 × stock of EcoRQ111 is added to the initial EC for 2 min, followed by addition of all four NTPs (0.5 mM). The chase reaction proceeds for 1–3 min depending on the length and properties of the template. More than 90% of the EC is expected to be blocked by EcoRQ111 under these conditions. The roadblock complex is washed with TB. Longer incubation with EcoRQ111 may lead to a noticeable cleavage of DNA at the EcoRI site because of a residual cutting activity of the enzyme. To remove the roadblock, beads are washed twice with TB containing 0.7 M KCl, followed by washing with TB or other buffers. The halted EC is now ready for further walking. Note that the position of the roadblocked EC is about 14 ± 1 bp upstream of the EcoRI site (GAATTC).

Applications

Many biochemical procedures (e.g., DNA and RNA footprinting, cross-linking) can be performed without releasing the EC from the beads. Exonuclease III and chemical DNA footprintings of the EC have been described in detail.[21,22] Walking can be used for generating specifically modified RNA transcripts. Many different NTP analogs can be utilized by E. coli RNAP and incorporated into the nascent RNA at specific positions during walking. The cross-linkable NTP analogs, such as 4-thio-UTP and 6-thio-GTP, are particularly useful to study the protein–nucleic acid interactions in the elongation complexes.[23,24] These reagents and conditions for their use are available from our laboratory upon request.

Acknowledgment

This work was supported by the National Institutes of Health (NIH).

[21] E. Nudler, A. Goldfarb, and M. Kashlev, Science 265, 793 (1994).
[22] E. Zaychikov, L. Denissova, and H. Heumann, Proc. Natl. Acad. Sci. USA 92, 1739 (1995).
[23] E. Nudler, I. Gusarov, E. Avetissova, M. Kozlov, and A. Goldfarb, Science 281, 424 (1998).
[24] N. Korzheva, A. Mustaev, M. Kozlov, A. Malhotra, V. Nikiforov, A. Goldfarb, and S. A. Darst, Science 289, 619 (2000).

[12] Stepwise Walking and Cross-Linking of RNA with Elongating T7 RNA Polymerase

By Younghee Sohn, Haihong Shen, and Changwon Kang

The single subunit 99-kDa RNA polymerase encoded by the bacterio-phage T7 genome accomplishes the entire process of transcription without any auxiliary factors. This polymerase can perform the basic features of transcription in a manner similar to the multisubunit RNA polymerases of prokaryotic and eukaryotic organisms. Transcription can be divided into three sequential stages: initiation, elongation, and termination. Transcription complexes exit initiation and enter elongation once nascent transcripts become long enough to stabilize the complexes. For multisubunit RNA polymerases, the composition of elongation complexes differs from that of initiation complexes. *Escherichia coli* sigma factors (or subunits) and eukaryotic initiation assembly factors are dissociated from the core enzymes during transition from initiation to elongation. Although there is no change in the composition of T7 RNA polymerase, significant structural changes occur during the transition.[1,2] Also, elongation complexes undergo structural changes during transition from elongation to termination.[3]

Elongation complexes are generally resistant to harsh conditions such as high salt concentration and high temperature, and they remain active through diverse purification processes and biochemical manipulations. This stability makes it possible to halt elongation complexes artificially at desired positions while maintaining their activity, a process referred to as stepwise walking.[4] This procedure provides a uniquely strong tool for studies on elongation and termination, such as monitoring changes in conformation and interactions of components.

For stepwise walking, RNA polymerases can be modified with 6-histidine tags at N or C termini, complexes purified using Ni^{2+}-agarose beads, and transcripts extended by successive cycles of limited elongation.[4,5] It is important to verify that the polymerase remains active after such modification, particularly for T7 RNA polymerase, because any C-terminal

[1] Y. W. Yin and T. A. Steitz, *Science* **298**, 1387 (2002).
[2] T. H. Tahirov, D. Temiakov, M. Anikin, V. Patlan, W. T. McAllister, D. G. Vassylyev, and S. Yokoyama, *Nature* **420**, 43 (2002).
[3] H. Song and C. Kang, *Genes Cells* **6**, 291 (2001).
[4] E. Nudler, E. Avetissova, V. Markovtsov, and A. Goldfarb, *Science* **273**, 211 (1996).
[5] D. Temiakov, P. E. Mentesana, K. Ma, A. Mustaev, S. Borukhov, and W. T. McAllister, *Proc. Natl. Acad. Sci. USA* **97**, 14109 (2000).

modification is detrimental to its activity. Unlike the procedure that involves modification of the key enzyme, the method described here uses modified DNA templates, which have biotins added at 5′ or 3′ ends, and complexes are then immobilized and purified using streptavidin-coated magnetic beads.[3,6] Recovery of elongation complexes in each walking cycle is higher in the biotin–streptavidin method of DNA immobilization than in the His–nickel method of polymerase immobilization. Furthermore, recovery is simpler in the method using magnetic beads than in the one using nonmagnetic beads, which requires centrifugation.

Elongation complexes obtained by stepwise walking can be subjected to various studies. For example, structural changes can be monitored in regard to the transcription bubble, and molecular interactions between DNA, RNA, polymerase and ribonucleotides, and elongation pathways leading to termination can be assessed.[3] Here we describe a method to elucidate changes in RNA–polymerase interactions.[6] A photocross-linker is incorporated into nascent RNA during stepwise walking of T7 RNA polymerase on DNA templates containing a sequence-specific terminator. After each step the complex is irradiated by ultraviolet (UV) light to cross-link the RNA with the polymerase. Chemical and enzymatic mapping of RNA cross-links on the polymerase reveals the path of RNA through elongation.

Biotinylated DNA Template

Phage T7 RNA polymerase transcribes virtually any sequence templates containing a T7 promoter. Using a primer that is biotinylated at the 5′ end, it is possible to synthesize biotinylated DNA templates using polymerase chain reaction (PCR) amplification of a promoter-containing region from an appropriate plasmid. The details of the PCR-based synthesis are as follows, with the reaction performed in 100 μl for 27 cycles.

10× PCR buffer	10 μl
(100 mM Tris–HCl, pH 8.3, 500 mM KCl, 15 mM MgCl$_2$)	
Deoxynucleotide mix (2.5 mM each)	8 μl
Template DNA	20 pmoles
Two primers (one biotinylated)	20 pmoles each
Taq DNA polymerase	5 units
Deionized water	up to 100 μl

[6] H. Shen and C. Kang, J. Biol. Chem. 276, 4080 (2001).

Immobilization of DNA Templates

Streptavidin-coated magnetic beads are used to isolate biotinylated DNA templates. Streptavidin is a 66-kDa protein consisting of four identical subunits, each containing a high affinity binding site for biotin ($K_D = 10^{-15}$ M). Compared with avidin, streptavidin has the same binding capacity for biotin but displays less nonspecific binding. Dynal MPC or other magnetic apparatus can be used to capture streptavidin-coated magnetic beads (complexes are completely immobilized within 30 s using Dynal MPC).

Before use, streptavidin-coated beads are washed with 2 volumes of buffer I (DEPC-treated 100 mM NaOH, 50 mM NaCl) and buffer II (DEPC-treated 100 mM NaCl) sequentially in order to remove ribonucleases. The beads are resuspended in 2× binding buffer (10 mM Tris–HCl, pH 7.5, 2 M NaCl, 1 mM EDTA) to a final concentration of 5 μg/μl and mixed with an equal volume of biotinylated DNA (the PCR product). Incubation is usually performed at room temperature for 20 min with gentle rotation or occasional tapping of the tube. However, the optimal incubation time and temperature depend on the length of bound nucleic acids, with DNA fragments larger than 1 kb requiring a longer incubation time (1 h) and a higher incubation temperature (43°). Streptavidin-bound biotinylated DNA is resuspended to a desired concentration with 1× binding buffer.

Purification of T7 RNA Polymerase

T7 RNA polymerase is commercially available, but it can be produced and purified at a high concentration ($\sim$ 100 units/μl) from *E. coli* transformed with a plasmid containing the gene. We constructed such a plasmid, called pSAT7, as follows: the 5.1 kb *Eco*RI/*Sal*I fragment of pAR1219 was inserted into the *Eco*RI/*Xmn*I site of pACYC184 after the *Sal*I- and *Xmn*I-cleaved ends were made blunt. The recombinant plasmid was reduced in size by *Sty*I digestion and self-ligation, generating pSAT7.

E. coli BL21 is transformed with pSAT7, and a single colony is inoculated in 3 ml LB broth containing 25 μg/ml tetracycline. After 12 h shaking incubation, 1 ml culture media is inoculated into 200 ml LB–tetracycline broth. Once the culture achieves $OD_{600} = 0.4$ at 37°, isopropyl-β-D-thiogalactopyranoside (IPTG) is added to a final concentration of 0.5 mM and incubation is continued for 4 h. Chilled and harvested cells are resuspended in 8 ml buffer A (20 mM Tris–HCl, pH 7.9, 1 mM dithiothreitol, 1 mM EDTA, 5% glycerol) containing 50 mM NaCl. Protease inhibitor phenylmethylsulfonyl fluoride (PMSF) can be added to a final concentration of 0.1 mM.

Resuspended cells are sonicated on ice intermittently over a 30-min period with a cycle of 1 min sonication followed by a 1-min interval. After the sample is clarified by centrifugation, the supernatant is applied to a phosphocellulose column (2 ml bed volume) preequilibrated with buffer A containing 50 mM NaCl. The column is washed with 200 ml buffer A containing 100 mM NaCl. Polymerase elution is accomplished by 4 ml buffer A containing 400 mM NaCl, and 200-μl aliquots are collected. The polymerase-containing fraction is identified using 8% SDS–PAGE. Polymerase is stored at 1 μg/μl in 50% glycerol.

Stepwise Walking of the T7 Transcription Complex

Stepwise walking is accomplished by a transcription reaction using a limited set of ribonucleotides, followed by purification of halted elongation complexes, and then another round of transcription with another limited set of ribonucleotides. The method described here uses a DNA template that contains a sequence-specific termination signal, in order to illustrate that stepwise walking is possible even near a terminator. The sequence of the sample template from the T7 promoter start site (+1G) is shown in Fig. 1. The underlined terminator sequence causes termination at the boxed G at +51 and consists of two modules. The conserved upstream module of ATCTGTT (+37 through +43) is responsible for pausing and conformation change, and the U-rich downstream module facilitates release of the transcript.[3,7]

Stepwise walking is performed as follows. First, the biotinylated template bound to streptavidin-coated beads is washed twice with 1× transcription buffer (40 mM Tris–HCl, pH 7.9, 6 mM MgCl$_2$, 100 mM KCl) and is then transcribed in the following manner:

1. Gently mix together the following components and incubate at room temperature for 5 min.

10× transcription buffer (pH 7.9)	4 μl
100 mM dithiothreitol	8 μl
T7 RNA polymerase	1000 units
DEPC-treated H$_2$O	to final volume of 40 μl

2. Add the following mixture and incubate at room temperature for 20 min with intermittent gentle mixing.

[7] Y.-S. Kwon and C. Kang, *J. Biol. Chem.* **274**, 29149 (1999).

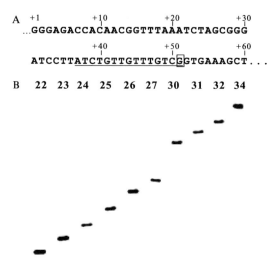

FIG. 1. Stepwise walking of transcription elongation complexes. (A) Sequence of the sample template downstream of the T7 promoter start site, +1G. The underlined terminator sequence causes termination at the boxed G at +51. (B) The nascent RNA transcript produced by stepwise walking. Transcripts were radiolabeled by incorporation of [α-^{32}P]UTP at +16 to +18 positions. The gel shows separation of transcripts, and the number of nucleotides are shown above the gel.

10× transcription buffer (pH 7.9)	4 μl
Ribonucleotide mixture stock	8 μl
(500 μM ATP, 500 μM GTP and 50 μM CTP)	
RNasin ribonuclease inhibitor	10 units
DEPC-treated H$_2$O	to final volume of 40 μl

The previous 80-μl reaction produces a nascent transcript of 15 nucleotides from the Fig. 1 template. This elongation complex has escaped from the unstable initiation stage and is engaged in stable conformation.

3. The reaction is stopped by trapping beads using a magnetic apparatus and washing transcription complexes five times with 200–400 μl 1× transcription buffer to remove any remaining NTPs and dissociated RNA and polymerase.

4. Resuspend the 15-mer RNA-containing complex beads in 80 μl 1× transcription buffer containing 0.33 μM radioactive [α-^{32}P]UTP (3000 Ci/mmol). Incubate at room temperature for 1 min to obtain a

radiolabeled 18-mer RNA complex, and then wash four times with 200–400 μl 1× transcription buffer.

5. Resuspend beads in 80 μl 1× transcription buffer containing 0.5 μM ribonucleotide required in the next position. In the case of the Fig. 1 template, a 1-min incubation at room temperature for incorporation of 3 ATP produces a 21-mer RNA complex. Wash complexes four times with 200–400 μl 1× transcription buffer.

6. Repeat step 5, but 40-s incubation is enough for single nucleotide incorporation of UTP in the case of the Fig. 1 template.

7. Repeat step 5 until the ternary complex reaches the desired position, with incubation time varying, depending on the number of nucleotides per walk.

Aliquots can be taken from the suspension solution in the last washing step. If the same amount of complex is to be taken as aliquots, the volume of washing solution needs to be proportionally reduced. When a 15-μl aliquot is taken out in each walking step, the last washing solution is reduced by 15–20 μl, taking into account some loss of sample during washing. The 15-μl aliquots are used for determination of size and purity of transcripts by gel electrophoresis, and for other experiments such as photocross-linking (see the following).

For gel electrophoresis, a 5-μl portion of the 15-μl aliquot is mixed with 10 μl gel-loading buffer (10 mM EDTA, 12 M urea, 0.1% bromophenol blue) prewarmed to 55°, and the mixture is heated for 1 min at 95° and then loaded on an 8 M urea–10% polyacrylamide gel. The results of stepwise walking on the sample template from 22-mer to 34-mer are shown in Fig. 1.

A 1-nucleotide walk needs 0.5 μM of a ribonucleotide and 40 s of time. However, jumping, or leaping-out, is required for long-range transcription. Three-ribonucleotide mixtures lacking one ribonucleotide are used for such jumping. The final concentration of each ribonucleotide is 0.5 μM, but the reaction time needs to be lengthened, depending on the number of nucleotides in a jumping step. A 2-min incubation is enough for a 5-nucleotide-long jump.

In the vicinity of a terminator, faster walking is required because ternary complexes become unstable. A 30-s incubation is suitable near termination. Complexes become very unstable two nucleotides before the termination site on the template. Thus the +49 complex on the template should not be washed, and serial addition of the next ribonucleotides without washing produces 50-mer and terminated 51-mer transcripts.[3]

If transcripts form a stable secondary structure, elongation complexes get paused, arrested, or dissociated and stepwise walking fails. For example, when the three base pairs AAA at +19 to +21 in the Fig. 1 template are replaced by CCC, the 27-mer RNA is mostly released from the complex. An RNA secondary structure prediction program would be helpful in this regard.

Cross-Linking between Transcripts and Elongating RNA Polymerase

Incorporation of a ribonucleotide cross-linker analogue into transcripts allows for cross-linking between the ribonucleotide and an RNA polymerase amino acid(s). The widely used 4-thio-uridine can be photo-cross-linked to virtually any amino acid. It can be incorporated into transcripts at a specific site by stepwise walking. The optimal condition is to use 1 μM 4-thio-uridine in the previous transcription buffer, but at pH 7.2.

To create cross-links, a 10-μl portion of the 15-μl complex-bead aliquot is placed on ice and irradiated with 360 nm UV light for 20 min using a UVGL-25 lamp.[8] Photocross-linked complexes are separated from non–cross-linked complexes by 8% SDS–PAGE. Before electrophoresis, strep-tavidin beads must be dissociated from biotinylated DNA by boiling in 0.1% SDS for at least 5 min. Alternatively, heating in 10 mM EDTA (pH 8.2) and 95% formamide at 65° for 5 min, or at 90° for 2 min, also breaks the conjugation. Separated streptavidin beads are removed by Dynal MPC before gel loading. After electrophoresis, the gel is dried and exposed to a phosphorimager screen. Measurement of radioactivity using a Storm 860 scanner (Amersham Bioscience, Piscataway, NJ) reveals the extent of photocross-linking in the elongation complexes.

Mapping the RNA Cross-Links in Elongating RNA Polymerase

The polypeptide region of T7 RNA polymerase that is photocross-linked to the RNA transcript of each elongation complex can be mapped by enzymatic and chemical proteolysis. Trypsin, hydroxylamine (HA), 2-nitro-5-thiocyanobenzoic acid (NTCB), N-chlorosuccinimide (NCS), and cyanogen bromide (CNBr) can be used to map the linked regions using digestion conditions described previously.[5,6] For mapping experiments, elongation complexes are obtained by stepwise walking in an 80-μl tran-scription reaction for each step, but without taking aliquots. The complex-beads are finally resuspended in 100 μl 1× transcription buffer, which is enough for 10 mapping experiments.

For coarse mapping, the polymerase is partially digested with trypsin and/or hydroxylamine, separately. Although trypsin can digest the T7 RNA polymerase into many small fragments, the polymerase free and

---

[8] V. Markovtsov, A. Mustaev, and A. Goldfarb, *Proc. Natl. Acad. Sci. USA* **93,** 3221 (1996).

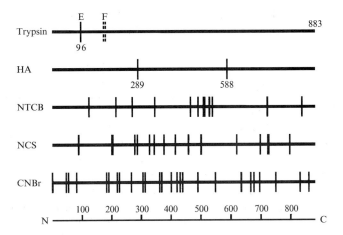

FIG. 2. Cleavage map of the T7 RNA polymerase. Each horizontal line represents the total 883 amino acids of the polymerase. Potential cleavage sites of trypsin, hydroxylamine (HA), 2-nitro-5-thiocyanobenzoic acid (NTCB), N-chlorosuccinimide (NCS), and cyanogen bromide (CNBr) are indicated by vertical bars. The initial trypsin cleavage sites of elongating and free polymerases are differentiated by E and F, respectively. Numbers at the bars indicate the C-terminal amino acids of cleavage fragments (bottom line represents 100 amino acid sections of the polymerase).

in an initiation complex is initially cleaved under mild conditions at the N-terminal one-fifth position (after Arg[173] and/or Lys[180]) into 20 kDa and 79 kDa fragments. However, the polymerase in an elongation complex is initially cleaved at the N-terminal one-tenth location (after Arg[96]) into 11 kDa and 88 kDa fragments (Fig. 2).[9] Thus the initial cleavage site of trypsin digestion can be changed upon conformation change of the T7 RNA polymerase. For partial digestion with trypsin, 10 μl complex-beads are added to 2 μl trypsin of 100 μg/μl dissolved in 10 mM Tris–HCl (pH 8.0) and 1 mM EDTA, and incubated at room temperature for 5 min. Digestion can be stopped by a trypsin inhibitor or by the previously mentioned heat treatment used for breaking conjugation. Digested fragments separated from the beads are analyzed by 10% SDS–PAGE and a phosphorimager.

Hydroxylamine cleaves peptide bonds between Asn and Gly. It cleaves T7 RNA polymerase only at two sites, but total digestion produces three polypeptides of approximately the same size (Fig. 2). When RNA is cross-linked to the N-terminal trypsin fragment, hydroxylamine digestion

[9] D. Temiakov, M. Anikin, and W. T. McAllister, *J. Biol. Chem.* **277**, 47035 (2002).

is not necessary. If RNA is cross-linked to the C-terminal trypsin fragment, partial cleavage of the fragment with hydroxylamine produces different sets of radiolabeled fragments, depending on which of three regions is linked to the RNA.[6] Complex-beads in 10 μl are recovered by a magnet and incubated in 30 μl 1 M hydroxylamine dissolved in 1 M NaCl at 43° for 15 min. Digested fragments separated from the beads are analyzed by 10% SDS–PAGE and a phosphorimager.

Fine mapping can be performed using an additional proteolytic agent(s). Denatured T7 RNA polymerase can be cleaved by NTCB at 12 Cys residues, by NCS at 17 Trp residues, and by CNBr at 26 Met residues (Fig. 2). Because the polymerase is cleaved into too many small fragments after extensive digestion, single-cut conditions should usually be used for mapping purposes. The procedure for using NTCB for single-cut digestion is as follows. Complex-beads in 10 μl are recovered by a magnet and resuspended in 10 mM Tris–HCl (pH 7.9), 10 mM dithio-threitol, and 1% SDS at room temperature for 30 min. After powdered NTCB is added to the suspension to a final concentration of 5 mM, 0.4 M NaOH is added to make pH 8.0, and incubation continues for 10 min at room temperature. It is good practice to determine the appropriate amount of NaOH with a non–cross-linked control sample before performing experiments with cross-linked samples. When more extensive digestion is needed, pH and reaction time are increased. Digested fragments separated from the beads are analyzed by 5–20% gradient SDS–PAGE and a phosphorimager.

When single-cut patterns are ambiguous, cross-linked fragments can be isolated and digested with another chemical. In-gel digestion is useful under these circumstances because it does not require elution of fragments from gels, thus reducing sample loss. The in-gel NCS digestion procedure is as follows:

1. Excise a gel slice containing the band of interest and wash the slice twice with 2 ml distilled water for 20 min.
2. Rinse the gel slice with equilibrium buffer (1 g of urea dissolved in 1 ml water and 1 ml acetic acid) for 20 min.
3. Incubate in equilibrium buffer containing 40 mM NCS for 1 h.
4. Wash the gel slice five times with 10 mM Tris–HCl (pH 8.0).
5. Soak in 1× SDS gel-loading buffer for 2 h with three changes of buffer.
6. Place the gel slice at the bottom of a well in a stacking gel and fill the well with 1× SDS gel-loading buffer.
7. Run the 12% SDS–PAGE and analyze the digestion products using a phosphorimager.

The in-gel CNBr digestion procedure is as follows:

1. Excise a gel slice containing the band of interest and place the slice in a microcentrifuge tube.
2. Add 150 μl CNBr solution (150 mM in 70% formic acid) and incubate overnight in the dark at room temperature with occasional agitation.
3. Remove the CNBr solution and wash the gel slice with deionized water.
4. Dry the gel slice using a Speedvac to remove remaining formic acid.
5. Follow procedures 5–7 described previously for NCS digestion.

Elongation complexes obtained by stepwise walking can be subjected to various chemical and enzymatic assays for functional and structural studies. For example, using the previous procedures, two-site contact of nascent transcript with elongating T7 RNA polymerase was elucidated and mapped, revealing an RNA exit path,[6] and transcription bubbles of elongating complexes were mapped near a sequence-specific terminator, revealing a conformation change a few base pairs before termination.[3]

Acknowledgments

This work was partially supported by grants from the Korea Science and Engineering Foundation (R01-1999-00111), the National Research Laboratory Program (M1-0104-00-0179), and the Brain Korea 21 Project.

[13] Characterization of Protein–Nucleic Acid Interactions that are Required for Transcription Processivity

By Evgeny Nudler, Ekaterina Avetissova, Nataliya Korzheva, and Arkady Mustaev

Generally, the ternary elongation complex (EC) of the nascent transcript, DNA template, and a cellular RNA polymerase (RNAP) is a highly stable structure that is resistant to dissociation under a variety of disrupting conditions. The majority of ECs can withstand incubation with high salt (e.g., 2 M KCl), high temperature (e.g., 70 °C), mild detergents (e.g., 1% sarcosyl), or nucleic acid competitors (e.g., 2 mg/ml heparin) for hours. This remarkable stability, however, occurs without strong specificity to any particular DNA or RNA site, because RNAP slides along the nucleic acids with a speed of 20–80 nt/s. Combination of these seemingly contradictory properties—strong nucleic acid binding and smooth sliding in one protein—ensures a high processivity of RNAP that is an ability to rapidly transcribe long stretches of DNA without premature termination. Recently published high-resolution structures of the elongation competent forms of bacterial RNAP and yeast RNAP II provide a comprehensive information about the localization of RNA and DNA in the enzyme.[1,2] High-resolution structural models of the EC show extensive contacts between nucleic acids and two largest subunits of RNAP[1–3]; however, they do not conclusively indicate which of those contacts actually contribute to the EC stability and processivity. To address this issue the biochemical approaches have been developed to functionally analyze RNAP–DNA and RNAP–RNA interactions. Results obtained by these approaches led to the structure/functional model of the EC, in which three major nucleic acid binding sites that hold the complex together were defined: the front DNA duplex binding site (DBS), RNA:DNA hybrid binding site (HBS), and single-stranded RNA binding site (RBS) (Fig. 1A).[4] Perturbations in each of these sites render the complex salt sensitive and less processive. Here we provide a detailed protocol for developing a minimal nucleic acid scaffold that defines each of three nucleic acid binding sites and integrates them in the fully stable EC (Fig. 1B).

Determination of the Minimal DNA Template Needed for EC Stability

To determine the minimal requirements for the DNA to sustain the fully stable and functional EC, two *in vitro* approaches were combined in manipulating with the EC: "walking" and "template switching." Walking (i.e., controlled stepwise transcription in solid phase) is described in detail in this volume.[5] Template switching utilizes the ability of the EC to perform end-to-end transposition from the primary, promoter-containing linear DNA template to another DNA molecule (the secondary template) without interrupting RNA synthesis.[6] After the switch, transcription continues on the secondary template, which can be a single-stranded, double-stranded, or partially double-stranded DNA oligonucleotide (Fig. 2). The template-switching phenomenon provides a powerful method to study

---

[1] A. L. Gnatt, P. Fu, J. Cramer, D. A. Bushnell, and R. D. Kornberg, *Science* **292,** 1876 (2001).

[2] S. A. Darst, *Curr. Opin. Struct. Biol.* **11,** 155 (2001).

[3] N. Korzheva, A. Mustaev, M. Kozlov, A. Malhotra, V. Nikiforov, A. Goldfarb, and S. A. Darst, *Science* **289,** 619 (2000).

[4] E. Nudler, *J. Mol. Biol.* **288,** 1 (1999).

[5] E. Nudler, I. Gusarov, and G. Bar-Nahum, *Methods Enzymol.* **371,** (2003).

[6] E. Nudler, E. Avetissova, V. Markovtsov, and A. Goldfarb, *Science* **273,** 211 (1996).

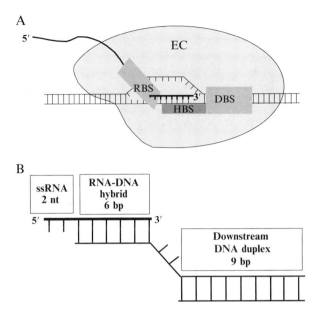

FIG. 1. The model of the elongation complex (EC). (A) Basic functional elements of the EC. DBS, downstream DNA duplex binding site; HBS, RNA:DNA hybrid binding site; RBS, single-stranded RNA binding site.[4] The catalytic site carries the 3′ end of the nascent transcript. (B) The minimal RNA–DNA scaffold of the EC.[10]

protein–DNA interaction in the EC because the portion of DNA normally protected by RNAP from the chemical and enzymatic attack can be changed at will in the secondary template. Template switching has been observed with other bacterial RNAP (E.Nudler, unpublished observations) and also mammalian RNAP II *in vitro* and *in vivo*.[7,8]

Template Switching

The transcribed sequence of the primary template is shown below:
ATCGAGAGGG$_{10}$ACACGGCGAA$_{20}$ TAGCCATCCC$_{30}$ AATC-GACAAC$_{40}$ GACACCCCGG$_{50}$GAAAA

It is designed to support the switch-over reaction of the EC stalled at position +34 upon addition of the ATP+CTP+GTP mix (Fig. 2A). *Escherichia coli* RNAP that is used in this protocol carries 6His-tag at the COOH terminus of β' subunit.[5] The detailed purification protocol for

[7] M. G. Izban, M. A. Parsons, and R. R. Sinden, *J. Biol. Chem.* **273**, 27009 (1998).

[8] E. S. Kandel and E. Nudler, *Mol. Cell* **10**, 1495 (2002).

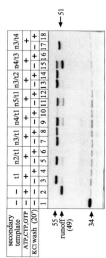

B

#	Secondary template	sequence	Elongation complex	Half-life (min) 500 mM KCl	Half-life (min) 5 mM KCl
1	n1/t1	5'-AGCGGATAACAATTTCACACAGGA 3'-TCGCCTATTGTTAAAGTGTGTCCT	+55	80	>120
2	t1	3'-TCGCCTATTGTTAAAGTGTGTCCT	+55	<1	>120
3	n2/t1	5'-TAACAATTT A 3'-TCGCCTATTGTTAAAGTGTGTCCT	+55	2	>120
4	n3/t1	5'-ACAATTTCA 3'-TCGCCTATTGTTAAAGTGTGTCCT	+55	60	>120
5	n4/t1	5'-AATTTCACA 3'-TCGCCTATTGTTAAAGTGTGTCCT	+55	8	>120
6	n5/t1	5'-TTTCACACA 3'-TCGCCTATTGTTAAAGTGTGTCCT	+55	<1	>120
7	n3/t2	5'-ACAATTTCA 3'-TCGCCTATTGTTAAAGT	+55	60	>120
8	n4/t3	5'-TAACAATTTCA 3'-TCGCCTATTGTTAAA	+55	1.5	>120
9	n3/t4	5'-ACAATTTCA 3'-TCATTGTTAAAGTGTGTCCT	+51	<1	30
10	n3/t5	5'-ACAATTCA 3'-TCGCATTGTTAAAGTGTGTCCT	+53	10	>120
11	t2	3'-TCGCCTATTGTTAAAGT	+55	<1	>120

A

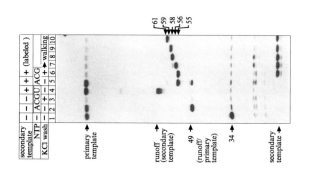

primary template

T7A1 GACACCCCGGAACA-3'
CTGTGGGGCCCTTGT-5'
+34   +49

secondary template

5'-AGCGGATAACAATTTCACACAGG A-3'
3'-TCGCCTATTGTTAAAGTGTGTCCT-5'
+55

the His-tagged enzyme, primary template DNA, and nucleotide triphosphates (NTPs) is described in Chapter 11 of this volume.[5] All reactions
are performed in siliconized Eppendorf plastic tubes (USA Scientific
Plastic, Ocala, FL).

*Step 1. Preparation of the "Ready to Jump" EC at the Primary
Template.* The initial EC stalled at position +20 (EC20) of the primary
template is prepared as follows: 1 pmol ($\sim$0.5 μl) of His-tagged RNAP,
2 pmol ($\sim$1 μl) of the primary template containing T7A1 promoter and
8 μl of transcription buffer (TB) (10 mM Tris–HCl, pH 7.9; 10 mM MgCl$_2$,
and 100 mM KCl) were incubated for 5 min at 37°, followed by addition of
1.5 μl of the 10$\times$ starting mix (100 μM ApUpC RNA primer [Oligos Etc,
Inc.], 250 μM ATP and GTP) and 3 μl of TB–preequilibrated Ni^{++}-NTA-
agarose (Qiagen, Chatsworth, CA) or TALON resin (Clontech, Palo Alto,
CA). After 5 min incubation at 37°, [α-^{32}P]CTP (3000 Ci/mmol; NEN Life
Sciences) is added to 0.3 μM at room temperature for 8 min, followed by
addition of CTP to 25 μM. The reaction proceeds for another 3 min at
room temperature, followed by two rounds of washing with "high-salt"
TB (1 mM KCl) and two rounds of standard TB washing. The standard
rounds of washing (i.e., pelleting and resuspending the beads) includes
centrifugation in a table-top microcentrifuge at its maximum speed for
3–5 s, removal of the supernatant leaving $\sim$50 μl of buffer covering the
beads, and resuspension in 1.5 ml of the appropriate TB. To obtain the
EC that is ready for the switch-over reaction, EC20 is walked to position
+34 in three successive steps:(1) ATP+CTP+GTP, (2) ATP+UTP+GTP,
+GTP, and (3) CTP+ATP+UTP. The final concentration of NTPs is
25 μM. Each step proceeds for 3 min at room temperature, followed by
four cycles of standard washing.

Fɪɢ. 2. The Minimal Template System. (A) A representative template switching
experiment. The autoradiogram demonstrates RNA products and end-labeled DNA
templates. The structure of the primary template (after position +34) and the secondary
template and shown in the bottom panel. EC34 is prepared on affinity beads as described in
text, washed, and chased in the presence of indicated sets of NTPs in the absence (lanes 2 and
3) or in the presence of the secondary template. Where indicated, the complexes were washed
with high-salt TB. The secondary template in the high-salt washed switch-over complex
(EC55, lane 6) was end-labeled with polynucleotide kinase. EC55 was then walked to
positions +56T, +58A, +59C, and +61A with subsets of NTPs (lanes 7–10). (B) Determination of the minimal template parameters required for EC stability. The summary table shows
the half-lives of the immobilized switch-over ECs in low and high salt (see text for details). In
all cases the decay followed first-order kinetics. A representative experiment for the 20-min
time point is shown in the bottom panel. Experimental details are as in (A).

Step 2. Preparation of the Secondary Templates. Gel-purified DNA oligonucleotides for the secondary template can be obtained from Oligos Etc., Inc. or another commercial source. For 10 reaction samples, 100 pmol each of the top- and bottom-strand oligos are mixed in a volume of 10 μl in "low-salt" TB5 (10 mM Tris–HCl, pH 7.9), 8 mM MgCl$_2$, and 5 mM KCl), heated for 30 s at 90° in a PCR machine, and then allowed to slowly ($\sim$10 min) cool down to room temperature while remaining in the heating block. The material would be then diluted to one-tenth its original concentration into the transcription reaction. The secondary templates are designed to support unidirectional switch-over reaction (i.e., the EC is allowed to continue transcription only from one side of the double-stranded molecule). For that purpose, the top strand has adenine residue at its 3′ terminus and the bottom strand carries the first adenine several nucleotides away from the 3′ terminus (Fig. 2). At the same time, the primary template has no adenine residues downstream from the position +34. Thus by withholding uridine triphosphate (UTP) from the chase reaction (ATP+CTP+GTP chase), the switch-over synthesis is restricted to the bottom strand of the secondary template.

Step 3. Preparation and Analysis of the Switch-over EC. The beads containing EC34 are washed with low-salt TB (5 mM KCl). The supernatant is removed, leaving 10 μl TB above the beads. The secondary template is added to the final concentration 5 pmol/μl, followed by addition of ATP+CTP+GTP to 100 μM. After 5 min of the chase reaction at room temperature, the beads are washed twice with high-salt TB and twice with TB. The resulting switch-over EC55 can be now walked along the secondary template according to its sequence (Fig. 2A). Upon washing with high-salt TB (500 mM KCl), the primary template dissociates from the beads. Also, the high-salt wash removes the EC population, which attempted to switch-over from another end of the secondary template and was halted at the run-off position (+49).

As mentioned previously, a series of secondary templates can be designed to determine various parameters of the DNA template that are involved in maintaining the processive EC. Various portions of the template and nontemplate strands upstream and downstream of the catalytic site can be deleted or modified. The secondary templates shown in Fig. 2B were prepared as described previously. In each case, the half-life of the switch-over EC can be determined at "high" (0.5 M KCl) and "low" (5 mM KCl) salt, thus discriminating ionic and nonionic contacts. To this end, the amount of retained [^{32}P] RNA is determined during time-course incubation of the switch-over EC in TB with various KCl concentrations, followed by washing the beads and resolving the recovered RNA products by the sequencing polyacrylamide gel electrophoresis (PAGE) (Fig. 2B, bottom panel). To measure the minimal requirement for the portion of the template strand normally base-paired with the nascent RNA in the transcription bubble, the adenine in the bottom strand of the secondary template is shifted toward the 3' terminus (Fig. 2B, lines 9, 10). Six nucleotides of the template strand just behind the catalytic site were shown to represent the shortest portion of DNA that renders the EC fully resistant to high salt. The EC becomes unstable even in low salt if less than four bases of the template strand were left behind the catalytic site (Fig. 2B, line 9). By shortening the bottom strand of the secondary template from the 5' end, one can determine the minimal portion of the template strand downstream of the catalytic center required for EC stability. This portion is 8–9 nt long (Fig. 2B, lines 7, 8). By assembling the partially double-stranded secondary templates, one can assess the role of the non-template strand in the complex stability. As Fig. 2B (line 1 versus line 7) shows, the portion of the nontemplate strand upstream of the catalytic site is dispensable for the complex stability. On the other hand, the downstream 8–9 nt of the nontemplate strand render the EC resistant to high salt. In summary, this analysis demonstrates that the minimal template that renders the *E. coli* EC fully stable is the partially double-stranded DNA positioned in the complex in such a way that the first six nucleotides were transcribed and the 9-bp DNA duplex is located just downstream of the catalytic center (Fig. 1B).

Analysis of EC Stability at the End of the DNA Template

Destabilization of the switch-over EC by shortening the duplex area of the secondary template to less than 8 bp could be due at least in part to spontaneous melting of short DNA before the switch-over reaction. To assess the role of the downstream DNA duplex in the complex stability more precisely, a complementary approach was developed, in which a series of ECs were obtained by walking close to the end of the primary template so that the front salt-resistant contacts could not exist (Fig. 3). The stability of these complexes can be analyzed in high-and low-salt TB as described previously. To obtain the EC near the physical end of the primary template, EC34 that was prepared as before is walked to positions +42, +45, +47, or +48. The resulted ECs will be located 7, 4, 2, or 1 bp upstream of the blunt end of the template, respectively. Each walking reaction is performed in low-salt TB at room temperature for 5 min. The final concentration of each NTP is 25 μM. Walking steps: GTP→wash→ATP+CTP (EC42)→wash→GTP (EC45) or ATP+GTP (EC47)→wash→CTP (EC48). The walking reaction is performed as described in Chapter 11 of this volume.[5]

Primary template		Elongation complex	Half-life (min)		#
	sequence		500mM KCl	5mM KCl	
	+34 *Sma* I	+42	>120	>120	1
P	GACACCCCGGG AACA-3′	+45	<1	10	2
	CTGTGGGGCCCTTGT-5′	+47	4	>120	3
	42 45 47	+48	3	>120	4
	48 49	+49	<1	4	5
		+42	>120	>120	6
		+45	~100	>120	7
C	GACACCCCGGGAACA **AGCGGATAACAATTTCACA** -3′	+47	>120	>120	8
	CTGTGGGGCCCTTGT **TCGCCTATTGTTAAAGTGT** -5′	+48	>120	>120	9
	42 45 47 48 49	+49	>120	>120	10

FIG. 3. Analysis of EC stability at the end of the primary template. The summary table shows the half-lives of the immobilized ECs in low and high salt (see text for details). The positions of ECs are indicated by arrows. P, primary template; C, control template.

Determination of the Minimal RNA Product Needed for EC Stability

To determine the minimal size of the RNA product that renders the EC fully stable, an experimental system was designed that allows analysis of ECs carrying transcripts shorter than 9 nt. Initiation complexes (IC) at the promoter that possess RNA of a similar length are not suitable for the minimal product system for several reasons: (1) unlike most ECs, the IC contains the initiation factor σ; (2) the IC does not hold the transcript tightly, but continuously releases RNA in the cycle of abortive initiation; (3) the IC is sensitive to salt, indicating that elongation-specific (processivity) interactions have not been established in this type of ternary complex.

To prepare the EC with the shortest possible RNA without disrupting the complex, the reaction of pyrophosphorolysis is utilized.[9] Processive pyrophosphorolysis drives a "normal" EC carrying transcripts of 11 nt or longer into a reverse reaction, resulting in progressive shortening of RNA. The basic experiment includes (1) the preparation of the initial EC bound to the Ni-NTA-agarose beads; (2) the removal of NTPs and traces of σ^{70} from the EC by high-salt washing; (3) treatment of the EC with pyrophosphate (PPi); and (4) challenging the PPi-treated EC with 1 M NaCl for various time intervals, followed by washing the beads. The retention of radioactive RNA in the PPi-treated population of EC is determined by PAGE analysis. The results of these experiments indicated that RNA products of less than 6 nt long cannot be retained in the EC in high-salt

[9] T. A. Rozovskaya, A. A. Chenchik, and R. Sh. Beabealashvilli, *FEBS Lett.* **137,** 100 (1982).

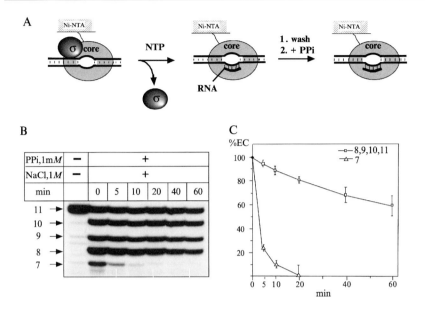

Fig. 4. The Minimal Product System. (A) A general scheme for generating the EC with the transcript shorter than 9 nt. Ni-NTA stands for His$_6$-tag affinity resin. (B) The decay profiles of EC7-11 generated by pyrophosphorolysis (PPi). (C) Graphic representation of data from (B).

TB.[10] The EC with RNA of 8 nt or longer are fully resistant to salt whereas the EC carrying 7-nt RNA displays intermediate salt stability (Fig. 4). The 5′-terminal seventh and eighth RNA bases may not be paired with the DNA template strand to render the EC fully stable. EC6–EC9 obtained by pyrophosphorolysis remain elongation competent.

Generation of EC6–EC11 and Assessing Their Stability

The initial EC11 carrying the radiolabeled transcript is prepared as follows: 1 pmol (~0.5 μl) of His-tagged RNAP, 2 pmol (~1 μl) of the T7A1 promoter template, and 8 μl of TB (10 mM Tris–HCl, pH 7.9, 10 mM MgCl$_2$, and 100 mM KCl) are incubated for 5 min at 37°, followed by addition of 1.5 μl of the 10× starting mix (100 μM ApUpC RNA primer, 0.6 μM [α-^{32}P] ATP (3000 Ci/mmol; NEN Life Sciences), 25 μM GTP, and 5 μl of TB–preequilibrated Ni^{++}-NTA-agarose (Qiagen). After 5 min incubation at 37°, the beads are washed several times with TB1000 (1 M KCl) and then with TB. Pyrophosphorolysis is performed at 37° for

[10] N. Korzheva, A. Mustaev, E. Nudler, V. Nikiforov, and A. Goldfarb, *Cold Spring Harb. Symp. Quant. Biol.* **63**, 337 (1998).

5 min by adding water solution of PPi to 1 mM, followed by washing the beads with TB. All kinetics are measured at 20° in TB1000. Electrophoretic analysis of the RNA samples is performed in 23% denaturing PAGE (20:3 acrylamide:N,N'-methylene bisacrylamide). Quantitation of radioactive RNA bands is carried out by PhosphoImager (Amersham, Piscataway, NJ) (Fig. 4).

To rule out a potential effect of the close proximity of the promoter, a similar PPi experiment can be performed on distal DNA sites. To prepare the EC with the short transcript (6–9 nt) stalled far from the promoter, the EC is treated with a ribonuclease to trim RNA up to the area of protection by RNAP, followed by the PPi reaction. For example, EC26 prepared by walking[5] is treated with 3000 units/ml RNase T1 (Roche Molecular Biochemicals, Indianapolis, IN) that cleaves the nascent transcript at 16 nt from the 3' terminus.[11] Next, EC26 is washed with TB and subjected to pyrophosphorolysis as described previously. Like the situation with EC11, in this case the minimal stable RNA product was 8 nt long as well.[10] Thus the minimal size of the RNA product that renders the *E. coli* EC fully stable is 8 nt and does not depend significantly on the sequence of DNA or RNA. Ribonucleotide residues at positions -7 and -8 serve as the critical determinants of EC stability and processivity because the 6-nt RNA dissociates immediately from the EC under high ionic strength conditions.

Superimposition of the minimal RNA product and the minimal DNA template yields the minimal nucleic acid scaffold of the EC as shown in Fig. 1B. It should be noted that this structure is inferred from minimization of either DNA or RNA in the context of the other component being full-sized. Additional experiments have demonstrated that such preassembled minimal scaffolds bind the RNAP core enzyme efficiently to form fully functional and stable ECs.[3,12]

Determination of the Minimal Size of the RNA:DNA Heteroduplex Needed for EC Stability

Although the minimal size of RNA required for EC stability is 8 nt, it does not necessarily mean that all bases in the minimal RNA should be hybridized with the DNA template strand in the transcription bubble. From the parameters of the minimal DNA template presented previously, it appears that the salt-stable EC is possible if the hybrid is only 6 bp long (Fig. 2B, line 7). Thus in the Minimal Template System, RNA does not need to be base-paired to DNA at -7, -8, and -9 positions to support fully stable ECs.

[11] N. Komissarova and M. Kashlev, *Proc. Natl. Acad. Sci. USA* **95,** 14699 (1998).
[12] K. Kuznedelov, N. Korzheva, A. Mustaev, and K. Severinov, *EMBO J.* **21,** 1369 (2002).

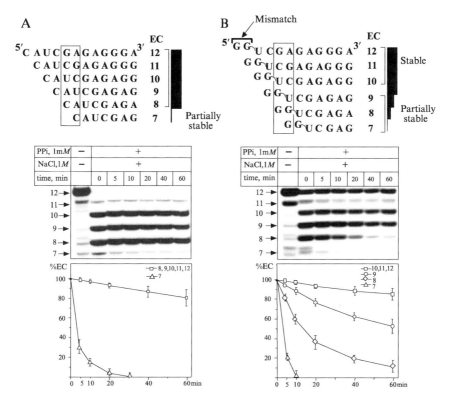

FIG. 5. EC stability as a function of the RNA:DNA hybrid length. Stability of transcription elongation complexes (TECs) containing matching (A) and mismatching (B) residues at RNA 5′ terminus. The ladders of RNAs generated by PPi are listed on the left. Positions −7 and −8 are boxed.

To address the role of hybrid base pairing in the Minimal Product System, a series of ECs can be prepared in which the transcript is only partially complementary to the template. This is accomplished by using synthetic RNA primers with mismatches at their 5′ termini to prepare the initial EC, followed by shortening of the transcripts in 3′ → 5′ direction by pyrophosphorolysis (Fig. 5). The ECs in this experiment are prepared as described previously, except that the RNA primer (GpGpUpC, 25 μM) in the starting mix has two mismatches at the 5′ terminus. The kinetics of dissociation of such complexes demonstrate that noncomplementary RNA nucleotides at −8 and −9 positions (Fig. 5B, line 9) stabilize the transcript more than thirty-fold as compared with the fully complementary 7-nt RNA control (Fig. 5A, line 7). By comparing stability of EC8, which RNA mismatches at −7 and −8 positions (Fig. 5B, line 8), with that of EC6 (Fig. 5A), one can conclude that the 5′ unpaired nucleotides are primarily responsible for the complex stabilization. The release of fully complementary 6-nt RNA occurs almost instantaneously under the specified conditions and could not be detected in the time frame of this experiment. To emphasize this finding, the minimal scaffold of Fig. 1B has the crucial RNA positions −7, −8, and −9 in a single-stranded form rather than as a part of the DNA:RNA hybrid.

It should be noted, however, that in the Minimal Product System, the presence of mismatched nucleotides at −8 and −9 positions do not stabilize the EC completely: perfectly hybridized 9 nt RNA is held ∼3 times stronger (Fig. 5A), reflecting some additional role of base pairing at the 5′ border of the hybrid in EC stability. This result is seemingly in disagreement with the conclusion derived from the Minimal Template System (Fig. 2, line 7), in which the presence of DNA bases at positions −7, −8, and −9 are dispensable for the salt-stable EC. There are two, not mutually exclusive, explanations that reconcile those findings. They reflect a principal difference between two systems. In the Minimal Template System, the EC carries excessive RNA, whereas in the Minimal Product System, the EC contains excessive DNA. The full-sized transcript is likely to be additionally stabilized through upstream interactions with the protein up to position −14 in the Minimal Template System.[6] In the case of the shortened transcripts in the Minimal Product System, base pairing at positions −7, −8, and −9 could compensate for the loss of the upstream RNA–protein contacts. The other explanation is that the upstream DNA duplex may exert a destabilizing effect on the complex, which is countered by RNA base pairing at positions −7, −8, and −9. In the Minimal Template System, such a destabilizing effect would not be detected because the upstream duplex does not exist.

The methods of DNA and RNA minimization described here can be used to measure the parameters of the nucleic acid scaffold for other bacterial and also eukaryotic RNAPs. This information is essential for understanding the mechanisms of transcription elongation and termination and their regulation by various intrinsic signals and protein factors.

Acknowledgments

This work was supported by the National Institutes of Health (E.N.).

[14] Strategies and Methods of Cross-Linking of RNA Polymerase Active Center

By ARKADV MUSTAEV, EUGENY ZAYCHIKOV, MIKHAIL GRACHEV,
MAXIM KOZLOV, KONSTANTIN SEVERINOV, VITALY EPSHTEIN,
NATALIYA KORZHEVA, OXANA BERESHCHENKO, VADIM MARKOVTSOV,
EUGENY LUKHTANOV, IGOR TSAREV, TATYANA MAXIMOVA, MIKHAIL
KASHLEV, IRINA BASS, VADIM NIKIFOROV, and ALEX GOLDFARB

RNA polymerase (RNAP) is one of the largest and complex enzymes in the living cell, which poses a big challenge in studies of its structure and function. Cross-linking as a complementary approach to X-ray crystallography has been successfully used to study the mechanism of RNA synthesis. A few main issues were addressed in these studies: (1) super-selective labeling and identification of the residues that contribute to the active center and its closest environment[1,2]; (2) spatial arrangement of priming substrate, antibiotic rifampicin, and template DNA in the active center, which revealed the mechanism for rifampicin inhibition of RNA synthesis;[3] (3) determination of the size of DNA–DNA heteroduplex in transcription elongation complex (TEC), which resolved a long-term controversy about the existence and functional role of the hybrid[4,5]; (4) construction of a high-resolution model of TEC by mapping of the nucleic acid–protein contacts in TEC and projection of these contacts on the structure of RNAP core enzyme[6,7]; (5) mapping of RNA contacts in the active center of transcription complex, which suggested the structural models for transcriptional arrest,[8] substrate entry to the active center, and RNAP translocation.[9] In

[1] M. A. Grachev, T. I. Kolocheva, E. A. Lukhtanov, and A. A. Mustaev, *Eur. J. Biochem.* **163,** 113 (1987).

[2] A. Mustaev, M. Kashlev, J. Lee, A. Polyakov, A. Lebedev, K. Zalenskaya, M. Grachev, A. Goldfarb, and V. Nikiforov, *J. Biol. Chem.* **266,** 23927 (1991).

[3] A. Mustaev, E. Zaychikov, K. Severinov, M. Kashlev, A. Polyakov, V. Nikiforov, and A. Goldfarb, *Proc. Natl. Acad. Sci. USA* **91,** 12036 (1994).

[4] E. Nudler, A. Mustaev, E. Lukhtanov, and A. Goldfarb, *Cell* **89,** 33 (1997).

[5] N. Korzheva, A. Mustaev, E. Nudler, V. Nikiforov, and A. Goldfarb, *in* "CSH Symposia for Quantitative Biology," Vol. 63, p. 337. Cold Spring Harbor Press, Cold Spring Harbor, New York, 1998.

[6] N. Korzheva, A. Mustaev, M. Kozlov, A. Malhotra, V. Nikiforov, A. Goldfarb, and S. A. Darst, *Science* **289,** 619 (2000).

[7] N. Korzheva and A. Mustaev, *Curr. Opin. Microbiol.* **4,** 119 (2001).

[8] V. Markovtsov, A. Mustaev, and A. Goldfarb, *Proc. Natl. Acad. Sci. USA* **93,** 3221 (1996).

[9] V. Epshtein, A. Mustaev, V. Markovtsov, O. Bereshchenko, V. Nikiforov, and A. Goldfarb, *Mol. Cell* **10,** (2002).

this article we present the basic strategies and methods used in some of these studies.

Labeling of the Priming Substrate Binding Site of the Active Center by Autocatalysis

The validity of affinity labeling depends on its selectivity, that is, on the ratio of specific to nonspecific modification (outside the binding site). It is hard to achieve high selectivity with large enzymes, especially when the binding constant of the ligand is low. It is not surprising, therefore, that numerous attempts to study the functional topography of RNAP substrate binding sites by means of traditional cross-linking have given relatively poor results.

In order to achieve the high selectivity of affinity labeling for RNA polymerase, we took advantage of "catalytic competence." This phenomenon reflects the ability of a substrate residue cross-linked at the active center of an enzyme to convert into a cross-linked product by the same enzyme molecule according to the normal mechanism of catalysis. This principle is illustrated later for RNAP[1,10] (Fig. 1A). At the first stage RNAP is treated in the binary complex with a promoter by affinity reagent, which is an analog of initiating substrate. This results in the cross-linking of affinity reagent residues both inside and outside the active center. At the second stage the modified enzyme is supplemented with the second radioactive substrate complementary to the next base of DNA template. Reagent residues tethered at the active center could be "catalytically competent" and elongated by radioactive nucleotide because of the catalytic action of the active center. Residues covalently bound outside the active center do not attach radioactivity at the second stage and remain "invisible" during subsequent analysis.

This method initially developed for *Escherichia coli* RNAP[1,10] has been extended to all kinds of NTP polymerizing enzymes, including representatives of eucaryotes,[11] procaryotes,[12] and phages[13] (for review see[12,14]).

[10] Y.-V. Smirnov, V. M. Lipkin, Y.-A. Ovchinnikov, M. A. Grachev, and A. A. Mustaev, *Bioorg. Khim.* (Russian), **7,** 1113 (1981).

[11] M. Riva, A. R. Schaeffner, A. Sentenac, G. Hartmann, A. A. Mustaev, and E. F. Zaychikov, *J. Biol. Chem.* **262,** 14377 (1987).

[12] G. R. Hartmann, C. Biebricher, S. J. Glaser, F. Gross, M. J. Katzamaeyer, A. J. Lindner, H. Mosig, H. P. Hasheuer, L. B. Rothman-Dennis, A. R. Shaffner, G. J. Schneider, K. O. Stetter, and M. Thomm, *Biol. Chem. Hoppe-Seyler* **369,** 775 (1988).

[13] T. G. Maximova, A. A. Mustaev, E. F. Zaychikov, D. L. Lyakhov, V. L. Tunitskaya, A. K. Akbarov, S. V. Luchin, V. O. Rechinsky, B. K. Chernov, and S. N. Kochetkov, *Eur. J. Biochem.* **195,** 841 (1991).

[14] M. A. Grachev, A. A. Mustaev, and F. F. Zaychikov, "Chemical Modification of Enzymes" (B. I. Kurganov, N. K. Nagradova, and O. I. Lavrik, eds.), p. 309. Nova Science Publishers, Inc., New York, 1997.

Unusually high selectivity of the labeling allows visualization of polymerase molecules in partially purified preparations[2] or even in crude cellular extracts.[15,16]

Autocatalytic Affinity Labeling of RNAP at the Active Center

A mixture (9 μl) containing 50 mM Tris–HCl, pH 8.5, 0.1 mM 2-mercaptoethanol, 50 mM NaCl, 10 mM MgCl$_2$, 2 pmol Bsp–1462 DNA fragment of T7 DNA containing promoters A$_0$–A$_3$, and 4 pmol RNAP is incubated for 5 min at 37°. 1 μl of 10 mM solution of one of the reagents I–V (Fig. 1D) is added. After incubation for 20 min at room temperature, the reaction mixtures containing reagents II–V are supplemented by 1 μl of immediately prepared 0.1 M NaBH$_4$ and the incubation continues for another 20 min. Then [α-^{32}P]UTP (3 μl of 1000–3000 Ci/mmol, 1 μCi/μl) is added to all the mixtures. In 20 min the mixtures are supplemented with 3 μl solution containing 5% SDS, 5% 2-mercaptoethanol, 50% glycerol, and 0.1% bromophenol blue, and after 30 min incubation at 37°, analyzed by (SDS)–PAGE. Labeled subunits are revealed by autoradiography.

RNA–Protein Cross-linking in the Active Center of Initiating Complex

As soon as during initiation short RNA products easily fall off the active center studies RNA–protein contacts therein (i.e., in the active center) is a complicated task. To overcome this difficulty the initiating Rif–GCU complex was constructed on the T7 A2 promoter using as a primer the chimerical compound Rif–GTP in which rifampicin is covalently linked to GTP.[3] Such compounds bind to RNAP bifunctionally so that Rif occupies its natural pocket, and the nucleotide enters the i site (Fig. 1B). Rif–GTP was used to prime the incorporation of the two next nucleotides. In the resulting ternary complex, the trinucleotide product is flexibly tethered to RNAP via the Rif moiety and serves to mimic the abortive product, with the difference that it is prevented from dissociating out of the enzyme.[3] The following is an example of efficient cross-linking in this system, which occurs at Methionine (Met)932 of β' subunit.

[15] O. Morozova, A. Mustaev, N. Belyavskaya, E. Zaychikov, E. Kvetkova, Yu. Wolf, and A. Pletnev, *FEBS Lett.* **277,** 75 (1990).
[16] V. Rait and F. Seifart, *in* "Nucleic Acid Symposium Series," Vol. 36, p. 162. Oxford University Press, Oxford, UK 1997.

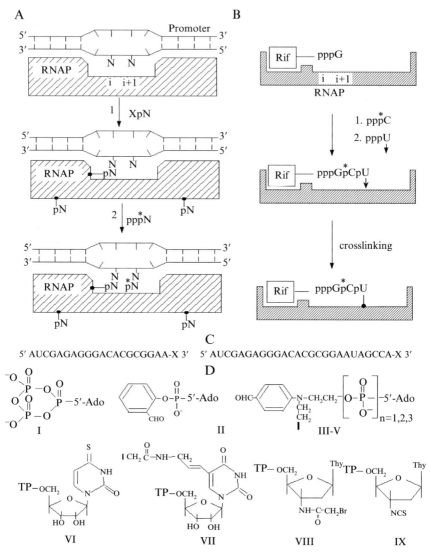

FIG. 1. Strategies used for cross-linking studies of RNAP active center. (A) Autocatalytic affinity labeling of RNAP in initiating complex. X-pN represents cross-linking derivative of initiating substrate; (*), radioactive phosphate. (B) Cross-linking in initiating complex. Rif symbolizes rifampicin residue tethered to GTP; arrow, reactive group; filled circle, a cross-link. (C) Sequence of RNA synthesized from T7A1 promoter. X-reactive nucleotide residue used for cross-linking in TEC. (D) Substrate analogs used in cross-linking studies. TP stands for triphosphate residue.

Cross-linking in Rif–GCU Complex[9]

Reaction mixture contains 4 pmol RNAP, 4 pmol T7A2 template, and 10 μM Rif–$(CH_2)_5$-GTP in suspension of 15 μl of Ni–NTA agarose and 15 μl of transcription buffer (TB 20 mM HEPES–HCl, pH 7.9, 50 mM NaCl, 10 mM MgCl$_2$). After 5 min incubation at 37° the mixture is washed with the same buffer (3 × 1 ml) and [α-^{32}P]CTP (3000 Ci/mmol) is added. The reaction is allowed to proceed for 10 min at 37° and washed (3 × 1 ml) by TB containing 2.5 mM MnCl$_2$ instead of 10 mM MgCl$_2$. The reagent VIII (Fig. 1D) is added to the final concentration of 200 μM. After 30 min min incubation at 37°, the reaction is quenched and cross-linked products analyzed as before.

RNA–protein Cross-linking in the Active Center of Elongation Complex[8,9]

To study the contacts of RNA in the active center of elongation complex, we obtained TECs containing 20 and 26 nt RNA (Fig. 1C) on T7 A1 promoter by "walking" protocol using RNA primers and an incomplete set of NTPs. Reactive derivatives of UTP (VI–IX; Fig. 1D) are used to introduce the cross-linking group at 3′ RNA terminus in the previous TECs. Derivative VI is photoreactive and can react with any residues, VII and VIII are alkylating compounds able to cross-link to nucleophylic residues, whereas IX forms stable adducts exclusively with lysine. As soon as the elongation complexes are stable, the cross-links therein are highly selective.

Cross-linking Protocol for TEC21[9]

Fifteen pmol of RNAP and T7A1 template are mixed with suspension of 15 μl preequilibrated Ni–NTA agarose and 15 μl of the transcription buffer, incubated for 5 min at 37°, and washed with the same buffer (3 × 1 ml). The volume is adjusted to 30 μl by TB, and AUC primer (10 μM) together with GTP and ATP (25 μM each) are added. The mixture is incubated at 37° for 5 min and washed again (3 × 1 ml) by the same buffer. ATP and GTP (25 μM each) are added with 2 μl [α-^{32}P]CTP (3000 Ci/mmol), incubated for 5 min at 20°; nonradioactive CTP (25 μM) is introduced, and incubation continues for another 2 min. The buffer is changed to TB containing 2.5 mM MnCl$_2$ instead of MgCl$_2$, the dUTP-cross-linking analogs VIII and IX are added to each mixture to final concentration 100 μM, and incubation continues for 30 min at 37°. The reaction is quenched and analyzed as before.

Mapping of Cross-linking Sites

To map the cross-linking sites, we used the method developed earlier in one of our labs.[17] Assume that a polypeptide is labeled by radioactive affinity adduct at a single site. Single-hit degradation of this polypeptide (e.g., at Met residues by CNBr) will generate the products, which possess C or N terminus of the original polypeptide. It is obvious that upon the cleavage at each particular site, only one product will contain radioactivity. Separation of the products according to their sizes by SDS–PAGE will give characteristic pattern of radioactive products, which depends on the position of the radioactive label on the polypeptide and distribution of Met residues in the sequence. Resulting patterns are compared with theoretical ones calculated under the assumption that the label resides close to either the C or N terminus of the polypeptide.

This is illustrated in Fig. 2A, where RNAP β subunit was cross-linked to either the initiating substrate or RNA product. In this case the distribution of CNBr degradation products on the autoradiogram is very close to the theoretically inferred C-terminal degradation pattern and does not match the N-terminal one. The position of the tag is located between CNBr cleavage sites, which give rise to the shortest labeled cleavage product and the next shortest product, which is not radioactive. Thus it is seen that in the presented cases the label resides within the intervals 951–1066 (reagent II), 1232–1242 (reagent I), and 1304–1315 (TEC14). Besides CNBr, other specific cleaving agents can be used:

1. Hydroxylamine—selectively cleaves Asn–Gly bonds at pH 9–10.
2. N-chlorosuccinimide (NCS) and N-bromosuccinimide (NBS)—at single-hit conditions perform cleavage at Trp residues.
3. Bromine—cleaves at Trp and Tyr with preference to Trp.
4. 2-nitro-5-thiocyanobenzoic acid (NTCBA)—cleaves at Cys residues.

Limited single-hit proteolysis at particular types of residues can also be used (Lys-C, Glu-C, Asp-N, trypsin, etc.) for mapping in combination with chemical degradation. As soon as the cleavage rates at a particular site differ significantly because of unequal accessibility to the protease (even in the presence of SDS) and different neighboring residues adjacent to the cleaved bond, the conditions of proteolysis must be carefully chosen to avoid multiple-hit degradation. An application of single-hit trypsinolysis for mapping is described.[17]

[17] M. Grachev, E. Lukhtanov, A. Mustaev, E. Zaychikov, M. Abdukajumov, I. Rabinov, V. Richter, Y. Skoblov, and P. Chistyakov, *Eur. J. Biochem.* **180,** (1989).

For the mapping, different cleavage strategies may be used either in parallel or consequently (e.g., cleaving the proteolytic fragment with CNBr[17] or single-hit cleavage of complete CNBr product by NH_2OH[14]). Supplementary data obtained by independent means (e.g., determination of the size of product of complete cleavage by SDS–PAGE[14] or mass spectrometry,[18] partial sequencing of cleavage products,[14] and their affinity purification or separation using His-tags[19]) can help in providing and refining the results obtained by single-hit cleavages.

There are a few requirements that must be obeyed to avoid the misinterpretation of the mapping results. (1) There must be equal susceptibility of the particular peptide bonds to the cleaving reagent to obtain the pattern representing all expected bands, which is achieved by denaturation of the protein before cleavage. (2) It is very important to distinguish between single- and multiple-hit products. To this end, the cleavage must be performed to a low extent when the main part of the protein remains uncleaved. Single-hit cleavage products must follow first-order time course, which can be judged by quantitation of the cleavage products. (3) Interpretation of the data may be complicated by the abnormal electrophoretic mobility of the cleavage products. For instance, three peptides (Fig. 2A, lane 6) corresponding to cleavage at Met 1107, 1119, and 1131 migrate as a single band, which can be resolved, although on a longer run (lane 10). It should be stressed, however, that the method is based on the analysis of the whole pattern of the bands rather than on determination of the exact sizes of the peptides from their electrophoretic mobility.

The interpretation of the degradation patterns is straightforward for those cases where the cross-linking site is located close to one of the termini, because the two sets of nested radioactive fragments corresponding to the NH_2 and COOH termini do not overlap on the autoradiogram. For cross-links located closer to the middle of the subunit, a mixture of NH_2- and COOH-degradation products has to be interpreted. As reference, degradation patterns of terminally labeled polypeptide can be used in this case. An example is given in Fig. 2C where the cross-link resides in the middle part of the β' subunit as revealed by single-hit CNBr cleavage (bottom panel). As a reference, CNBr degradation products of the β' subunit are labeled at the kinase site engineered at the C terminus.[9] Comparison of the patterns allows unequivocal positioning of the cross linking site between Met 747 and Met 821. This difficulty may be also avoided by

[18] K. Severinov, D. Fenyo, E. Severinova, A. Mustaev, B. Chait, A. Goldfarb, and S. A. Darst, *J. Biol. Chem.* **269**, 20826 (1994).

[19] K. Severinov, A. Mustaev, E. Severinova, M. Kozlov, S. A. Darst, and A. Goldfarb, *J. Biol. Chem.* **270**, 29428 (1995).

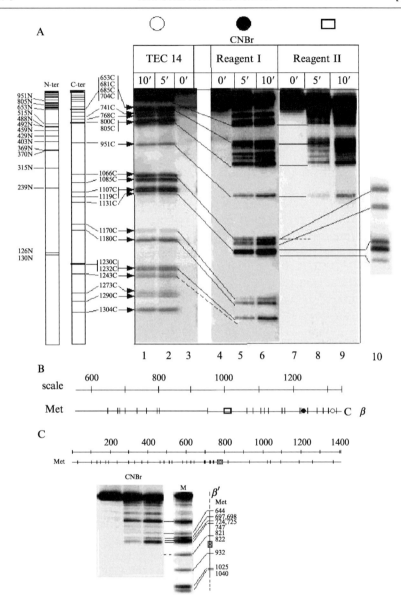

FIG. 2. Principle of the mapping of cross-linking sites. (A) *Right*: Single-hit CNBr degradation patterns of the cross-linked β subunit. *Left*: Theoretically calculated N- and C-terminal patterns of single-hit CNBr degradation products. The numbers indicate the position of Met residues in the sequence of β subunit, which give the corresponding cleavage product. Letters indicate the presence of C or N terminus of the original polypeptide in the cleavage product. Rectangle and filled and open circles here and in (B) symbolize the radioactive

precleaving of the labeled polypeptide (e.g., by limited proteolysis in non-denaturing conditions that give a limited number of products at the initial stage), followed by single-hit chemical degradation of purified proteolytic fragment.[14,17]

Single-Hit Degradation of Polypeptides at Particular Residues

Preparation of Starting Material

When only one subunit is labeled after the cross-link, the material can be used directly in degradation reactions. However, if more than one subunit is a target, each labeled polypeptide must be isolated and treated separately.

After SDS–PAGE separation of the labeled enzyme, the radioactive bands are excised from the gel and placed in Eppendorf tubes. After washing with water (2 × 1 ml) for 5 min with shacking, the gel pieces are crashed and soaked in ~0.6 ml of 0.03% SDS. After 1 h elution at 37° with shacking, the tubes are centrifuged and the eluate is carefully collected, avoiding the uptake of any gel pieces. The gel phase is additionally washed by 100 μl of 0.03% SDS, and the solutions are combined. The remaining gel and solution are checked by radioactivity monitor. If the recovery of radioactivity is less than 50%, the elution is continued. The lesser the mobility of the corresponding product in the separation gel, the more time is required for elution. The eluate is either freeze-dried or placed into Eppendorf tubes (200–300 μl portions), supplemented with 3 volumes of acetone, and kept at −20° for an hour. After centrifugation, the precipitate is dried under diminished pressure, dissolved in 1% SDS, and used in degradation reactions.

CNBr Degradation at Met Residues[17]

The reaction mixture after the cross-link to RNAP β or β' subunit is adjusted by water (if necessary) to 24 μl, supplemented with 3 μl of 10% SDS (if the radioactive polypeptide is purified as before, this stage is omitted), and kept at 37° for 30 min for denaturation. 1.5 μl of each 1 M HCl and 1 M CNBr in water are added. At time intervals 0, 5, and 10 min, 9-μl

cross-linking tag. *Far right*: Degradation products resolved on the longer gel. Solid lines link identical cleavage products; dashed lines indicate the position of the next cleavage product, which is not radioactive. (B) The scheme indicating the positions of Met residues in the C-terminal half of β subunit. (C) *Top*: Positions of Met residues in the sequence of β' subunit; *bottom*: CNBr degradation patterns of C-terminally labeled β' subunit (M) and of the same subunit cross-linked in initiation complex according to scheme B (Fig. 1) using reagent IX.

aliquots are withdrawn and quickly mixed with 3 μl of solution containing 5% 2-mercaptoethanol, 0.5 M Tris–HCl, pH 9.0, 0.2% bromophenol blue, and 50% glycerol. These mixtures are subjected to SDS–PAGE in a Läemmli system on 13% or 8–20% gradient gel.

NTCBA Degradation at Cys Residues[8,9]

To a solution (30 μl) of labeled polypeptide in 8 M urea and 50 mM Tris–HCl, pH 8.3, 1 μl of 150 mM 2-mercaptoethanol is added and the mixture is incubated for 30 min at 37°. 3 μl of 100 mM freshly prepared NTCBA in methanol is added, and the incubation continues at 37°. In 15 min the pH of the mixture is adjusted to 9–9.5 by 1 M NaOH. In 30 and 60 min the aliquots (10 μl) are withdrawn and mixed with 3 μl of solution containing 5% SDS, 50% glycerol, and 0.2% bromophenol blue. The samples are analyzed as before.

Hydroxylamine Degradation at Asn–Gly Sites

To the solution of labeled enzyme or its purified subunit SDS and 2-mercaptoethanol is added a final concentration of 1%. After incubation at 37° for 30 min an equal volume of 2 M NH$_2$OH/0.2 M Na$_2$CO$_3$ pH 10.0 (prepared from NH$_2$OH♦HCl by titration with NaOH to pH 10.0 and mixing with Na$_2$CO$_3$, pH 10.0) is added, and incubation continues at 37°. In 30 and 60 min, aliquots are withdrawn and mixed with 1:5 (v/v) solution of 50% glycerol and 0.1% bromophenol blue.

Cleavage by NBS and NCS at Trp Residues and by Br, at Trp and Tyr Residues

The labeled enzyme is denatured by 1% SDS as before and mixed with Na-formate, pH 4.0, to final concentration 0.1 M. The mixture is supplemented by freshly prepared water solutions of NCS, NBS, or Br$_2$ to final concentrations 3 and 10 mM (or 1 mM in the case of bromine) and kept for 10 min at 20°. 1:3 (v/v) of 5% 2-mercaptoethanol, 0.5 M Tris–HCl, pH 9.0, 0.2% bromophenol blue, and 50% glycerol is added, and the mixture is analyzed as before.

Detection and Mapping of Multiple Cross-linking Sites

As soon as the active sites are usually formed at the interface of different subunits or structural domains, multiple cross-linking sites could be expected for the reactive ligands. If the cross-link occurs at different subunits, this could easily be seen after separation of the labeled polypeptides. More complicated is the situation when two or more cross-linking sites reside within the same subunit. The following are a few examples of the detection and mapping of such sites.

Multiple cross-linking sites can be revealed by quantitative analysis of single-hit degradation products. Fig. 3 shows CNBr degradation pattern for cross-linked β subunit according to scheme B (Fig. 1) using reagent IX (lanes 1–3) and the reference degradation pattern (lane 4). In both cases the shortest degradation product corresponds to cleavage at Met 1232, which defines the cross-linking site in the region 1232–1243. Also, in both cases the intensity of the products on the upper part of the gel is fairly equal. However, in one case (lane 3) the products starting from 1066 and lower are much weaker than those in the reference pattern (lane 4). Indeed, quantitation of the corresponding products (B) shows that in one case the intensity of the weaker products is about three times less than in the reference. These

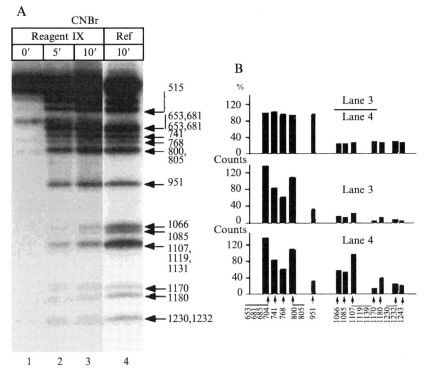

FIG. 3. Detection of the double cross-linking site by quantitative phosphoimagery. Limited CNBr degradation patterns of RNAP β subunit labeled according to Fig. 1C using reagent IX (A, lanes 1–3,) and reference degradation pattern for the β subunit labeled according to Fig. 1A using reagent I (lane 4). (B) Histograms of the degradation products.

data point to two cross-linking sites, one of which (major) resides in the region 951–1066 and the other (minor) in the region 1232–1243.

Multiple cross-linking sites also can be detected by limited proteolysis (Fig. 4). The major cross-link in the active center using derivatives III–V according to scheme A (Fig. 1) occurred to β subunit.[19] Limited trypsinolysis of the labeled enzyme (which is known to occur between the residues 903 and 904 of the β subunit[20]) gave two radioactive fragments of the expected size (Fig. 4A,B), which is indicative for at least two cross-linking sites. Using the enzyme containing insertion in β subunit (see Fig. 6C and Fig. 4A), which is much more sensitive to trypsin, allows quantitative cleavage of the modified subunit (Fig. 4C). Purified proteolytic fragments were used for further refining of the mapping.

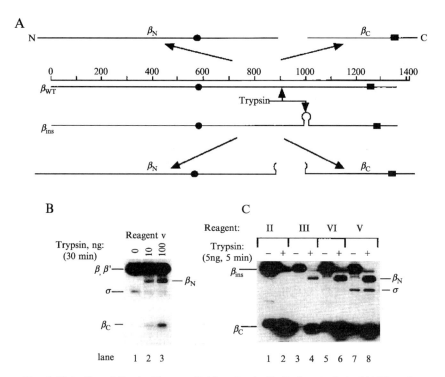

Fig. 4. Detection of the double cross-linking sites by limited proteolysis. (A) The scheme of tryptic cleavage of WT and insertion containing β subunit in the context of cross-linked initiating complex. Filled circle and box indicate the position of cross-linking tags. (B) Trypsinolysis of the cross-linked RNAP in the case of WT β subunit. (C) The same with β_{ins}.

[20] S. Borukhov, K. Severinov, M. Kashlev, A. Lebedev, I. Bass, G. C. Rowland, P. P. Lim, R. E. Glass, V. Nikiforov, and A. Goldfarb, *J. Biol. Chem.* **266,** 23921 (1991).

Another approach that appears to be very helpful for the mapping is based on the usage of functionally active enzymes assembled from the fragments of RNAP subunits.[21] A cross-link from 3'-RNA terminus using reagent VII (Fig. 1D) according to scheme B (Fig. 1) occurred to both β and β' subunit. CNBr degradation pattern of the purified β subunit was complex, suggesting two cross-linking sites in C- and N-terminal regions of the subunit. Indeed, when in the same experiment RNAP assembled from the individually expressed fragments of β subunit was used, three radioactive products were detected on the gel (Fig. 5B). One of them co-migrated with β' subunit, and the smaller two with the fragments of β subunit. CNBr degradation patterns of gel-purified smaller fragment of β subunit is shown in Fig. 5C and reveals the segment 1243–1273 as a cross-linking site.

Refining the Mapping

The precision of the described mapping depends on the distance between two cleavable sites flanking the cross-linking tag. The number of available cleavage reactions at specific residues is limited, which in unfortunate cases is a reason for low resolution of the mapping. The obvious way to increase that

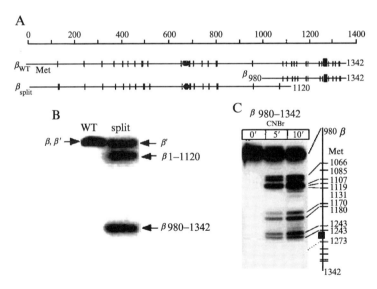

FIG. 5. Detection of the double cross-linking sites using RNAP containing split site in the β subunit. (A) Positions of Met residues in the intact β subunit and in the fragments of split β subunit used for cross-linking experiment. Rectangle and circle indicate the cross-linking tags. (B) Radioactive polypeptides cross-linked according to Fig. 1B with reagent VIII. (C) Single-hit CNBr degradation patterns of the purified fragment.

resolution is to engineer additional cleavage sites around the putative cross-link. This principle is illustrated in Fig. 6A. The cross-link at the i site of the active center by Lys-specific probe was mapped within the interval 1232–1243 of the β subunit.[17] There are two lysine residues in the segment—Lys1234 and Lys1242. To distinguish between those two residues, we introduced conservative substitution L1238M between them and used the mutant enzyme in the same experiment. Compared with wild type (WT) enzyme, single-hit CNBr cleavage showed one extra band migrating slightly below the cleavage product at Met 1232, which appears as a result of cleavage at Met 1238 (Fig. 6A). Therefore the target of cross-link is Lys1242.

Resolution of mapping can be increased by conservative substitution of the presumed cross-linking residues to nonreactive ones unable to produce the cross-link with the used affinity probe. In our work this principle has been successfully used in a few cases.[2,9] The environment of the catalytic center was probed by introducing the reactive UMP analogs VIII and IX (Fig. 1D) at RNA 3' terminus in the context of elongation complex[9] (Fig. 1C). Analog IX is Lys-specific, whereas VIII can potentially react with all nucleophylic amino acid residues. In both cases β' subunit was the main target of cross-link (Fig. 6B). Single-hit chemical cleavage allowed mapping of the cross-linking area 747–821 in the case of reagent IX. There are three Lys residues in this segment: Lys749, Lys781, and Lys789. Lys749 and Lys781 cannot be the targets, because for the probe to reach them, dramatic distortions of the TEC structure would be required. That Lys789 is the cross-link site was confirmed using the K789R substitution (Fig. 6B, lane 2), which led to the dramatic reduction of the cross-link. Further single-hit degradation of the residual cross-linking product showed that it is due to the cross-link to some other region of the subunit. The same effect is observed when M932L substitution was used in the experiment with reagent VIII (Fig. 6B, lane 4), suggesting that this residue was a target.

In some cases, when the level of the expression of mutant subunit is not very high, the reconstituted enzyme used for the previous purposes is contaminated with WT subunits from the host strain, complicating the interpretation. A new assay developed to overcome this complication is presented in Fig. 6C, D. Active center-directed cross-link using Lys-specific reagent II (Fig. 1D) according to scheme A has been mapped in the segment 1036–1066 of the β subunit by limited CNBr degradation and trypsinolysis.[17] This fragment contains four Lys residues. Lys1065 of this segment is highly conserved in evolution and therefore seems to be the most likely target of cross-link. Two substitutions were made, including

[21] K. Severinov, A. Mustaev, E. Severinova, I. Bass, M. Kashlev, R. Landick, V. Nikiforov, A. Goldfarb, and S. A. Darst, *Proc. Natl. Acad. Sci. USA* **92,** 4591 (1995).

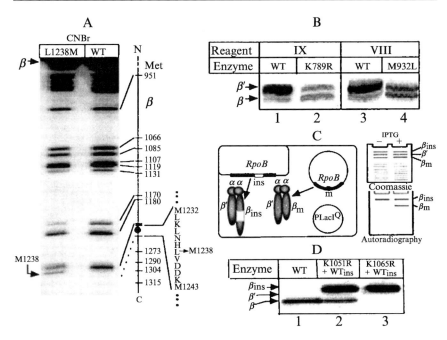

FIG. 6. Ways to increase precision of the mapping. (A) Engineering of the additional cleavage site between the potential targets. Limited CNBr degradation patterns of a wild type (WT) labeled RNAP and the enzyme containing L1238M substitution. B–D Conservative substitutions of the potential targets with nonreactive residues unable to produce the cross-link. (B) Cross-link from 3'-RNA terminus in TEC by reactive nucleotide analogs VIII and IX in TEC21. (C, *left*) Genetic system constructed to discriminate between the cross-links in the RNAP normal-sized mutant β subunit encoded by a plasmid, carrying a mutation (m) and WT chromosomal β$_{ins}$ subunit containing insertion in nonessential region of *RpoB* gene. PLacIQ, expression plasmid carrying a gene for *lac* repressor used to "turn off" the production of a plasmid encoded β subunit. (C, *right*) A typical picture of Coomassie-stained SDS–PAGE of partially purified cross-linked RNAP preparations containing cross-linking (+) and noncrosslinking β subunit and radioautography of the same gel. (D) Radioautography of cross-linked RNAP using reagent II according to Fig. 1A. Lane 1, WT RNAP with normal-sized β subunit; lane 2, partially purified RNAP from induced cells containing K1051R mutation in plasmid-borne β subunit; lane 3, the same but with K1065R substitution.

K1065R and K1051R as a control.[2] To determine whether the mutant β subunit in the context of holoenzyme would be labeled by using Lys-specific probe, β subunit expression plasmid containing each mutation (m) was placed into the host strain (Fig. 6C) harboring the insertion of 127 extra amino acids into the nonessential region of the β subunit. This insertion leads to elongated polypeptide β$_{ins}$ easily distinguishable from the normal-sized plasmid-encoded β subunit on SDS–PAGE (Fig. C, right panel).

Crude RNAP fraction was prepared from the IPTG induced cells and subjected to affinity labeling according to the protocol presented in Fig. 1A using reagent II. It is seen (Fig. 6D) that both chromosomal β_{ins} and plasmid encoded β subunit were labeled in the case of mutant K1051R, but only β_{ins} subunit was labeled when the plasmid carried the substitution K1065R. This highly suggests that the target of cross-link was indeed Lys1065.

Concluding Remarks

The autocatalytic affinity labeling described previously is simple and highly selective and has a wide applicability. Besides initiation, it has been used to map the product binding site of RNAP.[22,23] This method has been applied to more than 30 different NTP polymerases and allowed not only to get valuable information about the active sites, but also to address questions about their structure and function. In this way the contribution of the different RNAP subunits to the active center was determined[24] and the structural modules of the subunits that are able to assemble to a functional polymerase were defined.[21] Another example of the application of this approach is identification of the genes coding for replicases of tick-borne encephalitis virus.[15]

As for the described mapping technique, it has a number of important advantages:

1. It is very sensitive and fast. The whole experiment can be performed with picomole amount of an enzyme during one working day.

2. It may be used with unstable labels that do not survive standard procedures of mapping.

3. It is very simple and does not depend on any expensive equipment.

4. It is potentially general and could be used for mapping of not only affinity, but also other kinds of labels and particular sites (e.g., epitopes,[25,26] phosphorylation sites).

Acknowledgment

This work was supported by NIH grants GM49242 and GM30717 to A.G.

[22] T. Godovikova, M. Grachev, I. Kutyavin, I. Tsarev, V. Zarytova, and E. Zaychikov, *Eur. J. Biochem.* **166,** 611 (1987).

[23] I. Tsarev, A. Mustaev, E. Zaichikov, T. Alikina, A. Ven'iaminova, and M. Repkova, *Bioorg. Khim.* (Russian), **16,** 765 (1990).

[24] T. Naryshkina, A. Mustaev, S. A. Darst, and K. Severinov, *J. Biol. Chem.* **276,** 13308 (2001).

[25] V. Rar, E. Zaychikov, S. Vorobyev, I. Degtyarev, and S. Kochetkov, *Molecul. Biologia* (Russian), **25,** 1357 (1991).

[26] R. Burgess, T. Arthur, and B. Pietz, *Methods Enzymol.* **328,** 141 (2000).

[15] Using a *lac* Repressor Roadblock to Analyze the E. Coli Transcription Elongation Complex

By RODNEY A. KING, RANJAN SEN, and ROBERT A. WEISBERG

After binding to a promoter and initiating transcription of a short oligoribonucleotide, RNA polymerase (RNAP) undergoes a significant structural rearrangement.[1–5] The resulting transcription elongation complex (TEC) elongates the initial transcript until it reaches a terminator, at which point the complex dissociates. Analysis of the elementary steps of elongation and termination is demanding because it is difficult to obtain a population of TECs that are all in the same state. Several techniques have been used to overcome this problem. Visualization of individual molecules is a relatively new technique that has already provided important information,[6] but it requires additional apparatus and expertise. Biochemists have synchronized TECs by stopping them at a defined point on the template. Although it is not certain that all the halted TECs are in the same state or even in a state that is normally occupied during the elongation cycle, this technique has provided much useful information. Synchronization has been achieved in several ways. Perhaps the most commonly used method is "walking" the TEC along the template in steps by serial addition and removal of incomplete sets of nucleotide triphosphates (NTPs).[7,8] Another method, one that we describe here, is to block the progress of the TEC with a sequence-specific DNA binding protein.

The idea that a DNA binding protein can block the progression of RNA polymerase is rooted in the early genetic studies of the *lac* operon[9,10]

[1] C. L. Poglitsch, G. D. Meredith, A. L. Gnatt, G. J. Jensen, W. H. Chang, J. Fu, and R. D. Kornberg, *Cell* **98,** 791 (1999).

[2] J. Fu, A. L. Gnatt, D. A. Bushnell, G. J. Jensen, N. E. Thompson, R. R. Burgess, P. R. David, and R. D. Kornberg, *Cell* **98,** 799 (1999).

[3] K. S. Murakami, S. Masuda, and S. A. Darst, *Science* **296,** 1280 (2002).

[4] G. Zhang, E. A. Campbell, L. Minakhin, C. Richter, K. Severinov, and S. A. Darst, *Cell* **98,** 811 (1999).

[5] D. G. Vassylyev, S. Sekine, O. Laptenko, J. Lee, M. N. Vassylyeva, S. Borukhov, and S. Yokoyama, *Nature* **417,** 712 (2002).

[6] H. Yin, I. Artsimovitch, R. Landick, and J. Gelles, *Proc. Natl. Acad. Sci. USA* **96,** 13124 (1999).

[7] M. Kashlev, E. Martin, A. Polyakov, K. Severinov, V. Nikiforov, and A. Goldfarb, *Gene* **130,** 9 (1993).

[8] C. K. Surratt, S. C. Milan, and M. J. Chamberlin, *Proc. Natl. Acad. Sci. USA* **88,** 7983 (1991).

METHODS IN ENZYMOLOGY, VOL. 371

and in studies of *lac* operator containing constructs.[11–13] A number of other prokaryotic and eukaryotic site-specific DNA binding proteins have since been identified that are also capable of blocking the progression of TECs. Some examples are a noncleaving mutant of the EcoRI restriction enzyme,[14,15] polyoma virus T antigen,[16] the CCAAT box protein,[17] TTFI,[18] the replication terminator proteins of *E. coli* and *B. subtilis*,[19] and myco- phage L5 repressor.[20] We find that the Lac repressor-operator complex is particularly useful because it can be easily disrupted in physiological condi- tions, and it can be adapted for comparison of *in vivo* and *in vitro* conditions.

The *E. coli* Lac repressor is one of the best characterized sequence- specific DNA binding proteins. The first studies to address the mode of action of the repressor-operator complex on the TEC analyzed *lac* operon transcripts *in vivo* and transcripts generated from operator containing con- structs *in vivo* and *in vitro*.[21,22] The results suggested that the repressor- operator complex acts as an efficient transcription terminator. This conclusion was based in part on the inability to chase short transcripts into larger products upon prolonged incubation with IPTG, which disrupts the repressor-operator complex, and NTPs. Although the repressor-operator complex might, in some contexts, promote termination *in vivo*, the *in vitro* studies did not conclusively distinguish between true termination, which in- volves disassociation of the TEC, and stable RNAP arrest. Subsequent studies on the activity of elongating RNAP revealed that TECs can become stably arrested at certain template positions *in vitro* without dissociation from the template.[23] These complexes can be rescued through the action of transcript cleavage factors, whose activity was first described by Surratt and Chamberlin.[8] It is likely that the terminated transcripts observed in the

[9] W. S. Reznikoff, J. H. Miller, J. G. Scaife, and J. R. Beckwith, *J. Mol. Biol.* **43**, 201 (1969).

[10] D. H. Mitchell, W. S. Reznikoff, and J. R. Beckwith, *J. Mol. Biol.* **93**, 331 (1975).

[11] G. L. Herrin, Jr. and G. N. Bennett, *Gene* **32**, 349 (1984).

[12] M. Besse, B. Wilcken-Bergmann, and B. Muller-Hill, *EMBO J.* **5**, 1377 (1986).

[13] H. Horowitz and T. Platt, *Nucleic Acids Res.* **10**, 5447 (1982).

[14] P. A. Pavco and D. A. Steege, *Nucleic Acids Res.* **19**, 4639 (1991).

[15] P. A. Pavco and D. A. Steege, *J. Biol. Chem.* **265**, 9960 (1990).

[16] J. Bertin, N. A. Sunstrom, P. Jain, and N. H. Acheson, *Virology* **189**, 715 (1992).

[17] S. Connelly and J. L. Manley, *Mol. Cell Biol.* **9**, 5254 (1989).

[18] A. Kuhn, I. Bartsch, and I. Grummt, *Nature* **344**, 559 (1990).

[19] B. K. Mohanty, T. Sahoo, and D. Bastia, *EMBO J.* **15**, 2530 (1996).

[20] K. L. Brown, G. J. Sarkis, C. Wadsworth, and G. F. Hatfull, *EMBO J.* **16**, 5914 (1997).

[21] U. Deuschle, R. Gentz, and H. Bujard, *Proc. Natl. Acad. Sci. USA* **83**, 4134 (1986).

[22] M. A. Sellitti, P. A. Pavco, and D. A. Steege, *Proc. Natl. Acad. Sci. USA* **84**, 3199 (1987).

[23] K. M. Arndt and M. J. Chamberlin, *J. Mol. Biol.* **213**, 79 (1990).

early investigations of repressor-operator complexes were the result of RNAP arrest, not termination.

Although several studies have shown that the Lac repressor can decrease the expression of downstream genes independently of its effect on transcription initiation, there is considerable variation in the efficiency. Deuschle et al[21] reported a 90% decrease in a transcription unit that contains an isolated copy of the natural operator inserted upstream of a reporter gene. Steege and collaborators[22,24] reported repression efficiencies of 50% and 90%, depending on the construct that was assayed. Others have shown that the location of the operator relative to the transcription start site influences the degree of repression. Besse *et al.*[12] inserted synthetic symmetric operator sites at various distances downstream of a promoter and looked at the effects on the expression of a reporter gene. They found that the degree of repression of β-galactosidase expression decreased with increasing distance between the promoter and the operator, and suggested that repression depends on the position of the operator relative to the promoter and perhaps the relative positions of the repressor and polymerase on the DNA template.

Despite the variability of the efficiency of repression, there are several advantages of using *lac* repressor as a transcriptional roadblock. First, comparable studies often can be performed *in vivo* and *in vitro*. Second, the roadblock can be easily, gently, and efficiently removed by addition of the inducer IPTG. This feature distinguishes the *lac* repressor from other DNA binding proteins that have been used to trap elongating polymerases. Transcriptional roadblocks, such as the noncleaving Gln111 mutant of *Eco*RI, can be released by increasing the salt concentration. However, changing the ionic conditions may disrupt important interactions between RNAP and auxiliary factors or the nascent transcript. Third, RNAP walking by selective nucleotide deprivation, another widely used method of halting elongation, cannot always be readily reversed and becomes progressively more difficult to use as the distance of the arrest site from the start site of transcription increases.

Despite the advantages of *lac* repressor as a transcriptional roadblock, this experimental tool has seen only limited use in experiments on *E. coli* TECs, perhaps because of the earlier reports that its efficiency was variable and that it induced transcription termination. More recently, Rahmouni and collaborators[25,26] performed high-resolution mapping of *E. coli* TECs

[24] K. C. Cone, M. A. Sellitti, and D. A. Steege, *J. Biol. Chem.* **258,** 11296 (1983).

[25] F. Toulme, M. Guerin, N. Robichon, M. Leng, and A. R. Rahmouni, *EMBO J.* **18,** 5052 (1999).

[26] M. Guerin, M. Leng, and A. R. Rahmouni, *EMBO J.* **15,** 5397 (1996).

that were trapped by repressor-operator complexes formed *in vivo*. The roadblocked complexes were made in the presence of excess repressor and probed *in situ* with a variety of chemical and physical reagents. In addition, as described later, we have successfully exploited the properties of *lac* repressor to study *E. coli* RNAP elongation complexes that have been modified by a phage-encoded antiterminator RNA. Other investigators have used *lac* repressor to block the progression of the single subunit T3 and T7 RNAPs[27] and eukaryotic Pol II.[18,28–31] Although Deuschle *et al.*[28] suggested that the roadblock induces Pol II termination, the results reported by Reines and Mote[29] show that roadblocked complexes have not terminated because they can be reactivated by a transcript cleavage factor. Keen *et al.*[31,30] combined the use of *lac* repressor and a biotinylated template to isolate roadblocked elongation complexes and to probe their association with a transcription elongation factor.

We have used *lac* repressor to study the mechanism of transcription antitermination in HK022, a temperate phage that is related to λ.[32,33] After host RNAP transcribes HK022 antiterminator sequences, called "*put*" sites, it is modified so that it resists pausing and terminates less efficiently at intrinsic and *rho*-dependent terminators. Modification can be prevented by mutations of *put* or by mutations in the β' subunit of RNAP. This and other evidence suggested that antitermination requires sequence-specific interaction of nascent *put* transcripts with RNAP.[34,35] To demonstrate and characterize the putative interaction, we used *lac* repressor to block TECs well downstream of one of the *put* sites and isolated the roadblocked complexes. The properties of these complexes led us to conclude that *put* DNA is not required for antitermination after it is transcribed, that nascent *put* RNA associates specifically with the TEC during further translocation, and that persistent association is required for antitermination[32] (see the following).

[27] T. J. Giordano, U. Deuschle, H. Bujard, and W. T. McAllister, *Gene* **84,** 209 (1989).

[28] U. Deuschle, R. A. Hipskind, and H. Bujard, *Science* **248,** 480 (1990).

[29] D. Reines and J. Mote, Jr., *Proc. Natl. Acad. Sci. USA* **90,** 1917 (1993).

[30] N. J. Keen, M. J. Gait, and J. Karn, *Proc. Natl. Acad. Sci. USA* **93,** 2505 (1996).

[31] N. J. Keen, M. J. Churcher, and J. Karn, *EMBO J.* **16,** 5260 (1997).

[32] R. Sen, R. A. King, and R. A. Weisberg, *Mol. Cell* **7,** 993 (2001).

[33] R. A. Weisberg, M. E. Gottesman, R. W. Hendrix, and J. W. Little, *Annu. Rev. Genet.* **33,** 565 (1999).

[34] M. Clerget, D. J. Jin, and R. A. Weisberg, *J. Mol. Biol.* **248,** 768 (1995).

[35] R. A. King, S. Banik-Maiti, D. J. Jin, and R. A. Weisberg, *Cell* **87,** 893 (1996).

Effect of a *lac* Repressor Roadblock on Expression of a
Downstream Gene *in vivo*

We measured β-galactosidase production from a prophage containing a fusion of the HK022 P_L promoter and *putL* site to the *lac* operator and the structural genes of the *lac* operon in the presence and absence of *lac* repressor. Repressor was provided by a multicopy plasmid carrying the *lacIq* gene. We anticipated that *put* might help RNAP to overcome the transcriptional roadblock, and to test this possibility, we measured β-galactosidase production in a strain carrying the *rpoCY75N* mutation, which prevents *put*-mediated antitermination,[34] as well as in an *rpoC*+ strain. We found that repressor reduced the amount of β-galactosidase to 25% when antitermination was prevented (Table I, lines 1 and 2). A comparable fusion without the *lac* operator did not respond to *lac* repressor (data not shown). Antitermination decreased but did not completely prevent repression (Table I, lines 3 and 4). We conclude that repressor decreases transcript

TABLE I
EFFECT OF *lac* REPRESSOR ON EXPRESSION OF DOWNSTREAM *lacZ* GENE IN VIVO[a]

Strain	*lac* Repressor	*rpoC*	Steady State Activity	(+Repressor)/(−Repressor)
RAK353	No	*Y75N*	3389; 4163	25%; 24%
RAK350	Yes	*Y75N*	849; 1009	
RAK343	No	WT	6070; 6450	61%; 66%
RAK347	Yes	WT	3685; 4163	

[a] Beta-galactosidase was assayed as described[36] at two or three points during exponential growth of the cells in LB broth. The numbers in column 4 are the mean specific activities determined in two experiments, and the quotients (+Repressor)/(−Repressor) are given separately for each experiment (column 5). The standard errors of the mean were less than 10% in each experiment. The strains were single-copy lysogens of ΔX74[*lac*] or ΔX74[*lac*] *rpoCY75N*, as indicated in column 3. The prophage was a derivative of λ RS88[37] containing an insertion of the HK022 P_L promoter followed by the *putL* site, a *lac* operator, and the *lac* operon. Double-stranded DNA containing the wild type *lac* operator sequence (5'-TAAGCTTGTGGAATTGTGAGCGGATAACAATTTCA-CACCC) was made by annealing complementary oligosnucleotides. The DNA was phosphorylated and cloned into SmaI digested pRAK31.[35] This places the operator sequence approximately 150 bp downstream of the HK022 P_L start site. Strains containing *lac* repressor carried a plasmid with a *lacIq* gene.

[36] J. H. Miller, "Experiments in Molecular Genetics." Cold Spring Harbor Laboratory Press, Cold Spring Harbor, New York, 1972.
[37] R. W. Simons, F. Houman, and N. Kleckner, *Gene* **53,** 85 (1987).

elongation and that the block can be partially overcome by *put* modification of RNAP. *In vitro* experiments, presented next, suggest that the mechanism of *put* suppression of the *lac* repressor roadblock is indirect.

Effect of a *lac* Repressor Roadblock on Transcription *in vitro*

Our templates were linear DNA molecules produced by PCR. They contained a symmetric *lac* operator downstream of the HK022 P_L promoter and *putL* site and upstream of one or more transcription terminators.[32] A *lac* operator sequence[38] can be placed at any position of a template by incorporating its sequence in the appropriate primer, as indicated:

$$5' \frac{\text{GAATTGTGAGCGCTCACAATTC}}{\text{Symmetric } lac \text{ operator}}$$

$$- \frac{\text{NNNNNNNNNNNNNNNNNNNNNN}}{\text{Sequence complementary to the template}} \ldots 3'$$

An intrinsic transcription termination site can be included in the primer if desired. In the roadblocked complex, the 3'-end of the growing RNA chain is 3 to 4 nucleotides upstream of the operator sequence.

To maximize the recovery of roadblocked complexes and to minimize nonspecific inhibition of transcription by *lac* repressor, it is necessary to determine the optimal amount of repressor. Too little repressor decreases the amount of blocked TEC, and too much increases arrest at positions other than the *lac* operator, perhaps because of nonspecific binding of repressor (data not shown). In addition, it is necessary to optimize the time of elongation. We found that a significant fraction of TECs could read through the roadblock, and read-through increased with time (Fig. 1). In the experiment shown in Fig. 1, the *lac* operator was 180 bp from the transcription start, and we routinely allowed elongation to proceed for 2 minutes before isolating the complexes. Usually more than 50% of the transcripts had 3' ends just upstream of the *lac* operator. Clearly, the incubation time needed for optimal recovery would change if the location of the operator relative to the transcription start were changed.

The roadblocked TECs can be isolated free from most unincorporated NTPs and any unbound transcription factors by using biotinylated templates that are fixed to streptavidin-coated magnetic beads as described later. Upon addition of the *lac* operon inducer IPTG to the washed, immobilized TECs, we observed a slow and limited increase in transcript

---

[38] A. Simons, D. Tils, B. von Whicken-Bergmann, and B. Muller-Hill, *Proc. Natl. Acad. Sci. USA* **81**, 1624 (1984).

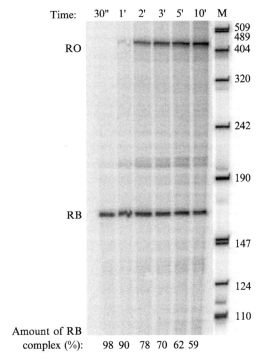

Time: 30" 1'   2'   3'   5' 10' M

RO

RB

Amount of RB
complex (%):   98 90 78 70 62 59

FIG. 1. Time course of transcription in the presence of *lac* repressor. The template was a linear molecule containing the HK022 P_L promoter, *putL* antitermination site, and a symmetric *lac* operator. At the indicated times after the initiation of transcription, 5 μl samples were withdrawn and analyzed as described in text. The concentrations of WT RNAP and DNA templates were 40 nm and 12 nm, respectively. The "M" lane contains labeled DNA size markers. The positions of the run-off ("RO") and roadblocked ("RB") transcripts are marked. The numbers below the lanes are (radioactivity in RB × 100)/([radioactivity in RB] + [counts in RO]). Estimates of radioactivity were divided by the number of CTP residues in the transcript.

length upon further incubation (data not shown). This probably reflects incomplete removal of NTPs from the bead matrix. When IPTG and NTPs were added, all the isolated TECs were rapidly chased to the end of the template (Fig. 2A). However, the kinetics and extent of chasing did vary with different templates (data not shown), and it is therefore necessary to test new templates. We have previously shown that nascent *put* RNA binds to a roadblocked wild type (WT) TEC and increases its ability to read through downstream terminators[32] (see the following). However, the proportion of TECs that were roadblocked by *lac* repressor and the

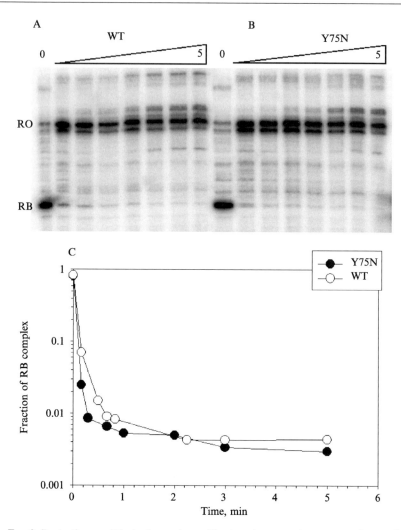

FIG. 2. Restarting roadblocked complexes. The template was the same as that used for the experiment of Fig. 1, and the bands formed by roadblocked and read-through transcripts are marked as before. We have no explanation of the band just below the run-off. (A) Immobilized, labeled, roadblocked TECs were formed according to the protocol described in text. After washing the complexes with transcription buffer, they were chased with 1 mM IPTG and 200 μm of each of the NTPs. Five-μl samples were withdrawn at intervals up to 5 min after addition of IPTG and NTPs and analyzed as described. (B) As in (A) except that RNAP containing the β'-Y75N substitution was used instead of WT RNAP. (C) The fraction of roadblocked complex after different times of chasing. Open circles, WT RNAP; closed circles, Y75N RNAP.

kinetics and extent of chasing were similar for WT and β'-Y75N RNAPs (Fig. 2A,B). Therefore *put* modification does not by itself allow RNAP to overcome the roadblock. To account for the ability of *put* modification to mitigate the effects of a *lac* repressor roadblock *in vivo* (Table I), we suggest that the modification interferes with the action of an elongation factor that dissociates or inactivates a roadblocked TEC, and that the factor is not present in the *in vitro* transcription reactions.

The following experiment provides an example of the manipulations that can be performed on a roadblocked TEC without compromising its activity. The interaction between nascent *put* RNA and the TEC could either be transient or stable. If the interaction is persistent, it could be the result of the intrinsically high affinity of the interacting partners for each other, or it could be favored by association of the 3'-end of the nascent transcript with the catalytic center of the TEC. This association tethers the transcript to the elongation complex and increases the effective local concentration of *put* RNA. To see if persistence of the *put* RNA–RNAP association is needed for antitermination and if tethering is important for persistence, we hybridized a oligodeoxyribonucleotide to the tether of a TEC that had been roadblocked 150 bp downstream of *put* and treated the complex with RNase H, which cleaves the RNA in RNA:DNA hybrids (Fig. 3A). We found that RNase H cleavage released most of the *putL* RNA, as shown by its separation from the template during fractionation on streptavidin-coated beads (Fig. 3B). This suggests that the *putL* RNA–RNAP interaction is intrinsically weak and persists only when the local concentration of *putL* RNA is high, as is the case when it is tethered to the elongation complex by the 3'-end of the nascent transcript. Cleavage also prevented antitermination when the TEC was restarted (Fig. 3C), showing that persistent association of *put* RNA with the TEC is required for antitermination.

Protocol for Roadblocking the TEC with *lac* Repressor

1. Incubate 35 μl of solution A at 37° for 10 min in a microcentrifuge tube to form the open initiation complex.

2. (Use this step only to immobilize a biotinylated initiation complex.) Add 35 μl of streptavidin-coated paramagnetic beads (Promega, www.promega.com) preequilibrated in T-buffer. Mix well and incubate for 10 min at 37°. Place the tube in a magnetic stand that accepts microcentrifuge tubes (Promega, www.promega.com). Carefully remove the supernatant liquid with a pipette. The complexes may be washed several times with

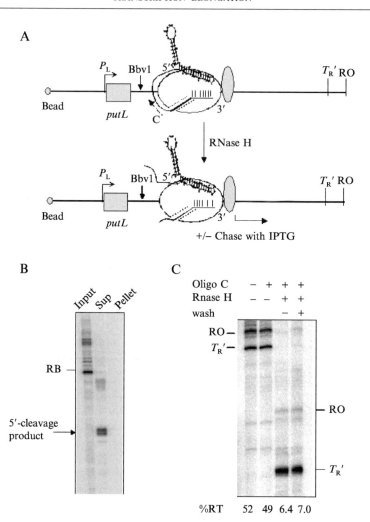

FIG. 3. Cleaving the tether between nascent *putL* RNA and the 3' end of the transcript (from Sen *et al.*[32]). (A) The cartoon shows the design of the experiment. The upstream end of the template (shaded circle) is bound to magnetic beads to facilitate the separation of released and retained RNA after cleavage of the tether at the position of oligonucleotide C by RNase H. The concentrations of DNA, RNAP, and *lac* repressor were 15 nm, 50 nm, and 40 nm, respectively. (B) The template of panel A was transcribed in the presence of [α-^{32}P]CTP and *lac* repressor. An aliquot was treated with oligonucleotide C and RNase H and separated into supernatant and pellet fractions. The RNA was fractionated in a denaturing gel and scanned. Input, Roadblocked transcript ("RB") before RNase H cleavage. Sup, Roadblocked transcript after RNase H cleavage and release from magnetic beads. Pellet, Roadblocked transcript cleaved by RNase H and retained on magnetic beads after three washes with T buffer (200 μl each wash). The 3' cleavage product is too small to be seen on

T-buffer to ensure the complete removal of unbound DNA and RNAP. After the final wash, suspend the complexes in 35 μl of T-buffer.

3. Add ~50 nm *lac* repressor to the open complexes, and incubate for 10 min at 37° to allow repressor binding. The exact concentration of repressor that gives the maximum yield of roadblocked complexes must be determined experimentally.

4. Add 35 μl of solution B, and incubate at 37° to form roadblocked TECs. The time of incubation that gives the maximum yield should be determined experimentally (Fig. 1).

5. To remove unincorporated NTPs from immobilized TECs, wash the complexes three times as described in step 2 with 200 μl T-buffer. This procedure can be performed in less than 10 min, and the complexes remain active for at least that long. The labeled transcripts can be analyzed at this point (step 6). Alternatively, the TECs can be probed in various ways before analysis.[26,32] To assess the effects of different manipulations on polymerase activity, restart transcription by adding IPTG (1mM) and the complete set of NTPs (200 μM each).

6. Reactions are stopped by phenol extraction and precipitated with ethanol. The precipitated products are then suspended in 10 μl of formamide loading buffer (Amersham-Pharmacia Biotech, Piscataway, NJ), 80% formamide) heated for 3 min in boiling water bath or temp block and fractionated in denaturing polyacrylamide gels. The gels are dried, exposed to a phosphoimager cassette, and scanned using a Molecular Dynamics "Storm" PhosphorImager. Data are analyzed with the associated ImageQuant software.

Solution A: T-buffer + 12 nm DNA + 50 nm RNAP. Our usual template was a PCR fragment containing the HK022 P_L promoter and a symmetric *lac* operator. Templates were amplified using standard protocols, and the amplified DNA was purified and concentrated on Centricon-100 (Ambion, www.ambion.com). We used a 5′-biotinylated upstream PCR primer (Biotechnologies, www.bioserve.com) to produce templates we wished to immobilize. RNA polymerase was obtained either from commercial sources (Epicentre, www.epicentre.com) or purified from an

---

this gel. (C) Unlabeled, stalled elongation complexes were chased with IPTG in the presence of [α-^{32}P]CTP and fractionated in a denaturing gel. The presence or absence of oligonucleotide C, RNase H, and washes of the magnetic beads with T-buffer are indicated above the scanned image, and the percent read-through (%RT) of the T_R' terminator is indicated below. The positions of the terminated (T_R') and run-off (RO) bands for uncleaved and cleaved transcripts are indicated to the left and right of the scanned image.

E. coli strain that expresses a plasmid-encoded RNAP β' subunit with a polyhistidine tag in its C-terminus.[39] RNAP stocks were 2 mg/ml ($\sim$4 μM).

Solution B: 200 μM of ATP, GTP, and UTP + 50 μM of CTP + 0.1 μM of [α-^{32}P]CTP (Amersham-Pharmacia Biotech) + 100 μg/ml heparin in T-buffer. This protocol labels the transcript between the start point and the roadblock. It is also possible to label the 5' end of the transcripts[40] or between the roadblock and the 3' end. For 3' labeling, include 4 unlabeled NTPs in solution B and 1 labeled and 3 unlabeled NTPs in the mixture added with IPTG (step 5). The latter protocol has the advantage that TECs that read through the roadblock or fail to restart will produce unlabeled transcripts.

T-buffer: We have used both chloride and glutamate containing T-buffers. Both worked equally well for making and manipulating the road-blocked complexes. Chloride T-buffer: 50 mM Tris–HCl, pH 8.0, + 10 mM MgCl + 100 mM KCl + 1 mM DTT + 100 μg/ml acetylated BSA (Ambion). Glutamate T-buffer: 20 mM Tris–glutamate, pH 8.0, + 10 mM Mg–glutamate + 50 mM K–glutamate + 1 mM DTT + 100 μg/ml acetylated BSA (Ambion).

Concluding Remarks

The ability to pause and restart TECs at preselected points on a template has been an invaluable tool for understanding transcript elongation and its control. Several studies have shown that the *lac* repressor is a versatile transcriptional roadblock *in vivo* and *in vitro*. Roadblocked TECs remain active and can be probed by a number of physical, chemical, and enzymatic methods. The exact state of the halted TECs depends on the sequences in the vicinity of the halt site,[41,42] and there is no guarantee that the roadblocked population will be homogeneous. Therefore results should be interpreted with caution.

[39] M. Kashlev, E. Nudler, K. Severinov, S. Borukhov, N. Komissarova, and A. Goldfarb, *Methods Enzymol.* **274,** 326 (1996).
[40] A. Das, M. Pal, J. G. Mena, W. Whalen, K. Wolska, R. Crossley, W. Rees, P. H. von Hippel, N. Costantino, D. Court, M. Mazzulla, A. S. Altieri, R. A. Byrd, S. Chattopadhyay, J. DeVito, and B. Ghosh, *Methods Enzymol.* **274,** 374 (1996).
[41] N. Komissarova and M. Kashlev, *Proc. Natl. Acad. Sci. USA* **94,** 1755 (1997).
[42] E. Nudler, A. Mustaev, E. Lukhtanov, and A. Goldfarb, *Cell* **89,** 33 (1997).

[16] Biochemical Assays of Gre Factors of *Thermus Thermophilus*

By OLEG LAPTENKO and SERGEI BORUKHOV

Prokaryotic transcript cleavage factors GreA and GreB affect the efficiency of transcription elongation *in vitro* and *in vivo* by stimulating the intrinsic nucleolytic activity of RNA polymerases (RNAPs).[1–5] The cleavage of nascent RNA is an evolutionarily conserved function among all multisubunit RNAPs.[4,6] It occurs in backtracked ternary complexes (TCs) of RNAP, typically 2–18 bases upstream from the 3′-terminus, followed by dissociation of the 3′-proximal RNA fragment and restart of transcription from newly generated 3′-terminus.[1,2,7,8] GreA induces hydrolysis of mostly dinucleotides and trinucleotides, whereas GreB induces cleavage of fragments of various lengths, depending on the extent of RNAP backtracking.[1,2,4] Biochemical studies of *Escherichia coli* Gre factors indicate that they are not nucleases but RNAP co-factors, which activate the same catalytic center involved in both RNA synthesis and RNA hydrolysis[4,6] reactions. The biological role of factor-induced endonucleolytic reaction includes: (1) the enhancement of transcription fidelity, by helping RNAP excise misincorporated nucleotides[9]; (2) suppression of transcriptional pausing and arrest by reactivation of RNAP during reversible and irreversible backtracking[1–5]; and (3) stimulation of RNAP promoter escape and transition from initiation to elongation stage of transcription by helping the catalytic center reengage with nascent RNA 3′-terminus during abortive synthesis.[10–12] The biological significance of Gre factors is underscored

[1] S. Borukhov, A. Polyakov, V. Nikiforov, and A. Goldfarb, *Proc. Natl. Acad. Sci. USA* **89,** 8899 (1992).

[2] S. Borukhov, V. Sagitov, and A. Goldfarb, *Cell* **72,** 459 (1993).

[3] M. T. Marr and J. W. Roberts, *Mol. Cell* **6,** 1275 (2000).

[4] S. M. Uptain, C. M. Kane, and M. J. Chamberlin, *Annu. Rev. Biochem.* **66,** 117 (1997).

[5] F. Toulme, C. Mosrin-Huaman, J. Sparkowski, A. Das, M. Leng, and A. R. Rahmouni, *EMBO J.* **19,** 6853 (2000).

[6] M. Orlova, J. Newlands, A. Das, A. Goldfarb, and S. Borukhov, *Proc. Natl. Acad. Sci. USA* **92,** 4596 (1995).

[7] N. Komissarova and M. Kashlev, *Proc. Natl. Acad. Sci. USA* **94,** 1766 (1997).

[8] E. Nudler, A. Mustaev, E. Lukhtanov, and A. Goldfarb, *Cell* **89,** 33 (1997).

[9] D. A. Erie, O. Hajiseyedjavadi, M. C. Young, and P. H. von Hippel, *Science* **262,** 867 (1993).

[10] G. H. Feng, D. N. Lee, D. Wang, C. L. Chan, and R. Landick, *J. Biol. Chem.* **269,** 22282 (1994).

[11] M. H. Hsu, N. V. Vo, and M. J. Chamberlin, *Proc. Natl. Acad. Sci. USA* **92,** 11588 (1995).

[12] A. Das, Unpublished data.

by the fact that Gre-homologs have been found in more than 60 bacterial organisms, including *Mycoplasma genitalium*, an organism with the smallest known genome,[13] the extreme thermophiles *Thermus aquaticus (Taq)*[14] and *Thermus thermophilus (Tth)*,[15] and *Dinoccocus radiodurans*,[16] an organism that survives high dosage of radiation.

All members of the Gre family are homologous polypeptides of ∼160 amino acids. They have similar structural organization and surface charge distribution, and are made of two domains: an N-terminal extended coiled-coil domain (NTD) and a C-terminal globular domain (CTD).[17–19] The NTD is responsible for the induction of type-specific nucleolytic activity by Gre factors, as well as read-through and antiarrest activities,[19,20] whereas CTD is responsible for the high-affinity binding of Gre to RNAP.[19–21] One of the distinct structural features of most Gre factors is the basic patch, a cluster of positively charged residues present on the side of the protein surface presumed to face RNAP in TC.[18] Basic patch residues are responsible for interaction of Gre factors with nascent RNA in TC and are required for efficient antiarrest activity.[17] The exact molecular mechanism by which Gre factors activate the catalytic center of RNAP and induce the RNA cleavage in TC is unknown. The current model of Gre action involves the following three essential steps: (1) the binding of Gre–CTD to RNAP near the secondary channel opening,[22] (2) the interaction of Gre–NTD's basic patch with the 3′-terminal portion of the transcript extruding from the secondary channel in backtracked TC, and (3) the activation of

[13] C. A. Hutchison, S. N. Peterson, S. R. Gill, R. T. Cline, O. White, C. M. Fraser, H. O. Smith, J. C. Venter, *Science* **286,** 2165 (1999).

[14] B. P. Hogan, T. Hartsch, D. A. Erie, *J. Biol. Chem.* **277,** 967 (2002).

[15] O. Laptenko, T. Hartsch, A. I. Bozhkov, S. Borukhov, *Biologicheskiy Vestnik* **4,** 3 (2000).

[16] O. White, J. A. Eisen, J. F. Heidelberg, E. K. Hickey, J. D. Peterson, R. J. Dodson, D. H. Haft, M. L. Gwinn, W. C. Nelson, D. L. Richardson, K. S. Moffat, H. Qin, L. Jiang, W. Pamphile, M. Crosby, M. Shen, J. J. Vamathevan, P. Lam, L. McDonald, T. Utterback, C. Zalewski, K. S. Makarova, L. Aravind, M. J. Daly, and C. M. Fraser, *Science* **286,** 1571 (1999).

[17] D. Kulish, J. Lee, I. Lomakin, B. Nowicka, A. Das, S. Darst, K. Normet, and S. Borukhov, *J. Biol. Chem.* **275,** 12789 (2000).

[18] C. E. Stebbins, S. Borukhov, M. Orlova, M. Polyakov, A. Goldfarb, and S. A. Darst, *Nature* **373,** 636 (1995).

[19] D. Koulich, M. Orlova, A. Malhotra, A. Sali, S. A. Darst, and S. Borukhov, *J. Biol. Chem.* **272,** 7201 (1997).

[20] D. Koulich, V. Nikiforov, and S. Borukhov, *J. Mol. Biol.* **276,** 379 (1998).

[21] A. Polyakov, C. Richter, A. Malhotra, D. Koulich, S. Borukhov, and S. A. Darst, *J. Mol. Biol.* **281,** 262 (1998).

[22] G. Zhang, E. A. Campbell, L. Minakhin, C. Richter, K. Severinov, and S. A. Darst, *Cell* **98,** 811 (1999).

RNAP's endonucleolytic activity via residue(s)-specific interaction between RNAP and Gre–NTD.

Until recently, *E. coli* GreA and GreB were the only Gre proteins studied in detail, both structurally and biochemically. Recently, two Gre homologs from *Thermus thermophilus*, GreA1[15] (GreA[14]) and GreA2[15] (Gfh1[14]), have been cloned and characterized independently in two laboratories.[14,15] GreA1 shows substantial sequence homology to both functional domains of *E. coli* GreA, whereas GreA2 displays limited homology, mostly to the CTD, and lacks several conserved elements characteristic for the NTD, most notably, the "GDLS(K)ENAEY" motif in the loop connecting two α-helices of NTD, which is essential for the induction of cleavage activity in TC.[23] Both Gre proteins bind specifically and competitively to *Tth* RNAP core and holoenzyme, but not to *E. coli* RNAP.[15] GreA1 stimulates RNA cleavage in TCs with specificity similar to that of *E. coli* GreA (cleavage and release of dinucleotides and trinucleotides) and generally facilitates transcription elongation.[15] In contrast to GreA1, GreA2 lacks any transcript cleavage activity; moreover, it inhibits GreA1 activity and the intrinsic (Gre-independent) nucleolytic activity of *Tth* RNAP,[14,15] as well as RNA synthesis and pyrophosphorolysis.[15] Thus GreA2 appears to function not just as an antagonist of GreA1, but also as a general inhibitor/repressor of RNAP. These activities of GreA2 suggest some implications regarding its possible biological role in the cell.

Most of the methods described earlier[24,25] for *in vitro* analysis of *E. coli* GreA and GreB can be applied to *Tth* GreA1, but not to GreA2. These include indirect Gre–RNAP binding assay,[25] transcript cleavage[24] and Gre–RNA photocross-linking[26] assays, and hydroxyl radical mapping of RNAP–Gre interactions.[26] Other assays, such as transcription read-through and antiarrest assays,[24] are not applicable because *Tth* RNAP (unlike *E. coli* RNAP) does not form arrested TCs during transcription on either T7A1 or *E. coli* rrnB P1 promoters.[15] Here we describe four new methods useful for biochemical and structure-functional studies of *Tth* Gre factors: (1) direct chromatographic assay for competitive binding of GreA1 and GreA2 to RNAP, (2) specific transcript cleavage (misincorporation–excision) assay for GreA1, (3) inhibition of RNA synthesis assay for GreA2, and (4) localized Fe^{2+}-induced hydroxyl radical mapping of GreA1 and GreA2 sites proximal to RNAP catalytic center.

[23] J. Lee, Unpublished data.
[24] S. Borukhov and A. Goldfarb, *Methods Enzymol.* **274,** 315 (1996).
[25] N. Loizos and S. A. Darst, *J. Biol. Chem.* **274,** 23378 (1999).
[26] S. Borukhov, O. Laptenko, and J. Lee, *Methods Enzymol.* **342,** 64 (2001).

Direct Gre–RNAP Binding Assay

The binding of Gre factors to RNAP can be visualized and quantified by a direct assay using native electrophoresis on PhastGels (Amersham-Pharmacia Biotech, Piscataway, NJ) of ^{32}P-labeled Gre protein, tagged with a phosphorylation site for heart muscle kinase (HMK), in complex with core RNAP (gel-shift assay).[25] This fast and simple method was first used to measure the binding affinity of *E. coli* GreB[25] and later applied for *E. coli* GreA[23] and *Tth* GreA1.[15] The apparent K_d values calculated for these factors by this method were ~87 nm, ~1 μm, and ~0.2 μm, respectively, and were in good agreement with those determined by other approaches.[19] However, despite its advantages, this method is not suitable for *Tth* GreA2 because of the higher hydrophobicity of this protein compared with *Tth* GreA1 or *E. coli* Gre factors. During electrophoresis, GreA2 tends to adsorb to the gel and plastic lining of the PhastGels, resulting in a visible smear on the autoradiogram and no discernible bands.

To complement the native gel-shift binding assay and to overcome some of its limitations, we developed a direct, semiquantitative Gre–RNAP binding assay based on the difference in the retention times of free [^{32}P]GreA2 or [^{32}P]GreA1 and the high molecular weight [^{32}P]Gre–RNAP complex during microanalytical size-exclusion high-performance liquid chromatography (HPLC).[19] The advantage of this method is that the protein samples (with and without carrier proteins and/or internal controls) can be applied over a wide range of concentrations. This assay allows us to estimate the binding affinity for both GreA1 and GreA2 and to demonstrate the competitive nature of their interactions with RNAP. The results of these experiments are illustrated in Fig. 1A and 1B, respectively. The major limitation of this method is that because of the multistep dilution/distribution nature of the gel-filtration process, the reliable detection and quantification of the complexes can be achieved only when the K_d values are less than 1–2 μM.

Procedure 1

Step 1. Tth Gre factors carrying N-terminal hexahistidine tag (6His) and site for HMK (LRRASV)[27] (6His-HMK-Gre) are purified as described[15] from *E. coli* strain BL21 (DE3)pLysS (Invitrogen, Carlsbad, CA) transformed with plasmids pET-N6HisHMK-GreA1 or pET-N6HisHMK-GreA2. Both plasmids were constructed by inserting the double-tagged *greA1* or *greA2* genes, which were PCR-amplified from *Tth* chromosomal DNA,[15] using appropriate primers, into NcoI/NdeI-linearized expression vector pET19b (Invitrogen).

[27] Z. Keiman, V. Naktinis, and M. O'Donnell, *Methods Enzymol.* **262,** 430 (1995).

A

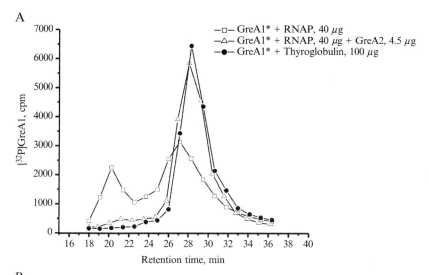

B

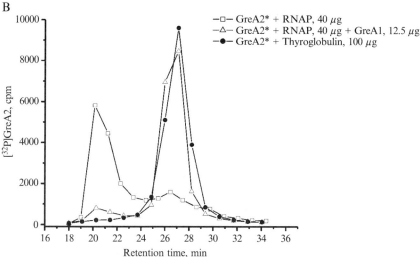

FIG. 1. Microanalytical size-exclusion HPLC of *Tth* Gre–RNAP complexes on Superdex 200HR 3 × 300 mm column. (A) Analysis of [^{32}P]GreA1–RNAP core complex and competitive displacement of [^{32}P]GreA1 by unlabeled GreA2. (B) Analysis of [^{32}P]GreA2–RNAP core complex and competitive displacement of [^{32}P]GreA2 by unlabeled GreA1.

Step 2. The purified Gre factors are radiolabeled in a 25-μl reaction containing 1.5 μg (3 μM) of double-tagged GreA1 or GreA2, 3 U of reconstituted HMK (P2645; Sigma, St. Louis MO), 0.3 μM [γ-^{32}P]ATP (3000 Ci/

mmol; ICN Pharmaceuticals, Costa Mesa, CA), 20 mM Tris–HCl, pH 7.9, 1 mM DTT, 0.1 M NaCl, and 10 mM MgCl$_2$. After the incubation (20 min at 30°), the radiolabeled proteins are supplemented with 10 mM EDTA and 0.5 mg/ml BSA and passed through QuickSpin G-50 desalting column (Roche Molecular Biochemicals, Indianapolis, IN) to remove free ATP and any remaining divalent metal ions. The column should be pre-equilibrated with buffer A (20 mM Tris–HCl, pH 7.9, 0.5 mM DTT, 5% glycerol, and 0.15 M NaCl). This procedure provides selective and specific phosphorylation and results in ~90% pure radiolabeled GreA1 and GreA2 (see Fig. 4). The typical yield of radiolabeled Gre is ~1 μg, 2.0–2.5 μCi. For all experiments, the radiolabeled Gre are diluted with the "cold" Gre proteins at different proportions to yield desirable specific radioactivity.

Step 3. Specific amounts of purified *Tth* RNAP core[28] (1–50 μg) are incubated for 20 min at 30°C with radiolabeled GreA1 or GreA2 (0.05–0.5 μg, 10^5 cpm) (with and without various amounts of competitors) in 20 μl of buffer A containing 0.5 mg/ml of BSA. In competition binding experiments the final concentration of radiolabeled Gre is 0.025 μg/μl (1.3 μM), *Tth* RNAP is 2 μg/μl (5.3 μM), and the concentrations of competitor Gre are 0.01–0.6 μg/μl (0.5–30 μM). In the control experiments, instead of *Tth* RNAP we used either thyroglobulin (Sigma) or an *E. coli* RNAP isolated from $greA^-$ $greB^-$ strain AD8571.[6] The samples (20 μl) are applied onto a microanalytical Superdex 200HR (3 × 30 mm) column (Amersham) and chromatographed at a flow rate of 70 μl/min in buffer A using an HPLC system HP1100 (Agilent Technologies, Palo Alto, CA). The fractions (1 drop, ~25 μl) are collected into 1.5-ml plastic tubes that contained 1 μl of 5% SDS (to prevent nonspecific adsorption of radiolabeled Gre to plastic). The fractions are incubated for 5 min at 70°, the radioactivity of a 5-μl aliquot is measured in liquid scintillation counter Beckman 6000IC (Beckman, Fullerton, CA), and the data are plotted as shown in Fig. 1. The chromatogram is integrated and the amount of Gre–RNAP complex is calculated based on the radioactivity in the area under the first radioactive peak with retention time ~20 min. The apparent K_d values for Gre–RNAP complexes are calculated graphically as described.[25]

Using this method, we estimated the apparent K_d values for GreA1 and GreA2 complexes with RNAP core to be ~0.4 μm and 0.05 μm, respectively. No complex formation was observed for GreA1 or GreA2 when *Tth* RNAP was substituted for thyroglobulin (with molecular mass of ~660 kDa) (see Fig. 1) or *E. coli* RNAP (data not shown).

[28] M. N. Vassylyeva, J. Lee, S. I. Sekine, O. Laptenko, S. Kuramitsu, T. Shibata, Y. Inoue, S. Borukhov, D. G. Vassylyev, S. Yokoyama, *Acta Crystallogr. D. Biol. Crystallogr.* **58,** 1497 (2002).

Specific Transcript Cleavage (Misincorporation–Excision) Assay

Gre factors were proposed to have a proofreading role in transcription.[9] It was suggested that when RNAP incorporates noncomplementary nucleotide into the growing end of the nascent RNA during transcription elongation, the enzyme may backtrack on the DNA template and become temporarily inactivated.[7,8] The backward translocation is accompanied by an extrusion of the 3'-terminal portion of RNA (which carries misincorporated nucleotide) from TC. This makes the backtracked TC susceptible to Gre, which induces the cleavage of the extruded RNA, resulting in the removal of misincorporated nucleotide and maintenance of transcription fidelity. According to this hypothesis, a TC that carries the correct RNA has a lower propensity to backtrack and therefore will be less susceptible to Gre. To test the ability of *Tth* GreA1 to discriminate between TCs with correct (perfectly matched) and incorrect (mismatched) RNA, and to induce RNA cleavage preferentially in the TCs carrying misincorporated nucleotide, we developed a specific transcript "misincorporation–excision" assay. For this assay, which is based on a modified protocol of Kashlev and co-workers,[29] we used two TCs, "correct" and "incorrect," both consisting of short promoterless DNA fragment, *Tth* RNAP core, and 17-meric RNA primer. In the correct TC, the 3'-terminal nine bases of RNA are fully complementary to the DNA template, whereas in the incorrect TC the DNA has a mismatch at the base corresponding to the RNA 3'-terminus (position -1). The extension of both TCs with GTP and $[\alpha^{32}P]$ATP results in the formation of 3'-terminally radiolabeled TCs: the correct (TC19) and the incorrect TC (TC19*), with a mismatch at position -3 of the RNA (Fig. 2A). As shown in Fig. 2B and 2C, in the presence of GreA1 the cleavage of RNA in incorrect TC19* occurs $\sim$4–5 times faster than in the correct TC19, supporting the initial hypothesis. At pH 7.6, in the absence of GreA1, the intrinsic cleavage reaction is clearly visible, although it occurs at a lower rate than in the presence of GreA1. Intriguingly, the intrinsic cleavage is significantly more efficient in TC19 than in TC19* (Fig. 2B and 2C), suggesting that mismatched hybrid renders the RNAP catalytic center less efficient in endonucleolytic function. Therefore in incorrect TCs, the effect of GreA1 is more dramatically manifest.

Procedure 2

Step 1. The RNA–DNA hybrid is prepared by incubation of 16 pmol of 17-meric RNA (AAUUGAGUAUCGAGAGG) with 8 pmol of 30-meric template DNA (DNA30 or DNA30*, see Fig. 2A) in 24 μl of standard

[29] I. Sidorenkov, N. Komissarova, M. Kashlev, *Mol. Cell* **2**, 55 (1998).

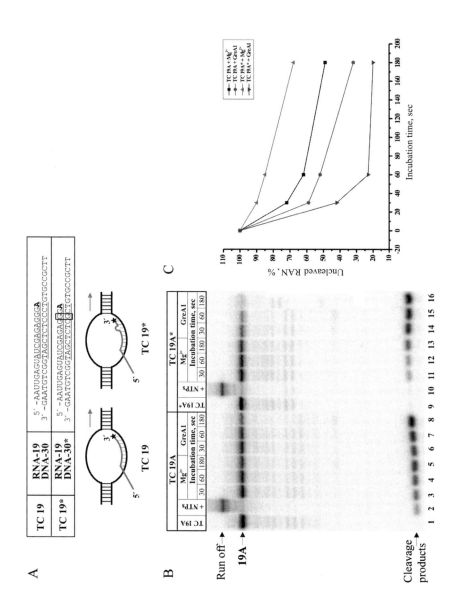

transcription buffer (STB: 40 mM Tris–HCl, pH 7.6, 40 mM NaCl, 1 mM DTT, 5% glycerol, and 10 mM MgCl$_2$) containing 0.5 mg/ml BSA for 5 min at 60°. After cooling the reaction mixture to room temperature for 10 min, it is supplemented with 2 μl of *Tth* RNAP core (16 pmol) in STB and incubated for 10 min at 30 °C. A solution of 160 pmol of 30-meric non-template DNA (complementary to template DNA30) in 12 μl of STB is added to the reaction mixture, followed by additional incubation for 10 min at 37 °C, resulting in formation of "cold" TC17 and TC17*.

Step 2. TCs are further extended by two nucleotides by addition of GTP and [α^{32}P]ATP (3000 Ci/mmol) in 12 μl of STB to a final concentration of 10 μm and 1 μm, respectively. The reaction mix is incubated for 10 min at 37 °C and terminated by addition of EDTA to a final concentration of 20 mM. The resulting radiolabeled TC19 and TC19* are purified by gel filtration on a Quick-Spin G-50 column (Roche Molecular Biochemicals) in STB lacking MgCl$_2$.

Step 3. 1 μl of the purified TCs are supplemented with 9 μl of STB with or without 30 ng of GreA1 and incubated at 37 ° for various time periods, as indicated in Fig. 2B. For the run-off transcription, 1 μl of TC19 and TC19* is incubated in 10 μl of STB containing 100 μM of all four (NTPs) for 10 min at 37°C. Reactions are terminated by addition of 1 μl of 0.5 M EDTA and 10 μl of electrophoresis loading buffer (90% v/v of formamide in 50 mM Tris–borate buffer, pH 8.3, containing 10 mM EDTA). The radioactive products of the cleavage reaction are separated by denaturing 23% polyacrylamide gel electrophoresis (PAGE) in the presence of 8 M urea, visualized by autoradiography and quantitated using PhosphorImager (Storm, Amersham, Piscataway, NJ).

Assays for Inhibitory Activity of GreA2

Tth GreA2 has been shown to be a competitive inhibitor of GreA1 in several transcript cleavage assays based on *E. coli* ribosomal *rrnB* P1, phage T7A1, and phage λ P$_R$ promoters.[14,15] These assays have previously

---

FIG. 2. Specific transcript cleavage (misincorporation–excision) assay. (A) Schematic representation of a synthetic promoterless DNA template system with partially hybridized 19-meric RNA primer used for preparation of "correct" TC19 and "incorrect" TC19*. The mismatched pair of nucleotide bases in TC19* is boxed. The complementary regions in 30-meric DNA template strand and in 19-meric RNA primer are underlined. The nontemplate strand is not shown. Bold letter symbolizes radioactive nucleotide at the RNA 3' terminus. (B) Preferential cleavage of "incorrect" TC19* by GreA1. An autoradiogram of urea–PAGE (23% [w/v]) analysis of the RNA cleavage products. The electrophoretic mobility for initial 19A RNA, run-off product, and the 3'-terminal cleavage product (most likely a trinucleotide, pGpGpA) is indicated. (C) Graphic representation of the time course of the cleavage reaction in TC19 and TC19* in the presence of Mg^{2+}, alone or together with GreA1 (see Procedure 2 for details).

been described in detail. To demonstrate that GreA2 is a general inhibitor of the catalytic activity of RNAP, we applied two modified standard assays performed on TCs generated on T7A1 promoter: the intrinsic transcript cleavage and run-off transcription assays (illustrated in Fig. 3A and 3B, respectively).

The cleavage assay consists of the incubation of TC carrying radio-labeled 21-meric RNA (TC21U) in the presence of Mg^{2+} ions, either alone or with GreA2 added to TC21U at various concentrations (Fig. 3A). Similar to *E. coli* RNAP, the intrinsic nucleolytic activity of *Tth* RNAP is pH-dependent and requires mild alkaline pH for induction of Gre-independent cleavage. However, compared with *E. coli* enzyme (pK 8.3[6]), its titration curve is more shallow, the pK is shifted toward acidic pH ($\sim$7.2[15]), and at pH below 6.8 the intrinsic activity of *Tth* RNAP is negligible. Thus for efficient cleavage of *Tth* TCs, we recommend that the pH of the reaction should be 7.5 or higher. As shown in Fig. 3A, GreA2 inhibits the intrinsic cleavage activity of *Tth* RNAP (compare lanes 2–6), but not that of *E. coli* RNAP (data not shown). Under given conditions, a 50% inhibition of transcript cleavage is achieved at GreA2 concentrations of $\sim$30 ng/μl ($\sim$1.5 μm).

The run-off transcription assay is a time course of the chase reaction when the starting TC carrying radiolabeled 20-meric RNA (TC20A) is incubated with four NTPs, either alone or in the presence of various amounts of GreA2 (Fig. 3B). Under chosen conditions of limited concentrations of NTPs (5 μM), *Tth* RNAP tends to pause significantly at positions 21U, 25C, and 27U, reaching the expected 50-mer run-off product only after 15 min incubation (lanes 1–4). The addition of GreA2 results in both the prolonged duration of the existing pauses and the appearance of new pause sites, most notably, around position 30G (lanes 2–15). From the pattern of intermediate RNA products, it appears that GreA2 stimulates pausing and strongly inhibits incorporation of NTPs and/or RNAP translocation along the DNA template. Typically, a 50% inhibition of RNA synthesis is achieved at GreA2 concentrations of $\sim$10 ng/μl ($\sim$0.5 μM) (data not shown).

Procedure 3

Step 1. The initial radiolabeled TC20A is prepared by incubation of 2 pmol of 375 bp-long DNA fragment bearing T7A1 promoter (end points from $-$325 to $+$50) with 4 μg (10 pmol) of *Tth* RNAP holoenzyme,[28] 1 mg/ml of BSA, 0.5 mM ApU (Sigma), 10 μM each of ATP, GTP, and CTP, and 0.5 μM [α-^{32}P]ATP (3000 Ci/mmol; ICN) in 50 μl of STB for 15 min at 60°. The reactions are stopped by addition of 2 μl of 0.5 M

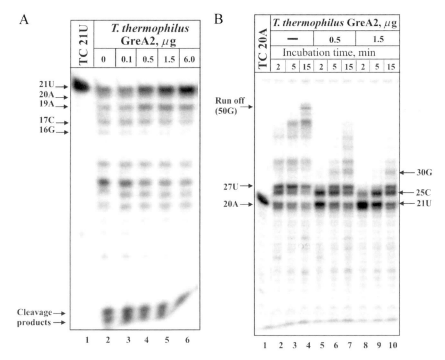

FIG. 3. Inhibitory activity of GreA2. (A) Intrinsic transcript cleavage assay. An autoradiogram of urea–PAGE (23% [w/v] analysis of the RNA cleavage products induced in TC21U at pH 7.5 in the presence and absence of GreA2. The electrophoretic mobility of the initial 21U RNA and some of the 5′-terminal RNA degradation products are indicated by arrows. The cleavage products visible in the bottom of the gel are most likely the 3′-proximal dinucleotides. (B) Run-off transcription assay. An autoradiogram of urea–PAGE (20% [w/v]) analysis of RNA extension reaction for the initial TC20A incubated with four NTPs in the absence or presence of GreA2. The mobility of the starting 20A RNA, the paused 21U, 25C, 27U, and 30G, and the run-off transcript (50G) is indicated by arrows (see Procedure 3 for details).

EDTA and kept on ice. The TC20A is further purified by gel filtration on a Quick-Spin G-50 column (Roche Molecular Biochemicals). The typical yield of TC20A is ~0.5 pmol (~25% based on the initial amount of DNA used in the reaction).

Step 2. The radiolabeled TC20A (0.2 pmol) is incubated in the presence of 10 μm UTP in 50 μl of STB at pH 7.0 for 5 min at 60°, and the resulting TC20U is supplemented with 1 mg/ml BSA and purified on a Quick-Spin G-50 column (Roche Molecular Biochemicals) as before.

Step 3. For the cleavage assay, 10 fmol of the resulting TC21U are incubated in 10 μl of STB alone, or in the presence of different amounts of

GreA2 (0.01–0.6 $\mu g/\mu l$, 0.5–30 μm) for 2, 5, and 15 min at 60°. Alternatively, for the run-off assay, 20 fmol of the initial radiolabeled TC20A are incubated in 10 μl of STB containing 5 μM NTPs alone, or in the presence of different amounts of GreA2 (0.01–0.6 $\mu g/\mu l$, 0.5–30 μm) for 2, 5, and 15 min at 60°. All reactions are terminated and analyzed by denaturing PAGE as described previously in Step 3 of Procedure 2.

Localized Fe^{2+}-Induced Hydroxyl Radical Mapping of GreA1 and GreA2 Sites Proximal to RNAP Catalytic Center

To identify the sites on Gre factors located near the enzyme's active center in Gre–RNAP complex, we applied Fe^{2+}-induced hydroxyl radical mapping analysis. This method is based on Fenton reaction,[30] and it exploits the property of chelated Fe^{2+} ion to generate highly reactive hydroxyl radicals in aqueous solutions in the presence of molecular oxygen and reducing agent. The hydroxyl radicals cause cleavages in biopolymers with a diffusion-limited rate in the distance range of 10–15 Å. By exchanging the catalytic Mg^{2+} ion in the active center of RNAP with Fe^{2+}, specific and highly localized cleavages of protein, DNA, and nascent RNA can be introduced in the immediate vicinity of the catalytic center.[31,32]

Recently we reported a method for detection of possible perturbations within the RNAP active center caused by interactions with Gre factors.[26] We have subjected *E. coli* RNAP to localized Fe^{2+}-induced hydroxyl radical mapping in both the presence and absence of GreA and GreB.[26] To visualize protein cleavages and facilitate mapping, we used N-terminally radiolabeled β' subunit. We have shown that the presence of Gre factors, especially GreB, results in specific and reproducible alterations of the β' cleavage pattern. These alterations include protection and enhancement of cleavage sites located mostly within the evolutionarily conserved regions G and F factors.[26] These regions comprise the buried part of the RNAP secondary channel located 10–15 Å away from the catalytic center factors,[33] and could serve as effective targets (direct or allosteric) for GreA and GreB action.

[30] M. A. Price and T. D. Tullius, *Methods Enzymol.* **212**, 194 (1992).

[31] E. Zaychikov, E. Martin, L. Denissova, M. Kozlov, V. Markovtsov, M. Kashlev, H. Heumann, V. Nikiforov, A. Goldfarb, and A. Mustaev, *Science* **273**, 107 (1996).

[32] A. Mustaev, M. Kozlov, V. Markovtsov, E. Zaychikov, L. Denissova, and A. Goldfarb, *Proc. Natl. Acad. Sci. USA* **94**, 6641 (1997).

[33] D. G. Vassylyev, S. I. Sekine, O. Laptenko, J. Lee, M. N. Vassylyeva, S. Borukhov, S. Yokoyama, *Nature* **417**, 712 (2002).

Assuming that Gre factors interact with RNAP in the vicinity of regions G and F at a distance of ~15 Å or less from the enzyme's catalytic center, we subjected *Tth* and *E. coli* Gre factors to localized Fe^{2+}-induced hydroxyl radical mapping in the presence and absence of cognate RNAPs. To visualize protein cleavages and facilitate mapping, we used N-terminally radiolabeled GreA1 and GreA2. The results of these experiments are shown in Fig. 4. Both *Tth* Gre factors undergo efficient hydroxyl radical cleavages only in the presence of *Tth* RNAP (lanes 3 and 6), but not *E. coli* RNAP (data not shown). Conversely, no cleavage of radiolabeled *E. coli* GreA was observed in the control reaction with *Tth* RNAP (lane 9), indicating that the hydroxyl radical cleavages in Gre factors are specific and reflect the complex formation with RNAP. Using chemical and enzymatic degradation of the radiolabeled Gre, we mapped the location of the cleavage site in both GreA1 and GreA2. In both proteins, the cleavage occurs within 6 residues of the tip of NTD (data not shown). We propose that the interactions of the NTD tip with regions F and G cause allosteric changes at the active center of RNAP, resulting in the activation of the dormant RNase activity of this enzyme.

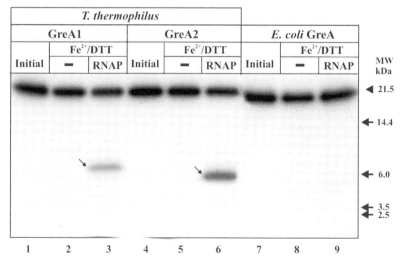

FIG. 4. Localized Fe^{2+}-induced hydroxyl radical mapping of *Tth* GreA1 and GreA2 in complex with RNAP. An autoradiogram of Tris–Tricine SDS gradient–PAGE (10–20%) shows the products of localized Fe^{2+}-induced hydroxyl radical cleavages (marked on the gel by arrows) of [^{32}P]GreA1 (lane 3) and [^{32}P]GreA2 (lane 6) in complex with *Tth* RNAP core. In the control reaction, N-terminally radiolabeled *E. coli* GreA is incubated with *Tth* RNAP (lane 9). Molecular weight standards (Mark 12; Invitrogen) are shown at right by arrowheads.

Procedure 4

Step 1. Purification and radiolabeling of 6His-HMK-GreA1 and 6His-HMK-GreA2 are performed as described in Steps 1 and 2 of Procedure 1. To remove free radioactive ATP and traces of divalent metal ions, the radiolabeled proteins are passed through Quick-Spin G-50 desalting column (Roche Molecular Biochemicals) preequilibrated with chelated[31] 20 mM Na–HEPES buffer, pH 7.5, containing 50 mM NaCl. The typical yield of radiolabeled GreA1 and GreA2 is 1 μg, 2–2.5 μCi. *Tth* and *E. coli* RNAPs (3 mg/ml), and BSA (6 mg/ml) are purified similarly.

Step 2. Fe^{2+}-mapping reactions are performed at 25° for 10 min in 10-μl reactions containing 20 ng of radiolabeled Gre factor (0.1 μm), 7 μg BSA (1 μm) with or without 2.5 μg of purified RNAP (0.6 μm), 20 mM Na–HEPES buffer, pH 7.5, and 0.2 M NaCl. After 5 min of preincubation at 25°, reactions are initiated by adding 1 μl of freshly made 500 μm $(NH_4)_2$-$Fe(SO_4)_2$ (Sigma) and 1 μl of 10 mM DTT to an 8 μl-mixture of $[^{32}P]$Gre with BSA alone or with RNAP. In the control experiments, inhibitors (10 mM $MgCl_2$ or 10 mM EDTA) should be included to verify the specificity of Fe^{2+}-induced cleavage reaction.

Step 3. Reactions are terminated by addition of 1 μl of 0.5 M EDTA and 10 μl of 2 × SDS-gel sample loading buffer containing 10% β-mercaptoethanol, followed by boiling for 2 min. The products of Fe^{2+}-induced cleavage reactions are resolved by Tris–Tricine/SDS 10–20% gradient PAGE (Invitrogen), visualized by autoradiography and analyzed by PhosphorImager (Storm Molecular Dynamics). Molecular weight markers for GreA1 and GreA2 are generated as described[19] by residue-specific cleavages of the radiolabeled Gre at Cys, Met, Trp, and Asp under limited cleavage conditions.

Acknowledgments

Authors are grateful to Dr. J. Lee for critical comments and help in preparation of the manuscript. The ongoing work in the laboratory is supported by the NIH grant GM54098-07 to S.B.

[17] Engineering of Elongation Complexes of Bacterial and Yeast RNA Polymerases

By Natalia Komissarova, Maria L. Kireeva, Jodi Becker, Igor Sidorenkov, and Mikhail Kashlev

Certain DNA sequences induce pausing, arrest, or termination of transcription[1] modulating catalytic activity and stability of the elongation complex (EC) between RNA polymerase (RNAP), template DNA, and nascent RNA.[2] The ECs of bacterial RNAP and eukaryotic RNAP II (Pol II) have similar structure, in which the enzyme covers 30–35 nucleotides (nt) of the double-stranded DNA containing ~12-nt melted segment called the transcription bubble.[2,3] Between 8–9 nt of the 3'-proximal RNA hybridize with the template DNA strand within the bubble.[4,5] The transcript exits RNAP at a distance of 14 nt from its 3' end;[6] therefore, 5–6 nt of single-stranded RNA upstream from the hybrid are located inside the enzyme.[7] Although RNAP forms multiple contacts with RNA and DNA within the protected regions, only RNA:DNA hybrid and 9–12 nt of the DNA duplex downstream from the RNA 3' end are needed for high stability of bacterial EC.[8] Surprisingly, the downstream DNA duplex is dispensable for stability of the EC formed by yeast Pol II.[9]

Elongation of RNA is accompanied by stepwise forward translocation of RNAP along the template. In addition, RNAP is capable of backward movement, which is induced by degradation of the transcript from the 3' end, either by pyrophosphorolysis[10] (a reaction reverse to nt addition) or by endonucleolytic cleavage stimulated by protein factors GreA and GreB.[11] Also, at certain DNA sequences RNAP moves backward along the RNA and DNA without any shortening of the transcript.[12] This translocation, or

[1] R. A. Mooney, I. Artsimovitch, and R. Landick, *J. Bacteriol.* **180,** 3265 (1998).
[2] P. H. von Hippel, *Science* **281,** 660 (1998).
[3] E. Zaychikov, L. Denissova, and H. Heumann, *Proc. Natl. Acad. Sci. USA* **92,** 1739 (1995).
[4] N. Korzheva, A. Mustaev, M. Kozlov, A. Malhotra, V. Nikiforov, A. Goldfarb, and S. A. Darst, *Science* **289,** 619 (2000).
[5] A. L. Gnatt, P. Cramer, J. Fu, D. A. Bushnell, and R. D. Kornberg, *Science* **292,** 1876 (2001).
[6] N. Komissarova and M. Kashlev, *Proc. Natl. Acad. Sci. USA* **95,** 14699 (1998).
[7] M. M. Hanna, E. Yuriev, J. Zhang, and D. L. Riggs, *Nucleic Acids Res.* **27,** 1369 (1999).
[8] I. Sidorenkov, N. Komissarova, and M. Kashlev, *Mol. Cell* **2,** 55 (1998).
[9] M. L. Kireeva, N. Komissarova, D. S. Waugh, and M. Kashlev, *J. Biol. Chem.* **275,** 6530 (2000).
[10] T. A. Rozovskaya, A. A. Chenchik, and R. Beabealashvilli, *FEBS Lett.* **137,** 100 (1982).
[11] S. Borukhov, V. Sagitov, and A. Goldfarb, *Cell* **72,** 459 (1993).
[12] N. Komissarova and M. Kashlev, *Proc. Natl. Acad. Sci. USA* **94,** 1755 (1997).

backtracking, disengages the RNAP catalytic center from the RNA 3' end and thereby inactivates, or arrests, the EC.[13] These two types of movement, along with dissociation of the EC in termination, are the principal processes that modulate the level of elongation.

To understand the mechanisms by which DNA sequences and protein factors regulate elongation, we need to identify the components of the EC that are targeted by regulatory signals. For this purpose, we have to systematically alter the protein, DNA, and RNA components of the EC. The recent resolution of RNAP structure allows targeted mutagenesis of separate domains in the enzyme.[5] It is also easy to modify the DNA template. The challenging task that remains is to address the role of the transcript sequence and structure. Generally, all modifications in the RNA should be encoded in the DNA, which by itself may affect the properties of the EC. In this chapter, we describe posttranscriptional modifications of the RNA in the EC, utilized to introduce changes into the RNA without altering the sequence of the DNA template and to address the effects of transcript composition on EC stability and catalytic activity.[6,9,14] Another set of experimental techniques is based on the reconstitution (assembly) of the EC from RNAP and synthetic RNA and DNA oligonucleotides.[8] We found this approach extremely useful for the introduction of mismatches into the RNA, and for obtaining ECs with RNAs shorter than 8 nt. Experiments with assembled ECs allowed us to address the specific roles of the RNA and the DNA within the regions protected by the enzyme, which could not be accomplished using promoter-initiated ECs.

Materials and Reagents

Buffers, Enzymes, and Reagents

Most reactions are performed in transcription buffer (TB): 20 mM Tris–HCl, pH 7.9, 40 mM KCl, 5 mM MgCl$_2$, 1 mM β-mercaptoethanol. In some cases, KCl concentration in TB varies from 5 to 1000 mM, and it is indicated in parentheses [e.g., TB(300) is TB containing 300 mM KCl]. Gel-loading buffer for denaturing PAAG contains 10 M urea, 50 mM EDTA, pH 7.9, 0.05% of bromophenol blue, and xylene cyanol. Gel-loading buffer for nondenaturing PAAG contains 50% (v/v) glycerol, 50 mM EDTA (pH 7.9), 0.25 mg/ml sheared salmon testes DNA (Sigma), 0.05% of bromophenol blue, and xylene cyanol. TGED for the assembled EC

---

[13] N. Komissarova and M. Kashlev, *J. Biol. Chem.* **272,** 15329 (1997).

[14] M. L. Kireeva, N. Komissarova, and M. Kashlev, *J. Mol. Biol.* **299,** 325 (2000).

purification contains 10 mM Tris–HCl, pH 7.9, 0.5 mM EDTA, 5% (v/v) glycerol, and 0.1 mM β-mercaptoethanol.

RNase T1 is from Boehringer Mannheim (Indianapolis, IN), and RNase A is from Sigma (St. Louis, MO). The stock solutions of RNase A and T1 can be stored for several months at $-20°$ and $+4°$, respectively. Prime RNase inhibitor is from 5Prime->3Prime (Boulder, CO), and RNase T1 inhibitor guanosine 2′-monophosphate is from Sigma. GreB protein is purified as described and stored at $+4°$.[15] Bacteriophage phage T4 polynucleotide kinase, T4 DNA ligase, and Quick Ligation kit are from New England Biolabs (Beverly, MA). NTPs are from Boehringer Mannheim, $[\alpha^{-32}P]$NTPs (3000 Ci/mmol) are from New England Nuclear (Boston, MA), and $[\gamma^{-32}P]$ATP (7000 Ci/mmol) is from ICN Pharmaceuticals (Costa Mesa, CA). All other chemicals are from Sigma or Fisher Scientific (Chicago, IL). Ni^{2+}-NTA agarose is from Qiagen (Chatsworth, CA), and streptavidin agarose is from Sigma.

RNA Polymerases

Hexahistidine-tagged *E. coli* RNAP is purified from RL721 strain (provided by Dr. R. Landick) as described[16] and stored at $-20°$. Hexahistidine-tagged Pol II from *S. cerevisiae* is purified as described in this volume[17] and stored at $-70°$ in small aliquots.

Promoter DNA Templates and RNA/DNA Oligonucleotides

Linear promoter templates 150–300 nt long are obtained by polymerase chain reaction (PCR) with nonphosphorylated primers and purified through agarose or nondenaturing PAAG. The left primer can carry biotin group at the 5′ end for immobilization of the EC on streptavidin agarose.

RNA and DNA oligonucleotides for PCR and EC assembly are from Oligos, Etc., Inc. (Wilsonville, OR), and are dissolved in H_2O to 1.5 mM, and stored at $-70°$. For routine use, the oligos are diluted to 15 μm with TB and stored at $-20°$ for 1–2 months. Some oligos require additional purification by denaturing polyacrylamide gel electrophoresis (PAGE) or high-performance liquid chromatography (HPLC).

---

[15] S. Borukhov and A. Goldfarb, *Methods Enzymol.* **274,** 315 (1996).

[16] M. Kashlev, E. Nudler, K. Severinov, S. Borukhov, N. Komissarova, and A. Goldfarb, *Methods Enzymol.* **274,** 326 (1996).

[17] M. L. Kireeva, L. Lubkowska, N. Komissarova, and M. Kashlev, *Methods Enzymol.* **370,** 138 (2003).

Equipment

Small 6-tube table-top centrifuge (Labnet Intl., Inc., Woodbridge, NJ); large orifice pipette tips and 1.5-ml nonstick Eppendorf tubes (USA Scientific Plastics., Ocala, FL) for handling complexes bound to Ni^{2+}-NTA agarose; Centricon-100 Filter Device (Millipore Corp., Bedford, MA); Superose 12 HR 10/30 column and 1-ml disposable Heparin column (Amersham-Pharmacia Biotech, Piscataway, NJ).

Promoter Initiation, Walking with RNAP, and 5′/3′ Truncation of Nascent RNA

Our approach is based on the immobilized *in vitro* transcription system,[9,16] which utilizes RNAP carrying a genetically attached six-histidine tag. RNAP and the EC can be immobilized on Ni^{2+}-NTA agarose beads through this tag. The presence of the tag and the immobilization do not affect the biochemical properties of RNAP and its EC. Alternatively, we immobilize the EC through the DNA on streptavidin agarose beads. For this purpose, a biotin group is introduced to the template.

The EC immobilization allows an easy exchange of the reaction components used for obtaining a specific EC and its analysis, including NTP subsets used for limited extension of the transcript (walking), enzymes (DNase, RNases, etc.) and chemicals (e.g., footprinting agents and pyrophosphate).

Obtaining Defined ECs from Promoters: Basic Protocol

Initiation of Transcription from Promoter with E. coli *RNAP*

All procedures are performed at $24°$, unless indicated otherwise.

1. In 5–20 μl of TB, incubate equimolar amounts (0.5–2 pmol) of RNAP holoenzyme and a linear DNA fragment containing a strong *E. coli* promoter for 5 min at $37°$. To obtain the halted EC, add a di-, tri-, or tetranucleotide RNA primer (corresponding to $-1/+3$ positions around the start site of transcription) to final concentration 20–100 μm, and a subset of NTPs to 50 μm. Transcription will stop before the template position that encodes the first nt missing from the NTP subset. The initially transcribed sequence should allow halting of the ECs at 10–30 nt downstream from the start site in the presence of a subset of NTPs in which one or two NTPs are missing.[16]

2. Wash Ni^{2+}-NTA agarose beads with TB by five cycles of centrifugation in a table-top low-speed microcentrifuge (5–10 s each cycle) and

resuspend in 1 ml of fresh buffer. To immobilize the EC halted downstream from the promoter, incubate it for 5 min with 20–40 μl of washed Ni^{2+}-NTA agarose (50% suspension, v/v). Wash the immobilized EC five times with TB to remove the excess RNA primer and NTPs. Leave 100 μl of TB in the tube after each centrifugation to prevent accidental withdrawal of the beads with the washing fluid. *Note*: (1) To achieve complete immobilization, increase the time of incubation of the EC with the beads to 30 min, with intense shaking, but beware of RNAP reading through the halting site. (2) If an excessive amount of RNAP is used for initiation, agarose beads may stick to the sides of the tube. In this case the RNAP-to-agarose ratio needs to be changed.

3. Alternatively, the EC can be immobilized on streptavidin agarose beads using a biotin group at the upstream end of the DNA. Handling of the beads and the immobilization of the EC is the same as described for Ni^{2+}-NTA agarose. The advantage of this method is that immobilization of the EC through the DNA separates true ECs from artificial binary complexes between RNAP and nascent RNA, which may form after dissociation of an unstable EC.[18]

4. The RNA and the DNA in the immobilized EC can be labeled, RNAP can be walked further downstream, and the complex can be treated with ribonucleases and other agents as described later. Most of the immobilized ECs (except those prone to arrest and dissociation) can be stored on ice for up to a week. Before using such a sample after the storage, briefly wash it with 1 ml of TB(1000) to remove the RNA and DNA from the dissociated ECs.

LABELING OF THE TRANSCRIPT IN THE EC. The transcript in the EC can be labeled in one of the following three ways.

1. To label an internal site in the RNA, first bring the volume of the Ni^{2+}-NTA agarose or streptavidin agarose suspension with the immobilized EC to minimum. Add the appropriate $[\alpha-^{32}P]NTP$ (one tenth of the suspension volume) for 10 min and wash the EC. To increase the efficiency of labeling, incorporate radioactive NTPs in several consecutive RNA positions by incubating the EC with a subset of $[\alpha-^{32}P]NTP$.

2. To label the RNA at the 5′ end, halt the immobilized EC more than 15 nt downstream from the start point of transcription to expose the 5′ end of the transcript out of RNAP. Incubate the EC with 2 μm $[\gamma-^{32}P]ATP$ and 10 U of phage T4 polynucleotide kinase for 10 min. 5′ ends of the template will also be labeled, which is useful for footprinting analysis of the EC.[8,9,14]

---

[18] M. Kashlev and N. Komissarova, *J. Biol. Chem.* **20,** 20 (2002).

3. Alternatively, prelabel the 5' end of the RNA primer before initiation. In 20 μl of TB, incubate the primer with 2 μm [γ-^{32}P]ATP and 10 U of T4 polynucleotide kinase for 10 min at 37°, followed by 20 min at 65° to inactivate the kinase. If ATP should be absent from the start reaction, purify the labeled RNA primer from nonincorporated [γ-^{32}P]ATP. In this case, because the concentration of the labeled primer in the initiation efficiency will be lower than the optimal 20–100 μm, the yield of the EC will decrease.

WALKING WITH THE IMMOBILIZED EC ALONG THE DNA. To transcribe the DNA in controlled steps, adjust the volume of the suspension to 100 μl with TB and add the appropriate subset of NTPs (10 μm each) for 3 min. Wash five times. Repeat as many times as needed with different NTP subsets.[6,12,13]

ANALYSIS OF EC ON DENATURING AND NONDENATURING PAAG.

1. Before loading samples on a denaturing PAAG, add an equal volume of gel-loading buffer and incubate the sample at 90° for 2 min. This procedure is used for the analysis of RNA and DNA in the EC immobilized on Ni^{2+}-NTA agarose and for the analysis of the RNA in the complexes immobilized on streptavidin agarose.

2. Before loading samples on a nondenaturing PAAG, elute the EC from Ni^{2+}-NTA agarose by a 10-min incubation with 100 mM imidazole, and add nondenaturing gel-loading buffer (one fifth of the volume).

Comments

1. The DNA sequence immediately downstream from the promoter (we routinely use A1 promoter from bacteriophage T7) tolerates some mutations with no effect on the EC formation. However, the highest yield of the specific EC on each sequence may require different concentrations of the primer and NTPs, and different incubation time. Generally, decreasing the volume of the initiation reaction results in a higher yield.

2. Longer RNAs are sometimes observed along with the product of the expected length after the initiation and during the walking. This is caused by reading through the halting site because of misincorporation in the RNA of an NTP, which is not encoded in the template, or because of cross-contamination of the NTPs. Also, a shorter-than-expected RNA may be obtained, which probably originates from the anomalous stable complexes with short RNAs containing σ subunit.[19]

[19] S. Borukhov, V. Sagitov, C. A. Josaitis, R. L. Gourse, and A. Goldfarb, *J. Biol. Chem.* **268,** 23477 (1993); B. Krummel and M. J. Chamberlin, *Biochemistry* **28,** 7829 (1989).

3. Sometimes, instead of a homogeneous transcript, multiple RNA species are produced after the RNA labeling with several radioactive NTPs. This may be caused by the low concentration of $[\alpha\text{-}^{32}P]$NTPs (3 μm stock produces 0.3 μm in the reaction). Thus at some positions RNA cannot be completely elongated. Subsequent incubation with the same subset of unlabeled NTPs taken in greater concentration should extend the shorter products to the expected length.

4. The RNA can be labeled internally with $[\alpha\text{-}^{32}P]$NTP at any step of walking. The efficiency of labeling and the homogeneity of the transcript may depend on the exact position where the radioactive nt is introduced. It is worth trying different template positions for labeling.

5. Pausing or arrest of transcription at certain sites in the course of walking may generate the shorter RNA species. Increased incubation time and higher NTP and KCl (up to 1 M) concentrations may help obtain a uniform RNA. Also, avoid stopping at the arrest-prone sites. The best walking strategy must be developed experimentally for each DNA sequence.[16]

6. The amount of initial EC can decrease during walking due to spontaneous dissociation of ECs at some template positions. Transcription through the "risky" sites without stopping should prevent the loss of the EC.

For a typical example of obtaining a defined EC and of walking along the DNA, see Fig. 1. Figure 1 also illustrates how to determine stability of an EC in the immobilized transcription system by separation of the samples into supernatant and pellet fractions, over timed intervals.

Truncation of Nascent RNA from 5' End in the EC with RNase T1 and RNase A

Treatment of the ECs with C/U-specific RNase A and G-specific RNase T1 allows alteration of the structure and length of the nascent transcript in the complexes. These RNases degrade the 5' end of the transcript to the "minimal" 14-nt fragment because the 3'-proximal region of the RNA is protected by RNAP. Therefore the site where the EC is halted in a course of walking defines the cleavage site in the RNA.[6,20] RNases and the digested 5'-proximal RNA can be washed off the immobilized EC, and walking with RNAP can be resumed. This method can be applied not only to promoter-initiated ECs but also to assembled ECs of *E. coli* RNAP and the yeast Pol II (see next section).

---

[20] N. Komissarova, J. Becker, S. Solter, M. L. Kireeva, and M. Kashlev, *Mol. Cell* **10,** 1151 (2002).

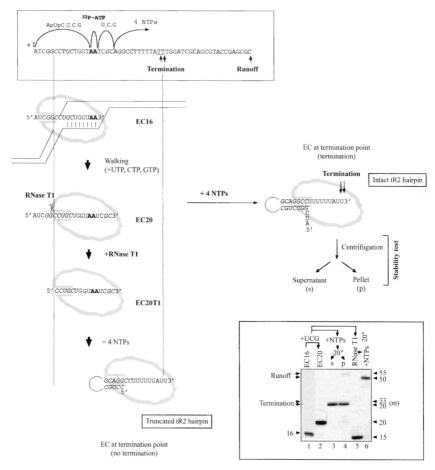

FIG. 1. Walking with immobilized *E. coli* RNAP and 5′ RNA truncation in EC. Transcription was carried out on 124-nt linear DNA fragment containing the A1 promoter of bacteriophage T7 and *rho*-independent tR2 terminator of bacteriophage lambda (the transcribed sequence of the nontemplate DNA strand and walking strategy are shown at top). First, the EC was halted at 14 nt downstream from the start point of transcription (EC14; the number indicates the length of the RNA). It was obtained by 5 min incubation of 2 pmol of the template with 2 pmol of RNAP in 5 μl of TB at 37°. Next, 50 μm ApUpC primer and 20 μm CTP, GTP, and UTP were added for 5 min at 37°. The complex was immobilized on 20 μl of prewashed Ni^{2+}-NTA agarose for 10 min and washed. The volume of the reaction was adjusted to 30 μl, and 30 μCi [α-^{32}P]ATP were added for 10 min to form EC16 (lane 1). The free label was removed by five washes. The upper left scheme symbolizes the structure of EC16. RNAP with the RNA exit channel is shown by a shaded oval, DNA is shown by thin black lines. In the RNA sequence the hairpin-forming part is italicized; positions of RNA labeling are bold. EC16 was walked to EC20 by incubation with 10 μM UTP, CTP, and GTP (lane 2). KCl concentration was increased to 300 m*M*, 100 μm of 4 NTPs were added for 20 s, the sample was quickly centrifuged, and half of the supernatant was withdrawn ("s" fraction).

Basic Protocol

1. To completely remove the 5′ region of the RNA in the EC, walk with RNAP to the position 14 nt downstream from the desired cleavage site. During the walking, label the RNA downstream from the intended cleavage site. Incubate the EC with 5000 U/ml of RNase T1, or 10 μg/ml of RNase A for 10 min in TB.

2. To remove RNase T1, wash the EC 10 times with TB; the removal of RNase A requires 1–2 additional washes with TB(1000). The EC containing the truncated transcript can be efficiently walked and analyzed after the wash.

3. Because residual amounts of the RNases remain in the sample even after the wash, special precautions should be taken when fixing the sample for loading on denaturing PAAG. For a 10-μl sample, first add 2 μl of phenol and vortex immediately for 2–3 s. The vortexing is important to generate instant denaturing conditions in the sample. Then combine the sample with an equal volume of denaturing gel-loading buffer.

4. Sometimes complete digestion of the 5′ part of the RNA is not required for the purpose of an experiment (e.g., probing of the nascent RNA folding into secondary structure behind RNAP). In this case, use a lower concentration of the RNase (50 U/ml RNase T1, 0.5 μg/ml RNase A), find the appropriate time experimentally (usually 1–20 min), and skip washing off the RNase. Decreasing the concentration of the RNase will minimize postdenaturation degradation (see the following) and simplify handling of a large number of samples.

Comments

1. RNase A, and to a lesser extent RNase T1, remain partially active in the presence of phenol, urea, or SDS[6] when RNA is already released from the denatured EC. To avoid the postdenaturation degradation of the RNA (which manifests itself by the appearance of 3′-proximal RNA fragments shorter than 14 nt), incubate the sample with 3 μl of ribonuclease inhibitor (Prime RNase Inhibitor for RNase A and 100 m*M* guanosine 2′-monophosphate for RNase T1) for 2–3 min at room temperature before adding phenol.

The "s" fraction and the remaining half of the supernatant, together with pellet ("p" fraction), were analyzed separately on PAAG (lanes 3 and 4). Note that all RNA products, terminated at tR2 signal, were released from the EC. Separately, EC20 was incubated in 20 μl with 5000 U/ml RNase T1 for 10 min, washed with TB 10 times (lane 5) and chased with 100 μm of 4 NTPs (lane 6). In EC20, only two G residues of the upstream arm of the hairpin are exposed for the cleavage with RNase T1. The 2-nt deletion of the upstream arm of the 7-nt long tR2 hairpin completely abolishes termination.[6] Reprinted with permission from N. Komissarova and M. Kashlev (1998).

2. In urea–PAGE, the mobility of short RNA species (2–10 nt) depends on their sequences. This sometimes complicates the determination of the cleavage site by mobility of the RNA in the gel. Instead, compare the cleavage products originated from the EC with the cleavage products derived from the ECs halted several nt upstream and downstream. In addition, treat the same EC with an RNase of different specificity.

3. Stable hairpins with 7–8-nt stems cannot be completely denatured in 8 M urea–PAAG. Folding into a secondary structure substantially increases the RNA mobility in the gel. When an RNase cleaves off a part of the hairpin, the secondary structure is destabilized. The cleaved RNA, instead of migrating faster in the gel (as it became shorter), migrates slower (as the hairpin is now denatured). This effect is especially noticeable with short (20–40 nt) RNA.[20]

4. Usually, the two RNases provide sufficient flexibility for the cleavage site selection. If needed, cleavage after U residues in the RNA can be achieved with RNase I, which degrades the transcript in the EC to 15–17 nt.[21] Two factors may extend the zone of RNA protection in the EC beyond 14 nt. (1) Formation of the secondary/tertiary RNA structure.[20] Addition of higher doses of the RNases, longer incubation time, and performing the reaction at 37° may help resolve the problem. Also, double-strand specific RNase V1 can be used as an alternative. (2) At some sites, RNAP backtracks at 2–15 nt along the RNA and DNA[12,13] and thus protects the upstream part of the transcript (its 3′ end can become exposed to the RNases instead). In this case, find a nearby position where RNAP does not backtrack.

A typical example of RNA truncation in the EC and the elongation of the truncated RNA are shown in Fig. 1.

Isolation of the EC with 8–10-nt RNA by Treatment with Pyrophosphate and GreB

Inorganic pyrophosphate (PP$_i$), or transcript cleavage factor GreB induces shortening of the nascent RNA from the 3′ end in the reactions of pyrophosphorolysis and internal transcript cleavage, respectively[10,11] The shortening is accompanied by RNAP retreat along the DNA. In the absence of secondary structures in the transcript, RNAP stops at the distance of 8–10-nt from the 5′ end of the RNA.[6] Because the 5′ end can be generated by the cleavage of the transcript with RNases, subsequent degradation

[21] S. Milan, L. D'Ari, and M. J. Chamberlin, *Biochemistry* **38**, 218 (1999).

of the cleaved RNA from the 3′ end produces ECs with 8–10 nt of RNA at any template sequence and any distance from the promoter.[6,20]

Basic Protocol (Fig. 2)

1. To truncate the RNA in the EC to 8–10 nt, halt RNAP at 14 nt downstream from the desired 5′ end of the 8–10-nt RNA. During the walking, label the RNA within the 8–10-nt region. Incubate the EC with RNase T1 or RNase A as described earlier and wash off the RNase.

2. Incubate for 10–20 min with 2.5 mM sodium pyrophosphate or potassium pyrophosphate (PP$_i$), or with 0.5 mg/ml GreB in TB. Wash five times with TB.

3. Confirm the identity of the 3′ ends in the EC by walking with subsets of NTPs, or by cleaving the 8–10-nt RNA purified by PAGE with different RNases.[6,20]

Comments

1. The efficiency of the RNA shortening from the 3′ end by pyrophosphorolysis varies for different template positions. In some ECs, only a fraction of the transcript reaches the minimal length, and the increase of the incubation time and PP$_i$–GreB concentration are not always effective. For some difficult tasks, such as the isolation of EC with 8-nt oligo-U RNA, this fraction constitutes less than 10% of the initial RNA.[6]

2. Because pyrophosphorolysis is the reverse reaction to elongation, at each template position, RNAP backtracking, without cleavage of the transcript, competes with pyrophosphorolysis and blocks RNA truncation from the 3′ end. For pyrophosphorolysis to occur, the RNAP active center must be juxtaposed with the 3′ end of the RNA. Also, some 3′-terminal nucleotides are intrinsically resistant to pyrophosphorolysis. If the problem emerges, combine pyrophosphorolysis with GreB treatment (which, on the contrary, efficiently cleaves the RNA in the backtracked EC), or use GreA factor, which has a different cleavage specificity.[11]

Assembly of Functional EC of *E. coli* RNAP and *S. cerevisiae* Pol II from Synthetic DNA and RNA Oligonucleotides

In this article, we present an alternative technique for obtaining ECs, which is based on the ability of RNA polymerases to bind the 3′ end of short RNA primer annealed to a single-strand template DNA oligo and to incorporate the nontemplate DNA oligo, fully complementary to the template DNA. The resulting assembled ECs are similar to the complexes obtained by promoter-specific initiation in all parameters that we tested,

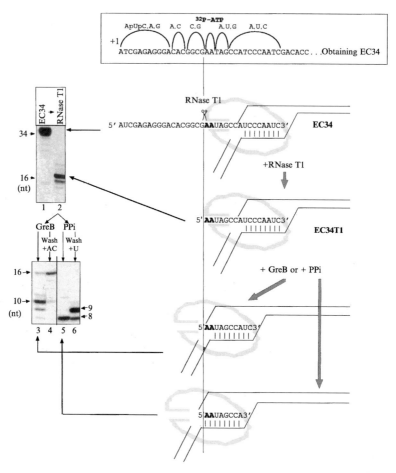

Fig. 2. Isolation of the EC containing 8–10-nt RNA. The transcribed sequence of the nontemplate DNA strand used in this experiment and the walking strategy employed to obtain EC34 labeled in 19A and 20A positions are shown at top. EC34 (lane 1) was treated with 5000 U/ml RNase T1 in 20 μl for 10 min and washed with TB 10 times. The 34-nt RNA was truncated to 16-nt (EC34T1, lane 2). EC34T1 was incubated with 0.5 mg/ml of GreB (lane 3) or 2.5 mM PP$_i$ (lane 5) for 30 min and washed. The cleavage with GreB produced the EC with 10-nt RNA, whereas pyrophosporolysis produced the shorter 8-nt product with the same 5' end. The identity of the 3' end in both complexes was confirmed by walking with subsets of NTPs that corresponded to the template sequence (lanes 4 and 6).

including all structural parameters (determined by footprinting assays), the ability to elongate the RNA according to the template sequence, and adequate recognition of termination signals in the DNA.[8,9]

This technique has many advantages over previously reported methods of EC formation. First, the EC can be obtained in great quantities suitable for crystallography and other physical methods of structural analysis. Second, this technique can easily produce ECs carrying RNA shorter than 8 nt, or RNA containing chemically modified or mismatched bases. Template and nontemplate DNA strands can be selectively modified as well. Third, previously reported promoter-independent methods of obtaining ECs were often complicated by formation of a long RNA:DNA hybrid in the course of transcription.[22] These methods include the use of a preformed mismatched "artificial bubble" in the DNA and the use of DNA templates that had a single-strand tail of 9–12 C residues.[5,22] Our technique utilizes a pair of fully complementary DNA oligos, which provides for successful displacement of the RNA from the hybrid with the DNA.[8,9] The ability to obtain authentic ECs from the core polymerase enzyme in the absence of transcription initiation factors is especially important for the studies of transcription by Pol II. Using this method, the specific promoter sequence becomes dispensable.

Assembly of EC with Histidine-Tagged Bacterial RNAP and Yeast Pol II and Immobilization of the EC on Ni^{2+}-NTA Agarose Beads

Basic Protocol

The principal steps in the protocol are shown in Fig. 3A.

1. *RNA/DNA labeling.* RNA and/or DNA oligos can be prelabeled at the 5′ end with bacteriophage T4 polynucleotide kinase before the assembly. 1.5 μm RNA or DNA oligo is incubated in TB with 2 μm of [γ-^{32}P]ATP and 3 U/μl of T4 kinase for 30 min at 37°, followed by 20 min at 65° to inactivate kinase. Normally, we do not purify labeled oligos from [γ-^{32}P]ATP. Alternatively, the RNA can be labeled with [α-^{32}P]NTP using catalytic activity of the enzyme in later steps of assembly.

2. *Formation of RNA:DNA hybrid.* Mix template DNA oligo and RNA oligo in equimolar amounts or using 2-fold excess of the RNA, and anneal the hybrid in a PCR machine programmed in the following way: 45° 5 min; 42°, 39°, 36°, 33°, 30°, 27°— 2 min each; 25°—10 min. The standard concentration of DNA and RNA is 1.5 μm. For shorter RNAs, oligos with mismatches, or oligos that form weaker base pairs, increase the concentration of the oligos up to 150 μm, and modify the annealing temperature accordingly.

[22] S. S. Daube and P. H. von Hippel, *Biochemistry* **33**, 340 (1994).

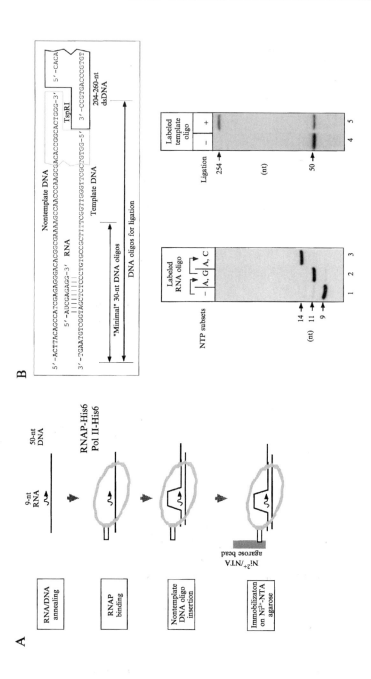

3. *RNAP binding.* Combine 10 μl (15 pmol) of the preannealed RNA:DNA hybrid with 1–2 pmol (0.5–1 μg) of hexahistidine-tagged *E. coli* RNAP[16] or Pol II[17] and incubate for 10 min at 24°. Both core and holoenzyme of bacterial RNAP can be used; the yield of EC is about the same.

4. *Incorporation of the nontemplate DNA strand.* Add 300 pmol of nontemplate DNA oligo and incubate for 10 min at room temperature or at 37°. Normally, these conditions lead to very efficient (more than 90% as determined by potassium permanganate footprinting) incorporation of the nontemplate DNA into EC. Yeast Pol II incorporates the nontemplate strand more easily than *E. coli* RNAP, and 2-fold molar excess may be sufficient. Later (steps 5 and 6) we describe the methods that allow the removal of the small admixture of the ECs, which have not properly incorporated nontemplate oligo.

5. *Immobilization.* Add 50 μl of washed Ni^{2+}-NTA agarose and incubate for 5–30 min at 24°, preferably with shaking. 50 μl of 50% (v/v) suspension of Ni^{2+}-NTA agarose beads bind up to ~1–2 μg of the assembled EC. Alternatively, the EC can be immobilized on streptavidin agarose beads through the biotin group at the 5' end of the DNA oligo. If the nontemplate DNA oligo is biotinylated, the EC that properly incorporated the nontemplate DNA strand will be purified from the admixture of the "single-stranded" intermediate described earlier. *Note:* RNAP can be immobilized on Ni^{2+}-NTA agarose first, before, or after the addition of the RNA:DNA hybrid.

Fig. 3. Assembly of the EC from hexahistidine-tagged *E.coli*/yeast RNA polymerases and synthetic RNA/DNA oligos. (A) Steps in the EC assembly. The oval shape and the solid and waved lines represent RNAP, DNA oligo, and RNA oligo, respectively. (B) Assembly and walking with the yeast EC and its ligation to a 204-nt DNA fragment. The sequence of the DNA and RNA oligos used for the assembly is shown at top. The walking and the positions of the resulting complexes are indicated above the nontemplate DNA sequence. The boxed areas show the 3' overhangs used for ligation of the DNA in the EC to the linear 204–260-nt DNAs containing TspRI cleavage site at one end. The EC was assembled from 1 pmol of the tagged Pol II with 9-nt RNA primer labeled at the 5' end by phosphorylation, 5'-phosphorylated 50-nt template DNA oligo, and 59-nt nontemplate DNA oligo (EC9, lane 1). EC9 was immobilized on Ni^{2+}-NTA agarose beads and walked to position +11 (lane 2) by incubation with 10 μm of ATP and GTP, and then to position +14 (lane 3) with 10 μm of ATP and CTP. Alternatively, the 59-nt nontemplate DNA oligo was labeled at the 5' end with [γ-^{32}P] ATP and assembled into the EC with the unlabeled RNA primer. Of DNA-labeled complex, 1 pmol was ligated to 1 pmol of the 204-bp DNA fragment. After washing with TB, the equal amount of the ligated (lane 5) and nonligated (lane 4) EC9 was analyzed in 6% urea–PAAG. The efficiency of ligation was ~50%. The DNA oligos of the shorter 30-nt length (indicated as the "minimal" oligos) can also be used for the assembly. From M. L. Kireeva, W. Walter, V. Tchernajenko, V. Bondarenko, M. Kashlev, and V. M. Studitsky, *Mol. Cell* 9, 541 (2002).

6. *Purification of the EC from the excess oligonucleotides.* Wash the immobilized EC three times with TB, wash briefly with 1 ml TB(1000), and wash with TB twice. The wash with TB(1000) removes DNA and RNA nonspecifically associated with RNAP. The resulting EC can be walked along the template (Fig. 3B), as well as modified and analyzed as described earlier for the promoter-initiated EC. The EC can be stored on ice for several days. However, the EC with a weak or short RNA:DNA hybrid may dissociate during storage. *Note:* Because the nontemplate strand is required for stability of the bacterial EC,[8] washing with 1 *M* KCl can be used to dissociate the "single-strand" intermediates, leaving only the fully assembled ECs. To utilize this property, increase the time of the EC incubation with 1 *M* KCl to 10–15 min. Because the stability of ECs differs depending on the sequence and length of the transcript, conditions for the selective dissociation of the "single-strand" ECs must be optimized for each DNA template.

Comments

1. *Selection of DNA and RNA oligos for the EC assembly.* We have used 30–70-nt DNA oligos and 3–20-nt RNA oligos. G/C-rich 9-nt RNA oligos give the best yield of the EC. ECs assembled with RNA oligos shorter than 8 nt, with A/U-rich oligos, or with RNA oligos longer than 10–12 nt may be unstable and can dissociate in 1 *M* KCl during the EC purification. DNA segments at both sides of the RNA:DNA hybrid in the assembled complex should be at least 10-nt long to form a stable transcription bubble in RNAP. Also, the length of the DNA downstream from the 3′ end of the RNA primer should be more than 10 nt to prevent arrest of the EC, which occurs at the end of the linear DNA.[23] DNA oligos longer than 70 nt are likely to form a secondary structure, which may interfere with the annealing of the RNA and may inhibit the incorporation of nontemplate DNA oligo.

The assembly method allows introduction of mismatches to the RNA:DNA hybrid, as well as having single-strand RNA "tails" not encoded in the DNA at the 5′ or 3′ ends of the RNA oligos.[8,14] Unpaired patches of variable length can be introduced to the DNA oligos as well.[6] Also, the nucleic acid "scaffold" in the EC can be assembled from more than three oligos. For instance, the nontemplate DNA strand can be assembled from two fragments with a nick or gap between them.[8]

2. *Partial dissociation of the EC upon incorporation of nontemplate DNA oligo (RNA displacement).* In standard assembly conditions, a substantial

[23] M. G. Izban, I. Samkurashvili, and D. S. Luse, *J. Biol. Chem.* **270,** 2290 (1995).

fraction (up to 50%) of the EC formed with the 9-nt RNA dissociates during incorporation of the nontemplate DNA oligo. The dissociation is greater when shorter than 9-nt or A/U-rich RNA oligos are used.

3. *The assembly efficiency.* The amount of RNAP that should be taken for assembly, as well as the amount of oligos, depends on the goal of the experiment and can vary widely. As little as 1 fmol of radioactively labeled RNA can be detected and quantified. However, for most experiments we assemble EC with 1–2 pmol of RNAP using at least 10-fold molar excess of the RNA:DNA hybrid. The yield of the EC depends on the concentration of RNA polymerase. If a high concentration of *E. coli* RNAP (above 1 mg/ml) is used, 95% of the enzyme forms catalytically active complex with the RNA:DNA hybrid. Because of the RNA displacement (see previous text), the yield of the EC drops to 20–80% after the nontemplate DNA strand is incorporated. The assembly efficiency also depends on the sequence and length of RNA and DNA oligos. The 9-nt RNA oligo and a pair of fully complementary 30–60-nt DNA oligos that we used to develop the assembly protocol are shown in Fig. 3B.

Ligation of a Long DNA Template to the Assembled EC

The length of DNA oligos that can be used for assembly is limited to 70–80 nt. However, many tasks require transcription of much longer DNA. ECs containing longer templates are obtained by ligation of the assembled complex to a double-strand DNA fragment. High efficiency of the ligation is achieved by using a 9-nt 3' end overhang (5'...CCCAGTGCC-3') of the nontemplate DNA oligo (Fig. 3B). It can be ligated to any linear DNA fragment produced by digestion with the TspRI restriction enzyme. For the DNA fragments that we tested (>200 nt long), the ligation occurred with ~50% efficiency.

The template DNA oligo must be phosphorylated at the 5' end to allow the ligation. The DNA between the 3' end of the RNA and the ligation site should be sufficiently long (>30 nt) to ensure that the downstream end of the template is not protected from ligation by RNAP. Positioning of RNAP just 22 nt from the TspRI site on the same template dramatically decreased the ligation efficiency.

Basic Protocol

1. Assemble the "double-strand" EC with 1–2 pmol of the bacterial RNAP or yeast Pol II and with DNA oligos producing the 9-nt overhang as described earlier, and immobilize the EC on Ni^{2+}-NTA agarose beads. Adjust the suspension volume to 90 μl with TB.

2. Add acetylated bovine serum albumin (BSA) to 0.5 mg/ml, polyethylene glycol (PEG)8000 to 1% (v/v), and adenosine triphosphate (ATP) to 100 μm. Add 100–200 ng of 200–250-nt DNA fragment (cut with TspRI at one end and blunt at the other) and 50 U of phage T4 DNA ligase, and incubate at 12° for 1 h. Alternatively, the ligation can be done using the Quick Ligation kit for 5 min at room temperature.

3. After ligation, wash the EC twice with 1 ml of TB, incubate in TB(1000) for 10 min, and again wash twice with TB to remove the obligated DNA fragment and other components of the ligation reaction.

Assembly and Purification of Large Quantities of the E. coli EC in Solution

Crystallographic or biophysical studies of transcription often require milligrams of EC. Moreover, the hexahistidine tag in the enzyme may interfere with crystallization, or it may affect some functions of RNAP, as has been shown for other proteins.[24] The following protocol was developed for large-scale assembly of EC with wild type (nontagged) E. coli RNAP. The homogeneous EC is purified from the admixture of free RNAP, the oligos, and the EC lacking the nontemplate DNA strand by combination of affinity and size-exclusion chromatography (Fig. 4).

1. *EC assembly.* Assemble EC from 1 mg (3 nmol) of RNAP, 15 nmol of RNA oligo, 8 nmol 30-nt template DNA oligo, and 160 nmol of the nontemplate DNA oligo in 1–2 ml of TB as described earlier. For the assembly, we used the 9-nt RNA and 30-nt DNA oligos shown in Fig. 3B.

2. *Purification of the EC from free RNAP on heparin column.* Dilute the assembled EC 2–3-fold with TGED containing 40 mM KCl. Load onto a 1-ml disposable heparin column and wash the column with 5 ml of TGED(40). Elute with gradient: 40 mM to 1 M KCl in TGED (0–35% of the gradient in 2 ml—peak that elutes contains admixture of the fully assembled EC and EC lacking the nontemplate DNA strand; 35–100% of the gradient in 1 ml—peak that elutes contains free RNAP).

3. Concentrate fractions containing the EC to 600–700 μl or less using Centricon-100.

4. Bring KCl concentration to 1 M to dissociate the unstable EC that did not incorporate the nontemplate DNA strand. Incubate at 24° for 20 min. At this step the properly assembled EC becomes contaminated with free RNAP and the RNA:DNA hybrid derived from the unstable EC.

---

[24] A. Goel, D. Colcher, J. S. Koo, B. J. Booth, G. Pavlinkova, and S. K. Batra, *Biochim. Biophys. Acta* **1523,** 13 (2000).

Content of the peaks

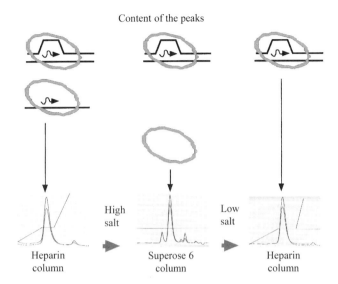

Fig. 4. Purification of the assembled bacterial EC in solution. The method involves large-scale assembly of the EC in solution with nontagged *E. coli* RNAP, followed by three-step purification by chromatography. All symbols are the same as in Fig. 3A. The cartoon on the top shows the content (top) and the position (bottom) of the major peaks eluted from the columns. The three-step purification allows the effective separation of the assembled EC from the free RNAP, nucleic acids, and partially assembled intermediate lacking the nontemplate strand.

5. *Purification of the EC from RNA:DNA hybrid (derived from dissociation of the unstable EC in high salt)*. Load the EC to Superose 12 HR 10/30 column preequilibrated with TGED containing 1 *M* KCl. The size-exclusion chromatography is performed in 1 *M* KCl to prevent spontaneous reassociation of RNAP with the RNA:DNA hybrid. The EC and free RNAP elute in the second peak, at about 10 ml of the buffer volume.

6. *Purification of the EC from free RNAP (derived from dissociation of the unstable EC in high salt)*. Dilute fractions containing EC with TGED to 100 m*M* KCl. Load on 1 ml disposable Heparin column and elute as before. This step removes free RNAP derived from dissociation of the single-stranded complex from the properly assembled double-stranded EC. Concentrate final fractions using Centricon-100 and dilute with TGED to 40 m*M* KCl.

[18] Assay of Transient State Kinetics of RNA Polymerase II Elongation

By Yuri A. Nedialkov, Xue Q. Gong, Yuki Yamaguchi, Hiroshi Handa, and Zachary F. Burton

Elongation of an RNA chain by human RNA polymerase II is a highly regulated process with implications for the control of gene expression and for the genesis of disease. Understanding how transcription elongation is regulated requires a full description of mechanism and a complete analysis of how that mechanism is controlled by protein factors. A powerful approach is the application of rapid quench-flow techniques, such as those pioneered to characterize analogous DNA polymerase mechanisms.[1,2] The primary advantage of transient state kinetic analysis is that this method allows tracking of elongation rates through formation of individual bonds with millisecond precision, which may allow real-time resolution of the fastest elongation rates for RNA polymerase II. Comparing the functional dynamics with the recently published structure of the yeast RNA polymerase II elongation complex[3] suggests adequate and testable elongation models.

We outline an approach for transient state kinetic analysis of elongation by human RNA polymerase II. In the previous detailed studies of DNA polymerase mechanisms, a complete kinetic analysis of bacteriophage T7 DNA polymerase elongation was presented in a set of classic papers.[4-6] For human RNA polymerase II, such a complete analysis may not yet be possible, but highly reliable rate data that are informative for inferring essential aspects of mechanism can be obtained. In the presence of elongation factors, specific steps that are targets of regulation can be identified.

Currently, this approach is most applicable to factors demonstrated to interact directly with the elongation complex to stimulate, repress, or edit RNA synthesis. Such factors include transcription factor IIF (TFIIF),[7-10]

[1] K. A. Johnson, *Methods Enzymol.* **249,** 38 (1995).

[2] K. A. Johnson, *Enzymes* **20,** 1 (1992).

[3] A. L. Gnatt, P. Cramer, J. Fu, D. A. Bushnell, and R. D. Kornberg, *Science* **292,** 1876 (2001).

[4] S. S. Patel, I. Wong, and K. A. Johnson, *Biochemistry* **30,** 511 (1991).

[5] I. Wong, S. S. Patel, and K. A. Johnson, *Biochemistry* **30,** 526 (1991).

[6] M. J. Donlin, S. S. Patel, and K. A. Johnson, *Biochemistry* **30,** 538 (1991).

[7] L. Lei, D. Ren, and Z. F. Burton, *Mol. Cell Biol.* **19,** 8372 (1999).

[8] S. Tan, T. Aso, R. C. Conaway, and J. W. Conaway, *J. Biol. Chem.* **269,** 25684 (1994).

[9] M. G. Izban and D. S. Luse, *J. Biol. Chem.* **267,** 13647 (1992).

[10] E. Bengal, O. Flores, A. Krauskopf, D. Reinberg, and Y. Aloni, *Mol. Cell Biol.* **11,** 1195 (1991).

the TFIIF-interacting component of the CTD phosphatase (FCP1),[11,12] DRB sensitivity-inducing factor (DSIF),[13–17] negative elongation factor (NELF),[14,15,17] ELL (a leukemia chromosome translocation partner),[18–20] SIII/elongin,[21–23] Cockayne syndrome type B protein (CSB),[24] hepatitis delta antigen (HDAg),[25] and SII/TFIIS.[26,27] So far, the method is validated by analyzing stimulation of RNA polymerase II by TFIIF and HDAg and inhibition by the mushroom toxin α-amanitin.[28–30]

Preparation of Precisely Stalled Elongation Complexes on Bead Templates

To adequately perform single-bond elongation studies, a method was required for stalling a bead-immobilized RNA polymerase II elongation complex at a defined template position.[7,12,14,31] To accomplish this goal,

[11] H. Cho, T. K. Kim, H. Mancebo, W. S. Lane, O. Flores, and D. Reinberg, *Genes Dev.* **13**, 1540 (1999).

[12] S. S. Mandal, H. Cho, S. Kim, K. Cabane, and D. Reinberg, *Mol. Cell Biol.* **22**, 1173 (2002).

[13] T. Wada, T. Takagi, Y. Yamaguchi, A. Ferdous, T. Imai, S. Hirose, S. Sugimoto, K. Yano, G. A. Hartzog, F. Winston, S. Buratowski, and H. Handa, *Genes Dev.* **12**, 343 (1998).

[14] D. B. Renner, Y. Yamaguchi, T. Wada, H. Handa, and D. H. Price, *J. Biol. Chem.* **276**, 42601 (2001).

[15] Y. Yamaguchi, T. Takagi, T. Wada, K. Yano, A. Furuya, S. Sugimoto, J. Hasegawa, and H. Handa, *Cell* **97**, 41 (1999).

[16] Y. Yamaguchi, T. Wada, T. D. Watanabe, T. Takagi, J. Hasegawa, and H. Handa, *J. Biol. Chem.* **274**, 8085 (1999).

[17] Y. Yamaguchi, N. Inukai, T. Narita, T. Wada, and H. Handa, *Mol. Cell Biol.* **22**, 2918 (2002).

[18] A. Shilatifard, D. Haque, R. C. Conaway, and J. W. Conaway, *J. Biol. Chem.* **272**, 22355 (1997).

[19] A. Shilatifard, D. R. Ruan, D. Haque, C. Florence, W. H. Schubach, J. W. Conaway, and R. C. Conaway, *Proc. Natl. Acad. Sci. USA* **94**, 3639 (1997).

[20] A. Shilatifard, W. S. Lane, K. W. Jackson, R. C. Conaway, and J. W. Conaway, *Science* **271**, 1873 (1996).

[21] T. Aso, W. S. Lane, J. W. Conaway, and R. C. Conaway, *Science* **269**, 1439 (1995).

[22] R. J. Moreland, J. S. Hanas, J. W. Conaway, and R. C. Conaway, *J. Biol. Chem.* **273**, 26610 (1998).

[23] J. N. Bradsher, S. Tan, H. J. McLaury, J. W. Conaway, and R. C. Conaway, *J. Biol. Chem.* **268**, 25594 (1993).

[24] C. P Selby and A. Sancar, *Proc. Natl. Acad. Sci. USA* **94**, 11205 (1997).

[25] Y. Yamaguchi, J. Filipovska, K. Yano, A. Furuya, N. Inukai, T. Narita, T. Wada, S. Sugimoto, M. M. Konarska, and H. Handa, *Science* **293**, 124 (2001).

[26] M. G. Izban and D. S. Luse, *J. Biol. Chem.* **268**, 12864 (1993).

[27] D. Reines, M. J. Chamberlin, and C. M. Kane, *J. Biol. Chem.* **264**, 10799 (1989).

[28] Y. A. Nedialkov, X. Q. Gong, S. L. Hovde, Y. Yamaguchi, H. Handa, J. H. Geiger, H. Yan, and Z. F. Burton, *J. Biol. Chem.* **278**, 18303 (2003).

[29] C. Zhang, H. Yan, and Z. F. Burton, *J. Biol. Chem.* **278,** (in press) (2003).

[30] J. D. Funk, Y. A. Nedialkov, D. Xu, and Z. F. Burton, *J. Biol. Chem.* 46998 (2002).

[31] J. A. Arias and W. S. Dynan, *J. Biol. Chem.* **264**, 3223 (1989).

DNA templates containing the adenovirus major late promoter were synthesized by the polymerase chain reaction using an upstream biotinylated primer and an unmodified downstream primer. DNA templates were immobilized at the highest achievable density on Promega (Madison, WI) MagneSphere beads, according to manufacturer's instructions.

The sequence downstream of the promoter was modified by addition of a 39 nucleotide (nt) CU cassette downstream of +1A (+1- ACTCTCTT -CCCCTTCTCTTTCCTTCTCTTCCCTCTCCTCC-+40-AAAGGCCTT-T-+50). An extract of human HeLa cell nuclei (4–6 μl per reaction) is the source of initiation factors.[32] HeLa cells are purchased from the National Cell Culture Center (Minneapolis, MN). All transcription reactions are at 25° in transcription buffer (12 mM HEPES, pH 7.9, 12% (w/v) glycerol, 0.12 mM EDTA, 0.12 mM EGTA, 1.2 mM DTT, and 0.003% NP-40). Buffers contain 60 mM KCl and 12 mM MgCl$_2$ unless otherwise indicated. Siliconized microfuge tubes are used for all steps because they minimize adsorption of paramagnetic beads to tube walls. Beads (4–6 μl of a 50% slurry per reaction) are incubated for 60 min with extract and then washed two times with transcription buffer to remove any contaminating nucleoside triphosphates (NTPs) and unbound proteins. Samples are processed in batch and then later divided for individual reactions. Transcription is initiated by addition of 10 μm dATP, 300 μm ApC, 20 μm UTP, and 1 μCi per reaction[α^{32}P]CTP for 10 min (20 μl solution per reaction). 20 μm CTP is then added and the reaction incubated for an additional 10 min to complete synthesis of C40 (the accurately initiated 40-nt transcript, which ends in a 3′-CMP). After C40 synthesis, complexes are washed in batch two times with 0.5 ml transcription buffer lacking MgCl$_2$ but containing 1% Sarkosyl and 0.5 M KCl. Complexes are then washed three times with transcription buffer containing 60 mM KCl and lacking MgCl$_2$.

Because RNA polymerase II is fully processive, and because of the methods we use for sample comparison, reactions need not contain identical molar quantities of elongation complex. Data among experiments, from different days, done with different bead and extract preparations, appear to be readily comparable. Other laboratories have used similar techniques to prepare elongation complexes for bench-top elongation experiments,[7,12,14] but with hand pipetting, the time resolution for starting and stopping a reaction is iimited to about 2–5 s, which is far too slow to observe single-bond formation under natural elongation conditions.

[32] D. J. Shapiro, P. A. Sharp, W. W. Wahli, and M. J. Keller, *DNA* **7**, 47 (1988).

Rapid Quench Experiments

To study elongation in real time, our laboratory began studies using the KinTek (Austin, TX) Rapid Chemical Quench-Flow (RQF-3) instrument, which allows reactions to be started and quenched within 0.002 s.[1,2] The RQF-3 is a relatively inexpensive instrument (about $30,000), which handles the small reaction volumes necessary for these experiments and which can be thought of as a rapid mixing device. Different experimental designs to analyze elongation through the sequence 40-CAAAGG-45 are shown in Fig. 1 and described in detail later. The left and right sample ports for the RQF-3 have a 15-μl volume. The standard RQF-3 mixing chamber has three syringes mounted under control of a powerful drive motor. Typically, elongation complexes (in our experiments, A43 complexes are formed by brief addition of ATP to C40 complexes) are loaded in transcription buffer into the left sample port (1), and NTP substrates in transcription buffer at twice their working concentration are loaded into the right sample port (2). Initiation of the reaction program activates the drive motor to push transcription buffer from syringes 1 and 2 and stop solution from syringe 3. Activation of the program causes equal volume mixing and then quenching after a delay $\geq$0.002 s (Fig. 1A).

Other experimental designs are possible with the 3-syringe instrument, such as pulse-chase reactions (Fig. 1B). In this design, after a precisely timed ATP pulse ($\geq$0.002 s), GTP is added to the reaction, and quenching occurs in the sample collection tube, which contains 0.5 M EDTA. Using the standard 3-syringe mixing chamber, however, the time of the GTP chase cannot be precisely controlled in the millisecond range, because the minimum time is limited to about 0.015 s by the flow rate and the dimensions of the exit line. A modified, 4-syringe mixing chamber is also available from KinTek that is better able to support the pulse-chase reaction design (Fig. 1C). Because, as we show later, pulse-chase strategies are helpful to analyze RNA polymerase II elongation, the 4-syringe mixing chamber may have wide utility. Our laboratory is just in the process of initiating studies with the 4-syringe chamber.

Erie and colleagues reported rapid quench experiments to analyze rates of elongation by *Escherichia coli* RNA polymerase,[33] and our initial studies were modeled after theirs. Our plan was to examine RNA polymerase II elongation in the presence and absence of TFIIF to learn how TFIIF stimulates elongation rate and suppresses pausing.[7–10,34] As these experiments progressed, we began studies on HDAg, an RNA polymerase II elongation factor encoded by hepatitis delta virus, a satellite of hepatitis B

[33] J. E. Foster, S. F. Holmes, and D. A. Erie, *Cell* **106,** 243 (2001).
[34] D. H. Price, A. E. Sluder, and A. L. Greenleaf, *Mol. Cell Biol.* **9,** 1465 (1989).

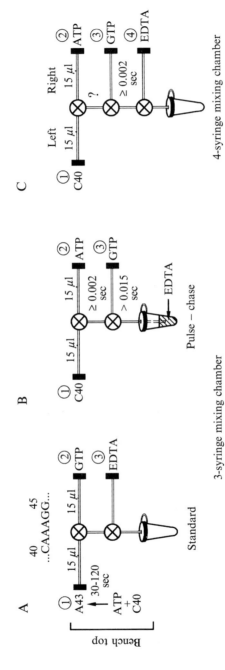

Fig. 1. Experimental designs using the KinTek RQF-3 for elongation through the sequence 40-CAAAGG-45. The left and right sample ports have a 15-μl volume. X indicates a rapid mixing event. 1, 2, 3, and 4 indicate syringes to drive solution (buffer, chase reagent, or EDTA stop solution) through a mixing chamber. (A) The running-start protocol (see Fig. 4) using the 3-syringe mixing chamber. (B) The pulse–chase protocol (see Fig. 3) using the 3-syringe mixing chamber. (C) The pulse–chase protocol using the 4-syringe mixing chamber. The question mark indicates that the precision of the pulse time is not yet determined.

virus.[25,29] So far, our approach has provided significant insight into RNA polymerase II mechanism and into inhibition by α-amanitin.[28,29] TFIIF and HDAg have complex effects on the bond addition cycle, and these factors are highly informative probes of the RNA polymerase II mechanism.

Elongation from C40

Using our standard template, the RNA sequence from C40 is 40-CAAAGGCCUUU-50, so adenosine monophosphate (AMP) is the next base added in the chain. To our dismay, however, when ATP was added to stalled C40 complexes, for the most part, elongation was very slow, indicating that few C40 complexes were initially poised on the forward elongation pathway (Fig. 2). In the absence of a stimulatory factor (Fig. 2A), a very slow elongation rate is observed, and only about 20% of C40 complexes are converted to longer products within 1 s. In the presence of TFIIF (Fig. 2B), elongation rates are significantly faster. With TFIIF, however, few C40 complexes are initially poised on the active elongation pathway, as indicated by the meager burst in the initial disappearance of C40 and the initial appearance of A41, A42, and A43. Very little evidence can be seen for rapid passage of complexes through the A41 and A42 positions. Slow accumulation of A43 is attributed primarily to the slow rate of disappearance of C40. With the viral elongation factor HDAg (Fig. 2C), however, about 25% of C40 complexes are initially poised on the rapid elongation pathway, as can be seen from the instantaneous burst in C40 disappearance and A43 appearance upon ATP addition. Passage of these complexes through the A41 and A42 positions is rapid, because most race through to the A43 position. For these highly poised C40 complexes, it appears that three bonds can be formed within 0.002 s (Fig. 2C, lane 2), because A43 is observed to accumulate from C40 at the earliest time points. Based on these experiments, studies of rapid elongation from C40 complexes prepared in this fashion may be possible in the presence of HDAg, but not in the absence of factor or in the presence of TFIIF. At present, the blocks to rapid elongation from the C40 template position are not fully understood.

Because of persistent problems measuring fast elongation rates from C40, we sought a method to commit a larger fraction of elongation complexes to rapid extension. Our strategy was to advance C40 complexes to A43 by addition of ATP. After a brief stall at A43, GTP was added, allowing extension to the G44 and G45 positions (Fig. 3). We refer to this protocol as a "running start." To perform these experiments, the RQF-3 was run in pulse–chase mode, as shown in Fig. 1B. In the absence of a stimulatory elongation factor (Fig. 3A), 100 μm ATP was added to C40 complexes and incubated for variable times from 0.5 to 480 s. 250 μm GTP was then

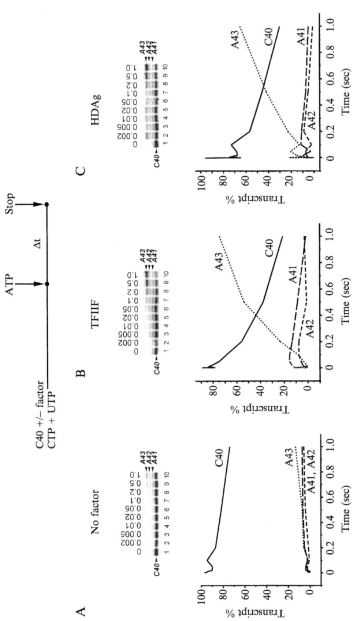

FIG. 2. Elongation from C40, in the absence of elongation factor (A), the presence of TFIIF (12 pmol per reaction) (B), or the presence of HDAg (77 pmol per reaction) (C). The reaction protocol is indicated above the figure. CTP and UTP were initially 20 μm. ATP was 100 μm. Times are indicated in seconds. For each experiment, gel data is shown above and corresponding PhosphoriMager quantitation below. There is little evidence for rapid elongation in the absence of factor or in the presence of TFIIF, although TFIIF does stimulate escape from C40. In the presence of HDAg, about 25% of C40 complexes are initially on the rapid elongation pathway. Electrophoresis was in 16% acrylamide gels (20:1 acrylamide:bisacrylamide), containing 50% w/v urea and run in 1× Tris–borate–EDTA buffer.

added and reactions quenched after 0.2 s, a time found sufficient to advance activated complexes. Using the running start, a significant fraction of elongation complexes are committed to forward synthesis. With pulse times 60 s or longer, most C40 complexes are converted to A43, and the system appears to approach a steady state from about 20 to 480 s. Throughout this interval, about 60% of A43 complexes remain strongly paused, and about 40% advance within 0.2 s, indicating that about 40% of A43 complexes can be committed to rapid elongation by the running-start protocol. In the absence of a stimulatory elongation factor, the C40 and A43 stops are precisely maintained (Fig. 3A, lanes 1–3). Furthermore, there is little evidence that RNA editing by backtracking and cleavage significantly affects partitioning between paused and activated elongation complexes within 480 s. For running-start assays in the absence of a stimulatory elongation factor, we have adopted an ATP pulse time of 120 s.

In the presence of TFIIF (Fig. 3B), the situation is quite different. First of all, there is a tendency to overrun the C40 and A43 transcription stops even when no ATP or GTP is added to the reaction (lanes 1, 4, and 5). Apparently, TFIIF induces RNA polymerase II to scavenge trace ATP and GTP that contaminate washed beads or reagent stocks. Background A41 synthesis from C40 (lane 1) and background G44 synthesis from A43 (lanes 4 and 5) probably represent accurate incorporation rather than misincorporation because the bands precisely co-migrate with the accurately synthesized products (compare with neighboring lanes). We do not believe that TFIIF preparations are significantly contaminated with ATP and GTP because mutant TFIIF carrying the deleterious I176A mutation in the RAP74 subunit and purified by the same method is incapable of scavenging trace ATP and GTP (data not shown). Therefore we attribute ATP and GTP scavenging to stimulation of RNA polymerase II by TFIIF. Because of the potential for background problems, the level of background G44 synthesis must be ascertained in each experiment with TFIIF, or misleading results will be obtained. In the experiment shown, the times of the 100-μm ATP pulse were varied from 0.25 to 60 s.

The time of the 200-μm GTP chase was about 0.015 s, the shortest possible chase time using the standard 3-syringe mixing chamber. C40 is mostly converted to A43 within 20 s. A steady state between paused and active complexes is established between about 20 and 60 s. About 55% of complexes appear to be committed to the forward synthesis pathway at the time of GTP addition, as indicated by their extension within 0.015 s. For running-start assays with TFIIF, a 30-s ATP pulse was adopted. There is little evidence of editing within 60 s, although, in the presence of TFIIF, editing is observed to affect the distribution of paused and active complexes after about 240 s (data not shown).

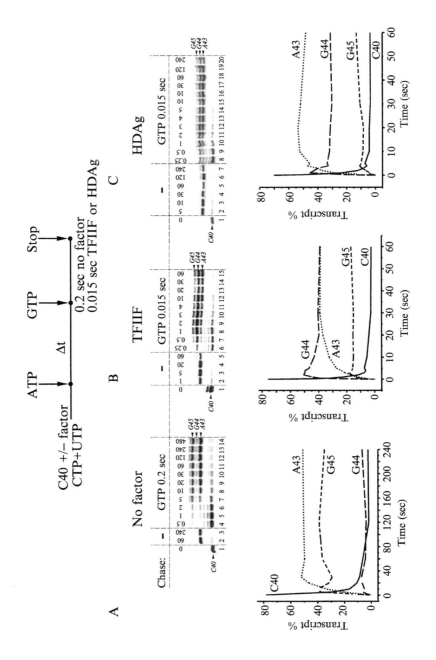

In the presence of HDAg, evidence was obtained for rapid elongation from C40 (Fig. 2), indicating differences in the modes of TFIIF and HDAg regulation of RNA polymerase II. In a running-start protocol with HDAg (Fig. 3C), as with TFIIF, there is a tendency to overrun transcription stops at C40 and A43 (lanes 1, 6, and 7). Because of the potential for background synthesis, controls with no GTP chase are always included in HDAg experiments. With HDAg, the steady state condition appears to be maintained with pulse times between about 10 to 240 s, and most C40 complexes are converted to A43 within 10 s. For routine experiments with HDAg, a 60-s ATP pulse has been adopted. There is little evidence for backtracking and cleavage within 240 s in the presence of HDAg.

The Running-Start, Two-Bond Assay

Based on experiments such as those described earlier and additional considerations described later, a "running-start, two-bond" protocol was adopted to analyze RNA polymerase II elongation (Fig. 4). Using the 40-CAAAGG-45 template, C40 complexes (10 μl in transcription buffer) were mixed with 100 μm ATP (added as 10 μl 200 μm ATP in transcription buffer) on the bench top. The ATP pulse time was 120 s in the absence of factor, 30 s in the presence of TFIIF, and 60 s in the presence of HDAg. During the short incubation with ATP, 15 μl of the reaction mix was rapidly injected into the left sample port of the KinTek RQF-3 and the valve closed and set in the "fire" position. Initiation of the reaction induces mixing of the sample with 15 μl GTP (at twice its working concentration before equal volume mixing) previously loaded into the right sample port. Reactions were quenched with 0.5 M EDTA at the times indicated. Beads were collected with a magnetic particle separator, supernatant was removed, and beads were resuspended in formamide gel loading dyes. Samples were electrophoresed in 16% polyacrylamide gels containing 50% (w/v) urea and analyzed using a Molecular Dynamics PhosphorImager (Amersham Biosciences, Piscataway, N. J.). Recovery of samples was sometimes variable, so, for each sample lane, the sum of all transcripts generated from A43 was determined as 100%.

FIG. 3. Running-start elongation protocol to test alternate ATP pulse times, in the absence of stimulatory factor (A), the presence of TFIIF (12 pmol per reaction) (B), or the presence of HDAg (77 pmol per reaction) (C). The reaction protocol is shown above the figure. CTP and UTP were initially 20 μm. ATP was initially 100 μm. GTP was at 250 μm (No factor or HDAg) or 200 μm (TFIIF). For each experiment, gel data is shown above and corresponding PhosphorImager quantitation below. In each experiment, a steady state condition is established in which the fraction of A43 complexes on the active elongation pathway and the pausing pathway remain balanced.

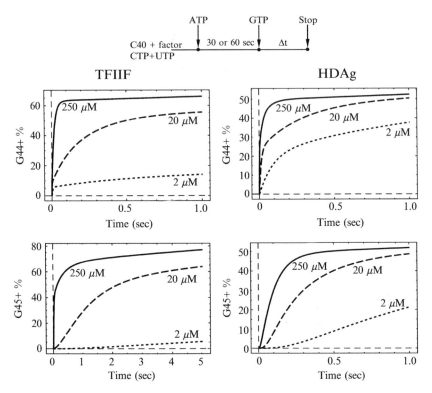

FIG. 4. Running-start, two-bond elongation protocol. The reaction scheme is shown above the figure. CTP and UTP were initially 20 μm. ATP was initially 100 μm. GTP was 250, 20, or 2 μm, as indicated. G44 and G45 synthesis rates (G44+% and G45+% to indicate the percent of G44 or G45+ all longer transcripts generated from A43) are shown in the presence of TFIIF (12 pmol per reaction) (left panels) or HDAg (77 pmol per reaction) (right panels). The curves shown are from a DYNAFIT[35] simulation of experimental data using an adequate kinetic model (not shown). Note that for the reaction in the presence of TFIIF, G45+% data is on a 5-s time scale. G45 synthesis is much more highly dependent on GTP in the presence of TFIIF than in the presence of HDAg.

NTP-Dependent Steps in Transcription

Studies with *E. coli* RNA polymerase indicate an unusual substrate NTP dependence for elongation, particularly at low NTP concentrations.[33] Based on these observations, we tested RNA polymerase II elongation in a running-start assay through the sequence 40-CAAAGG-45, varying substrate GTP concentration. Interestingly, the shapes of rate curves for G44

[35] P. Kuzmic, *Anal. Biochem.* **237,** 260 (1996).

and G45 synthesis are different (Fig. 4, compare upper and lower panels), indicating that distinct kinetic information is obtained from analysis of each bond (Fig. 4). Furthermore, G44 and G45 synthesis rate curves have different shapes in the presence of TFIIF or HDAg (Fig. 4, left and right panels), indicating that these factors, although they stimulate bulk elongation to a similar extent, do so according to different regulatory strategies. As shown for *E. coli* RNA polymerase,[33] human RNA polymerase II has at least two distinct rates on the forward elongation pathway (Fig. 4, top panels). Briefly stated, the manner in which changing GTP concentration bends the G44 synthesis rate curves indicates two distinct forms of A43 elongation complex with different requirements for GTP substrate. One form is more highly poised on the active synthesis pathway, and one form must first bind GTP and then undergo a conformational change to be converted to a catalytically competent state. Analysis of the G44 bond, therefore, provides evidence for a GTP "induced-fit" enzyme mechanism, in which prebinding of GTP induces a change in the elongation complex that stimulates subsequent phosphodiester bond formation. The conformational change associated with the induced-fit step, however, does not block access of the RNA polymerase II active site, because the most highly poised A43 complexes are observed to bind GTP and extend to the G44 position rapidly.[28] DNA polymerases have also been shown to elongate according to an induced-fit mechanism, but one in which the active site may close with each round of dNMP addition.[4,5] Rate curves for G45 synthesis have a distinctly sigmoidal shape, indicating a slow step after formation of the G44 bond that must occur before rapid synthesis of G45 can commence.[28–30] In the TFIIF-stimulated mechanism, the initial lag phase of the sigmoidal G45 synthesis rate curve is highly dependent on the GTP concentration, although such strong GTP dependence for G45 synthesis is not observed with HDAg (Fig. 4, compare left and right lower panels). The running-start assay, therefore, enabled us to measure fast elongation rates and obtain evidence for substrate GTP induced fit. Furthermore, analysis of G45 synthesis rates revealed a slow step in the elongation mechanism after G44 addition but before rapid synthesis of G45 could commence. Kinetic models for RNA polymerase II elongation that account for each of these observations and the known elongation steps of pausing, NTP-binding, translocation, chemistry, and pyrophosphate release will be published elsewhere.[28,29]

Conclusion

Transient state kinetic analysis of human RNA polymerase II elongation allows characterization of the forward synthesis mechanism with single-bond resolution. So far, using this approach, we have obtained evidence for distinct steps on the pausing and forward synthesis pathways and gained significant insight into substrate NTP-driven events in bond formation. The probable role of α-amanitin in RNA polymerase II inhibition has also been identified (not shown). Transient state kinetic analysis is further shown to have utility for analysis of two regulatory human elongation factors (Figs. 2, 3, and 4), one of which, HDAg, is associated with severe manifestations of hepatitis B infection.

Acknowledgments

This work was supported by a grant from the National Institutes of Health to Z.F.B. X.Q.G. was supported by the Michigan State University Development Foundation as part of the Gene Expression in Development and Disease initiative. Z.F.B. is a member of the Agricultural Experiment Station at Michigan State University.

[19]   Analysis of RNA Polymerase II Elongation *In Vitro*

By TODD E. ADAMSON, SARAH M. SHORE, and DAVID H. PRICE

Precursors to eukaryotic mRNAs are synthesized during the elongation phase of transcription by RNA polymerase II, and processing of these transcripts into mature mRNAs occurs while the polymerase is still engaged. Entry into productive elongation is regulated by initiation of transcription[1] and by a postinitiation elongation control event mediated by P-TEFb.[2] Exit from the elongation mode, termination, usually occurs close to the promoter (premature termination) or after the signal for 3' end processing of the mature transcript has been passed. RNA polymerase II does not act alone, but rather is influenced by a dynamic array of factors.[2,3] Most of the factors affecting RNA polymerase II elongation were discovered through the use of functional assays that provided a means of purifying factors that had observable effects on the inherent elongation properties of RNA polymerase II. In addition to the factors that affect and control

[1] A. Dvir, J. W. Conaway, and R. C. Conaway, *Curr. Opin. Genet. Dev.* **11,** 209 (2001).

[2] D. H. Price, *Mol. Cell Biol.* **20,** 2629 (2000).

[3] J. W. Conaway, A. Shilatifard, A. Dvir, and R. C. Conaway, *Trends Biochem. Sci.* **25,** 375 (2000).

elongation directly, the processing machinery required to generate the mature mRNA is thought to interact with and be influenced by the elongation complex.[4] The purpose of this article is to provide methods for studying elongation and associated events and to provide information useful in interpreting the results obtained.

To facilitate the discussion of elongation, a few basic definitions are provided. *Initiation* refers to the events that start with the formation of a preinitiation complex and end with promoter clearance. This includes the formation of the first phosphodiester bond, abortive initiation that repetitively generates transcripts between 2 and approximately 10 nt in length, and promoter escape that requires a helicase activity in TFIIH.[5] *Elongation* includes all subsequent transcript extension that occurs until dissociation of the polymerase, transcript, and template, in the *termination* event. During elongation the polymerase may *pause* for short or long periods of time at specific positions along the template, or it may *arrest* while still associated with the template and transcript. The only difference between a pause and an arrest is that a paused polymerase will resume elongation alone, and an arrested polymerase requires the action of a specific elongation factor, S-II, to resume elongation.[6,7] Polymerases can be *stalled* at any site by removal of nucleotides. Normally, RNA polymerase II encounters negative elongation factors shortly after initiation and requires the DRB-sensitive action of a cyclin-dependent kinase, P-TEFb, to enter *productive elongation*, during which full-length transcripts are generated.[2,8–10]

The best method to determine whether expression of a gene is controlled at the level of elongation is a nuclear run-on assay. Nuclei are isolated from cells, and the stalled polymerases are allowed to extend their nascent transcripts for a short time in the presence of labeled nucleoside triphosphate (NTPs).[11] The labeled transcripts are isolated and hybridized to filters or arrays containing DNA fragments spanning regions of genes of interest. To determine if elongation is controlled, nuclei are isolated from cells grown under two conditions of interest (for example, with and without hormone) and the ratio of RNA hybridizing to 5′ and 3′ probes in each set of nuclei are compared. If one condition results in strong 5′, but weak 3′ signal, and the other has more equal signal for both probes, then elongation

[4] N. J. Proudfoot, A. Furger, and M. J. Dye, *Cell* **108,** 501 (2002).
[5] J. Bradsher, F. Coin, and J. M. Egly, *J. Biol. Chem.* **275,** 2532 (2000).
[6] D. Reines, *J. Biol. Chem.* **267,** 3795 (1992).
[7] H. Guo and D. H. Price, *J. Biol. Chem.* **268,** 18762 (1993).
[8] N. F. Marshall and D. H. Price, *Mol. Cell. Biol.* **12,** 2078 (1992).
[9] N. F. Marshall and D. H. Price, *J. Biol. Chem.* **270,** 12335 (1995).
[10] J. Peng, Y. Zhu, J. T. Milton, and D. H. Price, *Genes Dev.* **12,** 755 (1998).
[11] W. F. Marzluff, *Methods Enzymol.* **181,** 30 (1990).

control is indicated. Many genes are controlled in this manner, such as HSP70[12,13] and c-fos.[14]

To identify factors that affect elongation or RNA processing, and to obtain details about their mechanism of action, soluble *in vitro* transcription systems have been and continue to be invaluable. Initially, dC-tailed templates[15] were used to study the properties of elongating RNA polymerase II,[16] and to identify elongation[17] and processing factors.[18] The advantage of dC-tailed templates was that they allowed initiation of RNA polymerase II without the auxiliary factors needed for promoter-containing templates. However, we have found that immobilized templates are more suitable for all types of studies for which dC-tailed templates were once used because they are easier to prepare, do not require purified RNA polymerase II, and never leave the transcript in heteroduplex with the template strand, as frequently happens with tailed templates.[16,19] The following sections describe methods useful for studying elongation by RNA polymerase II *in vitro* and provide illustrative examples of typical experiments.

Preparation of Nuclear Extract

The quality of the results obtained from *in vitro* transcription reactions is directly related to the nuclear extract used. The following method results in efficient extraction and concentration of RNA polymerase II, general transcription factors, and elongation control machinery. High-quality nuclear extracts can be prepared from a variety of cell lines or tissue types across many species. When making nuclear extracts derived from tissues or whole organisms, it is important to carefully separate the nuclei from the cytosol to reduce nuclease and protease activities released from lysosomes or associated degradative tissues. Please refer to an earlier article in *Methods in Enzymology* for an excellent description of the preparation of active nuclei from a variety of sources.[11] Stringent washing of isolated nuclei is generally not required when tissue culture cells are utilized. The following protocol has been used to make extracts from HeLa cells and *Drosophila* K_c cells. Tissue culture cells should be grown in suspension

[12] A. E. Rougvie and J. T. Lis, *Cell* **54,** 795 (1988).
[13] J. T. Lis, P. Mason, J. Peng, D. H. Price, and J. Werner, *Genes Dev.* **14,** 792 (2000).
[14] M. A. Collart, N. Tourkine, D. Belin, P. Vassalli, P. Jeanteur, and J. M. Blanchard, *Mol. Cell Biol.* **11,** 2826 (1991).
[15] T. R. Kadesch and M. J. Chamberlin, *J. Biol. Chem.* **257,** 5286 (1982).
[16] A. E. Sluder, D. H. Price, and A. L. Greenleaf, *J. Biol. Chem.* **263,** 9917 (1988).
[17] A. E. Sluder, A. L. Greenleaf, and D. H. Price, *J. Biol. Chem.* **264,** 8963 (1989).
[18] D. H. Price and C. S. Parker, *Cell* **38,** 423 (1984).
[19] C. M. Kane, *Biochemistry* **27,** 3187 (1988).

under conditions that allow the maximum growth rate. Cells should not be allowed to reach saturation at any time, and they should be at least one doubling time away from reaching saturation at time of extract preparation. Growth of *Drosophila* K_c cells and extract preparation was described in some detail earlier,[20] and the protocol provided here is very similar. We routinely use HeLa cells grown by the National Cell Culture Center (Minneapolis, MN) that are washed with PBS and shipped on wet ice. Although the following protocol is written for 10 liters of cells, it can be adapted to other amounts of starting material as long as the volumes are adjusted to maintain a similar concentration of nuclei. All steps are carried out at 0–1°. The washed cells are brought up to 20 ml with low-salt lysis buffer (25 mM HEPES, pH 7.6 150 mM NaCl, 0.1% NP-40, 0.1% PMSF saturated in isopropanol, 1 mM DTT) and homogenized in a Dounce homogenizer using the loose pestle for 20 to 40 strokes. Stroke rate should be as rapid as possible without causing cavitation on the up stroke or expulsion of the homogenate on the down stroke. Cell lysis is monitored using a phase contrast microscope and should be at least 95%. Nuclei are pelleted at 6,000 rpm ($6000g_{av}$) for 10 min in a JS13.1 (Beckman, Fullerton, CA) swinging bucket rotor or equivalent. The nuclear pellet is resuspended up to 20 ml in low-salt lysis buffer using a Dounce homogenizer, and the homogenate transferred to a 26-ml Oak Ridge (Oak Ridge, TN) tube. The volume of nuclei is calculated by subtracting the volume of buffer added from the total volume of nuclei and buffer. 4 M NH$_4$SO$_4$, pH 7.9, is added to achieve 360 mM (add $\frac{1}{10}$ total volume), and the contents of the tube are quickly mixed before being rotated end-over-end for 30 min in a Lab Quake (Barnstead-Thermolyne, Dubuque, IA) rotator. The chromatin is pelleted with a 45-min spin at 45,000 rpm ($150,000g_{av}$) in a Beckman (Fullerton, CA) 55.2 TI angle head rotor at 1°. The supernatant is transferred to a new Oak Ridge tube, and 0.25 g of solid NH$_4$SO$_4$ is added per ml of supernatant. After rotating end-over-end for 20 min, the tube is spun at 45,000 rpm for 15 min at 1° and the supernatant is removed from the pellet. The pellet is scraped from the tube in an amount of HGEDP (20 mM HEPES, pH 7.6, 15% glycerol, 0.1 mM EDTA, 1 mM DTT, 0.1% PMSF saturated in isopropanol) that is half of the nuclear volume, and the pelleted protein is forced to dissolve by homogenization with a small Dounce homogenizer. The extract is then dialyzed (14,000 molecular weight cut-off) against 100 volumes of 150 mM HGKEDP (HGEDP with 150 mM KCl) with frequent mixing of the contents in the bag. After 3 h the extract is removed from the bag and spun in multiple tubes in a microfuge at full speed for 10 min. Aliquots of the extract are pipetted into new tubes

---

[20] D. H. Price, A. E. Sluder, and A. L. Greenleaf, *J. Biol. Chem.* **262**, 3244 (1987).

and stored at $-80°$. The extract should be stable for many years and can be frozen and thawed several times without loss of activity.

Pulse–Chase Assay

The most convenient way to examine elongation *in vitro* is to force synchronized initiation of polymerases and then follow the progress of the elongating polymerases with time. These conditions are met by the pulse–chase assay. Preinitiation complexes are formed during a 10-min incubation (9 μl) containing 1.2 μl of HeLa nuclear extract (HNE), 20 μg/ml cytomegalovirus (CMV) promoter containing template DNA that has been linearized to produce a 548-nt run-off, 20 mM HEPES pH 7.6, 7 mM MgCl$_2$, and 60 mM KCl. All steps are normally carried out at room temperature ($20°$), but temperatures up to $37°$ can be used. The elongation rate is dependent on the temperature, with higher temperatures giving higher rates. RNasin (Promega, Madison, WI) can be added if desired, but is usually not necessary. If DRB is to be included to inhibit P-TEFb, it is added into the preincubation to 50 μm from a freshly diluted 500-μm solution made from a 10-mM stock in ethanol that is stored at $-80°$. For consistency, preincubations are set up for multiple reactions in one tube and then divided into individual tubes. For most experiments the preincubation time is not critical and can be increased if convenient. More dilute extracts require greater amounts of extract and longer preincubation time and may have different template DNA optima. Formation of the preinitiation complexes is very sensitive to salt, so it is important to determine the salt concentration of the starting extract using a conductivity meter. Addition of 3 μl of a pulse mixture containing 5 μCi-[α-^{32}P]CTP brings each reaction up to 500 μm GTP, UTP, and ATP while maintaining the same concentration of buffer, Mg, and salt. After 30 s, the reaction is stopped or, if elongation is to continue, a 3-μl chase mixture is added maintaining all components at the same level except that cold CTP is added to 1.2 mM to block further incorporation of label. The individual reactions are stopped by the addition of 200 μl of stop solution (100 mM Tris, pH 8.0, 100 mM NaCl, 10 mM EDTA, 1% Sarkosyl with 200 μg/ml low molecular weight Torula yeast RNA [Sigma, St. Louis, MO]) and extracted with 150 μl of water-saturated, neutralized phenol. 3 volumes of 95% ethanol with 0.5 M ammonium acetate are mixed into the aqueous phase, and the mixture is incubated at $-80°$ for 10 min before centrifugation for 10 min at full speed in a microfuge at $4°$. The pelleted RNA and the entire inside surface of the microfuge tube is washed with 70% ethanol, and the tube is recentrifuged for 3 min in the same orientation as the first spin. After careful removal of the wash solution, the pellet is air dried and dissolved in 7 μl RNA loading buffer (8 M urea, 0.25× TBE containing bromophenol blue and xylene cyanol) by heating at 70°, with frequent tapping of the tube for 5 min. The RNA is analyzed on a 0.8-mm-thick 6% polyacrylamide gel with 1× TBE buffer and 6 M urea in the gel and 1× TBE running buffer. Wells are cleared of urea, and samples are carefully loaded into the center lanes. The gel is run at a rate that allows significant warming of the gel plates (40–50°), and when the bromophenol blue reaches the bottom, the gel is soaked for about 15 min in water containing a small amount of ethidium bromide with rocking to soak out the urea and stain the nucleic acids before drying under vacuum. Autoradiography of the dried gel provides an image of the results, and quantitation is accomplished with a Packard InstantImager (Perkin Elmer, Woodbridge, Ontario) or a phosphoimager.

Figure 1 illustrates typical results from pulse/chase experiments. During the pulse, transcripts less than 25 nt in length are efficiently labeled (Fig. 1A, lane marked P). Because CTP was used as label, tRNAs were labeled by transfer of CCA to their 3′ ends. UTP or GTP can be used as label instead to bypass this artifact, but in some extracts other bands higher in the gel are artifactually labeled. When reactions were chased for different times, the labeled transcripts were extended, generating a pattern that reflects a combination of the elongation properties of the polymerase and the effect of elongation factors in the extract. A comparison of the results of similar reactions carried out in the presence of 50 μm DRB demonstrates that the effect of P-TEFb is to allow polymerases to enter productive elongation and reach the run-off transcript faster (Fig. 1A). If continuous labeling conditions had been used, the run-off would have been essentially the only transcript seen, because it is the longest and the most extensively labeled. Pulse/chase conditions allow comparison of amounts of short and long transcripts because each transcript is labeled equally at the 5′ end. To determine if a factor affects elongation versus initiation, an experiment like that in Fig. 1B, examining the function of human immunodeficiency virus (HIV) Tat protein, can be performed. The short transcripts produced from the HIV–LTR during a pulse with and without Tat were identical, indicating that Tat does not affect initiation (Fig. 1B, pulse). During the chase, the reaction containing Tat was able to produce more run-off transcript in the same period of time than the control reaction. Adding DRB to the reactions gave identical patterns with or without Tat. Consistent with the model of Tat function in which Tat recruits P-TEFb to the nascent HIV transcript,[21,22] the viral protein stimulated polymerases more than 70 nt in length (above the tRNAs) to reach run-off in a DRB-sensitive manner (Fig. 1B).

[21] Y. Zhu, T. Peéry, J. Peng, Y. Ramanathan, N. Marshall, T. Marshall, B. Amendt, M. B. Matthews, and D. H. Price, *Genes Dev.* **11,** 2622 (1997).

[22] M. E. Garber, P. Wei, V. N. Kewal Ramani, T. P. Mayall, C. H. Herrmann, A. P. Rice, D. R. Littman, and K. A. Jones, *Genes Dev.* **12,** 3512 (1998).

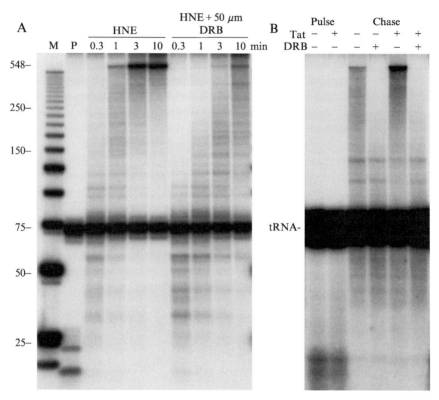

Fɪɢ. 1. (A) Pulse/chase assay using CMV template. Preinitiation complexes were formed in HNE. Initiation was accomplished during a 30-s pulse and was followed by a chase for the indicated times in the absence or presence of 50 μm DRB. M, 25-bp ladder; P, pulse only. (B) Tat transactivation on HIV–LTR template. A 30-s pulse with or without 30 ng of purified Tat protein[23] (first two lanes) was followed by a 10-min chase with 50 μm DRB and/or Tat as indicated.

Preparation of Immobilized Templates

Immobilization of the DNA template exponentially increases the number of reaction manipulations possible during transcription. The easiest way to attach DNA to a solid matrix is to generate the template using a polymerase chain reaction (PCR) reaction in which one of the primers is biotinylated. The resulting biotinylated DNA is bound to either agarose beads or plastic-coated paramagnetic particles that have streptavidin

[23] T. P. Cujec, H. Cho, E. Maldonando, J. Meyer, D. Reinberg, and B. M. Peterlin, *Mol. Cell Biol.* **17,** 1817 (1997).

covalently attached. Several parameters need to be considered in the design and synthesis of an immobilized template. The biotin should be located upstream of the promoter to allow polymerases that initiate to be able to efficiently run off the end of the template. The template should contain all known elements of the promoter required for efficient initiation *in vitro*. Practically, this means the biotin should be about 300 bp upstream of the start point of transcription. To facilitate examination of the elongation properties of the polymerase, the length of the transcribed region should be between 500 and 1000 bp. However, the shorter the DNA fragment, the more efficiently it will bind to the beads. We normally optimize each template primer set using Taq (Invitrogen, Carlsbad, CA) so that about 15 μg of DNA is generated in each 100-μl reaction. Products from 10 to 40 PCR reactions are pooled, ethanol precipitated, and dissolved into TE (10 mM Tris, pH 8.0, 1 mM EDTA). Unincorporated biotinylated primers can be removed using a 5-ml G-200 gel filtration column equilibrated and run in TE plus 500 mM NaCl or by using a DNA purification kit, UltraClean 15 (Mo Bio Laboratories, Carlsbad, CA). The purified DNA is incubated with the streptavidin-coated beads (paramagnetic Dynal Dynabeads M-280; Dynal Biotech; Lake Succ. NY) for about 1 h. It is important to carry out the binding in the presence of high salt (TE with 500 to 1000 mM NaCl) to minimize electrostatic repulsion of the negatively charged DNA molecules as they are concentrated on the beads. The capacity of the beads varies inversely with the length of the DNA, but usually we use 1 ml of bead slurry (10 mg of beads) for binding of 150 μg of DNA that is about 1000 bp long. Some DNA will not bind, possibly because of incompletely modified primers. The beads are washed with TE in high salt and then just TE for storage at 4° at 20 mg of beads/ml. If large amounts of biotinylated DNA are synthesized, it is best to bind only a portion to beads at a time for use over several months. The remaining DNA can be frozen at −80° until needed. It is important to assay from 1 to 8 μl of the immobilized template bead slurry using a standard pulse/chase assay to find the optimum amount for transcription (usually 2 μl of bead slurry). Large amounts of beads can be used by concentrating them using a magnet, removing the storage buffer, and then adding the preincubation mixture, ignoring the small bead volume.

Isolation of Elongation Complexes

Elongation complexes can be isolated using the immobilized template by exploiting the high stability of the RNA polymerase II ternary complex. Washing elongation complexes with 1 M KCl and 1% Sarkosyl strips all known factors from the polymerase and template. For isolation of early elongation complexes, a standard preincubation with extract as described earlier is set up with 1 to 20 reactions' worth of template and extract. The entire reaction is pulse labeled as before and after 30 s, EDTA is added to 20 mM to stop transcription. The beads are concentrated using a magnetic concentrator from Dynal and washed three times with 200 μl of buffer containing 20 mM HEPES, pH 7.6, 7 mM MgCl$_2$, 1 M KCl, and 1% Sarkosyl. Resuspension of the beads between each wash is accomplished by quickly removing the supernatant from the preceding wash, removing the tube from the magnet, and then pipetting the beads up and down several times with the new wash solution. The beads should not be left in association with the magnet too long or they will become difficult to pipet. After washing with high salt and detergent, the beads are washed several times in low-salt buffer (20 mM HEPES, pH 7.6, 7 mM MgCl$_2$, 60 mM KCl) containing 200 μg/ml bovine serum albumin (BSA). The elongation complexes are fairly stable, but we normally use them within about 30 min. After the initial isolation of the complexes, the transcripts can be extended by the addition of all four or selected subsets of NTPs along with 20 mM HEPES, pH 7.6, 7 mM MgCl$_2$, and 60 mM KCl to obtain transcripts of desired length. The rate of elongation of human RNA polymerase II without any elongation factors and at saturating levels of NTPs (500 μm) is 50 to 100 min at room temperature. The complexes can be reconcentrated and washed many times. However, during subsequent isolations at some template positions, a fraction of the polymerases fall into a conformation that is refractory to further elongation.

Elongation complexes can be isolated under less stringent conditions to look for the physical or functional association of factors. Washing elongation complexes directly after the pulse with low-salt buffer allows retention of the termination factor TTF2 because it binds tightly to the template.[24] Most elongation factors are not stably associated with the elongation complex and are functionally removed by washing even at low salt. Specific versus nonspecific association of factors with low-salt washed complexes is an important issue. We have been able to detect some factors, such as P-TEFb, after low-salt washing by Western blotting, but do not see any functional consequence of this association using a transcription assay. Nonspecific binding of many factors, including RNA polymerase II, to the beads and template can be a problem. Using a restriction enzyme to digest the template and release the elongation complex into the supernatant can help reduce nonspecifically associated proteins, but this method does not eliminate the nonspecific binding of RNA polymerase II to DNA or the nonspecific association of other factors with the template, polymerase, or

[24] Z. Xie and D. H. Price, *J. Biol. Chem.* **273,** 3771 (1998).

other associated factors. We believe that the best assay for association is one that has a functional readout.

Add-Back Assay

Isolated elongation complexes are very useful in add-back assays to determine if a crude extract or purified protein has an effect on elongation. dC-tailed templates were used for this purpose in the past, but as explained earlier, isolated elongation complexes have significant advantages, not the least of which is that the complexes can be reisolated after incubation with other proteins or mixtures of proteins. The effect of TFIIF on elongation is illustrated in Fig. 2A. Elongation complexes containing mostly transcripts less than 25 nt in length were isolated using a high-salt and Sarkosyl washing protocol and then allowed to elongate for 0.3, 1, 3, or 10 min with no addition or with recombinant TFIIF.[25] With no factors added back, the polymerases move steadily and slowly down the template and begin to reach the run-off site after 10 min. The band between 50 and 75 nt that appears after 10 min is an artifact caused by contamination of a nuclease and can be eliminated by including RNasin (Promega) in the elongation reaction. The 3- and 10-min lanes in the absence of TFIIF have a similar pattern to the 0.3- and 1-min lanes with TFIIF, demonstrating that the factor increases the elongation rate about 10-fold. The add-back assay is a sensitive method to screen for factors that affect elongation by increasing or decreasing either the maximal or average elongation rate. Also, subtle differences in the pattern of transcripts during elongation can be used to document effects of factors. If the pattern of transcripts does not show a steady increase in length during transcription in the presence of the tested factor, it could be because the factor has a contaminating RNAse or DNAse, or because the polymerase is terminating. All three possibilities can yield transcripts that are no longer associated with the beads. To test for the presence of RNAses, long transcripts such as those resulting from a 10-min chase can be incubated with the factor (or extract) in the presence of α-amanitin or absence of NTPs to stop transcription. Incubation of the material to be tested for elongation activity with DNA, followed by gel analysis of DNA, can be used to check for the presence of DNAses.

The factors required for RNA polymerase II elongation control can be added back to isolated elongation complexes. As shown in Fig. 2B, whole HNE can be added back to isolated elongation complexes and the pattern of transcripts generated in the absence or presence of DRB resembles that

---

[25] J. Peng, M. Liu, J. Marion, Y. Zhu, and D. H. Price, *Cold Spring Harbor Symp. Quant. Biol.* **63**, 365 (1998).

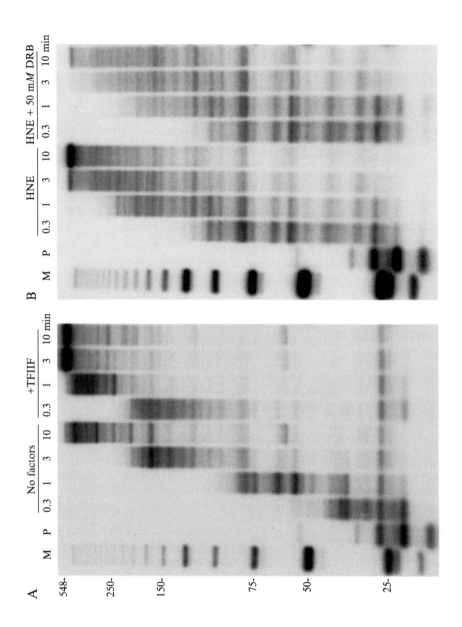

generated during normal transcription of the template (compare Fig. 2B with Fig. 1A). This indicates that both negative and positive factors needed for elongation control can function even when elongation is separated from initiation. The add-back assay has been used to develop a defined system of elongation control comprised of two purified negative factors NELF and DSIF and P-TEFb.[26]

Capping Assay and Coupling of RNA Processing Events to Transcription

Capping has recently been shown to be coupled to transcription such that guanylylation of transcripts in elongation complexes occurs about 100,000 times faster when the transcripts are present in an elongation complex.[27] Functional coupling was demonstrated using an immobilized template and an add-back assay. It was found that during normal transcription reactions *in vitro*, capping of the nascent transcripts longer than 25 nt occurs very rapidly. The $t_{1/2}$ for guanylylation was 15 seconds and for methylation of the 7 position was 90 seconds. The extent of guanylylation or methylation was determined using a quantitative antibody binding assay that preserves transcript length information (see Moteki and Price[27] for details). With this assay and the availability of recombinant human capping enzyme (Hce) and human cap methyl transferase (Hcm), it is now possible to generate isolated elongation complexes that contain transcripts that are completely capped (or completely uncapped) for use in add-back assays examining the effect of capping on the function of other factors. The add-back assay should be useful for examining other RNA processing events that may be functionally coupled to transcription, such as polyadenylation and splicing.

Acknowledgment

This work was supported by NIH grants GM35500 and AI43691 for D.H.P.

[26] D. B. Renner, Y. Yamaguchi, T. Wada, H. Handa, and D. H. Price, *J. Biol. Chem.* **276,** 42601 (2001).
[27] S. Moteki and D. H. Price, *Mol. Cell* **10,** 599 (2002).

FIG. 2. Elongation complexes were isolated using a high-salt and Sarkosyl wash protocol (lanes marked P) as described in text. M, 25-bp ladder. (A) Effect of TFIIF on isolated elongation complexes. Transcripts were extended in the presence of all four NTPs at 500 mM each in the absence or presence of TFIIF for the indicated times. (B) Reconstitution of elongation control on isolated elongation complexes. HNE or HNE and 50 μm DRB was added back to isolated elongation complexes, followed by transcript extension for the indicated times.

[20] Preparation and Assay of RNA Polymerase II Elongation Factors Elongin and ELL

By Stephanie E. Kong, Ali Shilatifard, Ronald C. Conaway, and Joan Weliky Conaway

Introduction

Eukaryotic mRNA synthesis is an elaborate biochemical process controlled by the action of a remarkably large collection of transcription factors that support accurate initiation and efficient elongation of transcripts by the multisubunit enzyme RNA polymerase II.[1] The early observation that purified RNA polymerase II is "slow" and unable to elongate transcripts *in vitro* at the observed rates of mRNA synthesis in cells of 2000 or more nucleotides per minute prompted biochemical searches for transcription factors that could stimulate the enzyme's elongation rate.[2] Among these transcription factors, Elongin and ELL were discovered and purified to homogeneity from rat liver by their abilities to stimulate the overall rate of elongation by RNA polymerase II that had either initiated transcription from promoters in reactions dependent on the general initiation factors TFIIB, TFIID, TFIIE, TFIIF, and TFIIH or initiated transcription from the ends of oligo(dC)-tailed DNA templates[3,4] in the absence of general initiation factors.

Elongin is a heterotrimeric complex composed of a transcriptionally active Elongin A subunit of ~800 amino acids and two smaller Elongin B and Elongin C subunits of ~100 amino acids.[3,5] Mammalian cells contain at least three distinct but structurally related Elongin A proteins designated Elongin A, Elongin A2, and Elongin A3,[6,7] all of which have been shown to bind specifically to Elongins B and C and to be capable of stimulating

[1] A. Shilatifard, *Biochemistry* **379,** 27 (1998).

[2] S. M. Uptain, C. M. Kane, and M. J. Chamberlin, *Annu. Rev. Biochem.* **66,** 117 (1997).

[3] J. N. Bradsher, K. W. Jackson, R. C. Conaway, and J. W. Conaway, *J. Biol. Chem.* **268,** 25587 (1993).

[4] A. Shilatifard, W. S. Lane, K. W. Jackson, R. C. Conaway, and J. W. Conaway, *Science* **271,** 1873 (1996).

[5] J. N. Bradsher, S. Tan, H.-J. McLaury, J. W. Conaway, and R. C. Conaway, *J. Biol. Chem.* **268,** 25594 (1993).

[6] T. Aso, K. Yamazaki, K. Amimoto, A. Kuroiwa, H. Higashi, Y. Matsuda, S. Kitajima, and M. Hatakeyama, *J. Biol. Chem.* **275,** 6546 (2000).

[7] K. Yamazaki, L. Guo, K. Sugahara, C. Zhang, H. Enzan, Y. Nakabeppu, S. Kitajima, and T. Aso, *J. Biol. Chem.* **277,** 26444 (2002).

the rate of elongation by RNA polymerase II *in vitro*.[6–9] In addition, a *Caenorhabditis elegans* Elongin A homolog has been shown to be capable of binding to mammalian Elongins B and C and stimulating the rate of elongation by mammalian RNA polymerase II *in vitro*.[9]

ELL, the founding member of the ELL family, is an ~700 amino acid protein.[10] The human *ELL* (eleven-nineteen lysine-rich in leukemia) gene on chromosome 19p13.1 was initially identified as a gene that undergoes frequent translocations with the *MLL* (mixed-lineage leukemia) gene on chromosome 11q23 in acute myeloid leukemia.[10] In mammalian cells, the ELL family includes at least three members designated ELL, ELL2, and ELL3, all of which are capable of stimulating the rate of elongation by RNA polymerase II *in vitro*.[4,11,12] In addition, a *Drosophila melanogaster* ELL homolog is capable of stimulating the rate of elongation by RNA polymerase II *in vitro*,[13] and mutations in the *D. melanogaster ELL* gene were recently found to impair selectively RNA polymerase II transcription of longer fly genes.[14]

Based on varied evidence from biochemical studies, Elongin and ELL appear to function similarly to increase the overall rate of transcript elongation by RNA polymerase II by a mechanism that involves suppression of transient pausing by polymerase at many sites along the DNA template.[4,15] Both Elongin and ELL appear to suppress RNA polymerase II pausing through direct interactions with polymerase by decreasing the time the enzyme spends in inactive conformations.[16] Next we describe methods for preparing recombinant Elongin and ELL and for assaying their activities in promoting rapid and efficient elongation by RNA polymerase II.

[8] T. Aso, W. S. Lane, J. W. Conaway, and R. C. Conaway, *Science* **269**, 1439 (1995).

[9] T. Aso, D. Haque, R. J. Barstead, R. C. Conaway, and J. W. Conaway, *EMBO J.* **15**, 5557 (1996).

[10] M. J. Thirman, D. A. Levitan, H. Kobayashi, M. C. Simon, and J. D. Rowley, *Proc. Natl. Acad. Sci. USA* **91**, 12110 (1994).

[11] A. Shilatifard, D. R. Duan, D. Haque, C. Florence, W. H. Schubach, J. W. Conaway, and R. C. Conaway, *Proc. Natl. Acad. Sci. USA* **94**, 3639 (1997).

[12] T. Miller, K. Williams, R. W. Johnstone, and A. Shilatifard, *J. Biol. Chem* **275**, 32052 (2001).

[13] M. Gerber, J. Ma, J. C. Eissenberg, and A. Shilatifard, *EMBO J.* **20**, 6404 (2001).

[14] J. C. Eissenberg, J. Ma, M. A. Gerber, A. Christensen, J. A. Kennison, and A. Shilatifard, *Proc. Natl. Acad. Sci. USA* **99**, 9894 (2002).

[15] R. J. Moreland, J. S. Hanas, J. W. Conaway, and R. C. Conaway, *J. Biol. Chem.* **273**, 26610 (1998).

[16] B. J. Elmendorf, A. Shilatifard, Q. Yan, J. W. Conaway, and R. C. Conaway, *J. Biol. Chem.* **276**, 23109 (2001).

Materials

Reagents

Unlabeled ultrapure ribonucleosides 5′-triphosphates and [α-^{32}P]CTP (>400 Ci/mmol) were purchased from Amersham Biosciences (Piscataway, NJ); recombinant RNAsin from Promega (Madison, WI); phenylmethylsulfonyl fluoride (PMSF) from Roche Molecular Biochemicals (Indianapolis, IN); and leupeptin and antipain from Sigma (St. Louis, MO). PMSF is dissolved in dimethylsulfoxide (DMSO) to 1 M and added to buffers immediately before use. Omnipur ammonium sulfate and glycerol were obtained from EM Science (Gibbstown, NJ). Isopropyl-β-D-thiogalacto-pyranoside (IPTG), polyvinyl alcohol (type II), and bovine serum albumin (BSA) (Fraction V) were obtained from Sigma, and Zwittergent ZC-8 came from Calbiochem (La Jolla, CA).

Chromatography Supplies

ProBond nickel resin was obtained from Invitrogen (Carlsbad, CA). TSK phenyl 5-PW and TSK SP 5-PW columns came from Tosoh Biosep (Montgomeryville, PA). All high-performance liquid chromatography (HPLC) was performed using either a Beckman (Fullerton, CA) System Gold or Amersham Biosciences SMART system chromatograph.

Buffers

Buffer A contains 40 mM HEPES–NaOH (pH 7.9), 1 mM EDTA, 1 mM DTT, and 10% (v/v) glycerol. Buffer B consists of 40 mM Tris–HCl (pH 7.9), 0.5 mM EDTA, 1 mM DTT, and 10% (v/v) glycerol. Buffer D contains 40 mM HEPES–NaOH (pH 7.9), 0.5 mM EDTA, 1 mM DTT, and 10% (v/v) glycerol. Buffer N contains 40 mM Tris–HCl (pH 7.9), 0.5 mM PMSF, 0.5 M NaCl, and 6 M guanidine–HCl. Buffer N$_2$ contains 20 mM HEPES–KOH (pH 7.9), 0.1 M KCl, 10% (v/v) glycerol, and 0.5 mM PMSF. Buffer R consists of 40 mM HEPES–NaOH (pH 7.9), 0.1 M KCl, 0.1 mM EDTA, 2 mM DTT, 50 μM ZnSO4, and 10% (v/v) glycerol.

Methods

Preparations of ELL

Although ELL can be purified from rat liver, it is simpler and more convenient to prepare recombinant ELL from *Escherichia coli* infected with an M13mpET bacteriophage vector[17] encoding ELL containing an N-terminal

6-histidine tag.[4] ELL expressed in *E. coli* is isolated from inclusion bodies and refolded.

Preparation of Recombinant ELL in E. coli

Expression of ELL in *E. coli* utilizes bacteriophage containing an N-terminal His-tagged ELL ORF. First, a 500 ml culture of *E. coli* strain JM109(DE3) is grown at 37° with gentle shaking in Luria broth supplemented with 2.5 mM MgCl$_2$ to an OD$_{600}$ of 0.3. Cells are then infected with bacteriophage to a multiplicity of infection (MOI) of 20. After 3.5 h of growth at 37°, cells are equilibrated to 30°, induced with 1 mM IPTG, and allowed to grow for an additional 12 h. Cells are centrifuged at 2000 $x g$ for 15 min at 4° and resuspended in 5 ml of ice-cold 50 mM Tris–HCl (pH 8.0) containing 6 M guanidine–HCl and 10 mM imidazole (pH 7.9). Recombinant ELL protein is then applied to a 1-ml ProBond Ni^{2+}-agarose column that has been equilibrated with Buffer N containing 10 mM imidazole (pH 7.9). The column is washed with buffer N$_2$ containing 40 mM imidazole (pH 7.9), and ELL is eluted with Buffer N$_2$ containing 300 mM imidazole (pH 7.9). Protein is then renatured by the addition of 2 vol of Buffer R and dialysis for 3 h at 4° against Buffer R without DTT.

Preparation of the Recombinant Elongin ABC Complex in E. coli

Two approaches for preparing the intact, transcriptionally active Elongin ABC complex from bacterially expressed proteins have proven successful. First, the Elongin ABC complex can be reconstituted by refolding its individually expressed and purified Elongin A, B, and C subunits, which are expressed in *E. coli* in insoluble form and are purified from inclusion bodies. Second, the Elongin ABC complex can be reconstituted by coexpression of its individual Elongin A, B, and C subunits in *E. coli*. The latter method is advantageous, because the complex is expressed in soluble form in bacteria expressing all three subunits.

Reconstitution of the Elongin ABC Complex by Refolding

Expression of the Elongin A, B, and C proteins in *E. coli* is accomplished using either an M13mpET bacteriophage expression system[17] or pRSET plasmid vectors. The Elongin A, B, and C proteins are expressed with N-terminal 6-histidine tags.

Using the M13mpET system, 100-ml cultures of *E. coli* strain JM109(DE3) (Promega) are grown to an OD$_{600}$ of 0.6 with gentle shaking

---

[17] S. Tan, R. C. Conaway, and J. W. Conaway, *BioTechniques* **16,** 824 (1994).

in LB medium containing 2.5 mM MgCl$_2$ 37°. Individual cultures are infected with M13mpET encoding Elongin A, B, or C at a MOI of 20. After a 2-h incubation at 37°, expression of Elongin proteins is induced by addition to cultures of IPTG to a final concentration of 0.4 mM, followed by an additional 2.5-h incubation.

Using pRSET plasmid expression vectors, 100-ml cultures of *E. coli* strain JM109(DE3) expressing Elongin A, B, or C are grown to an OD$_{600}$ of 0.3 with gentle agitation in LB medium containing 2.5 mM MgCl$_2$ at 37°. The expression of Elongin proteins is induced by addition of 0.4 mM IPTG to cultures, followed by an additional 2.5-h incubation.

All cultures are harvested by centrifugation at 2000 $x\,g$ for 10 min at 4°. The cell pellets are resuspended in 10 ml of ice-cold buffer containing 30 mM Tris–HCl (pH 8.0), 20% (w/v) sucrose, and 1 mM EDTA and kept on ice for 10 min. Cells are collected by centrifugation at 6000 $x\,g$ for 10 min and resuspended in 10 ml of ice-cold water. After 30 min, cells are collected by centrifugation at 8000 $x\,g$ for 10 min. After one cycle of freeze–thaw, inclusion bodies are collected by centrifugation at 100,000 $x\,g$ for 30 min at 4°. The resulting pellet is resuspended by homogenization with a Dounce homogenizer in 5 ml of Buffer N containing 10 mM imidazole (pH 7.9). The resulting suspension is clarified by centrifugation at 100,000 $x\,g$ for 30 min at 4°.

The 6-histidine-tagged Elongin A, B, and C proteins are purified from the supernatants of guanidine-solubilized inclusion bodies by Pro-Bond Ni^{2+}-agarose chromatography. Solubilized inclusion bodies from a 100-ml culture are applied to an ~1-ml ProBond column equilibrated in Buffer N containing 10 mM imidazole (pH 7.9). The ProBond column is washed with Buffer N$_2$ containing 100 mM KCl and 40 mM imidazole (pH 7.9) and eluted stepwise with Buffer N$_2$ containing 0.3 M imidazole (pH 7.9). The Elongin A, B, and C proteins can be stored at −80° for at least a year.

To reconstitute and purify the Elongin ABC complex, 500 μg Elongin A, 100 μg Elongin B, and 100 μg Elongin C from Ni^{2+}-agarose chromatography are mixed and diluted 5-fold with Buffer R. After a 90-min incubation on ice, the mixture is dialyzed against Buffer D until the conductivity is equivalent to Buffer D containing 0.1 M KCl and centrifuged at 60,000 $x\,g$ for 20 min at 4°. The supernatant is applied to a TSK SP 5-PW HPLC column (7.5 × 75 mm) preequilibrated in Buffer A containing 0.1 M KCl. The column is eluted at 1 ml/min with a 30-ml linear gradient in Buffer B from 0.1 M to 0.5 M KCl. One-ml fractions are collected. The Elongin ABC complex elutes between 0.35 and 0.4 M KCl. The TSK SP 5-PW fraction of Elongin ABC can be used in transcription assays; however, for some applications it may be necessary to purify the complex further to remove trace amounts of contaminating ribonuclease(s). To accomplish this, the TSK SP 5-PW fraction is diluted with an equal volume of Buffer A containing 3 M ammonium sulfate and applied to a TSK phenyl 5-PW column (7.5 × 75 mm) preequilibrated with Buffer A containing 1.5 M ammonium sulfate. The column is eluted at 1 ml/min with a 30-ml linear gradient in Buffer B from 1.5 M to 0 M ammonium sulfate. The purified Elongin ABC complex is stored in aliquots at −80°.

Reconstitution of the Elongin ABC Complex by Coexpression of the Elongin A, B, and C Subunits

Coexpression of the Elongin A, B, and C proteins in *E. coli* is accomplished by cotransformation of *E. coli* strain JM109(DE3) with a pRSET plasmid encoding Elongin A with an N-terminal 6-histidine tag and an ampicillin-resistance gene and with a pRSET plasmid encoding both Elongin B and Elongin C and a kanamycin-resistance gene. A 100-ml culture of *E. coli* strain JM109(DE3) expressing Elongin A, B, and C is grown to an OD_{600} of 0.3 with gentle shaking in LB medium containing 2.5 mM $MgCl_2$, 50 μg/ml ampicillin, and 50 μg/ml kanamycin at 37°. The expression of Elongin proteins is induced by addition of 0.4 mM IPTG to cultures, followed by an additional 2.5-h incubation. The culture is harvested by centrifugation at 2000 $x\ g$ for 10 min at 4°, and the cell pellets are resuspended in 10 ml of ice-cold buffer containing 30 mM Tris–HCl (pH 8.0), 20% (w/v) sucrose, and 1 mM EDTA. After 10 min, cells are collected by centrifugation at 6000 $x\ g$ for 10 min, resuspended in 10 ml of ice-cold water, kept on ice for 30 min, and collected by centrifugation at 8000 $x\ g$ for 10 min. After one cycle of freeze–thaw, the suspension is centrifuged at 100,000 $x\ g$ for 30 min at 4°.

The Elongin ABC complex is purified from the supernatant by consecutive Ni^{2+}-agarose and TSK SP 5-PW chromatography. The supernatant is applied to an ∼1-ml ProBond Ni^{2+}-agarose column equilibrated in Buffer N containing 10 mM imidazole (pH 7.9). The ProBond column is washed with Buffer N_2 containing 40 mM imidazole (pH 7.9) and eluted stepwise with Buffer N_2 containing 0.3 M imidazole (pH 7.9). Fractions containing the Elongin ABC complex are pooled, dialyzed against Buffer D to a conductivity that is equivalent to Buffer D containing 0.1 M KCl, and centrifuged at 60,000 $x\ g$ for 20 min at 4°. The supernatant is applied to a TSK SP 5-PW HPLC column (7.5 × 75 mm) preequilibrated in Buffer A containing 0.1 M KCl. The column is eluted at 1 ml/min with a 30-ml linear gradient in Buffer B from 0.1 M to 0.5 M KCl. One-ml fractions are collected. The Elongin ABC complex elutes between 0.35 and 0.4 M KCl. The Elongin ABC complex is stored in aliquots at −80°.

Assays of Elongin and ELL

Elongin and ELL can both stimulate the rate of elongation by purified RNA polymerase II *in vitro*. Two approaches for assaying Elongin and ELL transcription activities have proven successful. First, Elongin and ELL have both been shown to be capable of stimulating the rate of elongation by RNA polymerase II that has initiated transcription from a promoter in the presence of the general initiation factors TFIIB, TFIID, TFIIE, TFIIF, and TFIIH[4,5]; in this assay, Elongin and ELL are assayed for their abilities to increase the rate of accumulation of full-length run-off transcripts synthesized by RNA polymerase II from the adenovirus 2 major late (AdML) promoter. Second, Elongin and ELL have both been shown to be capable of stimulating the rate of elongation by RNA polymerase II on oligo(dC)-tailed DNA templates in the absence of auxiliary transcription factors[4,18]; in this assay, Elongin and ELL are assayed for their abilities to increase the rate of synthesis of long transcripts on linear duplex DNA templates that have been "tailed" at one end by terminal transferase addition of single-stranded oligo(dC).

The AdML Run-off Transcription Assay

RNA polymerase II preinitiation complexes are assembled at the AdML promoter by a 30-min preincubation at 28° of 60-μl reaction mixtures containing 20 mM HEPES–NaOH (pH 7.9), 20 mM Tris–HCl (pH 7.9), 60 mM KCl, 1 mM DTT, 0.5 mg/ml BSA, 2% (w/v) polyvinyl alcohol, 7% (v/v) glycerol, 6 U recombinant RNAsin, 100 ng of the EcoRI–NdeI fragment from the DNA template pDN–AdML,[19] 10 ng of recombinant TFIIB,[20] 10 ng of recombinant TFIIF,[17] 7 ng of recombinant TFIIE,[17] 40 ng of purified TFIIH (rat δ, fraction VI[21,22]), 50 ng of *Saccharomyces cerevisiae* TBP (Aca 44 fraction,[23] and 0.01 U of RNA polymerase II.[24] After assembly of the preinitiation complex, Elongin or ELL is titrated into reaction mixtures at levels of ~10 ng to 200 ng, and transcription is

[18] J. W. Conaway, J. N. Bradsher, S. Tan, and R. C. Conaway, *Cell. Mol. Biol. Res.* **39**, 323 (1993).

[19] R. C. Conaway and J. W. Conaway, *J. Biol. Chem.* **263**, 2962 (1988).

[20] A. Tsuboi, K. Conger, K. P. Garrett, R. C. Conaway, J. W. Conaway, and N. Arai, *Nucleic Acids Res.* **20**, 3250 (1992).

[21] R. C. Conaway, D. Reines, K. P. Garrett, W. Powell, and J. W. Conaway, *Methods Enzymol.* **273**, 194 (1996).

[22] R. C. Conaway and J. W. Conaway, *Proc. Natl. Acad. Sci. USA* **86**, 7356 (1989).

[23] J. W. Conaway, J. P. Hanley, K. P. Garrett, and R. C. Conaway, *J. Biol. Chem.* **266**, 7804 (1991).

[24] H. Serizawa, R. C. Conaway, and J. W. Conaway, *Proc. Natl. Acad. Sci. USA* **89**, 7476 (1992).

initiated by addition of 7 mM MgCl$_2$, 50 μm ATP, 1–2 μm UTP, 10 μm CTP, 50 μm GTP, and 10 μCi of [α-^{32}P]CTP. Transcription reactions are carried out at 28° for 10–30 min, and 10-μl aliquots are removed at regular intervals, mixed with an equal volume of 0.2 M Tris–HCl (pH 7.5), 25 mM EDTA, 0.3 M NaCl, and 2% (w/v) SDS, digested with 1 mg/ml proteinase K at room temperature, and ethanol precipitated with 20 μg of total yeast RNA as carrier. Run-off transcripts are suspended in 9 M urea, 0.025% (w/v) bromophenol blue, and 0.025% (w/v) xylene cyanol FF and analyzed by electrophoresis through 6% polyacrylamide, 7 M urea gels.

The Oligo(dC)-tailed DNA Template Assay

Elongin and ELL activities can be measured directly or in pulse–chase assays using the linear, oligo(dC)-tailed duplex DNA template pCpGR220 S/P/X.[25] In this template, the first nontemplate strand (dT) residues are 136, 137, and 138 nucleotides from the oligo(dC) tail; the next run of (dT) residues are located from ~240 to ~250 nucleotides from the oligo(dC) tail. In pulse–chase assays, 0.01 U of RNA polymerase II are mixed with 100 ng of the pCpGR220 S/P/X DNA template, and transcription, in which a radioactive transcript of ~135 nucleotides is synthesized, is carried out for 20 to 30 min at 28° in the presence of 20 mM HEPES–NaOH (pH 7.9), 20 mM Tris–HCl (pH 7.9), 2% (w/v) polyvinyl alcohol, 0.5 mg/ml BSA, 60 mM KCl, 7 mM MgCl$_2$, 0.2 mM DTT, 3% (v/v) glycerol, 3 units of recombinant RNAsin, 50 μM ATP, 50 μM GTP, 2 μM CTP, and 10 μCi [α-^{32}P]CTP. Elongin or ELL at levels of 10 ng to 200 ng are then added to reaction mixtures, and transcripts are chased into longer RNAs for 5 to 10 min by addition of 100 μM CTP and 2 μM UTP. Transcripts are suspended in 9 M urea, 0.025% (w/v) bromophenol blue, and 0.025% (w/v) xylene cyanol FF and analyzed by electrophoresis through 6% polyacrylamide, 7 M urea gels.

Acknowledgments

Work in the authors' laboratories is supported by National Institutes of Health grants R37 GM41628 (R.C.C. and J.W.C.) and RO1 CA089455 (A.S.).

[25] G. A. Rice, C. M. Kane, and M. J. Chamberlin, *Proc. Natl. Acad. Sci. USA* **88,** 4245 (1991).

[21] Use of RNA Yeast Polymerase II Mutants in Studying Transcription Elongation

By Daniel Reines

Studies of the elongation phase of RNA polymerase II transcription have benefited greatly from complementary genetic and biochemical studies in the budding yeast *Saccharomyces cerevisiae*. The tools of molecular biology have enabled the generation of mutant yeast strains and have enabled analysis of the biological consequences of the mutations on cell growth, viability, and transcript elongation. Preparative amounts of genetically altered RNA polymerase II have been purified and assayed for biochemical function using conventional chromatography or affinity separation.[1,2] The roster of the components of cellular machinery that participate in elongation has been expanded through the study of genetic interactions using a drug-sensitive phenotype in concert with demonstrated elongation defects in RNA polymerase II. As a result, proteins involved in DNA repair and chromatin structure (as well as others) have been linked to elongation.[3,4,5] The mechanistic basis of defects in mutant RNA polymerases can be appreciated in the context of a recently developed, atomic-level structural model of the RNA polymerase II elongation complex.[6]

History and Rationale

One of the earliest-identified and best-studied transcription elongation factors for RNA polymerase II is SII (also known as TFIIS).[7] Mutation of *DST1* (also known as *PPR2*), the non-essential gene encoding SII in yeast, leads to reduced growth rates in the presence of 6-azauracil (6AU).[8,9] 6-Azauracil is an inhibitor of IMP dehydrogenase (IMPDH), the rate-limiting

[1] P. A. Kolodziej and R. A. Young, *Methods Enzymol.* **198,** 508–519 (1991).
[2] A. M. Edwards, S. A. Darst, W. J. Feaver, N. E. Thompson, R. R. Burgess, and R. D. Kornberg, *Proc. Natl. Acad. Sci. USA* **87,** 2122–2126 (1990).
[3] G. A. Hartzog, T. Wada, H. Handa, and F. Winston, *Genes Dev.* **12,** 357–369 (1998).
[4] S.-K. Lee, S.-L. Yu, L. Prakash, and S. Prakash, *Mol. Cell. Biol.* **21,** 8651–8656 (2001).
[5] C. L. Denis, Y. C. Chiang, Y. Cui, and J. Chen, *Genetics* **158,** 627–634 (2001).
[6] A. L. Gnatt, P. Cramer, J. Fu, D. A. Bushnell, and R. D. Kornberg, *Science* **292,** 1876–1882 (2001).
[7] M. Wind and D. Reines, *Bioessays* **22,** 327–336 (2000).
[8] T. Nakanishi, A. Nakano, K. Nomura, K. Sekimizu, and S. Natori, *J. Biol. Chem.* **267,** 13200–13204 (1992).
[9] G. Exinger and F. Lacroute, *Curr. Genet.* **22,** 9–11 (1992).

enzyme in *de novo* GTP synthesis. Treatment of cells with 6AU results in depletion of intracellular nucleotide pools.[9] This cardinal finding has led to the use of 6AU-sensitivity as a phenotypic marker for a compromised elongation machinery in yeast. It has been suggested that depressed intracellular nucleotide pools stress the elongation machinery by reducing the levels of these RNA substrates below RNA polymerase's K_m.[10] When the concentration of nucleotides available to RNA polymerases is restricted *in vitro*, elongation rates are correspondingly reduced. RNA polymerase molecules that elongate slowly are prone to becoming arrested, a state reversed by SII.[7,11,12] Mutation and/or deletion of a number of RNA polymerase II subunit genes has also been shown to confer 6AU-sensitivity.[10,13,14,15] Following purification, biochemical defects were observed in the elongation behavior of mutant yeast RNA polymerase II.[13,15,16] Hence, it is attractive to think that due to its role as an IMPDH inhibitor, 6AU imposes on the cell an increased requirement for optimal elongation factor function.[10] Presumably, this requirement cannot be met by cells with mutated elongation factors or RNA polymerase II, which results in slowed growth.

Mycophenolic acid (MPA) has also become useful to assay for mutations in the elongation machinery. This drug is a well-characterized and specific inhibitor of IMPDH. There are detailed pharmacological studies in human cells and an atomic-level structural model of MPA in a complex with mammalian IMPDH.[17] Yeast strains lacking a functional SII protein are MPA sensitive, as are elongation-defective RNA polymerase II mutants.[9,18]

Thus, starting with mutations in genes encoding RNA polymerase II subunits, it is possible to assay for: (1) growth rate changes in the presence of drugs that inhibit nucleotide synthesis and (2) intrinsic or elongation

[10] J. Archambault, F. Lacroute, A. Ruet, and J. D. Friesen, *Mol. Cell. Biol.* **271,** 6866–6873 (1992).

[11] S. M. Uptain, C. M. Kane, and M. J. Chamberlin, *Annu. Rev. Biochem.* **66,** 117–172 (1997).

[12] W. Gu and D. Reines, *J. Biol. Chem.* **270,** 11238–11244 (1995).

[13] W. Powell and D. Reines, *J. Biol. Chem.* **271,** 6866–6873 (1996).

[14] A. Ishiguro, Y. Nogi, K. Hisatake, M. Muramatsu, and A. Ishihama, *Mol. Cell. Biol.* **20,** 1263–1270 (2000).

[15] S. A. Hemming, D. B. Jansma, P. F. Macgregor, A. Goryachev, J. D. Friesen, and A. M. Edwards, *J. Biol. Chem.* **275,** 35506–35511 (2000).

[16] J. Wu, D. E. Awrey, A. M. Edwards, J. Archambault, and J. D. Friesen, *Proc. Natl. Acad. Sci. USA* **93,** 11552–11557 (1996).

[17] M. D. Sintchak, M. A. Fleming, O. Futer, S. A. Raybuck, S. P. Chambers, P. R. Caron, M. A. Murcko and K. P. Wilson, *Cell* **85,** 921–930 (1996).

[18] J. C. Lennon, M. Wind, L. Saunders, M. B. Hock, and D. Reines, *Mol. Cell. Biol.* **18,** 5771–5779 (1998).

factor-related defects in elongation by mutant RNA polymerase II enzymes *in vitro*. Conversely, a collection of mutagenized yeast can be screened for drug-sensitive growth and candidates can be tested for a role in transcription elongation. This approach has been used to successfully screen for 6AU and MPA sensitive mutants in a collection of haploid yeast deleted for nonessential genes.[19,20]

This chapter addresses the application of 6AU- and MPA-sensitivity as a genetic assay and will provide procedures for the purification of RNA polymerase II for use in *in vitro* transcription assays.

Drug-Sensitive Growth Assays

6-Azauracil

One of the limitations of the 6AU-sensitivity assay is that cells must be tested on uracil-free media since uracil competes with and neutralizes the drug's effect. Therefore, the strain must be URA$^+$. Synthetic complete medium lacking uracil (SC-ura) is favored over rich medium such as YPD. Standard recipes for these media have been described.[21]

1. Media should be prepared, autoclaved, and allowed to cool to the touch.
2. Prepare a 5 mg/ml stock of 6-AU (Sigma, Catalog #A-1757) in water. The drug will take a number of hours to go into solution with constant stirring at 30°C. Sterilize by passing the fully dissolved solution through a 0.2 μm filter (Gelman, Product #4192).
3. Add the drug to media while stirring. A series of liquid cultures or plates can be made with varying amounts of the drug to assess sensitivity; effective concentrations range from 50–100 μg/ml (w/v). The drug is stable for days or weeks in media.

Mycophenolic acid

MPA is effective in rich medium or synthetic dropout medium. However, it is somewhat unstable, so it is recommended that plates be poured as close to use as possible and that stock solutions be made fresh prior to addition to liquid media.

[19] C. Desmoucelles, B. Pinson, C. Saint-Marc, and B. Daignan-Fornier, *J. Biol. Chem.* **277**, 27036–27044 (2002).
[20] L. Riles, R. Shaw, M. Johnston, and D. Reines (submitted).
[21] F. Sherman, *Meth. Enzymol.* **194**, 3–21 (1991).

1. Medium should be prepared, autoclaved, and allowed to cool to the touch.

2. Prepare a 15 mg/ml stock of MPA (Sigma, Catalog #M-5255) in dimethylsulfoxide and sterilize through 0.2 μm filters (Gelman, Product # 4192).

3. Add the drug to media with stirring. A series of liquid cultures or plates can be made with varying drug concentrations. Effective concentrations range from 5–30 μg/ml (w/v).

Cell Growth

Growth inhibition is a relative measure. At a high enough concentration, even wildtype cells show reduced growth rates for a drug such as MPA. Therefore, it is important to have a quantitative measure of this parameter. Strains to be compared should be obtained from a culture in the logarithmic stage of growth. Growth on solid medium can be gauged by diluting cells to an optical density (600 nm) of 0.001 ($\approx 5 \times 10^7$ cells/ml/OD). Ten μl of 4 to 10-fold serial dilutions from this stock are spotted onto agar plates for incubation at 30°C. For liquid cultures, the cell density is monitored and the time it takes for the number of cells in a liquid culture to double is calculated (slope of the logarithmic portion of the growth curve divided by log[2]). A typical strain with a disruption of the *DST1* gene shows a 1.4 to 1.9-fold increase in doubling time in SC-ura with 75 μg/ml 6AU.[18] Using doubling time as a gauge, synthetic effects of combining mutations can be readily observed.[18] It should be noted that the relative drug sensitivity of different strains can vary. Drug-sensitive cells should grow well when guanine is also included in the medium, demonstrating the specificity of the phenotype since this base can be converted into GTP by a salvage pathway, bypassing the pathway blocked by 6AU and MPA.

Induction of IMP Dehydrogenase

After exposing wildtype yeast to 6AU or MPA there is a robust transcriptional response of at least one of the four genes encoding IMPDH (*IMD2*; also known as *PUR5*).[22,23,24] Furthermore, this induction is compromised in a number of transcription elongation mutants.[22] Thus, measurements of *IMD2* induction using northern blotting have been an asset in understanding phenotypes of candidate elongation mutants.[25,26]

[22] R. J. Shaw and D. Reines, *Mol. Cell. Biol.* **20,** 7427–7437 (2000).

[23] R. J. Shaw, J. L. Wilson, K. T. Smith, and D. Reines, *J. Biol. Chem.* **276,** 32905–32916 (2001).

[24] M. Escobar-Henriques and B. Daignan-Fornier, *J. Biol. Chem.* **276,** 1523–1530 (2001).

RNA Polymerase II Purification and Transcription

Once a candidate RNA polymerase II mutant has been identified by the drug-sensitive phenotype, the enzyme can be readily purified and subjected to *in vitro* transcription assays.

Conventional Chromatography (after Edwards et al., *1990)*[2]

1. Cells are grown at $30°C$ in 6 L of rich medium (YPD), pelleted, washed with cold water, and resuspended in 0.2 volumes of 0.25 M Tris-HCl, pH 7.9, 5 mM EDTA, 2.5 mM DTT, 10 mM sodium pyrophosphate, 5% (v/v) dimethylsulfoxide, and 50% (v/v) glycerol, and frozen at $-80°C$.

2. Thawed cells are mixed at $4°C$ with 1.25 volumes glass beads and disrupted by fifteen 30–s bursts of a bead beater (Biospec Products, Bartlesville, OK) with 90–s of cooling between bursts.

3. The lysate is diluted into 0.6 volumes of buffer A (50 mM Tris-HCl, pH 7.9, 1 mM EDTA, 0.5 mM DTT, 10 mM NaF, 10 mM sodium pyrophosphate, 10% glycerol) and protease inhibitors (per liter: 10 mg of aprotinin, 320 mg of benzamidine, 1 mg of pepstatin, 10 mg of leupeptin, 5 mg L-1-chloro-3-[4-tosylamido]-4 phenyl-2 butanone, 24 mg of L-1-chloro-3-[4-tosylamido]-7 amino-2-heptanone hydrochloride, 174 mg of phenylmethylsulfonyl fluoride). Glass beads and cell debris are removed by centrifugation at 27,500 × g for 40 min at $4°C$. The supernatant is filtered through 3MM paper (Whatman, Catalog #3030917).

4. Apply the supernatant to a 40-ml heparin-Sepharose CL-6B column (Pharmacia Biotech Inc.). The column is washed with buffer A containing 0.1 M KCl and eluted with buffer A containing 0.6 M KCl. Peak protein fractions can be identified by spot immunoblotting with monoclonal antibody 8WG16[27] directed against the largest subunit of RNA polymerase II. Peak fractions are pooled and precipitated with 50% (w/v) saturated ammonium sulfate at $4°C$. This precipitate is collected by centrifugation at 34,000 × g for 30 min at $4°C$, dissolved in buffer B (40 mM Tris-HCl, pH 7.9, 0.5 mM EDTA, 1 mM DTT, 10% glycerol, 0.5 mM phenylmethylsulfonyl fluoride), and dialyzed versus buffer B containing 0.1 M KCl.

5. This protein is applied to a DEAE 5-PW column (TosoHaas; 75 × 7.5 mm, part #07164) using a Pharmacia FPLC system. Bound material is

[25] A. Ferdous, F. Gonzalez, L. Sun, T. Kodadek, and S. A. Johnston, *Mol. Cell.* **7**, 981–991 (2001).
[26] S. L. Squazzo, P. J. Costa, D. L. Lindstrom, K. E. Kumer, R. Simic, J. L. Jennings, A. J. Link, K. M. Arndt, and G. A. Hartzog, *EMBO J.* **21**, 1764–1774 (2002).
[27] N. E. Thompson and R. R. Burgess, *Methods Enzymol.* **274**, 513–526 (1996).

eluted with a 9-ml gradient from 0.1–0.5 M KCl at a flow rate of 0.6 ml/ min. Peak fractions are identified by immunoblotting and assayed for nucleotide incorporation using calf thymus or salmon sperm DNA as a template.[28] Nucleotide incorporation is typically greater than 90% α-amanitin-sensitive (100 μg/ml), and the specific activity ranges from 18–42 units/mg protein (1 unit = 1 nmol nucleotide incorporated per minute into an acid insoluble form). This procedure results in a preparation that is >10 to 20% pure using specific activity and silver staining of SDS gels.

Affinity Chromatography

The convenience of affinity purification is preferable to conventional chromatography for analyzing multiple mutants. If native (i.e., untagged) polymerase is required, affinity purification of RNA polymerase II using an immobilized monoclonal antibody (8WG16) against the carboxyl terminal domain of the largest subunit is effective[27] and has enabled the purification of milligram amounts of enzyme.[2,6] One limitation of this approach is the availability of large amounts of the monoclonal antibody.

If modified polymerase is acceptable, a high affinity ligand binding site can be added to a subunit using an appropriate genetically engineered strain. Recombinant plasmid encoding a subunit such as Rpb3p with a peptide extension recognized by a monoclonal antibody (12CA5), can be readily introduced into a yeast strain.[1] Similarly, a fusion of Rpb3p to glutathione-S-transferase can be used to purify the enzyme on glutathione agarose.[29,30] Alternatively, the tag can be added by introducing a piece of DNA encoding a ligand binding site into the end of a subunit's open reading frame in the chromosome.[31] Purification of such tandem affinity purification (TAP)-tagged multiprotein complexes is rapid and efficient and facilitates a large-scale analysis of mutant RNA polymerases. The TAP-tag can be engineered such that the tag "polypeptide" can be removed by a specific protease if native protein is desired.

For an analysis of elongation rates there are two useful experimental systems that do not require promoter-sequences, enabling the analysis of purified polymerase in the absence of general initiation factors. The first employs DNA duplexes extended at their 3′ ends with deoxycytidylate

[28] M. Sawadogo, A. Sentenac, and P. Fromageot, *J. Biol. Chem.* **255,** 12–15 (1980).

[29] E. M. Phizicky, M. R. Martzen, S. M. McCraith, S. L. Spinelli, F. Xing, N. P. Shull, C. Van Slyke, R. K. Montagne, F. M. Torres, S. Fields, E. J. Grayhack, *Methods Enzymol.* **350,** 546–559 (2002).

[30] J. Mote and D. Reines, unpublished results.

[31] O. Puig, F. Caspary, G. Rigaut, B. Rutz, E. Bouveret, E. Bragado-Nilsson, M. Wilm, and B. Seraphin, *Methods* **24,** 218–229 (2001).

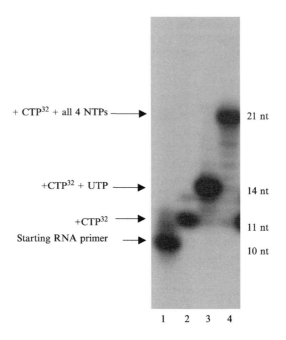

5' GTGGGGGCGCGTTCGTCCTCACTCTCTTCCTCTAGAGTCG3'   40-mer NON-template DNA strand

3' CACCCCCGCGCAAGCAGGAGTGAGAGAAGGAGATCTCAGC5'   40-mer Template DNA strand

                CACUCUCUUC   RNA 10-mer primer

                    +CT**P**^{32}

                CACUCUCUUC**C**   11-mer product

                    +CTP + UTP

                CACUCUCUUC**C**UCU   14-mer product

                    + all 4 NTPs

                CACUCUCUUC**C**UCUAGAGUCG   21-mer product

FIG. 1. Assembly and function of a synthetic elongation complex composed of purified yeast RNA polymerase II, DNA oligonucleotides, and an RNA oligonucleotide. The template strand DNA and 10-base RNA shown were annealed and mixed with 2 μg of purified yeast RNA polymerase II (DEAE fraction) as described.[34] The non-template DNA was annealed to the template strand and the resulting complexes were provided with 10 μCi of α-^{32}P-CTP (400 Ci/mmol) for 5 min at 22°C. An aliquot of RNA was prepared for gel electrophoresis (lane 2). The reaction was split and adjusted to 30 μM UTP (lane 3) or 30 μM each of ATP, (also known as, "tailed" templates).[13,32] Tailed templates can be used to study elongation on specific cloned DNA sequences of choice for template lengths of a few kilobases. The influence of elongation factors upon transcription rates can also be assayed. A detailed discussion of tailed templates can be found in an earlier volume in this series.[33]

Elongation complexes can also be assembled directly from two synthetic oligodeoxynucleotides that represent template and non-template strands, a small oligoribonucleotide "primer," and purified yeast RNA polymerase II.[34] This can be a very useful transcription system for quickly characterizing elongation by many tagged, affinity-purified mutant yeast RNA polymerases. One limitation of this method is that only relatively small lengths of template can be transcribed unless additional sequences are ligated onto pre-initiated complexes.[35] An advantage of the synthetic oligodeoxynucleotide templates over the use of tailed templates is that the nascent RNA is displaced from the template strand and the parameters of DNA rewinding and RNA-DNA melting are comparable to that seen for "normal" elongation complexes.[33] Elongation complexes formed on tailed templates may not displace the RNA chain, resulting in an extended RNA-DNA hybrid.[33] Purified yeast RNA polymerase II prepared using the conventional chromatographic procedure described can be "walked" stepwise down such a synthetic oligodeoxynucleotide template. As an example, wildtype enzyme was used to extend a 10-base oligoribonucleotide (Fig. 1, lane 1) by a single CMP residue in the presence of α-^{32}P-CTP (lane 2), by four bases in the presence of CTP and UTP (lane 3), or to completion in the presence of all four nucleoside triphosphates (lane 4). This system enables the study of the microscopic steps involved in nucleotide incorporation. In this example, DNA sequences representing the adenovirus major late promoter have been selected. To study more complex interactions, such as those between RNA polymerase II and the basal machinery, the mutant enzyme can be

[32] T. R. Kadesch and M. J. Chamberlin, *J. Biol. Chem.* **257**, 5286–5295 (1982).

[33] A. Edwards and C. M. Kane, *Meth. Enzymol.* **274**, 419–436 (1996).

[34] M. L. Kireeva, N. Komissarova, D. S. Waugh, and M. Kashlev, *J. Biol. Chem.* **275**, 6530–6536 (2000).

[35] M. L. Kireeva, W. Walter, V. Tchernajenko, V. Bondarenko, M. Kashlev, and V. M. Studitsky, *Mol. Cell.* **9**, 541–552 (2002).

CTP, GTP, and UTP (lane 4) and incubated at 22°C for 5 min before RNA was isolated from each for gel electrophoresis. The 15% (w/v) poyacrylamide/50% urea (w/v) gel shown was dried and exposed to XOMAT (Dupont) film. For reference purposes, lane 1 shows the starting 10-base oligoribonucleotide labeled with γ-^{32}P-ATP and polynucleotide kinase.

added to a reconstituted reaction in which RNA polymerase initiates transcription from a standard double-stranded promoter using recombinant general transcription factors.[36]

Acknowledgments

I thank John Mote, Randy Shaw, and Judy Wilson for excellent technical assistance. Work in the author's laboratory was funded by NIH grant GM46331.

[36] L. C. Myers, K. Leuther, D. A. Bushnell, C. M. Gustafsson, and R. D. Kornberg, *Methods* **12,** 212–216 (1997).

[22]   Acetylation of Human AP-Endonuclease 1, A Critical Enzyme in DNA Repair and Transcription Regulation

By KISHOR K. BHAKAT, SUK HOON YANG, and SANKAR MITRA

The 36 kD mammalian AP-endonuclease 1 (APE1), shown to be essential for survival of mouse embryos, plays a central role in DNA base excision repair.[1] It cleaves abasic (AP) sites in DNA to initiate the repair process.[2] It also removes 3' blocking groups at DNA single-strand breaks generated directly or indirectly by reactive oxygen species (ROS).[1–3] APE1 is a highly conserved protein, and present in species ranging from *E. coli* to mammals. However, mammalian APE1, with a unique *N*-terminal regions dispensable for its enzymatic activity, has two distinct, additional functions in transcription regulation.[4] It reductively activates many transcription factors including p53 and c-Jun, and thus plays a role in transcription regulation.[5–6] It was also shown to act directly as a *trans*-acting factor, which binds to negative Ca^{2+} response elements A and B (nCaRE-A and -B), and thereby downregulates parathyroid hormone expression in the presence of excess Ca^{2+}.[7] The fact that recombinant

[1] B. Demple and L. Harrison, *Annu. Rev. Biochem.* **63,** 915 (1994).
[2] P. W. Doetsch and R. P. Cunningham, *Mutat. Res.* **236,** 173 (1990).
[3] S. Mitra, T. K. Hazra, R. Roy, S. Ikeda, T. Biswas, J. Lock, I. Boldogh, and T. Izumi, *Mol. Cells* **7,** 305 (1997).
[4] T. Izumi and S. Mitra, *Carcinogenesis*, **19,** 525 (1998).
[5] S. Xanthoudakis, G. Miao, F. Wang, Y. C. Pan, and T. Curran, *Embo J.* **11,** 3323 (1992).
[6] L. Jayaraman, K. G. Murthy, C. Zhu, T. Curran, S. Xanthoudakis, and C. Prives, *Genes Dev.* **11,** 558 (1997).

APE1 by itself does not stably bind to either nCaRE-A or nCaRE-B suggests that APE1 is covalently modified, which would enhance its binding to nCaREs. In support of this, 2-D gel electrophoresis shows the presence of a second, minor species of APE1 *in vivo* with pI of about 8 compared to 8.6 for the recombinant APE1. It is also possible that APE1 does not bind directly to DNA but instead to other proteins present in the bound complex.[8] Subsequent studies indicate that APE1 is acetylated at Lys 6 or Lys 7 in the *N*-terminal region of the polypeptide. Furthermore, this acetylation is carried out by p300/CBP. The methods for identifying acetylated APE1 and characterizing its *in vivo* functions are described later.

In Vivo Acetylation of APE1

Identification of acetylated APE1 (AcAPE1) utilized FLAG-tagged APE1 in transiently-expressed human cells. Thus, acetylated FLAG-APE1 rather than endogenous AcAPE1 was isolated and characterized because of the ease of quantitative recovery of FLAG-tagged proteins as immunoprecipitates using a commercially available, anti-FLAG-monoclonal mouse antibody. In contrast, immunoprecipitation of endogenous APE1 with available antibodies is poor and variable at best.

Assay Procedure

1. Human colon carcinoma line HCT116 (p53-/-) cells were grown in McCoy 5A (Gibco Life Technologies) medium supplemented with 10% FBS, penicillin (100 units/ml) and streptomycin (100 μg/ml) on 100 mm petri dishes (50 to 60% confluent at 3 to 4 $\times$ 10^6 cells/dish) in a tissue culture incubator at 37 $^\circ$C with 5% CO_2.

2. Cells were transfected with FLAG-tagged APE1 expression plasmid (1 μg/dish) using LipofectAMINE 2000 (Gibco Life Technologies).

3. The cells, 40 h after transfection, were incubated in McCoy 5A medium containing 1 mCi/ml [^{3}H]-sodium acetate (5 Ci/mmol; NEN) for 1 h.

4. The cells were then washed twice with cold phosphate buffered saline (PBS), and lysed on plates with a cold lysis buffer containing 50 mM

[7] T. Okazaki, U. Chung, T. Nishishita, S. Ebisu, S. Usuda, S. Mishiro, S. Xanthoudakis, T. Igarashi, and E. Ogata, *J. Biol. Chem.* **269,** 27855 (1994).
[8] D. T. Kuninger, T. Izumi, J. Papaconstantinou, and S. Mitra, *Nucleic Acids Res.* **30,** 823 (2002).

Tris-HCl, pH 7.5, 150 mM NaCl, 1 mM EDTA, 1% Triton X-100, 10 mM sodium butyrate, and protease inhibitor cocktail (Roche).

5. The lysate was centrifuged (14,000 rpm, 15 min at 4°C) and the supernatant collected.

6. A suspension (60 μl) of anti-FLAG M2 monoclonal antibody linked to agarose beads was washed three times with cold TBS (50 mM Tris-HCL, pH 7.5, 150 mM NaCl) and collected by centrifugation at 10,000 rpm for 5 sec. The pellet was suspended in the supernatant (1 mg/ml) and incubated for 3 h at 4°C with gentle shaking.

7. The proteins bound to the immunobeads were eluted by gentle shaking with 100 μl TBS containing 300 ng/μl FLAG peptide (Sigma) for 30 min.

8. The eluate was concentrated by precipitation with an equal volume of acetone for 1 h at −20°C.

9. The precipitate, collected by centrifugation (14,000 rpm for 20 min) at 4°C, was dissolved in 30 μl SDS/PAGE loading buffer.

10. The proteins were separated in SDS/PAGE (12% polyacrylamide).

11. The gel was fixed for 1 h with a solution containing 40% methanol and 10% acetic acid combined with gentle shaking. Subsequently the gel was incubated with an enhancing solution (Amplify, Amersham) for 45 min to enhance the ^{3}H signal and dried for 1 h at 80°C. Autoradiography was performed at −70°C by exposing Hyperfilm-MP (Amersham) for 7 days (Fig. 1).

Acetylation of APE1 by p300/CBP

General Considerations

Acetylation, first discovered in the histones, is mediated by p300 and its close homolog CBP (CREB-binding protein), which are transcription co-activators and possess intrinsic acetyltransferase (HAT) activity.[9–11] Although p300/CBP are the major contributors of cellular HAT activity, such activity was subsequently identified in a number of other proteins. More recently, p300 was found to acetylate a number of *trans*-acting factors including p53 and some DNA metabolizing proteins (e.g., FEN1 and DNA polymerase β).[12–14] Thus, whether p300 is also responsible for acetylating APE1 was examined first.

[9] A. J. Bannister and T. Kouzarides, *Nature*, **384,** 641 (1996).

[10] R. Eckner, M. E. Ewen, D. Newsome, M. Gerdes, J. A. DeCaprio, J. B. Lawrence, and D. M. Livingston, *Genes Dev.* **8,** 869 (1994).

[11] R. H. Goodman and S. Smolik, *Genes Dev.* **14,** 1553 (2000).

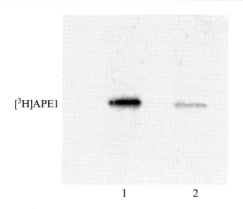

FIG. 1. Fluorogram of [³H] AcAPE1 labeled *in vivo* (lane 2) and *in vitro* with p300 HAT domain (lane 1).

Assay Procedure

1. Human colon carcinoma cells HCT116 (p53-/-) were grown in McCoy 5A on 100 mm petri dishes as before.

2. The cells were cotransfected with 1 μg of FLAG-tagged APE1 and p300 expression plasmid or FLAG-APE1 alone using LipofectAMINE 2000.

3. The cells were washed twice 36 h after transfection with cold PBS and lysed on plates by a cold lysis buffer as already described.

4. The lysates were centrifuged (14,000 rpm, 15 min at 4°C), and the supernatants collected.

5. FLAG-tagged APE1 was immunoprecipitated using anti-FLAG M2 monoclonal antibody (Sigma) and eluted from immunobeads using FLAG peptide as before.

6. The eluted proteins were separated in 12% polyacrylamide SDS/PAGE and transferred on to a PVDF membrane overnight at 35 v at 4°C using a transfer buffer (25 mM Tris-HCl, pH 8.3, 192 mM glycine, and 20% v/v methanol).

7. The membrane was immunoblotted using anti-acetylated lysine antibody (NEB) according to the manufacturer's instruction. The level of acetylated FLAG-APE1 in p300-transfected cells was found to be higher

[12] W. Gu and R. G. Roeder, *Cell*, **90,** 595 (1997).
[13] S. Hasan, M. Stucki, P. O. Hassa, R. Imhof, P. Gehrig, P. Hunziker, U. Hubscher, and M. O. Hottiger, *Mol. Cell*, **7,** 1221 (2001).
[14] S. Hasan, Ei-Andaloussi, U. Hardeland, P. O. Hassa, C. Burki, R. Imhof, P. Schar, and M. O. Hottiger, *Mol. Cell*, **10,** 1213 (2002).

than empty vector-transfected cells (data not shown). This strongly suggests that p300 is the major enzyme to acetylate APE1 *in vivo*.

Characterization of AcAPE1

General Considerations

Once it was shown that APE1 is acetylated *in vivo* by p300/CBP, it was important to identify the acetyl acceptor Lys residue(s). For this purpose, recombinant APE1 and its 20-mer *N*-terminal peptide were acetylated *in vitro* with recombinant p300 together with acetyl donor acetyl Coenzyme A.

Purification of Recombinant p300

The cDNA of human full-length p300 or its HAT domain was fused with the FLAG peptide-coding sequence at the *N*-terminus, and then inserted in a baculovirus expression plasmid.[15] Recombinant baculovirus was produced in Sf9 insect cells following established protocols. Recombinant p300 and p300 HAT domain were purified from virus-infected cells by immunoaffinity chromatography using anti-FLAG M2 antibody-bound beads as described earlier.

In Vitro *Acetylation of APE1*

Recombinant APE1 (0.5 mg) was incubated with immunopurified p300 HAT domain (5 μg) together with 2 mM acetylCoA (Sigma) in 500 μl of HAT buffer (50 mM Tris-HCl, pH 8.0, 0.1 mM EDTA, 10% v/v glycerol and 1 mM DTT, 10 mM Na-butyrate) at 30°C for 1 h. The incorporation of [^{3}H] acetyl group from acetyl CoA into APE1 as assayed by binding of the protein to P81 phosphocellulose filter, was linear for up to 90 min (data not shown). Acetylated APE1 was separated from the unmodified recombinant APE1 by FPLC on a 1 ml Mono-S column, using 150 mM to 500 mM NaCl gradient (25 ml) in a buffer containing 25 mM Tris-HCl, pH 7.5, 1 mM DTT, and 10% (v/v) glycerol. AcAPE1 and nonacetylated APE1 were eluted at 250 and 290 mM NaCl, respectively.

Identification of Acetylacceptor Lys Residues in AcAPE1

Purified AcAPE1 (3 μg) was subjected to cycles of Edman degradation for determining the sequence from the *N*-terminus. Although multiple Lys residues were present in APE1, only Lys 6 or Lys 7 in the sequence[1]

---

[15] D. Chakravarti, V. Ogryzko, H. Y. Kao, A. Nash, H. Chen, Y. Nakatani, and R. M. Evans, *Cell*, **96**, 393 (1999).

MPKRGKKGAVAEDGDELRTE[20] were found to be acetylated. Because no acetylation with p300 was observed in APE1 when 20 N-terminal residues were deleted (data not shown), it is clear that acetylacceptor Lys residues are localized exclusively in the N-terminal region. Mass spectroscopic analysis of the acetylated 20-mer synthetic peptide of the sequence showed the presence of a monoacetylated species with 42 mass units higher than of the unmodified peptide. The absence of diacetylated peptide suggests that APE1 could be acetylated either at Lys 6 or Lys 7 but not both.

Enhanced Binding of APE1 to nCaREs Due to Acetylation

General Considerations

The presence of AcAPE1 *in vivo* suggests strongly a physiological function of this modified species. The lack of effect of acetylation on the AP-endonuclease activity of APE1 suggests that acetylation primarily affects its *trans*-acting activity. APE1 was identified as one of the proteins bound to nCaREs.[7] It appeared likely that acetylation enhances the affinity of APE1 for the nCaRE complex which was examined as follows.

Electrophoretic Mobility Shift Assay with nCaRE-B

The nCaRE-B oligodeoxynucleotide of the following sequence, derived from the human PTH promoter,[7] was purchased from Integrated DNA Technologies.

5'-TTTTTGAGACACGGTCTCACTCTG-3'

The oligonucleotide (1–2 pmol) was labeled with ^{32}P by T4 polynucleotide kinase in the presence of 30 μCi [γ-^{32}P] ATP (3000 Ci/mmol) after incubation for 30 min at 37 °C. After removal of ATP and ADP by gel filtration through Sephadex G-25 (Amersham), the radio-labeled oligos were annealed with 1.2 molar equivalent of the complementary strand.

The nuclear extracts of HCT116 cells were prepared as described earlier.[16] The oligonucleotides were incubated with various nuclear extracts (10 μg), alone or supplemented with 50 or 100 ng of APE1 or AcAPE1, for 30 min at room temperature in a buffer containing 35 mM HEPES, pH 7.9, 1 mM MgCl$_2$, 0.5 mM EDTA, 0.5 mM DTT, 10% glycerol, and 0.5 μg poly (dI-dC), followed by electrophoresis in nondenaturing

[16] E. Schreiber, P. Matthias, M. M. Muller, and W. Schaffner, *Nucleic Acids Res.* **17,** 6419 (1989).

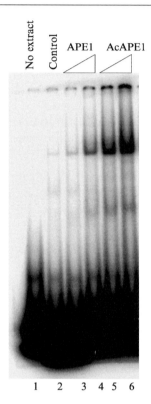

Fig. 2. AcAPE1 enhances binding of HCT 116 nuclear extract to nCaRE-B. Lane 1, free probe; extract alone (lane 2), or supplemented with 50 and 100 ng of either APE1 (lanes 3 and 4) or AcAPE1 (lanes 5 and 6).

polyacrylamide gel (5% polyacrylamide) using TBE buffer containing 45 mM Tris-borate, pH 8.3, and 1 mM EDTA.

Fig. 2 shows that exogenous AcAPE1 enhanced the binding of nuclear extract to nCaRE-B to a much larger extent than unmodified APE1 at the same concentration. It thus appears that a major cellular role of acetylation of APE1 is to channel this protein to transcriptional regulation.

Repression of Reporter Gene Expression Dependent on APE1 Acetylation

The gel mobility shift assay results predict that nCaRE-dependent gene expression would be affected by the acetylation status of APE1. This was tested independently by using a luciferase reporter whose expression is under the control of minimal SV40 promoter. The nCaRE-B sequence of the PTH gene was placed upstream of the promoter in an expression plasmid which was then introduced into BHK21 cells by LipofectAMINE 2000-mediated transfection as before.

1. Baby hamster kidney (BHK21:ATCC CCL-10) cells were grown to subconfluent level in 60 mm plates in minimal essential medium supplemented with 2 mM L-glutamine, 1 mM sodium pyruvate, 0.1 mM nonessential amino acids, 10% FBS, 100 units/ml penicillin, and 100 μg/ml streptomycin.

2. The cells were transfected with 1 μg SV40 promoter-reporter or SV 40-nCaRE-B promoter reporter together with 1 μg APE1 expression plasmid (encoding wild type or nonacetylable K6R/K7R mutant) or 1 μg of p300 expression plasmid along with 0.2 μg β-galactosidase expression plasmid PCMVβ (Clontech) using LipofectAMINE 2000 as before.

3. The cells were incubated for 8 h for transfection and then in fresh medium for another 36 hours.

4. The cells were then harvested and lysed with a reporter lysis buffer (Promega), and the luciferase activity of the extract was measured in a luminometer using a luciferase assay kit (Promega). Luciferase activity was normalized with coexpressed β-galactosidase measured with a β-galactosidase assay kit (Promega).

Table I shows the effect of p300 overexpression on luciferase expression in nCaRE-B-containing plasmid. It is evident from a comparison of the luciferase activity of control versus nCaRE-containing promoter-vector that the presence of nCaRE-B significantly reduced luciferase expression in cells overexpressing wild type APE1 or p300 or both but not K6R/K7R APE1 mutant. These results are consistent with the *in vitro* gel

TABLE I
IMPACT OF APE1 ACETYLATION ON EXPRESSION OF LUCIFERASE REPORTER *IN VIVO*

Cotransfecting plasmid	Normalized luciferase activity	
	SV40 promoter-reporter	SV40-nCaRE-B promoter-reporter
Control (empty vector)	1.00	0.4 + 0.05
APE1 (wild type)	1.1 $\pm$ 0.2	0.2 $\pm$ 0.06
K6R/K7R (mutant APE1)	1.1 $\pm$ 0.3	0.45 $\pm$ 0.1
p300	3.0 $\pm$ 0.5	0.7 $\pm$ 0.3
Wild type APE1 + p300	3.1 $\pm$ 0.6	0.35 $\pm$ 0.2

mobility shift results (see Fig. 2) and supports the hypothesis that acetylation of APE1 is critical for nCaRE-B-dependent gene regulation.

It should be noted that nCaRE-dependent gene expression occurs in the presence of Ca^{2+}. The Ca^{2+} effect is not shown in these experiments.

Concluding Remarks

This chapter has discussed an unexplored function of a critical DNA repair protein, APE1, as a *trans*-acting factor in mammalian cells. Furthermore, because APE1 was found to be acetylated *in vivo*, procedures have been described for identification and characterization of acetylated APE1. Finally, assays for examining the effect of acetylation of APE1 on its affinity to the negative Ca^{2+} responsive elements have been described. The *in vitro* binding studies were further validated in a reporter expression assay using transiently transfected cells. Many unanswered questions remain for future studies regarding the mechanism of activation of AcAPE1 including its interaction with distinct partners in transcription regulation that could be either positive and negative. The presence of nCaREs in the promoter of APE1 gene itself suggests that APE1 is autoregulated.[17]

Acknowledgments

We gratefully acknowledge the help of our colleague Istvan Boldogh for generating recombinant p300, and Tapas K. Hazra and Tadahide Izumi for providing many reagents. The research was supported by US Public Health Service grants ES08457 and CA53791 and NIEHS Center Grant ES06676.

[17] T. Izumi, W. D. Henner, and S. Mitra, *Biochemistry*, **35,** 14679 (1996).

[23] Characterization of Transcription-Repair Coupling Factors in *E. Coli* and Humans

By C. P. SELBY and AZIZ SANCAR

Elongation of RNA polymerase (RNAP) is blocked by the presence of DNA damage in the template strand. In *E. coli*, the transcription-repair coupling factor (TRCF), also called the Mfd protein, recognizes the stalled polymerase and performs two functions. First, it dissociates the stalled polymerase and nascent RNA from the template. Then, as a consequence of its affinity for the nucleotide excision repair enzyme (A)BC excinuclease, it delivers the repair enzyme to the damage site.[1] Consequently, lesions in the template strand of actively transcribed genes are repaired more rapidly than lesions elsewhere.[2] Elongation of RNAP may also be blocked by DNA-bound proteins[3]; such blocked polymerases are also a target for dissociation by TRCF.[4] This function of TRCF may modulate gene expression in cases where transcription is controlled by repressor binding to sequences within the transcription unit.[5] Finally, TRCF interacts with arrested, backtracked elongation complexes, and either dissociates them from the template or reverts them to actively transcribing complexes and thus may enhance elongation.[6]

The CSB (Cockayne's syndrome B) protein may be considered the closest mammalian counterpart to TRCF. Genetic analysis implicates CSB as having a role in transcription-coupled repair,[7] which is known to occur in mammalian cells.[8,9] TRCF (1148 amino acids) and CSB (1493 amino acids) are relatively large proteins[1,10]; both possess helicase motifs but lack helicase activity, and both bind to DNA and hydrolyse ATP.[11,12] Also, in vitro analysis has shown that CSB interacts with elongating RNA-PII and may enhance the rate of transcription by RNAPII,[12] and recent studies show that under certain circumstances TRCF may function as an elongation factor.[6] Accordingly, the biochemical characterization of Mfd and CSB has revealed several interesting similarities and differences between these two proteins. The purification and characterization of these two factors are detailed later.

[1] C. P. Selby and A. Sancar, *Science* **260,** 53 (1993).

[2] I. Mellon and P. C. Hanawalt, *Nature* **342,** 95 (1989).

[3] P. A. Pavco and D. A. Steege, *J. Biol. Chem.* **265,** 9960 (1990).

[4] C. P. Selby and A. Sancar, *Biol. Chem.* **270,** 4890 (1995b).

[5] J. M. Zalieckas, L. V. Wray, A. E. Ferson, and S. H. Fisher, *Molec. Microbiol.* **27,** 1031 (1998).

[6] J. S. Park, M. T. Marr, and J. W. Roberts, *Cell* **109,** 757 (2002).

[7] J. Venema, L. H. Mullenders, A. T. Natarajan, A. A. van Zeeland, and L. V. Mayne, *Proc. Natl. Acad. Sci. USA* **87,** 4707 (1990).

[8] V. A. Bohr, C. A. Smith, D. S. Okumoto, and P. C. Hanawalt, *Cell* **40,** 359 (1985).

[9] I. Mellon, G. Spivak, and P. C. Hanawalt, *Cell* **51,** 241 (1987).

[10] C. Troelstra, A. van Gool, J. de Wit, W. Vermeulen, D. Bootsma, and J. H. J. Hoeijmakers, *Cell* **71,** 939 (1992).

[11] C. P. Selby and A. Sancar, *J. Biol. Chem.* **272,** 1885 (1997a).

[12] C. P. Selby and A. Sancar, *Proc. Natl. Acad. Sci. USA* **94,** 11205 (1997b).

General Methods

Transcription-Repair Coupling Factor (TRCF)

DNA Constructs for Expression

pMFD19 contains the mfd gene, which encodes TRCF and upstream sequences cloned into the pIBI25 vector. High-level constitutive expression of the Mfd protein is observed in cells transformed with this construct,[1] as shown in Fig. 1A. Presumably this level of expression reflects control of expression from the *mfd* promoter on the multicopy plasmid. The *mfd* gene has also been subcloned into the pKK233-2 vector (pKKMFD), in which case expression is under control of the *trc* promoter and may be induced by addition of IPTG to a level comparable to that seen with pMFD19. A construct pMalMFD in which Mfd is fused to the C-terminus of Maltose Binding Protein (MBP) has been made using the pMalc2 vector (New England Biolabs) and has generated active Mfd protein after purification of the expressed protein with amylose resin.[13]

Protein Purification

CELL-FREE EXTRACT. Extract was prepared based on the procedure of Lu *et al.*[14] from DH5α cells transformed with pMFD19 grown in 1-liter batches of LB supplemented with 0.2% glucose to an A600 of about 1.0. Cells were pelleted and resuspended in 2 ml of 50 mM Tris.HCl, pH 7.6, 10% sucrose, then frozen in dry ice. Cells were thawed in water at room temperature, then placed on ice. To thawed cell suspensions were added dithiothreitol to 1.2 mM, KCl to 0.15 M, and lysozyme to 0.23 mg/ml. The suspension was manually shaken in a 37° water bath until suspensions warmed to 20°, then returned to ice. The suspension was pelleted by centrifugation at 50,000g for 15 min. To the supernatant 10 μl/ml of MgCl$_2$, 1 M was added, followed by 0.42 g of (NH$_4$)$_2$SO$_4$ per ml of supernatant. For the next 30 min, the suspension was kept on ice except for frequent vigorous mixing by hand. The suspension was centrifuged at 10,000 rpm for 30 min in a Sorvall RC-5B centrifuge. The pellet was resuspended in 0.3 ml of 25 mM Hepes pH 7.6, 0.1 mM EDTA, 100 mM KCl, and 2 mM dithiothreitol and dialysed against the same buffer for about 90 min. The yield was 30 to 100 mg protein in 1 ml.

DEAE AGAROSE (BioRad) CHROMATOGRAPHY. Purifications commonly begin with extract from about 12 L of cells. A 55 ml DEAE agarose column was equilibrated with 0.1 M KCl in buffer A (50 mM Tris HCl,

[13] C. P. Selby and A. Sancar, *J. Biol. Chem.* **270**, 4882 (1995a).

[14] A. L. Lu, S. Clark, and P. Modrich, *Proc. Natl. Acad. Sci. USA* **80**, 4639 (1983).

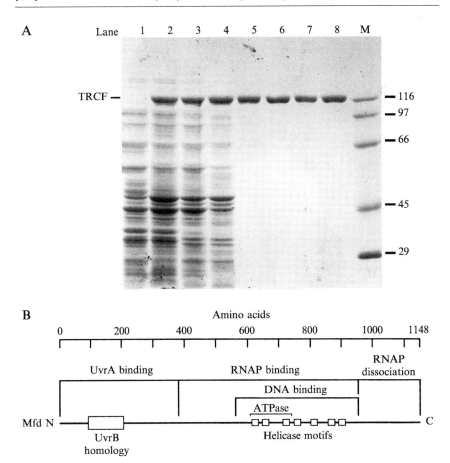

FIG. 1. TRCF (Mfd) protein from *E. coli*. (A) Purification. *E. coli* DH5α F' (lane 1, total cell lysate) was transformed with pMFD19 (lane 2, total cell lysate) and cell-free extract was prepared (lane 3). The peak flow-through fractions from DEAE biogel were pooled (lane 4) and applied to blue sepharose. Peak fractions obtained by isocratic elution were pooled (lane 5), concentrated, and resolved by size exclusion chromatography using AcA34. Peak fractions from the sizing column were pooled (lane 6) and further purified by two passes through heparin agarose (lanes 7 and 8). Fractions were resolved by SDS-polyacrylamide gel electrophoresis followed by staining with Coomassie blue. Reproduced with permission from Selby and Sancar.[1] Copyright (1993) AAAS. (B) Primary Structure. The line from N to C represents the protein and the line above locates the amino acids from 0 to 1148. Features evident from the sequence include a region of homology to UvrB protein (large box) and the seven helicase motifs region including motifs associated with hydrolysis of ATP, as indicated. Also indicated are regions associated with functional activities as described in the text. From Selby and Sancar.[13]

pH 7.5; 1 mM EDTA; 10 mM 2-mercaptoethanol; 20% glycerol). Extract was loaded onto the column by gravity. The column was then washed with 0.1 M KCl in buffer A. TRCF did not bind and fractions containing TRCF were identified by Coomassie blue staining of samples separated by SDS polyacrylamide gels.

BLUE SEPHAROSE (SIGMA) CHROMATOGRAPHY. Peak fractions of TRCF from the DEAE column were combined and applied to a 40-ml column of blue sepharose equilibrated with 0.1 M KCl in buffer A. The column was washed with the same buffer until TRCF eluted by isocratic elution after the flow-through fractions.

AcA34 GEL PERMEATION CHROMATOGRAPHY. Peak fractions from the blue sepharose column were concentrated to 2 to 5 ml and applied to an AcA34 column (2.5 × 80 cm) equilibrated with buffer A plus 0.3 M KCl. The column was developed with the same buffer and TRCF was located by SDS-PAGE.

HEPARIN AGAROSE (SIGMA) CHROMATOGRAPHY. Peak fractions from the sizing column were pooled and dialysed against 50 mM KCl in buffer A and then loaded onto a 13-ml heparin-agarose column that had been equilibrated with the same buffer. The column was developed with a 65-ml gradient of 0.05 to 0.3 M KCl in buffer A. TRCF eluted at approximately 0.2 M, and heparin agarose chromatography was repeated one more time. Peak fractions were pooled, aliquoted, and stored at −80 °C.

Substrates

DNA duplexes of known sequences and lengths containing promoters, defined DNA damage at a unique site, and a radiolabel were generated to study the mechanism of TRCF protein action. Two such substrates, each 138 bp in length, are shown in Fig. 2A.[1] Each of these two substrates was generated by annealing and then ligating a damage-containing oligomer 8 to 12 nucleotides in length with seven other oligomers. The damage in these substrates, indicated with a closed circle, is a psoralen-thymine monoadduct. Promoters for *E. coli* and T7 RNAPs are indicated and the ^{32}P-radiolabel at the 5′ end is indicated with an asterisk. An additional substrate was UV-irradiated pDR3274 plasmid. The plasmid was radiolabeled with 32-P on the template strand at the 3′ end of an EcoRI site located 200 bp upstream from the *tac* promoter. To study the effect of TRCF on a protein "roadblock" to elongation, we used the substrate pUNC211, which possesses an EcoRI site at +270 to +275 in a transcriptional unit controlled by the *tac* promoter.[15] A mutant EcoRI protein, E111Q, which

[15] D. C. Thomas, M. Levy, and A. Sancar, *J. Biol. Chem.* **260,** 9875 (1985).

binds to but does not cleave DNA,[16,17] was added to the template. The bound protein blocks elongation.[3]

In Vitro Assays for Transcription-Coupled Repair

Transcription-repair buffer contained 40 mM Hepes, pH 7.8, 50 mM KCl, 8 mM MgCl$_2$, 5 mM dithiothreitol, 4% glycerol, 100 μg/ml bovine serum albumin, 6% (w/v) polyethylene glycol 6000, 500 μM NAD, 2 mM ATP, 200 μM (each) CTP, GTP, and UTP, 40 μM (each) dATP, dGTP, dTTP, and 4 μM dCTP.

In the "repair synthesis assay," pDR3274 (covalently closed circle) was at about 1 nM (plasmid) and 5 μCi of [α-^{32}P] dCTP was included. Radiolabeled oligomers, when used, were also at about 1 nM. When extract was used, it was included at about 1.2 mg/ml (final). Reactions with the reconstituted system used 1.2 units/ml RNAP (Promega), 4 nM UvrA, 100 nM UvrB, 70 nM UvrC, 5 nM UvrD, 80 μ/ml DNA PolI (BRL), and 48 μ/ml T4 DNA ligase. Reactions of 25 μl were incubated for 25 min at 37°. DNA was then extracted with phenol and ether, then precipitated with ethanol using oyster glycogen (5–10 μg) as carrier. To analyze repair synthesis, reactions were digested with restriction enzymes and then precipitated. DNA was then resolved with sequencing gels and analyzed by autoradiography.

In the "incision assay," pDR3274 was linearized to generate an EcoRI-HindIII fragment about 5 kbp in length, which was labeled with ^{32}P at the 3′ end of the EcoRI site. In this fragment the label is located in the template strand 200 bp upstream from a *tac* promoter. The DNA was incubated in the transcription-repair buffer with RNA polymerase, UvrA, UvrB, and UvrC in the presence or absence of TRCF. Digestion products were extracted, precipitated, and resolved on a sequencing gel.

CSB/ERCC6 Protein

DNA Constructs for Expression

The excision repair cross complementation group 6 (ERCC6) gene encoding CSB was cloned by complementing a CHO mutant belonging to complementation group 6 of a UV-sensitive CHO mutant panel using a human cDNA library and was provided by Dr. J. H. J. Hoeijmakers (Erasmus University). The gene was subcloned into the p2Bac vector (Invitrogen) so that the gene is downstream of the p10 promoter. In this construct, pBacCSAB, the CSA gene is also present, downstream from

[16] K. King, S. J. Benkovic, and P. Modrich, *J. Biol. Chem.* **264,** 11807 (1989).
[17] D. J. Wright, K. King, and P. Modrich, *J. Biol. Chem.* **264,** 11816 (1989).

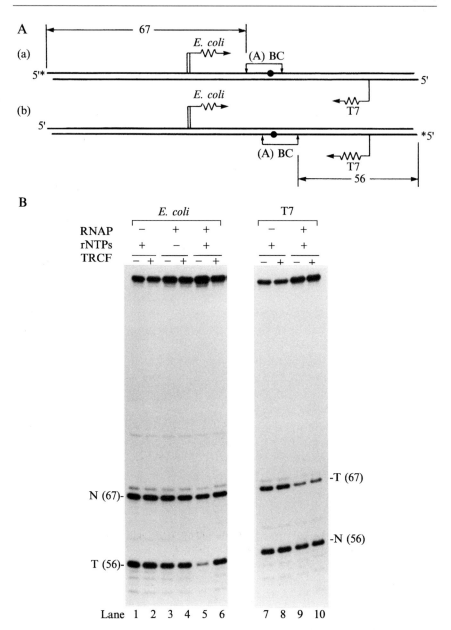

Fig. 2. TRCF overcomes transcription-dependent inhibition of repair. (A) Template-substrate. Two 138 bp duplexes were used, each having a unique psoralen adducted T-residue (closed circle) in the strand labeled at the 5' end with a ^{32}P (asterisk). Incision sites made by ABC excision nuclease with respect to the lesion and the radiolabel are indicated; incision of (a) gives a 67-mer labeled product and incision of (b) gives a 56-mer. In (a) the damage is in

the polyhedron promoter. Sf21 cells were transfected with pBacCSAB and recombinant virus was amplified and used to express CSB in Sf21 cells. Expression of the 168 kDa protein expressed in Sf21 cells was detectable by visualizing the band with Coomassie blue staining of total cellular proteins separated by SDS-PAGE.[11]

Protein Purification

Initial attempts to purify native CSB from human cells were unsuccessful due to extensive degradation. Attempts to express and purify full-length CSB and various regions of the CSB protein fused to MBP or Glutathione-S-Transferase (GST) in *E. coli* were also largely unsuccessful due to poor expression or degradation. An exception was the expression of the helicase motifs region, amino acids 528-1222. The GST-CSB (528-1222) was expressed and purified by glutathione sepharose affinity chromatography. The eluted protein, which was mostly insoluble, was resolved by SDS gel electrophoresis and pieces of the gel slice containing the fusion protein were injected into a rabbit to generate antibodies. Antibodies were affinity purified with an antibody affinity column consisting of MBP-CSB (528-1222) covalently linked to Sulfolink affinity gel (Pierce).

CELL-FREE EXTRACT. Five to seven liters of Sf21 cells were infected with recombinant baculovirus and incubated for 42–48 h at 26 °C. Cells were harvested by centrifugation, and cell-free extract was prepared by the method of Manley *et al.*[18] Approximately 0.8 g of protein in 100 ml of extract was prepared.

DE52 (WHATMAN) CHROMATOGRAPHY AND SP-SEPHAROSE (PHARMACIA) CHROMATOGRAPHY. An 82-ml DEAE precolumn was prepared and attached above a 34-ml column of SP-sepharose. Both were equilibrated with buffer E, 25 mM Hepes, pH 7.9, 12 mM $MgCl_2$, 0.5 mM EDTA, 2 mM EDTA, 17% glycerol (containing 100 mM KCl). Cell-free extract was applied to the DE52 column. CSB was in the flow-through but bound to the SP-sepharose column. The DE52 column was detached, and the SP- sepharose column was washed with 240 ml of buffer E containing 0.2 M KCl. CSB was then eluted with a linear gradient of 0.2 M to 0.8 M KCl in buffer E. CSB eluted at 0.3 to 0.4 M KCl.

---

[18] J. L. Manley, A. Fire, A. Cano, P. A. Sharp, and M. L Gefter, *Proc. Natl. Acad. Sci. USA* **77,** 3855 (1980).

---

the template strand downstream of an *E. coli* RNAP promoter and in (b) the lesion is in the template strand downstream of a T7 RNAP promoter. (B) Effect of transcription on incision. Substrates were incubated with the combinations of RNAPs, rNTPs and TRCF as indicated, and the products were resolved on an 8% sequencing gel. Reproduced with permission from Selby and Sancar.[1] Copyright (1993) AAAS.

ssDNA CELLULOSE (SIGMA) CHROMATOGRAPHY. The peak fractions from the SP-sepharose column (53 ml) were pooled and mixed with 3 volumes of buffer E containing no KCl and NP40 at 0.013%, and then applied to a DNA cellulose column of 11 ml equilibrated with buffer E containing 0.1 M KCl and 0.01% NP40. The column was washed with the same buffer and then eluted with a gradient of 0.1 M to 0.8 M KCl in buffer E with 0.01% NP40. CSB eluted at 0.2 to 0.3 M KCl.

DE52 AND Q-SEPHAROSE (PHARMACIA) CHROMATOGRAPHY. Peak fractions, about 10 ml, were diluted and mixed with buffer E containing no KCl to arrive at a final concentration of approximately 0.1 M KCl and 0.01% NP40. CSB was applied to a DE52 (4 ml) precolumn attached to a Q-sepharose (0.45 ml) column. The CSB flowed through the DE52 column and bound to the Q-sepharose column. The DE52 column was disconnected and CSB was eluted in a single step with 0.55 M KCl in buffer E with 0.01% NP40. Initially the protein eluted mostly in insoluble form at a nominal concentration of about 3 mg/ml, but subsequent, more dilute (1 mg/ml) fractions were completely soluble. Several hundred micrograms of soluble protein were aliquoted and stored at −80 °C.

Substrates

The pMLU112 template was used to generate human RNA polymerase II (RNAPII) stalled by nucleotide starvation. Omission of UTP from transcription reactions results in stalling of RNAPII at +112.[11] pPu192 was used to generate RNAPII stalled at a cyclobutane thymine-thymine dimer in the template strand. This plasmid possesses a dimer at +149–150 downstream from the transcription start site of the major late promoter, and a ^{32}P label is located in the damaged strand at the 13th phosphate 5′ to the dimer[12] This ^{32}P label was constructed by primer elongation of a 20-mer containing a cyclobutane thymine dimer and a single-stranded circular plasmid DNA with a complementary sequence.[19]

Results

Transcription-Repair Coupling Factor (TRCF) and Mutation Frequency Decline (Mfd)

Two lines of investigation conducted decades apart eventually converged to show that Mfd and TRCF were the same protein. First, the phenomenon of mutation frequency decline (Mfd) was characterized and an

[19] J. C. Huang, D. L. Svoboda, J. T. Reardon, and A. Sancar, *Proc. Natl. Acad. Sci. USA* **84,** 3664 (1992).

mfd- mutant was isolated.[20,21] Mutation frequency decline refers to a phenomenon in which certain incubation conditions following UV treatment of wild-type *E. coli* cells are associated with a decline in the observed mutation frequency. The decline was attributed to rapid repair of a thymine dimer which could occur at a site in the template strand of the DNA, which encodes the anticodon-encoding region of a Gln suppressor tRNA, and it was suggested that transcription in some way enhanced the repair of the template strand lesion and led to the mutation frequency decline.[22] Later, the phenomenon of transcription-stimulated repair of DNA damage located in the template DNA strand was found to occur in *E. coli* cells using an assay which directly measured repair in the template or coding strands of a *lacZ* promoter-driven gene.[2] Transcription-repair coupling was then demonstrated *in vitro* with extracts from *E. coli* cells (see later).[23] Transcription-repair coupling required a functional nucleotide excision-repair pathway and transcription by *E. coli* RNAP; using this *in vitro* assay, a protein named transcription-repair coupling factor (TRCF) was purified and the gene encoding TRCF was identified as the *mfd* gene responsible for MFD *in vivo*.[1,24]

Primary Structure of the TRCF

The sequence of TRCF,[1] represented in Fig. 1B, revealed a region of homology to UvrB in its amino terminus, and a region of homology to RecG protein in the middle. The UvrB subunit of the (A)BC excision nuclease binds to the UvrA subunit and it was subsequently shown that TRCF also binds to UvrA through its amino terminus.[13] The region of sequence similarity with RecG contains the seven motifs commonly found in helicases. RecG is involved in recombination,[25] and the sequence similarity between the two proteins may be associated with the fact that they both interact with DNA. Mutation of *mfd* does not affect recombinational repair or conjugational recombination in *E. coli*,[26] but it apparently affects both transcription-coupled repair and recombination in *Bacillis subtilis*.[27] A potential leucine zipper motif identified in the *C*-terminus of TRCF[1] probably does not exist as such since a corresponding motif was not seen in the TRCF from *B. subtilis*.[27]

[20] E. M. Witkin, *Bioessays* **16**, 437 (1994).
[21] B. H. Li, A. Ebbert, and R. Bockrath, *J. Mol. Biol.* **294**, 35 (1999).
[22] R. C. Bockrath and J. E. Palmer, *Mol. Gen. Genet.* **156**, 133 (1977).
[23] C. P. Selby and A. Sancar, *Proc. Natl. Acad. Sci. USA* **88**, 8232 (1991).
[24] C. P. Selby, E. M. Witkin, and A. Sancar, *Proc. Natl. Acad. Sci. USA* **88**, 11574 (1991).
[25] M. C. Whitby, S. D. Vincent, and R. G. Lloyd, *EMBO J.* **13**, 5220 (1994).
[26] G. J. Sharples and L. M. Corbett, *Microbiol.* **143**, 690 (1997).
[27] S. Ayora, F. Rojo, N. Ogasawara, S. Nakai, and J. C. Alonso, *J. Mol. Biol.* **256**, 301 (1996).

Transcription Inhibition of Repair and TRCF Overcoming of Inhibition

The recombinant TRCF was purified to near homogeneity as shown in Fig. 1A. The template and/or substrates shown in Fig. 2A were made to examine its activity. The two 138mers have a psoralen monoadduct located in either the template (T) or nontranscribed (N) coding strand downstream from *E. coli* or T7 RNAP promoters. Each 138mer is radiolabeled at the 5′ end of the damaged strand to enable detection of incision of the damaged strand by ABC excinuclease. ABC excinuclease, which results from sequential and partly overlapping functions of UvrA, UvrB, and UvrC proteins, initiates nucleotide excision repair by cleaving the damage on both sides of the lesion,[28] as indicated in Fig. 2A. The repair assay results shown in Fig. 2B show that transcription by either T7 or *E. coli* RNAP results in inhibition of repair of the downstream psoralen but only when the psoralen is located in the template strand. Addition of TRCF to the reconstituted reaction had no effect on the inhibition by T7 RNAP; however, TRCF specifically reversed the inhibition seen with *E. coli* RNAP, suggesting that TRCF may remove the repair-inhibiting RNA polymerase stalled at the lesion.[1,29]

TRCF Removal of RNAP Stalled at Damage in the Template

The disposition of the RNAPs during the repair reaction was examined by DNaseI footprinting. Using the same conditions as described earlier, transcription was conducted with T7 and *E. coli* RNAPs using the 138 mers with the damage in the template strand. Then, the DNA was subjected to partial hydrolysis by DNaseI under standard conditions to obtain a DNaseI footprint of RNAP in the presence and absence of TRCF. The footprints in Fig. 3 showed that both polymerases formed stable elongation complexes stalled at the psoralen adducted thymine. Thus, the stalled elongation complexes inhibit repair. TRCF removed the *E. coli* RNAP footprint but not the T7 RNAP footprint. Thus, TRCF overcomes repair inhibition by elongating *E. coli* RNAP by removing the stable, stalled elongation complex from the damage site.[1] Similar results were obtained by Ayora *et al.*[27] with RNA polymerase and TRCF from *B. subtilis*.

Transcription-Stimulated Repair by TRCF: Repair Synthesis Assay

The results in Figs. 2 and 3 demonstrated roles for TRCF in transcription and repair but failed to demonstrate stimulation of repair as seen in living cells. Stimulation was seen *in vitro* when using a repair synthesis assay

[28] A. Sancar, *Ann. Rev. Biochem.* **65,** 43 (1996).
[29] C. P. Selby and A. Sancar, *J. Biol. Chem.* **265,** 21330 (1990).

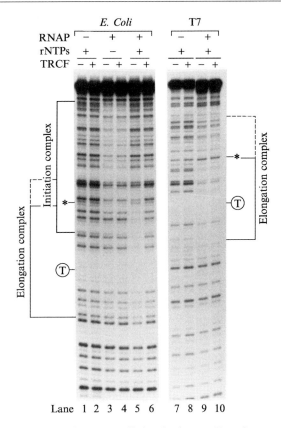

FIG. 3. TRCF removes *E. coli* RNAP stalled at the damage. Reactions were prepared as in Figure 2B except ABC excinuclease was omitted. Instead, reactions were subjected to DNaseI footprinting. Products were separated on an 8% sequencing gel. Reproduced with permission from Selby and Sancar.[1] Copyright (1993) AAAS.

which employed a UV-irradiated plasmid template-substrate. In nucleotide excision repair, dual incisions of the DNA strand on either side of the damage are made by (A)BC excinuclease, followed by synthesis of a repair patch by DNA polymerase I and sealing of the resulting nick by DNA ligase.[28] The repair synthesis assay detects incorporation of radiolabeled nucleotide into repair patches. Results of the repair synthesis assay are shown in Fig. 4A. In this assay, following repair synthesis the plasmid is digested with restriction enzymes and DNA strands are separated on a sequencing gel. The gel resolves the transcribed/template (T) and nontranscribed/coding (N) strands of 299 and 337 bp fragments from the strongly transcribed *tac*-controlled transcriptional unit, and 465 and 560 bp fragments

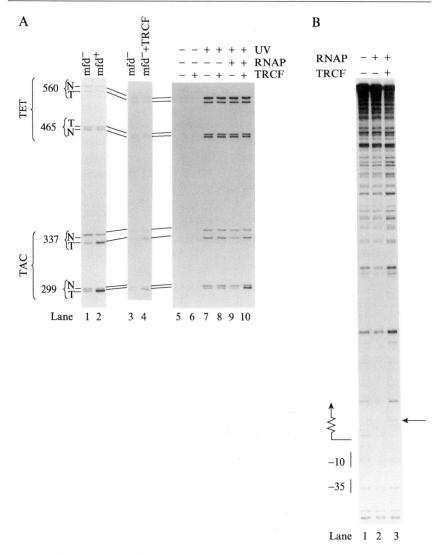

FIG. 4. Stimulation of repair by RNAP and TRCF. The template-substrate was pDR3274, which possesses the *tac* promoter. Panel (A) shows the results of a repair synthesis assay. Incorporation of radiolabeled nucleotide into repair patches in the UV-irradiated plasmid is stimulated in the transcribed (T) as compared to the nontranscribed (N) coding strand of fragments from the strongly transcribed *tac* transcriptional unit when transcription and repair occur and TRCF is present. For the assay in panel (B), the plasmid was labeled at the 3′ end of the template strand 200 bp upstream from the promoter. The −10, −35, and transcription start site of the promoter are indicated. In this assay, dual incision (repair) by ABC excinuclease gives rise to the series of bands corresponding to fragments extending from the 3′ incision to the radiolabel. The horizontal arrow indicates in lane 3 the first site in the transcriptional unit in which TRCF mediates transcription-stimulated repair. Reproduced with permission from Selby and Sancar.[1] Copyright (1993) AAAS.

from the weakly transcribed *tet* gene. The left panel of Fig. 4A shows results obtained with cell-free extracts: the wild-type extract shows transcription-stimulated repair specifically in the transcribed strand of the *tac* transcriptional unit, and this repair is absent in the extract from *mfd-* cells. The center panel shows the lack of transcription-stimulated repair in extract from *mfd-* cells and its restoration by adding TRCF. The right panel shows that transcription-repair coupling may be reconstituted with purified proteins proving that TRCF is necessary and sufficient for coupling transcription to repair in *E. coli*.

Transcription-Stimulation of Repair by TRCF: Incision Assay

In the incision assay with the 138-bp oligonucleotides, repair inhibition by stalled *E. coli* RNAP was overcome by TRCF but stimulation of repair did not occur (see Fig. 2). We considered the possibility that the failure to stimulate repair was because the 138-mers lacked sufficient length to serve as transcription-repair substrates. Therefore a plasmid was tested as template-substrate in a reconstituted transcription-repair incision assay. The pDR3274 plasmid possesses a *tac* promoter[15] and was labeled with ^{32}P at the 3′ end of the template strand 200 bp upstream from the transcription start site. The plasmid was damaged with UV light, which causes dimerization of adjacent pyrimidine residues with varying efficiency. Incubation of the damaged DNA with the excision nuclease produces the series of bands shown in Figure 4B, lane 1, each of which corresponds to an incision product extending from the radiolabel at the 3′ end of the template/substrate to the 3′ incision made by the excision nuclease 4 nucleotides 3′ to the photoproduct. Transcription with *E. coli* RNAP inhibited repair, as seen in lane 2, and addition of TRCF in lane 3 produced an enhanced rate of repair compared to lane 1. Additional controls showed that TRCF had no effect on repair in the absence of transcription, and transcription-repair coupling mediated by TRCF was also observed with different DNA damages. Thus, with substrates of sufficient length, TRCF overcomes the inhibitory effect of stalled RNAP and causes faster rates of repair at these sites relative to lesions in the coding strand or in non-transcribed DNA.

TRCF Binding to UvrA and RNAP

To understand the mechanism of stimulation of excision repair at sites of stalled RNAP we investigated the interactions of TRCF with the excision nuclease and RNAP by conducting affinity chromatography using a resin to which TRCF was attached. RNAP bound to TRCF; core polymerase bound more than holoenzyme.[13] UvrA, the subunit of ABC excision nuclease that binds to DNA damage, bound more strongly to the TRCF resin than RNAP, and UvrA bound to TRCF even in the presence of cell-free extract. UvrB failed to bind.[1] It was concluded that by virtue of its affinity to UvrA and RNAP, TRCF interacts with the elongation complex and causes dissociation of the RNAP stalled at the damage site and in a concerted manner delivers UvrA to the damage to initiate repair. A model depicting the role of TRCF is shown in Fig. 5. Since the damage recognition

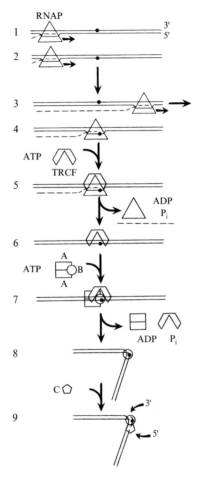

FIG. 5. Model for transcription-repair coupling in *E. coli*. Dotted line, RNA; closed circle, DNA damage; A, UvrA; B, UvrB; C, UvrC. The lesion in the template (2) blocks RNAP in (4) but the lesion in the coding strand (1) does not (3). The blocked polymerase and UvrA of the A2B1 complex are recognized by TRCF which delivers UvrB to the lesion from the A2B1 complex as it removes the blocked polymerase in a sequential (as shown here) or concerted reaction (4 through 8). Reproduced with permission from Selby and Sancar.[1] Copyright (1993) AAAS.

step is the rate-limiting step in excision repair, the process of transcription-repair coupling produces an enhanced rate of repair compared to repair of lesions at sites that do not block RNAP.[1]

Removal of Stalled RNAP by TRCF

The capacity of TRCF to interact with a stalled RNAP is not limited to RNAP blocked at DNA damage. As shown in Fig. 6, TRCF removes RNAP stalled by a DNA-bound protein roadblock to transcription. TRCF also dissociates RNAP stalled by nucleotide starvation.[13] Thus TRCF may have roles in biochemical processes other than repair. In *B. subtilis*, mutation in *mfd* partially relieves carbon catabolite repression of genes when the repressor binding element (cre) is located downstream of the transcription start site. Repression was not effected when the repressor binding element was located upstream.[5] These results are consistent with the

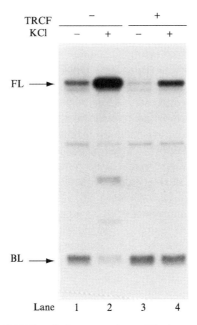

Fig. 6. TRCF removes RNAP stalled at a protein roadblock to transcription. Transcription was from the *tac* promoter in the template pUNC211. The plasmid was linearized with PvuII to give a 465-nt runoff transcript (FL). EcoRI-E111Q was added to reactions; it binds but does not cleave the EcoRI site at +270–275. The mutant EcoRI blocked transcription (BL). TRCF was added as indicated and then KCl was added to 1 M to remove mutant EcoRI and allow remaining blocked complexes to resume elongation (lane 4). From Selby and Sancar.[4]

finding that TRCF dissociates RNAP stalled in elongation mode but has no effect on binding of RNAP to the promoter. Similar results on the effect of TRCF on an RNAP roadblock by a DNA-bound protein were obtained with the well-characterized nut-Nun system of lambdoid phages.[30,31] Phage HK022 excludes superinfecting lambda. The 109 amino acid Nun protein, which is expressed from a HK022 prophage, binds nascent phage lambda RNA at the *nut* sites and induces termination and transcript release. In contrast, Nun arrests transcription on lambda DNA templates in a purified *in vitro* system. The evidence indicates that TRCF dissociates Nun-arrested RNAP. First, Nun activity is enhanced by an *mfd*-null mutation. Second, expression of nut RNA titrates Nun in the mutant host, allowing superinfecting lambda to form plaques. Finally, addition of TRCF and ATP efficiently releases a Nun-arrested transcription complex *in vitro*.[32]

Removal of Arrested RNAP by TRCF

An additional situation in which RNAP becomes stalled has been characterized. In certain situations RNAP becomes stalled in an arrested state in which polymerization ceases since the active site has "backtracked" from the 3' end of the transcript. It was found that GreA and GreB may cause resumption of elongation after inducing cleavage of several nucleotides from the 3' end of the transcript. This cleavage presumably places the 3' end of the transcript in proximity to the active site.[33,34] It has been found that TRCF can also interact with the arrested, backtracked RNAP.[6] When presented with a particular backtracked complex, TRCF, in the absence of nucleotides, removed the arrested complex. In the presence of nucleotides TRCF induces resumption of transcription and thus promotes elongation. However, unlike GreA and GreB, TRCF induces resumption of transcription without cleavage of the 3' end of the transcript. It was suggested that TRCF induces forward translocation of the polymerase so that the active site is repositioned with respect to the 3' end of the transcript so as to allow polymerization. In this scenario, it is this forward translocation induced by TRCF, which, in the absence of NTPs, results in polymerase being forced off of the template. This scenario could be applied to the mechanism of dissociation of RNAP stalled by a protein roadblock and by DNA damage. That is, the blocked polymerase, when pushed forward by TRCF, comes off the template. With respect to physiological relevance, in situations

[30] R. S. Watnick, S. C. Herring, A. G. Palmer, and M. E. Gottesman, *Gen. Dev.* **14,** 731 (2000).
[31] R. S. Watnick and M. E. Gottesman, *Science* **286,** 2337 (1999).
[32] R. S. Washburn, Y. Wang, and M. E. Gottesman *J. Mol. Biol.* **329,** 655 (2003).
[33] S. Borukhov and A. Goldfarb, *Meth. Enzymol.* **274,** 315 (1996).
[34] N. Komissarova and M. Kashlev, *Proc. Natl. Acad. Sci. USA* **94,** 1755 (1997).

where arresting is a significant impediment to transcription, the net effect of TRCF could be to act as an elongation factor.[6]

Substrate and Cofactor Requirements of Transcription-Repair
  Coupling Factor

TRCF binds to DNA in an ATP-dependent manner. It hydrolyses ATP, and ATP hydrolysis is associated with dissociation of TRCF from DNA. ATP hydrolysis is also necessary for the dissociation of stalled RNAP by TRCF.[13] Transcription-stimulated repair begins at lesions located at +15 in the transcription unit, suggesting that the transition of RNAP from the transcription initiation to transcription elongation mode is necessary for productive interaction with TRCF, even though TRCF is capable of interacting with RNAP in the absence of DNA.[4] Similarly, it has been found that dissociation of a stalled RNAP by TRCF is inhibited by the presence of the sigma subunit,[6] which is involved in initiation of transcription. Transcription-stimulated repair requires the presence of approximately 87 bp of DNA downstream of the lesion.[4] In contrast, removal of RNAP stalled by nucleotide starvation requires no more than 3 to 4 nucleotides downstream from the stalled polymerase.[6] The additional DNA required for transcription-stimulated repair may be imposed by the UvrA protein which binds to damaged and undamaged DNA.[28] Although TRCF possesses the so-called helicase motifs, it has no helicase activity towards various DNA and DNA:RNA duplexes.[4] Structure-function analysis of TRCF constructs has revealed the organization of domains shown in Figure 1B.[13]

CSB/ERCC6 Protein

Structure and Function of CSB

Cockayne's syndrome is associated with general growth failure in humans and at the cellular level by abnormally slow recovery of RNA and DNA synthesis following UV irradiation.[35] Repair kinetics analysis showed that CS cells are defective in transcription-coupled repair. Complementation analysis indicates that defects in one of two proteins, CSA and CSB, are associated with the syndrome. CSA, a 44 kDa protein, is a member of the WD-repeat family, which is known to have regulatory but not enzymatic activity.[36,37] CSB exhibits similarities to TRCF. It is

[35] L. V. Mayne and A. R. Lehmann, *Cancer Res.* **42**, 1473 (1982).
[36] K. A. Henning, L. Li, N. Iyer, L. D. McDaniel, M. S. Reagan, R. Legerski, R. A. Schultz, M. Stefanini, A. R. Lehmann, L. V. Mayne, and E. C. Friedberg, *Cell* **82**, 555 (1995).
[37] E. J. Neer, C. J. Schmidt, R. Nambudripad, and T. F. Smith, *Nature* **371**, 297 (1994).

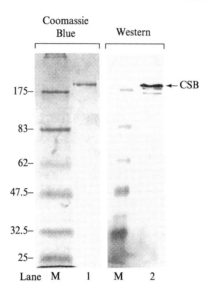

FIG. 7. CSB protein purified from Sf21 cells. The protein demonstrates anomalous mobility during SDS-polyacrylamide gel electrophoresis in that the 168 kDa protein migrates more slowly than the 175 kDa prestained molecular weight marker. The protein from Sf21 cells had the same mobility as native human CSB. From Selby and Sancar.[11]

relatively large, 168 kDa, and possesses the "helicase motifs."[10] CSB was cloned by complementation of a mildly UV-sensitive CHO mutant belonging to complementation group 6 of a collection of UV-sensitive CHO mutant cell lines. Hence it is also referred to as the ERCC6 (Excision Repair Cross Complementing group 6) gene. Although it has been reported that CSA and CSB interact with one another,[38] studies with cell-free extracts showed that the native and recombinant CSA and CSB proteins do not copurify.[11,39]

Failure of CSB to Dissociate Stalled RNAPII

Purification of recombinant CSB from baculovirus-infected cells has been a useful route to obtain active CSB.[11] The purified protein is shown in Fig. 7. Purified CSB hydrolyses ATP and hydrolysis is stimulated by

[38] N. Iyer, M. S. Reagan, K. J. Wu, B. Canagarajah, and E. C. Friedberg, *Biochemistry* **35,** 2157 (1996).
[39] A. J. van Gool, E. Citterio, S. Rademakers, R. van Os, W. Vermeulen, A. Constantinou, J. M. Egly, D. Bootsma, and J. H. J. Hoeijmakers, *EMBO J.* **16,** 5955 (1997).

DNA. In addition, CSB binds to DNA in the presence and absence of ATP or ATP-γ-S. However, CSB failed to remove RNAPII stalled by nucleotide starvation. Figure 8 shows that stalled elongation complexes were readily chased to runoff products both in the presence and absence of CSB.

Effect of CSB on Elongating and Stalled RNAPII

In contrast to its failure to remove stalled RNAPII from the template, CSB did interact with RNAPII in solution in a pull-down assay.[12] Furthermore, interactions of CSB with RNAPII in a ternary complex were detected in four different assays. As shown in Fig. 9A, after incubating

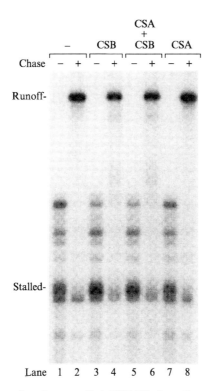

Fig. 8. CSB does not dissociate a stalled RNAPII elongation complex. pMLU112 was transcribed in the absence of UTP, to generate a stalled elongation complex at +112. The 112 nt RNA product is labeled "Stalled." Digestion with PvuII then generated an end such that further elongation would generate a discrete product, labeled "Runoff." Then, as indicated, elongation complexes were incubated with CS protein, followed by chasing with UTP and cold CTP to dilute the radiolabel. RNA products were separated on a 5% sequencing gel. From Selby and Sancar.[11]

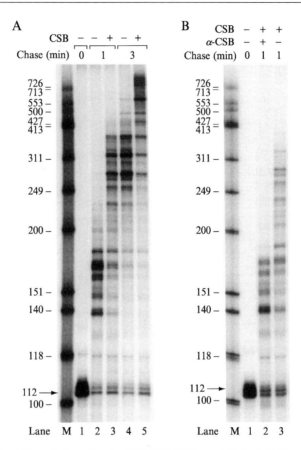

FIG. 9. Stimulation of elongation by CSB. RNAPII was stalled at +112 on the template pMLU112 as described in the legend to Fig. 7. After adding CSB, reactions were chased for 0 to 3 minutes as indicated. RNA was separated with a 6% sequencing gel. Panel (A) demonstrates results of a time course and panel (B) demonstrates results obtained after immunoprecipitating the CSB preparation with anti-CSB antibodies. From Selby and Sancar.[12]

stalled RNAPII with CSB, its rate of elongation during a subsequent chase was faster than in the absence of CSB. The enhancement of elongation was inhibited by addition of anti-CSB antibodies, as shown in Fig. 9B. Pausing occurs when elongating RNAPII reaches the attenuation site in the adenovirus major late promoter,[40] and CSB-stimulated transcription past this site (not shown). RNAPII forms a stable ($t_{1/2}$ is about 20 h) stalled elongation

[40] D. K. Weist, D. Wang, and D. K. Hawley, *J. Biol. Chem.* **267,** 7733 (1992).

complex at the site of the lesion,[41] in which case the major transcript ends one base before the 3'T of the dimer. Upon addition of CSB, RNAPII added an additional nucleotide directly across from the 3'T of the dimer. Finally, CSB negated the transcript-shortening effect of TFIIS on RNAPII stalled at either a dimer[42] or by nucleotide starvation.[12] Thus each of these four different endpoints showed a tendency for CSB to enhance elongation by RNAPII.

Cyclobutane thymine dimer is a strong block to RNAPII[12,42–44] and consistent with CSB functioning as an elongation factor, *in vivo* studies with a yeast strain mutated in the CSB homolog RAD26 have clearly shown a serious transcriptional defect in the mutant strain even in the absence of DNA damage.[45,46]

"Passive" Model for Human Transcription-Coupled Repair

How might transcription-stimulated repair occur without removal of stalled RNAPII from the DNA damage? In contrast to bacteria, the DNA of higher organisms is packaged in chromatin, which inhibits excision repair[47,48] even at the nucleosomal level of compaction[49,50] and hence the status of chromatin structure during transcription may be an important modulator of the rate of repair. In contrast to the prokaryotic RNAP, human RNAPII stalled at a lesion does not inhibit repair of the lesion: The transcription-blocking lesion is repaired at the same rate as the lesion in naked DNA.[41] Thus, by becoming stalled at a lesion present in chromatin, RNAPII may preserve the "open" chromatin conformation characteristic of transcribed genes and thus may prevent inhibition of its repair by chromatin, and in this way rapid repair may be targeted to transcription-blocking lesions located in the template strand. On the other hand, CSB was shown to interact with human repair proteins XPA and XPG and with transcription-repair protein TFIIH and an active role of CSB in coupled repair cannot be ruled out.[11,38]

[41] C. P. Selby, R. Drapkin, D. Reinberg, and A. Sancar, *Nucleic Acids Res.* **25,** 787 (1997).

[42] B. A. Donahue, S. Yin, J. S. Taylor, D. Reines, and P. C. Hanawalt, *Proc. Natl. Acad. Sci. USA* **91,** 8502 (1994).

[43] S. Tornaletti, B. A. Donahue, D. Reines, and P. C. Hanawalt, *J. Biol. Chem.* **272,** 31719 (1997).

[44] S. Tornaletti and P. C. Hanawalt, *Biochimie* **81,** 139 (1999)

[45] S. N. Guzder, Y. Habraken, P. Sung, L. Prakash, and S. Prakash, *J. Biol. Chem.* **271,** 18314 (1996).

[46] S. K. Lee, S. L. Yu, L. Prakash, and S. Prakash, *Mol. Cell Biol.* **21,** 8651 (2001).

[47] Z. Wang, W. Xiaohua, and E. C. Friedberg, *J. Biol. Chem.* **266,** 22472 (1991).

[48] K. Sugasawa, C. Masutani, and F. Hanaoka, *J. Biol. Chem.* **268,** 9098 (1993).

[49] R. Hara, J. Mo, and A. Sancar, *Mol. Cell Biol.* **20,** 9173 (2000).

[50] R. Hara and A. Sancar, *Mol. Cell Biol.* **22,** 6779 (2002).

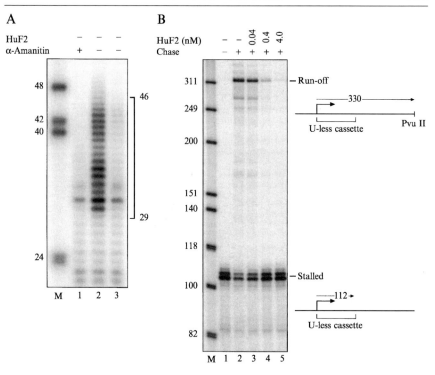

Fig. 10. Transcription release factor 2 (HuF2) dissociates RNAPII stalled at a cyclobutane thymine dimer in the template strand. (A) Template pPu192 possesses in the template strand both a cyclobutane dimer at +149–150 and a ^{32}P radiolabel at the 13th phosphodiester bond 5′ to the dimer. Transcription by RNAPII results in a stable, stalled elongation complex that protects radiolabeled fragments 29–46 nt in length from digestion with DNaseI, as shown in lane 2. Lane 1 is a negative control in which transcription in the presence of all transcription factors was inhibited with α-amanitin at 20 μg/ml. Human factor 2(HuF2) dissociated the polymerase, resulting in complete digestion with DNaseI (lane 3). (B) Stalled elongation complexes were prepared as in Figure 8 and then HuF2 was added in the amounts indicated. The ternary complexes were then chased, and RNA was analyzed as in Fig. 8. M, size markers. From Hara et al.[53]

Dissociation of Stalled RNAPII by Transcription Factor 2

Even though CSB or the CSA/CSB combination do not displace RNA-PII stalled at a lesion, human cells do possess a factor which does so very efficiently. It was found that human transcription factor 2 (HuF2), a negative transcription elongation factor,[51,52] efficiently releases RNAPII stalled at a thymine dimer or stalled by nucleotide starvation.[53] Figure 10A shows a

[51] N. F. Marshall and D. H. Price, Mol. Cell Biol. **12,** 2078 (1992).
[52] Z. Xie and D. H. Price, J. Biol. Chem. **271,** 11043 (1996).

DNaseI footprint of RNAPII stalled at a thymine dimer located in the template strand. The presence of a radiolabel adjacent to the dimer allows visualization of the 29–46 nt region protected by the bound polymerase. Addition of HuF2 causes release of the stalled polymerase and the radiolabeled plasmid is no longer protected against digestion by DNaseI. Similarly, HuF2 prevented RNAPII stalled by nucleotide starvation from subsequent elongation, presumably by removing the ternary complex from the template (Fig. 10B).[53] These findings suggest that even though the mammalian structural homolog of bacterial TRCF does not dissociate RNAPII stalled at a lesion, these complexes could be rapidly and efficiently disrupted by human Factor 2 and hence transcription-coupled repair associated with release of stalled RNAPII may occur in mammalian cells.

Comparison of TRCF and CSB

Table I summarizes the comparisons between the *E. coli* TRCF and human CSB proteins. A better perspective for relating these properties with the similarities and differences between the reaction mechanisms of these proteins will become possible only when the mechanism for human transcription-repair coupling and the molecular basis for Cockayne's

TABLE I
PROPERTIES OF TRCF AND CSB PROTEINS

Property	TRCF	CSB
Mutant phenotype	Mfd-	Cockayne's
UV sensitivity of mutant	moderate	moderate
No. of amino acids	1148	1493
Size (kDa)	130	168
Helicase motifs	yes	yes
Helicase activity	no	no
DNA binding	yes	yes
ATPase	yes	yes
Interacting repair proteins	UvrA	TFIIH, XPA, XPG
Cognate RNAP	*E. coli* RNAP	human RNAPII[*]
RNAP interacting:		
In solution	yes	yes
Elongation complex	yes	yes
Dissociate ternary complex	yes	no
Transcription-repair coupling		
In vivo	yes	yes
In vitro	yes	ND

ND, not done.
[*] A recent report indicates enhanced repair rate of RNA PolI transcribed rRNA genes in mammalian cells.

syndrome are understood. The mechanism for transcription-repair coupling in humans may be quite different than that in *E. coli* for several reasons. These include the presence of chromatin in humans and its effect on repair, as well as the finding that the RNAP from *E. coli* stalled at damage inhibits repair, but stalled human RNAPII does not. In addition, TRCF removes stalled RNAP, but CSB does not. Accordingly, the roles of these two proteins in transcription-coupled repair will likely exhibit differences. However, several intriguing similarities between TRCF and CSB have been revealed, notably, their associated phenotypes, their ATPase activities, their interactions with their respective RNAPs, their ability to "push" the blocked polymerase forward, and their binding to DNA and to transcription and repair proteins.

Acknowledgments

This work is dedicated to Richard Bockrath, Philip Hanawalt, and Evelyn Witkin for their pioneering work on Mfd, strand-specific mutagenesis, and strand-specific repair. Our research was supported by the NIH grant GM32833.

---

[53] R. Hara, C. P. Selby, M. Liu, D. H. Price, and A. Sancar, *J. Biol. Chem.* **274**, 24779 (1999).

[24] Techniques to Analyze the HIV-1 Tat and TAR RNA-Dependent Recruitment and Activation of the Cyclin T1:CDK9 (P-TEFb) Transcription Elongation Factor

By NATHAN GOMES, MITCHELL E. GARBER, and KATHERINE A. JONES

The human immunodeficiency virus (HIV-1)-encoded Tat protein has provided a unique paradigm in our understanding of the process of regulated transcription elongation by RNA polymerase II (RNAPII). Tat functions primarily as a virus promoter-selective inducer of transcription elongation in cytokine-activated T cells and macrophages.[1-3] Tat regulates a step in early transcription elongation that is marked by hyphosphorylation of the *C*-terminal domain (CTD) of the largest subunit of RNAPII and associates tightly in nuclear extracts with a CTD kinase (TAK, *Tat*

[1] M. Emerman and M. H. Malim, *Science* **280** (1998) 1880–1884.
[2] J. Karn, *J. Mol. Biol.* **293** (1999) 235–254.
[3] M. E. Garber and K. A. Jones, *Curr. Opin. Immunol.* **11** (1999) 460–465.

*a*ssociated *k*inase).[4] More recently, the catalytic subunit of TAK was identified as CDK9,[5–7] catalytic subunit of the positive transcription elongation
factor, P-TEFb.[8] Tat is directed specifically to the HIV-1 promoter through
the TAR element, which folds into an RNA hairpin at the 5' end of the
emerging viral transcript (Fig. 1). Work from our laboratory and others
has shown that Tat interacts directly with human cyclin T1 (CycT1),[9–13]
which is a regulatory subunit of the nuclear CDK9 subunit of
transcription elongation factor the positive P-TEFb[14] CDK9 has been
shown to phosphorylate the CTD of RNAPII, and previous work has
established that both CDK9 and the RNAPII CTD are essential for Tat-
mediated transactivation.[1,3,8] Interestingly, studies with purified recombinant proteins have shown that Tat binds cooperatively with CycT1 to TAR
RNA, and that this complex binds to TAR RNA in a loop- and bulge
sequence–specific manner.[9] Thus, through its ability to interact with CycT1,
Tat can recruit the P-TEFb complex to nascent viral RNA transcripts. In
addition, binding of Tat to P-TEFb strongly activates its CTD kinase activity, resulting in hyperphosphorylation of the RNAPII CTD and efficient
elongation of viral transcription.[15] At least *in vitro*, Tat also strongly alters
the specificity of CDK9 to phosphorylate both Ser-2 and Ser-5 in the CTD
heptapeptide repeats,[15] whereas P-TEFb principally phosphorylates the
Ser-2 position in the absence of Tat.[8] Tat binds CycT1 in a zinc-dependent
manner through a Cys-rich activation domain, and contacts a region of
CycT1 overlapping the cyclin box but on the opposite face to that through
which CycT1 binds CDK9.[10] Although other cyclin partners for CDK9
have been isolated from HeLa cells, only CycT1 is able to interact with
Tat. Similarly, the murine homolog of CycT1 binds relatively weakly to

[4] C. H. Herrmann and A. P. Rice, *J. Virol.* **69** (1995) 1612–1620.
[5] H. S. Mancebo., G. Lee, J. Flygare, J. Tomassini, P. Luu, Y. Zhu, J. Peng, C. Blau, D. Hazuda, D. Price, and O. Flores, *Genes Dev.* **11** (1997) 2633–2644.
[6] Y. Zhu, T. Peery, J. Peng, Y. Ramanathan, N. Marshall, *et al. Genes Dev.* **11** (1997) 2633–2644.
[7] M. D. Gold, X. Yang, C. H. Herrmann, and A. D. Rice, *J. Virol.* **72** (1998) 4448–4453.
[8] D. H. Price, *Mol. Cell. Biol.* **20** (2000) 2629–2634.
[9] P. Wei, M. E. Garber, S. M. Fang, W. H. Fischer, and K. A. Jones, *Cell* **92** (1998) 451–462.
[10] M. E. Garber, P. Wei, V. N KewalRamani, T. P. Mayall, C. H. Herrmann, A. P. Rice, D. R. Littman, and K. A. Jones, *Genes Dev.* **12** (1998) 3512–3527.
[11] M. E. Garber, P. Wei, and K. A. Jones, *Cold Spring Harbor Symp. Quant. Biol.* **63** (1998) 371–380.
[12] P. D. Bieniasz, T. A. Grdina, H. P. Bogerd, and B. R. Cullen, *EMBO J.* **17** (1998) 7056–7065.
[13] K. Fujinaga, T. Cujec, J. Peng, J. Garriga, D. H. Price, X. Grana, and B. M. Peterlin, *J. Virol.* (1998) 7154–7159.
[14] J. Peng, Y. Zhu, J. T. Milton, and D. H. Price, *Genes Dev.* **12** (1998) 755–762.

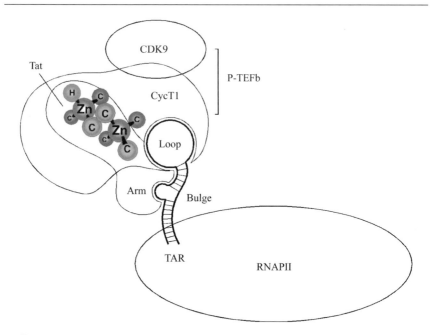

Fig. 1. Schematic diagram depicting the zinc-dependent interaction between Tat and Cyclin T1 (*CycT1*) and the loop– and bulge sequence–specific interactions that recruit the Tat:P-TEFb complex to nascent TAR RNA. In addition to recruiting CDK9 to the transcript early in transcription elongation, Tat strongly activates the CTD kinase activity of P-TEFb in cells. TAR RNA-binding studies with recombinant P-TEFb subunits suggest that the complex must undergo autophosphorylation in the presence of Tat in order to bind avidly to TAR RNA (see Fig. 4), indicating that the proteins bound to the RNA may be highly phosphorylated. It has been proposed that additional post-translational modifications, including acetylation or ubiquitination, may play a role in the later steps of elongation or in the release of the P-TEFb complex from TAR RNA, and the biochemical techniques described here can be readily adapted to address these mechanistic questions. (See color insert.)

Tat, and the cooperative binding to TAR RNA is greatly diminished.[10,12] These differences arise from the lack of conservation of residue Cys261 in the cyclin proteins that fail to be recognized by Tat.

Recent studies have shown that P-TEFb activity can be regulated in many ways *in vivo*, both transcriptionally and post-transcriptionally.[16,17,17a,18,18a]

[15] M. E. Garber, T. P. Mayall, E. M. Suess, J. Meisenhelder, N. E. Thompson, and K. A. Jones, *Mol. Cell. Biol.* **20** (2000) 6958–6969.

[16] C. H. Herrmann, R. G. Carroll, P. Wei, K. A. Jones, and A. P. Rice, *J. Virol.* **72** (1998) 9881–9888.

[17] R. Ghose, L. Y. Liou, C. H. Herrmann, and A. P. Rice, *J. Virol.* **75** (2001) 11336–11343.

Thus both mRNA and protein levels for the CycT1:CDK9 proteins are in-duced upon PMA activation of quiescent primary T cells, and protein levels and kinase activity are also subject to cytokine control in quiescent CD4+ T cells.[16,17,17a] Binding of Tat to P-TEFb in cells clearly activates the resident protein kinase activity of CDK9 strongly, and this effect of Tat is likely to play a significant role in the transcription activation mechanism. P-TEFb has been shown biochemically to reside in large and small complexes, and Tat interacts preferentially with the latter complex.[18,18a,19] P-TEFb in the larger complex is also associated with 7SK RNA, potentially in an inactive or storage form of the kinase, although the role of 7SK RNA interaction remains largely an open question at the present time. Over-expression of a dominant-negative form of CycT1 predisposes T cells to undergo apop-tosis,[20] indicating that P-TEFb is an anti-apoptotic factor in lymphoid cells. Importantly, CDK9 is also a primary target for flavopiridol, which also down-regulates genes involved in apoptosis.[8]

In this chapter, we describe the *in vitro* transcription and TAR RNA-binding protocols that we have used to study Tat transactivation of the HIV-1 promoter. In addition, we present a TAR RNA-affinity selection procedure that can be used to study the activity of the Tat complex with native P-TEFb in nuclear extracts.[11] Finally, we outline the GST-Tat protein-affinity selection protocol that we used to identify CycT1 as the direct binding partner of Tat in cells[9]; see Figs. 2 and 3 and the protocol used to examine binding to TAR (Fig. 4). These approaches continue to be useful to analyze the P-TEFb complex *in vitro* and *in vivo*.

Although recent studies have highlighted the role of chromatin modifi-cation and remodeling in controlling the processivity of transcription elongation, HIV-1 Tat controls the phosphorylation of the CTD in a step that occurs efficiently and accurately *in vitro* on nonchromatin templates. Tat activation in HeLa nuclear extracts requires only a minimal promoter containing the Sp1-binding sites, the TATA box, and distinctive HIV-1 ini-tiator sequence, as well as the TAR RNA element. The most commonly used procedures for monitoring elongation are run-off assays using tem-plates with G-less cassettes or riboprobe protection assays, and most of the published procedures are very sensitive methods to study regulation of transcription by Tat. Although there has been considerable progress identifying the various human factors necessary for transcription elonga-tion, at the present time there is no well-characterized completely purified

[17a] L. Liou, C. H. Herrmann, and A. P. Rice, *J. Virol.* (2002) 10579–10587.

[18] Y. W. Fong and Q. Zhou, *Mol. Cell. Biol.* **20** (2000) 5897–5907.

[18a] V. T. Nguyen, T. Kiss, A. A. Michels, and O. Bensaude, *Nature* **414** (2001) 322–325.

[19] Z. Yang, Q. Zhu, K. Luo, and Q. Zhou, *Nature* **414** (2001) 317–322.

[20] S. M. Foskett, R. Ghose, D. N. Tang, D. E. Lewis, and A. P. Rice, *J. Virol.* **75** (2001) 1220–1228.

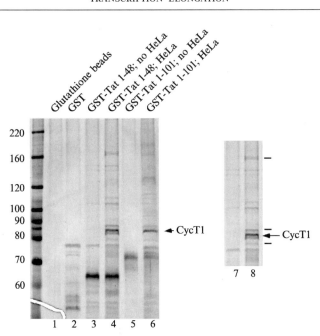

FIG. 2. Identification of human Cyclin T1(CycT1) as the nuclear protein-binding partner of HIV-1 Tat in GST-protein pull-down assays. The top of each lane indicates the GST protein that was coupled to the beads used in each experiment. The reactions shown in lanes 1, 2, 4, and 6 were incubated with crude HeLa nuclear extract, and the tightly associated proteins retained after stringent washing were analyzed by boiling the beads and analyzing the proteins following SDS-PAGE and silver staining. The arrow indicates native CycT1 that is recovered from beads containing the GST-Tat(aa1-48) activation domain or the full-length GST-Tat(aa1-101) proteins in lanes 4 and 6, respectively. The panel at the right shows an insert from lanes 2 and 3 where additional nuclear proteins associated with Tat are indicated with dashes. Other proteins in the lower half of the gel are either nuclear GST-binding proteins or proteins contaminating the GST-Tat preparations (such as dimers). The GST-Tat fusion proteins themselves have been run off of the bottom of the gel in this experiment.

reconstituted transcription system for elongation. Our protocols for preparing HeLa nuclear extract and recombinant Tat proteins have been reported elsewhere, and in this paper we review the methods we use to: (1) study Tat transactivation by "run-off" *in vitro* transcription assays; (2) monitor binding of recombinant Tat and CycT1:CDK9 to TAR RNA in gel mobility shift assays; (3) analyze binding of Tat and native HeLa CycT1:CDK9 in extracts using an RNA-tethered beads assay, and (4) analyze and identify the nuclear proteins that associate with Tat in extracts using a simple and very sensitive protein: protein affinity purification technique.

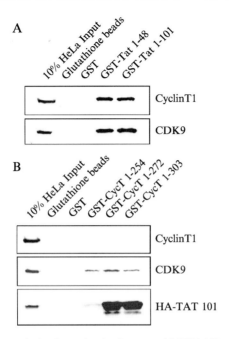

FIG. 3. Western blot analysis of proteins that interact with HIV-1 Tat or Cyclin T1 (Cyc T1) in GST pull-down assays from HeLa nuclear extract. (A) Both GST-Tat activation domain (aa1-48) and the full-length GST-Tat (aa1-101) interact with CycT1 and CDK9 proteins in nuclear extract. Note that no binding to the GST-coupled beads is observed under the stringent conditions used in these assays. (B) Western blot analysis of the ability of different truncated CycT1 proteins to bind to CDK9 and to HA-tagged Tat (aa1-101) in nuclear extracts.

The *In Vitro* Transcription Elongation Assay

Preparation of Recombinant HIV-1 Tat Protein

A single 50-ml overnight culture of GST-Tat (BL21 DE3) is diluted with L-broth to 500 ml, the bacteria are grown to a density of 0.9 A_{600} and induced with 0.1 mM IPTG for 7 h at 30°. Cells are resuspended in lysis buffer (1X PBS containing 10 mM DTT, 5 mM EDTA, 1 mM PMSF, 2 mg/ml benzamidine, 2 μg/ml pepstatin A, 8 μg/ml leupeptin, 20 μg/ml aprotinin, 40 μg/ml soybean trypsin inhibitor, and 200 μg/ml lysozyme), sonicated for 30 sec, and frozen and thawed once. The cell lysate is clarified by centrifugation at 4° in an SW41 rotor for 30 min at 30,000 RPM. The supernatant is bound to 500 μl glutathione beads for

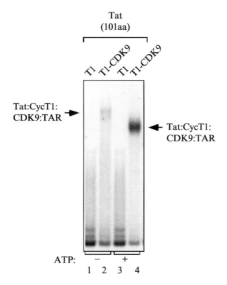

FIG. 4. Analysis of the binding of recombinant purified Tat, Cyclin T1 and CDK9 to TAR RNA in gel mobility shift experiments. Reactions were carried out in the absence (lanes 1, 2) or presence of 1 mM ATP (lanes 3, 4). The arrow indicates the migration position of Tat:CycT1:CDK9:TAR ternary complex, which migrates differently in the gel, depending on whether or not the proteins have undergone CDK9 autophosphorylation.

1 h at 4°, washed three times in 1X PBS containing 10 mM DTT and 0.1 mM PMSF, rinsed once in Tat elution buffer (50 mM Tris-HCl, pH 7.5, 100 mM NaCl, 10 mM DTT, and 0.1 mM PMSF) and eluted in Tat elution buffer containing 10 mM glutathione. Glycerol is added to 10% (final concentration), and the Tat protein is stored at −80° prior to use. The GST peptide can be removed by thrombin cleavage at room temperature for 2 h before freezing.

In Vitro *Transcription Assay with HIV-1 Tat*

There are several different procedures to monitor Tat activation of HIV-1 transcription, and all work comparably well, though the following protocol may be the simplest. In general, the HIV-1 promoter template is most strongly activated *in vitro* by either the HIV-1 or HIV-2 Tat proteins (up to 50-fold) because of its unusual forked TAR RNA structure that has two stem-loop structures. However, strong activation of up to 30-35–fold can also be observed for the HIV-1 promoter. Tat did not stimulate transcription from a mutant HIV-2 promoter that lacks the cis-acting TAR element ($\square$TAR), indicating that Tat activity requires the TAR RNA element in this assay. The relative inability of HIV-2 Tat to activate transcription from the HIV-1 promoter is consistent with previous *in vivo* findings and has been attributed to differences in the ARM regions of the two Tat proteins that weaken the binding of HIV-2 Tat to the single HIV-1 TAR RNA.

Solutions

Each transcription reaction (30 μl, final volume) contains the following components:

9 μl TM buffer containing 160 mM KCl
4–8 μl HeLa nuclear extract (8–12 μg/μl)
1.2 μl poly (dl-dC) 0.5 mg/ml
1.2 μl poly (rl-rC) 0.5 mg/ml
1 μl linearized pHIV2-CAT DNA template (150 ng); cut with Ncol
1 μl linearized pHIV2Δ34-CAT DNA template (100 ng); cut with Ncol
1.5 μl 20 $\times$ rNTP-Mix (50 mM DTT; 2 mM EDTA; 0.2 M phosphocreatine; 12 mM rATP; 12 mM rGTP; 12 mM rCTP; 0.8 mM rUTP)
0.2–0.4 μl [α-32P]-rUTP (800 Ci/$m$mol, 20 μCi/μl)

Procedure

1. Incubate the mixture just described for 30 min at $30°$.
2. Terminate the transcription reactions by adding 100 μl of stop solution per reaction (1% SDS; 20 mM EDTA; 0.1 mM NaCl; 100 μg/μl yeast t-RNA).
3. Extract the mixture once with phenol:chloroform (1:1) and once with chloroform.
4. Add 0.5 volume of ammonium acetate and 2.5 volumes of ethanol to the supernatant (for 120 μl supernatant, use 60 μl 7.5 mM NH$_4$Ac and 300 μl 100% ethanol) and allow the RNA to precipitate for 30 min on ice.
5. After spinning for 30 min at $4°$, the supernatant is removed and the pellet is air-dried.
6. The pellet is dissolved in 8 to 10 μl loading dye (98% deionized formamide; 10 mM EDTA; 0.04% bromphenol blue; 0.04% xylene cyanol).
7. Electrophoresis is carried out on a 6% denaturing polyacrylamide gel (41 cm $\times$ 20 cm $\times$ 0.4 mm) at 40 W constant power for 2 to 3 h.
8. After electrophoresis, the gel is dried and visualized by autoradiography.

TAR RNA-Affinity Selection Procedures to Isolate Complexes
  Containing Recombinant HIV-1 Tat and the Native
  CycT1:CDK9 in HeLa Nuclear Extracts

The Tat-associated kinase (TAK) complex can be readily isolated from HeLa nuclear extracts by affinity selection with GST-Tat (aa1-48) following the procedure of Herrmann and Rice.[4] Our modifications to this approach, and the scaled-up procedure necessary to purify hCycT1 from HeLa nuclear extracts (see Fig. 2) are outlined below. We have also used a reiterative affinity selection procedure in which the Tat:P-TEFb complex is first purified from crude HeLa nuclear extracts by TAR RNA affinity chromatography, as outlined below, and then subjected to GST-Tat affinity selection.[11] In this experiment, Tat is incubated with HeLa nuclear extract, and the Tat:P-TEFb complex is isolated by the GST-Tat affinity selection procedure described below. CDK9 and CycT1 are visualized by incubation of the beads with [γ-32P] ATP, which permits autophosphorylation of the kinase and adventious labeling of the cyclin T1 and Tat proteins, as well as other unidentified high-molecular weight factors. We and others have shown that P-TEFb does not associate with mutant Tat proteins carrying substitutions in the Cys-rich activation domain.[10,12] If the Tat:P-TEFb complex is first purified by incubation with wild-type TAR RNA-coupled beads, the P-TEFb complex is more highly purified and can be recovered efficiently following GST-Tat affinity selection. Importantly, the Tat:P-TEFb complex in nuclear extracts does not associate with loop mutant TAR RNA beads, indicating that this assay is highly selective.[11] Through this approach we showed that the native HeLa P-TEFb does not bind to TAR RNA in the absence of Tat, and that the nuclear Tat:P-TEFb complex, like the complex formed between recombinant CycT1 and Tat binds to TAR RNA in a loop-specific manner. Details of these procedures follow.

HIV-1 TAR RNA Affinity Chromatography

  Solutions

    Tat kinase buffer (50 mM Tris-HCl, pH 7.5, 10 mM MgCl$_2$, and
      5 mM DTT)
    Binding buffer (50 mM Tris-HCl, pH 8.0, 0.5% NP-40, 5 mM DTT)
    Tat elution buffer (50 mM Tris-HCl, pH 7.5, 100 mM NaCl, 10 mM
      DTT, and 0.1 mM PMSF)

  Procedure

    Large-scale synthesis (400 μl reaction vol) of biotinylated TAR RNA was performed with Sp6 RNA polymerase as described[5] except that reactions contained 100 nmole unmodified rUTP (1.0 μl of 0.1M) and 120 nmole biotin-rUTP (12 μl of a 10-mM stock). This results is an approximately 85% incorporation of biotin-rUTP and no more than one biotin per TAR molecule. The RNAs were labeled to low specific-activity by the addition of 10 μCi of [α-32P] rUTP to the synthesis reaction in order to quantitate the efficiency of coupling of TAR RNA to the beads, and the extent of RNA release from the beads upon incubation with crude nuclear extract.

For the batch purification of P-TEFb by TAR RNA affinity chromatography, the following components were assembed, per reaction:

   40 μl NE (500 μg protein)
   12 μl TM buffer containing 100 mM KCl and 0.2% NP-40
   11 μl Tat elution buffer
   1 μl GST-cleaved Tat (200 ng)
   6.8 μl ddW

1. Incubate 30' at 30°. Add 20 μl 0.1 M NaCl and 28 μl ddW. Allow to stand at 0° for an additional 40 min. Spin to pellet.

2. Bind 10 μg biotin-TAR RNA to 25 μl streptavidin beads in 1X SSC. The TAR RNA-bound beads were equilibrated and resuspended in 80 μ 0.5× TM containing 75 mM KCl and 0.05% NP-40.

3. A 120-μl nuclear extract mix was added to a slurry (80 μl volume) of TAR RNA-coupled beads. Rotate in a silanized tube for approximately 14 h at 4°. Roughly 90% of the TAR RNA remained bound to the beads during the 14-h incubation as measured by radioactivity.

4. Wash in 120 μl of 0.5× TM buffer containing 75 mM KCl and 0.05% NP-40. To elute, add 120 μl TM buffer containing 1M KCl and 0.05% NP-40, incubate on ice with gentle agitation for 30 minutes, and spin to pellet the beads.

5. For the flowthrough and wash fractions, dilute the 120 μl aliquot of the supernatant fraction to 600 μl with a binding buffer containing 0.25% gelatin. For the 1M elution, dilute to 600 μl with a binding buffer containing 120 mM NaCl and 0.25% gelatin. Add roughly 2 μl crude Tat polyclonal serum and rotate at 4° for 1 h.

6. Protein A sepharose beads (15 μl), equilibrated in binding buffer, were added and incubated with the antibody/antigen complex at 4° for 1 h. Beads were washed three times in binding buffer containing 0.03% SDS, once in Tat kinase buffer; the kinase reaction was then initiated with the addition of 25 μl of Tat kinase buffer containing 2.5 mM MnCl$_2$, 1 μM cold ATP and 10 μCi [γ-32P] ATP, as described previously.[7,8] Following a 1-h incubation at room temperature, beads were boiled in 50 μl 1X SDS-PAGE

sample buffer, the proteins were separated by 10% SDS-PAGE, and the gel was visualized by autoradiography.

RNA Gel Mobility Shift Analysis of Tat:CycT1:CDK9 Complexes

Solutions

5X binding buffer
50 mM Tris-HCl, pH 8
1 mM EDTA
30% glycerol
1mM DTT (add fresh)

Reactions are 15 μL final volume

3 μL 5X binding buffer
4 μL ddW
1 μL labelled RNA probe; 3-8 ng/μL; >50 cps per μL
1 μL KCl (dilution appropriate for 100 mM final concentration)
2 μL competitor RNA as required; see below for example reaction
4 μL recombinant protein fraction

Protocol

1. Mix binding buffer, water, and labeled TAR RNA probe.
2. Add competitor RNA to reactions on ice.
3. Add protein fractions to the reactions.
4. Incubate 20 min on ice.
5. Load directly onto on a 6% Tris-glycine gel, and run at 4° for 3 h at 25 mAmps constant.

Gel mobility shift experiments (15 μl) are carried out in an RNA-binding buffer (RBB) containing 135 mM KCl, 1 μg of poly(rl-rC), 15 ng of HIV-1 TAR RNA probe, and HeLa total RNA (600 ng). HIV-1 TAR RNA (nucleotides 1 to 80) was uniformly labeled *in vitro* using a linearized template and T7 RNA polymerase. Where indicated, ATP was added to CDK9 prebound to bacterially expressed CycT1(1-303), and HIV Tat and complex formation on TAR RNA was allowed to proceed for 30 min at 30°. Reaction products are separated on a pre-run 4–6% Tris-glycine polyacrylamide gel.

Protocol for Purification of Native Cyclin T1 from Nuclear Extracts by GST-Tat Pull-Down Affinity Selection

In this procedure, a crude HeLa nuclear extract is incubated in batch with GST-Tat bound to glutathione beads and washed extensively to purify the P-TEFb subunits.

1. Ten μl NE (10 mg/ml) was diluted 1:1 with TM0.1M KCl and pre-cleared by successive incubation with: (1) 15 μl glutathione-S-Sepharose beads and (2) 15 μl GST (or mutant GST-Tat)–coupled beads. Incubations were carried out at 4° for 30 to 60 min per step.

2. Two μg of purified GST or GST-Tat proteins (proteins had been eluted from beads with glutathione, but not dialyzed) were incubated with 15 μl glutathione beads (as a 50% slurry) in EBC buffer (50 mM Tris-HCl, pH 8, 120 mM NaCl, 0.5% NP-40) at 4° for 15 to 30 min. The beads were washed twice with 150 μl EBC buffer containing 5 mM DTT and 0.075% SDS, and then washed once with EBC buffer containing 5 mM DTT.

3. GST-Tat beads were incubated with pre-cleared NE at 4° for 60 min. The complexes were washed three times with 10-20x (bead vol) EBC buffer containing 5 mM DTT and 0.03% SDS, and then washed once with Tat kinase buffer (TKB; 50 mM Tris-HCl, pH 7.5, 5 mM MnCl$_2$, and 5 mM DTT).

4. The washed beads were analyzed by SDS-PAGE (after boiling beads in sample buffer) and also by *in vitro* kinase assays:

A 50-μl reaction containing TKB, 5 μCi of [α-32P] ATP, 2 μM ATP and the bead complexes were incubated for 30 min at room temperature. The complexes were washed twice with TKB buffer and boiled in Laemmli buffer before analysis on SDS-PAGE.

It was important to optimize the efficiency of coupling of GST-Tat proteins to the beads and to minimize contamination of the preparation by bacterial proteins, because these could obscure the detection of nuclear proteins that bind Tat.

Following the GST-pulldown reaction and Coomassie staining, proteins bands can be excised, destained in acetonitrile and ammonium bicarbonate buffer and subjected to in-gel tryptic digestion. A portion of the concentrated tryptic fragments are analyzed by MALDI-TOF with Reflectron to identify the proteins by comparison with existing databases. A scale-up of this procedure 80-fold yielded around 1 μg of CycT1 (86 kDa), which was run out on 7 to 8 lanes of a preparative SDS-PAGE for protein microsequencing.

Concluding Remarks

The mechanism used by HIV-1 Tat to recruit the RNAPII P-TEFb complex to the HIV-1 promoter in activated cells provides a detailed and unique paradigm of regulated transcription elongation. Recent studies have highlighted the activation of the P-TEFb complex, and induced expression of CycT1 in activated T cells and macrophages suggest that Tat evolved through a need to induce viral gene expression rapidly upon T cell activation. In this paper we reviewed the biochemical approaches we use to analyze the mechanism of Tat activation, in particular the protein-affinity approach we used originally to identify Cyclin T1 as the direct binding partner of Tat. The cooperative binding of Tat and P-TEFb to TAR RNA provides a unique method to recruit the kinase to the RNAPII large subunit, but of equal importance is the finding that Tat strongly activates the CTD kinase activity of P-TEFb. When coupled to enzymatic assays of kinase activity, the approaches described here may yield much useful information about the regulation of CDK9 kinase activity in the cell. Future work is needed to define how the kinase activity of CDK9 is inhibited until the particular stage of elongation where it is needed, and whether interaction of P-TEFb with other cellular proteins may regulate its activity *in vivo*. Although some cellular gene targets of CDK9 have been identified, and its role as an anti-apoptotic, flavopiridol-sensitive kinase regulated in T cells have begun to be elucidated, much remains to be learned of the role of this kinase in the cell and the mechanisms that may govern its recruitment to cellular genes. Thus far no cellular genes have been identified that use a TAR-like mechanism to attract P-TEFb, and consequently it is not clear whether the Tat mechanism is unique or general. The RNA-binding assay using TAR-coupled beads that we describe here is perhaps the most robust approach to analyze the native P-TEFb complex bound with Tat on TAR. As specific regulatory cofactors and modifications of the Tat: P-TEFb complex are identified, these biochemical approaches should continue to provide important details of the mechanism of HIV-1 Tat trans-activation.

Acknowledgments

We thank current and previous members of the Jones Laboratory for their continuing contributions to the development of procedures to analyze the Tat transactivation mechanism. The research in our laboratory on HIV-1 Tat and the P-TEFb complex is funded by the NIH and the California Universitywide Task Force on AIDS.

Section II

Transcription Termination

[25] Assay of Intrinsic Transcript Termination by E. Coli RNA Polymerase on Single-Stranded and Double-Stranded DNA Templates

By Susan M. Uptain

The final step in the transcription cycle, termination, occurs when the RNA polymerase releases the nascent RNA transcript and dissociates from the DNA template. Termination is irreversible; any subsequent transcription requires reinitiation at a promoter with formation of another RNA transcript. In prokaryotes, termination sites are often targets for regulation of gene expression[1,2]; in humans, regulation of termination plays an important role in human development and disease (for review see Ref. 3). Although much progress has been made,[2] the exact mechanism of transcript termination remains unclear.

There are two types of terminators in *Escherichia coli*. Factor-dependent terminators depend on the action of ancillary factors. Because the predominant termination factor in *E. coli* is a hexameric protein called Rho, these terminators are often called Rho-dependent. Rho mediates transcript release in an adenosine triphosphate (ATP)-dependent manner signaled by sequences that are weakly conserved.[2,4] Intrinsic terminators are recognized by the polymerase *in vitro* without the aid of additional factors, and they are commonly referred to as Rho-independent. Recognition of these terminators by RNA polymerase usually depends on formation of an RNA hairpin followed by a run of seven or eight U residues, although not all RNA hairpins lead to termination and the U-rich region is frequently interrupted with other nucleotides. RNA transcript release usually occurs seven to nine nucleotides downstream from the base of the RNA hairpin stem *in vitro*.[2,5,6]

The purpose of this article is to describe the principal techniques used for biochemical characterization of transcript termination. The major focus will be on intrinsic termination by *E. coli* RNA polymerase on

[1] T. M. Henkin and C. Yanofsky, *Bioessays* **24**, 700 (2002).
[2] E. Nudler and M. E. Gottesman, *Genes Cells* **7**, 755 (2002).
[3] S. M. Uptain, C. M. Kane, and M. J. Chamberlin, *in* "Annual Review of Biochemistry" (C. C. Richardson, ed.), Vol. 66, p. 117. Annual Reviews, Inc., Palo Alto, Calif., 1997.
[4] J. P. Richardson, *Biochim. Biophys. Acta* **1577**, 251 (2002).
[5] T. M. Henkin, *Ann. Rev. Genet.* **30**, 35 (1996).
[6] P. H. von Hippel, *Science* **281**, 660 (1998).

METHODS IN ENZYMOLOGY, VOL. 371

double-stranded (dsDNA) and single-stranded (ssDNA) DNA templates[7]; however, these methods can be easily adapted to study transcript termination of other RNA polymerases and factor-dependent termination, or extended to study aspects of transcript elongation.

Methods

Overview of Purification of Histidine-Tagged E. coli RNA Polymerase

E. coli RNA polymerase, which has a hexahistidine tag at the carboxyl-terminus of the β subunit, is purified from E. coli strain RL721 (kindly provided by R. Landick, University of Wisconsin, Madison). Cell extracts are prepared and are fractionated with polyethylenimine (polymin P) by the method of Burgess and Jendrisak.[8] The RNA polymerase is then precipitated from the polymin P eluate by addition of 35 g/100 ml ammonium sulfate. The resulting pellet is extracted twice with equal volumes of 2.0 M ammonium sulfate, and then RNA polymerase is eluted from the pellet with 1.6 M ammonium sulfate. This eluate is fractionated by chromatography on Ni^{2+}-NTA agarose. Subsequently, holoenzyme, which contains α, β, β', and σ subunits, is separated from core RNA polymerase, which lacks the σ subunit, by phosphocellulose chromatography.[9] The holoenzyme preparations obtained using this method typically contain 60% to 90% active RNA polymerase relative to the amount of protein added, as determined by the quantitative assay for RNA polymerases[10] and are estimated to be more than 95% pure, judging from Coomassie Brilliant Blue staining of SDS–PAGE gels. The following procedure is designed for a large-scale preparation using 200 g of frozen E. coli cell pellet; however, it may be scaled up or down as needed.

Buffers

Buffer composition: TGED [50 mM Tris–HCl, pH 7.9, 5% v/v glycerol, 2 mM EDTA, pH 8.0, 0.1 mM dithiothreitol (DTT)]; TGβ [10 mM Tris–HCl, pH 7.9, 5% v/v glycerol, 7 mM beta-mercaptoethanol (βME)]; P$_{50}$ [40 mM potassium phosphate, pH 8.0, 50% v/v glycerol, 0.1 mM EDTA, 1 mM DTT].

---

[7] S. M. Uptain and M. J. Chamberlin, *Proc. Natl. Acad. Sci. USA* **94**, 13548 (1997).

[8] R. R. Burgess and J. J. Jendrisak, *Biochemistry* **14**, 4634 (1975).

[9] N. Gonzalez, J. Wiggs, and M. J. Chamberlin, *Arch. Biochem. Biophys.* **182**, 404 (1977).

[10] M. J. Chamberlin, W. C. Nierman, J. Wiggs, and N. Neff, *J. Biol. Chem.* **254**, 10061 (1979).

His₆-Tagged RNA Polymerase Purification

A. *Cell Lysis*

 1. Break up 200 g of frozen *E. coli* cells into small pieces (<2 cm) with a hammer at 4°. To further break up cells, grind cells in a warm blender without buffer using short bursts. Add a small volume (usually 50–100 ml) of TGED buffer with 233 mM NaCl and blend until homogeneous. We recommend using a blender fitted with a water jacket and water pump. The temperature of the jacket may be modified as needed simply by circulating water of the appropriate temperature. Alternatively, temperature can be adjusted by placing the blender top in a water bath.
 2. Add the remaining TGED + 233 mM NaCl to cells: 3 ml per gram of cells, keeping the temperature at about 4°.
 3. Add lysozyme (Sigma, St. Louis, MO) to 130 μg/ml.
 4. Blend on low speed 2–3 min, avoiding vortex. Add phenylmethylsulfonyl fluoride (PMSF) to 23 μg/ml as the cells are breaking up. Blend until the cells are completely suspended and the temperature is 2–5°.
 5. Incubate 20 min at 4°, to allow lysozyme to disrupt the cell walls.
 6. Warm to 10° and add freshly prepared 4% w/v sodium deoxycholate to 0.05% w/v, to lyse the cells.
 7. Blend 30 s at low speed.
 8. Incubate 20 min at 8–12°.
 9. Blend 30 s on high to shear the DNA. Extract should be noticeably less viscous.
10. Add 4 ml TGED + 0.2 M NaCl per gram of cells.
11. Blend 30 s on high speed.
12. Centrifuge the lysate in a Sorvall (Asheville, NC) GS-3 rotor or equivalent at 4°, 8000 rpm (10,816g), 45 min.
13. Retain the supernatant fluid (8 K) and discard the pellets.

B. *Polymin P Precipitation*

1. Add 10% (w/v) polymin P (Sigma), pH 7.9, dropwise on ice with stirring to a final concentration of typically 0.3–0.35% (w/v). The precise concentration will vary with individual batches of polymin P and it should be predetermined with a small-scale trial.
2. Incubate 30 min on ice at 4° with stirring.
3. Sediment the precipitant in a Sorvall GS-3 rotor at 4°, 8000 rpm (10,816g) 45 min. The supernatant fluid should be clear and yellow; however, if it is cloudy, try sedimenting longer or harder.

4. Pour off the supernatant fluid and carefully transfer the pellets to a cold blender. Add 4 ml TGED + 0.5 M NaCl per gram of cells.

5. Blend very slowly for about 5–10 min until homogeneous, using a glass pestle to break up the pellet if necessary. Avoid vortexing and foaming.

6. Sediment the precipitant in a Sorvall GS-3 rotor, 6000 rpm (6,084g), 30 min.

7. Pour off the supernatant fraction and transfer the pellets to a cold blender. Blend carefully until homogeneous with 2 ml of TGED + 1 M NaCl per gram of cells.

8. Sediment the precipitant in a Sorvall GS-3 rotor, 7000 rpm (8,281g), 30 min.

9. Remove supernatant fluid to a chilled beaker (1.0 M extract) and measure its volume.

C. Ammonium Sulfate (AmSO₄) Precipitation and Back Extraction

1. Add 35 g of freshly ground (NH₄)₂ SO₄ per 100 ml of the 1.0 M extract slowly with stirring on ice. The pH should be 7.9; if not, adjust with NaOH.

2. Incubate with stirring at least 30 min on ice or overnight.

3. Sediment the precipitant 45 min at 10,000 rpm (11,952g) in a Sorvall SS-34 rotor at 4°.

4. Remove the supernatant fluid and resuspend the pellets at 4° using a glass pestle with an equal volume of TGED + 2.0 M (NH₄)₂SO₄ to eliminate any remaining DNA and polymin P.

5. Sediment the precipitant 30 min at 10,000 rpm (11,952g) in a Sorvall SS-34 rotor at 4°.

6. Repeat steps 4–5.

7. Extract pellets with TGED + 1.6 M (NH₄)₂SO₄ using half the volume of the 1.0-M extract (see step B9) to minimize column load volume. RNA polymerase is soluble to at least 5 mg/ml.

8. Sediment the precipitant 45 min at 10,000 rpm (11,952g) in Sorvall SS-34 rotor at 4°.

9. Save supernatant fluid (1.6 M AmSO₄ eluate).

10. Precipitate proteins from the 1.6-M AmSO₄ eluate with 2.0 M (NH₄)₂ SO₄ by adding the appropriate amount of freshly ground (NH₄)₂ SO₄ slowly, on ice with stirring.

11. If needed, incubate overnight at 4° before proceeding.

D. Ni²⁺-NTA Agarose Chromatography

1. Pour a 25-ml Ni²⁺-NTA agarose column (Qiagen, Chatsworth, CA) and equilibrate the resin with 10 bed volumes of TGβ + 1 mM imidazole, 0.1 M NaCl.

2. Sediment the 1.6-M AmSO$_4$ eluate precipitant 45 min at 10,000 rpm (11,952g) in a Sorvall SS-34 rotor at 4°.
3. Resuspend the pellets in the smallest possible volume of TGβ + 1 mM imidazole, usually 50–80 ml for a 200-g preparation.
4. Load the column slowly, 3–4 column volumes per hour, and collect 5 ml fractions.
5. Wash the column with TGβ + 1 mM imidazole, 0.1 M NaCl until the absorbance 280 of the output is within 0.01 of A$_{280}$ of TGβ + 1 mM imidazole, 0.1 M NaCl buffer.
6. Wash the column with TGβ + 20 mM imidazole, 0.1 M NaCl.
7. Elute RNA polymerase with TGβ + 100 mM imidazole, 0.1 M NaCl buffer.
8. Pool the fractions with RNA polymerase activity,[10] and dialyze them into P$_{50}$ buffer. Dialyzed protein may be stored frozen or at −20° as needed.

E. Phosphocellulose Chromatography

The purified RNA polymerase eluted from the Ni^{2+}-NTA agarose is a mixture of holoenzyme, which contains the σ subunit, and core enzyme, which lacks the σ subunit. These two catalytically active forms can be separated using phosphocellulose chromatography,[9] or using other ion-exchange resins such as Mono-Q.[11] Fractions containing the RNA polymerase are adjusted to 10 mM MgCl$_2$ and are stored at −20° to −80°.

Solid Phase Transcription and Transcript Termination in Vitro

To study intrinsic termination of E. coli RNA polymerase, we generate 5′ biotinylated linear dsDNA templates by polymerase chain reaction (PCR) using one 5′ biotinylated oligonucleotide and one unmodified oligonucleotide. The template contains a promoter and an intrinsic termination site. Transcription is initiated using a dinucleotide or trinucleotide to prime transcript initiation and a limited subset of nucleoside triphosphate (NTP) substrates, resulting in a stable RNA polymerase elongation complex halted at a defined template site[12] downstream of the transcriptional start site, but upstream of the termination site. Often, it is desirable to move the elongation complex further downstream by sequentially adding and removing different subsets of nucleotides. Transcript termination occurs at the terminator when all four nucleotides are added and the halted RNA polymerase resumes transcript elongation. Because linear DNA templates are used, those elongation complexes that do not release at the intrinsic

[11] D. A. Hager, D. J. Jin, and R. R. Burgess, *Biochemistry* **29**, 7890 (1990).
[12] J. R. Levin and M. J. Chamberlin, *J. Mol. Biol.* **196**, 61 (1987).

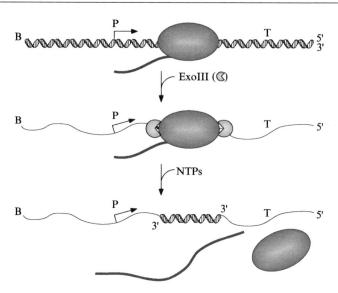

Fig. 1. Diagram of experimental design for dsDNA and ssDNA templates. Transcription is initiated at the promoter (P, bent rightward arrow), and the elongation complexes are halted proximal to the termination site (T) on dsDNA, PCR-generated templates that are biotinylated (B) at one of the 5′ ends. Next, the halted elongation complexes (filled ellipse) are digested with ExoIII (filled, cut-out circles), and then ATP, GTP, CTP, and UTP are added, allowing the elongation complex to resume transcript elongation. The 5′ and 3′ ends of the DNA (black lines) are as indicated. Upon recognizing the intrinsic termination site, the RNA polymerase releases the completed transcript (red line) and dissociates from the DNA template. (See color insert.)

termination signal are expected to run off the end of the remaining DNA template. Transcript termination can be studied using either dsDNA or ssDNA templates. To generate ssDNA templates, *E. coli* Exonuclease III (ExoIII), a processive 3′ to 5′ dsDNA-specific exonuclease,[13] is added to digest from the 3′ ends the nontemplate and template strands not protected by the halted elongation complex (Fig. 1).[7] Once ExoIII is removed, the efficiency of termination (%TE) at the single-stranded intrinsic termination site can be compared to % TE on dsDNA templates.

We developed two solid-phase transcription methods to prepare transcript elongation complexes halted at a distinct position along the DNA template.[7] One involves immobilization of the biotinylated DNA template to paramagnetic particles coated with streptavidin. We are experienced

[13] S. G. Rogers and B. Weiss, *Methods Enzymol.* **65,** 201 (1980).

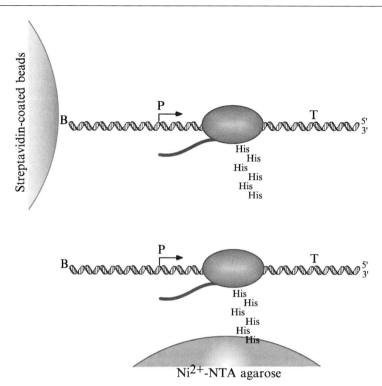

FIG. 2. Diagram of two solid-state transcription methods. There are two methods to perform transcription in the solid phase. RNA polymerase elongation complexes (filled ellipse) can be bound to Ni^{2+}-NTA agarose (blue half-circle) via its hexahistidine tag (His), or the 5′ biotinylated (B) DNA template can be bound to streptavidin-coated paramagnetic beads (yellow half-circle). (See color insert.)

using M-280 streptavidin Dynabeads (Dynal Biotech, Lake Success, NY). The other method involves chelating His_6-tagged RNA polymerase to Ni^{2+}-NTA agarose (Fig. 2). When Dynabeads are used, the 5′ biotinylated DNA template is bound essentially irreversibly to magnetic beads by streptavidin. We use templates in which the promoter is at least 100 nucleotides away from the position of the biotinylated end, primarily because occasionally transcription from promoters proximal to the 5′ biotinylated end bound to streptavidin dissociates the biotinylated DNA from the paramagnetic beads.[14] Templates should be designed such that RNA polymerase will have moved entirely downstream of the remaining dsDNA

[14] K. Fujita and J. Silver, *Biotechniques* **14,** 608 (1993).

region upon transcribing the termination signal on the ssDNA template. Biotinylated DNA templates are bound to streptavidin-coated paramagnetic beads (3–4 pmol DNA template/100 μg beads) using conditions suggested by the manufacturer. The bead-bound DNAs are then used as templates for *in vitro* transcription reactions. When Ni-NTA agarose is used, His_6-tagged RNA polymerase is reversibly bound to beads before being used for transcription.

Both solid-state methods use similar procedures, but there are some important differences. Unincorporated NTPs can be removed from halted elongation complexes by pelleting the resin either with a magnetic field (Dynabeads) or by gentle centrifugation at 2000xg (Ni-NTA agarose) and removing the supernatant fraction. The resin may then be resuspended in any buffer of choice. By repeating washes of the resin, it is possible to thoroughly remove unincorporated NTPs, dramatically reducing undesired transcriptional read-through encountered with other methods that rely on gel filtration. In our experience, solid-phase methods are preferable to gel filtration methods because they are less stringent and indisputably quicker, thereby reducing the proportion of elongation complexes that become inactivated while removing the unincorporated NTPs (S. M. U, unpublished observation). We routinely use buffers that lack $MgCl_2$ during the washes to minimize any read-through to downstream positions along the DNA. Moreover, no difference was observed in the RNA products between reactions in the solid phase compared with reactions in solution, even regardless of the direction of transcription relative to the position of the streptavidin-coated paramagnetic bead (S. M. U, unpublished observation). One caveat when using M-280 Dynabeads is that this resin is hydrophobic, which causes the resin to stick to pipette tips and tube walls and interferes with quantitative recovery of beads during the washing steps. Silicification of tubes and tips exacerbates this problem. Our solution has been to use 500 μg/ml acetylated bovine serum albumin (BSA) (Ac-BSA; preparation of Ac-BSA from Ref. 9) in all buffers when using these beads. Another solution, which we have not tested yet, is to use less hydrophobic streptavidin-coated beads, such as M-270 (Dynal Biotech).

Buffers

Buffer composition: general transcription buffer, TKG-$B_{40}M_4$ [44 mM Tris–HCl, pH 8.0, 30 mM KCl, 4 mM $MgCl_2$, 2% (v/v) glycerol, 40 μg/ml of acetylated BSA (Ac-BSA; preparation of Ac-BSA from Ref. 9)], 1 × TBE buffer (89 mM Tris base, 89 mM boric acid, 2.5 mM EDTA), and urea–SDS load buffer [10 M urea, 0.5% (w/v) SDS, 1 × TBE,

10 mM EDTA, 0.025% (w/v) xylene cyanol, 0.025% (w/v) bromophenol blue]. The variable components of the general transcription buffer are 0, 1, or 4 mM MgCl$_2$ (M$_1$ or M$_4$), 0 or 350 mM NaCl (N), and 40 or 500 μg/ml ml Ac-BSA (B$_{40}$ or B$_{500}$).

Protocol

To form open promoter complexes (EP$_o$, binary complexes between the RNA polymerase and the promoter that contain a ssDNA transcription bubble), 40 nM His$_6$-tagged RNA polymerase is incubated with 20 nM Dynabead-bound DNA template for 10 min at 30° in TGK-B$_{500}$M$_4$. Alternatively, 40 nM His$_6$-RNA polymerase is bound to 10 μl of preequilibrated 1:1 Ni^{2+}-NTA agarose slurry per 4 pmol RNA polymerase in the presence of 80 nM 5' biotinylated DNA template in TGK-B$_{40}$M$_4$ for 15 min at room temperature. Once EP$_o$ complexes are formed, transcription is initiated by adding the appropriate nucleotides for 5 min at 30° such that the RNA polymerase halts transcription at a defined position downstream of the promoter region. A typical reaction volume is 100 μl and contains 2 to 10 μM of high performance liquid chromatography (HPLC)-grade NTPs (Amersham-Pharmacia Biosciences, Piscataway, NJ). Depending on the DNA sequence 10–15 nucleotides downstream of the transcriptional start site, dinucleotides, or trinucleotides may be necessary. We typically use 50 to 250 μM dinucleotides (Sigma) or 250 μM trinucleotides (Oligos, Etc., Wilsonville, OR). To facilitate RNA detection, all reactions also include one [α^{32}P] radiolabeled NTP (Amersham-Pharmacia Biosciences) with a specific activity of 50,000 cpm/pmol. If desired, the elongation complex can be moved further downstream by sequentially adding (30° for 5 min) and removing different subsets of NTPs. Unincorporated NTPs can be removed by either pelleting the resin with a magnetic field (Dynabeads) or else by gentle centrifugation at 2000g (Ni^{2+}-NTA agarose), withdrawing the supernatant fluid, and resuspending the beads in an appropriate buffer. When Dynabeads are used, complexes are washed twice with 3 volumes of TGK-B$_{500}$ and then are resuspended in 1 volume of TGK-B$_{500}$M$_4$. When Ni^{2+}-NTA agarose is used, the complexes are washed five times with 1 ml of TGK-B$_{40}$ and then are resuspended in 1 volume of TGK-B$_{40}$M$_4$.

Divide the reaction volume in half and use one half for analyzing transcript termination on dsDNA and the other for termination on ssDNA. To generate single-stranded template downstream of the halted RNA polymerase, digest the DNA with Exo III (Roche Applied Science, Indianapolis, IN).

ExoIII Digestion

1. Immediately before digestion with ExoIII, residual EP_o are disrupted by high-salt treatment (either $TGK-B_{40}N$ or $TGK-B_{500}N$). Remove an aliquot of the undigested template for analysis of digestion efficiency.
2. ExoIII digestion is usually performed at 2000–5000 U/ml, in $TGK-B_{40}M_1$ or $TGK-B_{40}M_4$ for 5 min at either $30°$ or $37°$.
3. Quench digestion by washing the elongation complexes with $TGK-B_{40}$ or $TGK-B_{40}M_4$. Remove an aliquot of the digested template for analysis of ExoIII efficiency.

Termination Assay

1. Disrupt any residual EP_o complexes by washing elongation complexes with high-salt buffer, either $TGK-B_{40}N$ (Ni^{2+}-NTA agarose) or $TGK-B_{500}N$ (Dynabeads). Add 20 μg/ml rifampicin (Sigma) to prevent reinitiation and nonspecific transcription. If using Ni^{2+}-NTA agarose, also include 0.8 mg/ml yeast Torula RNA (Sigma) to minimize RNA polymerase binding to RNA in a binary complex and thereby remaining in the pellet, despite having been initially released from the ternary complex.
2. Remove an aliquot of the total reaction for analysis later.
3. Enable the halted elongation complex to resume elongation by adding 500 μM ATP, UTP, CTP, and GTP and incubate for 10 min at 30–$42°$ (chase reaction). The chase reactions are performed in either $TGK-B_{40}M_1$ or $TGK-B_{40}M_4$. (Lowering the Mg^{2+} concentration can enhance the efficiency of termination.[15])
4. Separate the supernatant fluid, containing released RNA, from the pellet, which contains RNAs associated with elongation complexes.
5. Terminated RNA released into the supernatant fluid was precipitated by addition of ammonium acetate to 1 M, yeast Torula RNA to 40 μg/ml, and ethanol to 70% (v/v).
6. Individual aliquots of the transcription reactions are stopped by the addition of an equal volume of urea–SDS load buffer, heated to $95°$ for 3 min, chilled on ice, and then loaded directly onto a 7 M urea, 15% acrylamide [38:2 acrylamide:bis(acrylamide)] gel. Electrophoresis is performed at 1800 volts.

The termination efficiency (%TE) can be calculated for each terminator on ssDNA or dsDNA templates using a Storm PhosphorImager system and ImageQuant software (Amersham-Pharmacia Biosciences). The calculation of termination efficiency is based on quantifying RNA longer than expected for termination (run-off) compared with the amount of RNA released from the polymerase at the termination site.[15] $\% TE = (R/ORC) \times 100\%$, where R is release RNAs, O is run-off RNAs, and C is unreleased RNAs at the termination site.

One notable difference between transcript elongation on dsDNA templates and on ssDNA templates is that a substantial fraction of complexes can fail to reach the termination site on ssDNA templates. These shorter RNAs are found in the pellet fraction rather than the supernatant fluid, suggesting that these RNAs are part of paused or arrested elongation complexes.[7] The fraction of these prematurely blocked complexes decreases as the distance between the elongation complex and the terminator is shortened. Increasing the temperature during the chase reaction (i.e., from $30°$ to $42°$) also decreases the fraction of these blocked complexes. We believe that these elongation blocks are caused in part by DNA secondary structure impeding the progress of the RNA polymerase.

Occasionally, some of the elongation complexes may not resume elongation after treatment with ExoIII. The mechanism of this inactivation is unknown, but others have reported similar observations.[16,17] To distinguish released RNA transcripts from RNA transcripts associated with inactive elongation complexes, we take advantage of the properties of the solid-phase transcription system (Fig. 2). We separate any remaining bead-bound ternary complexes from the RNA transcripts released into the supernatant fluid and then load the RNAs present in the pellet and the supernatant fluid in separate lanes for denaturing electrophoresis. Inactive complexes are indicated by the presence of RNA transcript in the pellet; release complexes are indicated by the presence of the RNA transcript in the supernatant fluid.

Use of Chemiluminescence or Fluorescence rather than Radioisotopes for Detection of ExoIII Digestion

When using ExoIII to generate ssDNA templates, it is important to determine the efficiency and extent of DNA digestion, as a control. The 5' biotinylated DNAs in these reactions readily facilitates determining the extent of ExoIII digestion using chemiluminescent or fluorescent detection. We resolve the DNA fragments before and after ExoIII treatment on a 6% urea–PAGE gel and transfer the DNA fragments from the gel to Hybond N^+ nylon membrane (Amersham-Pharmacia Biosciences) by capillary

[15] R. Reynolds, R. M. Bermudez-Cruz, and M. J. Chamberlin, *J. Mol. Biol.* **224,** 31 (1992).
[16] D. Wang, T. I. Meier, C. L. Chan, G. Feng, D. N. Lee, and R. Landick, *Cell* **81,** 341 (1995).
[17] E. Nudler, M. Kashlev, V. Nikiforov, and A. Goldfarb, *Cell* **81,** 351 (1995).

transfer (1–2 h). The DNA fragments are ultraviolet (UV) cross-linked to the membrane (120 mJ/cm^2). Detection of the biotinylated DNA is facilitated by binding a streptavidin–alkaline phosphatase conjugate to it, followed by exposure to the alkaline phosphatase substrate, such as CSPD or CDP-Star (Applied Biosystems, Foster City, CA). Upon dephosphorylation by alkaline phosphatase, the substrate decomposes, generating visible light, which can be detected using scientific imaging film (Hyperfilm ECL; Amersham-Pharmacia Biosciences). A typical exposure time is 10 min, between 4 to 6 h after addition of CSPD. Alternatively, a fluorescent alkaline phosphatase substrate, such as Attophos (Roche Applied Science) or ECF substrate (Amersham-Pharmacia Biosciences), can be used together with a fluorescence scanner for quantitative detection.

If desired, sequence-specific markers can be generated by chemical sequencing of biotinylated DNA templates, as previously described[18] with the following modifications: Purine reactions (G and G+A) are performed with the DNA template bound to streptavidin-coated magnetic beads. The G reaction is quenched by addition of 1 M 2-mercaptoethanol (βme), and the G+A reaction is quenched by addition of 80 μl of distilled water. Washing of the modified DNAs is accomplished by pelleting the magnetic beads with a magnetic field and by removing the supernatant. The beads are then resuspended in 20 mM NaCl. Wash the beads three times, then resuspend them in 50 μl of distilled H$_2$O, to which 50 μl of freshly distilled 2 M piperidine is added. Incubate the modified DNAs at 90° for 30 min. Lyophilize the samples, wash the beads twice with distilled H$_2$O, and suspend in 9 M urea/0.1% SDS loading buffer. Sequencing ladders can be stored at −20° for prolonged periods. Immediately before loading on a gel, the ladders are heated to 95° for 5 min to ensure that the biotinylated DNAs are not bound to the streptavidin-coated magnetic beads.

Pyrimidine reactions: C+T and C reactions were not performed using magnetic beads because hydrazine disrupts the streptavidin–biotin interaction. A T-ladder can be generated using biotinylated DNAs bound to paramagnetic beads if KMnO$_4$ is used. Denature the DNA template in the presence of 0.1 N NaOH for 10 min at room temperature and then withdraw the supernatant fluid, which contains the nonbiotinylated strand. The streptavidin-bound ssDNA is then washed once with 0.1 N NaOH and twice with 10 mM Tris–HCl and 20 mM NaCl. The ssDNA is resuspended in 2 mM KMnO$_4$ and incubated for 1 min at 20°. The reaction is quenched by addition of βME to 1.0 M. The beads are then washed and heated with piperidine as described earlier. As an alternative to the purine Maxam–Gilbert

[18] A. M. Maxam and W. Gilbert, *Proc. Natl. Acad. Sci. USA* **74,** 560 (1977).

reactions, an A-ladder can also be generated by substituting a saturated solution of diethylpyrocarbonate (DEPC) for $KMnO_4$.

There are many advantages to using chemiluminescence instead of radioisotopes for detection. Perhaps the most obvious is that it minimizes exposure to radioactivity and the markers never radioactively decay. Chemiluminescence detection is quite sensitive; in control experiments, as little as 0.1 fmol biotin was detected (S. M. U, unpublished observation). Use of biotinylated templates provides an attractive alternative to using T_4 polynucleotide kinase and $[\gamma^{32}P]ATP$ to radiolabel one of two oligonucleotides required to amplify a linear DNA template by PCR.[19] Whereas nearly every DNA will be biotinylated in our method, the percentage of radioactively labeled templates can vary tremendously because the efficiency of T_4 kinase depends unpredictably on the sequence of the oligonucleotide to be labeled.

Acknowledgments

These methods were devised in the laboratory of Michael J. Chamberlin, University of California, Berkeley, CA. S. M. U. would like to thank M. J. Chamberlin, C. M. Kane, L. Hsu, L. D'Ari, and the members of the Chamberlin and Kane labs for their technical help and advice.

[19] R. Higuchi, B. Krummel, and R. K. Saiki, *Nucleic Acids Res.* **16,** 7351 (1988).

[26] Bacteriophage HK022 Nun Protein: A Specific Transcription Termination Factor that Excludes Bacteriophage λ

By Hyeong C. Kim and Max E. Gottesman

Prophage HK022 excludes superinfecting phage λ by terminating transcription on the lambda chromosome.[1] This exclusion is promoted by the 109 amino acid HK022 Nun protein. Nun carries an arginine-rich RNA-binding motif (ARM) at its N terminus. The ARM binds to a stem-loop structure, BOXB, in the nascent transcripts of the λ *pL* and *pR* operons.[2,3]

[1] J. Robert, S. B. Sloan, R. A. Weisberg, M. E. Gottesman, R. Robledo, and D. Harbrecht, *Cell* **51,** 483 (1987).
[2] J. Oberto, R. A. Weisberg, and M. E. Gottesman, *J. Mol. Biol.* **207,** 675 (1989).

Binding to BOXB brings Nun into proximity to elongating RNA polymerase (RNAP). Nun forms a termination-prone complex with RNAP and four *Escherichia coli* host factors, NusA, NusB, NusE (S10), and NusG.[4,5] Within this complex, the Nun C terminus interacts with RNAP as well as with the λ DNA template.[6–8] Nun induces termination at sites just promoter-distal to *nutL* and *nutR*.[1,9,10] *In vitro*, Nun alone causes transcription arrest on λ DNA template, rather than termination.[5] The four host factors stimulate Nun-mediated transcription arrest and lessen the need for high Nun concentrations.[5] True termination (i.e., the release of RNAP and transcript from the DNA template) is induced by the host MFD protein.[11] Recently it has been found that Nun can also prevent λ growth by repressing translation of the λ N protein.[12]

Purification of Nun

The Nun overproducer plasmid pT7NunII is a pET-21d(+) derivative that carries *nun* between the plasmid *Nco*I and *Hind*III sites. The *nun* gene is under the control of the T7 promoter, which is in turn controlled by an adjacent *lac* operator and a *lac*I gene. Overexpression of Nun from pT7NunII is achieved in *E. coli* strain BL21(DE3) pLysS, which contains a chromosomal copy of the T7 RNA polymerase gene under the control of the *lacUV* promoter. Basal Nun expression is highly down-regulated by the T7 lysozyme encoded by the pLysS plasmid. *E. coli* carrying the pT7NunII plasmid is maintained in LB–ampicillin.

Induction

Inoculate LB (10 ml) containing 50 μg/ml ampicillin with an isolated colony of BL21(DE3) pLysS/pT7NunII and grow in each of two 50-ml flasks with shaking for 12 to 14 h at 37°. These stationary phase cultures

[3] S. Chattopadhyay, S. C. Hung, A. C. Stuart, A. G. III Palmer, J. Garcia-Mena, A. Das, and M. E. Gottesman, *Proc. Natl. Acad. Sci. USA* **92**, 12131 (1995).

[4] J. Modgridge, T. F. Mah, and J. Greenblatt, *Genes Dev.* **9**, 2831 (1995).

[5] S. C. Hung and M. E. Gottesman, *J. Mol. Biol.* **247**, 428 (1995).

[6] R. S. Watnick and M. E. Gottesman, *Proc. Natl. Acad. Sci. USA* **95**, 1546 (1998).

[7] R. S. Watnick and M. E. Gottesman, *Science* **286**, 2337 (1999).

[8] R. S. Watnick, S. C. Herring, A. G. III Palmer, and M. E. Gottesman, *Genes Dev.* **14**, 731 (2000).

[9] S. B. Sloan and R. A. Weisberg, *Proc. Natl. Acad. Sci. USA* **90**, 9842 (1993).

[10] S. C. Hung and M. E. Gottesman, *Genes Dev.* **11**, 2670 (1997).

[11] R. Washburn, Y. S. Wang, and M. E. Gottesman, *J. Mol. Biol.* **329**, 655 (2003).

[12] H. C. Kim, J. Zhou, H. R. Wilson, G. Mogilnitskiy, D. L. Court, and M. E. Gottesman, *Proc. Natl. Acad. Sci. USA* **100**, 5308 (2003).

are then added to 1 liter of LB–ampicillin in a 2-liter flask and grown with vigorous shaking to an OD_{600} of 0.4 to 0.5 at 37°. Isopropyl-β-D-thiogalac-topyranoside (IPTG) is then added to 1 mM, and growth is continued for an additional 4 to 5 h at 32°. Chill culture in ice and harvest cells by centrifuging at 4000 rpm for 30 min in a Sorvall RC3-B rotor. Freeze cell pellet in a dry ice–ethanol bath, and store at $-70°$.

Purification

Most of the Nun protein in the induced culture is soluble, and it is readily purified from a supernatant fraction by two chromatographic steps (Fig. 1). The following procedures are carried out on ice or at 4°.

Soluble Extract. After thawing on ice, resuspend the cell pellet with 30 ml buffer A [50 mM MES, pH 6.0, 0.2 M NaCl, 2 mM EDTA, 1 mM DTT, 1 mM ABESF (4-(2-Aminoethyl)benzenesulfonylfluoride, HCl)] and obtain a homogeneous suspension by repeated pipetting. Aliquot 6 ml of the lysate into five separate 15-ml tubes. Sonicate the lysate to shear DNA at a power level 1.5 of 10 and 15% duty cycle using Branson Sonifier (Branson Ultrasonics Corp., Danbury, CT) 250, for 4 min in the first round and 2 min in the second round per tube. Avoid overvigorous sonication, which yields a mainly truncated form of Nun (T-Nun) deleted for the last 13 C-terminal amino acids. T-Nun is defective in arresting transcription *in vitro* (Fig. 2). Remove cell debris by centrifugation of the lysate (12,000 rpm for 30 min in a Sorvall RC2-B rotor).

SP-Sepharose FF Column Chromatography. Equilibrate a 20-ml SP-Sepharose Fast Flow (Amersham, Piscataway, NJ) column with 100 ml of buffer B [50 mM MES, pH 6.0, 0.2 M NaCl, 2 mM EDTA, 1 mM DTT] at a rate of 5 ml/min. Apply supernatant to the SP-Sepharose column at a rate of 3 ml/min. Wash the column with 50 ml of the same buffer at a rate of 2.5 ml/min. Elute Nun with 50 ml of buffer C [50 mM MES, pH 6.0, 0.6 M NaCl, 2 mM EDTA, 1 mM DTT] at a rate of 2 ml/min. Collect the first 10-ml fraction and the following six 5-ml fractions successively. Most of the Nun protein elutes in fractions 3–7; major contaminants elute before Nun. Pool fractions containing Nun and concentrate to 7–8 ml using a Centriprep (Amicon, Beverly, MA) centrifugal filter device with a YM-10 membrane. Dialyze the Nun preparation against 1 liter of buffer D [50 mM Na–phosphate, pH 7.0] using a Slide-A-Lyzer (Pierce, Rockford, IL) dialysis cassette with 10,000 MWCO overnight. Centrifuge the dialyzate at 15,000 rpm for 30 min in a Sorvall RC2-B rotor and collect the clear supernatant.

Mono S Chromatography. Load supernatant onto a 1-ml Mono S HR 5/ 5 column (Amersham, Piscataway, NJ) equilibrated with buffer D. Wash

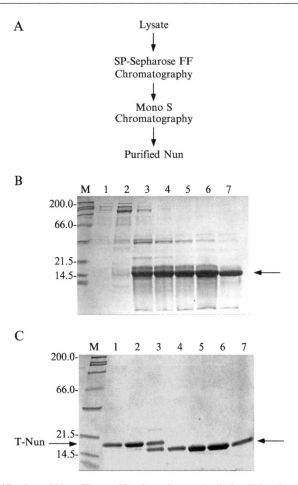

FIG. 1. Purification of Nun. The purification scheme. Analysis of Nun fractions from SP-Sepharose FF purification on SDS–PAGE (15%). Lane 1 is the first 10 ml collected. Lanes 2–7 represent subsequent 5-ml fractions. Analysis of various Mono S fractions at the last step of purification. Lanes 1–7 represent some of the 500-μl fractions of 0.37–0.42 M sodium chloride gradient. The truncated form of Nun (T-Nun) elutes before full-length Nun and runs more slowly on SDS–PAGE.

the column with 5 ml of buffer E [50 mM Na–phosphate, pH 7.0, 0.2 M NaCl]. Elute Nun with a 20-ml NaCl gradient (0.2–0.55 M in buffer D), followed by a 2.5-ml NaCl gradient (0.55–1 M) and collect 500-μl fractions. Most of the Nun protein elutes between 0.37–0.42 M NaCl. Pool the purest Nun fractions and dialyze against 4 liters of buffer D overnight as before.

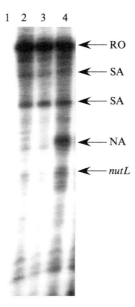

Fig. 2. Nun-mediated transcription arrest *in vitro*. RO, Run-off transcript; SA, spontaneously-arrested transcript; NA, major Nun-arrested transcript.

Centrifuge the dialyzate (15,000 rpm for 30 min in a Sorvall RC2-B rotor) and quick-freeze in convenient aliquots on dry ice before storing at −70°.

This protocol yields about 3 mg Nun per g cell pellet. The preparation is free of significant RNase and DNase activities and is active in transcription arrest *in vitro*.

Assay of Nun Protein

In vitro transcription is performed with template derived from a *Hind*III-digested λ genome that contains the *pL–nutL* segment (35,253–35,718) (Figure 2). 10 nM holo RNAP is incubated with 20 nM template at 32° for 10 min. Transcription elongation is initiated with 5 nM each rNTP, with (Lane 4) or without 200 nM Nun (Lane 2), and continued for 20 min at 32°. Lane 1 is without RNAP. Lane 3 shows transcription with 200 nM T-Nun. Reaction is stopped by the addition of 187.5 mM Na–acetate, pH 5.2, and 31.25 mM EDTA. The transcript is then phenol-chloroform extracted and ethanol precipitated. Purified transcripts are loaded onto 6% 8 M urea sequencing gel and run at 24 W for 2 h.

[27] Assay of Transcription Termination by Ribosomal Protein L4

By Janice M. Zengel and Lasse Lindahl

Although transcription initiation is a common stage for regulation of gene expression, a variety of mechanisms have been discovered for regulating transcription at the level of elongation/termination.

We have been investigating an example of regulation exerted at the level of transcription termination. Ribosomal protein (r-protein) L4 of *Escherichia coli* is a component of the 50S ribosomal subunit. However, this protein also regulates expression of its own transcription unit, the S10 operon, which encodes L4 and 10 other r-proteins. L4 accomplishes this regulation by acting at two levels. First, it regulates translation by blocking initiation of translation at the most proximal gene of the operon.[1,2] Second, L4 induces RNA polymerase to terminate prematurely at a specific site within the 172-base leader of the S10 operon.[3–5] Both types of regulation contribute to autogenous control of the S10 operon, but the mechanisms are independent and require overlapping but distinct determinants within the untranslated leader RNA.[2,6] We have focused on the role of L4 in regulating transcription termination.

The best proof for regulation of transcription by termination is differential changes in rates of transcription within the transcription unit. That is, if transcription is terminated at an attenuator in response to a regulatory cue, the transcription rate downstream of the attenuator will be reduced, but the transcription rate upstream of the attenuator will not. Previously described methods for measuring transcription rates of specific genes include pulse labeling the RNA and then characterizing the labeled RNA by comparing hybridization to DNA complementary to the upstream and downstream sequences, using either filters or nuclease protection experiments.[4,6,7] Northern analysis is also possible, but it will only work for

[1] J. L. Yates and M. Nomura, *Cell* **21,** 517 (1980).

[2] L. P. Freedman, J. M. Zengel, R. H. Archer, and L. Lindahl, *Proc. Natl. Acad. Sci. USA* **84,** 6516 (1987).

[3] J. M. Zengel, D. Mueckl, and L. Lindahl, *Cell* **21,** 523 (1980).

[4] L. Lindahl, R. Archer, and J. M. Zengel, *Cell* **33,** 241 (1983).

[5] J. M. Zengel and L. Lindahl, *Proc. Natl. Acad. Sci. USA* **87,** 2675 (1990).

[6] J. M. Zengel and L. Lindahl, *J. Mol. Biol.* **213,** 67 (1990).

[7] J. M. Zengel, J. R. McCormick, R. H. Archer, and L. Lindahl, *in* "Ribosomes and Protein Synthesis: A Practical Approach" (G. Spedding, ed.), p. 213. Oxford University Press, New York, 1990.

transcripts with a lifetime long enough to allow detectable amounts of specific transcripts to accumulate. For example, the 140–145 nt prematurely terminated transcript from the S10 operon can be detected among pulse-labeled transcripts, but not in purified bulk RNA.[4,6] The reason for this is that in pulse-labeled transcripts, each radioactive RNA is represented according to its rate of synthesis, but in bulk RNA, each sequence is represented according to accumulation and therefore depends not only on the synthesis rate, but also on the degradation rate.

Although *in vivo* pulse labeling of RNA is the most direct way of characterizing transcription regulation, the methods are cumbersome and time consuming. We describe here two other approaches, reporter genes and *in vitro* transcription, which we have used to characterize attenuation control of the S10 ribosomal protein operon in *E. coli* and other bacteria.

In Vivo Assay of L4-Mediated Transcription Termination Using Reporter Genes

Construction of Reporter Genes

The use of reporter genes usually assumes that transcription rates can be deduced from the rate of accumulation of a protein product. However, the outcome of the measurement depends not only on accumulation of transcript, but also on the rate of its translation. That is, if the number of protein molecules produced from each mRNA synthesized changes, then these changes in translation efficiency will contribute to the amount of protein synthesized. This possibility is clearly an issue with the S10 operon, because we know it is regulated at both the level of transcription attenuation and the level of translation initiation.

To study regulation of the S10 operon, we have utilized two types of reporter plasmid. In one, the reporter gene is a fusion of the S10 gene, the first gene in the operon, with the *lacZ* gene coding for β-galactosidase. Because the construct contains the entire S10 leader sequence, including the S10 translation initiation site, synthesis of reporter protein from this construct is regulated not only by the transcription attenuator, but also by the translation regulation governing the translation start at the S10 gene (Fig. 1A).[6,8] In the other reporter, the S10 ribosome binding site is eliminated and replaced with a spacer followed by the genuine *lacZ* leader and translation start site (Fig. 1B).[6] Synthesis of the reporter protein from this plasmid is only regulated by L4-mediated attenuation.

[8] L. P. Freedman, J. M. Zengel, and L. Lindahl, *J. Mol. Biol.* **185,** 701 (1985).

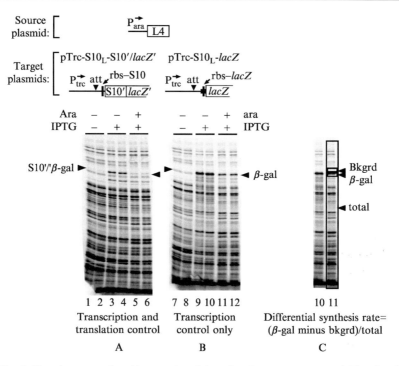

FIG. 1. Use of reporter plasmids to analyze L4-mediated autogenous control. The plasmids used for analysis of L4-mediated autogenous control are shown at top. Para, Arabinose promoter, inducible by addition of arabinose to 0.25%; Ptrc, trc promoter, inducible by addition of IPTG to 1–2 mM; att, site of L4-mediated termination; rbs–S10, Shine–Dalgarno sequence for S10 gene; rbs–lacZ, Shine–Dalgarno sequence for *lacZ* gene. (A) and (B) illustrate typical results from pulse-labeling cells carrying the indicated plasmids with [^{35}S]methionine before and 10 min after the addition of IPTG and arabinose. (C) illustrates the strategy for quantitating the effect of L4 on expression of the reporter gene. Bkgrd, Background.

One potential complication with reporter gene constructs is that the constitutive and often very high level of expression of the reporter transcript and/or protein can sometimes inhibit the growth of the cells. This may result in selection of mutations in the reporter construct. Indeed, that was our observation when we used the constitutive S10 operon promoter for expression of reporter genes. We have therefore found it advantageous to replace the natural promoter with an inducible promoter, such as the lactose, *trc*, or arabinose promoter.[9–11]

[9] X. Li, L. Lindahl, Y. Sha, and J. M. Zengel, *J. Bacteriol.* **179,** 7046 (1997).

Expression of a reporter gene can be quantified by measuring accumulation of protein or enzyme activity. However, a much more sensitive method is to determine the rate of protein synthesis by pulse labeling with a radioactive amino acid (usually [^{35}S]methionine). The pulse-labeling technique allows instantaneous measurements of regulatory responses (see Fig. 1), whereas the accumulation measurements require measurements of protein or enzyme activity over a lengthy period to ensure an accurate rate measurement. Furthermore, it is our experience that some fusion proteins are not stable and their turnover rate may be sensitive to the growth state of the culture, introducing additional possible errors in the accumulation measurements.

Basic [^{35}S]Methionine Labeling Protocol

Cultures should be grown in defined medium with no methionine in order for the uptake of radioactive methionine to be successful. Aliquots of the main culture are labeled before any perturbation; additional samples of the culture are labeled after each change in the growth conditions. For example, using a culture with a plasmid expressing the S10 leader and reporter protein expressed from the *trc* promoter, and a second plasmid expressing L4 from the arabinose promoter, the experiment consists of the following steps:

1. Inoculate the appropriate defined medium lacking methionine with an aliquot of an overnight culture that has been grown in the same medium. The initial density should be about about 10^7 cells per ml. Our growth medium is typically AB minimal,[12] with glycerol (0.4% v/v) as the carbon source. We also include a mix of all amino acids except methionine (at 20 μg/ml each) to encourage more rapid growth of the cells, and the appropriate antibiotics to ensure maintenance of the target and source plasmids.

   Note on growth medium: It is essential that the cells used to inoculate the working culture are grown in the same medium as will be used for the labeling experiment. If cells are grown in a "rich" medium such as LB, they will have a very long lag after switching to the defined medium. Furthermore, methionine will be transferred to the working culture and reduce uptake of the radioactive methionine.

[10] T. Allen, P. Shen, L. Samsel, R. Liu, L. Lindahl, and J. M. Zengel, *J. Bacteriol.* **181,** 6124 (1999).

[11] J. M. Zengel, Y. Sha, and L. Lindahl, *RNA* **8,** 572 (2002).

[12] D. J. Clark and O. Maaløe, *J. Mol. Biol.* **23,** 99 (1967).

Note on measuring cell density: Whether a spectrophotometer or a Klett colorimeter is used, it is necessary to calibrate the relationship between absorption (or OD) reading and cell density as it varies with geometry of the instrument, species, and cell growth rate. For frame of reference, an OD_{450} reading of 1 in a Hitachi (Schaumburg, IL) U-1100 spectrophotometer with a 10-mm cuvette corresponds to about 2×10^8 cells per ml of an *E. coli* culture.

2. Grow the cells in a shaking water bath at 37° to a density of about 0.5–1×10^8 cells per ml. Transfer a 100-μl aliquot to a separate tube (we use a 5-ml Falcon tube) prewarmed at 37° for several minutes and containing 2–5 μCi [^{35}S]methionine (approximately 1000 Ci/mmol). Continue shaking. This amount of methionine should be completely taken up within 30 s. When the exogenous radioactive methionine is exhausted, internally synthesized methionine will effectively be used for a chase of radioactivity into complete peptide chains. We allow this "chase" to continue for about 90 s (see step 3), but depending on length and stability of reporter protein, this chase period may have to be varied (under standard growth conditions, the translation rate is about 15 amino acids per second).

3. Two min after addition of the radioactive methionine, add 200 μl of "1.6 × sample buffer" for SDS gel electrophoresis to the tube; incubate at 90–95° for 2 min. Transfer the sample to ice until the experiment is complete, then store at −20° until the gel is run (Fig. 1, lanes 1 and 2, and 7 and 8).

4. To induce expression of the reporter gene, under control of the *lac* or *trc* promoter, transfer 1 ml aliquot from the master culture to a prewarmed small Falcon tube; at $t = 0$ min, add isopropyl-β-D-thiogalactopyranoside (IPTG) to 2 mM (4 μl of 0.5 M IPTG). Continue shaking at 37°.

5. Immediately after addition of IPTG, remove 0.5 ml of IPTG-induced cells to another (prewarmed) Falcon tube. To induce expression of the regulatory protein (r-protein L4), at $t = 1$ min, add arabinose to 0.25% (12.5 μl of 10% arabinose).

6. At $t = 10$ min, remove 100 μl aliquot from IPTG-induced cells and label as described in steps 2–3 (Fig. 1, lanes 3 and 4, and 9 and 10).

7. At $t = 11$ min, remove 100 μl aliquot from IPTG- and arabinose-induced cells and label as described in steps 2–3 (Fig. 1 lanes 5 and 6, and 11 and 12).

Note on stability of samples: Proteins will be degraded if they are kept for longer periods at room temperature, or if they are thawed and refrozen more than a couple of times. If a sample must be used on multiple days, we recommend that aliquots be made before storage and to limit refreezings of each aliquot to two or three times.

Labeling Protocol for Test of Protein Stability

The basic labeling protocol can be modified to test the stability of the reporter protein or the regulatory protein. A larger mix of culture and [^{35}S]methionine is made up, and at various times during the chase, aliquots of this mix are transferred to tubes containing 2 vol of 1.6 × SDS sample buffer, heated to 90–95° for 2 min, and stored. We typically label 0.5 ml of exponentially growing cells with 15 μCi [^{35}S]methionine and remove 100 μl aliquot at times ranging from 2 min to 30 min. If a protein is less stable than the average cell protein, analysis by gel electrophoresis will show that the corresponding radioactive band becomes weaker relative to other bands.

SDS–Polyacrylamide Gel Electrophoresis

We use the protocol of Laemmli[13] as our basic electrophoresis protocol, but several parameters must be optimized to obtain satisfactory resolution. One of these parameters is the acrylamide concentration. For example, a 7.5% gel will resolve β-galactosidase (135 kD/monomer) but a 12.5% gel is required to resolve r-protein L4 (22.1 kD). A second parameter is the ratio between acrylamide and bisacrylamide, which in turn determines the degree of cross-linking in the gel. We have found that reducing the acrylamide:bisacrylamide ratio from 30:0.8 to 30:0.2 in the separating gel can improve resolution of larger proteins, such as β-galactosidase. However, the stacking gel should be made with the standard 30:0.8 mix, because the combination of low acrylamide concentration and low cross-linking can turn the stacking gel into disorderly slime.

Preparing and Running an SDS–Polyacrylamide Gel

Warning: acrylamide is a neurotoxin. Consider purchasing ready-made solutions. Ready-made buffers are also available. For a detailed description of pouring and running an SDS–polyacrylamide gel, see Ref. 14. When the tracking dye reaches the bottom of the gel, turn off the power, carefully transfer gel to filter paper (e.g., Whatman 3MM) (Whatman Group, Maidstone, Kent, U.K.) and dry on a vacuum gel dryer.

---

[13] U. K. Laemmli, *Nature* **227,** 680 (1970).
[14] J. Sambrook and D. W. Russell, "Molecular Cloning: A Laboratory Manual," ed. 3. Cold Spring Harbor Laboratory Press, Cold Spring Harbor, N. Y. 2001.

Imaging and Quantification

Place the completely dried gel on a phosphorimager screen or X-ray film for an appropriate time (usually between overnight and a couple of days). Develop film or scan phosphorimager screen. Identify bands of interest by comparing runs of strains with and without expression or oversynthesis of relevant proteins (see Fig. 1). Quantification is done most conveniently with a phosphorimager, but can also be performed by cutting out suitable pieces of the gel and counting in scintillation counter. The strategy for quantification is illustrated in Fig. 1C. We prefer to do manual background corrections, because it is difficult to evaluate what is going on in the automatic black box correction in software packages.

We use differential rate of synthesis to evaluate regulation. The differential rate is defined as the rate of synthesis of protein I over the total rate of synthesis $[(dPi/dt)/(dP/dt) = dPi/dP$, where Pi is a specific protein, P is total protein and t is time], that is, the ratio between radioactivity in a given protein and radioactivity in total protein. The latter is determined from the total amount of radioactivity in a lane. This parameter is not sensitive to changes in the overall rate of protein synthesis and variations in pipetting.

In Vitro Assay of L4-Mediated Transcription Termination

Although much characterization of the regulation can be performed through *in vivo* analysis of mutations in the regulatory protein and its target, more detailed mechanistic insight requires *in vitro* strategies. We have gained significant understanding of the mechanism of L4-mediated transcription control of the S10 operon by using a cell-free transcription system. This system reproduces the *in vivo* effect of r-protein L4 on transcription: in the presence of RNA polymerase and an ancillary protein, NusA, purified r-protein L4 stimulates the termination of transcription at the site in the S10 leader that *in vivo* studies had identified as the site of L4-stimulated premature termination (Fig. 2).[5,6] The principal approaches used in our *in vitro* experimentation have included kinetic analysis of regulation, mutational analysis of the mRNA target,[5,15] disruption of nascent RNA structures with oligonucleotides,[15] and competition for L4 by addition of other RNAs.[5,16] Furthermore, purification and restart of the transcription complex in the presence or absence of its ligands allowed us to order the NusA- and L4-dependent steps in the regulation.[17]

[15] Y. Sha, L. Lindahl, and J. M. Zengel, *J. Mol. Biol.* **245,** 486 (1995).
[16] J. M. Zengel and L. Lindahl, *Nucleic Acids Res.* **21,** 2429 (1993).
[17] Y. Sha, L. Lindahl, and J. M. Zengel, *J. Mol. Biol.* **245,** 474 (1995).

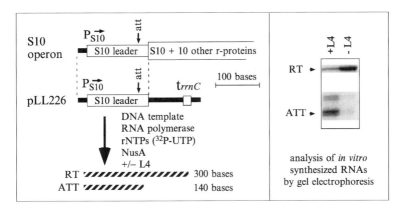

FIG. 2. Analysis of L4 effect on *in vitro* transcription of the S10 operon. The plasmid used for *in vitro* transcription analysis and the scheme for the reaction are shown at left. Typical results are shown at right. RT, Read-through transcripts; ATT, attenuated transcripts.

We use a transcription reaction with purified RNA polymerase, but it may be advantageous to use crude cell extracts for preliminary experiments, because the regulation studied may require ancillary factor(s). In fact, our initial *in vitro* transcription experiments were performed with RNA polymerase that we later found was contaminated with NusA, an RNA polymerase auxiliary protein enhancing the polymerase response to pause and termination signals.[5] Had we used completely pure RNA polymerase, our initial attempts to observe L4-mediated transcription attenuation would have failed, because we now know that the L4 effect is completely dependent on the presence of NusA.[5]

Purification of DNA Templates

Most of our transcription assays have been performed with intact supercoiled plasmid DNAs carrying the S10 promoter and leader (wild type and mutant derivatives) upstream of a synthetic *rrnC* terminator. These plasmids, pLL226 and its derivatives, direct the synthesis of attenuated and read-through transcripts, which migrate to a region of a denaturing gel that is devoid of transcripts initiated from other promoters (Fig. 2).[5] Plasmid DNA can be purified using traditional cesium chloride banding procedures. Plasmid DNA can also be prepared using a Qiagen (Chatsworth, CA) kit, but the column elution must be followed by a phenol/chloroform/isoamylalcohol extraction to remove traces of RNase. We have also used polymerase chain reaction (PCR) products as template.

Purification of Proteins

E. coli RNA polymerase is available commercially from sources such as Epicentre (Madison, WI) although there are published procedures for purification of the enzyme using standard chromatography strategies. Moreover, purification of the enzyme is relatively simple using genetically engineered strains carrying His$_6$-tagged subunits of RNA polymerase.[18,19]

Because L4-mediated transcription control requires NusA, we have also purified that protein. Initially we used an overproducing strain carrying the wild type NusA gene, and purified the protein according to the procedure of Schmidt and Chamberlin.[20] However, we later cloned an N-terminal His$_6$-tagged version of NusA under control of an IPTG-inducible promoter and purified active protein using immobilized metal affinity chromatography (IMAC).

L4 protein is more difficult to purify, because the protein is highly prone to aggregation and precipitation. Most of our experiments have been done with L4 received from other labs, especially the Nierhaus lab (Max Planck Institute, Berlin, Germany). However, recently we have been able to engineer mutant versions of the protein that can be purified by IMAC (Zengel and Lindahl, manuscript in preparation).

Design of the Transcription Reaction

Our *in vitro* transcription studies have shown that L4-mediated transcription termination is preceded by a NusA-stimulated pause of the RNA polymerase at the site of termination.[21] In order to study the pattern of transcription pauses, a single wave of RNA polymerases moving synchronously on the DNA template must be created. To accomplish this, transcription initiation complexes are first formed by incubating the DNA template with RNA polymerase and the ribonucleoside triphosphates needed to make the first few nucleotides of the RNA strand. In the case of the S10 transcript, the transcript begins pppGpGpCpUpUp. . . . Hence, transcription complexes are formed in the presence of GTP and CTP, while no ATP and UTP are present. Subsequent addition of ATP and UTP commences the elongation phase of the reaction. To ensure that no more initiation complexes are formed, further initiation can be blocked by simultaneous

[18] M. Kashlev, E. Martin, A. Polyakov, K. Severinov, V. Nikiforov, and A. Goldfarb, *Gene* **130,** 9 (1993).

[19] H. Tang, Y. Kim, K. Severinov, A. Goldfarb, and R. H. Ebright, *Methods Enzymol.* **273,** 130 (1996).

[20] M. C. Schmidt and M. J. Chamberlin, *Biochemistry* **23,** 197 (1984).

[21] J. M. Zengel and L. Lindahl, *Genes Dev.* **6,** 2655 (1992).

addition of heparin or rifampicin. After this, stopping aliquots of the reaction at various times monitors the progress of transcription.

During our experiments with the S10 promoter, we discovered that the kinetics of the transcription reaction is not changed by the addition of rifampicin. Thus it appears that under our assay conditions, there is little or no release and restart of transcription at the S10 promoter. We have not investigated if this property is specific to the DNA template we have used, so we would recommend the use of rifampicin or heparin to ensure a single wave of synchronous RNA polymerases until the effect of rifampicin or heparin on the detailed transcription kinetics has been characterized.

The efficiency and lifetime of RNA polymerase pausing and/or termination can be influenced by the concentration of the nucleotides to be incorporated at the pause site. For example, in the S10 operon, transcription of upstream U-patches occur without any detectable pausing, but transcription of the Us in the termination region is much slower, causing transient accumulation of paused transcripts. This phenomenon is visible only at low (below 10 μM) UTP concentrations.[17]

Transcription pausing is also sensitive to ligands of RNA polymerase. In the case of the *E. coli* S10 operon, a stable pause is dependent on the addition of NusA to the reaction. At 100 μM UTP, a pause with a half-life of 1–2 min is seen when NusA is present in about 4-fold molar excess over the RNA polymerase. Pausing at the S10 attenuator is further enhanced by r-protein L4, but only in the presence of NusA.[21]

Recipes for in Vitro *Transcription*

$\underline{10 \times TME}$: 200 m$M$ Tris–acetate, pH 7.9/40 mM Mg–acetate/1 mM EDTA

$\underline{1.5\ M\ KGlu}$: L-glutamate, potassium salt

$\underline{50\ mM\ EDTA + carrierRNA}$: contains 0.25 mg/ml yeast RNA (Turola yeast ribonucleic acid type IV from Sigma [St. Louis, MO], extracted with phenol and chloroform-isoamylalcohol as described,[7] or purchased from a source such as Ambion, Inc. [Austin, TX])

$\underline{10\ mM\ DTT}$: dithiothreitol

$\underline{1\ mg/ml\ BSA}$: diluted from stock of RNase-free bovine serum albumin

$\underline{10\ mM\ CTP/10\ mM\ GTPmix}$: diluted from ultrapure 100 mM cytosine 5'-triphosphate, sodium salt and guanosine 5'-triphosphate, sodium salt purchased from Pharmacia

$\underline{100\ mM\ ATP}$: ultrapure adenosine 5'-triphosphate, sodium salt from Pharmacia AB (Uppsala, Sweden)

20 mM UTP: diluted from 100 mM ultrapure uridine 5′-triphosphate
   from Pharmacia

[^{32}P]UTP: Uridine-5′-[α-^{32}P]triphosphate, triethylammonium salt,
   ~3000 Ci/mmol, 10 mCi/ml (Amersham Biosciences Corporation,
   Piscataway, NJ)

Rifampicin : 150 μg/ml, freshly dissolved in diethylpyrocarbonate
   (DEPC) treated water

10.5 M Ammonium Acetate

Gel Loading Buffer : 95% Formamide/20 mM di-sodium EDTA/
   0.05% bromophenol blue/0.05% xylene cyanol

Single-Round Transcription Reaction (40 μl Final Volume)

1. Prepare a reaction mix containing RNA polymerase, plasmid or
   other DNA template, and NusA and/or L4 where desired:

Per 40 μl reaction :	vol.	(final concentration)
10 × TME	4 μl	(20 mM Tris–acetate/4 mM MgAc/0.1 mM EDTA)
1.5 M KGlu	2.7 μl	(100 mM)
10 mM DTT	0.8 μl	(0.2 mM)
1 mg/ml BSA	2 μl	(25 μg/ml)
10 mM CTP/GTP	2 μl	(500 μM each)
RNA polymerase		(20 nM)
DNA		(20 nM)
NusA (optional)		(80 nM)
L4 (optional)		(160 nM)
H$_2$O		to bring volume to 36 μl

2. Incubate for 10 min at 37° to allow initiation and synthesis of
   pppGpGpC.

3. Meanwhile, prepare Elongation Solution containing rifampicin,
   ATP, and [^{32}P]UTP by mixing (per 40 μl reaction):

Rifampicin	2.7 μl
100 mM ATP	0.2 μl
20 mM UTP	0.2 μl
[α-^{32}P]UTP	0.4–1 μl
H$_2$O	if necessary to bring volume to 4 μl

4. At $t = 10$ min, start a single round of elongation by adding 4 μl
   Elongation Solution (rifampicin to 10 μg/ml, ATP to 500 μM, UTP
   to 100 μM and 4–10 μCi [α-^{32}P]UTP).

5. At the desired time (0.2 to 10 min after the start of elongation), stop reaction by addition of 40 μl 50 mM EDTA plus carrier RNA. Place on ice.
6. Add 16 μl 10.5 M ammonium acetate and 250 μl 100% ethanol.
7. Store 2 h to overnight at $-20°$ or 15–20 min at $-80°$.
8. Centrifuge for 15 min at 13,000 rpm in refrigerated microfuge. Aspirate off and discard the supernatant. Wash with 100 μl ice-cold 70% v/v ethanol, repeat centrifugation, and discard supernatant. Dry pellet in Speedvac.
9. Dissolve pellet in 10 μl Gel Loading Buffer. Load 5–10 μl on a urea–acrylamide (8%) gel in 0.5 × or 1 × TBE buffer (90 mM Tris, 90 mM boric acid, 2.5 mM EDTA, pH 8.1–8.3). For details, see Ref. 14.

Kinetic Analysis of the NusA and L4 Effects on Transcription

Using the synchronized transcription assay, we were able to show by late addition of L4 that the r-protein can stabilize the paused transcription complex even if it is not added to the elongation reaction until the RNA polymerase, in the presence of NusA, has already reached the attenuation site, as late as 2 min after the start of elongation.[21] Similarly, late addition of NusA to paused transcription complexes formed at 5 μM UTP enhanced the pause.[17]

Purification of Paused Transcription Complexes and Restart of Transcription

L4-mediated NusA-dependent transcription termination can be characterized in more detail by separating paused transcription complexes from the NTP substrates and ligands by gel filtration.[17] We have isolated transcription complexes made at 5–100 μM UTP in the presence or absence of NusA and/or L4. The regulatory ligands could then be added and transcription restarted by addition of NTPs. These experiments showed that complexes formed in the presence of NusA (or NusA and L4) when restarted maintained the influence of the regulatory ligand(s).

1. Pass the transcription reaction through a column of Sepharose 4B (cross-linked) previously equilibrated with 20 mM Tris–acetate (pH 7.9), 4 mM Mg–acetate, 0.1 mM EDTA, 130 mM KCl, 10 μg/ml rifampicin or heparin, 50 μg/ml BSA, and 0.1 mM DTT.
2. Fractions containing the transcription complex, with attached nascent RNA chains, are located by counting aliquots of each fraction and then pooled.

3. The fractions can be stored for at least 24 h at 2–4°.
4. The transcription reaction is resumed by addition of all nucleoside triphosphates (NTPs) (0.1–1 mM of each) in the presence or absence of combinations of NusA and L4.

Assay for Release of Terminated Transcripts

Kinetic analysis of the *in vivo* transcription of the S10 leader during the autogenous response shows that the RNA polymerase truly terminates (i.e., dissociates from the DNA template). However, in our *in vitro* transcription system, release of the transcription apparatus from the DNA template is not as efficient, even in the presence of L4. We use two approaches to analyze transcript release. First, we test the effect of raising the UTP concentration on the accumulation of short (attenuated) transcripts.[21] Transcripts are synthesized in reactions containing 5 or 100 μM UTP. At suitable times after the start of elongation (e.g., 6 min), UTP is added to an aliquot to a final concentration of 100 μM (for 5-μM reactions) or 1 mM (for 100-μM reactions). Transcription is allowed to continue for another 2 to 5 min. Transcripts that are not released will be chased into read-through transcripts because the pausing is very sensitive to the UTP concentration. The abundance of attenuated/paused versus read-through transcripts in the presence and absence of the UTP "chase" is then compared.

Another technique for analyzing paused versus released transcripts is nitrocellulose filtration. RNA molecules stably bound to RNA polymerase are retained on the filter, whereas terminated/released molecules are found in the flow-through fraction.[17]

Transcript Release Filtration Assay

1. Prepare a standard transcription reaction, 80 μl total volume.
2. At the appropriate time after start of elongation (e.g., 5 min), quench the reaction with EDTA (to 10 mM).
3. Prepare one 40-μl aliquot as for a typical transcription reaction (see previous text).
4. Pass the remaining 40 μl through a nitrocellulose filter (20 mm, BA85). Collect pass-through, and concentrate by ethanol precipitation.
5. Wash the filter two times with 100 μl Wash Solution containing 250 mM KCl, 50 μg yeast carrier RNA, and 10 mM EDTA.
6. Cut filter into slices, and place in microcentrifuge tube containing 200 μl 10 mM EDTA and 40 μg yeast carrier RNA. Boil for 5 min. Transfer aqueous fraction to new tube and concentrate RNA by ethanol precipitation.

RNA Competition Experiments

An essential part of the model for autogenous regulation of ribosomal protein synthesis is that the regulatory r-protein binds to similar targets on both its own messenger and its target on rRNA. This principle was demonstrated for L4 regulation of the S10 operon by adding rRNA to the transcription reaction: addition of 23S rRNA abrogated the L4 stimulation of premature termination. By adding specific fragments of the 23S rRNA we were able to map the specific site for the initial binding of L4 to the rRNA.[16]

Procedure: Perform a standard single-round transcription reaction, except that 60–120 nM of rRNA is added 1 min before addition of Elongation Solution. The rRNA is made by transcription of suitable templates with phage RNA polymerases (e.g., T7, T3, or similar RNA polymerase).

DNA Competition Experiments

Competition experiments can also be performed with oligonucleotides complementary to a specific region of the transcript. These oligonucleotides are added to a final concentration of 5 μM just before the start of transcription elongation. The oligonucleotides can serve to block binding sites for the L4 and NusA protein ligands and can probably also affect the folding of the nascent transcript.[15]

[28] Analysis of the Intrinsic Transcription Termination Mechanism and Its Control

By Evgeny Nudler and Ivan Gusarov

Intrinsic terminators are specific signals in the bacterial genome leading to rapid dissociation of transcription elongation complexes (EC). A canonical intrinsic terminator consists of a GC reach inverted repeat, followed by a stretch of five or more T bases in the nontemplate strand. Upon transcription of this signal, several U residues and a strong RNA hairpin appear at the 3′ end of the nascent transcript. Intrinsic terminators specify the ends of most bacterial and bacteriophage genes, allowing the RNA polymerase (RNAP) to release DNA and RNA and begin the next round of transcription. In addition, intrinsic terminators frequently reside in the beginning of a gene in an operon and are directly involved in gene regulation. Although the intrinsic terminators terminate transcription *in vitro* without assistance

Copyright 2003, Elsevier Inc.
All rights reserved.
0076-6879/03 $35.00

by any additional factors, hence the name "intrinsic," their efficiency is tightly regulated *in vivo*.[1]

The major technical challenge in studying the mechanism of termination is a short-lived nature of intermediates of this process. Under normal salt conditions (e.g., 100 mM KCl, 10 mM MgCl$_2$) inactivation and dissociation of the EC occur almost instantaneously, precluding conventional biochemical analysis. Here we describe approaches that permit dissecting the termination process into several steps and identifying individual roles of the terminator elements, the hairpin and T-stretch, in each step. These approaches also determine which step of the termination process is a target for regulation by termination or antitermination factors.

The Minimal Termination and Antitermination System

To study the basic mechanism of intrinsic termination and its control we use an *in vitro* reconstituted system that contains *Escherichia coli* RNAP, a linear DNA template with a strong promoter and the intrinsic terminator, and pure elongation factors, NusA and N. *E. coli* NusA protein works as a general termination factor, which significantly increases the efficiency of many intrinsic terminators *in vitro* and *in vivo*.[1] Phage λ N protein is an antitermination factor, which inhibits intrinsic termination and also Rho-dependent termination *in vitro* and *in vivo*.[2] The physiological role of N is to activate the early operons of phage λ by allowing the EC to read through multiple terminators at intergenic sites. Processive N antitermination *in vivo* requires several *E. coli* elongation factors (NusA, B, G, and E) and also a cis-acting RNA sequence, the *nut* site, which is composed of the boxB hairpin and an evolutionary conserved element, boxA.[1-3] *In vitro*, however, if provided at higher concentrations, N alone causes antitermination, suggesting that the very mechanism of antitermination lies in the direct interactions between N and the EC.[4,5] NusA significantly enhances the inhibitory effect of N on termination *in vitro*.[4-6] This observation implies that N induces fundamental changes in the way NusA interacts with the EC, converting it into the antitermination factor.

[1] E. Nudler and M. E. Gottesman, *Genes Cells* **7**, 755 (2002).
[2] A. Das, *Annu. Rev. Biochem.* **62**, 893 (1993).
[3] S. W. Mason, J. Li, and J. Greenblatt, *J. Biol. Chem.* **267**, 19418 (1992).
[4] W. A. Rees, S. E. Weitzel, T. D. Yager, A. Das, and P. H. von Hippel, *Proc. Natl. Acad. Sci. USA* **93**, 342 (1996).
[5] I. Gusarov and E. Nudler, *Cell* **107**, 437 (2001).
[6] W. Whalen, B. Ghosh, and A. Das, *Proc. Natl. Acad. Sci. USA* **85**, 2494 (1988).

```
                    TCCAgA TCCCgAAAAT TTATCAAAAA

       gAgTATTgAC TTAAAgTCTA ACCTATAggA TACTTACAgC
       ┌─►           -35                      -10
       CATCgAgAgg gACACggCgA ATAgCCATCC CAATCgAACA
       +1
                                          ▼
       ggCCTgCTgg TAATCgCAgg CCTTTTTATT TggATCCCCg
           ──►
```

Fig. 1. DNA sequence of the A1-tR2 template. −10 and −35 regions and the start (+1) site of the T7A1 promoter are indicated. The inverted repeat of the tR2 hairpin is shown by underlined arrows. The major termination point is indicated by the vertical arrowhead.

Utilization of the minimal *in vitro* system that contains only essential components for monitoring termination and antitermination greatly simplifies the biochemical analysis of these processes.

1. *The DNA template.* The complete sequence of the "master" template, called A1-tR2, is shown in Fig. 1. The template is designed to support efficient "walking" of the EC along the terminator (i.e., controlled stepwise transcription in solid phase)[7,8] which allows one to prepare and analyze homogenous ECs stalled within the terminator sequence. Important characteristics of the template to be considered are: (1) A relatively short length (∼200 bp) of the template. This makes it ideal for high-yield polymerase chain reaction (PCR) amplification, one-step PCR-generated mutagenesis, and also footprinting analysis. (2) T7A1 promoter. This is the strongest promoter to be utilized by *E. coli* RNAP *in vitro*, which generates the highest yield of the EC.(3) The initial transcribed sequence that permits one-step preparation of stable EC stalled at position +20 (EC20).(4) A short distance between the promoter and terminator (40 bp), which greatly facilitates walking of the EC to the desired position of the terminator.(5) tR2 terminator. Phage λ tR2 is a terminator of choice for a number of reasons. First, this is a canonical intrinsic terminator that has been characterized in detail *in vitro* and is known to be potentiated by NusA and suppressed by N *in vivo*.[9–11] Second, under standard *in vitro* conditions (see the following) the efficiency of tR2 is about 50%, leaving a significant margin for both up- and down-regulation. Third, tR2 is well suited for the biochemical analysis involving walking and stalling the EC at desired positions within the terminator sequence. The hairpin of tR2 is

[7] M. Kashlev, E. Nudler, K. Severinov, S. Borukhov, N. Komissarova, and A. Goldfarb, *Methods Enzymol.* **274,** 326 (1996).

[8] E. Nudler, I. Gusarov, and G. Bar-Nahum, *Methods Enzymol.* **371,** 160 (2003).

[9] K. S. Wilson and P. H. von Hippel, *Proc. Natl. Acad. Sci. USA* **92,** 8793 (1995).

[10] E. Nudler, M. Kashlev, V. Nikiforov, and A. Goldfarb, *Cell* **81,** 357 (1995).

[11] I. Gusarov and E. Nudler, *Mol. Cell* **3,** 495 (1999).

relatively small, and the T-stretch is naturally interrupted by A between fifth and seventh T (Fig. 1).

T7-tR2 is obtained by PCR from pENtR2 plasmid using the upstream primer 5'-TCCAGATCCCGAAAATTTATC-3' and the gel purified downstream primer 5'-CGGGGATCCAAATAAAAAggCC-3'. To generate the mutant tR2 templates, the downstream primer varies depending on a particular substitution to be made in the terminator. A typical amplification reaction (50 μl) includes 41.5 μl H$_2$O, 5 μl 10 × Pwo reaction buffer (Roche Molecular Biochemicals, Indianapolis, IN), 1 μl dNTP (1 mM stock), 1 μl of each oligo (50 pmol/μl stock), and 0.5 μl Pwo DNA polymerase (Roche Molecular Biochemicals). The PCR reaction is programmed as follows: 95° 45 s, 55° 45 s, 72° 45 s (27 cycles). Bulk mixture is usually prepared for 20 PCR reactions and then distributed in 50-μl aliquots in 250 μl thin-walled plastic tubes (Perkin Elmer, Shelton, CT). The amplified product is resolved in a 1.5% (w/v) agarose gel (BMA, Rockland, ME), followed by electroelution, extraction with phenol-chloroform, and precipitation with ethanol. The DNA pellet is diluted in 50 μl TE (10 mM Tris–HCl, pH 7.9, and 1 mM EDTA). The final DNA concentration is ~2 pmol/μl as determined by spectrophotometry.

2. *RNAP.* The highly pure *E. coli* RNAP holoenzyme carrying the 6His-tag at the COOH terminus of the β' subunit is used in the minimal termination/antitermination system. The detailed protocol for RNAP purification is described in Chapter [11].[8] RNAP isolated by this method is highly active, so that at least 30% of RNAP molecules are elongation competent.

3. *NusA protein.* To purify *E. coli* NusA using the pMS7 overproducing plasmid,[12] the protocol described by Das *et al.* is utilized without modifications.[13] The stock NusA solution (~300 μm) is stored at −20° and contains 200 mM potassium acetate (KAc), 10 mM Tris–HCl, pH 7.9, 0.1 mM EDTA, 1 mM DTT, and 40% (v/v) glycerol. The activity of NusA was measured in a single-round termination assay using the A1-tR2 template (see the following).

4. *N protein.* λN is induced and purified according to Das *et al.* with modifications of later steps.[13] N protein in the induced culture is mostly in inclusion bodies. As a good substrate for La protease, N is highly unstable in wild type cells. Hence, it is important to use a *lon*-deficient strain for N induction (e.g., BL21[DE3] from Stratagene, LaJolla, CA). Highly pure N is obtained by solubilizing the inclusion bodies in 8 M guanidinium hydrochloride, followed by one-step chromatography. All procedures are carried out at 4°.

[12] M. C. Schmidt and M. J. Chamberlin, *J. Mol. Biol.* **195,** 809 (1987).
[13] A. Das, M. Pal, J. G. Mena, *et al.*, *Methods Enzymol.* **274,** 374 (1996).

2 g of frozen cell pellet is resuspended in 30 ml buffer A [20 mM Tris–HCl, pH 7.5, 10% (v/v) glycerol, 1 mM DTT, 0.5 mM PMSF] with 0.2% sodium deoxycholate. Homogenous suspension rests on ice for 10 min. Cells are sonicated using Sonic Dismembrator 60 (Fisher Scientific, Chicago, IL) in 30 pulses 10 times on the maximum output, followed by 30 rest. The lysate is centrifuged for 15 min at 25,000×g, and supernatant is discarded. Sonication is repeated two times. For the last round of sonication, 0.5% Triton X100 is added. Next, the pellet is dissolved in buffer B (20 mM Tris–HCl, pH 7.9, 0.1 mM EDTA, 1 mM β-mercaptoethanol), containing 8 M guanidinium hydrochloride and centrifuged for 20 min at 25,000 × g. The clear supernatant is then dialyzed against buffer B. Aggregates are removed from the dialysate by centrifugation for 20 min at 25,000 × g. The clarified supernatant is applied to the 5-ml MonoS column (Amersham-Pharmacia Biosciences, Piscataway, NJ) preequilibrated, and washed with 10 ml of buffer C (20 mM Tris–HCl, pH 7.9, 50 mM NaCl, 0.1 mM EDTA, 1 mM β-mercaptoethanol, 5% (v/v) glycerol). The bound proteins are eluted with a linear gradient of 0.05–1 M NaCl in buffer C at the rate 0.3 ml/min. Highly pure N protein is in fractions with about 0.25 M NaCl. The collected fractions are dialyzed against buffer C and concentrated on the Amicon (Millipore, Bedford, MA) filter concentrator (Centricon YM-3). Glycerol is added to 40%, and the protein is stored at $-70°$. About 3 mg of N per 1 g of wet cell pellet is expected from this protocol. The N prep is virtually free of nuclease activities and is highly active in the single-round antitermination assay.

5. *Single-round (anti)termination assay*. The following assay is designed for a rapid assessment of the intrinsic termination efficiency and also the activity of termination/antitermination factors. In a standard reaction (for 10–15 samples), 0.5 pmol ($\sim$0.5 μl) of His-tagged RNAP and 1 pmol ($\sim$1 μl) of A1-tR2 template are mixed in 8 μl TB (10 mM Tris–HCl, pH 7.9, 10 mM MgCl$_2$, and 100 mM KCl) and incubated for 5 min at 37° to form an open promoter complex. 1 μl of the 10× starting mix (100 μm CpApUpC RNA primer [Oligos Etc., Inc., Wilsonville, OR], 250 μm ATP and GTP) and 1 μl [α-^{32}P]CTP [3 μm, 3000 Ci/mmol; NEN Life Science, Shelton, CT] are added for another 5 min at 37°, followed by addition of CTP to 25 μm for another 3 min. This generates [α-^{32}P]EC stalled at position +20 (EC20). All following reactions are performed at room temperature. EC20 is diluted with 100 μl buffer, which is either a standard TB or TBAc (50 mM potassium acetate (Ac), 5 mM MgAc, 20 mM Tris–Ac, pH 7.6). TBAc is preferred when the activity of N is tested. Rifampicin (Sigma, St. Louis, MO) is added to 10 μm to prevent reinitiation by RNAP. EC20 is then distributed in 10-μl aliquots in 10 Eppendorf tubes.

Aliquots can be stored at $+4°$ for 12 h. Elongation factors (usually 1 μl) are added before the chase reaction for 2 min at room temperature. The chase reaction is performed by adding 1 μl of $10\times$ NTP mix (ATP+CTP+GTP+UTP, 1 mM) for 5 min. The reaction is stopped by 2 vol of the hot loading mix (12 M urea, 20 mM EDTA, 10 mM Tris–HCl, pH 7.5, 0.0125% each of BPB and XC). Samples are heated at $100°$ for 30 s and immediately loaded on denaturing sequencing gel [0.5-mm thick; 10% acrylamide:bisacrylamide (29:1)] that has been prerun for at least 10 min and is hot. The remainder of the samples is frozen at $-70°$ for additional gels. Radioactive RNA bands are quantitated using the PhosphoImager and software from Molecular Dynamics (Piscataway, NJ). To calculate the efficiency of termination, the amount of radioactivity in a termination band is divided by the total radioactivity present in that and all read-through bands. Termination efficiency (%T) depends on monovalent and divalent salt concentration, NTP concentration, temperature, and the presence of elongation factors. Under specified conditions (50 mM KAc, 5 mM MgAc, 20 mM Tris–Ac, pH 7.6, 250 μm 4 NTPs, $25°$) %T at the A1-tR2 template is ~50%.[5] NusA (0.1 μm) increases %T to ~70%. N (0.5 μm) or N (0.2 μm)+NusA (0.1 μm) decrease %T to ~30%.[5]

Obtaining and Analyzing the Trapped Intermediate in Termination

The trapped complex (TC) represents a unique configuration of the EC that occurs exclusively at the intrinsic termination points.[11] The TC can be isolated as a long-lived intermediate under low ionic strength conditions. RNAP is inactive in the TC while carrying both the DNA template and the nascent transcript (Fig. 2). Unlike other inactive forms of the EC, such as pauses or arrests, the TC is irreversibly inactive and extremely unstable. Inactivation and loss of stability depend strictly on the termination hairpin folding at the proper distance (7–9 nt) from the catalytic site of RNAP.[11] The TC is likely to represent an intermediate of the intrinsic termination process, which uncouples its inactivation and dissociation steps. An ability to obtain the intact TC suitable for conventional biochemical analysis provides an opportunity for monitoring the structural transitions that occur in the complex during termination in real time. The following protocols are designed for obtaining the TC at the tR2 terminator.

1. *Two-step preparation of the TC at the A1-tR2 template.* All reactions are performed in siliconized 1.5-ml plastic tubes (USA Scientific Plastics, Ocalar, FL). In the standard reaction (for 10–15 samples), 1 pmol (~0.5 μl) of His-tagged RNAP, 2 pmol (~1 μl) of the T7-tR2 template and 8 μl of TB (10 mM Tris–HCl pH 7.9, 10 mM MgCl$_2$, and 100 mM

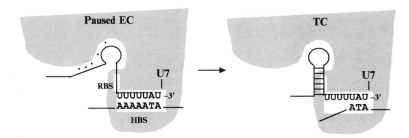

Fig. 2. The termination mechanism. Transition from a paused EC to a termination (trapped) complex (TC) occurs as a result of hairpin formation that irreversibly inactivates the complex. The folding hairpin breaks protein–RNA contacts in the single-stranded RNA binding site (RBS) and partially unwinds the RNA:DNA hybrid in the hybrid binding site (HBS).[11]

KCl) are mixed and incubated for 5 min at 37° to form an open promoter complex. 1.5 μl of the 10× starting mix (100 μm CpApUpC RNA primer, 250 μm ATP and GTP) and 5 μl of TB-preequilibrated Ni^{++}-NTA agarose beads (Qiagen, Chatsworth, CA) are added for 8 min at 37° with slight agitation. The beads are washed four times with 1.5 ml TB and resuspended in 10 μl TB containing 0.3 μm [α-^{32}P]CTP (3000 Ci/mmol; NEN Life Sciences) and ATP + GTP (25 μm). The reaction proceeds for 8 min at room temperature, followed by two rounds of washing with "high-salt" TB (1 mM KCl) and two rounds of standard TB washing. The resulted EC stalled at position +20 (EC20) is expected to count 3000–5000 cpm, if the bottom of the tube was placed next to the membrane of the Geiger counter (RPI, Inc., Mt. Prospect, IL) The standard round of washing (i.e., pelleting and resuspension of the beads) includes 3–5 s centrifugation in a microcentrifuge, removal of supernatant leaving ~50 μl TB covering the beads, and resuspension in 1.5 ml of appropriate TB.

To follow the fate of DNA during the experiment, the template can be radiolabled at its 5' ends. For this, 0.5 μl [γ-^{32}P]ATP (7000 Ci/mM; ICN, Costa Mesa, CA and 5 U (0.5 μl) of T4 polynucleotide kinase (New England Biolabs, Beverly, MA)) are added to a 10 μl of TB containing 2 pmol of the A1-tR2 template for 15 min at 37°. The T4 polynucleotide kinase is inactivated at 65° (10 min) before the template is used in preparation of the initial EC. The radioactive material retained on beads after the first round of washing is expected to count 2000–4000 cpm, which is indicative for efficient DNA labeling.

To prepare the TC, an aliquot of EC32 is washed twice with TB5 (10 mM Tris–HCl, pH 7.9, 8 mM MgCl$_2$, and 5 mM KCl) and chased to the terminator by adding 4 NTPs (ATP, CTP, GTP, and UTP—100 μm). The reaction proceeds for 5 min at room temperature, followed by four rounds of washing with TB5. Under these conditions, much of the radiolabled RNA and DNA remain on beads. The amount of TC that does not dissociate after the first wash with TB5 varies with the particular preparation of RNAP within the range from 40% to 95% and constitutes ~75% on average. Challenge of TC with a higher-salt TB (e.g., 40 mM KCl) or heparin (1 mg/ml) causes rapid complex dissociation. This control is useful to assess the amount of stable EC arrested at the termination point. A fraction of EC becomes arrested at the termination point because of backtracking of RNAP at the T-stretch.[10,11] This reaction competes with the hairpin formation and termination process (see the following). The amount of arrested EC at tR2 constitutes 5–10% of the terminated fraction.[10]

To confirm that the retained RNA is positioned in the RNAP as a part of the TC rather than being nonspecifically bound to the beads or RNAP, the GreB transcript cleavage factor can be used. GreB potentiates the intrinsic RNA endonuclease activity in the catalytic center of backtracked RNAP.[14] Sensitivity to GreB reflects a probability of the RNAP for backtracking at a particular DNA position.[15] The TC is about 10^3 times more resistant to GreB than the arrested EC,[11] meaning that at the moment of GreB addition, TC was not backtracked for more than 1 nt and its RNA was specifically positioned in the RNAP to be accessible for cleavage. GreB protein is purified as described by Borukhov and Goldfarb.[14] GreB is added to an aliquot of TC (10 μl) to the final concentration of 1 μm for 5 min at 37°.

2. *Preparation of the TC by walking to the tR2 terminator.* As mentioned earlier, the TC provides an opportunity to study the termination process in real time, that is, probing the protein–nucleic acid interactions in the complex after the termination hairpin has been formed. Walking to the termination point incorporates the cross-linkable analogs of NTP into certain positions of the terminator RNA and thus directly monitors structural changes induced by the hairpin in three functional RNA binding sites of RNAP: the catalytic site, the RNA:DNA hybrid binding site (HBS), and the single-stranded RNA binding site (RBS). The latter is located just upstream of the ~7–8 bp hybrid (Fig. 2, see [13] in this volume).[16] To prepare the TC and also the ECs within the terminator sequence, a standard walking procedure is utilized,[8] with some modifications at the later steps. The derivative of the A1-tR2 template is used with the T→C substitution within the T-stretch sequence TTTTTA*T*CT (the termination point is underlined) that allows halting the EC at the termination point. Formation of the initial EC20 immobilized on Ni^{++}-

[14] S. Borukhov and A. Goldfarb, *Methods Enzymol.* **274**, 315 (1996).

[15] E. Nudler, A. Mustaev, E. Lukhtanov, and A. Goldfarb, *Cell* **89**, 33 (1997).

NTA agarose is carried out as described previously. From this point, the successive walking steps are performed to move the EC through the terminator: AUG(EC23)→ AUC(EC36)→ AGC(EC43)→ UGC(EC50)→ AUC(EC54)→ AGC(EC61). All walking steps are done for 3 min at room temperature using 25 μm NTPs. To prepare the TC (EC68), EC61 is washed with TB5 for the last "AU" step. The walking steps can be modified to meet the particular aim of the experiment. For example, the hairpin can be gradually weakened by substituting one or more G with inosine (I) at specific positions in the hairpin's stem. Also, cross-linkable 4-thio-UTP and 6-thio-GTP can be inserted at the desired positions within the termination sequence to probe the protein–RNA contacts at the terminator. These cross-linkable analogs do not interfere significantly with the hairpin structure and termination.[11] NTP analogs are used for walking at concentrations: 15 μm ITP (Roche Molecular Biochemicals), 40 μm 4-thio-UTP and 6-thio-GTP. 4-thio-UTP and 6-thio-GTP are synthesized from the corresponding nucleosides (Sigma) and purified as described by Hoard and Ott.[17] UV inducible cross-linking and mapping of the cross-linking sites in RNAP are described in detail by Mustaev et al.[18] These analogs are available upon request.

Another cross-linkable analog (U•), an alkylating derivative of UTP containing 3-[3'-(N,N-bis-2-iodoethyl)amino-4'-formylphenyl]propionate moiety attached to position 5 of the pyrimidine ring through an aminoallyl spacer,[15] can be used to directly determine the change in the length of the RNA:DNA hybrid during transition from EC to TC.[11] The T-stretch of the A1-tR2 template needs to be modified to incorporate U• into a single position within the A:U hybrid at the terminator. For example, the TTT**ATA**TC derivative would allow insertion of U• at the −3 position (in bold) counting from the termination point (underlined); the TA**TAT**TTC derivative makes it possible to insert U• at the −5 position. These substitutions do not preclude termination at tR2. To incorporate U•, the EC is washed with TBMn (10 mM HEPES, pH 7.2, 100 mM KCl, and 2 mM MnCl$_2$). U• is added to the final concentration of 40 μm for 30 min at 25°, followed by multiple washing with TB5 and walking. To induce cross-linking, a freshly prepared aqueous solution of NaBH$_4$ (4 mg/ml; Aldrich, St. Louis, MO) is added to 0.4 mg/ml final concentration for 20 min at room temperature.

The DNA oligo, 5'-CAGCAGGCCTGTTC-3', which is complementary to the first half of the termination hairpin, can be used to suppress hairpin

[16] E. Nudler, E. Avetissova, N. Korzheva, and A. Mustaev, *Methods Enzymol.* **371,** 179 (2003).

[17] D. E. Hoard and D. G. Ott, *J. Am. Chem. Soc.* **87,** 1785 (1965).

[18] A. Mustaev et al., *Methods Enzymol.* **371,** (2003).

folding and obtain the active EC at the termination point by walking.[11] The oligo (1 μl, 50 pmol/μl) is added to the 10-μl aliquot of EC54 for 5 min, followed by washing with TB5 and walking to the termination point. Once annealed to the RNA, the oligo remains in the complex at least for two walking steps in TB or TB5 at 25°. This technique permits a direct comparison of the EC and TC that carry the same RNA sequence. Using this approach, we have shown that the RNA:DNA cross-linking pattern differs dramatically between the TC and the corresponding EC. For instance, cross-linking from RNA to DNA can be detected at the -5 position only in the active EC, but not the TC. On the other hand, RNA:DNA cross-linking at the -3 position is detectable in both types of complexes.[11] These results demonstrate directly that formation of the termination hairpin leads to unwinding of the upstream portion of the hybrid in the transcription bubble (Fig. 2). They also serve as an internal control, indicating that RNA and DNA are positioned in the same RNAP molecule in the TC.

Functional Characterization of the Terminator Elements

The A1-tR2 template can be easily manipulated to disable partially or completely the T-stretch or hairpin elements of the terminator. All changes in the template can be introduced by a single PCR reaction using a modified downstream oligo that overlaps with the termination sequence. PCR conditions and template purification are described previously. The mutant templates are used to determine the individual role of the T-stretch and hairpin in pausing at the terminator, complex destabilization, and responsiveness to termination/antitermination factors.

Pausing at the Termination Point: Detection and
   Functional Analysis

Pausing at the tR2 termination point is determined by the downstream portion of its T-stretch.[11] Pausing provides an additional time for the hairpin to form and is required for efficient termination under physiologic chase conditions (i.e., 5–10 mM Mg^{++}, NTPs >100 μm).[11] The tR2 hairpin, in the absence of the T-stretch, does not pause the EC at the termination point (E. Nudler and I. Gusarov, unpublished observations). To study pausing at the termination point, the A1-tR2 template is modified in such a way that three G at positions -24, -28, -29 in the hairpin stem (counting from the T7 termination point) are substituted for C. These substitutions completely eliminate hairpin formation and termination.[11] Pausing is detected in a time course single-round transcription experiment. The initial

EC20 is prepared in solution as described earlier. The sample of EC20 is diluted with TB and distributed in 20-μl aliquots to the wells of a 96-well plastic tray. To resume transcription, 4 μl of TB containing all four NTPs is added to the well to a final concentration of 1 mM for 2, 4, 6, 8, 10, 12, and 60 s. Transcription reaction is stopped by adding 2 vol of the hot loading mix (12 M urea, 20 mM EDTA, 10 mM Tris-HCl, pH 7.5, 0.0125% each of BPB and XC), then transferred to Eppendorf tubes and heated at 80° for 1 min in water bath before loading onto 8% polyacrylamide gel electrophoresis (PAGE) (19:1 acrylamid:bisacrylamid, 7 M urea, 0.5X TBE). To minimize the error attributed to the short-time chase reactions, at least 10 separate time trials are processed as independent samples for each short-time point and then combined in a single tube for resolving by PAGE. Under the described chase conditions, the major pause sites at the hairpin deficient A1-tR2 template correspond to the T7, T8, and T9 positions of the T-stretch (i.e., coincide with the termination points). The half-life of the pause is determined by estimating an apparent duration of a paused band on a radiogram of the sequencing PAGE. After 6 s, most of the EC reaches the run-off. Those EC that are stuck at the T-stretch pause sites for more than 5 s are considered as backtracked (arrested) EC (see the following). The radioactivity belonging to the transiently or permanently arrested band is not taken into account for calculating the pause half-life.

To determine how the effect on pausing at the termination point by elongation/termination factors affects the termination efficiency at tR2 or other intrinsic terminators, two calibrating plots need to be generated. One determines the termination efficiency (%T) as a function of NTP concentration (Fig. 3A). The native A1-tR2 template is used in this case in a standard single-round assay, in which the initial EC20 is chased to the terminator by adding NTPs to 0.1, 0.2, 0.3, 0.4, 0.5, and 1 mM. The other plot determines the apparent duration of the pause at the point of termination as a function of NTP concentration (Fig. 3B). The pause decay profiles in this case are nonlinear because of progressive backtracking of the EC that occurs at the end of T runs.[15] RNAP backtracking depends on NTP concentration and duration of the initial pause. Because progressive backtracking induced by the T-stretch and termination are mutually exclusive events,[10] it is important to consider only the early, close to linear, portion of each curve (marked by tangent lines in Fig. 3B) to calculate the approximate half-life of the pause ($t_{1/2}$). By combining the information from both plots, it is possible to plot the %T as a function of pause $t_{1/2}$ at the termination point (in this case T7, Fig. 3C). Using this plot, one can determine the kinetic component in the termination regulation (i.e., the extent by which the modulation of pausing by individual factors contributes to termination). For example, at 1 mM NTPs, NusA (100 nm) increases pause $t_{1/2}$

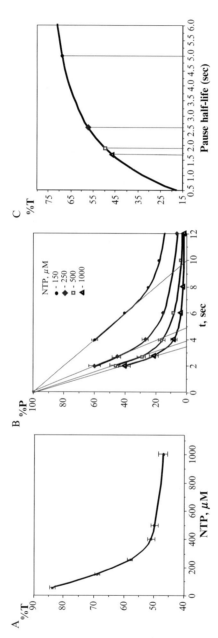

Fig. 3. Assessment of the pause contribution to the termination efficiency. (A) tR2 termination as a function of NTP concentration. The reaction conditions are described in text. (B) Pausing at the termination point (T7 position). The A1-tR2 template, carrying point substitutions in the upstream half of the tR2 hairpin that prevented hairpin folding, is used (see text). Pausing (%P) is calculated as a fraction of [^{32}P] elongated transcripts interrupted at the T7 position of the T-stretch at the indicated times. The fast components of the T7 pause decay curves are shown by tangent lines and used in further analysis. Pause curves generated at different NTP concentrations are indicated by corresponding geometric marks. (C) Efficiency of tR2 termination as a function of pausing. The curve combines the results of (A) and (B).

at the termination point of tR2 by ~15%. According to graph C, this should translate into an ~8.5% increase of %T. Because the actual increase of %T by NusA under the same chase conditions is 43%,[5] it is concluded that the increase of pause dwell time by NusA accounts for ~30% of its stimulation effect on termination under specified conditions. A similar analysis has been performed for the N protein,[5] and it can be done for any other elongation factor.

The "Slow Termination" Assay

To exclude the kinetic component (pausing) from the analysis of termination and to study the "mechanistic" component of the termination process in real time, the slow termination approach has been developed.[5] This approach utilizes a modified A1-tR2 template carrying point substitutions within the T-stretch, rendering the RNA:DNA hybrid strong enough to delay the hairpin folding. Using the methods described previously, we have demonstrated that the ability of the termination hairpin to rapidly fold at the 7–9 nt distance from the catalytic site of RNAP and terminate transcription strictly depends on melting of at least four A:U hybrid base pairs adjacent to the hairpin's stem.[11] Such a direct competition between the hairpin and the upstream portion of the hybrid in the transcription bubble occurs even if the hairpin does not include any Us (e.g., in case of tR2). Substitution of U1 through U5 with hybrid stabilizing guanines prevents hairpin formation, whereas weakening of this hybrid by inosines has an opposite effect.[11] Figure 4 shows the A1-tR2^{mut2} template that has two hybrid-stabilizing substitutions (T2→G and T3→G). The EC68 at the termination point is prepared by walking as described earlier except that all steps are done in TB. EC68 is washed four times with TBAc before measuring the rate of termination. The time course begins by adding the chase mix (CTP, UTP, and GTP) to 100 μm for 20 s to extend the U7 transcript for the next 5 nt. The radioactive RNA products are analyzed by 8% sequencing gel as described earlier. %T is calculated by dividing the amount of radioactivity in the unchaseable (terminated U7) band to the combined radioactivity in the read-through and terminated bands. To verify that the unchaseable band is indeed the result of termination and not a result of arrest formation, the equal aliquot of beads is washed with TBAc after the chase reaction. Most of the radioactivity belonging to the terminated band should disappear, while the read-through band stays the same. Small fraction of radioactivity (~5–10%) in the U7 band that remains after the chase and wash is considered to be an arrested fraction. The backtracked nature of this fraction may be further confirmed by probing with GreB. Incubation with 1 μg/ml GreB for 5 min at room temperature should be

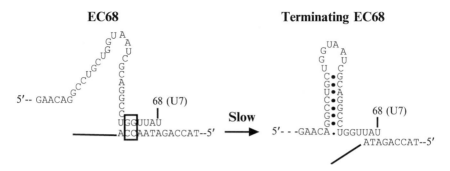

Fɪɢ. 4. The "slow termination" approach. Immobilized EC is walked to the termination point at position +68 (T7), followed by TB washing to remove unincorporated NTP. The A1-tR2 template carrying two hybrid stabilizing substitutions (in box) that prevent rapid hairpin folding and termination is used. The half-life of EC68 under specified conditions (see text) is ~20 min.

sufficient to cleave the 3' portion of RNA in the arrested fraction and completely reactivate it. The arrested fraction should not be accounted in calculating %T. Under specified conditions, it takes about 20 min to terminate half of the EC68. To test how an elongation factor may change the rate of termination in this assay, it should be added to the washed EC68 shortly before the chase reaction has been initiated. For example, NusA (100 nm) accelerates the rate of termination of EC68 by a factor of approximately 3. N protein (0.5 μm) has an opposite effect and slows down the termination rate significantly.[5] Individually, NusB, NusE, and NusG (100 nm) do not have any measurable effect on the termination rate (I. G. and E. N., unpublished observations).

Because pausing at the termination point is not relevant in this assay (the rate of termination is in the range of minutes), it is useful to quickly assess whether a certain factor affects termination through a direct interference with the hairpin formation. Apart from pausing, there are two mechanistic ways for controlling hairpin formation. A factor can either change the stability of the hybrid that competes with the hairpin formation, or it can affect protein–RNA interactions in the EC that interfere with hairpin formation. Analysis of protein–RNA and RNA–DNA contacts in the TC and EC at the terminator using highly specific cross-linking probes as described earlier can directly address these issues.

Acknowledgments

This work was supported by the National Institutes of Health.

[29] Rho's Role in Transcription Attenuation in the *tna* Operon of *E. Coli*

By FENG GONG and CHARLES YANOFSKY

Introduction

Organisms exploit virtually every feature of transcription and translation in achieving efficient regulation of gene expression. In many bacterial species the transcribed leader regions of specific operons contain regulated sites of transcription termination. Depending upon the appropriate signal received, transcription will be terminated at such a site or be allowed to continue into the structural genes of the operon. Transcription attenuation is the term used to describe this mechanism of gene regulation.[1] Transcription attenuation is advantageous as a regulatory strategy because it allows cellular components and events that do not participate directly in transcription initiation decisions to influence gene expression. Thus RNA sequences and structures play a decisive role in the process of transcription attenuation. RNA sequences can be specified that form competing, alternative RNA structures, one of which functions as a terminator, and the alternate as an antiterminator. These permit or prevent transcription termination, respectively. RNA involvement also allows translation of a short coding region and ribosome movement to serve as the basis for the decision whether or not to terminate transcription. The mechanisms of transcription attenuation used to regulate gene expression vary greatly.[2,3] Transcription attenuation is a common regulatory strategy, and genome sequence scans for likely attenuation-regulated leader regions suggest that as many as 5–10% of the operons of many bacterial species are regulated in this manner.[4]

Most microorganisms employ two mechanisms of terminating transcription, so-called intrinsic termination and factor-dependent termination.[5–9] In the former mechanism, the transcribing RNA polymerase synthesizes

[1] C. Yanofsky, *J. Bacteriol.* **182,** 1 (2000).

[2] P. Gollnick and P. Babitzke, *Biochim. Biophys. Acta* **1577,** 240 (2002).

[3] T. M. Henkin and C. Yanofsky, *Bioessays* **24,** 700 (2002).

[4] E. Merino and C. Yanofsky, *in* "*Bacillus subtilis* and its Closest Relatives: from Genes to Cells" (A. L. Sonenshein, Ed.), p. 323. ASM Press, Washington, D.C., 2002.

[5] E. Nudler and M. E. Gottesman, *Genes Cells* **7,** 755 (2002).

[6] T. Platt, *Mol. Microbiol.* **11,** 983 (1994).

[7] J. Richardson, *Biochim. Biophys. Acta* **1577,** 251 (2002).

[8] P. H. von Hippel, W. A. Rees, and K. S. Wilson, *Nucleic Acids Symp. Ser.* **33,** 1 (1995).

[9] J. P. Richardson, *Crit. Rev. Biochem. Mol. Biol.* **28,** 1 (1993).

a segment of RNA that can fold to form a stable hairpin structure, followed by a run of Us. This RNA structure/segment is recognized by the transcribing polymerase as a termination signal, and, when synthesized, directs polymerase to terminate, releasing polymerase, template, and transcript. Many organisms also form a specific RNA binding protein that functions as a termination factor. The factor uses features of a transcript as a binding site, and it then scans the transcript in the 3′ direction, searching for a paused polymerase. If it contacts a paused polymerase, it directs that polymerase to terminate transcription. With either type of termination event, specific proteins exist that can interact with defined transcript sequences and prevent or facilitate termination.[10–12]

The most widely studied termination factor is the Rho protein of *Escherichia coli*. Much is known about its structure and mechanism of action.[7] The most thoroughly studied example of Rho action concerns its participation in preventing early gene expression in bacteriophage lambda, in *E. coli*.[7] A second well-studied bacterial operon that is regulated by Rho-mediated transcription termination is the tryptophanase (*tna*) operon of *E. coli*.[13,14] In this article we describe the *in vitro* methods that have been used to study Rho action in mediating transcription attenuation in the leader region of this operon. We also describe methods allowing experimental analysis of the role of the *tna* operon leader peptide in influencing whether or not Rho-mediated transcription termination will occur.

The *tna* operon of *E. coli* contains two major structural genes, *tnaA* and *tnaB*.[15] The *tna* promoter is separated from the *tnaA* structural gene by a 319-nucleotide leader region. Expression of the *tna* operon of *E. coli* is regulated both by catabolite repression and tryptophan-induced transcription antitermination. Catabolite repression regulates transcription initiation, whereas the presence of excess tryptophan induces antitermination at Rho factor-dependent termination sites in the leader region of the operon.[13] The leader region encodes a 24-residue peptide, TnaC, that is essential for induction.[16–18] A Rho utilization site (*rut*) is located in the vicinity of *tnaC* stop codon, UGA.[19] Using an *E. coli* S-30 system, as described in this article, it was demonstrated that Rho-dependent transcription

[10] J. Stulke, *Arch. Microbiol.* **177**, 433 (2002).
[11] R. A. Weisberg and M. E. Gottesman, *J. Bacteriol.* **181**, 359 (1999).
[12] T. M. Henkin, *Curr. Opin. Microbiol.* **3**, 149 (2000).
[13] V. Stewart, R. Landick, and C. Yanofsky, *J. Bacteriol.* **166**, 217 (1986).
[14] C. Yanofsky, *Annu. Rev. Biochem.* **70**, 1 (2001).
[15] V. Stewart and C. Yanofsky, *J. Bacteriol.* **164**, 731 (1985).
[16] V. Stewart and C. Yanofsky, *J. Bacteriol.* **167**, 383 (1986).
[17] F. Gong and C. Yanofsky, *Science* **297**, 1864 (2002).
[18] F. Gong and C. Yanofsky, *J. Biol. Chem.* **276**, 1974 (2001).
[19] P. Gollnick and C. Yanofsky, *J. Bacteriol.* **172**, 3100 (1990).

termination in the leader region of this operon is prevented by inducing levels of tryptophan. Detailed studies using this cell-free system allowed a determination of the mechanism of tryptophan induction.[18,20,21]

The following stock solutions are used in this protocol. If not specified, all reagents are commercially available from Sigma (St. Louis, MO).

1. Circularized DNA template, 0.5 mg/ml in RNase-free H_2O
2. 5 mM amino acid mix (-Met, -Trp)
3. 5 mM NTPs (Diluted from 100 mM stock solutions from Boehringer Mannheim high-performance liquid chromatography (HPLC) purified)
4. 5 mM CTP and GTP (HPLC-purified, Boehringer Mannheim, Germany)
5. 2.5× S-30 reaction mix (without amino acids and NTPs), 87.5 mM Tris–acetate, pH 8.0, 500 mM potassium glutamate (should be optimized), 20 mM magnesium acetate (should be optimized), 75 mM ammonium acetate, 5 mM DTT, 5 mM ATP, 50 mM phosphoenol pyruvate (trisodium salt), 0.25 mg/ml $E.$ $coli$ tRNA, 87.5 mg/ml polyethylene glycol (molecular weight 8000), 0.75 U/ml pyruvate kinase, 50 μg/ml folinic acid, 2.5 mM cAMP, 67.5 μg/ml TPN (optional), 67.5 μg/ml pyridoxine hydrochloride (optional), 67.5 μg/ml FAD (optional)
6. RNase-free H_2O (DEPC-treated)
7. [^{35}S]-methionine (Perkin Elmer, Boston, MA, 1000 Ci/mmol, 10 mCi/ml)
8. [α-^{33}P]-UTP (Perkin Elmer, 1000-3000 Ci/mmol, 10 mCi/ml)
9. Tricine–SDS sample buffer, pH 6.8[22]
10. RNA loading buffer for a urea–PAGE denaturing gel[23]

Preparation of Active S-30 Extracts, and Standard Reaction Conditions and Procedures

S-30 extracts were prepared according to Zubay,[24] using a derivative of the $E.$ $coli$ A19 RNaseI$^-$ strain that was $trpR^-$ Δ $lacZ$ Δ $trpEA2$ $tnaA2bgl$::Tn10. The protein synthetic activity of an S-30 reaction is highly dependent on the concentrations of Mg^{2+}, K^+, and the S-30 extract in the

[20] F. Gong, K. Ito, Y. Nakamura, and C. Yanofsky, $Proc.$ $Natl.$ $Acad.$ $Sci.$ USA **98,** 8997 (2001).
[21] F. Gong and C. Yanofsky, $J.$ $Biol.$ $Chem.$ **277,** 17095 (2002).
[22] H. Schagger and G. von Jagow, $Anal.$ $Biochem.$ **166,** 368 (1987).
[23] J. Sambrook and D. W. Russell, "Molecular Cloning: A Laboratory Manual." Cold Spring Harbor Laboratory Press, Cold Spring Harbor, N.Y., 2001.
[24] G. Zubay, $Annu.$ $Rev.$ $Genet.$ **7,** 267 (1973).

reaction mixture. Optimization of Mg^{2+} and K^+ concentrations was performed as described by Lesley.[25]

A standard coupled transcription/translation reaction mixture (with a total volume of 50 μl) contains 35 mM Tris–acetate, pH 8.0, 10 mM magnesium acetate, 200 mM potassium glutamate, 30 mM ammonium acetate, 2 mM DTT, 2 mM ATP, 0.2 mM each of CTP, UTP, and GTP, 20 mM phosphoenol pyruvate (trisodium salt), 0.25 mg/ml E. coli tRNA, 0.3 U/ml pyruvate kinase, 35 mg/ml polyethylene glycol (molecular weight 8000), 20 μg/ml folinic acid, 1 mM cyclic AMP, 200 μm each of 19 amino acids, with varying amounts of added L-tryptophan, and approximately 2 μg DNA template. All reactions are performed at 37°. To monitor newly synthesized mRNA transcripts, cold UTP is replaced by 20 μCi [α-^{33}P]-UTP.

A typical procedure used in labeling newly synthesized protein by incorporation of [^{35}S]-methionine is shown as follows. Combine reagents on ice in the following order, in a reaction volume of 50 μl.

4 μl	∼2 μg DNA template in RNase-free H_2O
20 μl	2.5 × reaction mix (without amino acids and NTPs)
15 μl	S-30 extract
2 μl	5 mM 18 amino acids (-Met, -Trp)
2 μl	5 mM NTPs
2 μl	[^{35}S]-methionine (10 mCi/ml)
2 μl	Trp (vary the concentration of the stock solution to change the level of tryptophan added in the reaction)
3 μl	RNase-free H_2O

Mix the components briefly and incubate the reaction mixture at 37° for 30 min. Using this protocol, the NTPs or amino acids can be readily substituted by their radiolabeled forms. Optional additions are inhibitors of proteinases and/or RNases.

Construction of Circular DNA Templates Suitable for
in Vitro Analysis

Because of the presence of exonucleases in most S-30 preparations, a linear DNA template would not be stable, thus it could not be used. Intact plasmid DNA purified by banding twice in a cesium chloride–ethidium bromide gradient is a very efficient template to direct protein synthesis, provided transcription initiation is limited to a single promoter. The simplest source of template, however, is a circularized polymerase chain reaction (PCR) fragment bearing only the promoter and genetic region of interest. For tna operon studies, to facilitate clean transcription and translation,

[25] S. A. Lesley, in "Methods in Molecular Biology: in Vitro Transcription and Translation Protocols" (M. J. Tymms, Ed.), Vol. 37, p. 265. Humana Press, Inc., Totowa, N.J., 1995.

PCR was performed to amplify a DNA fragment bearing only the intact *tna* promoter, the *tna* region to bp +306 (relative to *tna* transcription initiation site), followed by an inserted *rpoBC* transcription terminator sequence. *Eco*RI sites were introduced at both the 5' and 3' ends of this fragment. The fragment was digested with *Eco*RI and then circularized using T4 DNA ligase, resulting in CF-tna+306rpoBC"t". Typically, ~2 μg CF-tna+306rpoBC"t" was used in each 50-μl S-30 reaction. The only transcripts expected from this template would be those initiated at the *tna* promoter.

Measurement of *in Vitro* Rho-Dependent Termination and its Prevention

In CF-tna+306rpoBC"t" template-directed S-30 preparations, the transcripts produced that escape Rho-dependent termination should be terminated at the *rpoBC* terminator, yielding a ~430-nucleotide read-through transcript. To measure Rho-dependent transcription termination when it occurs in the leader region of the *tna* operon, $[\alpha$-^{33}P]-UTP was used to label newly synthesized transcripts and transcription was analyzed closely by using the single-round transcription approach.[26] Using this procedure, the reaction mixture (100 μl volume) was incubated for 10 min at 37° without added CTP and UTP. Then a prewarmed mixture of 40 μCi of $[\alpha$-^{33}P]-UTP, 200 μm CTP, and 200 μg/ml rifampicin (an inhibitor of transcription initiation) was added. Samples were removed at various intervals therafter, extracted with phenol, and analyzed by electrophoresis on a 6% polyacrylamide–7 M urea gel. Results of a typical experiment are presented in Fig. 1.

Bicyclomycin is an antibiotic that specifically inhibits Rho action.[27] Rho-dependent transcription termination in the leader region of the *tna* operon can be readily demonstrated by comparing the products of two reactions, one with bicyclomycin addition (50 μg/ml), and the second without this antibiotic Fig. 1.

Rho-dependent transcription termination can also be prevented by the presence of inducing levels of tryptophan; for example, the mRNA banding pattern closely resembles the pattern observed in the presence of bicyclomycin (Fig. 1). One difference is observed in the presence of added tryptophan: a ~120-nucleotide RNA doublet band appears near the bottom of the gel. This doublet is presumably produced by RNase degradation of the ribosome-bearing read-through transcript. By testing templates analogous to CF-tna+306rpoBC"t" that bear appropriate *tnaC* mutations, the crucial role of translation of *tnaC* in tryptophan induction was demonstrated (data not shown).

---

[26] M. E. Winkler and C. Yanofsky, *Biochemistry* **20**, 3738 (1981).
[27] A. Zwiefka, H. Kohn, and W. R. Widger, *Biochemistry* **32**, 3564 (1993).

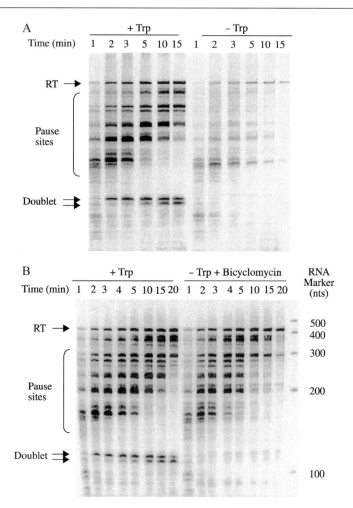

FIG. 1. Single-round transcription analyses examining the effects of added tryptophan and bicyclomycin on Rho-dependent transcription termination. (A) Single-round transcription analyses (coupled with translation) were performed in an S-30 system. A circularized fragment CF-tna+306rpoBC"t" bearing only the *tna* promoter and its leader region was used as template. S-30 reactions (100 μl) without CTP or UTP, in the absence or presence of tryptophan, were incubated at 37° for 10 min, then 50 μCi of [α-^{33}P]-UTP, 200 μm CTP, and 200 μg/ml rifampicin were added together to the reaction mixture. Samples (10 μl) were taken at indicated time points, the reaction stopped by phenol extraction, and the product loaded onto an RNA gel. The read-through transcript (RT) and the RNA doublet observed only in the presence of tryptophan are marked by arrows. (B) The effect of addition of the Rho inhibitor bicyclomycin to a reaction mixture incubated in the absence of tryptophan. Single-round transcription assays (coupled to translation) were performed as described in (A). Control reactions (+Trp) are shown at left. Reproduced from Ref. 18, with permission of American Society for Biochemistry and Molecular Biology.

Measuring Ribosome Stalling at the *tnaC* Stop Codon and
Peptidyl-tRNA Accumulation

CF-tna+306rpoBC"t"-directed cell-free transcription and translation
(with 20 μCi [^{35}S]-methionine, labeling for 30 min) reactions were stopped
by acetone precipitation, centrifuged, and the pellets boiled in $1\times$ SDS gel
loading buffer for 3 min. Samples were separated electrophoretically by
10% Tricine–SDS–PAGE.[22] In the presence of tryptophan, the labeled
TnaC peptide appeared as two species, free and as a peptidyl-tRNA. The
identity of TnaC was confirmed by selective labeling with specific radioactive amino acids (Fig. 2). The 25-kDa molecule band was identified as
TnaC–peptidyl-tRNA^{Pro} by RNase digestion, proteinase K treatment,
and RT–PCR using the following protocol.

The 25-kDa molecule band was localized by autoradiography, excised
from the gel, and recovered by immersion in a low pH buffer (20 mM
Tris–HCl, pH 6.8, 50 mM NaCl) overnight at $4°$. For RNase digestion
and proteinase K treatment, the recovered 25-kDa molecules were incubated for 10 min at $37°$ with 10 μg of RNase A per ml or 50 μg proteinase
K per ml, mixed with an equal volume of $2\times$ SDS gel loading buffer, and
separated on a 10% Tricine–SDS protein gel (Fig. 3).

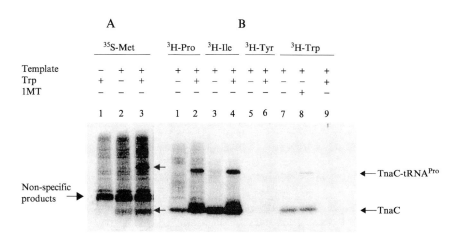

FIG. 2. Detection of TnaC peptide synthesis. CF-tna+306rpoBC"t" was used as template.
(A) [^{35}S]Methionine-labeled products produced in the presence and absence of added
tryptophan. A control sample, without template, is shown in lane 1. Arrows mark the
positions of TnaC, TnaC–tRNA^{Pro}, and the nonspecific products. (B) Selective ^{3}H labeling in
the presence or absence of added tryptophan to confirm the identity of TnaC. In the S-30
system the TnaC peptide is labile. Reproduced from Ref. 18, with permission of American
Society for Biochemistry and Molecular Biology.

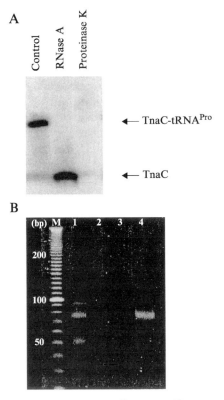

Fig. 3. Identification of TnaC–peptidyl-tRNA^{Pro}. (A) A [^{35}S]methionine-labeled ∼25-kDa band was treated with RNase A or proteinase K. RNase A digestion shifts the ∼25-kDa band to the TnaC band position. (B) RT–PCR identification of tRNA^{Pro}. *E. coli* total tRNA (Sigma) was used as the control RT–PCR template (positive control, lane 1). The 25-kDa TnaC–tRNA band was used as a RT–PCR template (lane 4). An S-30 reaction performed without the CF-tna+306rpoBC"t" template was also loaded onto the same gel, and the corresponding band was recovered as a control and used for RT–PCR (lane 2). A reaction mixture similar to the one used for lane 4, but without RT, was loaded in lane 3. A 10-bp DNA ladder (lane M, Life Technologies, Inc., Carlsbad, CA) is shown. Reproduced from Gong, F. and Yanofsky, C. (2001), *J. Biol. Chem.* **276,** 1974, with permission of American Society for Biochemistry and Molecular Biology.

Identification of the tRNA of the Peptidyl-tRNA in the Stalled Ribosome

The recovered 25-kDa molecules were used as RT–PCR templates with an ACCESS RT–PCR kit (Promega, Madison, WI). Two oligos specific for tRNA$_2$^{Pro} (tRNA A^{Pro}-Plus: 5′-CGGCACGTAGCGCAGCCTGGTAGC-3′; tRNA^{Pro}-Minus: 5′-TGGTCGGCACGAGAGGATTTGAAC-3′) were used. An S-30 reaction without the CF-tna+306rpoBC"t" DNA template was also loaded on the same gel, and the corresponding band was recovered as a negative control for RT–PCR. An *E. coli* total tRNA mixture was used as the positive control for RT–PCR. A control reaction without RT was also performed. The results of using this procedure are shown in Fig. 3.

Other Features and Events That Influence Rho-Dependent Termination

The concentration of DNA template used in the S-30 system should be optimized. We found that addition of too high a concentration of DNA template resulted in very high read-through expression of the *tna* operon. Because addition of purified Rho lowered this high basal level expression, it appears that Rho becomes limiting in our cell-free system. When a high concentration of *tna* transcript was produced *in vitro*, the *rut* binding site probably sequestered much of the available Rho. Another parameter that is crucial is the concentrations of NTPs in the S-30 reaction mixture. We found that the presence of high concentrations of the NTPs in an S-30 reaction mixture could result in inefficient Rho-dependent termination, probably because pausing of the elongation complex at strategic positions was too short to facilitate the interaction of RNA polymerase with Rho factor.

Comments and Variations

Various *E. coli* strains have been described as sources of S-30 preparations[25]; S-30s from some strains offer the advantage that they will permit the use of linear DNA templates. The *E. coli* strain we used for our S-30 preparations was selected because we wanted to avoid having background levels of β-galactosidase or the TrpE protein. In our studies it was crucial to optimize the Mg^{2+} and K^+ concentrations and the amount of S-30 extract in the reaction mixture, because it was necessary to use a highly active S-30 preparation to demonstrate the regulatory features of the *tna* operon. We used a plasmid bearing the β-galactosidase coding sequence under the control of the *tna* promoter to optimize performance. The amount of Rho protein in our S-30 preparations was normally sufficient, but, as mentioned, the DNA template concentration had to be optimized to avoid Rho-titration. The single-round transcription approach was crucial to analysis of Rho-dependent termination in the *tna* operon leader region. It is impractical to try to synchronize transcription initiation in the S-30 system by omitting one or two of the NTPs, because S-30 preparations contain low levels of the NTPs. Information about Rho's structure and mechanism of action can be found in a recent review article.[7]

[30] Role of RNA Structure in Transcription Attenuation in *Bacillus subtilis*: The *trpEDCFBA* Operon as a Model System

By Paul Babitzke, Janell Schaak, Alexander V. Yakhnin, and Philip C. Bevilacqua

Transcription attenuation mechanisms modulate the extent of transcription read-through past termination signals, thereby regulating expression of the downstream genes. In transcription attenuation the nascent transcript generally folds into one of two overlapping RNA secondary structures termed the antiterminator and terminator. Because the sequences that comprise the antiterminator structure precede those involved in formation of the terminator hairpin, the default situation is transcription read-through. Efficient termination requires the action of a regulatory molecule that prevents the antiterminator structure from forming. RNA polymerase pausing also participates in several attenuation mechanisms.[1] We describe methods to examine the role that RNA structure plays in transcription attenuation, using our studies of the *Bacillus subtilis* *trpEDCFBA* operon as a model system.

The untranslated *trp* leader transcript can form four RNA secondary structures that participate in the attenuation mechanism: the 5′ stem-loop, the pause hairpin, the antiterminator, and the terminator.[2] When activated by tryptophan, the *trp* RNA-binding attenuation protein (TRAP) binds to 11 (G/U)AG repeats that overlap the 5′ portion of the antiterminator and prevents formation of this structure. As a consequence, formation of the overlapping terminator promotes transcription termination in the leader region. In the absence of TRAP binding, the antiterminator forms and transcription continues into the *trp* genes. In addition, TRAP interaction with the 5′ stem-loop increases its affinity for *trp* leader RNA (Fig. 1).[2] Finally, a pause hairpin promotes RNA polymerase pausing, which in turn provides additional time for TRAP binding.[3]

[1] P. Gollnick and P. Babitzke, *Biochim. Biophys. Acta* **1577,** 240 (2002).

[2] P. Babitzke and P. Gollnick, *J. Bacteriol.* **183,** 5795 (2001).

[3] A. V. Yakhnin and P. Babitzke, *Proc. Natl. Acad. Sci. USA* **99,** 11067 (2002).

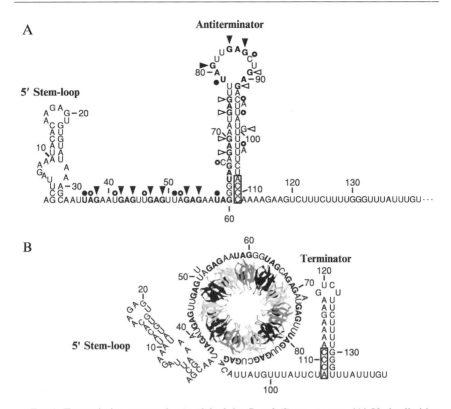

FIG. 1. Transcription attenuation model of the *B. subtilis trp* operon. (A) Under limiting tryptophan conditions, antiterminator formation prevents formation of the overlapping terminator, resulting in transcription read-through into the *trp* structural genes. Positions of cleavage from secondary structure mapping experiments are superimposed on the secondary structure. RNase Tl cleavage sites are marked by triangles, while Pb^{2+} cleavage sites are represented by circles. Filled symbols are sites of strong cleavage, while open symbols are sites of moderate or weak cleavage. (B) Under excess tryptophan conditions, TRAP is activated. TRAP binding prevents formation of the antiterminator, thereby allowing formation of the terminator, resulting in termination of transcription in the *trp* leader region. NusA-stimulated RNA polymerase pausing at U107 allows additional time for TRAP to bind. The (G/U)AG repeats are indicated in bold. The mutually exclusive antiterminator and terminator structures overlap by four nucleotides (boxed).

Prediction of RNA Secondary Structure

RNA secondary structure can be predicted by comparative sequence analysis if reliable secondary structures already exist for several homologues, or it can be predicted by folding programs such as mfold, which rely on free energy minimization.[4,5] These programs determine thermodynamically stable structures using experimentally determined nearest-neighbor thermodynamic parameters.[5] Results from chemical and enzymatic structure mapping experiments can be entered into mfold to further constrain the folding. Addition of only a few constraints greatly improves the success of predictions.[5] Directions for entering constraints are described in detail on the mfold web site (http://www.bioinfo.rpi.edu/applications/mfold/old/rna/) and will not be discussed here.

Biochemical Determination of RNA Secondary Structure

Enzymatic and Chemical Probes

Reagents that cleave the RNA backbone are particularly useful for RNA structure mapping because 5′ end-labeled RNA can be used to detect the site of cleavage directly. Commonly used single-strand-specific nucleases used to probe RNA structure include RNase T1 (G residues), RNase T2 (A $>$C, U, or G), RNase U2 (A $>$G $\geq$C $>$U), and RNase A (C and U). In addition, RNase V1 cleaves double-stranded and stacked residues. Chemicals can also be used to probe RNA secondary structure. Divalent and trivalent cations, as well as metal complexes that cleave single- and double-stranded residues, are also available. For example, Pb^{2+} cleaves single-stranded residues and in our hands has a preference for pyrimidines. Because of the small size and high reactivity of this reagent, only residues that are strongly cleaved in the presence of Pb^{2+} are considered to be single-stranded.

Additional chemicals are known that modify the bases on their Watson–Crick or Hoogsteen faces. Dimethylsulfate (DMS), 1-cyclohexyl-3-(2-morpholinoethyl) carbodiimide metho-p-toluene sulfonate (CMCT), and β-ethoxy-ketobutyraldehyde (kethoxal) modify the Watson–Crick face, whereas diethylpyrocarbonate (DEPC) can be used to probe the Hoogsteen face. The modifications to the basepairing face can be detected directly by reverse transcription using an end-labeled DNA primer, whereas those to the Hoogsteen face require aniline cleavage and ethanol precipitation before reverse transcription. We focus on the use of RNase T1 and Pb^{2+} in the following section. Further details of RNA structure mapping probes are available elsewhere.[6]

The concentration of each reagent that results in less than one cleavage or chemical modification event per transcript must be determined to

[4] M. Zuker, *Nucleic Acids Res.* **31,** 1 (2003).
[5] D. H. Mathews, J. Sabina, M. Zuker, and D. H. Turner, *J. Mol. Biol.* **288,** 911 (1999).
[6] C. Brunel and P. Romby, *Methods Enzymol.* **318,** 3 (2000).

minimize secondary cleavage or modification of refolded RNA. This is best determined by carrying out a concentration or time titration. The conditions reported here are those determined for *B. subtilis trp* leader RNA containing nucleotides +1 to +112.

Preparation of 5' End-Labeled RNA

All solutions used for *in vitro* RNA manipulation that do not contain primary amines should be treated with DEPC (0.1%, v/v) to remove RNases before autoclaving. RNA generated by *in vitro* run-off transcription using the MEGAScript kit (Ambion, Austin, TX) is dephosphorylated with calf intestinal alkaline phosphatase (CIP). Each 200-μl reaction contains 2 μm RNA and 20 U of CIP and is incubated at 37° for 1 h. The sample is then extracted once with phenol and twice with chloroform, percipitated with ethanol, and resuspended in 16 μl of TE (10 mM Tris–HCl, pH 8.0, 1 mM EDTA, pH 8.0) to give a final concentration of $\approx$ 25 μm. The RNA is subsequently 5' end-labeled with polynucleotide kinase (New England Biolabs, Beverly, MA) and excess [γ-^{32}P]ATP (6000 Ci/mmol) at 37° for 1 h. The labeled transcript is mixed with an equal volume of loading buffer (20 mM EDTA, pH 8.0, 0.3 mg/ml bromophenol blue and xylene cyanol in 95% formamide) and gel purified. The RNA is eluted from a gel slice by shaking overnight at 37° in buffer containing 10 mM Tris–HCl, pH 8.0, 1 mM EDTA, pH 8.0, 250 mM NaCl, and 0.2% sodium dodecyl sulfate. The RNA is then extracted once with phenol and three times with chloroform, concentrated by ethanol precipitation, and resuspended in TE. The RNA concentration is determined by scintillation counting or spectrophotometrically. The RNA is denatured at 90° for 1 min in TE and allowed to cool to room temperature for 10 min to obtain a conformationally homogenous population of RNA. To allow for proper folding under physiologic salt conditions, the RNA used for structure mapping is then renatured in TK buffer (40 mM Tris–HCl, pH 8.0, 100 mM KCl) at 50° for 10 min, and allowed to cool to room temperature for 10 min min before further manipulation.

Enzymatic and Chemical Probing

Aerosol-resistant tips should be used for structure mapping experiments. Each 5-μl reaction contains 4 nm 5' end-labeled RNA in TK buffer. Samples are incubated at 37° for 10 min before the addition of the mapping agent. RNase T1 (Life Technologies, Rockville, MD) is serially diluted in water to a concentration of 0.05 U/μl while lead (II) acetate (Sigma, St. Louis, MO) is dissolved in water, filter sterilized, and serially diluted to a concentration of 5 mM. One μl of each reagent is added, and the reactions are carried out at 37° for 10 min. These concentrations should be optimized for each transcript. RNase T1 reactions are stopped by adding an equal volume of 0.1 × TBE loading buffer (0.1 × TBE, 20 mM EDTA, pH 8.0, 0.1% xylene cyanol and 0.025% bromophenol blue in 95% formamide) and placed immediately on dry ice. 1 × TBE is 100 mM Tris, 83 mM boric acid, and 0.1 mM EDTA. Pb^{2+} reactions are stopped by adding 2 vol of 0.1 × TBE loading buffer and placed on ice. Control reactions containing all components except the mapping reagent are carried out to detect any background cleavage.

For the generation of RNA ladders, 5′ end-labeled RNA is cleaved using RNase T1 under denaturing conditions and by using general base hydrolysis.[7] To generate a G ladder, 3 μl of 10 nm RNA is added to 7 μl of denaturing buffer (10 M urea, 30 mM sodium citrate, pH 3.5, 1.5 mM EDTA, and 0.04% bromophenol blue and xylene cyanol). RNase T1 is serially diluted in water to 1 U/μl, and 1 μl is added to the reaction mixture. Reaction mixtures are incubated at 50° for 15 min and then placed on dry ice. To generate a limited base hydrolysis ladder, equal volumes of 10 nm 5′ end-labeled RNA and 2 × hydrolysis buffer (100 mM sodium carbonate buffer, pH 9.0, 2 mM EDTA, pH 8.0) are mixed together and incubated at 90° for 3 min. Reactions are stopped by adding an equal volume of 0.1 × TBE loading buffer and placed on ice.

RNA samples are fractionated through a 6% polyacrylamide (29:1 acrylamide:bisacrylamide)/8.3 M urea sequencing gel in 1 × TBE. Gels are prerun for at least 1 h before loading and are run at 100 W to obtain sharp bands. Samples from each structure mapping reaction, as well as the RNase T1 and base hydrolysis ladders, are quantified by scintillation counting so that an equal number of counts can be loaded per lane ($\geq$10000 cpm). Gels are transferred to 3 MM Whatman paper (Whatman, Clifton, NJ) and dried, and radioactive bands are detected by phosphorimagery.

An example of RNase T1 and Pb^{2+} cleavage of the *B. subtilis trp* leader RNA containing nucleotides +1 to +112 is shown in Fig. 2. To determine the intensity of cleavage, each structure mapping lane, as well as the control lane, is quantified using ImageQuant software (Amersham, Piscataway, NJ). Background cleavage of the control lane is subtracted from the structure mapping lanes. The amount of cleavage in each lane is normalized by dividing through by the most-cleaved position. Strong RNase T1 cleavage is observed for G residues between +33 and +59, as well as G81, G84, and G86, demonstrating that these residues are single-stranded. Moderate to weak cleavage of the other G residues between +60 and +100

[7] H. Donis-Keller, A. M. Maxam, and W. Gilbert, *Nucleic Acids Res.* **4,** 2527 (1977).

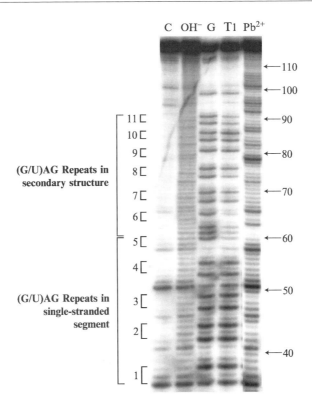

FIG. 2. Secondary structure mapping of the antiterminator from the *B. subtilis trp* leader. 5′-³²P-labeled *trp* leader RNA containing nucleotides +1 to +112 was subjected to partial RNase T1 digestion and Pb^{2+} cleavage to probe the secondary structure. C, Control without RNase T1 or lead; OH⁻, alkaline hydrolysis ladder; G, RNase T1 ladder; T1, RNase T1 mapping; Pb^{2+}, lead (II) acetate mapping. Important secondary structural features are given on the left side of the gel, including the number of triplet repeats.

is consistent with these residues being involved in secondary structure. Pb^{2+} cleaved four residues strongly (U36, U51, U58, and U79), indicating that these residues are single-stranded. Moderate to weak cleavage of other residues between +33 and +59, as well as C66, C87, C93, U96, and U102, is consistent with these residues being single-stranded at least a fraction of the time. Three of these residues are across from weakly cleaved G residues, indicating that some breathing occurs in this secondary structure, or that this structure is in equilibrium with another minor conformation (not shown). Residues that were strongly cleaved by RNase T1 or Pb^{2+} were constrained as single-stranded in the mfold prediction. The resulting mfold structure is pictured in Fig. 1A with the sites of cleavage superimposed onto the structure. The experimental constraints resulted in a substantially different secondary structure than the one predicted without constraints.

In vivo Demonstration That RNA Structure Participates in Attenuation

Once formation of an RNA secondary structure is verified biochemically, it is necessary to establish that the structure participates in transcription attenuation. This can be accomplished through the use of genetic and biochemical approaches. The investigator should be aware that it can be difficult to demonstrate that a secondary structure participates in an attenuation mechanism *in vivo*. Because the antiterminator and terminator structures overlap one another, care must be taken so that mutations only disrupt the intended structure and that the mutant transcript does not form a new structure.

We carried out a mutational analysis of the 5' stem-loop, which contains a 3-bp lower stem, a 5×2 asymmetric internal loop, a 6-bp upper stem, and a hexaloop (Fig. 1 and Table I).[8] A polymerase chain reaction (PCR)-based method is useful for deleting DNA sequences corresponding to any RNA structure. Primers are designed such that the 5' nucleotides of the primer pair border the region to be deleted, which in our case corresponded to $+2$ and $+33$ of the *trp* leader region. The use of VENT (New England Biolabs) or Pfu (Stratagene, LaJolla, CA) polymerase is

TABLE I
EFFECT OF 5' STEM-LOOP MUTATIONS ON *TRP* OPERON EXPRESSION

5' Stem-loop Mutation	β-Gal activity		β-Gal ratio ($-$Trp/$+$Trp)
	$+$Trp	$-$Trp	
Wild type	0.2 ± 0.1	28 ± 4	140
Δ5'S-L	44 ± 2	371 ± 49	8.4
U5A	8 ± 3	178 ± 37	22
G7A	31 ± 6	350 ± 19	11
A19U	0.9 ± 0.2	70 ± 9	78
C15G	111 ± 30	340 ± 2	3.1
G22C	100 ± 16	376 ± 9	3.8
C15G/G22C	2.0 ± 0.7	237 ± 9.3	119

Modified from S. Sudershana, H. Du, M. Mahalanabis, and P. Babitzke, *J. Bacteriol.* **181,** 5742 (1999).

[8] H. Du, A. V. Yakhnin, S. Dharmaraj, and P. Babitzke, *J. Bacteriol.* **182,** 1819 (2000).

recommended because both of these enzymes have low error rates and do not add nontemplated nucleotides to the 3′ end. The resulting blunt-ended, linear PCR product containing the entire plasmid with the exception of the deleted region is gel purified, phosphorylated with T4 polynucleotide kinase, and subsequently self-ligated and transformed into *Escherichia coli*. Site-directed mutations predicted to only disrupt the primary nucleotide sequence, destabilize the secondary structure, or restore the secondary structure by engineering compensatory changes are introduced using the Quick Change mutagenesis kit (Stratagene). DNA fragments containing each *trp* promoter and wild-type or mutant leader region are then separately subcloned into the ptrpBG1-PLK polylinker, resulting in the generation of *trpE′–′lacZ* translational fusions. ptrpBG1-PLK is designed to allow integration of the *trp* promoter and leader region into the *amyE* locus of the *B. subtilis* chromosome.[9] Recombinant plasmids are linearized with *Pst*I, and integration of each gene fusion is accomplished by linear transformation of naturally competent *B. subtilis*.

To examine the effect of each mutation on *trp* operon expression, cells are grown in minimal medium in the absence or presence of 200 μm L-tryptophan to late exponential phase and subsequently assayed for β-galactosidase activity.[10] The effect of exogenous tryptophan on expression is assessed from the ratio of expression in the absence of tryptophan to expression in the presence of tryptophan ($-$Trp/$+$Trp ratio). Deletion or disruption of the 5′ stem-loop results in higher expression levels and the reduced ability to regulate expression in response to tryptophan (Table I). The compensatory changes predicted to restore the secondary structure of the 5′ stem-loop resulted in partial restoration of wild-type expression levels (Table I).

In Vitro Demonstration That RNA Structure Participates in Attenuation

Two different *in vitro* transcription assays are generally used to determine if an RNA structure participates in transcription attenuation. Multi-round transcription assays have the advantage of their relative simplicity, whereas synchronized single-round transcription assays are more difficult to perform but have the advantage of allowing a more detailed experimental analysis. In both cases, a source of *B. subtilis* RNA polymerase and other regulatory factors is required. The majority of our transcription experiments have been carried out with purified vegetative (σ^A) RNA

---

[9] E. Merino, P. Babitzke, and C. Yanofsky, *J. Bacteriol.* **177,** 6362 (1995).
[10] S. Sudershana, H. Du, M. Mahalanabis, and P. Babitzke, *J. Bacteriol.* **181,** 5742 (1999).

polymerase that we obtained from Michael Chamberlin. A His-tagged version of RNA polymerase[11] gives similar results. We also used purified TRAP[12] and His-tagged NusA in our studies. DNA restriction fragments from plasmids or PCR-generated fragments are most commonly used for templates. Templates between 250 and 750 bp have been used routinely in our laboratory.

Multi-round in vitro *Transcription Assay*

Templates for transcription are prepared by restriction digestion of plasmid DNA containing the desired promoter and leader region. DNA fragments are gel purified, extracted with an equal volume of phenol/chloroform, and ethanol precipitated. DNA pellets are suspended in TE and stored at $-20°$. Reaction mixtures contain 2 μl of 5 ×RNA polymerase buffer (200 mM Tris–HCl, pH 8.0, 600 mM KCl, 20 mM MgCl$_2$, 50 mM β-mercaptoethanol, 20 mM spermidine, and 12.5% glycerol), 1 μl of a 10 × NTP mix (2.7 mM ATP, 0.7 mM CTP, 1.1 mM GTP, and 1.4 mM UTP), 0.6 μl [α-^{32}P]UTP (3000 Ci/mmol), 1 μl of 10 × DNA template (200 nm), 0.8 μl (1 μg) RNA polymerase diluted in 50 mM Tris–HCl, pH 8.0, 1 mM DTT, and water to a final 10-μl reaction volume. RNA polymerase buffer, NTPs, DNA template, and water for all reactions are mixed together and aliquoted into separate tubes. Reactions are initiated by adding RNA polymerase. Transcription reactions are carried out for 30 min at 30° and stopped by adding an equal volume of loading buffer (0.37% EDTA, pH 8.0, 0.3% bromophenol blue and xylene cyanol in 95% formamide). Reaction mixtures are heated to 90° for 3 min to denature the RNA and then placed on ice. RNA samples are fractionated by electrophoresing through a 6% polyacrylamide gel containing 7 M urea in 1 × TBE buffer that has been prerun for at least 10 min. Gels are transferred to 3 MM Whatman paper and dried, and radioactive bands are visualized by phosphorimagery. Bands are quantified using ImageQuant software (Amersham). This assay was modified to include TRAP (0.3 μm) and L-tryptophan (1 mM) to promote termination.[13] In these instances, TRAP and L-tryptophan were added before RNA polymerase (Fig. 3).

DNA oligonucleotide competition experiments are used to demonstrate that an RNA structure functions as predicted. For example, oligonucleotides complementary to the top (nucleotides 70–84, oligo A) or the base (nucleotides 55–69, oligo B) of the antiterminator were used to demonstrate that this structure prevents formation of the terminator.

---

[11] Y. Qi and M. Hulett, *Mol. Microbiol.* **28,** 1187 (1998).
[12] A. V. Yakhnin, J. J. Trimble, C. R. Chiaro, and P. Babitzke, *J. Biol. Chem.* **275,** 4519 (2000).
[13] P. Babitzke and C. Yanofsky, *Proc. Natl. Acad. Sci. USA* **90,** 133 (1993).

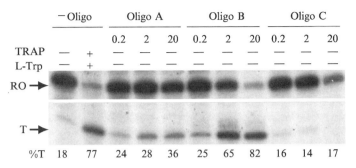

Fig. 3. Multi-round transcription attenuation assay demonstrating that the antiterminator structure functions *in vitro*. Competitor oligonucleotides were added at the indicated concentration (μm). −, No TRAP or no L-tryptophan added; +, TRAP (0.3 μm) or L-tryptophan (1 mM) was present. Positions of run-off (RO) and terminated (T) transcripts are shown. The molar percent of terminated transcripts is shown at the bottom of each lane. Only fragments of the gel containing bands of interest are shown. Modified from P. Babitzke and C. Yanofsky, *Proc. Natl. Acad. Sci. USA* **90**, 133 (1993).

Oligos are added before the addition of RNA polymerase (Fig. 3). Disruption of the base of the antiterminator structure by oligo B, which is the RNA segment that overlaps the terminator hairpin, was capable of replacing the requirement of tryptophan-activated TRAP for promoting termination.[13]

Single-round in vitro *Transcription Assay*

A two-step single-round *in vitro* transcription assay is used to study RNA polymerase elongation. This method is particularly useful to examine RNA polymerase pausing.[14,15] In the first step, transcription is synchronized by halting RNA polymerase at an early position by withholding one NTP. In the case of the *B. subtilis trp* leader, the absence of CTP makes transcription initiation dependent on an RNA primer that contains the first three nucleotides of the *trp* transcript and halts elongation after incorporation of A12 (Fig. 1). In the second step, elongation of the halted complexes is resumed by the addition of all four NTPs together with heparin (an inhibitor of reinitiation) and L-tryptophan. TRAP and/or NusA proteins, when present, are included in the second extension step.[3]

Reagents

DNA templates for *in vitro* transcription are obtained by PCR amplification of the region of interest. We used a template containing the *B. subtilis trp* promoter and leader region (−78 to +212 relative to the start

[14] R. Landick, D. Wang, and C. L. Chan, *Methods Enzymol.* **274**, 335 (1996).
[15] R. Reynolds, R. M. Bermudes-Cruz, and M. J. Chamberlin, *J. Mol. Biol.* **224**, 31 (1992).

of transcription). PCR fragments are extracted by phenol/chloroform and precipitated with spermine.[14] NTPs (Amersham) and 3'-dNTPs (TriLink BioTechnologies, San Diego, CA) obtained as 100 mM and 10 mM stock solutions, respectively, are stored as aliquots at $-80°$. Aerosol-resistant tips are used to avoid cross-contamination of NTPs.[14] RNA polymerase, TRAP, and NusA are stored in 50% glycerol (v/v) at $-20°$. All other stock solutions are stored at $-80°$ except where noted. The following stock solutions are used in this protocol:

> 10 × DNA template (0.2 μm) in 1/2 × TE (5 mM Tris–HCl, pH 8.0, 0.5 mM EDTA, pH 8.0) is stored at $-20°$.
>
> B. subtilis RNA polymerase (3.2 μm).
>
> B. subtilis N-terminally His$_6$-tagged NusA (30 μm).
>
> B. subtilis TRAP (24 μm).
>
> 6 × transcription buffer (TB) (240 mM Tris–HCl, pH 8.0, 24 mM MgCl$_2$, 0.6 mM EDTA. 24% trehalose, and 30 mM DTT).
>
> 150 μm ApGpC trinucleotide (Dharmacon Research, Boulder, Co), corresponding to the first three nucleotides of the trp transcript.
>
> 10 × initiating NTP mix (20 μm GTP, 80 μm ATP and UTP. CTP is omitted to halt elongation complexes).
>
> 5 × extension solution in 1 × TB (0.5 mg/ml heparin, 5 mM L-tryptophan, 25 μm ATP, 750 μm each GTP, CTP, and UTP).
>
> 5 × RNA sequencing mixtures in 1 × TB (0.5 mg/ml heparin, 50 μm GTP, 750 μm each ATP, CTP, and UTP, and one of four 3'-dNTPs at the concentration equal to that of the corresponding NTP).
>
> 6 × chase solution in 1 × TB (3mM each ATP, CTP, GTP, and UTP).
>
> [α-^{32}P]GTP (3000 Ci/mmol) NEN Life Science.
>
> Stop solution (20 mM EDTA, pH 8.0, 0.2% SDS, 0.3 mg/ml bromophenol blue and xylene cyanol in 95% formamide). Store at $-20°$. Before beginning the experiment, aliquot the stop solution into microfuge tubes.

Procedure

Combine reagents on ice in the following order:

> Water to 60 μl total in the initiation/halted elongation reaction.
>
> 10 μl of 6 × TB.
>
> 6 μl of 10 × initiating NTP mix.
>
> 6 μl of 150 μm ApGpC.
>
> 6 μl of DNA template.
>
> 3 μl of RNA polymerase.
>
> 6 μl of [α-^{32}P]GTP.

Incubate at $37°$ for 10 min to form halted elongation complexes. Place the reaction on ice. The halted complex can be stored for at least for 2 h at $0°$.

Make fresh dilutions of TRAP and/or NusA in 1 × TB to a concentration of 10 μm. Combine the following reagents on ice:

3 μl of 5 × extension solution.

1.5 μl of 10 μm TRAP (if desired, otherwise 1 × TBE).

1.5 μl of 10 μm NusA (if desired, otherwise 1 × TBE).

Prewarm the extension mixture for 5 min at 25°. Start the extension reaction by adding 9 μl of the halted elongation complex. Remove 2 μl at each preselected time and add to an equal volume of stop solution. After collecting all desired samples, mix the last 4 μl aliquot with 0.8 μl of 6 × chase solution and incubate for an additional 5 min at 37° before adding the stop solution. This final chased sample will detect run-off, terminated, and arrested transcripts.

Dilute 5 × sequencing mixtures (A, C, G, and U) 2-fold with 1 × TB. RNA sequencing reactions are started by adding 1.2 μl of the halted elongation complex to 0.8 μl of the 2.5 × A, C, G, and U sequencing mixtures. Incubate for 20 min at room temperature or 5 min at 37° before adding an equal volume of stop solution. To accurately map pause sites, sequencing reactions can be stopped earlier when the pause complexes are most pronounced.

Heat the samples to 80° for 4 min before loading on a 5% polyacrylamide sequencing gel containing 8 M urea in 1 × TBE buffer that has been prerun for at least 10 min. 1 × TBE buffer is 89 mM Tris, 89 mM boric acid, 2 mM EDTA, pH 8.3. Alternatively, sequencing gels in 1 × TTE buffer allows substantially more accurate comparison of the paused or terminated transcripts with the sequencing ladder. 1 × TTE buffer is 89 mM Tris, 40 mM taurine, and 1 mM EDTA (there is no need to adjust pH). After electrophoresis, the gel is transferred to a 3 MM Whatman paper and dried, and radioactive bands are detected by phosphorimagery. The relative amount of each RNA band is determined using ImageQuant software (Amersham). Calculate the pause half-lives and efficiencies (the fraction of RNA polymerase molecules that pause at a particular site) by plotting pause band intensities versus time using KaleidaGraph (Synergy Software, Reading, PA) and Microsoft Excel (Microsoft, Redmond, WA) software as described.[14] Calculate the efficiency of termination as a fraction of a terminated transcript versus the sum of all terminated and read-through (run-off) transcripts.

An example of a single-round transcription reaction is shown in Fig. 4. NusA stimulates pausing at U107 and U144. Oligonucleotide competition experiments indicate that both of these pausing events are hairpin-dependent (not shown). The pause at U107, the nucleotide that just precedes the overlap between the antiterminator and terminator structures, provides additional time for TRAP to bind to the nascent transcript and promote

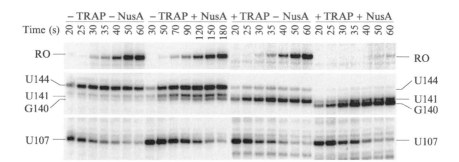

Fig. 4. Single-round transcription attenuation assay showing that TRAP interferes with NusA-stimulated U107 pausing. −, No TRAP or no NusA added; +, TRAP (1 μm) or NusA (1 μm) was present. Reactions were stopped at the times indicated above each lane. Positions of paused (U107 and U144), terminated (G140 and U141) and run-off (RO) transcripts are indicated. Only fragments of the gel containing bands of interest are shown. Modified from A. V. Yakhnin and P. Babitzke, *Proc. Natl. Acad. Sci. USA* **99,** 11067 (2002).

termination. Both TRAP and NusA stimulate termination by themselves and in combination shift the primary termination point from U141 to G140.[3]

Comments

This procedure using *B. subtilis* RNA polymerase is based on the method previously described for the single-round transcription with *E. coli* RNA polymerase.[14] In contrast to the *E. coli* enzyme, primer-dependent transcription initiation using *B. subtilis* RNA polymerase is highly sensitive to salt; concentrations as low as 20 mM NaCl or KCl dramatically inhibit initiation. Thus monovalent salts have been omitted from the transcription buffers. We also found that the *B. subtilis* enzyme efficiently initiates transcription at primer concentrations that are 10- to 100-fold lower than that routinely used with *E. coli* RNA polymerase.[3,14] In general, halted *E. coli* RNA polymerase elongation complexes must contain transcripts $\geq$10 nucleotides to be stable.[14] Similarly, *B. subtilis* RNA polymerase forms a stable halted complex with a 12-nucleotide transcript, but not with a 9-nucleotide transcript. Finally, comparison of paused or terminated transcripts with an RNA sequencing ladder generated with 3'-dNTPs is somewhat compromised if TBE-buffered sequencing gels are used. 1 × TTE buffer allows the precise localization of paused and terminated transcripts.

[31]   TRAP–RNA Interactions Involved in Regulating Transcription Attenuation of the *Bacillus subtilis trp* Operon

By PAUL GOLLNICK

In *Bacillus subtilis* and several related bacilli, transcription of the tryptophan biosynthetic (*trpEDCFBA*) operon is regulated by an attenuation mechanism involving two alternative RNA secondary structures that form in a 203-nt leader region in the nascent mRNA transcript upstream of the structural genes (Fig. 1).[1,2] In the presence of excess tryptophan, an RNA binding protein called TRAP (*trp* *R*NA-binding *A*ttenuation *P*rotein) binds to a specific target in the leader RNA and facilitates formation of a transcription terminator RNA structure, which induces RNA polymerase to halt transcription before the structural genes. Under conditions of limiting tryptophan, TRAP does not bind to the *trp* leader RNA, allowing it to form an alternative antiterminator structure that prevents the terminator from forming and thus permits transcription to continue into the structural genes.

TRAP is composed of eleven identical subunits (75 amino acids each) arranged in a symmetric ring (Fig. 1).[3] The secondary structure of the protein consists mainly of 11 7-stranded antiparallel β-sheets, each of which contains 4 β-strands from one subunit and 3 β-strands from the adjacent subunit. TRAP is activated to bind RNA by binding up to 11 molecules of L-tryptophan in pockets formed between adjacent subunits.[3–5] Each bound tryptophan interacts with amino acid residues on both adjacent subunits. Tryptophan-activated TRAP binds to a target site in the *trp* leader transcript that consists of 11 GAG or UAG repeats separated by two or three nonconserved nucleotides (Fig. 1).[3,6] The crystal structure of TRAP complexed with an RNA containing 11 GAG repeats shows that the single-stranded RNA wraps entirely around the outer perimeter of the protein ring.[7] The specificity of binding is generated by interactions (mostly

[1] P. Babitzke and P. Gollnick, *J. Bacteriol.* **183,** 5795 (2001).

[2] P. Gollnick, *Mol. Micro.* **11,** 991 (1994).

[3] A. A. Antson, J. B. Otridge, A. M. Brzozowski, E. J. Dodson, G. G. Dodson, K. S. Wilson, T. M. Smith, M. Yang, T. Kurecki, and P. Gollnick, *Nature* **374,** 693 (1995).

[4] P. T. X. Li and P. Gollnick, *J. Biol. Chem.* **277,** 35567 (2002).

[5] P. Babitzke and C. Yanofsky, *J. Biol. Chem.* **270,** 12452 (1995).

[6] P. Babitzke, J. T. Stults, S. J. Shire, and C. Yanofsky, *J. Biol. Chem.* **269,** 16597 (1994).

[7] A. A. Antson, E. J. Dodson, G. G. Dodson, R. B. Greaves, X.-P. Chen, and P. Gollnick, *Nature* **401,** 235 (1999).

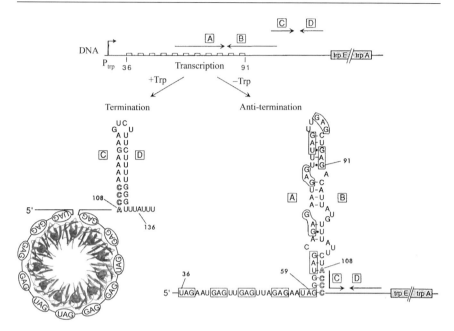

Fig. 1. Model of transcription attenuation of the *B. subtilis trp* operon. Numbers indicate the residue positions relative to the start of transcription. The large boxed letters designate the complementary strands of the terminator (C and D) and antiterminator (A and B) RNA structures. Nucleotides 108–111 overlap between the antiterminator and terminator structures and are shown as outlined letters. The TRAP protein is shown as a ribbon diagram with the 11 subunits in different colors, and the bound RNA is shown encircling TRAP, as seen in the crystal structure. The GAG and UAG repeats involved in TRAP binding are shown in ovals surrounding the protein and are also outlined in the sequence of the antiterminator structure. (See color insert.)

hydrogen bonds) between the bases of the RNA and amino acid residues on the protein. This structure suggests a model for the mechanism by which TRAP binding to *trp* leader RNA serves to alter its secondary structure and thus control attenuation of the *trp* operon (Fig. 1). This article describes the methods that have been used to characterize the interaction of TRAP with RNA and to study how these interactions are involved in regulating transcription attenuation of the *trp* operon.

Expression and Purification of TRAP

TRAP is encoded by *mtrB*, the second gene of the *mtrAB* operon.[8] *MtrA* encodes GTP cyclohydrolase I, an enzyme involved in folic acid biosynthesis.[9] Both *B. subtilis*[9–12] and *Bacillus stearothermophilus*[13] TRAP have been expressed in *Escherichia coli* using the T7 expression system first developed by Tabor and Richardson.[14] The *mtrB* gene was cloned into plasmids that contain a T7 promoter, such as pTZ18 (USB, Cleveland, OH) and pBlueScript (Stratagene, LaJolla, CA). With these plasmids, expression of TRAP relies on the natural *mtrB* ribosome binding site. Expression was done in *E. coli* strains such as SG62052/pGP1–2, which contains the gene for T7 RNA polymerase under control of the lambda P_L promoter on the plasmid pGP1–2, as well as a gene encoding a temperature-sensitive (CI857) lambda repressor to regulate P_L.[14] Cells are grown in T7 expression medium (2% tryptone, 1% yeast extract, 0.5% NaCl, 0.2% glycerol, and 50 mM potassium phosphate, pH 8.0) at 30° until they reach mid-log phase ($A_{600} \approx 0.8$). The cells are then heat shocked in a 42° water bath with gentle shaking for 15–20 min to inactivate the lambda repressor and induce expression of T7 RNA polymerase to drive transcription of the *mtrB* gene to produce TRAP. Growth is then continued for 4–6 h before the cells are harvested by centrifugation. TRAP is expressed to approximately 10% of the soluble protein in these cells.[10] Subsequent improvements to this system have included using plasmids with optimized *E. coli* ribosome binding sites such as pET9a and pET17b (Novagen, Madison, WI), as well as using the *E. coli* strain BL21(DE3). In BL21(DE3) the T7 RNA polymerase gene is under control of the *lac* promoter/repressor system, and hence expression is induced by adding 1 mM isopropyl-β-D-thiogalactopyranoside (IPTG) when the cells reach mid-log phase growth, which is more convenient than heat shock induction. Using these systems, TRAP is expressed to 20–40% of soluble protein.

Large amounts of *B. subtilis* TRAP can easily be purified in 1 day using ammonium sulfate precipitation, followed by hydrophobic chromatography on phenyl agarose.[10] Typically it is convenient to grow 4×1 L of cells as described earlier, collect the cells by centrifugation, and freeze the pellets at −20°. The frozen cells are later resuspended in lysis buffer (100 mM K_2HPO_4, 50 mM KCL, 1 mM EDTA, 160 μg/ml PMSF, and 1 mM benzamidine) on ice using approximately 10 ml buffer per L of cell

[8] P. Gollnick, S. Ishino, M. I. Kuroda, D. Henner, and C. Yanofsky, *Proc. Natl. Acad. Sci. USA* **87,** 8726 (1990).

[9] P. Babitzke, P. Gollnick, and C. Yanofsky, *J. Bacteriol.* **174,** 2059 (1992).

[10] A. A. Antson, A. M. Brozozowski, E. J. Dodson, Z. Dauter, K. S. Wilson, T. Kurecki, and P. Gollnick, *J. Mol. Biol.* **244,** 1 (1994).

[11] J. Otridge and P. Gollnick, *Proc. Natl. Acad. Sci. USA* **90,** 128 (1993).

[12] P. Babitzke and C. Yanofsky, *Proc. Natl. Acad. Sci. USA* **90,** 133 (1993).

[13] X. -P. Chen, A. A. Antson, M. Yang, C. Baumann, P. Li, E. J. Dodson, G. G. Dodson, and P. Gollnick, *J. Mol. Biol.* **289,** 1003 (1999).

[14] S. Tabor and C. C. Richardson, *Proc. Natl. Acad. Sci. USA* **82,** 1074 (1985).

culture and are then broken by two passages through a French pressure cell (1200 lb/in^2). The lysate is cleared by centrifugation at 30,000 × g for 20 min at 5°, and then an equal volume of saturated (NH$_4$)$_2$SO$_4$ is added slowly on ice, with mixing, to yield a final value of 50% saturation. After mixing for 20 min on ice, the precipitate is removed by centrifugation at 30,0000 × g for 20 min and the supernatant is applied directly onto a 2.5 cm × 10 cm column of phenyl agarose (Sigma, St. Louis, MO) at 5°, equilibrated with 25 mM Tris–HCl, pH 8.5, 50% saturated with (NH$_4$)$_2$SO$_4$. The column is then washed successively with 25 mM Tris–HCl, pH 8.5, buffer that is 50% saturated with (NH$_4$)$_2$SO$_4$, followed by the same buffer 15% saturated with (NH$_4$)$_2$SO$_4$, and then with the buffer alone. Each time washing is done until the A$_{280}$ of the column eluate is below 0.1. TRAP is then eluted from the column with 25% ethylene glycol in 25 mM Tris–HCl, pH 8.5. TRAP purified by this method is greater than 95% pure as judged by SDS–PAGE.[10] If this is not adequate, TRAP can be further purified using an Econo Pac Q (BioRad, Hercules, CA) ion exchange cartridge column. The sample is loaded on the column in 25 mM Tris–HCl, pH 8.0, and eluted using a gradient from 0 to 1.0 M NaCl in the same buffer, with TRAP eluting at approximately 0.6 M NaCl. In all cases, the pools containing TRAP are concentrated using Ultrafree-15 centrifugal filter devices (Millipore, Billerica, MA), then dialyzed against lysis buffer with 10% glycerol and frozen in aliquots at −70°. Yields are 25–50 mg of TRAP per L of cell culture.

   B. stearothermophilus TRAP does not bind to phenyl agarose, and thus an alternative method is used to purify this protein.[13] Cell growth, expression, harvesting, and breaking the cells are performed as described earlier for *B. subtilis* TRAP. After the lysate is cleared at 30,000 × g for 20 min, it is heated at 75° for 20 min. The precipitated *E. coli* proteins are then removed by centrifugation at 30,000 × g for 20 min. The supernatant is dialyzed overnight at room temperature against 50 mM Tris–HCl, pH 8.0, and then loaded on a Q-sepharose column (1.5 × 15 cm) equilibrated in the same buffer at room temperature. *B. stearothermophilus* TRAP does not bind to this column and is found in the flow-through, whereas the remaining contaminating *E. coli* proteins bind to the column. A high-salt wash (1.0 M NaCl) can be used to elute the *E. coli* proteins to regenerate the column. The TRAP-containing fractions are pooled, concentrated, dialyzed at room temperature, and stored as described earlier for *B. subtilis* TRAP.

   B. stearothermophilus TRAP purified by this method retains 4–6 molecules of bound tryptophan per TRAP 11mer, even after extensive dialysis versus lysate buffer for up to 72 h at room temperature. In cases where it is necessary to completely remove tryptophan, this can be done by denaturing the TRAP 11mer into unfolded monomers using ≥3.0 M guanidine

HCl (GdnHCl), followed by dialysis against GdnHCl. Upon removing the denaturant by dialysis against lysate buffer, TRAP spontaneously refolds into fully functional 11mers.[15]

Mutant TRAP proteins with single amino acid substitutions have been used extensively for structure/function studies.[4,13,15–17] Surprisingly, many seemingly innocuous amino acid substitutions in TRAP result in dramatic changes in the purification properties of the mutant proteins on phenyl agarose or Q-sepharose. Hence, to purify small amounts (up to 5 mg) of these proteins, we raised antibodies against both *B. subtilis* and *B. stearothermophilus* TRAP in rabbits and linked the affinity-purified antibodies to CnBr-activated Sepharose 4B to create affinity columns. Mutant TRAP proteins are expressed, the cells lysed, and the lysates cleared as described earlier. Standard purification protocols for immunoaffinity purification are followed,[18] and TRAP is eluted with a low-pH wash (0.1 M glycine, pH 2.5).

Several approaches have been used to create site-directed mutations in the gene for TRAP. One widely used technique is a two-step polymerase chain reaction (PCR) method to amplify the cloned *mtrB* gene using mutagenic oligonucleotides combined with oligonucleotide primers that bind to the plasmid vector.[16,19] PCR reactions contain: 10 mM Tris–HCl, pH 8.3, 50 mM KCl, 15 mM MgCl$_2$, 0.01% gelatin, 200 μm dNTPs, 1 ng plasmid DNA template, and 200 ng of each oligonucleotide primer. The PCR program consists of 94° for 1.5 min, followed by 25 cycles of 50° for 2 min, 60° for 5 min, and 94° for 1 min, and then finally 60° for 10 min. The DNA template contains the *mtrB* gene cloned into pBlueScript (Stratagene). In the first round of PCR, two reactions are performed. One reaction contains the mutagenic oligonucleotide combined with the Universal Reverse primer (anneals to sequences in the plasmid downstream of the *mtrB* gene). The second PCR reaction combines an oligonucleotide complementary to the 3′ end of *mtrB* with the Universal −20 primer (binds to the plasmid upstream of the *mtrB* gene insert). The two PCR products are purified by agarose gel electrophoresis using low melting temperature agarose (FMC SeaPlaque, Rockland, ME), mixed together in PCR buffer (described earlier) and denatured by boiling for 10 min. The overlapping regions between two PCR products are allowed to anneal by slowly cooling

[15] P. T. X. Li, D. J. Scott, and P. Gollnick, *J. Biol. Chem.* **277,** 11838 (2002).
[16] M. Yang, X.-P. Chen, K. Millitello, R. Hoffman, B. Fernandez, C. Baumann, and P. Gollnick, *J. Mol. Biol.* **270,** 696 (1997).
[17] A. V. Yakhnin, J. J. Trimble, C. R. Chiaro, and P. Babitzke, *J. Biol. Chem.* **275,** 4519–4524 (2000).
[18] E. Harlow and D. Lane, "Antibodies, a Laboratory Manual." Cold Spring Harbor Laboratory Press, Cold Spring Harbor, NY, 1988.
[19] E. Merino, J. Osuna, F. Bolivar, and X. Soberon, *BioTechniques* **12,** 508 (1992).

the mixture to room temperature, followed by extension with the Klenow fragment of DNA polymerase (5 U) in PCR buffer at 30° for 15 min. The hybrid DNA fragment is then used as template for another PCR reaction using the Universal −20 and Reverse primers to amplify the entire *mtrB* containing DNA fragment. The resulting PCR product, which contains 50% wild-type and 50% mutant *mtrB* sequences, is then subcloned into pBlueScript, and individual clones are sequenced using Sequenase (USB) to identify the desired mutant gene.

Recently a second method, termed the "Quick Change" mutagenesis procedure, (Stratagene) has been used to create mutations in *mtrB*. In this approach, two complementary mutagenic primers are used to amplify the entire plasmid (pBlueScript, pET9a or pET17b) containing the *mtrB* gene, followed by restriction digestion with *Nde*I to specifically eliminate the plasmid template (this enzyme does not digest the unmodified DNA resulting from PCR amplification). This method has the advantage of high efficiency (in our hands, approximately 80%), as well as not having to subclone the resulting PCR fragment after mutagenesis.

Assays of RNA Binding to TRAP

Several assays have been used to examine the interaction of TRAP with RNA. These assays use ^{32}P-labeled transcripts containing either the *trp* leader sequence or artificial TRAP binding sites. Labeled transcripts are prepared by *in vitro* transcription using T7 RNA polymerase and [α-^{32}P]UTP. Either linearized plasmids or PCR products can be used as templates. Standard conditions for *in vitro* transcription are 40 mM Tris–HCl, pH 8.0, 8 mM MgCl$_2$, 2 mM spermidine, 25 mM NaCl, 1 mM DTT, 0.5 mM GTP, ATP, and CTP, 0.025 mM UTP, 100 μCi of [α-^{32}P]UTP, 1 μg of linearized DNA template, and 50 U of T7 RNA polymerase (Invitrogen, Carlsbad, CA). Labeled RNAs are gel purified from denaturing polyacrylamide gels (1 × TBE with 7 M urea) by a crush and soak procedure. After exposing the gel to either X-ray film or a phosphorimager screen, the bands corresponding to the desired transcript are cut out of the gel with a sterile razor blade. The gel slice is crushed to a fine consistency in a 1.5-ml microcentrifuge tube using an RNase-free plastic pestle (Kontes, Vineland, NJ), then 0.4 ml of elution buffer consisting of 0.5 M NaCl, 0.1 M Tris–HCl, pH 8.0, and 20 mM EDTA is added and the sample is vigorously vortexed. The RNA is eluted for 30 min at room temperature and then extracted with phenol:chloroform, followed by chloroform. 10 μg yeast tRNA is added as a carrier, and the nucleic acid is precipitated with ethanol.

TRAP binding to *trp* leader RNA was first demonstrated using mobility shift gels (Fig. 2),[11] and this method has also recently been used to further

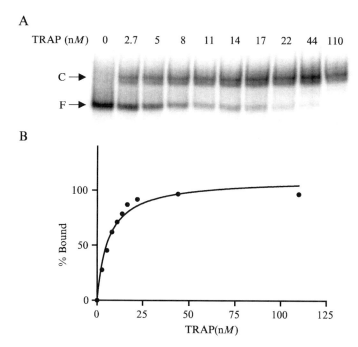

FIG. 2. (A) RNA mobility shift gel assay of TRAP binding to RNA. ^{32}P-labeled $(GAGUU)_{11}$ RNA, containing 11 tandem repeats of the sequence GAGUU, was incubated with 0 to 110 nm TRAP in the presence of 1 mM L-tryptophan and electrophoresed on a 6% native polyacrylamide gel. F indicates free (unbound) RNA, and C indicates complexed (bound) RNA. (B) Equilibrium binding curve of TRAP to $(GAGUU)_{11}$ RNA. Phosphorimager analysis was used to determine the fraction-bound (C) and unbound (F) RNA in each reaction. The curve is fit to the equation $Y = Bmax*X/(K_d + X)$ using the program Prism 3.0 (GraphPad software) to determine a K_d of 7 nM.

characterize the TRAP binding site in the leader RNA.[20] This assay has the advantage that the complexes formed between TRAP and the RNA are visualized as shifted bands in the gel, and hence it is easy to obtain information about the number of complexes formed and the specificity of the complexes. A single shifted band suggests a specific complex, whereas a smear suggests nonspecific binding. However, this assay has the disadvantages of being slow, having difficulty processing large numbers of samples, and providing a somewhat poor estimate of the binding equilibrium.[21] Standard mobility shift reactions for TRAP binding to RNA contain 12 mM HEPES

[20] H. Du, A. V. Yakhnin, D. Subramanian, and P. Babitzke, *J. Bacteriol.* **182,** 1819 (2000).
[21] D. Lane, P. Prentki, and M. Chandler, *Microbiol. Rev.* **56,** 509 (1992).

and 4 mM Tris, pH 8.0, 60 mM NaCl, 4 mM MgCl$_2$, 10 μg yeast tRNA, 1 mM tryptophan, 1 $\times$ 10^5 cpm (10 fmol) of labeled RNA, and various amounts of TRAP in a 20-μl reaction. The reactions are incubated until equilibrium is reached, usually 30 min, at the desired temperature, typically 37°, and then glycerol is added to 10%. The samples are then loaded onto a 6% polyacrylamide gel (79:1 acrylamide: bisacrylamide) in 0.25 $\times$ TBE (25 mM Tris–borate, pH 8.3, 0.5 mM EDTA) and 10% glycerol while the gel is running at 300 V at 5°. After the samples have entered the gel ($\approx$15 min), the voltage is lowered to 150 V and the gel is run overnight. It is important to preelectrophorese the gels for 4–10 h at 150 V before loading the samples. After the gels are run, they are fixed in 10% acetic acid:10% methanol for 20 min and then dried. The radioactive bands are quantified by phosphorimage analysis (ImageQuant software; Molecular Dynamics). Analysis of the percent of the RNA shifted (in complex with TRAP) as a function of protein concentration allows determination of the affinity (K_d) of TRAP for the RNA (Fig. 2; see the following section for data analysis).

Nitrocellulose Filter Binding Assay

One of the quickest and most reliable methods to quantitatively analyze TRAP binding to RNA is the nitrocellulose filter binding assay.[4,15,22,23] This technique involves incubating TRAP and radiolabeled RNA, followed by filtration (and washing) through a nitrocellulose filter. Nitrocellulose (0.45 μm pore size; Advantec MFS, Vineland, NJ) binds strongly to proteins, but does not bind RNA. Radiolabeled RNA retained on the filter therefore indicates that the nucleic acid is bound to the protein. Filter retention of the radiolabeled RNA is examined as a function of protein concentration, and the data are used to derive a K_d for each TRAP–RNA interaction. The advantages of this method are that it is quick, it requires small amounts of radiolabeled nucleic acid, it allows processing large numbers of samples, and analysis of the data is simple (see the following). The disadvantage of this method is that it is somewhat difficult to distinguish specific from nonspecific binding.

A typical filter binding experiment involves 12 0.1-ml reactions, each containing filter binding buffer (250 mM potassium glutamate and 16 mM HEPES, pH 8.0), 1 mM L-tryptophan, 1 fmol (approximately 1 $\times$ 10^4 dpm) of ^{32}P-labeled RNA, and progressively increasing concentrations of protein. The reactions are set up in 96 well plates and incubated at

[22] C. Baumann, J. Otridge, and P. Gollnick, *J. Biol. Chem.* **271**, 12269 (1996).
[23] C. Baumann, S. Xirasagar, and P. Gollnick, *J. Biol. Chem.* **272**, 19863 (1997).

37° (or any other desired temperature) until binding has reached equilibrium. Some TRAP–RNA complexes require incubation for as long as 1 h to reach equilibrium, although in most cases 15–20 min is adequate. The first tube contains no protein and is a measure of background RNA retention on the filter. The range of protein concentrations used in the remaining reactions varies depending on the affinity of TRAP for the particular RNA. To accurately determine a binding constant, five tubes should contain protein concentrations below the K_d and six should have concentrations above the K_d, with binding saturation achieved by the tenth or eleventh reaction.

A recent improvement of the filter binding assay involves using a minifold microsample filtration manifold (Schleicher and Shuell, Keene, NH), which is a 96-well vacuum dot blot system, as well as using two filter membranes simultaneously.[24] In this system a nitrocellulose filter (BA83; Schleicher and Shuell) is placed on top of a Hybond N+ filter (Amersham-Pharmacia Biosciences, Piscataway, NJ) in the filter manifold. Assays are performed as described earlier, and then 80 μl of the 100 μl of samples are filtered concurrently through both membranes. The TRAP-bound RNA is retained on the nitrocellulose, and the unbound RNA binds to the Hybond membrane. By phosphorimager analysis of both membranes, the percent of the RNA bound to TRAP is easily determined for each reaction. The advantage of this method is that measuring both the bound and unbound RNA to determine the fraction bound for each sample eliminates loading errors during filtration.

For either RNA binding assay described previously, the data are anyalyzed to generate a best-fit binding curve to determine the dissociation constant (K_d) by using a nonlinear least squares fitting algorithm. We use Prism 3.0 (GraphPad Software, San Diego, CA) to fit the data to the following equation:

$$Y = \text{Bmax} * X/(K_d + X)$$

This equation describes ligand binding to a receptor following the law of mass action. X is the total protein concentration in each reaction, Bmax is the maximal binding at saturation, and K_d is the ligand concentration necessary for half-maximal binding. Using the filter binding assay, the apparent K_d for *B. subtilis* TRAP binding to *trp* leader RNA has been measured to be 0.12 nm at 37°,[22] whereas several estimates of this value using mobility shift gels are significantly higher.[11,20]

---

[24] I. Wong and T. M. Lohman, *Proc. Natl. Acad. Sci. USA* **90**, 5428 (1993).

Tryptophan Binding Assays

Equilibrium dialysis was first used to demonstrate that tryptophan binds to TRAP using an assay.[3] This simple method involves placing TRAP on one side of a dialysis membrane and $[^{14}C]$L-tryptophan on the other side and then allowing dialysis to proceed until equilibrium is reached. After incubating for 16–24 h in an equilibrium microvolume dialyzer (EMD101; Hoefer, San Francisco, CA), samples are analyzed by scintillation counting to determine the concentration of tryptophan on both sides of the membrane. The concentration of tryptophan on the side without TRAP corresponds to free tryptophan ([Trp]f), whereas the total tryptophan concentration on the side with TRAP corresponds to free tryptophan plus tryptophan bound to the protein ([Trp]b). Hence, by subtracting the [Trp]f from the total [Trp] in the presence of TRAP, the [Trp]b can be determined. Tryptophan binding to TRAP is positively cooperative, and therefore the data are fit to the Hill equation:

$$[Trp]_b = \frac{a \times ([Trp]_f/S_{0.5})^n}{1 + ([Trp]_f/S_{0.5})^n}$$

where a is the saturation level of bound tryptophan ([Trp]b); $S_{0.5}$ is defined as the concentration of free tryptophan, [Trp]f, at which the [Trp]b reaches 50% of saturation, and is used to describe the binding affinity. The Hill coefficient, n, is used to describe the cooperativity of binding: n is 1.0 in the case of no cooperativity, >1.0 for positive cooperativity and <1.0 for negative cooperativity. The disadvantages of this assay are that it is slow and rather limited in the number of samples that can be analyzed at one time.

Recently Li and Gollnick have developed an assay to measure tryptophan binding to TRAP using circular dichroism (CD) spectroscopy.[4] This assay takes advantage of changes in the CD spectrum of TRAP that occur when tryptophan binds to the protein. In the absence of tryptophan, the CD spectrum of TRAP is typical of a protein composed predominantly of β-sheet secondary structure, showing a negative peak near 215 nm (Fig. 3A). In the presence of tryptophan, a new positive peak appears near 225 nm. This spectral change does not occur for TRAP mutants that do not bind tryptophan. Hence tryptophan binding can be assayed by examining the changes in CD_{225} ($\Delta\theta$) as a function of tryptophan concentration (Fig. 3B). At each tryptophan concentration, the spectrum of free tryptophan is subtracted and values of $\Delta\theta$ are normalized based on the maximal CD signal change at saturation to give $\%\Delta\theta$. The data for $\%\Delta\theta$ are fit to the Hill equation (as before) to yield an apparent $S_{0.5}$ of 24 μM and Hill coefficient (n) of 1.5 for wild-type TRAP at 37°. Both values are

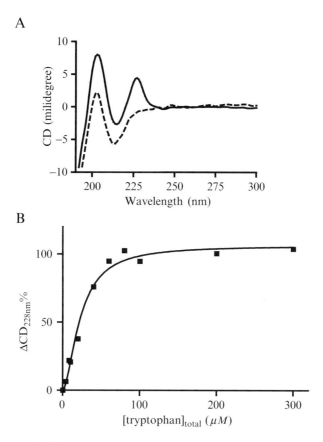

FIG. 3. Tryptophan binding to TRAP as measured by circular dichroism spectroscopy. (A) CD spectra of 12 μM TRAP in 50 mM sodium phosphate, pH 8.0 (dashed line) and with addition of 100 μM L-tryptophan (solid line) at 37°. Free tryptophan background was subtracted from the spectrum. (B) Equilibrium binding curve of tryptophan to TRAP (12 μM). The changes in CD at 228 nm upon addition of tryptophan are plotted and the curve fit to the equation $Y = Bmax *X/(K_d + X)$ using the program Prism 3.0 (GraphPad software).

similar to those derived previously from equilibrium dialysis of TRAP at 5° ($S_{0.5} = 5$–$10 \mu M$ and $n = 1.5 - 2.0$).

In Vitro Transcription Attenuation

To further study the function of TRAP in regulating transcription of the *trp* operon, an *in vitro* transcription attenuation assay was developed.[11,12] In this assay, *B. subtilis* RNA polymerase is used to transcribe a DNA template containing the *B. subtilis trp* leader sequence and the fraction of the transcripts that end at the attenuator, as opposed to those that read through, is determined. Standard transcription assays contain 20 mM Tris–HCl, pH 8.0, 20 mM NaCl, 10 mM MgCl$_2$, 14 mM β-mercaptoethanol, 0.1 mM EDTA, 0.5 mM L-tryptophan, 2.7 mM ATP, 1.1 mM GTP, 0.7 mM CTP, 1.4 mM UTP, 10 μCi [α-^{32}P]UTP, 40 nm DNA template, 50 nm *B. subtilis* RNA polymerase, and various amounts of purified TRAP protein. After incubation at 37° for 30 min, the reactions are stopped by addition of loading dye (7 M urea, 0.1% bromophenol blue) and run on a 6% denaturing (7 M urea) polyacrylamide gel. The gels are examined by phosphorimager analysis. Transcription of the *trp* leader region results in two major transcripts corresponding to the run-off transcript (320 nt) and a shorter product (138 nt) resulting from termination at the *trp* attenuator (Fig. 4). In the absence of TRAP, 5–10% termination at the attenuator is observed. Addition of increasing amounts of tryptophan-activated TRAP increases termination to 80–90%. This assay can be used to assess the affects of amino acid substitutions in TRAP on the ability of the protein to induce attenuation or to examine the effects of base substitutions in the *trp* leader region on transcription attenuation. Babitzke and Yanofsky[12] extended this assay to include the use of oligonucleotides complementary to various portions of the *trp* leader transcript during transcription in the absence of TRAP. Using this approach, they demonstrated that formation of the antiterminator structure is responsible for transcriptional readthrough and that base-pairing of residues at the base of the antiterminator (those which overlap with the terminator) is most critical.

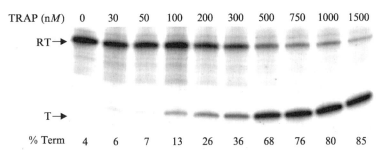

FIG. 4. *In vitro* attenuation assay for TRAP-induced transcription termination in the *trp* leader region. Denaturing (7 M urea) 8% polyacrylamide gel electrophoresis of *in vitro* transcription with *B. subtilis* RNA polymerase using a DNA template containing the *B. subtilis trp* promoter, leader and first 381 nt of *trpE* in the presence of [α-^{32}P]UTP. Transcription was performed in the absence or presence of the indicated concentrations of TRAP. The percentage termination (%Term), as determined by phosphorimager quantitation of the terminated transcripts (T), which are truncated at the attenuator, and the full-length run-off transcripts (RT) are indicated at the bottom of each lane.

In Vivo Gene Regulation

Both transcriptional and translational gene fusions containing the *trp* promoter and leader region fused to *lacZ* have been used extensively to examine TRAP function *in vivo*.[10,16,25–27] By examining regulation of β-galactosidase expression from these fusions in response to tryptophan in wild-type and *mtrB* mutant *B. subtilis* strains, the role of TRAP in regulation can be studied. The gene fusions are created in one of several integrative plasmids for *B. subtilis* derived from pTrpBG1.[28] In this plasmid, the *lac* fusion and an antibiotic resistance marker are placed between two portions of the *amyE* gene. The linearized plasmid is transformed into *B. subtilis* using natural competence,[29] and transformants are selected on minimal agar plates[30] supplemented with 0.2% acid-hydrolyzed casein, 0.2% glucose, 10 μg/ml L-arginine, 50 μg/ml 5-bromo-4-chloro-3-indolyl-β-D-galactopyranoside (X-gal), and the appropriate antibiotic. Because the plasmid does not contain a functional origin of replication for *B. subtilis*, transformants result only from homologous recombination between the *amyE* regions on the plasmid and the genome. Recombination results in disruption of the *amyE* gene and integration of the *trp–lac* fusion into the chromosome. The *amyE*[-] phenotype can be confirmed on starch plates, which, upon addition of iodine, show a clear halo around wild-type colonies but not around *amyE*[-] colonies.[31]

To assay β-galactosidase, cells are grown in liquid medium containing minimal salts[30] with 0.2% glucose, 0.2% acid-hydrolyzed casein, in the absence or presence of 50 μg/ml L-tryptophan at 37° until the optical density of the culture at 600 nm reaches between 0.4 and 0.8. 1.5 ml of cell culture is transferred to a microfuge tube and spun for 30 s to pellet the cells. The cells are washed once with ice-cold TE buffer and then resuspended in 1.5 ml of Z buffer (100 mM sodium phosphate, pH 7.0, 10 mM KCl, 1 mM MgSO$_4$, and 50 mM β-mercaptoethanol). 0.5 ml of the resuspended cells is transferred to another microfuge tube. The remaining 1.0 ml of the sample is stored on ice to read the A$_{600}$ value (see the following). 5 μl of freshly prepared 10 mg/ml lysozyme (Sigma) is added to the 0.5 ml of cells and incubated at 37° for 5 min, then 5 μl of Triton X-100 is added to lyse

[25] H. Yaknin, J. E. Babiarz, A. V. Yakhnin, and P. Babitzke, *J. Bacteriol.* **183,** 5918 (2001).
[26] M. Yang, A. de Saizieu, A. P. G. M. van Loon, and P. Gollnick, *J. Bacteriol.* **177,** 4272 (1995).
[27] M. I. Kuroda, D. Henner, and C. Yanofsky, *J. Bacteriol.* **170,** 3080 (1988).
[28] H. Shimotsu and D. J. Henner, *Gene* **43,** 85 (1986).
[29] C. Anagnostopoulos and J. Spizizen, *J. Bacteriol.* **81,** 741 (1961).
[30] H. J. Vogel and D. M. Bonner, *J. Biol. Chem.* **218,** 97 (1956).
[31] H. Shimotsu, M. I. Kuroda, C. Yanofsky, and D. J. Henner, *J. Bacteriol.* **166,** 461 (1986).

the cells. 0.1 ml of the lysed cells is transferred to a glass culture tube containing 0.9 ml of Z buffer. The reactions are started by adding 0.2 ml of 4 mg/ml ONPG, and each reaction is incubated at 30° until it turns light yellow, at which point it is stopped by the addition of 0.5 ml of 1 M Na_2CO_3 and the time is recorded. The absorbance at 420 nm and 550 nm of each assay is measured, as well as the absorbance at 600 nm of the saved sample of cells described earlier. The following formula is used to calculate the Miller units of β-galactosidase:

$$1000 \times (A_{420} - 1.75 \times A_{550})/t \times vA_{600}$$

where t is the time of the assay in min, and v is the volume of cells assayed.

Acknowledgments

This work was supported by grants from the National Science Foundation (MCB-9982652) and from the National Institutes of Health (GM62750). I wish to thank all the previous and current members of my laboratory who contributed to developing these methods. In particular, I thank Mirela Milescu, Xiufeng Li, and Pan Li for contributing figures.

[32] Analyzing Transcription Antitermination in Lambdoid Phages Encoding Toxin Genes

By MELODY N. NEELY and DAVID I. FRIEDMAN

Background

Phage λ, with its fellow lambdoid phages,[1] has for many years steadfastly served as a model system for studying many basic cellular processes. These include the nature of repressors and operators and their interactions, transcription elongation, morphogenesis, generalized and site-specific recombination, and replication, to name a few.[2] More recently the study of lambdoid phages has taken a more practical tack with the discovery that members of this phage family contribute to the pathogenic behavior of certain strains of *Escherichia coli*.[3-7] These lambdoid phages carry the genes encoding Shiga toxin (Stx A and B): most commonly, the genes encoding

[1] A. Campbell, *Ann. Rev. Microbiol.* **48**, 193 (1994).
[2] R. W. Hendrix, J. W. Roberts, F. W. Stahl, and R. A. Weisberg, *Lambda II*, Cold Spring Harbor Laboratory Press, Cold Spring Harbor, NY, 1983.
[3] H. W. Smith, P. Green, and Z. Parsell, *J. Gen. Microbiol.* **129**, 3121 (1983).

the two subunits of either the *stx1* or *stx2* genes. The two best-studied *stx*-carrying phages are Stx1 phage H-19B and Stx2 phage 933W.[8]

Shiga Toxin

The *stx* genes were named because their gene products are closely related to a toxin first described as a virulence factor in *Shigella dysenteria*.[9] Stx is a typical *AB* toxin.[10,11] The holotoxin has one *A* and five *B* subunits. The *A* subunit is the active component, whereas the five *B* subunits form the pentameric structure that recognizes the cellular receptor and serves as the port of entry of the *A* subunit. Stx is then transported in a retrograde pathway from endosomes through the Golgi body, where it is postulated that the *B* fragment dissociates from the *B* fragment. The *A* fragment is then thought to be transported to the endoplasmic reticulum and from there to the cytoplasm.[12] The *A* subunit is an *N*-glycosylase that blocks protein synthesis by removing a specific adenine residue in the 28S ribosomal RNA.[11]

E. coli strains expressing *stx* genes, STEC, can be highly pathogenic, with Stx playing a significant role in the virulence.[13,14] Infection with enterohemorrhagic *E. coli* (EHEC), such as O157:H7 and 026:H19, can lead to serious sequelae that include bloody diarrhea, hemolytic uremic syndrome, and death. A number of studies have shown that the *stx* genes in STEC are carried in partial or complete genomes of lambdoid phages.[8] Moreover, the *stx* genes are found in relatively the same position within the phage genome, downstream of the promoter controlling expression of late phage genes (Fig. 1) and upstream of the genes encoding lysis functions.[15] This common arrangement can occur even though *stx* and associated genes

---

[4] A. D. O'Brien, J. W. Newland, S. F. Miller, R. K. Holmes, H. W. Smith, and S. B. Formal, *Science* **226,** 694 (1984).

[5] S. M. Scotland, H. R. Smith, G. A. Willshaw, and B. Rowe, *Lancet* **2,** 216 (1983).

[6] J. W. Newland, N. A. Strockbine, S. F. Miller, A. D. O'Brien, and R. K. Holmes, *Science* **230,** 179 (1985).

[7] W. C. Hollifield, E. N. Kaplan, and H. V. Huang, *Mol. Gen. Genet.* **210,** 248 (1987).

[8] P. L. Wagner and M. K. Waldor, *Infect. Immun.* **70,** 3985 (2002).

[9] A. D. O'Brien, G. D. LaVeck, M. R. Thompson, and S. B. Formal, *J. Infect. Dis.* **146,** 763 (1982).

[10] A. D. O'Brien and R. K. Holmes, *Microbiol. Rev.* **51,** 206 (1987).

[11] W. K. Acheson, A. Donohue-Rolfe, and G. T. Keusch, *in* "Sourcebook of Bacterial Protein Toxins," (J. E. Alouf and J. H. Freer, eds.), p. 415. Academic Press, San Diego, 1991.

[12] L. Johannes and B. Goud, *Trends Cell Biol.* **8,** 158 (1998).

[13] J. B. Kaper and A. D. O'Brien, *Escherichia coli 0157, H7, and Other Shigatoxin-producing E. coli Strains.* American Society for Microbiology, Washington, DC, 1998.

[14] V. L. Tesh and A. D. O'Brien, *Mol. Microbiol.* **5,** 1817 (1991).

[15] M. Bitzan, H. Karch, H. Altrogge, J. Strehlau, and F. Blaker, *Pediatr. Infect. Dis. J.* **7,** 128 (1988).

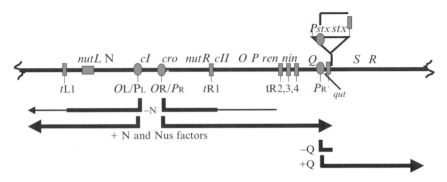

Fig. 1. Generic map of the early regulatory region of lambdoid phages based primarily on studies with λ (not drawn to scale). Shown are locations of relevant genes, promoters (*P*), operators (*O*), terminators (*t*), and the relative position of *stx* genes in *stx*-carrying phages. Shown below are the transcription patterns from promoters *P*L and *P*R in the presence and absence of N and *P*R′ in the presence and absence of Q. Shown above is the putative transcription pattern of the *stx* genes initiating at the *stx*-associated promoters.

show significant differences. For example, 933W not only differs from H-19B in having *stx2* genes rather than *stx1* genes, but it also has upstream tRNA genes encoding rare amino acids not present in H-19B.[16,17] Interestingly, *stx2* contains a number of the rare amino acid codons calling for amino acids corresponding to the tRNAs in the phage genome. Although both the *stx1* and *stx2* genes have associated promoters, only the promoter associated with *stx1* is iron regulated.[18]

The observation that the position of the *stx* genes downstream of phage late promoters *P*R′ (Fig. 1) is highly conserved suggested that there is a selective advantage for this placement. One plausible scenario is that transcription from *P*R′ is an important contributor to Stx expression. Transcription from *P*R′ is the last event in a regulatory cascade leading to expression of the late phage genes.[19] Thus if transcription from *P*R′ does play a significant role in Stx expression, the earlier elements of the cascade would by necessity be contributors to the process. Because a number of studies, primarily from our laboratory and that of Waldor,[20,21] have indicated that transcription from *P*R′ does make a contribution to Stx

[16] G. Plunkett III, D. J. Rose, T. J. Durfee, and F. R. Blattner, *J. Bacteriol.* **181,** 1767 (1999).

[17] M. N. Neely and D. I. Friedman, *Gene* **223,** 105 (1998).

[18] S. B. Calderwood and J. J. Mekalanos, *J. Bacteriol.* **169,** 4759 (1987).

[19] D. I. Friedman and D. L. Court, *Curr. Opin. Microbiol.* **4,** 201 (2001).

[20] P. L. Wagner, M. N. Neely, X. Zhang, D. W. Acheson, M. K. Waldor, and D. I. Friedman, *J. Bacteriol.* **183,** 2081 (2001).

[21] M. N. Neely and D. I. Friedman, *Mol. Microbiol.* **28,** 1255 (1998).

expression, it is necessary to understand the regulatory scheme of lambdoid phages to have a full understanding of the factor contributing to Stx expression. Although variations of this scheme diverge to some extent from that identified for the prototypical λ,[22] 933W and H-19B appear to conform without significant variation to this scheme.

The Lambdoid Family of Phages

λ is one member of a large family of related phages.[1,23] These phages share a similar genetic organization, but within that organization the individual genes may vary. For example, coliphage λ and *Salmonella* phage P22 both have two genes encoding replication functions, as well as their origins of replication, at relatively the same positions on their genomes. However, the P22 18 and 12 gene products are functionally different from the O and P gene products of λ.[24] Although the genes encoding the repressors and the cognate operator sites are at nearly the same relative positions on their genomes, each repressor is specific for its cognate operator. That said, we note that some lambdoid phages share the same repressors and operators while differing in other genes (e.g., λ–HK97 and P21–P22). In short, the lambdoid phages appear to be mosaics constructed from a large pool of homologous genes. In addition to genes essential for phage activities, some lambdoid phages have acquired other genes that when expressed may be useful to the bacterial host.[8,25] The *stx* genes in phages found in STEC strains fit this class of genes.

The Approach

We now outline procedures for analyzing the antitermination mechanisms in lambdoid phages, with a focus on our studies characterizing how these mechanisms contribute to Stx expression in phage H-19B.

Analysis: N-mediated Antitermination

Four methods of analysis were used to determine whether H-19B has an antitermination system similar to that of λ and assess its specific mechanism of action: (1) DNA sequence determination of regions of the H-19B

[22] M. N. Neely and D. I. Friedman, *Mol. Microbiol.* **38,** 1074–1085 (2000).
[23] A. Campbell and D. Botstein, *in* "Lambda II," (R. W. Hendrix, J. W. Roberts, F. W. Stahl, and R. A. Weisberg, eds.), p. 365. Cold Spring Harbor Laboratory Press, Cold Spring Harbor, NY, 1983.
[24] S. Wickner, *J. Biol. Chem.* **259,** 14038 (1984).
[25] F. A. Miao and S. I. Miller, *Proc. Natl. Acad. Sci. USA* **96,** 9452 (1999).

genome that may play a role in antitermination based on studies of λ N-mediated antitermination, (2) growth of H-19B in *E. coli* mutants defective for λ antitermination, (3) construction and analysis of H-19B phage mutants, and (4) reporter constructs designed to study antitermination mediated by the H-19B N protein.

Using the DNA Sequence to Identify Components of the N Antitermination System

Comparison of approximately 17 kb of H-19B genome sequence with sequences from other members of the lambdoid phage family revealed that H-19B shares the typical genome organization of lambdoid phages.[17] As observed with other lambdoid phages, H-19B is a genetic mosaic constructed from a pool of genes shared with a number of different lambdoid phages.[26,27] Thus H-19B has scattered regions of significant homology with genomes of different lambdoid phages. Additionally, there are regions in the H-19B genome with no obvious homology to any identified sequences in lambdoid phages.[17] Because, to a first approximation, the gene arrangement is conserved among lambdoid phage genomes, positive identification of one gene can be used as a landmark to locate likely positions of linked genes.

The *c*I gene, encoding repressor, was chosen as the primary landmark for this genomic analysis. The *c*I genes of lambdoid phages encode the repressor. Thus these *c*I genes are functionally identifiable because, when expressed, they provide the host bacterium immunity from infection by phages that have that particular repressor gene.[28] Based on previous analysis of the H-19B genome,[29] we determined the approximate location of the *c*I gene and used that as the basis for cloning a fragment that confers specific immunity to infection by H-19B, the hallmark of a *c*I gene. By analogy to the positioning of the λ *N* gene relative to the λ *c*I gene,[30] a putative H-19B *N* gene was identified. This open reading frame contains an N-terminal arginine-rich motif similar to those found in N proteins of other lambdoid phages.[31,32] A putative H-19B *c*II gene was also identified at the same position as the λ *c*II gene relative to the position of *c*I. This putative *c*II gene

[26] D. Botstein, *Ann. N. Y. Acad. Sci.* **354**, 484 (1980).

[27] J. Oberto, S. B. Sloan, and R. A. Weisberg, *Nucleic Acids Res.* **22**, 354 (1994).

[28] M. Ptashne, *The Genetic Switch*, Cell Press, Blackwell Scientific Publications, Cambridge, Mass., 1992.

[29] A. Huang, J. Friesen, and J. L. Brunton, *J. Bacteriol.* **169**, 4308 (1987).

[30] D. L. Daniels, J. L. Schroeder, W. Szybalski, F. Sanger, and F. R. Blattner, *in* "Lambda II," (R. W. Hendrix, J. W. Roberts, F. W. Stahl, and R. A. Weisberg, eds.), p. 469. Cold Spring Harbor Laboratory Press, Cold Spring Harbor, NY, 1983.

[31] N. C. Franklin, *J. Mol. Biol.* **231**, 343 (1993).

[32] D. Lazinski, E. Grzadzielska, and A. Das, *Cell* **59**, 207 (1989).

has significant homology to the *c*II genes of λ and P21,[33] information that was available at the time of that study. Later studies showed that the *c*II gene of H-19B also has homology to the *c*II genes of HK97 and HK022.[34]

Locating the *N* and *c*II genes provided the landmarks for identifying the *nut* sequences. The *nut* regions of λ (Fig. 1), as well as other lambdoid phages, are located immediately upstream of the *N* gene (*nutL*) and immediately upstream of the *c*II gene (*nutR*). Sequences resembling *nut* sites could be identified on the H-19B genome at the expected locations relative to the putative *N* and *c*II genes.[17] Both of these putative *nut* sequences contain regions of hyphenated dyad symmetry that are typical of a *boxB*.[32,35–37] Although sequences resembling *boxA* sites were identifiable in both the H-19B *nutL* and *nutR*, these *boxA* sequences not only differed in substantive ways from the consensus *boxA* sequence, but also differed from each other (Fig. 2). The highly conserved C nucleotide in the fifth position of the consensus *boxA* sequence (CGCTCTTTA), required for effective λ N-mediated antitermination,[38,39] is an A in the H-19B *nutR* (CGCTATTTT). Although the H-19 *nutR* has this conserved C, it is missing the run of Ts at the 3' end of *boxA*, as well as the highly conserved T nucleotide at the fourth position of the conserved *boxA* (CGCACTACT). The run of Ts has also been demonstrated to be important for N-mediated antitermination in λ.[40]

The spacer region between *boxA* and *boxB* has been shown to be a functionally important component of the λ *nut* sequence.[41–43] Although the spacer regions of previously analyzed *nut* regions varied from 8 to 14 nucleotides, the spacer regions in H-19B *nut* regions were significantly larger; 22 nucleotides in *nutR* and 42 nucleotides in *nutL* (Fig. 2).[22] Moreover, the spacer regions in both *nut* sites contain regions of dyad symmetry that could result in additional stem-loop structures in the NUT RNA. The

[33] Y. S. Ho and M. Rosenberg, *in* "The Bacteriophages," (R. Calendar, ed.), p. 725. Plenum Press, New York, 1988.

[34] R. J. Juhala, M. E. Ford, R. L. Duda, A. Youlton, G. F. Hatfull, and R. W. Hendrix, *J. Mol. Biol.* **299,** 27 (2000).

[35] J. S. Salstrom and W. Szybalski, *J. Mol. Biol.* **124,** 195 (1978).

[36] E. R. Olson, C. S. Tomich, and D. I. Friedman, *J. Mol. Biol.* **180,** 1053 (1984).

[37] N. C. Franklin, *J. Mol. Biol.* **181,** 75 (1985).

[38] R. Robledo, M. E. Gottesman, and R. A. Weisberg, *J. Mol. Biol.* **212,** 635 (1990).

[39] T. A. Patterson, Z. Zhang, T. Baker, L. L. Johnson, D. I. Friedman, and D. L. Court, *J. Mol. Biol.* **236,** 217 (1994).

[40] D. I. Friedman, E. R. Olson, L. L. Johnson, D. Alessi, and M. G. Craven, *Genes & Dev.* **4,** 2210 (1990).

[41] S. W. Peltz, A. L. Brown, N. Hasan, A. J. Podhajska, and W. Szybalski, *Science* **228,** 91 (1985).

[42] M. Zuber, T. A. Patterson, and D. L. Court, *Proc. Natl. Acad. Sci. USA* **84,** 4514 (1987).

[43] J. H. Doelling and N. C. Franklin, *Nucleic Acids Res.* **17,** 5565 (1989).

	boxA	spacer sequence	*boxB*
λ *nutR*	CGCTCTTAC	ACATTCCA	GCCCT GAAAA AGGGC
λ *nutL*	CGCTCTTAA	AAATTAA	GCCCT GAAGA AGGGC
H-19B *nutR*	CGCTATTTT	CACAATGGACATTCGTCCTACG	TCGCT GACAA AGCGA
H-19B *nutL*	CGCACTACT	CACCAGGGCGGTGATATACAACGATTCGAATATGAATCTACG	GCGCT GACAA AGCGC
Concensus *boxA*	CGCTCTTTA		

FIG. 2. Alignment of the *nut* sites of H-19B and λ. The *nut* sites are divided into their three identified components, *boxA*, spacer, or *boxB*. The arrows underneath a sequence denote regions of hyphenated dyad symmetry that potentially could form stem-loop structures in the RNA. The BOXB regions of λ have been shown to form a stem-loop structure in the RNA. See text for details.

NUT-N system of H-19B has been identified in two other lambdoid phages, HK97[34] and 933W.[16]

This summary of studies employing genome analysis of H-19B to identify the *N* gene and *nut* sites emphasizes the important role careful scanning of the sequence can play in initiating analysis of the N antitermination system in lambdoid phages.

Use of E. coli and H-19B Mutants to Analyze N-mediated Antitermination

A number of *E. coli* mutants that were identified because of their failure to support N action have proved invaluable in the characterization of the action of the H-19B N protein.[44,45] Four loci were identified by this mutational analysis: *nusA*, *nusB*, *rpsJ* (*nusE*), and *nusG*. Subsequent *in vitro* studies with purified proteins provided evidence that these proteins are directly involved in λ N action.[46,47] Together these genetic and *in vitro* studies provided insight into the interactions of N and the Nus proteins with the NUT RNA. N and NusA interact with the BOXB hairpin, and there is evidence that NusB and S10 (the product of the *rpsJ* gene and also referred to as NusE) interact with BOXA.[48,49] An *E. coli* derivative with the

[44] D. I. Friedman and L. S. Baron, *Virology* **58,** 141 (1974).

[45] Y. Zhou, J. J. Filter, D. L. Court, M. E. Gottesman, and D. I. Friedman, *J. Bacteriol.* **184,** 3416 (2002).

[46] A. Das and K. Wolska, *Cell* **38,** 165 (1984).

[47] S. W. Mason, J. Li, and J. Greenblatt, *J. Biol. Chem.* **267,** 19418 (1992).

[48] J. R. Nodwell and J. Greenblatt, *Cell* **72,** 261 (1993).

[49] D. L. Court, T. A. Patterson, N. Baker, N. Costantino, C. Mao, and D. I. Friedman, *J. Bacteriol.* **177,** 2589 (1995).

rpoAD305E mutation, an allele of the gene encoding the α subunit of RNA Pol, is also part of the collection of *E. coli* mutants that can be used in the analysis.[50] This *rpoA* allele suppresses the inhibitory effect of the *nusA1* mutation on λ N action. It was proposed that the suppressive effect of the *rpoAD305E* mutation is due to inhibition of an interaction of the α subunit with a putative inhibitor.

In addition to the *E. coli* mutants, the H-19B*nin* mutant proved useful in the analysis. The 2.8 kb *nin5* deletion in λ removes a region of at least three transcription terminators from the λ genome.[51] Because of their strategic location, deletion of these terminators permits phage growth to be independent of a requirement for N action.[52] H-19B*nin* was constructed by deleting a region of the H-19B genome analogous to the *nin* region of λ.[22] Presumably this deletion removes a similar collection of transcription terminators.

To construct the H-19B*nin* deletion, two fragments were amplified using polymerase chain reaction (PCR) with the H-19B genome serving as the template (Fig. 1). One fragment includes the *ren* gene, located upstream of the putative *nin* region. The other includes the *Q* gene, located downstream of the *nin* region. Using PCR splicing by overlap extension (SOE),[53] these two fragments served as templates for generating a contiguous fragment with the *ren* and *Q* genes intact. However, the fragment was deleted for sequences between *ren* and *Q* (Fig. 1), sequences encompassing the putative *nin* region. The composite fragment deleted for the *nin* region was cloned into the pBR322 vector, creating pH-19Bnin. H-19B phage was grown on an *E. coli* strain containing pH-19Bnin to allow recombination between the phage and the plasmid-based cloned fragments. Because deletion of the terminators in the *nin* region of λ allow N-independent growth, it was expected that a similar deletion in the H-19B genome would allow N-independent growth of that phage. Putative H-19B derivatives with *nin* deletions were isolated as phage-forming plaques on the *nusA1* host at 42°. The presence of the deletion was confirmed by PCR.

Studies with *nus* mutants provided insight into the nature of the interactions at the H-19B NUT site[22] (Table I). H-19B, like λ, fails to grow in *E. coli* carrying the *nusA1* mutation, whereas H-19B*nin*, like λnin, grow in the *nusA1* mutant. These observations provide compelling *in vivo* evidence that H-19B requires NusA for effective action of an N-like antitermination system. However, unlike λ, H-19B grows in *E. coli* and

[50] A. T. Schauer, S. W. Cheng, C. Zheng, L. St. Pierre, D. Alessi, D. Hidayetoglu, N. Constantino, D. Court, and D. Friedman, *Mol. Microbiol.* **21,** 839 (1996).

[51] S. C. Cheng, D. L. Court, and D. I. Friedman, *Genetics* **140,** 875 (1995).

[52] D. Court and K. Sato, *Virology* **39,** 348 (1969).

[53] R. M. Horton, H. D. Hunt, S. N. Ho, J. K. Pullen, and L. R. Pease, *Gene* **77,** 61 (1989).

TABLE I
GROWTH OF λ AND H-19B ON *E. COLI* STAINS WITH *nus* MUTATIONS

			Efficiency of plating[a]		
Strain	Temperature	Relevant genotype	λ	H-19B	H-19B *nin*
K95	42	*nusA1*	<0.0001	<0.0001	1
K450	40	*nusB5*	<0.0001	1	ND
K7554	40	*nusB::IS10*	<0.0001	1	ND
K551	40	*nusE71*	<0.0001	1	ND
K2049	40	*nusB5/nusE71*	<0.0001	1	ND
K4069	40	*rpoAD305E*	1	1	ND
K4047	42	*rpoAD305E/nusA1*	1	<0.0001	ND

[a] Efficiency of plating is calculated as the ratio of the phage titre on a lawn of the stain with the mutant genotype divided by the phage titre on a lawn of the parent *E. coli* strain wild type at the locus being tested.
ND, Not done.

has any one of the following mutations: *nusB5*, Δ*nusB*, or *nusE71* (an *rpsJ* mutation), as well as an *E. coli nusB5–nusE(rpsJ)71* double mutant (Table I). As discussed, NusB and S10 (NusE) appear to work through BOXA. Thus this finding is consistent with the sequencing data showing that H-19B has *boxA* sequences that, when compared with the consensus, can be described as degenerate. The sequences are apparently nonfunctional.

Studies with an RNA Pol mutant also contributed to the analysis. The *rpoAD305E* allele, encoding a mutant α subunit, suppresses the defect caused by the *nusA1* mutation in support of λ N action.[50] λ fails to form plaques on a lawn of the single *nusA1* mutant at 42°, but does on a lawn of the double *nusA1–rpoAD305E* mutant. H-19B fails to form plaques on lawns formed from bacteria having either the single *nusA1* or double *nusA1–rpoAD305E* mutations at 42° (Table I). The suppressor effect of the mutant α subunit is thought to be associated with an interaction at BOXA. The results with H-19B are consistent with this idea, because H-19B *nut* regions have degenerate *boxA* sequences. If the mutant works through BOXA, then it would be expected that a degenerate BOXA sequence would be unable to support the suppressor activity of the mutant α subunit.

These studies provide evidence of the utility of characterized phage and host mutants in obtaining important insight into the mechanism of a phage-encoded antitermination system.

Use of Reporter Constructs to Analyze N-mediated Antitermination

A two-plasmid reporter system was employed to assess H-19B NUT interaction with its cognate N *in vivo*, one based on the fusions used in the characterization of the action of λ N.[54,55] The reporter plasmid has an insert containing the *lac* promoter (*Plac*), the H-19B *nutR* site, a three-terminator cassette, and the *lacZ* reporter gene (Fig. 3). The terminator cassette contains λ $t_{R'}$ and the *rrn* T1 and T2 Rho-independent terminators.[56] The second plasmid, compatible with the reporter plasmid, has an insert containing the H-19B *N* gene cloned downstream of *Plac* (Fig. 3). N-mediated transcription antitermination can be quantified by measuring levels of expression of the reporter gene that is located distal to the terminator cassette.

The specificity of the H-19B N protein for its cognate NUT site was assessed using a similar reporter system that has the NUT site of λ rather

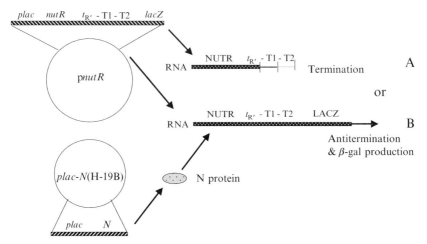

FIG. 3. Vector system for testing N-mediated antitermination. Levels of expression of β-galactosidase from the plasmid construct with the *Plac* promoter, a *nut* site, a terminator cassette, and a *lacZ* reporter gene provide a measure of the efficiency of antitermination. (A) In the absence of N, termination occurs within the terminator cassette, resulting in low levels of β-galactosidase. (B) In the presence of N, expressed from the second plasmid from the *Plac* promoter, read-through of the terminator cassette occurs, resulting in high levels of β-galactosidase.

[54] S. Adhya, M. Gottesman, and B. De Crombrugghe, *Proc. Natl. Acad. Sci. USA* **71,** 2534 (1974).

[55] N. C. Franklin, *J. Mol. Biol.* **89,** 33 (1974).

[56] R. A. King, S. Banik-Maiti, D. J. Jin, and R. A. Weisberg, *Cell* **87,** 893 (1996).

than that of H-19B. H-19B N expressed from a plasmid, unlike similarly expressed λ N, did not facilitate read-through of the downstream terminator. To test for specificity of the NUT site for its N, the N proteins of λ and P22 were tested. Expression of either of these N proteins from a plasmid, unlike a similarly expressed H-19B N, failed to facilitate terminator read-through from the H-19B reporter plasmid (Table II).

Use of Reporter Constructs to Analyze the nut Region

To determine whether the stem-loop structure suggested by the hyphenated dyad symmetry in the putative *boxB* sequence is required for antitermination activity, the sequence encoding the ascending arm was changed to eliminate the dyad symmetry and thus the potential hairpin structure in the RNA.[22] The resulting construct failed to support antitermination when supplied with H-19B N (Table II). This finding demonstrates the importance of this element of the *nut* region for H19B N action.

The functional significance of what were described as degenerate *boxA* sites was also assessed using this system. As discussed, the *boxA* sequences of both *nutR* and *nutL* diverge from the consensus *boxA* and differ from each other. The results of experiments testing growth of H-19B on *E. coli nus* mutants suggested that H-19B does not require a *boxA* sequence. In these experiments, variants of the reporter plasmid were used that had

TABLE II
SEQUENCE SPECIFICITY OF N PROTEINS FOR WILD-TYPE AND VARIANT H-19B NUT SITES

Relevant insert in reporter plasmid	Source of N	%Read-through[a]
nutR-wt	H-19B	76 +/−4.2[b]
nutR-wt	None	2 +/− 0.3
nutR-wt	λ	2.5 +/− 0.1
nutR-wt	P22	1.5 +/− 0.3
nutR/ΔboxA[c]	H-19B	72 +/− 7.5
nutR/boxB13[d]	H-19B	1.7 +/− 0.2

[a] Percentage read-through was determined from the levels of β-galactosidase expressed from the *Plac-nutR-term-lacZ* plasmid constructs. The indicated values were all normalized to the level of β-galactosidase synthesized from a *Plac-nutR-lacZ* construct deleted for all terminators. This value was set at 100% read-through ($\sim$10,000 Miller units).

[b] $n = 5$.

[c] Complete deletion of the *boxA* sequence.

[d] The ascending arm of the *boxB* dyad symmetry was changed from TCGCT to AGCGA.

H-19B *nut* sites with altered *boxA* sequences previously identified in λ as defective in support of N-mediated antitermination. Using the method of PCR SOEing,[57] the *boxA* site of H-19B *nutR* was specifically mutated without changing surrounding sequences. Reporter plasmids with these changes failed to exhibit any significant reduction in H-19B N-mediated antitermination. Moreover, deletion of the *boxA* sequence from the H-19B *nutR* site had little effect on N-mediated terminator readthrough (Table II). These results confirmed the conclusions drawn from both the sequence comparison and studies with *E. coli* mutants that the *boxA* sequence plays little if any role in H-19B N action.

The reporter constructs were also used to ask specific questions about the function of the extended H-19B spacer region (Figure 4). There was precedent for considering the role of the spacer region since previous experiments with the λ *nut* site showed that mutations in the spacer region decrease the effectiveness of N-mediated antitermination.[43] Further suggesting functional significance was the observation that the extended sequence in the H-19B spacer region with the hyphenated dyad symmetry are conserved in HK97[34] and 933W,[16] phages with the same N-NUT system as H-19B. Based on these considerations, the effect of changes in the spacer region and the significance of the dyad symmetry found in the spacer regions of the H-19B *nutR* regions were examined.[22] Changes in one arm of the dyad symmetry in the spacer sequence that would eliminate the hairpin structure in the RNA inactivated the NUT region, demonstrating that the hairpin structure was important for N action at NUT (Figure 4). However, deletion of the dyad symmetry had only minimal effects on N-mediated antitermination, suggesting that the role for the dyad symmetry was to allow formation of the hairpin structure in the RNA. Further support of this idea came from studies using a reporter plasmid in which both arms of the dyad symmetry were changed without affecting the potential for hairpin structure in the RNA. These changes had little effect on the effectiveness of N-mediated antitermination.

These results suggest that the stem-loop region may act to bring together important regulatory sequences either within the spacer region or to bring an upstream sequence, not likely to be BOXA, in closer contact to the *boxB* sequence. Furthermore, it suggested that the sequence composing the dyad symmetry does not play a direct role in complex formation. If this were the case, sequences on either side of the stem-loop would only serve to withdraw sequences from the run of nucleotides forming the spacer RNA. This idea was supported by results of an experiment with a variant in which the nucleotides in each arm were completely changed

[57] S. N. Ho, H. D. Hunt, R. M. Horton, J. K. Pullen, and L. R. Pcase, *Gene* **77,** 51 (1989).

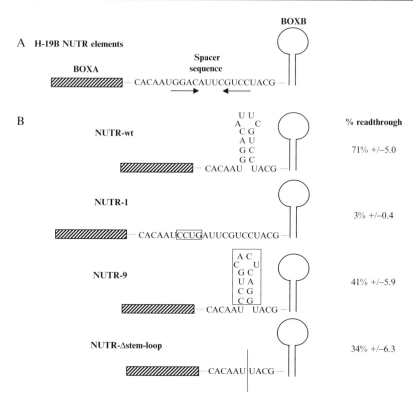

FIG. 4. Sequences and potential structures of H-19B NUTR RNA sites with variant spacer regions. (A) The wild-type H-19B NUTR site. Arrows under the spacer sequence indicate a region of hyphenated dyad symmetry with the potential to form a stem-loop structure in the RNA. (B) Shown are the predicted stem-loop structures in the RNA, spacer regions, and BOXB elements of the H-19B NUT site variants tested. The nucleotide sequence is shown only for the spacer region because both BOXA and BOXB are the same as wild type in all constructs. Transcription antitermination, determined as percentage read-through of terminators in the presence of N, is indicated at the right of each construct. Terminator read-through was determined by measuring β-galactosidase expression from the *Plac-nutR-term-lacZ* plasmid constructs in the presence of N as described in Fig. 3. The values were normalized to the level of β-galactosidase expressed from a *Plac-nutR-wt-lacZ* construct without the terminators; this reading was set for comparison as 100% read-through (∼10,000 Miller units). In the absence of the N-expressing plasmid, all constructs showed less than 3% read-through. Boxed sequences identify changed nucleotides. The perpendicular line in NUTR-Δstem-loop indicates the position of the deleted nucleotides.

but the dyad symmetry was maintained. In this case, antitermination in the presence of N was only slightly reduced. Reflecting this idea, the name "reducer" was given to this region of dyad symmetry.

This brief discussion was designed to provide an indication of the utility of the plasmid reporter system in analyzing NUT sites. In addition to demonstrating similarities in N-NUT systems (e.g., N action at BOXB), it also revealed important differences from other *nut* regions (e.g., formation of the reducer structure in the spacer region and a lack of a requirement for BOXA).

Analysis: Q-mediated Antitermination

Analysis of H-19B Q-mediated antitermination involved similar methods as those employed in analysis of N-mediated antitermination: (1) sequence determination of the late regulatory region of H-19B, which includes the *Q* gene, the *P*R′ promoter with its associated *qut* site, and the $t_{R′}$ terminator, as well as *stx* and lysis genes, (2) phage mutants, and (3) reporter–terminator constructs to measure factors required for antitermination at downstream terminators.

Use of the H-19B DNA Sequence to Identify Components of the Q Antitermination System

Using the putative *nin* region as a landmark, an open reading frame of 144 codons downstream of the *nin* region with homology to the bacteriophage 21 Q protein (41% identity)[58] was identified in the H-19B genome.[21] Associated with the *Q* gene were sequences corresponding to a *P*R′ promoter, a *qut* site, and a $t_{R′}$ terminator. Comparison of the 933W sequence with the regions of H-19B that have been sequenced showed that the 933W sequence in this region is nearly identical to that of H-19B. Although differing significantly in sequence, the *stx* genes of these two phages are similarly located on their respective phage genomes, downstream of $t_{R′}$.

Use of Reporter Constructs to Analyze Q-mediated Antitermination

To confirm that the putative H-19B Q protein acts with its cognate *P*R′–*qut* site to promote transcription antitermination, a plasmid reporter system similar to the two-plasmid reporter system used to measure N-mediated antitermination was constructed.[21] One plasmid has a fragment with the putative H-19B *P*R′–*qut* site and the downstream $t_{R′}$ terminator upstream of the *lacZ* reporter gene (Fig. 5). The second plasmid contains the H-19B *Q* gene downstream of *Plac* (Fig. 5). In *E. coli* with just the reporter plasmid, only a small, although measurable, amount of β-galactosidase was expressed. When both plasmids were present, significantly higher amounts

[58] H. C. Guo, M. Kainz, and J. W. Roberts, *J. Bacteriol.* **173,** 1554 (1991).

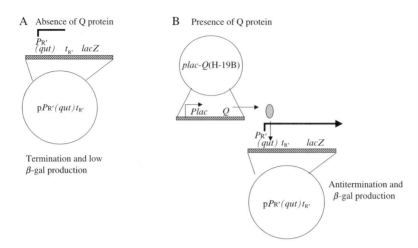

FIG. 5. Vector system for testing Q-mediated antitermination. Antitermination was assessed by the level of expression of β-galactosidase from the plasmid construct with the $P_{R'}$ promoter, the *qut* site, the $t_{R'}$ terminator, and the *lacZ* reporter gene. (A) In the absence of Q, the majority of transcripts will terminate at the $t_{R'}$ terminator, resulting in low levels of β-galactosidase. (B) When Q, under the control of the *Plac* promoter, is expressed from the second plasmid, high levels of β-galactosidase were observed, showing terminator read-through.

of β-galactosidase were expressed. When the second plasmid contained the λ Q gene rather than the H19B Q gene, only small amounts of β-galactosidase, similar to those seen in the absence of the H-19B Q protein, were expressed. To demonstrate that the action of Q was directed toward relieving termination, a variant of the reporter plasmid that was missing $t_{R'}$ was shown to express high levels of β-galactosidase.

The specificity of the H-19B Q protein was assessed by determining if it could act at the λ *qut* site to modify transcription initiating at λ $P_{R'}$. The plasmids discussed earlier with a Q gene from either H-19B or λ were tested for the ability to complement a λ Q^- phage (λ $Qam_{73}\text{-}am_{501}$).[59] Because the Q genes are transcribed from *Plac*, expression of the Q proteins was induced by the addition of isopropyl-β-o-thiogalactopyranoside (IPTG). Complementation was observed with the plasmid expressing λ Q, but not with the plasmid expressing H-19B Q. Hence the H-19B Q cannot interact with the λ *qut* site to support Q-mediated antitermination.[21] The observation that the Q expressed from each of these plasmids

[59] I. Herskowitz and E. Signer, *Cold Spr. Harb. Symp. Quant. Biol.* **35**, 355 (1970).

functions with its cognate *qut* site provides direct evidence that the plasmids express active Q proteins.

Q Action and stx Gene Expression

The location of the *stx* genes, in both the H-19B and 933W genomes downstream of the $P_{R'}$-*qut*-$t_{R'}$ region (Fig. 1), suggested that the $P_{R'}$ promoters might contribute to Stx expression. This would occur by Q-mediated antitermination of transcription initiating at $P_{R'}$ during the lytic growth of the phage. The H-19B Q expression plasmid provided a means to directly address this question. The plasmid was transformed into an *E. coli* that carries an H-19B prophage. Q could thus be supplied *in trans* to the prophage, and measurement of Stx levels provided a direct assay of the ability of Q to direct Stx expression. Using an uninduced lysogen permits assessment of the role of the $P_{R'}$ promoter in the absence of replication that would occur during phage induction. In this way, gene copy number does not influence the analysis. Because the *Q* gene in the plasmid is transcribed from *Plac*, maximal Q expression is obtained when bacteria are grown in the presence of IPTG.

The following procedure was used in the assessment of the effect of Q expression on Stx production. IPTG was added to exponentially growing cultures, and the cultures were allowed to express for 1 h. Cells were lysed by sonication, and total protein concentration was measured. Enzyme immunoassays[60] were used to measure Stx levels.

Stx expression from lysogens carrying either an H-19B or a 933W prophage was measured both in the presence and absence of Q. The plasmid expressing H-19B Q could be used to test Q action with both prophages, because 933W has the same Q-$P_{R'}$-*qut* sequence as H-19B. However, 933W, unlike H-19B, has the *stx2* genes. Enhanced toxin expression by both lysogens was observed when the H-19B Q protein was provided *in trans*.[21] However, the increases in Stx expressed in the presence of Q by the H-19B and 933W prophages over that observed in the absence of Q were very different. There was a 17-fold increase in expression of Stx2 from the 933W prophage, whereas there was only a 4-fold increase in expression of Stx1 from the H-19B prophage.[21] When considering these results, it should be kept in mind that although 933W and H-19B carry the same Q-$P_{R'}$-*qut*-$t_{R'}$ sequences, the *stx* promoters, as well as the region between $t_{R'}$ and the toxin genes, are very different. Some or all of these differences may explain the differences in the effects of Q on Stx production.

[60] A. Donohue-Rolfe, D. W. Acheson, A. V. Kane, and G. T. Keusch, *Infect. Immun.* **57**, 3888 (1989).

H-19BΔQ-P$_R'$ Phage

An H-19B derivative with a deletion extending from the distal part of Q through P_R' was used to assess the role of the P_R' promoter in the expression and release of Stx during phage induction. The ΔQ–P$_R'$ phage was constructed in a stepwise process. First, a ~900-bp fragment encompassing the late gene regulatory region with an 189-bp deletion that includes the 3' end of the Q gene and the P_R' (Fig. 1) sequence was synthesized by PCR SOEing[53] using H-19B DNA as template. Second, the ΔQ–P$_R'$ fragment was cloned into the suicide vector pCVD442[61], creating pΔQ–P$_R'$. This vector will only replicate autonomously in a host that has an expressable *pir* gene (e.g., *E. coli* strain SM10λpir). Moreover, the vector carries a *sacB* gene, which confers sensitivity to sucrose and can be used for counter selection. Third, the suicide plasmid with the cloned fragment was introduced into the H-19B lysogen by conjugation with the *E. coli* strain SM10λpir. Fourth, stable transformants were obtained by selecting for resistance to ampicillin. Because the plasmid cannot replicate autonomously in the recipient bacterium, to be maintained it would have to be integrated into the chromosome, presumably at the site of homology within the H-19B prophage. Fifth, haploid segregants were selected as sucrose-resistant colonies[61] and screened by PCR for the Q–P$_R'$ deletion. The presence of the deletion in the prophage was confirmed by DNA sequencing.

Induction of the H-19BΔQ-P$_R'$ prophage was employed to assess the role of the Q–P$_R'$ region in H-19B development. A log phase culture of an H-19BΔQ–P$_R'$ lysogen was treated with mitomycin C, and cell growth was monitored by optical density. Induction by similar treatment of a wild-type H-19B lysogen resulted in a drop in optical density that was first apparent approximately 2 h after addition of mitomycin C. Presumably this drop in OD resulted from the expression of phage-encoded lysis functions.[62] In contrast, induction of the H-19BΔQ–P$_R'$ lysogen did not result in any decrease in optical density, even after overnight incubation, indicating that phage lysis functions were not expressed even though the genes were present.[63]

Phage Induction and Stx Release

The contribution of phage-encoded functions to Stx release was also assessed using prophage induction. In this study the clinical isolate H19, which has an H-19B prophage,[63] was used. Previous studies showed that

---

[61] M. S. Donnenberg and J. B. Kaper, *Infect. Immun.* **59,** 4310 (1991).

[62] R. Young, *Microbiol. Rev.* **56,** 430 (1992).

[63] P. L. Wagner, J. Livny, M. N. Neely, D. W. Acheson, D. I. Friedman, and M. K. Waldor, *Mol. Microbiol.* **44,** 957 (2002).

the *stx1* promoter is regulated by iron through the action of the iron-dependent Fur transcriptional repressor; expression is derepressed in low iron.[18,64,65] Total production of Stx was measured under two growth conditions, phage-inducing conditions (full-iron medium made 10 μg/ml in mitomycin C) and Stx-inducing conditions (low-iron medium). Low-iron medium is syncase broth,[66] and full iron is syncase broth made 0.5 μg/ml in FeCl$_3$.[63] Although the total amounts of Stx produced under these two conditions were about the same, the distributions between supernatants and cells were very different. Under complete inducing conditions (mitomycin C and full iron), Stx was found almost entirely in the supernatant. Under low-iron conditions, Stx was found almost entirely within the bacteria. According to these results, regardless of whether phage induction is the major cause of its production, Stx release appears to be dependent on production of phage lysis functions (i.e., phage induction). This idea is attractive because no other mechanism for Stx release has yet been identified.

Analysis of Promoters Directly Involved in Stx Production

To assess the contribution of the *P*R′ promoter to Stx production *Pstx*,[18] the other promoter capable of directing transcription of *stx*, was removed. To this end, an H-19B prophage deleted for the *Pstx* was constructed, creating a lysogen with the H-19BΔ*Pstx* prophage. Based on the λ paradigm, this should result in *P*R′ being the primary promoter directing Stx expression in this lysogen. Stx expression by the H-19BΔ*Pstx* and H-19BΔQ–*P*R′ lysogens were compared to assess the contribution of *P*R′ to Stx expression.

To promote both phage- and toxin-inducing conditions, cultures of lysogens with mutant prophages and a control lysogen with wild-type H-19B were grown in low-iron medium supplemented with mitomycin C. The H-19BΔ*Pstx* lysogen produced levels of Stx essentially the same as those produced by the wild-type H-19B lysogen. Stx production had to be assessed differently for the ΔQ–*P*R′ prophage, because the loss of Q-modified transcription from *P*R′ results in a failure of late gene expression that includes the genes encoding lysis functions (see the following). Therefore Stx expression by this lysogen was compared with that from a lysogen

[64] S. de Grandis, J. Ginsberg, M. Toone, S. Climie, J. Friesen, and J. Brunton, *J. Bacteriol.* **169,** 4313 (1987).

[65] S. B. Calderwood, F. Auclair, A. Donohue-Rolfe, G. T. Keusch, and J. J. Mekalanos, *Proc. Natl. Acad. Sci. USA* **84,** 4364 (1987).

[66] R. A. Finkelstein, P. Atthasampunna, M. Chulasamaya, and P. Charunmethee, *J. Immunol.* **96,** 440 (1966).

carrying an H-19B prophage with an intact Q–P_R' region but defective in lysis because the phage has a mutation in the S lysis gene.[63] Induction of both lysogens resulted in essentially the same levels of Stx expression. In this way, use of lysogens with the promoter-deleted prophages permitted an assessment of the relative contributions of the two promoters to Stx production. Depending on the physiologic conditions, both promoters in H-19B appear to make substantial contributions to Stx expression.

These studies demonstrate how lysogens can be used to study mutations that render the phage defective in lytic growth. Here, deleted prophages permitted an assessment of the relative contributions of the two promoters to Stx production.

The nature of the mutations involved must be taken into consideration when considering the results of these experiments. Most importantly, the ΔQ–P_R' eliminates transcription of late phage genes even when $Pstx$ is active. This follows from the nature of Q-modified transcription from P_R', which reads through downstream transcription terminators. Transcription from $Pstx$ does not read through terminators. Therefore lysis functions located downstream of the stx genes with an intervening terminator are not expressed by transcription initiating at $Pstx$. This means that in the presence of mitomycin C the H-19BΔQ–P_R' lysogens will not express lysis functions and will continue to synthesize Stx for much longer than the wild-type or H-19B$\Delta Pstx$ lysogens.

The results with H-19B were in marked contrast to those observed with an *E. coli* 0157:H7 clinical isolate, 1:361, that carries the Stx2-encoding Φ361 prophage. This phage has the same Q, P_R', and qut site as H-19B and 933W, as well as the $stx2A$ and B genes with 933W.[16] However, a Q–P_R' deletion in the Φ361 prophage, constructed in the same manner as the ΔQ–P_R' in H-19B, rendered this phage incapable of producing Shiga toxin.[20] Little toxin is produced under noninducing conditions from the wild-type Φ361 prophage, whereas large amounts are produced when the prophage is induced with mitomycin C. In contrast, the Φ361 ΔQ–P_R' prophage was incapable of producing toxin in either condition. This suggests that for Stx2-producing prophages the P_R' promoter is critical for Shiga toxin 2 production. These results were confirmed by studies of Φ361 infection in mice in which Stx2 levels were measured in fecal samples. A 30-fold lower concentration of Stx2 was measured in the fecal samples of mice infected with the Φ361 ΔQ–P_R' lysogen compared with the 1:361 wild-type lysogen. This observation confirmed the idea that *in vivo*, as *in vitro*, Stx 2 expression from the Φ361 prophage is largely dependent on the P_R' promoter through prophage induction.[20]

Some Considerations

The close linkage of *stx* genes in association with phage regulatory sequences is highly conserved in most of the studied Stx producing bacteria.[67] In their study of the *stx*-flanking regions of 39 Shiga toxin-producing *E. coli* strains, including those that carry variant *stx* genes, Unkmeir and Schmidt[67] found that in 39 *E. coli* genomes with *stx* genes, 35 had the *stx* genes located between a *Q*-like antiterminator gene and an *S*-like lysis gene. Three of the four strains that did not have a *Q*-like gene had upstream sequences homologous to those found upstream of *Q* in the 933W genome. In addition, they analyzed eight *E. coli* strains carrying the variant Shiga toxin genes, stx_c and stx_{2f}, and found that all eight contained *S*-like lysis genes downstream of the *stx* genes. This high conservation of phage genes associated with the *stx* genes is even more significant when it is taken into consideration that the six stx_{2f} *E. coli* strains were isolated from pigeons. Analysis of the *S. dysenteriae* genome in the region of the Shiga toxin genes identified a lysis gene cassette downstream of the *stx* genes homologous to the sequences from H-19B and 933W.[68,69] Although a Q-like regulatory region upstream of the *stx* genes was not identified, an IS element, which may carry its own regulatory sequences, was identified upstream of the *stx* genes. The *stx*-encoding bacteriophage of *Shigella sonnei*, Ø 7888, has a *Q*-like antiterminator gene, as well as $P_{R'}$ and $t_{R'}$ upstream of the *stx* genes and a lysis cassette. The lysis cassette, as well as what appear to be $P_{R'}$ and $t_{R'}$ sequences, are over 95% homologous to those from H-19B and 933W.[70] The conservation of the toxin genes associated with lysis genes in a diverse group of bacteriophages isolated from different species suggests that there has been selective pressure for maintaining the linkage of the toxin to the lysis cassette. Furthermore, the ability of STEC carrying phages to infect and lysogenize various species of gram-negative organisms illustrates a scenario for recombination events that result in the vast diversity of *stx*-encoding lambdoid phages.

Numerous virulence factors have been found to reside on mobile genetic elements, allowing gene transfer from one organism to another and in this way contributing to the pathogenicity of the new host.[71] The first identified were those encoded on genomes of temperate bacteriophages.[72] The studies of *stx*-encoding phages have uncovered

[67] A. Unkmeir and H. Schmidt, *Infect. Immun.* **68,** 4856 (2000).
[68] M. A. McDonough and J. R. Butterton, *Mol. Microbiol.* **34,** 1058 (1999).
[69] S. Mizutani, N. Nakazono, and Y. Sugino, *DNA Research* **6,** 141 (1999).
[70] E. Strauch, R. Lurz, and L. Beutin, *Infect. Immun.* **69,** 7588 (2001).
[71] J. Hacker and J. B. Kaper, *Curr. Topics Microb. Immunol.* **264,** 157–175 (2002).

another role for phages carrying genes encoding virulence factors: contribution to the regulation of their expression and release.[21] The number of identified phage-encoded virulence factors has greatly expanded since the early reports of phage with toxin genes.[8,73] Thus an understanding of the role of phages in both the transfer and expression of virulence factors should continue to be an important component of the study of bacterial pathogenesis.

Acknowledgments

Jonathan Livny is thanked for helpful suggestions. Work by the authors was supported by grant AI43023 from the National Institutes of Health.

---

[72] L. Barksdale and S. B. Arden, *Ann. Rev. Microbiol.* **28,** 265 (1974).
[73] M. K. Waldor, *Trends Microbiol.* **6,** 295 (1998).

[33] Genetic and Biochemical Strategies to Elucidate the Architecture and Targets of a Processive Transcription Antiterminator from Bacteriophage Lambda

By Asis Das, Jaime Garcia Mena, Nandan Jana, David Lazinski, Gregory Michaud, Sibani Sengupta, and Zuo Zhang

Introduction

Antitermination of transcription is a gene activation mechanism by which cells regulate genes in response to environmental and developmental cues. A variety of molecular mechanisms enable RNA polymerase (RNAP) to read through a termination signal in bacteria.[1] Many of these antitermination mechanisms are terminator-specific in that the regulation is directed to a single terminator. By contrast, processive antiterminators of transcription act on the transcription machinery itself so that the modified polymerase can override many termination signals along its course.[2–6] The

[1] T. M. Henkin, *Ann. Rev. Genet.* 30:35–57 (1996).
[2] S. Adhya, M. Gottesman, B. De Crombrugghe, *Proc. Natl. Acad. Sci. USA* **71,** 2534–2538 (1974).
[3] S. Barik, B. Ghosh W. Whalen, D. Lazinski, and A. Das, *Cell* **50,** 885–99 (1987).
[4] R. J. Horwitz, J. Li, and J. Greenblatt, *Cell* **51,** 631–641 (1987).

archetype of these antiterminators is the *E. coli* phage lambda N gene product, a small (107aa) basic protein that binds three targets: RNAP, NusA (a loosely associated subunit of RNAP encoded by *E. coli*), and an RNA hairpin encoded by lambda *nutL* and *nutR* sites.[7–10] A previous article had described the methods of overproduction and purification of N protein and the host factors as well as the methods for *in vitro* transcription studies and RNA-binding studies used to characterize N.[11] Here, we focus on the strategies involved and the genetic and biochemical methods used to elucidate the architecture of N protein and its interaction with the various targets.

Transcription Antitermination in Lambdoid Phages

The N protein is vital for the development of lambda, the most extensively studied member among a small family of temperate phages that infect *E. coli* and its close relative, *Salmonella*.[12,13] The genome of a lambdoid phage contains three large operons, of which two are expressed early after infection (Fig. 1). During lysogenic response, a lambdoid phage establishes the expression a repressor protein (cI) that binds to specific operator sequences adjoining the early promoters and thereby limits early operon transcription (Fig. 1). During the lytic growth, however, the repressor is not made, and the expression of early operons is followed by the activation of the late operon. Because numerous terminators precede the distal genes of the early operons (Fig. 1), an active antitermination mechanism is required to allow optimal expression of the distal genes for lytic growth. In all but one of the lambdoid phages,[6] the product of the first gene of the leftward operon, N, mediates this early operon antitermination. Encoded within the early rightward operon are the phage DNA replication functions. This operon also encodes late gene activator, Q, whose expression is entirely dependent on N (Fig. 1). The Q protein is a transcription

[5] J. W. Roberts, *Cell* **52**, 5–6 (1988).

[6] R. A. Weisberg and M. E. Gottesman, *J. Bacteriol.* **181**, 359–367 (1999).

[7] A. Das, *J. Bacteriol.* **174**, 6711–6716 (1992).

[8] J. Greenblatt, J. R. Nodwell, and S. W. Mason, *Nature* **364**, 401–406 (1993).

[9] D. I. Friedman and D. L. Court, *Mol. Microbiol.* **18**, 191–200 (1995).

[10] D. I. Friedman and D. L. Court, *Curr. Opin. Microbiol.* **4**, 201–207 (2001).

[11] A. Das, M. Pal, J. G. Mena, W. Whalen, K. Wolska, R. Crossley, W. Rees, P. H. von Hippel, N. Costantino, D. Court, M. Mazzulla, A. S. Altieri, R. A. Byrd, S. Chattopadhyay, J. DeVito, and B. Ghosh, *Methods Enzymol.* **274**, 374–402 (1996).

[12] D. I. Friedman, E. R. Olson, C. Georgopoulos, K. Tilly, I. Herskowitz, and F. Banuett, *Microbiol. Rev.* **48**, 299–325 (1984).

[13] R. J. Juhala, M. E. Ford, R. L. Duda, A. Youlton, G. F. Hatfull, and R. W. Hendrix, *J. Mol. Biol.* **299**, 27–51 (2000).

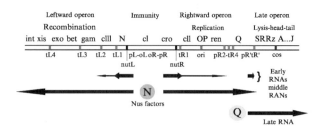

Fig. 1. Transcription antitermination in bacteriophage lambda. The diagram shows the genetic map of lambda (not drawn to scale), outlining the three operons that are regulated by N and Q. Relevant genes are shown on top, and the location of the *nut* sites and terminators are indicated at the bottom of the map. The origin of replication (ori) and the cleavage site for DNA packaging (cos) are shown. Transcripts relevant to the lytic program of lambda development are drawn below as arrows, reflecting direction of transcription, the start and stop sites, as well as the temporal order in which they are made. Some basal transcription of clll and cll-OP-ren occurs without the aid of N, as indicated, owing to inefficient termination at tL1 and tR1, respectively, both of which are Rho-dependent terminators. The terminators tL2 and tR2 are intrinsic terminators that are stimulated by NusA, while termination at tL4 is strongly enhanced by NusA. (See color insert.)

antiterminator, like N.[5] However, it binds DNA,[14] rather than RNA, to modify RNAP in the late operon, which encodes not only the head and tail proteins of the phage but also functions to package DNA and cause cell lysis.

A remarkable feature of antitermination in different lambdoid phages is their demonstrated genome specificity: N and Q proteins act preferentially on the cognate genomes to turn on the early and the late operons, respectively. Obviously, this specificity is due to sequence-specific interaction between the antiterminator and the cognate RNA (in case of N) or the DNA (in case of Q). Despite this disparity in the nature of their recognition signals, however, N and Q act by a common mechanism: each antiterminator binds and transforms RNAP at a promoter-proximal site so that termination is suppressed downstream, at many sites.[3,15] Many of these terminators act on RNAP directly and hence are classified as intrinsic terminators, while others require Rho factor or NusA for function. Notably, N and Q suppress each type of terminator. Exactly how the two proteins achieve this has yet to be understood.

An N-recognition site (*nut*) is present in each of the two early operons subject to regulation by N (Fig. 1). The *nutL* site lies upstream of N, while *nutR* precedes the tR1 terminator. The nut site is made of two functional elements (Fig. 2). The 5′ element (BoxA) is an ~12nt single-stranded

[14] W. S. Yarnell and J. W. Roberts, *Cell* **69,** 1181–1189 (1992).
[15] E. J. Grayhack, X. J. Yang, L. F. Lau, J. W. Roberts, *Cell* **42,** 259–269 (1985).

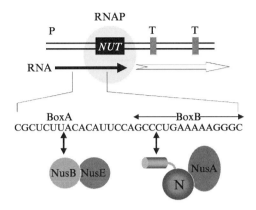

FIG. 2. Function of the *nut* site. Sequence of the lambda *nutR* RNA is shown along with the RNA-protein interactions involving BoxA and BoxB. The arrows above BoxB identify the hairpin stem, whose 5′ arm binds the ARM helix of N. (See color insert.)

RNA that is conserved among the *nut* sites of many lambdoid phages, including P21, P22 and H19B.[10,16] The BoxA RNA binds a heterodimer of two *E. coli* proteins, NusB and the S10 ribosomal protein or NusE.[17,18] The other element (BoxB) is a small interrupted palindrome that forms the unique N-binding RNA hairpin.[19,20] The lambda BoxB hairpin folds as a GNRA tetraloop structure that forms a ternary complex with NusA and the amino terminal domain of N.[21–23]

In addition to the *nut* site, several host factors (NusA, NusB, NusE and NusG) influence N activity *in vivo*.[11] *In vitro*, NusA alone suffices for N to transform RNAP at the *nut* site[24]; the additional factors are needed for the persistence of the antitermination complex over time and distance.[25,26]

[16] N. C. Franklin, *J. Mol. Biol.* **181,** 75–84 (1985).

[17] J. R. Nodwell and J. Greenblatt, *Cell* **72,** 261–268 (1993).

[18] H. Luttgen, R. Robelek, R. Muhlberger, T. Diercks, S. C. Schuster, P. Kohler, H. Kessler, A. Bacher, G. Richter, *J. Mol. Biol.* **316,** 875–885 (2002).

[19] D. Lazinski, E. Grzadzielska, and A. Das, *Cell* **59,** 207–218 (1989).

[20] S. Chattopadhyay, J. Garcia-Mena, J. DeVito, K. Wolska, and A. Das, *Proc. Natl. Acad. Sci. USA* **92,** 4061–4065 (1995).

[21] J. Mogridge, T. F. Mah, and J. Greenblatt, *Genes Dev.* **9,** 2831–2845 (1995).

[22] P. Legault, J. Li, J. Mogridge, L. E. Kay, and J. Greenblatt, *Cell* **93,** 289–299 (1998).

[23] M. Scharpf, H. Sticht, K. Schweimer, M. Boehm, S. Hoffmann, and P. Rosch, *Eur. J. Biochem.* **267,** 2397–2408 (2000).

[24] W. Whalen, B. Ghosh, and A. Das, *Proc. Natl. Acad. Sci. USA* **85,** 2494–2498 (1988).

[25] S. W. Mason, J. Li, J. Greenblatt, *J. Biol. Chem.* **267,** 19418–19426 (1992).

[26] J. DeVito and A. Das, *Proc. Natl. Acad. Sci. USA* **91,** 8660–8664 (1994).

NusA, NusE and NusG each bind RNAP.[8] NusA enhances transcription pausing and NusG can suppress a pause, while no effect of NusE on transcription on its own has been reported. Whether these activities of NusA and NusG play a role in N-mediated antitermination is not known.

The Domain Architecture and Targets of N

Experimental Strategies

A combination of genetic and biochemical experiments have helped to reveal that the lambda N protein has an architecture that is typical of many transcription activators (Fig. 3). N is made of two autonomous domains: an arginine-rich domain in the amino terminus binds the NUT RNA hairpin in the absence of any other protein, and the carboxy terminus binds NusA and RNAP, individually, in the absence of NUT RNA.

RNA-Binding Domain. Genetic evidence that the N proteins contain an autonomous *nut* site recognition domain came from the *in vivo* analysis of hybrid proteins.[19] Having the DNA sequences of the N genes and the *nut* sites from three members of the lambdoid phage family,[16] chimeric N genes were constructed in which either the amino or the carboxy terminus of one N protein (lambda) was swapped with the respective regions from

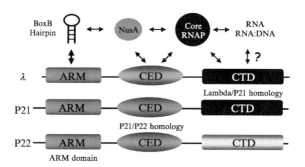

FIG. 3. Architecture and targets of N. A schematic representation illustrates the demonstrated function of the three domains of lambda N protein (top) and the similarities and differences among N homologues from P21 and P22. Note that P21 and P22 ARM domain is extended at the amino terminus. The central domain (CED) of P21 and P22 N proteins are very similar; however, lambda N CED is quite different. Lambda and P21 N proteins share a structurally similar and functionally equivalent carboxy terminal domain (CTD). Although P22 N differs from lambda and P21 N proteins in the CTD, it does show strong similarity in this domain with several other N homologues, including that from the Shiga-toxin phage H19B. The possibility that the CTD of lambda N may bind a nucleic acid to influence antitermination is indicated with a question mark. (See color insert.)

another N protein (P21 or P22) (Fig. 4). When cloned in an expression vector, the resultant hybrid N proteins not only proved to be active in antitermination in vivo but they displayed genome specificity (see Procedures 1–3). The specificity of the hybrid N protein was directed toward the *nut* site of the phage from which the amino terminus was derived, thus

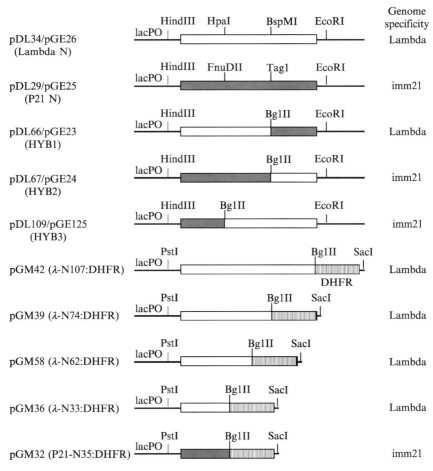

FIG. 4. *Nut* site recognition domain of N defined by hybrid and fusion proteins. The key hybrids between lambda and P21 N used to localize the BoxB–binding domain to the amino domain are shown along with the restriction sites used to create the hybrids. As indicated by the names of constructs, two plasmids were made for each hybrid, in the pUC19 and the p15A-lacIq vectors, respectively.[19] Also included in the figure are the DHFR constructs that demonstrated that the amino domain of lambda N is sufficient to bind BoxB *in vivo* and that its central domain binds both NusA and RNAP.

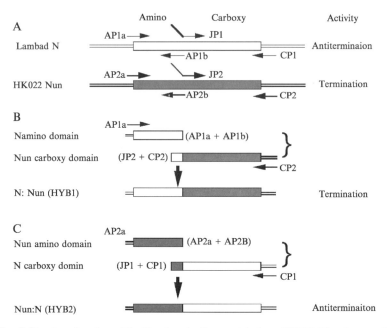

Fig. 5. Structure-function of the Nun termination protein from HK022. The diagram shows how a set of two reciprocal hybrids between N and Nun were made by using PCR methods. Details are described in the text. Note that Nun is larger than N; the two genes are drawn in the same scale for simplicity. Note also that the amino and the carboxy terminal primers contain restriction sites suitable for cloning of the hybrid genes in an expression vector.[27]

assigning unequivocally the *nut* site recognition function to the amino domain of N, which contains a conserved arginine-rich motif (ARM).[19] The same conclusion was reached from studies of hybrids made between lambda N protein and the HK022 phage Nun protein, which also contains the ARM motif at its amino terminus and causes transcription to terminate downstream from the lambda *nut* sites.[27] A hybrid N:Nun protein with the amino domain from N caused termination while a hybrid Nun:N having the amino domain of Nun caused antitermination (Fig. 5). Thus, like N, the amino domain of Nun must bind NUT RNA, while its carboxy terminus would signal RNAP. These conclusions have been confirmed by biochemical studies.[28,29]

[27] K. S. Henthorn and D. I. Friedman, *J. Mol. Biol.* **257,** 9–20 (1996).
[28] S. Chattopadhyay, S. C. Hung, A. C. Stuart, A. G. Palmer 3rd, J. Garcia-Mena, A. Das, and M. E. Gottesman, *Proc. Natl. Acad. Sci. USA* **92,** 12131–12135 (1995).
[29] R. S. Watnick, S. C. Herring, A. G. Palmer 3rd, and M. E. Gottesman, *Genes Dev.* **14,** 731–739 (2000).

Negative Dominant N Mutants. Genetic evidence that the amino domain of N (the arginine-rich motif) is sufficient to bind NUT RNA was obtained by constructing truncated forms of the lambda N gene and determining the negative dominant phenotype of the resultant N-derived peptides (Procedure 2). As said above, the N gene is essential for the lytic growth of lambda. Thus a lambda mutant phage that bears a nonsense mutation in its N gene can form plaques on a nonsense-suppressor host but not on an otherwise wild-type (non-suppressor) host (Fig. 6). The cloned N gene, when expressed in a non-suppressor host, enables plaque formation by the N mutant phage (i.e., phage complementation). Now consider that a domain of N that binds RNAP is deleted from the carboxy terminus of N. If the truncated peptide is stably expressed in *E. coli*, and its amino domain is sufficient to bind NUT RNA, then the peptide should be a competitive inhibitor of the wild type N protein (i.e., negative dominance).

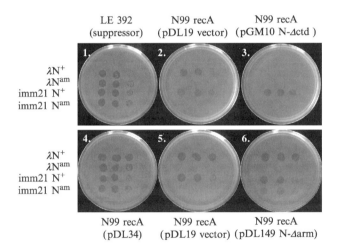

FIG. 6. Activity of lambda N protein and peptides on phage growth: Complementation and dominance tests. Serial dilutions of four phage stocks were spotted on the lawns of the various strains indicated, induced with IPTG. The two amber mutant phages (lambda and imm21), which produce plaques on the suppressor host LE392 (plate 1), fail to plate on the non-suppressor host (plates 2 and 5). pDL34 overproduces lambda N. The fact that it complements each of the two amber mutant (plate 4) demonstrates the loss of genome specificity described previously.[19] pDL149 is the overproducer of the ΔARM N mutant; it fails to complement the amber mutant of lambda demonstrateing the essential role of the ARM domain. However, the fact that pDL149 complements the imm21 mutant shows that the ARM domain is dispensable for antitermination, per se, hence establishing the autonomy of the remainder of the N protein. The CTD deletion mutant of N, expressed by pGM10, not only fails to complement each amber mutant (Plate 3), it blocks the growth of wild-type lambda. This dominance is attributed to the ability of the peptide to bind the *nut* site and sequester it from the wild type protein.

A host cell expressing this peptide would not only fail to plate the N mutant phage, reflecting a defect of the peptide in antitermination, but it should also fail to plate the wild-type phage that encodes an intact N gene, reflecting the ability of the peptide to compete with wild type N. This is indeed true (Fig. 6). An amino-terminal peptide derived from lambda N, as small as 33 amino acids, can block lambda growth completely. This block to wild-type phage growth is due to the inhibition of N function, since the peptide has no effect on a terminator-deletion mutant of lambda, λ*nin*, which can grow without N.[30] The peptide is also specific as an N-inhibitor, since it has no effect on a hybrid phage, λ*imm21*, which encodes the P21 N gene and the P21 *nut* sites in place of lambda (Fig. 6).

NusA/RNAP Binding Domains. Several pieces of genetic evidence hinted that the lambda N protein interacts with NusA and RNA polymerase.[11] First, suppressor mutations in N (called punA) have been isolated, which permits lambda growth in the *nusA1* mutant host.[31] Second, silent mutations in N (called mar) have been isolated that affect lambda growth in hosts with specific *rpoB* (RNAP beta subunit) mutations (called ron).[32] Curiously, the punA and mar mutations all localize to the central domain of N (Fig. 3).[33] To directly test the interaction of N with NusA/RNAP in vitro, and to localize the binding domains in N, a variety of fusion proteins have been engineered that contain GST, Hexa-His or DHFR tags (Fig. 7).[34] By coupling these fusion proteins to appropriate agarose beads and then testing the ability of these coupled beads to retain purified ligand proteins (Procedures 4 and 5), it has been possible to demonstrate that indeed the N protein forms binary complexes with NusA and core RNAP.

Next, by using fusion proteins that contain various parts of N, the NusA binding domain has been mapped to the central region of the protein, the same domain in which the punA substitution mutations have been mapped that allow N function in both *nusA* and *nusE* mutant hosts. In an independent approach using fluorescently labeled N protein and peptides, the central domain of N was shown to be responsible for binding NusA.[35] Consistent with the location of the mar mutations in this same region, the central domain of N binds the RNAP core as well (see later). This is not the only domain to bind RNAP, however, as the carboxy terminus of N binds

[30] D. Court and K. Sato, *Virology* **39,** 348–352 (1969).

[31] D. I. Friedman, A. T. Schauer, M. R. Baumann, L. S. Baron, and S. L. Adhya, *Proc. Natl. Acad. Sci. USA* **78,** 115–118 (1981).

[32] A. Ghysen and M. Pironio, *J. Mol. Biol.* **65,** 259–272 (1972).

[33] N. C. Franklin, *J. Mol. Biol.* **181,** 85–91 (1985).

[34] J. Mogridge, P. Legault, J. Li, M. D. Van Oene, L. E. Kay, and J. Greenblatt, *Mol. Cell* **1,** 265–275 (1998).

[35] M. R. Van Gilst and P. H. von Hippel, *J. Mol. Biol.* **274,** 160–73 (1997).

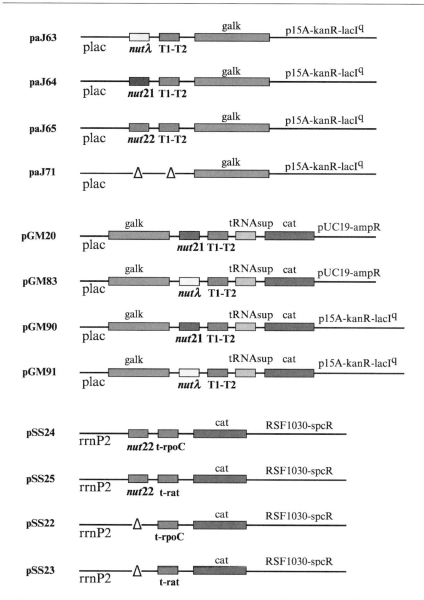

FIG. 7. A collection of antitermination reporter plasmids. The pAJ plasmids were derived from constructs made previously.[19] The rest of the plasmids shown represent a new generation of the reporters constructed in this laboratory.[36,37] The trpoC and t-rat (rho gene attenuator) terminators have been cloned from plasmids provided by C. Squires.[38] (See color insert.)

[36] G. A. Michaud, Ph. D. thesis, University of Connecticut, Storrs, 1998.

core RNAP with a similar affinity.[34] Together, these experiments indicate a bipartite structure of the RNAP-binding domain of N. The target of N in NusA has been localized to its carboxy terminus, away from the S1 and KH domains implicated in RNA binding.[39] The targets of N in RNAP has not yet been disclosed. As mentioned above, several mutations in RNAP affect N activity, including *groN* and *nusC*;[11] however, the amino acids altered by these mutations remain to be discerned.

Autonomy of the "Antitermination-Induction" Domain

The activity and specificity of the N hybrid proteins suggest that the antitermination-induction domain of N functions independently of the RNA-binding domain. Two pieces of genetic evidence support this conclusion (Fig. 6). First, an excess of the N protein can cause antitermination *in vitro* without Nus factors or a *nut* site in the template.[26,40] Second, a deletion mutant of N devoid of the ARM domain can cause antitermination *in vivo* under certain conditions.[41] The ΔARM deletion mutant of N does fail to complement the lambda N amber mutant, even when the ARM-less N protein is overproduced in the cell. However, on overproduction, the ARM-less N mutant complements the P21/P22 N amber mutant phages,[41] each of which contains an optimal boxA sequence in its nut sites.[42] Because the amber mutation in P21 N truncates the protein at the sixth tryptophan residue (TGG to TAG),[36] an amber peptide from the phage could not be involved in RNA binding and trans-complementation. Evidently, the lambda N protein contains an autonomous RNAP signaling domain that can function without binding NUT RNA. We suspect that the tight complex of an optimal BoxA sequence with NusB-S10[43] can somehow recruit the truncated N protein to the *nut* site without the aid of N-BoxB interaction.

Experimental Procedures

Procedure 1: Construction of Hybrids, Fusions, and Truncated Forms of N

Although PCR is now routinely used to generate hybrids, fusions, and truncated derivatives of a protein under study, part of the studies on N reviewed above was done by conventional cloning techniques, before PCR came of age. We describe both of these approaches briefly.

[37] S. Sengupta, Ph D. thesis, University of Connecticut, Storrs, 2000.

[38] B. Albrechtsen, C. L. Squires, S. Li, and C. Squires, *J. Mol. Biol.* **213**, 123–134 (1990).

[39] T. F. Mah, J. Li, A. R. Davidson, and J. Greenblatt, *Mol. Microbiol.* **34**, 523–537 (1997).

[40] W. A. Rees, S. E. Weitzel T. D. Yager A. Das, and P. H. von Hippel, *Proc. Natl. Acad. Sci. USA* **93**, 342–6 (1996).

[41] D. W. Lazinski Ph. D. thesis, University of Connecticut, Storrs, 1989.

[42] D. I. Friedman E. R. Olson, L. L. Johnson, D. Alessi, and M. G. Craven, *Genes Dev.* **4**, 2210–2222 (1990).

REGULATED EXPRESSION VECTORS. A series of vectors have been constructed thus far that allow regulated expression of N for genetic tests and also overproduction for purification and biochemical analysis (Fig. 4). These constructs utilize the *lac, tac* or the *trc* promoter, regulated either by a resident *lacl*^{q} repressor gene or a compatible plasmid vector supplying *lacl*^{q}. The vectors with the *tac/trc* promoter are efficient in overproducing the desired protein. Note, however, that the physiological tests of N protein function with these vectors are complicated by the limited ability to keep the expression to nontoxic levels.

CONSTRUCTION OF HYBRID PROTEINS. The generation of hybrid genes without the aid of PCR required that restriction sites were conveniently located at domain boundaries within each of the N homologs, as was the case for lambda and P21 N genes (Fig. 4). Therefore it was possible to generate a limited number of hybrids in three steps. First, the wild-type N genes were cloned in the same expression vector (Fig. 4); second, linkers with a unique restriction site (BgIII) were inserted at the domain boundaries by using the available sites (Hpal and BspMl in lambda N and FnuDll and Taq1 in P21 N); third, the amino and the carboxy terminal fragments were swapped to generate in-frame fusions at the two boundaries (Fig. 4). Although this approach was successful in localizing the RNA binding domains of the N homologues conclusively, attempts to produce active hybrids that substitute the ARM domain of P21 with that from lambda had failed. This could be due to many reasons, including aberrant protein structure and instability.

PCR-mediated joining of two genes is obviously the best approach to create hybrid proteins with many junction points, which was used to obtain active hybrids between N and Nun.[27] In this approach, also suitable for making fusions to a desired tag, a set of eight primer oligonucleotides is used to create two reciprocal hybrids (Fig. 5). As depicted in the figure, the amino terminus of each gene is amplified with two sets of amino terminal primers (AP1a + AP1b and AP2a + AP2b) that correspond to the beginning and the end of the amino domain of each of the two genes. The carboxy terminus of each gene is amplified by combining a specific junction primer (JP1 or JP2) and a gene-specific carboxy terminal primer (CP1 or CP2). Each junction primer's 5' end corresponds to the end of the amino domain of one gene while the 3' end corresponds to the beginning of the carboxy domain of the other gene. For instance, JP1 oligonucleotide could encode 7 amino acids of Nun and 7 amino acids of N, in that order, while JP2 would correspond to 7 amino acids of N and 7 amino acids of Nun, as shown in Fig. 5. The PCR is done in two steps. First, the amino and the carboxy domains of each gene are amplified in four separate

---

[43] J. Mogridge, T. F. Mah, and J. Greenblatt, *J. Biol. Chem.* **273,** 4143–4148 (1998).

reactions. Subsequently, the PCR products are combined in a second PCR reaction to extend the amino and the carboxy domains via homology at the fusion junction, resulting in DNA products that have the two domains fused. These products are amplified by including in the same reaction specific amino and carboxy terminal primers corresponding to the two hybrid constructs (AP1a + CP2 and AP2a + CP1).

CONSTRUCTION OF FUSIONS. For biochemical studies, the N protein has been fused to a number of tags at either the amino or the carboxy domain. Plasmids expressing DHFR fusion proteins (Fig. 4) were engineered by combining PCR amplified N and DHFR fragments with the vector backbone of pDL19, a derivative of the high copy Amp^R plasmid pUC19 that contained an altered multiple cloning site. The *E. coli* dihydrofolate reductase (DHFR) coding region was amplified from pLP5, a gift from L. Shulman, by using appropriate primers. Owing to their high copy number, the plasmids express the DHFR fusion proteins from the *lac* promoter at high levels in a variety of *E. coli* strains without induction by IPTG. For toxic fusions, this expression is reduced to marginal levels by introducing into these strains a compatible plasmid (p15A-lacI^q-Kan^R). A high level of expression is induced by IPTG when desired.

A second series of N protein fusions to His-tag and DHFR was generated in a modified derivative of the vector pTRC99. In it, the pT7 promoter derived from phage T7 follows the *trc* promoter. The resident *lacI*^{q} gene in this vector controls both promoters via *lac* operator sites adjoining each promoter. This series of plasmids all encode a His-tag, a FLAG-tag and a peptide for phosphorylation by the Heart Muscle Kinase (HMK), in that order, followed by the N gene. A series of truncations and internal deletions of the N gene was then constructed using standard PCR methods. Finally, the DHFR gene was fused to the carboxy terminus of these N gene constructs. Because the N protein is subject to severe proteolysis in *E. coli*, the fusions with a single tag often yield a mixture of peptides, the full-length, as well as distinct, cleavage products. Thus, tags were added at both the amino and the carboxy domain to ensure the purification of the intact protein free of cleavage products. Furthermore, the DHFR tag and the FLAG tag would permit binding studies with other proteins (NusA/RNAP) that are tagged with His. The HMK tag permits the *in vitro* labeling of the N protein with radioactive phosphate to aid binding studies and to identify and characterize the products of cross-linking to the various targets of N.

Procedure 2: Phage Complementation Assay: Sterile Media, Plates, and Solutions

LB MEDIUM. Use 10 g tryptone, 10 g yeast extract, 5 g NaCl per liter of distilled water.

TBMM MEDIUM. Use 10 g tryptone, 5 g NaCl per liter of distilled water. After autoclaving, add maltose to 0.2%, and $MgSO_4$ to 10 mM. A 20% maltose stock is made sterile by filtration; sterile 1 M $MgSO_4$ stock is made by autoclaving.

TB AGAR PLATES. Use 10 g tryptone, 5 g NaCl, 11 g Difco agar per liter of distilled water. Let the media cool to about 60° before pouring plates. Store plates sealed at 4° to reduce dehydration; fresh plates (1 to 2 weeks) are ideal for lambda plating experiments.

TB TOP AGAR. Use 10 g tryptone, 5 g NaCl, 7 g agar per liter of distilled water. Typically, 100-ml stocks are stored in round bottles. When needed, heat the bottle in a microwave to melt agar completely (do not overboil) and place the bottle in a block heater to bring agar temperature to 48°.

TMG SOLUTION. Use 1.2 g Tris base, 2.46 g $MgSO_4$ $7H_2O$ and 0.1 g gelatin (Sigma) per liter of distilled water. Adjust pH to 7.4, heat to dissolve gelatin, dispense in 100-ml glass bottles and autoclave.

PREPARATION AND TITRATION OF PHAGE STOCKS. Each phage stock should be prepared from single plaques produced on a permissive host. To obtain single plaques, grow an overnight TBmm culture (5 ml) of the permissive host from a single colony (C600 or LE392 for the amber mutant phages and N99 for others). Starting with a stock of the phage, make serial dilutions in TMG. Mix 0.1 ml of these dilutions with 0.2 ml of the host cell culture and incubate at 37° for 10 minutes. To this infection mixture, add 2.5 ml Top Agar, mix by mild vortex, and pour immediately onto a TB agar plate, taking care to spread agar uniformly. Let the agar solidify on bench top, and then incubate the plates overnight at 37°.

To grow a new stock, pick a single plaque (the agar underneath the clear area) with a sterile micropipette and add to 0.25 ml of a fresh, saturated TBmm culture of the host and incubate at 37° for 15 min. Transfer infected culture to a 250-ml flask with 25 ml of LB containing 5 mM $CaCl_2$ and shake for 5 to 6 h. The culture will become turbid within a few hours and then clear, indicating an efficient lysis. Add 0.2 ml chloroform, shake for 5 minutes, and then let the flask sit on the bench top for 10 min. Obtain a clear supernatant of the lysate after centrifugation (5000 rpm for 10 min), and save several aliquots at 4°.

To titer the phage stock, follow the procedure to obtain single plaques, detailed above. Samples of serial dilution tested in triplicate provide a dependable titer of the stock. For the amber mutants, the dilutions of the stock should also be tested on a non-permissive host (N99) to ensure that the efficiency of plating (EOP) is reduced to 10^{-4}, or less, in that host compared to the permissive host (LE392).

DETERMINATION OF EOP. To test various N gene constructs for antitermination activity, the ability of the respective N gene to complement an N amber mutant is tested in a non-suppressor host (Fig. 6). To avoid instability associated with toxic plasmids, N99 is transformed freshly with the desired N expression plasmids and a culture of the transformant is grown in TBmm. Following the method described above for obtaining single plaques, serial dilutions of the phage stocks are tittered on LE392, N99, and N99 with various plasmids. The efficiency of plating of the amber mutant is then determined by comparing the number of plaques produced on an N-plasmid bearing N99 versus that produced on LE392. Plasmid pDL34 encoding the wild-type N gene of lambda complements the lambda N amber mutant at 100% efficiency, meaning that a stock of the N amber mutant produces the same number of plaques on LE392 and N99(pDL34). By comparison, pGM39 that encodes a truncated lambda N gene (aa 1–74) fused to DHFR produces plaques of the lambda N amber mutant at an efficiency of 0.01% or less. Combined with the evidence that this truncated peptide is stably expressed in N99, the reduced EOP indicates the essential role of the carboxy terminus in N protein function.

REGULATION OF N EXPRESSION. The expression of the N gene from the plasmid vector is repressed with the aid of lac repressor. Therefore, test cultures are plated with 0.02 to 0.2 mM IPTG in the top agar to induce N expression to the desired level. It should be noted that there is often a basal level of N protein expression from these recombinant plasmids, without IPTG-induction, that is sufficient to allow complementation. An important point to remember is that N is toxic to *E. coli*, especially at temperatures above 40°. Therefore, tests for phage complementation must maintain the level of N expression below that which inhibits bacterial growth.

PLAQUE SIZE. Some N mutations under study may show little or no effect on EOP, yet it may be possible to distinguish a modest decrease in activity based on the character of plaques. Thus it is important to distinguish the size of the plaques produced in an EOP experiment. When a complementation test displays smaller-than-expected plaques consistently, the result means that the activity of the cloned N gene under test is compromised. This can then be quantified by reporter assays (below), quantitative RT-PCR experiments, or by determining the phage burst size in a single infection cycle, which is the most sensitive and quantitative measure of complementation.[41]

SPOT TESTS. A spot test of serial dilutions is the most convenient way to obtain rough estimates of EOP of several phages on a large number of hosts and at the same time obtain a visual record of such results, as shown in Fig. 6. To perform a spot test, use 96-well plates to prepare serial dilutions of the desired phage stocks. Grow fresh, overnight cultures of the host cells to be tested. Mix 0.2-ml aliquots of the host cell cultures with 2.5 ml Top Agar and pour onto TB-agar plates. Let the agar solidify. Identify plates with markings at the bottom to indicate the specific host lawns they carry as well as specific phages that are to be spotted. Next, use a multi-channel pipette to spot about 5 μl of the serial dilutions. Let the spots dry on the bench top, and incubate the plates overnight. In most cases, it should be possible to count the number of plaques emerging from the highest dilution spot; done in duplicates or triplicates, EOP can be estimated from these numbers. Pictures should be taken as rapidly as possible to record data.

TEST FOR NEGATIVE DOMINANCE. Negative dominance is conveniently determined by the spot test just described. If a mutant interferes strongly with the activity of the wild-type protein, the spot test would show a defect in phage growth by orders of magnitude, as was the case for truncated lambda N mutants (Fig. 6). A less-severe interference could be reflected in reduced plaque size, which is best documented by the measurements of EOP and confirmed by the reduction in burst size in a single cycle infection experiment.

Procedure 3: Quantitative Assay of Antitermination with a Dual Plasmid System

Although the phage complementation assay is extremely convenient to monitor the physiological activity and specificity of N protein mutants, this method alone is inadequate to provide a complete picture of how a mutation affects N activity. Therefore, reporter plasmids have been engineered to quantify the various parameters of N-mediated antitermination (Fig. 7).

One set of reporters of antitermination utilizes the *E. coli galK* (galactokinase) gene placed under the control of the *lac* promoter. Between the promoter and the *galK* gene, different terminators have been placed to limit *galK* expression. Next, a *nut* site is inserted upstream from the terminators so that N function can be monitored. This set of reporter plasmids includes different *nut* site cassettes (both natural or synthetic), and those in which BoxA and BoxB are mutated. Finally, the reporter fusion has been engineered in colE1 (AmpR), p15A (KanR) and RSF1030 (SpcR) replicons so that they may be combined conveniently with compatible N-supplier plasmids.

REPORTERS WITH THE *GALK* GENE. The *galK* reporter system is suitable not only for visual analysis and genetic screens (Fig. 8) but also for quantitative measurements of antitermination by a convenient radiochemical assay (Fig. 9). The same is true for reporters that are made of the *lacZ* (beta-galactosidase) gene from *E. coli* or the *cat* (chloramphenicol acetyl transferase) gene from transposon Tn9 (Fig. 8). With either *galK* or *lacZ* reporter, positive expression is indicated by the formation of red colonies on MacConkey agar plates containing galactose or lactose, respectively (Fig. 8). By contrast, TB-TTC-agar plates with the respective sugar are

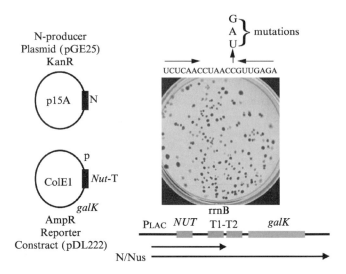

FIG. 8. A two-plasmid system for monitoring processive antitermination *in vivo:* screening of *nut* site mutants. One of several available two-plasmid systems is shown here, depicting the basic features of the plasmids, the *galK* reporter construct and the phenotype displayed on MacConkey-galactose agar plates containing 0.1 mM IPTG. The plate displays the colonies of a *galK* mutant strain (N99 *recA::Tn10*) with the P21 N plasmid, pGE25, that was transformed with the mutagenized clones of the reporter plasmid pDL222. Indicated at the top of the P21 BoxB RNA sequence are three substitutions that were introduced by oligonucleotide directed cassette mutagenesis.[18] That certain substitution(s) of a specific loop nucleotide has abolished the *nut* site function is revealed by the colorless nature of some of the colonies. The red colonies in the background, on the other hand, indicate that one or more substitutions of the same nucleotide have little effect on *nut* site function. (See color insert.)

used to display red colonies that do not express the reporter gene. In genetic screens, the latter type of plates is used to identify rare red colonies (mutants that do not express the reporter) among the background of colorless colonies (parental strain expressing the reporter).

INDICATOR AGAR PLATES. MacConkey agar plates are made by dissolving 40 g Bacto-MacConkey agar base from Difco in 950 ml dH$_2$O, and after autoclaving, supplementing the medium with 50 ml of a sterile solution of 20% galactose plus any required antibiotic. TB-TTC-agar plates are made by dissolving in 950 ml dH$_2$O 10 g tryptone, 10 g agar, 5 g NaCl, 25 mg 2,3,5-triphenyl-2H-tetrazolium chloride and adding the necessary antibiotics plus a 50 ml solution of 20% galactose, after autoclave.

GALACTOKINASE ASSAYS. The galactokinase assay consists of the following steps: (a) growth and induction of cultures; (b) toluene treatment to obtain permeabilized cells; (c) incubation of the cell suspension with

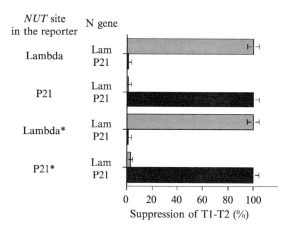

FIG. 9. Quantitative measurement of N activity and specificity by the two-plasmid system. The *galK* mutant strain N99 recA::Tn10 was transformed with KanR plasmids pGE25 or pGE26 (see Figure 1) plus an AmpR reporter plasmid that contained the indicated *nut* site. Lambda* and P21* indicate synthetic *nut* sites made of an optimal BoxA sequence (CGCTCTTTAACA) plus BoxB from lambda and P21, respectively.[19] The relative levels of galactokinase expressed after induction with IPTG were determined from the different cultures by the procedure outlined in the text.

a reaction mixture containing radioactive galactose; and (d) filtration to isolate the reaction product, galactose-1-phosphate.[2]

For a typical assay, 1-ml cultures are harvested in Eppendorf microtubes, and cells are resuspended in a lysis buffer (40 mM Tris-HCl, pH 7.5, 1 mM DTT) made fresh. Next, 500-μl aliquots are mixed with 50 μl toluene, vortexed, incubated at 37° for 10 min and the tubes are kept on ice. Reactions (100 μl) are set up in microtubes containing a reaction cocktail made with the following components. Mix 1: 0.8 M Tris-HCl, pH 7.9, 40 mM MgCl$_2$, 25 mM ATP; Mix 2: 80 mM NaF, 12.5 mM DTT; Mix 3: 20 mM D-galactose, 200 μCi/ml ^{3}H-D-galactose. For twenty reactions, 50 μl each, a complete reaction mixture is first prepared by combining 114 μl of Mix 1, 86 μl of Mix 2 and 100 μl of Mix 3. Next, 15 μl aliquots of this mixture are dispensed in microtubes containing 5 to 25 μl of the cell suspension plus dH$_2$O to make up the volume to 50 μl. Reactions are done at 37° for 30 min min and terminated by spotting 40-μl aliquots on Whatman DE81 filter discs. After air drying, the discs are washed several times with high quality deionized dH$_2$O by placing discs on a filter paper in a large funnel. The discs are dried again, and the radioactivity of galactose-1-phosphate retained on the discs is counted with an appropriate scintillation cocktail. For a measure of the relative levels of galactokinase expressed in the various strains, the volume and concentration of the culture aliquots should be adjusted such that no more than 20% of the substrate is converted to the product. One unit of galactokinase is presented as pmoles of ^{3}H-galactose-1-phosphate synthesized per minute per milliliter of culture ($A_{600} = 1$).

REPORTERS WITH THE *CAT* GENE. The *cat* gene is a useful reporter in that the rate of growth of cells in culture (or of colonies on plates) with varying amounts of chloramphenicol can reflect differential expression of the reporter with various mutants whose activity is compromised to different degrees. Wild-type *E. coli* is sensitive to 1 to 2 μg/ml chloramphenicol. Depending on the reporter system and the plasmid copy number, cells may be resistant to as much as 200-fold more chloramphenicol. This large window is quite convenient for phenotypic grouping of defective mutants and revertants. Results from the phenotypic screens are then extended by direct radiochemical measurements of the level of chloramphenicol acetyl transferase,[44] or by quantitative RT-PCR of the *cat* mRNA.

CAT ASSAYS. A typical CAT assay protocol is as follows. Freshly transformed cells are grown to saturation in LB containing the desired antibiotics. To obtain samples for the CAT assay, the culture is diluted into fresh medium to $A_{600} = 0.1$ and grown at 37° till the culture reaches about 0.5 A_{600}; IPTG is added to the dilution medium if induction is desired. One-milliliter aliquots of the culture are pelleted in Eppendorf microtubes, and the cell pellet is resuspended in 0.1 M Tris-HCl, pH 8, and kept on ice. For CAT assays, 0.1-ml aliquots of the samples are mixed with 0.1 ml of 2× lysis buffer (8 mM each of DTT, EDTA and Tris-HCl, pH8). Cells are permeabilized by treating samples with 4 μl toluene; after vigorous vortexing, the tubes are incubated at 30° for 30 minutes and returned to ice. The toluene-permeabilized cultures are next diluted 5- to 20-fold with the lysis buffer, and undiluted aliquots are frozen for additional future measurements. Next, 90-μl aliquots of a CAT reaction cocktail (0.1 M Tris-HCl, pH8, 4 mM DTT, 0.5 mM n-butyryl CoA, 20 μM chloramphenicol plus 0.1 μCi of 3H-chloramphenicol) are dispensed into Eppendorf tubes and preincubated at 37° for 5 minutes before 10 μl of culture samples is added. After mixing, the tubes are further incubated for 10 to 30 minutes. The reactions are terminated with 200 μl of TMPD-mixed xylenes (2:1) and vigorous vortexing. The organic phase containing the radioactive product is then separated from the reaction mixture by Eppendorf centrifugation for 5 minutes. Duplicate aliquots (50 μl each) are then removed, mixed with a scintillation cocktail (Fisher) and counted. Aliquots of the untreated reaction mixture dried onto a filter paper are also counted in similar fashion to determine the input radioactivity. One unit of CAT is presented as pmoles of ^{3}H-butyryl-chloramphenicol synthesized per

[44] B. Seed and J. Y. Sheen, *Gene* **67,** 271–7.

minute per milliliter of culture ($A_{600} = 1$). The various dilutions of the culture sample and the different incubation periods for each sample ensure that a set of reactions for each test culture represents the linear range of the assay (less than 20% substrate converted into the product); these values are then used to obtain a mean.

Procedure 4: Expression and Purification of Fusion Proteins

EXPRESSION. To obtain cells with optimal amounts of the fusion protein, it is crucial that freshly transformed strains are used. Our typical experiments use the *recA::Tn10* derivative of the *E. coli* strain N99 from the NIH collection. To improve the yield of some particularly labile fusion proteins, we used a protease defective strain (*lon::Tn10* derivative of N99) successfully. Freshly prepared LB-agar plates with ampicillin (50 μg/ml) are used to isolate transformants, and several thousand transformants obtained after overnight incubation are resuspended from the plates with 5 ml LB by using a sterile glass spreader and used directly to inoculate 2 L of LB medium containing 50 μg/ml ampicillin. After overnight growth at 37°, saturated cultures are centrifuged, the pellets resuspended thoroughly in 20 ml of Buffer CP (10 mM K-phosphate [pH 7.5], 50 mM Kcl, 5% glycerol, 0.1 mM EDTA, 1 mM PMSF, 1 mM DTT) and the suspension frozen at −80° until use.

CELL-FREE SUPERNATANT. To prepare a cell-free supernatant with the recombinant protein, the frozen pellet is thawed on ice, and a homogeneous suspension is treated with Lysozyme (250 μg/ml) at 4° for 45 min. The cell debris is then removed by a low-speed centrifugation (30,000 g for 20 min in a Beckman centrifuge). The resultant S30 supernatant is then processed by column chromatography, as detailed later.

PURIFICATION OF DHFR FUSION PROTEINS. The S30 supernatant is loaded onto a 100 ml DE52 (Whatman) column equilibrated with Buffer CP. The column is then washed with 100 mL of Buffer D (Buffer C containing 100 mM KCL), and the bound proteins are then eluted by washing the column with 200 ml Buffer E (Buffer C with 250 mM KCL). Peak protein-containing fractions (with ~90% of the fusion protein, as can be confirmed by SDS-PAGE analysis of individual fractions) are pooled and loaded directly onto a 5-ml methotrexate-agarose (Sigma) column that is equilibrated with Buffer F (a pH 8 version of Buffer C having 1 M KCL). The bound DHFR fusion protein is then recovered by slow washing of the column with 100 ml of Buffer F containing 5 mM folic acid (Sigma). Peak protein–containing fractions are then analyzed by SDS-PAGE, and the most pure fractions are pooled and concentrated by ultrafiltration with Amicon filters. The concentrated pool is then dialyzed extensively against Buffer S (10 mM K-phosphate [pH8], 100 mM KCl, 0.1 mM EDTA, 5%

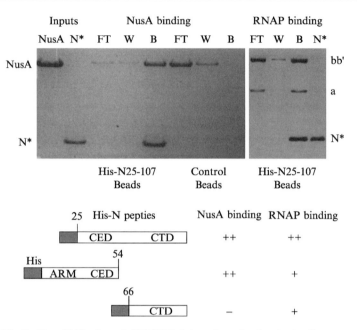

Fig. 10. Probing N-NusA and N-RNAP interactions by bead-retention assays. The SDS-PAGE gels show the binding of NusA and core RNAP to the His-tagged lambda N peptide (ΔARM). Lanes are marked as follows: FT, unbound, W, wash, and B, SDS-eluted fractions. Input lanes included show NusA and the His-N peptides, respectively. A control experiment with NTA beads is shown for NusA; the same experiment with RNAP shows <10% retention by NTA beads. Note also that NusA and RNAP do not bind to the ARM domain fused to DHFR.

glycerol, and 10 mM DTT) or chromatographed again with a DE52 column to remove the bound folate.[45] When desired, the pooled fraction is further purified by FPLC with a Mono-S column, separating some minor contaminants from the fusion protein.

Purification of fusion proteins with His-tag is done by procedures described by Qiagen, the supplier of vectors and nickel-agarose beads.

Procedure 5: Coupled-Bead Retention Assays

BEAD COUPLING. To couple N-DHFR fusion proteins to agarose beads, MTX-Agarose (Sigma) equilibrated in Buffer F is incubated with a desired amount of the protein solution to yield sub-saturated coupling of the beads. Batches of protein-coupled beads are conveniently processed in a micro column and washed thoroughly with Buffer F (about 10 × bead volume)

---

[45] P. T. Rajagopalan, S. Lutz, and S. J. Benkovic, *Biochemistry* **41,** 12618–12628 (2002).

followed by Buffer S (10 × bead volume). The coupled beads may be kept for several days in ice for a series of binding studies.

BINDING ASSAYS. Aliquots of the coupled beads (100 μl) are mixed with 200 μl samples of the ligand to be tested (NusA or RNAP), diluted in Buffer B (20 mM Tris-HCl [pH 7.5], 0.1 M NaCl, 5% glycerol, 1 mM beta-mercaptoethanol). The mixture is incubated at room temperature for 10 min, and the protein-bound beads are then isolated by centrifugation for 5 min. The supernatant (flow-through) is removed quantitatively with a micropipette, taking proper precaution so that neither the pellet is disturbed nor the beads are contaminated with residual supernatant. The pellet is then suspended in 400 μl of Buffer B and centrifuged again to collect the wash fraction. Finally, the bound ligand is eluted together with the receptor DHFR protein by washing the beads with 400 μl of Buffer B containing 2% SDS. Protein in each fraction is recovered quantitatively by TCA precipitation (adding an equal volume of 18% cold TCA containing 0.1 mg/ml deoxycholate), collection of the resultant precipitate by centrifugation, and finally, suspension in 30 μl of sample buffer (0.1 M NaOH, 62 mM Tris-HCl [pH 6.8], 2% glycerol, 0.005% CBB, and 5% beta-mercaptoethanol). Protein aliquots (10 μl) are then analyzed by 12% SDS-PAGE. Gels from a representative binding experiment are presented in Fig. 10.

The Qiagen supplier has extensive protocols established for bead-binding experiments with His-tagged proteins with use of nickel-agarose beads. The protocol described above is simply modified with buffers suitable for nickel beads. To the extent possible, attention should be given to avoid chelators (EDTA) and reducing agents (DTT or beta-mercaptoethanol), which affect nickel-coupling.

Concluding Remarks

Although the lambda N protein has served as an important model for understanding the basic mechanisms of processive antitermination, the homologous proteins from lambda's cousins show considerable variation in primary structures. Like the studies of different repressors and activators, a molecular genetic analysis of some of these N protein homologs (from P21 and Phi-80, for example) promises to reveal interesting variations in terms of the basic mechanisms by which RNA polymerase is transformed to a termination-resistant state. At the same time, these studies would provide a better understanding of the structure-function of RNA polymerase itself.

Acknowledgment

This work was supported by NIH grant GM28946 to A. D.

[34] Assay of Transcription Antitermination by Proteins of the CspA Family

By Sangita Phadtare, Konstantin Severinov, and Masayori Inouye

When an exponentially growing culture of *Escherichia coli* is shifted from 37° to 15°, there is a transient arrest of cell growth. During this period a number of genes called *cold-shock genes* are induced; however, general protein synthesis is severely inhibited. Synthesis of CspA protein is dramatically induced immediately after the temperature downshift, and this protein becomes one of the most abundant proteins in the cell.[1] Eventually the synthesis of CspA decreases as cells become acclimated to low temperature and growth resumes. *E. coli* has nine homologous proteins, CspA to CspI. Out of these, CspA, CspB, CspG and CspI are coldshock inducible, while CspC and CspE are constitutively produced at 37°. CspD is induced by nutritional deprivation.[2,3] *E. coli* cells harboring double or triple deletions of genes coding for coldshock inducible CspA homologs ($\Delta cspA\Delta cspB$, $\Delta cspA\Delta cspG$, $\Delta cspB\Delta cspG$, $\Delta cspA\Delta cspI$ or $\Delta cspA\Delta cspB\Delta cspG$) are not cold-sensitive. In a triple deletion strain $\Delta cspA\Delta cspB\Delta cspG$, CspE is overproduced at low temperatures.[4] This observation suggests that the functions of the CspA family members overlap and that they are able to substitute for each other during cold acclimation. Indeed, a quadruple deletion strain ($\Delta cspA\Delta cspB\Delta cspG\Delta cspE$) of *E. coli* exhibits cold sensitivity at 15°, which can be complemented by overproduction of any one of CspA homologs except CspD.[4] For *Bacillus subtilis* it has been shown that at least one of its three CspA homolog proteins (*viz*. CspB, CspC or CspD) is essential for cell viability under optimal growth conditions as well as during cold shock.[5]

CspA binds RNA without apparent sequence specificity and with low binding affinity. It destabilizes the secondary structures in RNA, hence it was described as an RNA chaperone.[6] Since secondary structures in RNA become more stable at lower temperatures, the nucleic acid melting by CspA may be crucial for efficient translation at low temperatures,

---

[1] P. G. Jones, R. A. VanBogelen, and F. C. Neidhardt, *J. Bacteriol.* **169,** 2092 (1987).

[2] S. Phadtare, J. Alsina, and M. Inouye, *Curr. Opin. Microbiol.* **2,** 175 (1999).

[3] S. Phadtare, K. Yamanaka, and M. Inouye, The Cold Shock Response in The Bacterial Stress Responses, (Storz, G and Hengge-Aronis, R., ed.), ASM Press, Washington DC pp. 33 (2000).

[4] B. Xia, H. Ke, and M. Inouye, *Mol. Microbiol.* **40,** 179 (2001).

[5] P. Graumann, T. M. Wendrich, M. H. Weber, K. Schroder, and M. A. Marahiel, *Mol. Microbiol.* **25,** 741 (1997).

[6] W. Jiang, Y. Hou, and M. Inouye, *J. Biol. Chem.* **272,** 196 (1997).

because keeping RNA in a linear form is an essential prerequisite for efficient initiation of translation. Nucleic acid melting was also demonstrated for *E. coli* CspE, and it is likely that most CspA homologs are able to function as RNA chaperones.

Hanna and Liu[7] demonstrated the interaction between CspE and the nascent RNA in transcription elongation complexes, suggesting that this protein is involved in the regulation of transcription. Indeed, the same authors reported that purified CspE interfered with Q-mediated transcription antitermination. We showed that CspA, CspE and CspC decreased transcription termination at several intrinsic terminators and also affected transcription pausing.[8] With single amino acid substitution mutants of CspE, it has been shown that its RNA-melting activity is essential for transcription antitermination function.[9] The cumulative data suggest that CspE affects transcription termination by preventing the formation of the nascent RNA secondary structures, most likely the formation of the stem-loop structure at the terminator. Thus the mechanism of Csp-induced antitermination appears to be similar to the recently deduced mechanism of transcription antitermination by λ N protein.[10] At least some of promoter-distal genes induced during cold acclimation are activated through a transcription antitermination mechanism.[8] Since the loss of nucleic acid melting activity of CspE leads to its inability to support cold acclimation,[9] it is possible that transcription antitermination through RNA-melting activity is an essential function of Csp proteins at low temperature.

CspA homologs are widely distributed in prokaryotes, and CspA is also homologous to the cold-shock domain (CSD) in human Y-box protein YB-1 and the Y-box proteins from other eukaryotes.[11,12] In addition, a number of proteins that are not similar to Csp proteins in sequence but assume a homologous structural fold are known. S1 domains of PNPase and translation initiation factor IF1 have significant structural similarity with CspA.[13,14] The growth and the sporulation defects observed in *cspB/cspC* double deletion mutant of *B. subtilis* are cured by expression of the translation initiation factor IF1 from *E. coli*.[15] This suggests that IF1 and CspA

[7] M. M. Hanna and K. Liu, *J. Mol. Biol.* **282,** 227 (1998).

[8] W. Bae, B. Xia, M. Inouye, and K. Severinov, *Proc. Natl. Acad. Sci. USA* **97,** 7784 (2000).

[9] S. Phadtare, M. Inouye, and K. Severinov, *J. Biol. Chem.* **277,** 7239 (2002).

[10] I. Gusarov and E. Nudler, *Cell* **107,** 437 (2001).

[11] A. P. Wolffe, *Bioessays* **16,** 245 (1994).

[12] P. L. Graumann and M. A. Marahiel, *Trends Biochem. Sci.* **23,** 286 (1998).

[13] M. Sette, P. van Tilborg, R. Spurio, R. Kaptein, M. Paci, C. O. Gualerzi, and R. Boelens, *Embo. J.* **16,** 1436 (1997).

[14] M. Bycroft, T. J. Hubbard, M. Proctor, S. M. Freund, and A. G. Murzin, *Cell* **88,** 235 (1997).

[15] M. H. Weber, C. L. Beckering, and M. A. Marahiel, *J. Bacteriol.* **183,** 7381 (2001).

homologs have partially overlapping cellular function(s), including possibly RNA chaperone activity.

In this chapter, we present protocols developed to study transcription antitermination and related biochemical functions of *E. coli* CspE protein. The methods described can be easily adapted to study other Csp proteins as well as unrelated proteins known or suspected to interact with RNA and affect transcription termination efficiency.

Expression and Purification of CspE

To test the effect of CspE on transcription antitermination *in vitro*, pure proteins free of RNase contamination must be obtained at high concentration.[8,9] The open-reading frame of *cspE* gene was cloned in pET11a vector (Novagen, Madison, WI) by using the *Nde*I and *Bam*HI sites. The pET11 vector contains a T7 promoter under the control of the *lac* operator and allows high and regulated levels of overproduction of CspE. The *cspE* expression plasmid is transformed in *E. coli* BL21 (DE3) cells (Novagen, Madison, WI). Transformed cells are grown overnight at 37° on LB plates (Luria broth plates; tryptone, 1%; yeast extract, 0.5%; NaCl, 0.5%; 1 mM, MgSO$_4$ and 1.5% agar) containing ampicillin (50 μg ml^{-1}). A single colony is inoculated in 10 ml M9 medium with casamino acids (dibasic sodium phosphate, 1.32%; monobasic potassium phosphate, 0.3%; sodium chloride, 0.05%; ammonium chloride, 0.1% casamino acids, 0.2%, MgSO$_4$, 0.02%; tryptophan, 0.005%, glucose, 0.4%, thiamine 0.0002%) and ampicillin (50 μg ml^{-1}) and incubated overnight at 37° with shaking. The resulting culture is used as an inoculum for 1 liter of M9 medium with casamino acids in a 5-liter flask and incubated on a rotary shaker at 37° until the culture reaches an early logarithmic phase (2 to 3 h or until OD$_{600}$ reaches ~0.5). CspE overproduction is induced by the addition of 1 mM IPTG (isopropyl β-D thiogalactopyranoside) for 2 h. Cells are harvested by centrifugation at 4,500 × g for 30 min at 4° and are suspended in 25 ml of 20 mM Tris-HCl, pH 8.0. The cells are lysed by 2 to 3 passes through a French press at 1000 psi, and the lysate is centrifuged at 10,000 × g for 10 min to remove cell debris. The supernatant is subjected to ultracentrifugation at 40,000 × g for 2 h at 4°. CspE and most other proteins are precipitated from the supernatant obtained after the ultracentrifugation step with 60% saturated ammonium sulfate. The protein is collected by centrifugation at 12,000 × g for 15 min at 4° and is dissolved in 10 ml 20 mM Tris-HCl, pH 7.5, and dialyzed against 4 × 1 liter changes of the same buffer. The dialysate is loaded onto a 50-ml Q-Sepharose Fast Flow (Amersham Pharmacia Biotech, catalog number 17-0510-01) column equilibrated with 20 mM Tris-HCl, pH 7.5. The column is developed isocratically with the same buffer. Fractions (15 ml each) are collected and checked by 17.5% SDS-PAGE. CspE-containing fractions are pooled and dialyzed against two 1-liter changes of 10 mM potassium phosphate buffer, pH 7.0. The dialysate is loaded on a 15-ml hydroxyapatite column (BioRad HT gel) equilibrated with 10 mM potassium phosphate buffer, pH 7.0. After loading, the column is washed with 30 ml of the same buffer, and pure CspE is eluted with 1 M NaCl in the same buffer. Fractions (5 ml each) are collected, and the homogeneity of the protein is established after SDS-PAGE by silver staining. Protein concentration is determined by the Bradford method with BioRad protein assay dye. The protein is concentrated on a Centricon (Millipore) centrifugal concentrator with an Mw cutoff of 3000 to a concentration of approximately 1 mg/ml, aliquoted, and stored at $-80°$. Repeated freezethawing of the protein should be avoided. Typically 10 mg of pure CspE can be obtained from 1 L of an induced cell culture.

Procedures for purification of Csps other than CspE have also been published.[16,17] In the protocol presented above, CspE does not bind to Q-Sepharose. The protocol has to be modified by including a salt gradient for CspA homologs that interact with Q-sepharose, such as CspB and CspC.[17] An additional S-Sepharose column chromatography step is also used in addition to Q-Sepharose and hydroxyapatite column chromatography to obtain highly pure CspC and CspA. This chromatography is carried out by using 10 mM potassium phosphate buffer, pH 7.0, as above, and a gradient of 100 to 150 mM salt is used for elution of protein from the column.

Transcription Antitermination *In Vivo*

An *E. coli* strain RL211 is used to demonstrate *in vivo* transcription antitermination activity of CspA homologs.[8,9] This strain contains the *cat* gene preceded by a strong ρ-independent *trpL* terminator (Fig. 1A) and is therefore sensitive to chloramphenicol. When transcription termination at *trpL* is reduced, the *cat* gene is expressed at higher levels and the cells become resistant to chloramphenicol.[18] Plasmid expressing the *cspE* gene from an IPTG-inducible *E. coli* RNA polymerase *lac* promoter is used to test the effect of CspE on transcription antitermination *in vivo*. The plasmid is based on the pINIII expression vector.[19] The multiple cloning sites

[16] S. Chatterjee, W. Jiang, S. D. Emerson, and M. Inouye, *J. Biochem. (Tokyo)* **114,** 663 (1993).
[17] S. Phadtare and M. Inouye, *Mol. Microbiol.* **33,** 1004 (1999).
[18] R. Landick, J. Stewart, and D. N. Lee, *Genes Dev.* **4,** 1623 (1990).
[19] M. Inouye, Multipurpose expression cloning vehicles in *Escherichia coli*, New York Academic Press, NY (1983).

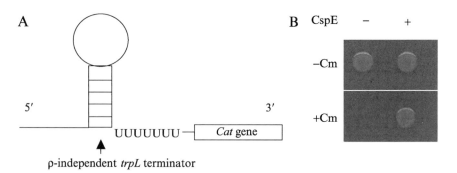

FIG. 1. Effect of CspE on transcription antitermination *in vivo*. A. Schematic representation of the *cat* gene cassette positioned downstream of the *trpL* terminator in the RL211 *E. coli* cells.[18] B. RL211 *E. coli* cells are transformed with pINIII vector alone as a control or with pINIII containing cloned *cspE* and are spotted on LB plates containing 50 μg/ml ampicillin, 1 m*M* IPTG and with and without 30 μg/ml chloramphenicol. The results of an overnight cell growth (without chloramphenicol) and after incubation of 2 overnights (with chloramphenicol) at 37° are presented.

of pET and pINIII plasmids allow direct subcloning of *csp* genes from one plasmid backbone to another.

The RL211 cells are transformed with pINIII-*cspE* or control pINIII and grown in LB broth at 37° to OD_{600} nm of 1. The culture (3 μl) is spotted on LB plates containing ampicillin (50 μg ml^{-1}), 1 m*M* IPTG, with and without chloramphenicol (30 μg ml^{-1}). The plates are incubated at 37°, and results are recorded after overnight incubation for the plates without chloramphenicol and after incubation for 2 to 3 overnights for plates containing chloramphenicol. For reproducible results, it is advisable to spot rather than streak bacterial cultures on plates, as streaking sometimes gives irreproducible results. Further, RL211 cells transformed with pINIII-*cspE* vector appear to be unable to form colonies on LB-ampicillin-chloramphenicol medium when plated directly after transformation. Therefore, picking up colonies from an LB-ampicillin master plate, growing them in liquid culture, and then spotting as described here is recommended when high-throughput genetic screens are conducted with this system.[9] Fig. 1B shows the results of a typical plate assay. As expected, both cells exhibit growth in the absence of chloramphenicol; however, only the ones overproducing CspE are able to grow in the presence of chloramphenicol.

In vivo transcription antitermination induced by Csp overproduction can also be shown directly by using Northern blotting. The products of the *metY-rpsO* operon genes *nusA*, *infB*, *rbfA*, and *pnp* are known to be induced at cold shock. We have shown that overexpression of CspA homologs induces transcription of prompter-distal genes of the *metY-rpsO* operon even at $37°$.[8] The genes of the *metY-rpsO* operon are separated by multiple transcription terminators located between *metY* and *yhbC*, *infB* and *rbfA*, and *rpsO* and *pnp*. CspA homologs reduce transcription termination on at least some of these terminators *in vitro*.[8] The protocol suitable for monitoring *csp*-dependent transcription of promoter distal genes of the *metY-rpsO* operon is presented below. *E. coli* JM83 strain [F-*ara*Δ (*lac-proAB*) *rpsL*(*str'*)] cells transformed with pINIII-*cspE* plamsid or pINIII as control are grown at $37°$ in 10 ml M9 medium containing casamino acids till OD_{600} nm of 0.3 to 0.4 is reached. IPTG (1 m*M*) is added and after 60 min, the RNA is isolated from the culture by the hot phenol method.[20] Before RNA isolation, samples of bacterial cultures are removed for SDS-PAGE analysis to ascertain that overproduction of plasmid-borne Csp did occur. Northern blot analysis is carried out by standard procedures.[21] Twenty micrograms of total RNA are loaded per lane on a 1.4% agarose gel with 2.2 M formaldehyde. After electrophoresis, RNA is transferred from the gels to GeneScreen membranes (New England Nuclear, Boston, MA) with a PosiBlot pressure blotter (Stratagene, La Jolla, CA). After transfer, the membranes are fixed by baking them at $80°$ for 2 h in a vacuum oven and hybridized with labeled DNA fragments at $42°$ overnight. Hybridization probes are generated by asymmetric PCR amplification of the antisense strands of *metY*, *nusA*, and *rbfA* in the presence of $[\alpha^{32}P]dCTP$ by using chromosomal DNA as template.[8]

Transcription Antitermination *In Vitro*

A DNA template containing the T7 A1 promoter fused to the λ tR2 terminator was described previously[22] and is routinely used in our laboratories, but other promoter fragments fused to different terminators can also be used. The 324-bp T7 A1-tR2 template is prepared by PCR by using "template 1" of Nudler *et al.*[23] and purified by using a Qiagen plasmid spin miniprep kit. Histidine-tagged RNAP σ^{70} holoenzyme is prepared by *in vitro* reconstitution or purified from *E. coli* cells containing a genomic copy of *rpoC* genetically fused to hexahistidine tag.[24,25] Elongation complexes

[20] P. Sarmientos, J. E. Sylvester, S. Contente, and M. Cashel, *Cell* **32,** 1337 (1983).

[21] J. Sambrook, E. F. Fritsch, and T. Maniatis, Molecular cloning: A Laboratory Manual, Cold Spring Harbor Lab. Press, Plainview, NY (1989).

[22] N. Zakharova, I. Bass, E. Arsenieva, V. Nikiforov, and K. Severinov, *J. Biol. Chem.* **273,** 24912 (1998).

[23] E. Nudler, M. Kashlev, V. Nikiforov, and A. Goldfarb, *Cell* **81,** 351 (1995).

[24] H. Tang, Y. Kim, K. Severinov, A. Goldfarb, and R. H. Ebright, *Methods Enzymol.* **273,** 130 (1996).

stalled at position +20 (EC^{20}) are prepared in 50 μl of transcription buffer containing 25 μl of Ni^{2+}-nitrilotriacetic acid agarose (Quiagen), 20 nM of template DNA fragment, 40 nM His-tagged RNAP, and 0.5 mM ApU primer. The standard transcription buffer contains 40 mM KCl, 40 mM Tris·HCl, pH 7.9, and 10 mM $MgCl_2$. Reaction mixtures are incubated for 15 min at 37° to form open promoter complexes and then transferred to room temperature. Fifty micromoles of ATP and GTP, and 2.5 μM [α-^{32}P]CTP (300 Ci/mmol) are added and reactions are allowed to proceed for 10 minutes. Ni-NTA agarose beads are next washed five times with 1.5 ml of transcription buffer and left in a small (40 μl) volume. Aliquots of Ni-NTA agarose beads suspension with bound and washed EC^{20} are transferred into fresh Eppendorf tubes, supplemented with Csp proteins (20 μM) and NTPs (250 μM). The reactions (total reaction volume 10 μl) are incubated for 10 min at room temperature. Reactions are terminated by the addition of 20 mM EDTA, and heparin (10 mg/ml) is then added to reactions followed by brief (2 to 3 min) incubation at room temperature. This step appears to be necessary to avoid nonspecific retardation of RNA by Csp proteins during electrophoretic separation of reaction products. Seven microliters of formamide-containing loading buffer is added, and reactions are boiled for 4 min and cooled on ice. Products are analyzed by denaturing urea-PAGE (7 M urea/10% polyacrylamide, 19:1). Before loading, the amount of radioactivity in each sample is determined by liquid scintillation counter, and an equal amount of radioactivity is loaded per each lane of the gel. After electrophoresis, reaction products are visualized and quantified using phosphorimager analysis. Fig. 2A shows transcription antitermination caused by CspE *in vitro*. The mean readthrough efficiency values (RE), defined as the ratio of the run-off transcripts to the total (terminated and run-off) transcripts produced is calculated. In the experiment shown in Fig. 2B, mean RE values obtained from three independent experiments are presented. As can be seen, RE was less than 30% in the control reaction in the absence of CspE, while in the presence of CspE it increased to 55%.

Properties of CspA Homologs Important for Transcription Antitermination

In *E. coli*, the factor-independent transcription terminator signal is encoded in the nascent RNA and consists of two essential elements, a stem loop structure followed by a stretch of U residues. CspA homologs appear

---

[25] H. Tang, K. Severinov, A. Goldfarb, and R. H. Ebright, *Proc. Natl. Acad. Sci. USA* **92,** 4902 (1995).

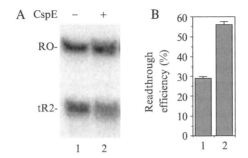

FIG. 2. Effect of CspE on transcription antitermination *in vitro*. A. The *in vitro* transcription assays are carried out as described in the text. The products are analyzed by urea-PAGE (7 M urea/10% polyacrylamide). RO and tR2 indicate the runoff and the tR2-terminated transcripts, respectively. Lane 1 is the control reaction without CspE and lane 2 is reaction carried out with CspE. B. The readthrough efficiency was calculated as described in text. The mean values calculated from three independent experiments are shown for each lane.

to target the stem-loop structure and thus decrease termination of transcription. Impairment of either the RNA binding or RNA melting activity can affect transcription antitermination function of CspA homologs. Biochemical assays have been designed that allow one to quantitatively determine these activities.

Checking RNA Binding Activity of CspE. E. coli CspA and its homologs studied so far bind RNA and single-stranded DNA.[6,17] While CspA binds nucleic acids without apparent sequence specificity, CspB, CspC and CspE prefer, respectively, Poly U/T, AGGGAGGGA and AU/AT-rich sequences as RNA binding substrates.[17] We use an AU-rich 88-nt RNA for checking the RNA binding activity of CspE,[17] but other RNAs should work provided they are long enough, since Csps require a certain minimum length of nucleic acids for efficient binding: at least 74-nt for CspA and at least 45-nt for CspB and CspE. To obtain an RNA substrate, an insert containing a T7 RNA polymerase promoter and the sequence of interest is cloned in the pUC19 plasmid. The resultant plasmid is used as a template for preparation of specific RNA fragments. Plasmid is prepared by using a Qiagen plamid spin miniprep kit. T7 RNA polymerase reaction is carried out in the presence of 5 μl [α-^{32}P]UTP. Total volume of the reaction is 30 μl, and it contains 0.5 μg template DNA, 0.2 mM NTPs (ATP, GTP, CTP, UTP), 1.5 μl T7 RNA polymerase (Boehringer Mannheim), and 3 μl T7 RNA polymerase buffer (10×). The reaction is carried out at 37° for 2 h. RNase free DNaseI (2 μl) (Boehringer Mannheim) is added, and the reactions are incubated at 37° for 15 more min. The RNA is purified by phenol/chloroform extraction, precipitated with ethanol and washed with 70% ethanol, and RNA substrate is dissolved in deionized water. Structural integrity of the RNA is checked by urea-acrylamide gel electrophoresis (8% acrylamide, 19:1), and a single radioactive band of the expected size should be obtained. The RNA substrate prepared as above may not be completely free of unincorporated nucleotides, including α-[^{32}P]UTP, which is measured by TCA precipitation. Equal amounts of the RNA substrate preparation are spotted on two nitrocellulose filters (Schleicher and Schuell filter type 0.45 μM HA 024). One filter is counted directly in a liquid scintillation counter, and the other is treated with 10% trichloroacetic acid on ice for 5 min, followed by washing with ethanol on ice for 5 min. The filter paper is dried and the amount of retained radioactivity is measured. The two values are compared to calculate percentage of radioactive RNA in the RNA substrate preparation. A good RNA substrate should have more than 90% of radioactivity in the incorporated form. If the incorporation is less than that, RNA should be reprecipitated with ethanol and washed with 70% ethanol. RNA concentration is estimated from the known specific activity of [α-^{32}P] UTP used in *in vitro* transcription reaction.

The CspE-RNA-binding assay is carried out in a 15-μl reaction mixture containing binding buffer (10 mM Tris-HCl buffer, pH 8.0, containing 1 mM EDTA, 10 mM KCl, 7.4% glycerol). Variable amounts of CspE (0–2 × 10^{-6} M) are incubated with constant amounts (50 fmoles) of RNA. Reactions are allowed to proceed for 20 min on ice. Individual reactions are then passed through a nitrocellulose filter (Schleicher and Schuell) to separate protein-bound RNA from free RNA (protein-bound RNA is retained on the filter). The filters are washed thoroughly to remove unbound RNA/DNA. The RNA bound to protein is retained on the filter. Radioactivity retained on the filter is measured by liquid scintillation counter. About 1% of the input radioactivity is detected as background in the absence of any protein in the reaction mixture. This background is subtracted from the measured amounts to get specific binding values. The apparent K_d value of the binding reaction is defined as the concentration of protein at which half of maximum binding is obtained.[26] For CspE, the K_d value is around 0.3 × 10^{-6} M and the maximum binding is about 53%.[17] It has been shown that CspE binds to RNA and ssDNA with similar affinity and specifity. Alternatively, RNA binding by CspE can be assayed by native gel retardation assay. The details of gel running are described in next section. However, in several cases the shifted bands representing RNA-CspE complexes can be smeary and difficult to quantify, presumably

[26] Y. Kajita, J. Nakayama, M. Aizawa, and F. Ishikawa, *J. Biol. Chem.* **270**, 22167 (1995).

because of the instability of the RNA-protein complexes under the electrophoresis conditions used.

Checking Nucleic Acid Melting Activity of CspE. We have shown that
the ability of CspE to cause transcription antitermination is a manifestation
of its ability to melt the secondary structures in nucleic acids.[9] There are
two methods by which the nucleic acid melting activity of CspA homologs
can be tested, as follows.

The RNA Chaperone Assay. To test the RNA melting activity of CspE,
a fragment of RNA containing secondary structure is used as substrate.
E. coli CspA mRNA has a long 159-base 5'UTR that forms extensive secondary structure. A DNA fragment corresponding to *cspA* 5'UTR positions
from +1 to +142 is fused to a T7 RNAP promoter,[6] and the substrate
RNA is produced by *in vitro* transcription in the presence of $[\alpha^{32}P]$ CTP
in a manner described for preparing the RNA substrate for testing RNA
binding by CspE, above. This is the substrate used in the assay described
here. ^{32}P-labeled RNA substrates (1×10^4 cpm) are incubated for 15 min
on ice in reaction buffer (total reaction volume 15 μl) containing 10 mM
Tris-HCl (pH 8.0), 1 mM EDTA, and 50 mM KCl, 7.4% glycerol in the
presence and absence of 27 μM CspE. RNase T1 (or RNase A) (Boehringer Mannheim) (25 to 100 units) is added to the reaction mixtures and the
reactions are kept on ice for 10 min and then run on a non-denaturing 8%
acrylamide (19:1) gel. The running buffer is $1 \times$ TBE (50 mM Tris-HCl,
1 mM boric acid). Electrophoresis is carried out at 130 V for 3.5 h at 4°.
The gel is dried and RNase digestion products are visualized by autoradiography. RNase T1 or RNase A cannot digest double-stranded molecules.
In the absence of CspA homologs, RNase T1 (or RNase A) digests the
RNA substrate only at target residues that are present in single-stranded
loops separating double-stranded regions of the secondary structure
(G residues are attacked by RNase T1 and U/C residues by RNase A)
giving rise to distinct, stable digestion products that correspond to segments of RNA that are double-stranded. In the presence of CspA homologs, double-stranded RNA is destabilized and the entire RNA becomes
susceptible to RNase attack and thus RNase-resistant digestion products
disappear.

Testing the Nucleic Acid Melting By Using Molecular Beacon System.
As mentioned above, CspE has similar binding affinity and specificity for
RNA and single-stranded DNA. Therefore its nucleic acid melting activity
can also be tested by using single-stranded DNA strands containing
fluorescent label (molecular beacon) at one end as described here. The
molecular beacon system used here was designed by Tyagi and Kramer.[27]

---

[27] S. Tyagi and F. R. Kramer, *Nat. Biotechnol.* **14,** 303 (1996).

The molecular beacon is a single-stranded nucleic acid molecule that forms a stable stem-loop structure. A fluorescent moiety is attached to the end of one arm of the DNA and a nonfluorescent quenching moiety is attached to the end of the other arm. The stem keeps these two moieties in close proximity to each other, resulting in efficient quenching of the fluorophore fluorescence. When a protein "opens up" the hairpin loop structure, the arms of the beacon move apart, causing the fluorophore and the quencher to move away from each other as well. Since the flourophore is no longer in close proximity to the quencher, its fluoresence will increase. Several companies manufacture molecular beacons (web site: http://www.molecular-beacons.org/company.html). The beacon used in our studies is a 82-nucleotide long hairpin-shaped molecule labeled with a fluorophore and quencher, tetramethyl rhodamine-AGGGTTCTTTGTGGTGTTTTTAT-CTGTGCTTCCCTATGCAC CGCCGACGACAGTCGCTAACCTCT-CGCTAAGAACCCT-DABCYL. It is made by ligating the two oligos: 5′-half-tetramethyl rhodamine-AGGGTTCTTTGTGGTGTTTTTATCT-GTGCTTCCCTATGCAC and 3′-half- CGCCGACGACAGTCGCTAA-CCTCTCGCTCAAGAACCCT-DABCYL. As a splint for ligation, the following oligonucleotide is used: 5′- TAGCGACTGTCGTCGGCGGTGCA-TAGGGAAGCACAGATAAAA-3′. The ligated full-length product is purified on a 10% acrylamide (19:1) gel containing 7 M urea. The product is eluted in GE buffer (10 mM Tris·HCl, pH 7.5, 3 mM EDTA, 400 mM NaCl) and precipitated with ethanol. A long beacon (82-nt) is used because E. coli CspA homologs do not bind well to shorter nucleic acid.[6]

Fluorescence measurements are performed on an LS-5B spectrofluorometer (Perkin Elmer) with 1 cm path length QS cuvettes (Hellma). The temperature of the cuvette can be controlled by a circulating bath. The CspE nucleic acid–melting assays are carried out at room temperature unless specified otherwise. The excitation and emission wavelengths used are 555 and 575 nm, respectively. Change in the fluorescence of a 100 μl solution of 32 nM molecular beacon dissolved in 20 mM Tris–HCl, pH 7.5, containing 1 mM MgCl$_2$ is monitored as CspE (1.5 μg) is added. The reactions are carried out at room temperature. The addition of the wild-type CspE results in the stably increased (4-fold) beacon fluorescence (Fig. 3). CspE alone does not show any fluorescence in a control experiment carried out without the molecular beacon. This experiment demonstrates that CspE can melt the secondary structures in nucleic acids. We created a mutant of CspE (CspE-H32R) that is deficient in transcription antitermination and in cold acclimation.[9] This mutant shows reduced nucleic acid melting as indicated by reduced level (20% fluorescence as compared with the wild-type CspE) of fluorescence achieved on addition of the mutant protein to the beacon reaction (Fig. 3). This result confirms

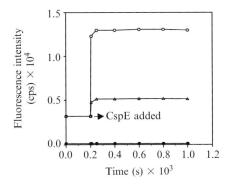

FIG. 3. Nucleic acid melting activity of CspE and its mutant using the molecular beacon system. The excitation and emission wavelengths used are 555 and 575 nm, respectively. The fluorescence of an 82-nt long molecular beacon is monitored as wild-type CspE or its mutant protein is added, as described in the text. Time points at which the proteins are added is shown with arrow. Fluorescence of CspE itself was checked by adding CspE in the buffer without the molecular beacon. Wild-type CspE, open circles; wild-type CspE added in buffer without molecular beacon, closed circles; CspEH32R, open triangles.

that nucleic acid melting activity of CspE is essential to antiterminate transcription and for its function in cold acclimation.

A control experiment to test the effect of degradation of CspE by trypsin on fluorescence can also be performed. Reaction temperature is adjusted to 37° and trypsin (CspE: trypsin ratio of 1:50) is added to the reaction mixture after the maximum fluorescence is achieved with CspE. As CspE is degraded by trypsin, a slow decrease in the fluorescence is seen as the two arms of the beacon come back together bringing back the quencher in close proximity of the fluorophore.

Comments

In this chapter we have mainly focused on analysis of ability of an RNA chaperone protein such as CspE to destabilize secondary structures in RNA and cause transcription antitermination. We have discussed both *in vivo* and *in vitro* assays as well as *in vivo* implications of such an activity. These methods should prove to be useful for testing proteins that exhibit functional or structural similarity to CspA.

[35]   Assay of Antitermination of Ribosomal RNA Transcription

By Catherine L. Squires, Ciarán Condon, and Hyuk-Kyu Seoh

The crystal structures of RNA polymerase of several different origins are beginning to reveal how the various subunits of both prokaryotic and eukaryotic RNA polymerases interact with each other, their DNA templates, and RNA products in transcription.[1–5] Questions as to how RNA polymerase functions in the elongation process and in termination are also being addressed.[6,7] However, exactly how auxiliary molecules modify RNA polymerases during elongation to alter their transcription properties is not yet well understood. Yet such modified transcription complexes are capable of reading through terminators and thereby dramatically influencing gene expression in a variety of systems.[8–10] The ribosomal RNA (rRNA) operons provide a specific example of an antitermination (AT) system in which RNA polymerase is modified by auxiliary factors that permit it to read through Rho-dependent transcription terminators. Because under fast growth conditions the majority of *E. coli* RNA polymerases are transcribing rRNA operons, it can be considered that most RNA polymerases in the cell are in fact complexed with supplementary proteins. The exact function of these proteins and how they interact with RNA polymerase is unknown. The rRNA antitermination signal consists of a sequence in the RNA, the most important feature of which is a sequence known as BoxA, GCTCTTTAACAA.[11,12] Transcription of this

[1] K. S. Murakami, S. Masuda, E. A. Campbell, O. Muzzin, and S. A. Darst, *Science* **296,** 1285 (2002).

[2] D. G. Vassylyev, S. Sekine, O. Laptenko, J. Lee, M. N. Vassylyeva, S. Borukhov, and S. Yokoyama, *Nature* **417,** 712 (2002).

[3] G. Zhang, E. A. Campbell, L. Minakhin, C. Richter, K. Severinov, and S. A. Darst, *Cell* **98,** 811 (1999).

[4] A. L. Gnatt, P. Cramer, J. Fu, D. A. Bushnell, and R. D. Kornberg, *Science* **292,** 1876 (2001).

[5] P. Cramer, D. A. Bushnell, and R. D. Kornberg, *Science* **292,** 1863 (2001).

[6] I. Artsimovitch, and R. Landick, *Proc. Natl. Acad. Sci. USA* **97,** 7090 (2000).

[7] E. Nudler and M. E. Gottesman, *Genes Cells.* **7,** 755 (2002)

[8] J. Devito and A. Das, *Proc. Natl. Acad. Sci. USA* **91,** 8660 (1994).

[9] J. W. Roberts, W. Yarnell, E. Bartlett, J. Guo, M. Marr, D. C. Ko, H. Sun, and C. W. Roberts, *Cold Spring Harbor Symp. Quant. Bio.* **63,** 319 (1998).

[10] R. Sen, R. A. King, N. Mzhavia, P. L. Madsen, and R. A. Weisberg, *Mol. Microbiol* **46,** 215 (2002).

[11] S. Li, C. L. Squires, and C. Squires, *Cell* **38,** 851 (1984).

[12] K. Berg, C. L. Squires, and C. Squires, *J. Mol. Biol.* **209,** 345 (1989).

sequence results in the formation of a transcription complex that is dramatically altered in two ways: (1) The rate of transcription increases from ~40 nuc/sec to ~85 nuc/sec; and (2) the complex becomes resistant to Rho-dependent termination. Both alterations represent striking changes from the normal transcription process.[13–16]

Previous *in vitro* studies with the rRNA AT system demonstrated an absolute requirement for the antitermination protein NusB, a protein also known to be necessary for the bacteriophage lambda N antitermination system.[17,18] Other studies have shown an association of NusA, NusE, and NusG, also involved in the lambda N antitermination system, with rRNA antitermination.[14,15,19] Recent studies showing that ribosomal protein (r-protein) S4 has properties similar to NusA also implicate S4 in rRNA antitermination.[20] The purification of these various factors, the preparation of an S100 cell extract—which permits isolation of antiterminated transcription complexes—and an *in vitro* transcription assay designed to study rRNA antitermination are described below. The methods described can be used to address the contribution of various protein factors to the Rho-dependent termination read-through abilities of RNA polymerase modified by the rRNA antitermination system.

Purification of Antitermination Factors and EcoRI*

The following are the purification protocols for the Nus factors, r-proteins S10 (NusE) and S4, and mutant restriction endonuclease EcoRI E111Q. The mutant EcoRI (EcoRI*) is used in the isolation of antitermination complexes (later). The procedures are straightforward protein purification protocols that use special production strains of *E. coli* for each protein.

NusA

The gene encoding wild-type NusA from *E. coli* was cloned into the pET-21b expression vector (Novagen, Madison, WI) and over-expressed in *E. coli* strain BL21 (λDE3). Cells were grown overnight in 25 ml of YT (8 g/liter bactotryptone; 5 g/liter yeast extract; 5 g/liter NaCl) + 0.5%

[13] U. Vogel and K. F. Jensen, *J. Biol. Chem.* **270,** 18335 (1995).
[14] U. Vogel and K. F. Jensen, *J. Biol. Chem.* **272,** 12265 (1997).
[15] M. Zellars and C. L. Squires, *Mol. Microbiol.* **32,** 1296 (1999).
[16] B. Albrechtsen, C. L. Squires, S. Li, and C. Squires, *J. Mol. Biol.* **213,** 123 (1990).
[17] C. L. Squires, J. Greenblatt, J. Li, C. Condon, and C. Squires, *Proc. Natl. Acad. Sci. USA* **90,** 970 (1993).
[18] J. Mogridge, T. F. Mah, and J. Greenblatt, *Genes Dev.* **9,** 2831 (1995).
[19] J. R. Nodwell and J. Greenblatt, *Cell* **72,** 261 (1993).
[20] M. Torres, C. Condon, J.-M. Balada, C. Squires, and C. L. Squires, *EMBO J.* **20,** 3811 (2001).

glucose + 100 μg/ml of ampicillin at 37°. The overnight culture was diluted into 2.5 liters of YT + 0.2% glucose + 100 μg/ml ampicillin and incubated at 37° with shaking. When the culture reached an O.D._{600} = 0.6, IPTG was added to 1 mM for induction of NusA protein synthesis. After 2 hours of induction, cells were harvested by centrifugation (8000 rpm for 10 min; JA-14 Beckman rotor) and washed once with 10 mM Tris/HCl, pH 8.0. Cell pellets were quick frozen and stored at $-80°$.

The frozen cell pellets were thawed on ice and resuspended in 25 ml Buffer A (20 mM Na$_2$HPO$_4$, pH 7.0; 10 mM MgCl$_2$; 1 mM DTT; 1 mM PMSF; 5% glycerol) and passed through a French Pressure cell twice (15,000 psi). DNase I (1 μg/ml) was added to the total lysate and the mix was incubated on ice for 15 min. The cell lysate was clarified by centrifugation at 15,000 rpm for 30 min (JA-20 rotor), and the supernatant was collected and passed through a 0.22 μm syringe filter. The resulting filtrate was loaded onto fast flow Q-Sepharose (Amersham Pharmacia Biotech Inc, Piscataway, NJ) packed into a Bio-Rad glass column (35 ml, 1.5 × 20 cm) and equilibrated with Buffer A with a Biologic FPLC system (Bio-Rad, Hercules, CA). The column was washed with 210 ml of 250 mM NaCl in Buffer A at the rate of 2 ml/min. The remaining proteins were eluted with a 250 mM to 500 mM NaCl gradient in Buffer A, 250 ml total volume. The NusA protein eluted at around 350 mM NaCl. NusA-containing fractions were identified by running fraction samples on a 12% SDS-PAGE gel, then pooled and stored at $-80°$. The purity of NusA protein was determined to be more than 95% by SDS-PAGE. The yield was about 100 mg of NusA per 2.5 liters of culture, estimated by UV$_{280}$.

NusB

NusB was purified according to the method of Swindle *et al.*, 1988.[21] J. Greenblatt kindly provided the *E. coli* strain JA510, expressing NusB from plasmid pLC28 under control of the λ c*I*857 repressor.

JA510 was grown overnight in 500 ml of M9 minimal medium (6 g/liter Na$_2$HPO$_4$; 3 g/liter KH$_2$PO$_4$; 0.5 g/liter NaCl; 1 g/liter NH$_4$Cl; 0.1 mM CaCl$_2$; 1 mM MgSO$_4$; 0.2% glucose) + 0.2% casamino acids + 20 μg/ml tryptophan + 20 μg/ml chloramphenicol at 30°. This culture was used to inoculate 10 liters of the same media (500 ml per 2 liter flask), also at 30°. Cells were grown with vigorous shaking to an OD$_{550}$ of 0.9 and then shifted to 42° by raising the temperature of the incubator, a process that took 30 min. The cells were incubated at 42° for a further 2 h for induction of

[21] J. Swindle, M. Zylicz, C. Georgopoulos, J. Li, and J. Greenblatt, *J. Biol. Chem.* **263,** 10229 (1988).

NusB protein. The cells were harvested by centrifugation (6000 rpm for 15 min at 4°; Beckman JA-14 rotor). The pellet of cell paste (39 g) in the centrifuge bottle was quick frozen in an ethanol-dry ice bath and stored frozen at −80°.

The 39 g of frozen cells were thawed on ice, washed in 45 ml of Buffer B1 (10 mM Tris/HCl, pH7.6; 14 mM MgCl$_2$; 1 mM DTT), centrifuged (6000 rpm for 15 min at 4°; Beckman JA-20 rotor), and resuspended in 20 ml of the same buffer. The resuspended cells were sonicated with fifty 30-second bursts of a Branson Sonifier (∼350 watts; Branson Co., Danbury, CT), with the macroprobe on maximum power. The lysate was cooled for 1.5 min between sonic bursts in an ethanol-dry ice bath. The sonicate was centrifuged (13,000 rpm for 30 min; JA-20 rotor) to remove cell debris. The 110 ml of supernatant collected was then centrifuged (48,000 rpm for 2.5 hrs; Ti50 rotor). In a 4° cold room, ammonium sulfate was added to the supernatant to 30% saturation. When all of the ammonium sulfate was dissolved, the preparation was allowed to stand 1 hour without stirring. It was then centrifuged (10,000 rpm for 10 min; JA-20 rotor) and the supernatant collected. More ammonium sulfate was added to achieve a final saturation of 60%. When the ammonium sulfate was totally dissolved, the preparation was allowed to stand 3 hours without stirring. The sample was then centrifuged (22,000 rpm for 20 mins; 50Ti rotor) the supernatant discarded and the pellets put on ice overnight.

The pellet from each tube was resuspended in Buffer B1 + 0.1 M NaCl, then pooled. The tubes were rinsed a second time with the same buffer, yielding a total sample volume of 15 ml. Chromatography, ammonium sulfate fractionation, and dialysis were all carried out in a 4° cold room. A concentrated solution of Dextran blue in water (200 μl; Sigma, St. Louis, MO) was added to the sample just before it was loaded onto a 100 cm × 4.3 cm column containing 1.35 liters of G75 superfine (Amersham Pharmacia Biotech Inc, Piscataway, NJ) equilibrated in Buffer B1 + 0.1 M NaCl. The column was run at a flow rate of 34 ml/h and 7-ml fractions were collected. Aliquots (6 μl) of every third fraction from 66 to 134 were run on a 14% PAGE gel and analyzed for a prominent band running at 15 to 16 Kda. The NusB protein eluted in fractions 116 to 131. These fractions were pooled, and ammonium sulfate was added to 80% saturation. The sample was allowed to stand for 3 hs. It was then centrifuged (10,000 rpm for 20 min; JA-20 rotor). The pellet was resuspended in 10 ml of Buffer B1 + 0.1 M NaCl + 200 μl Dextran blue and reapplied to the G75 column. The column was again run at a flow rate of 34 ml/h, and 7-ml fractions were collected. Aliquots of 6 μl were sampled from every fraction from 77 to 112 and run on a 14% PAGE gel. Fractions 100 to 111 were pooled, ammonium sulfate added to 80% saturation, and the sample allowed to sit for 3 hs. It was then centrifuged

(10,000 rpm for 20 min; JA-20 rotor) and the pellet dissolved in 7 ml Buffer B2 (50 mM HEPES/HCl, pH 7.5; 60 mM NaCl; 1 mM DTT) and dialyzed against 1 liter of Buffer B2 overnight, with one change.

The dialyzed sample was centrifuged (10,000 rpm for 10 min; JA-20 rotor) to remove any precipitated material. It was then loaded on an 11 ml P11 (Whatman, Maidstone, England) column that had been equilibrated in Buffer B2. The column was run at 6 ml/h in the same buffer, and 3.5-ml fractions were collected. Aliquots of 6 μl were sampled from every fraction and run on a 14% PAGE gel. Fractions 24 to 27 were pooled as NusB pool I; 23, 28 and 29 as NusB pool II; and 30 and 31 as NusB pool III. The pooled fractions were dialyzed against NusB storage buffer (10 mM HEPES/HCl, pH 7.9; 100 mM NaCl; 0.1 mM EDTA; 0.1 mM DTT; 50% glycerol) for 18 hours with one change. This concentrated the samples approximately 3-fold. The final concentration of NusB recovered was 18 mg of NusB from 39 g cells in a total of 6.5 ml storage buffer from the three pools. The activity of the pool I NusB was confirmed by its ability to facilitate Nun termination in an *in vitro* test system with Nus factors A, E, and G.[22]

NusE

The purification of NusE (ribosomal protein S10) protein was carried out by using a modification of the S8 purification method by Wu *et al.*, 1993.[23] The wild-type NusE (S10) gene was cloned into a pT7 expression vector and over-expressed in *E. coli* strain BL21 (λDE3)/pLysS. BL21 (λDE3)/pLysS/pNusE8 was grown overnight in 50 ml of 2xYT (16 g/liter tryptone; 10 g/liter yeast extract; 10 g/liter NaCl) + 0.2% glucose + 100 μg/ml ampicillin + 30 μg/ml chloramphenicol at 37°. The overnight culture was used to reinoculate 3 liters of 2 × YT (0.2% glucose + 100 μg/ml ampicillin + 30 μg/ml chloramphenicol). Cells were allowed to grow to an OD$_{600}$ of 0.5 and were then induced with 1 mM IPTG for production of NusE protein. After 3 hours of further incubation at 37°, cells were harvested by centrifugation (8000 rpm for 10 min, Beckman JA-14 rotor) and washed once with 10 mM Tris/HCl, pH 8.0. All subsequent centrifugation steps were carried out in a JA-20 rotor at 4°.

Cell pellets were resuspended in 30 ml of NusE lysis buffer (10 mM Tris/HCl, pH 7.8; 10 mM MgCl$_2$; 2 mM DTT; 2 mM PMSF; 1 mM Benzamidine) and passed through a French Press twice (16,000 psi), and then centrifuged (500 rpm for 5 min) to remove cell debris and unbroken

[22] S. C. Hung and M. E. Gottesman, *J. Mol. Biol.* **247**, 428 (1995).
[23] H. Wu, I. Wower and R. Zimmermann, *Biochemistry* **32**, 4761 (1993).

cells. The cell lysate was clarified by centrifugation at 15,000 rpm for 30 min. The supernatant (40 ml) was pooled and dialyzed overnight against 3 liters of Buffer E1 (50 mM NaOAc, pH 5.6; 1 mM DTT; 0.1 mM PMSF; 1 mM Benzamidine; 6M urea) with 2 changes. The dialyzed lysate was centrifuged at 16,000 rpm for 30 min to remove any precipitated proteins. The supernatant was loaded onto a fast flow SP-Sepharose (Amersham Pharmacia Biotech Inc, Piscataway, NJ) packed in a Bio-Rad glass column (1.5 cm × 17 cm) and equilibrated with Buffer E2 (Buffer E1 without Benzamidine) with a Biologic FPLC (Bio-Rad, Hercules, CA) at the rate of 1.5 ml/min. The column was washed with 100 ml of 150 mM LiCl$_2$ in Buffer E2. The remaining proteins were eluted with a 150 mM to 550 mM LiCl$_2$ gradient in Buffer E2. NusE containing fractions were identified by running eluted fraction samples on a 15% SDS-PAGE gel. The NusE protein, a major band of around 11 kD, was eluted at 170 mM LiCl$_2$ as a broad peak. NusE containing fractions were pooled and ammonium sulfate was added to 60% saturation and incubated for 4 hours with gentle stirring. The precipitate was collected by centrifugation at 20,000 rpm for 30 min and then kept at −80°.

The precipitate was dissolved in 5 ml of Buffer E3 (20 mM KPO$_4$, pH 7.4; 1 mM DTT; 0.1 mM PMSF; 6 M urea), and dialyzed against 1.5 liters of Buffer E3 overnight, with 2 changes. The dialyzed sample was centrifuged at 13,000 rpm for 30 min and applied onto a 5 ml Hi-Trap heparin column (Amersham Pharmacia Biotech Inc, Piscataway, NJ) equilibrated with Buffer E3. After washing with 40 ml of Buffer E3, a 5 mM to 350 mM KCl gradient in Buffer E3 was applied at the rate of 1.5 ml/min. NusE eluted around 150 mM KCl. NusE fractions were pooled (20 ml) and concentrated by Centricon 10 (Amicon Inc, Beverly, MA) to 3 ml.

Further purification was carried out by Gel-filtration. The concentrated NusE pool was loaded onto a 150 ml G-75 Sephadex column (Amersham Pharmacia Biotech Inc, Piscataway, NJ). Buffer E3 was applied at the rate of 0.5 ml/ml. The final NusE containing fractions were pooled and stored at −80°.

NusG

The purification of NusG is based on that described in Washburn et al., 1996.[24] In brief, SS392 ompT harboring a NusG expression plasmid was grown in LB (10 g/liter tryptone; 5 g/liter yeast extract; 5 g/liter NaCl; 1 g/liter glucose) + 50 μg/ml ampicillin at 32° to an A$_{550}$ of 1.0. The temperature was shifted to 42° and maintained for 2 hs. The cells were harvested

---

[24] R. S. Washburn, D. J. Jin, and B. L. Stitt, J. Mol. Biol. **260,** 347 (1996).

and resuspended in 3.2 ml of NusG lysis buffer (50 mM Tris/HCl, pH 7.6; 5% glycerol; 2 mM EDTA; 0.1 mM DTT; 0.23 M NaCl) per gram of cells. The suspension was sonicated with 20-second pulses at full power three times (Heat Systems, W 185 sonicator), and the cell debris was removed by centrifugation at 7000 rpm for 10 min in an SS34 rotor (Sorvall).

A 60-μl sample of 10% (v/v) polyethyleneimine (pH 8.0, polymin P; Sigma Chemical Co, St. Louis, MO) was added drop-wise to each 1 ml of supernatant. After 10 min of stirring, the mixture was clarified by centrifugation at 7000 rpm for 10 min in an SS34 rotor. Ammonium sulfate was added to the supernatant to 45% saturation, and the precipitate was collected by centrifugation. A 20-ml sample of 35% ammonium sulfate saturated TAGED buffer (20 mM Tris/acetate, pH 8.3; 5% glycerol; 1 mM EDTA; 0.1 mM DTT) was added to the precipitate. The dissolved proteins were applied onto a Phenyl Sepharose FPLC column (1.6 cm $\times$ 10 cm; Amersham Pharmacia Biotech Inc, Piscataway, NJ) previously equilibrated with 35% saturated ammonium sulfate in TAGED buffer. After the column was washed with 45 ml of equilibration buffer, a descending linear gradient from 35% saturated ammonium sulfate to 0% ammonium sulfate in TAGED buffer was applied to the column. NusG-containing fractions were eluted at the end of the gradient. PAGE gels (15%) were run to monitor the purification process and a band of 20.4 kDa. The NusG fractions were pooled and diluted 2-fold with TAGED buffer. The sample was loaded onto a Q-Sepharose FPLC column (2.5 cm $\times$ 10 cm) equilibrated with TAGED buffer. The column was washed with 60 to 80 ml of TAGED buffer and then a linear gradient of 0 to 5% sodium sulfate was applied at 1 ml/min for 25 min. The gradient was held at 5% until the major NusG-containing peak was eluted. The final NusG-containing fractions were collected and dialyzed against 500 ml of NusG storage buffer (10 mM Tris/HCl, pH 8.0; 50% glycerol; 0.1 mM EDTA; 0.1 mM DTT; 0.1 M NaCl) at 4°, with 2 changes. The sample was stored at $-20°$. The recovery was 2 mg NusG per g of cell paste.[24]

Ribosomal Protein S4

The purification of r-protein S4 was carried out by a previously described method with minor modifications.[25] An S4 overproducing plasmid, pET-S4, was obtained from David Draper.[25] *E. coli* strain BL21 (λDE3) harboring pET-S4 was grown in 2 $\times$ YT medium with 100 μg/ml of ampicillin at 37°. Cells were induced with 1 mM IPTG after cell density reached an OD$_{600}$ of 0.8. The culture was then allowed to grow for another

---

[25] A.-M. Baker and D. E. Draper, *J. Biol. Chem.* **270**, 22939 (1995).

3 hours. After the cells were harvested, cell pellets were washed once with S4 lysis buffer (20 mM MES, pH 5.5; 1 mM EDTA; 0.5 mM DTT; 0.1 mM PMSF) and resuspended in 20 ml of the same buffer. The suspension was then passed through a French Press (15,000 psi) twice. The total lysate was centrifuged at 11,000 rpm for 15 min. Unless otherwise noted, a Beckman JA-20 rotor was used for all centrifugation steps described below.

S4 protein was extracted from the pellet of cell debris with 20 ml of S4 Buffer (S4 lysis buffer + 6 M urea) for 1 hour and centrifuged at 11,000 rpm for 30 min. The extracted sample was loaded onto a 30 ml column of SP-Sepharose (Amersham Pharmacia Biotech Inc, Piscataway, NJ) at the rate of 1 ml/ml with a Biologic FPLC system (Bio-Rad, Hercules, CA). S4 protein was eluted with a 0 to 0.4 M KCl gradient in S4 Buffer with 240 ml of buffer at the rate of 1 ml/min. The fractions were analyzed by running samples on a 12% SDS-PAGE gel, and the S4 protein was identified as a prominent band around 25 kDa. S4-containing fractions were pooled and dialyzed against S4 storage buffer (30 mM Tris/HCl, pH 7.6; 350 mM KCl; 0.5 mM DTT). Dialyzed samples were stored at $-80°$.

Purification of EcoRI*

The purification of EcoRI* was an adaptation of the protocol in Cheng *et al.*, 1984.[26] *E. coli* strain M5248, expressing EcoRI* (Glu111->Gln) from plasmid pQEM111, under control of the $\lambda cI857$ repressor, was grown overnight in 50 mls 2×YT + 200 μg/ml ampicillin at 30°. This culture was used to inoculate 12 liters 2×YT + 0.5% glucose + 0.3% Na_2HPO_4 in a New Brunswick fermenter, also at 30°. Aeration was at 12 liters/min at 15 psi. Stirring was at 300 rpm. Cells were grown to OD_{600} of 1.0 and then shifted to 42°, a process that took 30 min. Cells were incubated at 42° for a further 2.5 h for induction of EcoRI* expression. Cells were harvested by centrifugation (4000 rpm for 15 min; HG-41 swinging bucket rotor), washed in 2L M9 salts (6 g/liter Na_2HPO_4; 3 g/liter KH_2PO_4; 0.5 g/liter NaCl; 1 g/liter NH_4Cl) and recentrifuged. A 75 g sample of cell paste was stored frozen at $-80°$.

The 75 g frozen cell paste was resuspended in 100 ml EcoRI* lysis buffer (20 mM KPO$_4$, pH 7.4; 15 mM β-mercaptoethanol; 1 mM EDTA; 1 μg/ml DNase I) with the help of a Waring blender. Cells were lysed by two passages through a French Pressure cell at 19,000 psi. The cell lysate (146 ml) was clarified by centrifugation at 12,000 rpm for 30 min in an SS34 Sorvall rotor. Ammonium sulfate was added to 60% saturation over a 15-min period and incubated on ice for 1 hour. The mixture was divided into four 50-ml tubes and centrifuged for 20 min at 10,000 rpm (SS34 rotor). Each

[26] S. C. Cheng, R. Kim, K. King, S. H. Kim, and P. Modrich, *J. Biol. Chem.* **259,** 11571 (1984).

pellet was resuspended in 20 ml RI Buffer (20 mM KPO$_4$, pH 6.8; 5 mM β-mercaptoethanol; 0.5 mM EDTA; 10% glycerol) and pooled. The mixture (143 ml) was dialyzed overnight against 2 liters RI Buffer + 200 mM KCl with one change. The volume increased to 170 ml and had an estimated protein concentration of 17 mg/ml. This was loaded onto a 400 ml Phosphocellulose P11 (Whatman, Maidstone, England) in an XK50 column (5 cm × 30 cm; Amersham Pharmacia Biotech Inc, Piscataway, NJ) equilibrated with RI Buffer + 200 mM KCl at 0.5 ml/min with a Waters 650E FPLC apparatus (Waters, Milford, MA). The column was washed overnight at 1 ml/min and eluted with a 200 to 700 mM KCl gradient in RI Buffer in 5 h at the same speed. Fractions (3 ml) were collected. EcoRI* eluted in a narrow peak at around 580 mM KCl; the EcoRI methylase, which is also overproduced by the same plasmid, eluted in a much wider peak at 410 mM KCl. The EcoRI* peak fractions were pooled (45 ml) and dialyzed against 2 liters RI Buffer + 200 mM KCl overnight. This fraction was applied to and eluted from a 5 ml HiTrap heparin column (Amersham Pharmacia Biotech Inc, Piscataway, NJ) in 3 × 15 ml aliquots. The column was equilibrated in RI Buffer + 200 mM KCl and eluted with a 200 to 850 mM KCl gradient in RI Buffer in 50 min at 0.5 ml/min. Fractions (0.5 ml) were collected in Eppendorf tubes. EcoRI* eluted at 730 mM KCl. The peak fractions from the three HiTrap Heparin runs were pooled (12.5 ml) and diluted in RI Buffer to reduce the KCl concentration to 200 mM. This was loaded on a MonoS 10/10 HR column (Pharmacia) equilibrated with RI Buffer + 200 mM KCl and eluted with a 200 to 600 mM KCl gradient in RI Buffer in 40 min at 1 ml/min. Fractions (1 ml) were collected in Eppendorf tubes. EcoRI* eluted at 260 mM KCl. The peak fractions were pooled and dialyzed against EcoRI* storage buffer (20 mM KPO$_4$, pH 7.4; 400 mM KCl; 0.2 mM DTT; 1 mM EDTA; 50% glycerol). The final yield was about 18 mg EcoRI* per 75 g of cells.

Preparation of AD7333 S100 Extracts

The method is a modification of that reported by Goda and Greenblatt, 1985.[27] A 20 ml overnight culture of E. coli strain AD7333 (rna, Δrnb, recBC, sbc, nadA::Tn10, galK; gift of W. Whalen and A. Das) was used to inoculate 1 liter of Z medium (500 ml per flask in a 2 liter flask). To make 1 liter of Z-medium, add 5.6 g KH$_2$PO$_4$, 28.9 g K$_2$HPO$_4$ and 10 g yeast extract. Thiamine and glucose are added at 10 mg/liter and 1%, respectively, after autoclaving is performed.[28] The cultures were shaken vigorously at 30° until the A$_{550}$ = 4.0. The flasks were then immediately

[27] Y. Goda and J. Greenblatt, Nucleic Acids Res. **13**, 2569 (1985).
[28] H.-Z. Chen and G. Zubay, Methods Enz. **101**, 674 (1983).

swirled in an ice/water bath to quickly cool the cultures (or mixed with ice). The cultures were centrifuged (5000 rpm for 10 min; Beckman JA-20 rotor) and the supernatant discarded. The cell pellets were washed in 200 ml ice-cold Buffer X1 (0.01 M Tris/acetate, pH 7.8; 0.06 M KCl; 0.014 M MgAc; 6 mM β-mercaptoethanol) and centrifuged again. The cells were washed a second time in 100 ml ice-cold Buffer X1, centrifuged (10,000 rpm for 10 min; Beckman JA-20 rotor), and the wash discarded. Each cell pellet was then weighed. (Pellets can be quick frozen on dry ice and stored at $-80°$ for future use).

Next, the cell pellet was ground vigorously with 1.5 g alumina (Type 305; Sigma, St. Louis, MO) per g of cells. (The mortar and pestle must be pre-chilled and the mortar put on ice as the grinding is performed. It is important to grind the cells until they are efficiently broken. After a snapping sound is heard, cells are ground for another 3 min). The alumina/cell paste was resuspended in Buffer X2 (0.01 M Tris/acetate, pH 7.8; 0.06 M KCl; 0.014 M MgAc; 1 mM DTT), 1.5 ml per g of cells. The slurry was then centrifuged (16,000 rpm for 30 min in a Beckman J-20 rotor at $4°$) and the supernatant collected and adjusted to a KCl concentration of 300 mM. The samples were centrifuged (40,000 rpm for 2 h; Beckman 50Ti rotor) and the supernatant was removed and put into dialysis tubing (Spectra-Por3; 3500 Da cut-off) and dialyzed against 1 liter of Buffer X3 (0.01 M Tris/acetate, pH 7.8; 0.06 M KAc; 0.014 M MgAc; 1 mM DTT) for 5 h. The sample was then collected, distributed as 50 μl aliquots in Eppendorf tubes, quick frozen in an ethanol-dry ice bath and stored at $-80°$. Note that antitermination activity is lost if the samples are thawed more than once or twice. When samples are to be used, thaw, add $CaCl_2$ to 1 mM and micrococcal nuclease to 500 units/ml (Boehringer Mannheim, Indianapolis, IN). Incubate 30 min at room temperature to destroy all nucleic acids in the extract. After the 30 min, add EGTA to 2 mM to inactivate the micrococcal nuclease and store the extract on ice. It is then ready for use in *in vitro* reaction mixtures.

In Vitro Transcription Reactions

The *in vitro* transcription templates were designed to provide a strong promoter, the *rrn* boxA AT sequence in either the forward or backward (control) orientation, a strong Rho-dependent terminator, and an EcoRI restriction site a convenient distance from the start of transcription (Fig. 1). The plasmids contained the *rrnG* operon P2 promoter, the *trpt'* Rho-dependent terminator,[11,29] and the gene encoding chloramphenicol acetyltransferase, *cat*, which also allows assay of antitermination *in vivo*.

[29] J. L. Galloway and T. Platt, *J. Biol. Chem.* **263,** 1761 (1988).

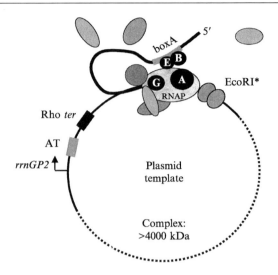

Fig. 1. Ribosomal RNA antitermination complexes. Plasmid templates (dotted line) contain the *rrnGP2* promoter with an AT motif, the *trp t'* Rho-dependent terminator (Rho-*ter*) and an EcoRI restriction site. The portion of the plasmid amplified by PCR for small-scale *in vitro* transcription assays on linear templates is shown as a solid line. The coordinates of the fragment containing the *rrnG P2* promoter are −88 to +3, relative to the initiating nucleotide. The *AT* sequence corresponds to coordinates +4 to +64 of the transcript originating from *rrnG P2*. On transcription of the AT sequence, the antitermination complex begins to assemble around the *boxA* RNA and is postulated to bind to RNA polymerase (RNAP) by the NusE/NusB heterodimer. The four known Nus factors (A, B, E, G) known to be part of the antitermination complex are shown as black spheres. Additional unknown antitermination factors provided by the S100 extract are shown as differently patterned gray ovals and spheres. The antitermination complex containing RNAP, with its transcript (curved black line) and associated proteins, is shown "roadblocked" at the site of EcoRI* mutant dimer binding. The estimated size of the known components of this RNA/DNA/protein complex is >4000 kDa.

The basic plasmid backbone was derived from pKK232-8.[30] An EcoRI restriction endonuclease site within the *cat* gene provided the binding site for the mutant EcoRI protein, EcoRI*. This protein can bind to its specific recognition site on the DNA but cannot cleave it. Bound EcoRI* is a potent inhibitor of RNA polymerase translocation, essentially blocking all complexes that are able to transcribe past the *trpt'* terminator and into the *cat* gene from going further or switching templates.[31,32] This transcription block provides uniformly sized read-through transcription products for analysis as well as a practical way of isolating the entire transcription complex by using a sizing column (see next section).

[30] J. Brosius, *Gene* **27**, 151 (1984).
[31] P. Pavco and D. Steege, *J. Biol. Chem.* **265**, 9960 (1990).
[32] E. Nudler, E. Avetissova, V. Markovtsov, and A. Goldfarb, *Science* **273**, 207 (1996).

In vitro transcription reactions are set up as follows. Two primers (upstream primer 5' TGA AAA TCT CGT CGA AGC TCG GCG G 3', downstream primer 5' CGG ATT TGA ACG TTG CGA AGC AAC G 3') are used to synthesize templates by PCR from pHBA17,[16] pRATT-1,[17] or the control plasmid, pCAB.[20] The first plasmid contains the *rrnG P2* promoter (−88 to +3, relative to transcription start) and a 250 nt fragment containing the *trp t'* Rho-dependent terminator. The second plasmid is identical to pHBA17 except it has a *Cla*I-*Bam*HI fragment containing the *rrnG* antiterminator *boxBAC* (+4 to +64, relative to transcription start) cloned between the *Cla*I-*Bam*HI sites. Plasmid pCAB is identical to pRATT-1 except it has the 61 nt *boxBAC* sequence inserted in the reverse orientation downstream of *rrnG P2*. The *in vitro* transcription reactions are performed in two steps: the formation of a post-initiation complex paused at nucleotide +8, followed by an elongation step. The 12.5 μl paused complex mix contains 50 to 100 ng of the PCR fragment; 41.8 nM of RNA polymerase from Epicentre (Madison, WI) or 20 nM of RNA polymerase from Sigma; 62 nM of EcoRI*; 100 μM of CpC (Sigma, St. Louis, MO); 2 μM of CTP, GTP and ATP (Amersham Pharmacia Biotech Inc, Piscataway, NJ); 12 units of RNasin (Promega, Madison, WI); and 5 μCi of $[\alpha^{32}P]$GTP (NEN Life Science Products, Boston, MA) in Transcription Buffer (20 mM Tris/glutamate, pH 8.0; 5 mM magnesium glutamate; 100 mM potassium glutamate; 5% glycerol; 1 mM DTT).[8] Practically, the paused complex mix is prepared as follows (Fig. 2). A mix (Step 1A) containing the initiating di-nucleotide, CpC, the three nucleotide triphosphates (CTP, GTP, ATP), RNasin and Transcription Buffer is prepared and split equally in two tubes. To the first is added template and EcoRI* (Step 1B). This mix is incubated for 10 min at 30° to allow EcoRI* to bind to the template. To the second tube is added α-P^{32}-GTP, RNA polymerase and antitermination factors (Step 1C). The amount of Nus factors or r-proteins added to the reactions is typically in the 250 to 750 nM range. Step 1C is then combined with 1B and incubated for 6 min at 30° to form the +8 paused complex. The 12.5 μl elongation mix (Step 2) contains 100 μM each UTP, GTP, and CTP, 4 mM ATP, 20 ng/μl rifampicin and 44.8 nM Rho hexamers in Transcription Buffer. The elongation mix is added to the paused complex and the reactions are incubated at 30° for 5 min. Reactions are stopped by addition of 100 μl of a solution containing 0.1 M sodium acetate (pH 5.2), 0.4% sodium dodecylsulfate, and 1.3 mg of carrier yeast tRNA per ml. The samples are extracted with phenol/chloroform, ethanol-precipitated, and resuspended in 10 μl of 32% formamide, 6.7 mM EDTA, 0.03% bromophenol blue, 0.03% xylene cyanol, heated at 95° for 2 min and then loaded on 5% polyacrylamide/7 M urea gels. After electrophoresis, gels are dried, scanned by PhosphorImager (Storm; Molecular Dynamics), and then the desired bands quantified by the ImageQuant program.

E.coli IVT protocol

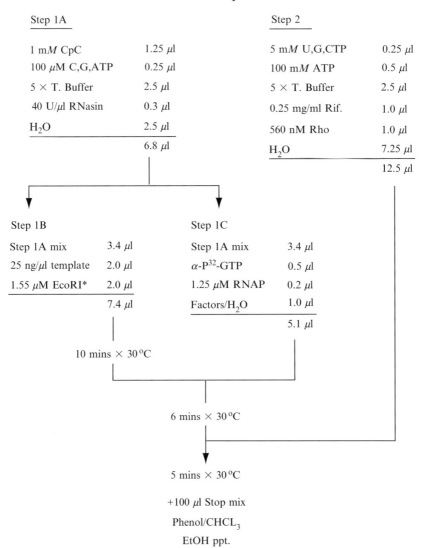

Step 1A

1 mM CpC	1.25 μl
100 μM C,G,ATP	0.25 μl
5 × T. Buffer	2.5 μl
40 U/μl RNasin	0.3 μl
H₂O	2.5 μl
	6.8 μl

Step 2

5 mM U,G,CTP	0.25 μl
100 mM ATP	0.5 μl
5 × T. Buffer	2.5 μl
0.25 mg/ml Rif.	1.0 μl
560 nM Rho	1.0 μl
H₂O	7.25 μl
	12.5 μl

Step 1B

Step 1A mix	3.4 μl
25 ng/μl template	2.0 μl
1.55 μM EcoRI*	2.0 μl
	7.4 μl

Step 1C

Step 1A mix	3.4 μl
α-P³²-GTP	0.5 μl
1.25 μM RNAP	0.2 μl
Factors/H₂O	1.0 μl
	5.1 μl

10 mins × 30°C

6 mins × 30°C

5 mins × 30°C

+100 μl Stop mix

Phenol/CHCL₃

EtOH ppt.

FIG. 2. *In vitro* transcription protocol. For complete description, see text. T. Buffer = Transcription Buffer.

Isolation of Antitermination Complexes

The scheme we developed for isolating antitermination complexes was based on the observations that[1] efficient read-through of the *trpt'* terminator could be achieved by addition of an S100 extract to the *in vitro* transcription system[17] and that[2] the EcoRI* block to transcription elongation provides an excellent way to stabilize stalled transcription complexes.[31] We reasoned that the EcoRI*-stalled antitermination complex should contain the proteins that allowed it to efficiently read-through the *trpt'* Rho-dependent terminator and that such a large complex could be separated from all other components of the reaction by size fractionation. Based on the molecular weights of the proteins known to associate with RNA polymerase, we estimated the size of the stalled antitermination complex to be at least 4000 kDa and chose an appropriate sizing gel accordingly. We also scaled up the reaction volume 50-fold and separately processed ten reactions to obtain enough material to analyze on an SDS-PAGE gel and then extract protein bands for identification by N-terminal sequencing. Using this method, we identified the r-proteins S2, S4, L1, L3, L4, and L13—as well as H-NS—in the antiterminated transcription complexes, in addition to the known protein factors that we had added to the reactions.[20]

Antitermination complexes were isolated on a circular pRATT plasmid template in a scaled-up transcription reaction.[17] Reactions were done slightly differently from above. A typical 500 μl +8 paused complex mix contained 100 μg of plasmid DNA, 300 mM potassium glutamate, 2.5 μM cold GTP, 25 μCi of [α^{32}P]GTP, 260 nM of Sigma RNA polymerase, 400 units RNasin, 18 nM Rho hexamers, 80 nM EcoRI* dimers, GTP and CTP (40 μM GTP and 140 μM CTP) in 0KCl buffer (20 mM Tris/acetate, pH 7.9; 0.1 mM EDTA; 1 mM DTT; 4 mM magnesium acetate). This mix was allowed incubate 10 min at 37°C to allow EcoRI* to bind. The 500 μl elongation mix contained ATP, UTP (1.16 mM ATP and 140 μM UTP), 366 nM NusA, 839 nM NusB, 860 nM NusE, 3.43 μM NusG and AD7333 S100 extract in 0KCl buffer. The elongation mix was preheated 3 min at 37° before being added to the paused complex mix. The reaction was allowed to continue 6 min at 37°. The combined transcription mix (1 ml) was added directly to an 11 ml Ultrogel A4 sizing column (Sigma, St. Louis, MO), drained to the top of the resin (Fig. 3). The mix was allowed to run in and was washed in with 1 ml transcription buffer. The top of the column was then filled with 5 ml transcription buffer, and 500 μl fractions were collected in Eppendorf tubes. Radioactivity incorporated into transcription complexes was followed by using a Geiger counter. Two peaks of radioactivity were observed, the first corresponding to labeled GTP incorporated into newly synthesized mRNA in the antitermination

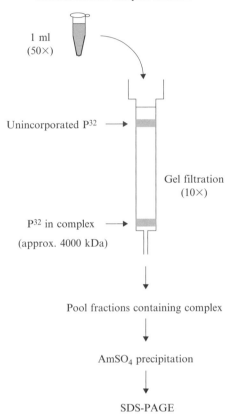

Antitermination complex Isolation

1 ml
(50×)

Unincorporated P³² ⟶

Gel filtration
(10×)

P³² in complex ⟶

(approx. 4000 kDa)

Pool fractions containing complex

AmSO₄ precipitation

SDS-PAGE

FIG. 3. Isolation of antitermination complexes by gel filtration. For full description, see text.

complex, "roadblocked" at the EcoRI* site of the template DNA. The second, much-larger peak corresponded to unincorporated GTP. Less than 5% of the radioactivity was incorporated under these conditions. The experiment was repeated 9 times. The peak fractions corresponding to the antitermination complex were pooled from the 10 experiments (total volume was 27.5 mls). The OD_{260} and OD_{280} were 0.72 per ml and 0.59 per ml, respectively. The proteins of this complex were precipitated by slow addition of 6.4 g of finely ground ammonium sulfate (80% saturation), incubation on ice for 3 h without stirring, and centrifugation at 15,000 rpm for 20 min (Sorvall SS34 rotor). The pellet was resuspended in 1 ml H_2O and dialyzed overnight at 4° against 2 liters H_2O, with 3 changes. The dialysate was dried

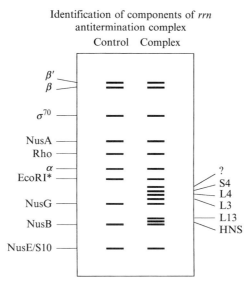

Fig. 4. Identification of components of rRNA antitermination complex. A sample of known proteins added to the reaction mix is shown in the left lane. α,β,β', and σ^{70} are the different sub-units of RNA polymerase. The proteins recovered in the antitermination complex are shown in the right lane. The presence of r-proteins S2 and L1, which co-migrate, was deduced from a mixed sequence.

to completion in a SpeedVac (Savant, Farmingdale, NY). The sample was resuspended in sample buffer and one quarter was run on a 14% polyacrylamide/SDS gel, alongside a lane containing a mixture of all of the known components added to the transcription reaction (i.e., RNA polymerase, Nus factors, EcoRI* and Rho and stained with Coomassie blue). The remaining three quarters of the sample was run on a separate gel, blotted to an Immobilon-P membrane (Millipore, Bedford, MA), and stained in 0.2% amido black. New bands present in the antitermination complex were cut out for N-terminal sequencing. A schematic of such a gel is shown in Fig. 4.

While the procedure outlined above was used to successfully isolate ribosomal RNA antitermination complexes and identify the different proteins present, this method of using EcoRI* as a roadblock to RNA polymerase can be adapted to any situation in which the stabilization of transcription complex might be useful. Although it has not yet been shown whether eukaryotic RNA polymerases can also be blocked by EcoRI*, this system may lend itself to the identification of some of the components of the many different kinds of transcription complexes that have been described in higher organisms.

Section III

Chromatin

[36]   Purification of Elongating RNA Polymerase II and Other Factors from Yeast Chromatin

By JESPER Q. SVEJSTRUP, THODORIS G. PETRAKIS, and JANE FELLOWS

Introduction

Biochemical isolation and characterization of multi-subunit protein complexes from tissues or cells derived from a variety of organisms has led to profound insight into the basic mechanisms of transcription and other nucleic acid–associated processes. In almost all cases, however, the DNA-binding proteins—or proteins involved in processes occurring on DNA—are deliberately purified from cleared extracts, devoid of nucleic acids. Soluble protein is salt-extracted from DNA before cell debris and large chromatin fragments are discarded after centrifugation, and little attention is then paid to proteins that remain tightly chromatin-associated in the pellet fraction. It is, however, an obvious possibility that different versions of the same protein exist as both weakly and strongly DNA-associated species, respectively, and that isolation of the easily extracted version gives only part of the information that can be obtained by the protein's biochemical characterization. Two examples of the potential value of investigating whether a different form of a protein exists in chromatin have recently been described in our laboratory. First, the hyper-phosphorylated, tightly chromatin-associated, elongating form of RNA polymerase II was isolated and shown to be a novel holoenzyme,[1] fundamentally different from the one involved in transcript initiation. Second, the transcription-repair coupling factor Rad26 was found to exist in similar amounts in the soluble and the tightly chromatin-associated fraction, and when purified from these sources it turned out to exist in different forms.[2] The form purified from the soluble fraction did not appear to be complexed with any other factor, whereas chromatin-associated Rad26 was associated with another protein, Def1. These examples suggest that purification from chromatin might shed new light on the function of a variety of DNA-related factors. Here, we describe procedures for the identification and purification of tightly chromatin-associated factors with particular emphasis on elongating RNA polymerase II.

[1] G. Otero, J. Fellows, Y. Li, T. De Bizemont, A. M. G. Dirac, C. M. Gustafsson, H. Erdjument-Bromage, P. Tempst, and J. Q. Svejstrup *Mol. Cell* **3,** 109 (1999).

[2] E. C. Woudstra, C. Gilbert, J. Fellows, L. Jansen, J. Brouwer, H. Erdjument-Bromage, P. Tempst, and J. Q. Svejstrup *Nature* **415,** 929 (2002).

Identification of Chromatin-Associated Protein *In Vitro*

Identification of tightly chromatin-associated proteins *in vitro* can be done in small scale by the use of a procedure that has been adapted from the work of Diffley and co-workers, who used it to show that DNA replication proteins such as MCM, Cdc6, and ORC are bound to chromatin.[3,4] This procedure can be used for quickly establishing if a protein of interest partitions to both the soluble and the chromatin-associated fraction *in vitro*.

Materials and Reagents

PIPES (Sigma), 1 M stock solution; pH adjusted with potassium hydroxide

Dithiotreitol (DTT), 1 M stock solution

Lyticase (L-5763, Sigma. 8000 units/mg protein)

Sorbitol (Sigma), 1 M stock in water, sterile-filtered

Triton X-100 (Sigma T-8787), 20% stock solution

YPD[5], sterile-autoclaved

Potassium acetate, 5 M stock, adjusted to pH 7.8 with potassium hydroxide, sterile-filtered

Magnesium acetate, 1 M stock

Protease inhibitors ($100\times$ stock solution in ethanol):

Phenymethyl-sulfonyl fluoride (PMSF) (Sigma), 100 mM

Leupeptin (Sigma), 1 mg/ml

Pepstatin A (Sigma), 100 μg/ml

Benzamidine HCl (Sigma), 1 M

Lysis Buffer: 0.4 M sorbitol, 150 mM potassium acetate, 2 mM magnesium acetate, 20 mM PIPES/KOH, pH 6.8, 1 mM PMSF, 10 μg/ml leupeptin, 1 μg/ml pepstatin A, 10 mM benzamidine HCl.

Procedure

The procedure below is for a 25-ml culture at 10^7 cells/ml.

1. Harvest and re-suspend the cells in 6.25 ml 100 mM PIPES/KOH, pH 9.4, 10 mM DTT. Incubate at 30° for 10 min with agitation.

2. Spin down cells and re-suspend in 2.5 ml YPD containing 0.6 M sorbitol, 25 mM Tris-Cl, pH 7.5. Add 50 μl 25 mg/ml lyticase.

---

[3] Donovan, J. Harwood, L. S. Drury, and J. F. Diffley *PNAS* **94,** 5611 (1997).

[4] C. Desdouets, C. Santocanale, L. S. Drury, G. Perkins, M. Foiani, P. Plevani, and J. F. Diffley *EMBO J.* **17,** 4139 (1998).

[5] F. Sherman, (1991) *in* "Guide to yeast genetics and molecular biology" (Guthrie, C., and Fink, G. R., eds.), Vol. 194, pp. 3, Academic Press, Inc., San Diego.

3. Incubate at $30°$ for approximately 15 min with agitation (until cells are 100% spheroplasted—as judged by microscopy after lysis in 1% Triton-X-100).

4. Spin 1000 RPM in bench top centrifuge 3 min to pellet cells. Re-suspend in 2.5 ml YPD containing 0.7 M sorbitol, 25 mM Tris-Cl, pH 7.5.

5. Incubate at $30°$ for 20 min with agitation. Then spin 1000 RPM in bench top centrifuge for 3 min to pellet cells.

6. Wash pellet three times with 1 ml lysis buffer, each time spinning cells out as above.

7. Re-suspend the pellet in 400–500 μl of lysis buffer. Determine the protein concentration by Bradford assay. Then adjust the volume of the suspension so that the protein concentration is about 5 mg/ml.

8. Add 10% Triton X-100 so that the final concentration is 1%. Mix gently. Examine cell lysis microscopically—should be close to 100%.

9. Remove 90 μl of the lysed material and add to 30 μl of 3× SDS loading buffer for "whole cell extract" control.

10. Remove 100 μl of the lysed material and spin in microcentrifuge at 14,000 RPM for 15 min at $4°$C. All of the chromatin will be in the pellet fraction while >90–95% of the protein should be in the supernatant.

11. The total volume of the supernatant usually turns out to be around 95 μl, but this needs to be measured each time. Add 30 μl of 3× SDS-loading buffer to the supernatant.

12. Wash the chromatin pellet once with 100 μl of the lysis buffer. Re-suspend the chromatin pellet in the same volume as the supernatant (usually ca. 95 μl; see above) and add 30 μl of 3× SDS-loading buffer. This generates three samples: whole cell extract, soluble extract, and low-salt chromatin. A substantial fraction of almost any DNA-binding protein will be found in the chromatin fraction under these conditions.

13. Remove another 100 μl of the lysed starting material and add 7.5 μl 5 M potassium acetate (KOAc), pH 7 (500 mM final KOAc concentration). Incubate 15–30 min at $4°$.

14. Spin in microcentrifuge at 14,000 RPM for 15 min at $4°$. Again, all of the chromatin will be in the pellet fraction. Treat samples as above (steps 11–12), making sure to end with equal volumes of soluble extract and *salt-stable* chromatin fraction.

15. Check for the presence of the protein of interest by Western blotting.

If the protein of interest is found in the salt-stable chromatin fraction, it is of relevance to ensure that this is in fact due to its binding to nucleic acids. This is best tested by also doing a separation (in 500 mM KOAc; step 14) with extensive sonication of the chromatin prior to spinning. We've found that the chromatin resulting from the protocol is often quite resistant to DNAse digestion but that it can be solubilized quite well into soluble 300–800 bp fragments by sonication. After sonication and spinning, a larger proportion of the protein should now be found in the supernatant with the sheared DNA.

Note on gel running: After the samples have been boiled prior to gel loading, spin at 14,000 RPM for 1 min in a microcentrifuge, and vortex well. Then re-spin, and load 5–10 μl of the soluble material. These steps minimize loading errors, making it easier to ascertain the precise proportion of the protein that can be detected in the different fractions.

Extraction and Purification of Elongating RNA Polymerase II and Other Chromatin-Associated Proteins

Once it has been established that a protein of interest is found in the salt-stable chromatin fraction, the protein can be purified from a larger quantity of cells. The chromatin isolation and purification procedure described below is based on at least 10 to 20, typically 50 to 80 liters of yeast culture (10^{11}–10^{13} cells) and breaking cells by bead-beating. Bead-beating shears DNA into semi-soluble chromatin fragments that can be separated from soluble protein as well as cell debris by ultra-centrifugation in a Ti45 rotor. Others and we have failed to reproduce this separation by using other procedures for making extracts, or scaling down so that other rotors than Ti45 can be used. This is probably because DNA shearing and the indicated centrifugation forces are required to obtain "solubilized" chromatin. Starting from such sheared, solubilized chromatin fragments thus avoids the contamination from cell debris and organelles in the pellet fraction and also allows efficient separation from non-DNA bound material.

Preparation of Yeast Whole Cell Extracts (WCE)

Materials and Reagents

> Bead-beater (BioSpec products Inc., Cat. No. 1107900)
> Glass beads (BioSpec products)
> Potassium acetate, 5 M stock, adjusted to pH 7.8 with potassium hydroxide, sterile-filtered
> Beckman Ultracentifuge, Ti45 rotor and tubes
> DNAse I (Stratagene)
> RNAse A (Sigma)
> Polyethyl-enimine (Sigma P-3143), 20% stock solution in water adjusted to pH 7.5 with HCl

3× lysis buffer: 60% glycerol, 450 mM Tris-Acetate, pH 7.8, 150 mM KOAc, 3 mM EDTA, 15 mM DTT, 1 mM PMSF, 10 μg/ml leupeptin, 1 μg/ml pepstatin A, 10 mM benzamidine HCl, 5 mM sodium fluoride, 5 mM sodium pyrophosphate, 10 mM sodium phosphate

Procedure

1. Yeast paste from the strain of choice is redissolved in the appropriate volume of 3× lysis-buffer (1 g yeast paste is taken to correspond to 1 ml). If starting from frozen yeast in lysis buffer (preferable), thaw just enough to make it possible to pour into a pre-cooled 450 ml bead-beater chamber; 200–300 ml per chamber (ideally 250 ml). Add glass-beads (0.5 mm diameter) to almost fill chamber and stir well to dissolve and to let air escape. Fill up with beads, enough to spill a little when the bead-beater propeller is inserted. This is important to avoid air being trapped in the chamber.

2. Assemble the bead-beater and fill the coat with ice-salt mixture. Firmly press the ice-salt mix around the whole chamber, before pouring cold water over to make good cooling connection from ice to chamber.

3. Bead-beat in cycles of 0.5 min ON/1.5 min OFF. We usually do a total of 35 to 45 min. Make sure that the ice doesn't melt away. Refill if necessary.

4. After the last beating, leave for 5 min to cool in chamber. Then invert, disassemble, and leave on ice for another 5 min to allow beads to settle. Pour off the thick yeast protein extract into pre-cooled glass beaker. Avoid getting too many beads (not easy). Use a 25-ml pipette to drain beads in chamber for the remaining extract: press against bottom of chamber to avoid getting beads stuck in pipette. Wash beads in the chamber twice with 25 ml 1 × lysis buffer to extract all the protein.

5. Spin extract in 250- or 500-ml centrifuge tubes for 30 min at 8000 to 9000 rpm in a pre-cooled rotor to remove unbroken cells and remaining glass beads. Pour into pre-cooled measuring cylinder. Measure volume.

6. Transfer to pre-cooled glass-beaker and slowly add 1/10 volume 5 M KOAc, pH 7.8 while stirring at 4°. Stir for 30 min. This step removes "weakly chromatin-bound proteins" by raising the salt concentration to more than 500 mM.

7. Spin 42,000 rpm in Ti45 rotor for 90 min at 4°.

8. Remove the upper clear supernatant (60% to 80% of the total volume) with 25-ml glass pipette, taking great care not to touch the brownish, murky chromatin layer above the gray pellet.

9. Carefully pour off the chromatin layer, leaving behind the solid pellet (mostly cell debris). Wash/dilute the chromatin layer with 3 volumes

1 × lysis buffer containing 500 mM KOAc (or simply pour chromatin layer into fresh Ti45 tube and fill with this buffer) and re-spin as in 7.

10. Remove and discard the supernatant above the murky layer. Chromatin at the bottom of the tube is often "collapsed" and partly forms a firm whitish/clear pellet. Re-dissolve the pellet in the murky liquid remaining. To obtain a homogenous solution, the chromatin solution can be homogenized with a few strokes in a 50-ml Dounce homogenizer. Typical yields range from 20 to 60 ml of very concentrated chromatin from six Ti45 tubes of starting whole cell extract material. A few microliters of material fractionated on an agarose gel and stained with ethidium bromide is sufficient to be convinced that there are very large amounts of nucleic acids in the fraction. The presence of the protein of interest should be checked by Western blotting. However, please note that the large amount of nucleic acids will affect gel running and often makes detection of non-abundant proteins difficult.

This fraction now forms the basis from which to purify strongly chromatin-associated proteins. But first, the proteins have to be extracted and the nucleic acids removed. For the extraction of most proteins, it is sufficient to simply raise the salt concentration by addition of ammonium sulfate. However, in the case of elongating RNAPII we found it helpful to first add nucleases and incubate at 4° for 30 to 45 min before adding salt. The advantage of not having to add nucleases is that nucleic acids are much easier to remove after high-salt extraction of the strongly bound proteins.

1. Add DNAse and RNAse to the concentrated chromatin fraction (10 U/ml and 5 μg/ml of DNAse I and RNAse A, respectively) and incubate 30 to 90 min at 4°.

2. Add ammonium sulfate to 1 M final concentration (as powder, or as cold, saturated solution, pH 7.5). Incubate 45 min, stirring at 4°. Please note that some proteins may precipitate at this ammonium sulfate concentration (approximately 25% of saturation).

3. Add 20% polyethyl-enimine to 0.5% final concentration. Incubate 45 min, stirring at 4°.

4. Spin 42,000 rpm in Ti45 rotor for 90 min at 4°. DNA is in the pellet fraction; extracted proteins are in the supernatant. The protein fraction can now be dialyzed prior to purification of the protein of interest.

Purification of Elongating RNA Polymerase II

Proteins released from chromatin can be purified conventionally or via affinity chromatography. We have found that for purification of elongating RNA polymerase II holoenzyme, conventional chromatography with the procedure below has the advantage of gradually co-concentrating components of the fragile holoenzyme, and we therefore prefer this procedure over others.

Materials and Reagents

> HEPES (Sigma), 2 M stock, adjusted to pH 7.6 with potassium hydroxide
>
> Tris (ICN Biomedicals, Ohio), 2 M stock, adjusted to pH 7.8 with acetic acid
>
> Di-potassium hydrogen orthophosphate (BDH laboratory supplies, Poole, UK), 1 M stock solution
>
> Potassium di-hydrogen orthophosphate (BDH laboratory supplies, Poole, UK), 1 M stock solution
>
> 8A7 monoclonal antibody (BD Pharmingen).
>
> 8WG16 monoclonal antibody (BabCo)
>
> Bio-Rex-70 (Bio-Rad)
>
> DEAE Sephacel (Amersham Pharmacia)
>
> HTP hydroxylapatite (Bio-Rad)
>
> Mono Q HR 5/5 (Amersham Pharmacia)
>
> Buffer A: 40 mM HEPES, pH 7.6, 10% glycerol, 1 mM EDTA, 5 mM sodium fluoride, 5 mM sodium pyrophosphate, 10 mM sodium phosphate
>
> Buffer B: 20% glycerol, 20 mM Tris-Acetate, pH 7.8, 1 mM DTT, 1 mM EDTA, 1 mM PMSF, 10 μg/ml leupeptin, 1 μg/ml pepstatin A, 10 mM benzamidine HCl, 5 mM sodium fluoride, 5 mM sodium pyrophosphate, 10 mM sodium phosphate
>
> Buffer C: 20% glycerol, 100 mM KOAc, pH 7.5–7.8, 1 mM DTT, 0.1 mM EDTA, 0.01% NP-40, 1 mM PMSF, 10 μg/ml leupeptin, 1 μg/ml pepstatin A, 10 mM benzamidine HCl, 5 mM sodium fluoride, 5 mM sodium pyrophosphate (buffer C is buffered by potassium phosphate, which is adjusted to the correct pH by mixing di-potassium hydrogen orthophosphate and potassium di-hydrogen orthophosphate[6]

Procedure

1. Dialyze protein fraction against buffer A containing 25 mM KOAc (A-25) until the conductivity of the fraction is less than that of buffer A-300 (buffer A with 300 mM KOAc). Then adjust the salt concentration of the fraction to that of A-150 by dilution in buffer A-0.

---

[6] J. Sambrook, E. F. Fritsch, and T. Maniatis, Molecular Cloning. A Laboratory Manual, Cold Spring Harbor Laboratory Press, Plainview, NY (1989).

2. Fractionate on Bio-Rex-70 (1 ml resin per 20 mg protein). After protein loading, wash with 1 column volume (CV) buffer A-150, then elute by consecutive step-wise elution with 2 CV buffer A-300, 2 CV buffer A-600, and 2 CV buffer A-1200, collecting 1/10 CV fractions. RNAPII will elute in approximately half a column volume in A-600.

3. Collect fractions containing RNAPII and dialyze fraction against buffer B-0 for 3 to 4 hours and dilute in buffer B-0 till the conductivity is less than that of buffer B-100 (100 mM KOAc).

4. Fractionate on DEAE Sephacel (1 ml resin per 20 mg protein), equilibrated in buffer B-100. After protein loading, wash with 1 column volume (CV) buffer B-100, then elute by consecutive step-wise elution with 2 CV buffer B-200, 2 CV buffer B-550, and 2 CV buffer B-1000, collecting 1/10 CV fractions. RNAPII will elute in B-550.

5. Collect fractions containing RNAPII and fractionate directly on HTP hydroxylapatite (1 ml resin per 5 mg protein), equilibrated in buffer C containing 10 mM potassium phosphate, pH 7.7 (Buffer C-10). After loading, wash with 1 CV of same buffer, before resolving the column with a 10 CV gradient from 10 to 300 mM potassium phosphate, pH 7.7, in buffer C. Resolve the column at a speed of no more than 1–1.5 CV/hour for best resolution. Collect 1/4 column volume fractions. Locate RNAPII by Western blotting with phosphorylation-specific RNAPII antibodies (8A7 monoclonal antibody, BD Pharmingen).

6. Collect fractions containing RNAPII and fractionate on Mono Q HR 5/5 (20–50 mg protein loaded per run) equilibrated in buffer B-300. After protein loading, wash with 1 column volume (CV) buffer B-300, and then elute with 10 CV gradient from B-300 to B-1200. Elongating RNAPII will elute at around 1 M KOAc.

Check the ratio of phosphorylated versus hypo-phosphorylated RNA-PII by comparing the reactivity of the crude chromatin fraction and the highly purified protein to 8WG16 (BabCo) (non-phosphorylated CTD) and 8A7 (BD Pharmingen) (phosphorylated CTD). The purified protein can be dialyzed to lower salt (100 mM KOAc final concentration) and stored at −80°. It is quite stable and can be frozen/thawed several times without loss of activity.

[37]   Chromatin Assembly *In Vitro* with Purified Recombinant ACF and NAP-1

By Dmitry V. Fyodorov and James T. Kadonaga

To study eukaryotic transcriptional mechanisms in vitro, it is important to analyze gene regulatory sequences in the context of chromatin. Here we describe the ATP-dependent assembly of chromatin by using completely purified components. This system uses chromatin assembly factors ACF and NAP-1 in conjunction with purified core histones for the assembly of extended periodic arrays of nucleosomes. We additionally describe the assembly of chromatin that contains the linker histone H1. This histone H1-containing chromatin resembles bulk native chromatin in metazoans.

In the eukaryotic nucleus, transcription, as well as DNA replication, recombination, and repair occur in the context of chromatin.[1] Therefore, regulatory mechanisms of DNA-utilizing nuclear processes would ideally be studied with chromatin. To study chromatin-dependent transcriptional regulation in vitro, a variety of methods have been employed for the reconstitution of chromatin.[2] The principal goal of these techniques has been to reconstitute periodic arrays of nucleosomes from purified DNA and core histones. A number of methods can yield randomly distributed nucleosomes from defined purified components. These methods are based on histone-DNA co-folding during dialysis/dilution of high-salt buffers[3] or histone transfer from negatively charged (polyanion) complexes.[4,5] Until recently, the assembly of periodic nucleosome arrays could only be achieved with crude cell extracts.[6–8] Each of these approaches has advantages and shortcomings, which are discussed elsewhere.[2] In this chapter, we describe the methodology for the ATP-dependent assembly of periodic nucleosome arrays by using purified recombinant chromatin assembly factors.

The assembly of DNA into chromatin is a critical step in the duplication and maintenance of the eukaryotic genome. Chromatin assembly is an

[1] A. P. Wolffe, "Chromatin: Structure and Function." Academic Press, San Diego (1995).

[2] D. V. Fyodorov and M. L. Levenstein, *in* "Current Protocols in Molecular Biology," p. 21.7.1. John Wiley & Sons, New York (2002).

[3] R. D. Camerini-Otero, B. Sollner-Webb, and G. Felsenfeld, *Cell* **8,** 333 (1976).

[4] A. Stein, *Methods Enzymol.* **170,** 585 (1989).

[5] D. Rhodes and R. A. Laskey, *Methods Enzymol.* **170,** 575 (1989).

[6] G. C. Glikin, I. Ruberti, and A. Worcel, *Cell* **37,** 33 (1984).

[7] P. B. Becker and C. Wu, *Mol. Cell Biol.* **12,** 2241 (1992).

[8] M. Bulger and J. T. Kadonaga, *Methods Mol. Genet.* **5,** 241 (1994).

ATP-dependent process and is mediated by the concerted action of core histone chaperones and ATP-dependent chromatin assembly machines.[9] NAP-1, CAF-1, ASF-1, nucleoplasmin and other histone chaperones play an essential role by escorting histones to the sites of nucleosome assembly.[10] ATP-utilizing factors mediate the energy-dependent steps of the chromatin assembly reaction, which are the deposition of core histone octamers onto the DNA and the organization of assembled nucleosomes into periodic arrays. ACF (*ATP-utilizing chromatin assembly and remodeling factor*) was identified on the basis of its ability to mediate the assembly of extended periodic nucleosome arrays in conjunction with NAP-1.[11] ACF can also catalyze the mobilization of nucleosomes[11,12] and is therefore a chromatin remodeling factor as well as a chromatin assembly factor. ACF consists of two subunits, the ISWI ATPase and a polypeptide termed Acf1.[13] Both subunits are essential for the full chromatin assembly activity of ACF. Recombinant ACF and NAP-1 can be synthesized by using baculovirus vectors and purified by affinity and conventional chromatography. Together with purified native or recombinant core histones,[14] ACF and NAP-1 constitute the chromatin assembly system that is discussed here.

Chromatin templates assembled with this purified recombinant system have been used for studies of transcription[15] and DNA recombination[16] and repair.[17] This *in vitro*-assembled chromatin possesses properties that are similar to those of bulk native chromatin and should be generally useful for the biochemical analysis of transcriptional regulation and other DNA-utilizing processes in the eukaryotic nucleus.

Purification of Chromatin Assembly Factors

Reagents and Equipment

In addition to common laboratory equipment (centrifuges, tissue culture equipment, etc.), reagents, and consumables (plasticware, dialysis

[9] T. Ito, J. K. Tyler, and J. T. Kadonaga, *Genes Cells* **2**, 593 (1997).

[10] J. A. Mello and G. Almouzni, *Curr. Opin. Gen. Dev.* **11**, 136 (2001).

[11] T. Ito, M. Bulger, M. J. Pazin, R. Kobayashi, and J. T. Kadonaga, *Cell* **90**, 145 (1997).

[12] A. Eberharter, S. Ferrari, G. Längst, T. Straub, A. Imhof, P. Varga-Weisz, M. Wilm, and P. B. Becker, *EMBO J.* **20**, 3781 (2001).

[13] T. Ito, M. E. Levenstein, D. V. Fyodorov, A. L. Kutach, R. Kobayashi, and J. T. Kadonaga, *Genes Dev.* **13**, 1529 (1999).

[14] M. E. Levenstein and J. T. Kadonaga, *J. Biol. Chem.* **277**, 8749 (2002).

[15] W. Jiang, S. K. Nordeen, and J. T. Kadonaga, *J. Biol. Chem.* **275**, 39819 (2000).

[16] V. Alexiadis and J. T. Kadonaga, *Genes Dev.* **16**, 2767 (2002).

[17] K. Ura, M. Araki, H. Saeki, C. Masutani, T. Ito, S. Iwai, T. Mizukoshi, Y. Kaneda, and F. Hanaoka, *EMBO J.* **20**, 2004 (2001).

tubing, Wheaton Dounce homogenizers, etc.), the following should be provided: log phase Sf9 cells and culture media (Pharmingen); high-titer baculoviruses for expression of 6His-NAP-1, Acf1-FLAG, and ISWI (Orbigene); Ni-NTA agarose resin (Qiagen); imidazole; FPLC; Source 15Q resin (Amersham Biosciences); FLAG-M2 resin and FLAG peptide (Sigma); recombinant human insulin (Roche, Cat. 1376497); assorted protease inhibitors (e.g., Sigma); bovine serum albumin mass standard (Pierce, Cat. 23209); pET-NDH6 bacterial expression construct encoding, *Drosophila* topoisomerase I catalytic domain (available by request).

Drosophila *NAP-1*

Sf9 cells are infected with the baculovirus expressing N-terminally histidine-tagged NAP-1. The chaperone protein is purified by affinity chromatography through the 6His tag, followed by anion exchange chromatography on Source 15Q resin. Note that the second step removes an inhibitor and is essential to obtain active NAP-1. This protocol was originally described by Fyodorov and Kadonaga.[18]

Log phase Sf9 cells are diluted to 1×10^6 cells/ml in a spinner flask and infected with 6His-NAP-1 baculovirus at a multiplicity of infection of at least 5. At 72 hours subsequent to infection, the cells are harvested and washed with ice-cold phosphate-buffered saline (PBS). All subsequent steps are performed on ice or in the cold room. The cell pellet is resuspended in the lysis buffer (1/40 of the cell culture volume) containing 50 mM sodium phosphate (pH 7.6), 500 mM NaCl, 20 mM imidazole, 15% (v/v) glycerol, 0.01% (v/v) NP-40, 10 mM β-glycerophosphate, 0.2 mM PMSF, and 0.5 mM benzamidine. The cells are homogenized by sonication (or douncing). The lysate is cleared by centrifugation (10 min at 15,000 g) and mixed with Ni-NTA resin slurry (2 ml per 1 liter Sf9 cell culture volume).

After incubation on a rocking platform or a spinning wheel from 3 h to overnight, the resin is pelleted by centrifugation (2 min at 500 g), washed twice with the lysis buffer and twice with the wash buffer (lysis buffer containing 100 mM, rather than 500 mM NaCl) and eluted by 4 successive elution steps with elution buffer (wash buffer plus 480 mM imidazole). The volume of the buffer for each elution step is equal to twice the volume of Ni-NTA resin slurry (i.e., 4 ml per 1 liter Sf9 cell culture volume). The eluted protein is pooled and dialyzed (10,000 molecular weight cutoff) twice for 2 h against 4 liters of buffer that contains 25 mM potassium HEPES, pH 7.6, 1 mM EDTA, 10% (v/v) glycerol, 100 mM NaCl, 0.01%

[18] D. V. Fyodorov and J. T. Kadonaga, *Mol. Cell. Biol.* **22**, 6344 (2002).

(v/v) NP-40, 10 mM β-glycerophosphate, 1.0 mM DTT, 0.2 mM PMSF and once for 2 h against 4 liters NAP-1 buffer (10 mM potassium HEPES, pH 7.6, 10 mM KCl, 1.5 mM MgCl$_2$, 0.5 mM EGTA, 10% glycerol, 0.01% NP-40, 10 mM β-glycerophosphate, 1.0 mM DTT, 0.2 mM PMSF) plus 100 mM NaCl. Precipitated material is removed by centrifugation (20 min at 20,000 g). The soluble protein is quantitated by SDS polyacrylamide gel electrophoresis on an 8% gel and Coomassie staining along with bovine serum albumin mass standard. The expected yield at this step is 2 to 10 mg protein from 1 liter of Sf9 cell culture. It should be at least 90% homogenous.

By using the FPLC, NAP-1 is further purified on Source 15Q chromatographic resin (1 ml of packed resin per 5 mg nickel-purified NAP-1). The column is equilibrated and the sample is injected in the NAP-1 buffer plus 100 mM NaCl, washed with 10 column volumes of NAP-1 buffer plus 200 mM NaCl and eluted with a 20 column volume gradient of NAP-1 buffer containing from 200 to 500 mM NaCl. NAP-1-containing material elutes in two peaks, between 250 and 350 mM NaCl. The column fractions are analyzed by SDS polyacrylamide gel electrophoresis on an 18% gel and Coomassie staining. The first peak contains a 14-kDa contaminating band (approximately 2% of the total protein by mass) in addition to the 56-kDa NAP-1 band. This fraction is inactive in chromatin assembly reactions. Fractions from the second peak, which are free from contaminating bands, are pooled and dialyzed against 2 changes of 2 liters NAP-1 Buffer plus 100 mM NaCl for a total of 4 h. The protein is quantitated by SDS polyacrylamide gel electrophoresis along with bovine serum albumin mass standard and stored at $-80°$. The expected yield is 1 to 3 mg NAP-1 protein from 1 liter Sf9 cell culture (Fig. 1).

Drosophila ACF

Drosophila ACF is prepared by co-expression of C-terminally FLAG-tagged Acf1 subunit and untagged ISWI subunit in baculovirus. The complex is purified in one step by FLAG affinity chromatography. This procedure typically results in purification of a stoichiometric complex of Acf1 and ISWI polypeptides (185 and 140 kDa, respectively). This protocol is a modification of the original method described by Ito *et al.*[13]

Log phase Sf9 cells are plated at 1.5 to 2.0×10^4 cells/cm^2 and infected with both Acf1-FLAG and ISWI baculoviruses at multiplicity of infection of at least 5. At 44 to 46 hours subsequent to infection, the cells are harvested and washed with ice-cold PBS. All subsequent steps are performed on ice or in the cold room. The cell pellet is resuspended in the lysis buffer (1 ml per 150 mm cell culture plate) that contains 20 mM Tris-HCl (pH

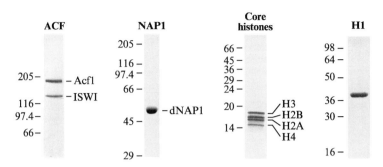

Fig. 1. Components of the purified recombinant chromatin assembly system. Purified recombinant ACF, purified recombinant NAP-1, purified native *Drosophila* core histones, and purified native *Drosophila* histone H1 were analyzed by SDS-polyacrylamide gel electrophoresis and stained by Coomassie brilliant blue R-250. The sizes of molecular mass markers (in kDa) are indicated. Adapted with permission from W. Jiang, S. K. Nordeen, and J. T. Kadonaga, *J. Biol. Chem.* **275,** 39819 (2000).

7.9), 500 mM NaCl, 20% (v/v) glycerol, 4 mM MgCl$_2$, 0.4 mM EDTA, 2 mM DTT, 20 mM β-glycerophosphate, 0.4 mM PMSF, 1.0 mM benzamidine-HCl, 4 μg/ml leupeptin and 2 μg/ml aprotinin. The cells are homogenized by douncing. The lysate is cleared by centrifugation (10 min at 15,000 g), diluted 1:1 (v/v) with a buffer that contains 20 mM Tris-HCl (pH 7.9), 10% (v/v) glycerol and 0.02% (v/v) NP-40 and mixed with FLAG-M2 resin slurry (20 μl per 150 mm cell culture plate).

After incubation on a rocking platform for 3 to 4 hours, the resin is pelleted by centrifugation (2 min at 500 g) and washed 4 times with the wash buffer that contains 20 mM Tris-HCl (pH 7.9), 150 mM NaCl, 15% (v/v) glycerol, 2 mM MgCl$_2$, 0.2 mM EDTA, 0.01% (v/v) NP-40, 1 mM DTT, 10 mM β-glycerophosphate, 0.2 mM PMSF, 0.5 mM benzamidine-HCl, 2 μg/ml leupeptin and 1 μg/ml aprotinin. The protein is eluted by 4 successive elution steps with the elution buffer (wash buffer containing 0.4 mg/ml recombinant insulin and 0.4 mg/ml FLAG peptide). The volume of elution buffer for each step is equal to 0.4 to 0.5 volumes of the FLAG resin used (i.e., 8 to 10 μl per 150 mm plate). The eluted protein is pooled and analyzed by SDS polyacrylamide gel electrophoresis on a 6% gel and Coomassie staining along with bovine serum albumin mass standard. Typical yield is 5 to 10 μg ACF per 150 mm plate of Sf9 cells (Fig. 1). The concentration of ACF can be conveniently expressed in units (1 unit of ACF equals 22 femtomoles of purified recombinant protein). The protein is stored in small aliquots at $-80°$. Five or more cycles of freezing and thawing should be avoided.

Drosophila *Topoisomerase I Catalytic Domain (ND423)*

An N-terminally truncated form[19] of *Drosophila* topoisomerase I (termed ND423) with a C-terminal 6His tag is expressed in the pET bacterial expression system (Novagen). The protein is purified by nickel affinity chromatography. The topoisomerase activity is necessary only for chromatin assembly on relaxed circular DNA or for chromatin analyses based on DNA supercoiling (see below). The bacterially expressed active topoisomerase I fragment is an economical alternative to commercially available enzymes.

The plasmid (pET-NDH6) is transformed in BL21(DE3) cells. Bacterial culture (LB containing 50 μg/ml kanamycin) is grown to A_{600} ~0.5 and induced with 0.42 mM IPTG. The cells are harvested after induction for 5 h at 30° and sonicated in a buffer containing 50 mM sodium phosphate (pH 7.0), 500 mM NaCl, 15% (v/v) glycerol, 15 mM imidazole, 0.1% (v/v) NP-40, 0.2 mM PMSF, 0.5 mM benzamidine and 10 mM sodium metabisulfite in a total volume that equals 1/50 to 1/20 of the original culture volume. The lysate is cleared by centrifugation (10 min at 15,000 g) and mixed with Ni-NTA resin slurry (4 ml per 1 liter original culture volume).

After incubation on a rocking platform or a spinning wheel for 3 h, the resin is loaded on a disposable 10-ml polypropylene column (BioRad, Cat. 731–1550) and washed 3 times with 10 column volumes of sonication buffer. The protein is eluted with 3 column volumes of elution buffer (sonication buffer plus 485 mM imidazole). The eluate is dialyzed for 2 h against 2 liters of dialysis buffer (10 mM potassium HEPES [pH 7.6], 0.1 mM EDTA, 10% glycerol, 50 mM NaCl, 0.01% NP-40, 0.5 mM DTT, 0.2 mM PMSF, 0.5 mM benzamidine, 5 mM sodium metabisulfite) and for 2 h against 1 liter storage buffer (similar to the dialysis buffer but containing 50% glycerol, 10 mM β-mercaptoethanol instead of DTT, and 1 μg/ml leupeptin instead of sodium metabisulfite). Precipitate is removed by centrifugation (5 min at 15,000 g), and soluble protein is quantitated as described above and stored at −80°. The typical yield is 3 to 4 mg ND423 protein per 1 liter cell culture (in 2.0 to 2.5 ml total volume). Note that this protein is extremely active and should be used as a 100-fold dilution in the storage buffer containing 0.2 mg/ml recombinant insulin. A 1 μl sample of the working 1:100 dilution of ND423 is sufficient to completely relax 10 μg of a 3.2-kbp plasmid after 10 min incubation at 30° in a 100 μl reaction in 1× topoisomerase buffer (50 mM Tris-HCl, pH 7.5, 10 mM MgCl$_2$, 0.1 mM EDTA, 50 μg/ml bovine serum albumin, 0.5 mM DTT; this buffer is

[19] W. L. Shaiu and T. S. Hsieh, *Mol. Cell Biol.* **18**, 4358 (1998).

typically prepared as a $10\times$ stock). The working dilution should be stored without freezing at $-20°$.

Native or Recombinant Core and Linker Histones

The recombinant assembly system is compatible with core histones purified from various species including *Drosophila*, human, and yeast (unpublished results). A number of protocols are available to purify native core histones from these sources.[2,20,21] Alternatively, *Drosophila* or *Xenopus* recombinant core histones can be expressed and purified from bacteria.[14,22] The protocols in this chapter are based on the use of native *Drosophila* core histones (Fig. 1). However, recombinant histones appear to function in a comparable fashion without significant changes to the protocol. Additionally, linker histones can be incorporated in the assembly system if desired. A method for purification of the native *Drosophila* or human linker histone H1 is published elsewhere.[23]

Chromatin Assembly with Plasmid DNA

The reaction conditions below have been optimized for the assembly of nucleosome arrays on circular plasmid DNA of 3 to 5 kbp. However, this protocol will work for shorter or longer DNA templates, circular or linear. For instance, it has been tested for chromatin assembly on a 23 kbp plasmid, a 1 kbp restriction fragment, and linear wild-type λ DNA (48 kbp). The reaction can be broken down into 3 steps: (1) pre-binding of core histones with the chaperone, NAP-1; (2) relaxation of the plasmid DNA (optional); and (3) ATP-dependent nucleosome assembly.

Reagents and Equipment

The following reagents are used: double CsCl-purified plasmid DNA, 0.3 to 2.0 mg/ml in TE; purified native or recombinant core histones, 0.1 to 2.0 mg/ml; recombinant NAP-1, 0.5 to 4.0 mg/ml; recombinant *Drosophila* ACF, 0.003 to 0.3 mg/ml (0.5 to 50 units/μl); recombinant topoisomerase I or its catalytic domain, ND423, 1:100 working dilution (see previous section); ACF dilution buffer (identical to ACF elution buffer, with or without the FLAG peptide); 0.5 M phosphocreatine (Sigma, Cat.

[20] P. E. Prevelige and G. D. Fasman, *Biochemistry* **26,** 2944 (1987).
[21] W. F. Brandt, K. Patterson, and C. von Holt, *Eur. J. Biochem.* **110,** 67 (1980).
[22] K. Luger, T. J. Rechsteiner, A. J. Flaus, M. M. Waye, and T. J. Richmond, *J. Mol. Biol.* **272,** 301 (1997).
[23] G. E. Croston, L. M. Lira, and J. T. Kadonaga, *Prot. Exp. Purif.* **2,** 162 (1991).

P-7936) in 20 m*M* HEPES, pH 7.6; creatine phosphokinase (Sigma, Cat. C-3755); solution of 5% polyvinyl alcohol (PvOH) and 5% polyethylene glycol (PEG) 8,000 (Sigma, Cat. P-8136 and P-2139) in HEG (described below).

Preincubation of NAP-1 and Core Histones

In a siliconized 1.5 ml plastic microcentrifuge tube, the following components are combined: 1.4 μg bovine serum albumin (Pierce, Cat. 23209), 1.76 μg NAP-1 (5:1 mass ratio to core histones), 0.353 μg core histones, 11.7 μl 300 m*M* KCl and 14 μl PvOH+PEG solution. The volume is made up to 56.6 μl with HEG buffer (25 m*M* potassium HEPES [pH 7.6], 0.1 m*M* EDTA, 10% glycerol). When recombinant core histones are used, the HEG buffer is replaced with HEG$^+$ buffer, which additionally contains 0.01% (v/v) NP-40. The mixture is vortexed gently and incubated on ice for 20 to 60 min. This step prevents subsequent non-specific aggregation of positively charged histones with ACF and/or DNA. When several assembly reactions are performed simultaneously, it is advisable to prepare the master (NH) mix of NAP-1 and core histones to avoid errors in pipetting small volumes of the core histone solution.

ATP Regeneration System and Relaxed DNA

During the incubation of core histones with NAP-1, the master ATP and Mg^{2+} (AM) mix is prepared by combining 3 μl 0.5 M ATP with 30 μl 0.5 M phosphocreatine, 16.5 μl double-distilled water and 25 μl 100 m*M* MgCl$_2$ at room temperature. A 0.5 μl sample of creatine phosphokinase solution is added immediately before use. (Creatine phosphokinase solution is prepared at 5 mg/ml in 10 m*M* potassium phosphate [pH 7.0], 50 m*M* NaCl and 50% [v/v] glycerol and stored at $-80°$ in 2 μl aliquots.) The volume of the AM mix can be scaled up if more than seven standard assembly reactions are performed simultaneously. The AM mix contains ATP and ATP regeneration system. The regeneration system is optional if chromatin assembly is performed for less than 2 h with less than 5 units of ACF.

Relaxed DNA template is prepared by combining 3.53 μg plasmid DNA with 2 μl topoisomerase (ND423 working dilution) in 1 × topoisome-topoisomerase buffer (see above) in a total volume of 20 μl. The DNA is relaxed for 10 min at 30° and kept at room temperature until ready to be used. The relaxation of the DNA is an optional step. The chromatin can be efficiently assembled on linear or supercoiled DNA templates. However, in the absence of topoisomerase I, the requirements for DNA quality are significantly more stringent (for instance, supercoiled DNA has to be purified by two rounds of CsCl banding). With the relaxation step,

DNA prepared by many standard methods (e.g., Qiagen plasmid prep) is suitable.

During the assembly of chromatin with a covalently closed circular DNA, positive superhelical tension is generated as a consequence of the wrapping of DNA around the histone octamers. To relieve this torsional stress, topoisomerase is included in the assembly reactions. A higher concentration of topoisomerase is needed to relax chromatin relative to DNA. Hence, a relatively high concentration of topoisomerase is included in the DNA relaxation reaction.

ATP-dependent Nucleosome Assembly and Spacing

ACF is diluted to a final concentration of 0.5 to 10 units/μl in dilution buffer and is kept on ice. The 56.6 μl NH mix is equilibrated to room temperature in a water bath. The following components are added in the indicated order (the reaction tube is gently vortexed *immediately* after the addition of every component): 1 μl ACF, 10.5 μl AM master mix, and 2 μl relaxed DNA (0.353 μg). The final concentrations of ATP, MgCl$_2$, and KCl in the reaction are 3 mM, 5 mM, and 50 mM, respectively. The assembly is allowed to proceed for up to 2.5 h at 27°C or at room temperature, and assembled chromatin is analyzed as outlined in the next section.

Methods for the Analysis of Chromatin Assembly

The assembled chromatin can be used directly in downstream assay reactions.[11,14–17] Certain applications (e.g., electron microscopy studies) may require further purification of the chromatin. The nucleosome arrays can be purified from chromatin assembly factors by sucrose gradient sedimentation or gel filtration.[8] One of the protocols for chromatin purification involves stripping non-histone proteins with 0.05% Sarkosyl and a subsequent gel filtration step.[24] The quality of assembled chromatin can be evaluated by a number of methods that are detailed below.

Reagents and Equipment

Micrococcal nuclease (Sigma, Cat. N-5386), 200 units/ml in 5 mM sodium phosphate (pH 7.0), 2.5 μM CaCl$_2$; RNase A (Sigma, Cat. R-5503), 10 mg/ml in water, boiled; proteinase K (Sigma, Cat. P-6556), 2.5 mg/ml in water; 123 bp and 1 kbp DNA ladder (Life Technologies).

---

[24] T. Tsukiyama and C. Wu, *Cell* **83**, 1011 (1995).

Micrococcal Nuclease Digestion Assay

Micrococcal nuclease preferentially digests DNA in internucleosomal (linker) regions of the chromatin. Thus, partial digestion with the nuclease should reveal periodic spacing of assembled nucleosomes. After chromatin is partially digested, it is deproteinized. The resulting DNA fragments are resolved by agarose gel electrophoresis and visualized by ethidium bromide staining. This technique should reveal the quality of assembled chromatin, as judged by the number of polynucleosome particles that can be recognized in the micrococcal ladder. In addition, it can be used to determine the nucleosome repeat length of reconstituted chromatin.

Two micrococcal nuclease dilutions, 1:500 and 1:1,500, are prepared in buffer R (10 mM potassium HEPES [pH 7.6], 10 mM KCl, 1.5 mM MgCl$_2$, 0.5 mM EGTA and 10% [v/v] glycerol). The standard 70-μl chromatin assembly reaction is divided into two equal aliquots. A 10 mM CaCl$_2$ stock solution is added to each aliquot to the final concentration of 2 mM. Each half-reaction aliquot is digested with 5 μl of each of the micrococcal nuclease dilutions for exactly 10 min at room temperature. The digestion is stopped by adding EDTA to 50 mM. (At this point, the samples can be optionally treated with 10 μg RNase A for 5 min at room temperature to digest contaminating RNA.) Each sample is mixed with 100 μl glycogen stop buffer (20 mM EDTA [pH 8.0], 200 mM NaCl, 1% [w/v] SDS and 0.25 mg/ml glycogen) and 5 μl proteinase K stock solution (2.5 mg/ml). Chromatin assembly factors and histones are digested for 30 min at 37°. The DNA is extracted once with equal volume of phenol-chloroform and then precipitated with ethanol. The precipitated DNA is resuspended in 5 μl loading buffer and run on a 1.2% agarose gel in 1 × TBE until the bromphenol blue dye reaches the bottom quarter of the gel. A 123 bp DNA ladder (0.5 μg per lane) is used as the size marker. The gel is stained for 15 min with 2 gel volumes of 0.75 μg/ml ethidium bromide in water and destained for 1 hour in several changes of water.

The result of a typical chromatin assembly experiment in conjunction with the micrococcal nuclease assay is shown in Fig. 2A. Please note the appearance of at least 8 distinct sharp nucleosome bands from the bottom of the gel. The nucleosome repeat length is approximately 165 bp. Additionally, this experiment demonstrates that ACF, NAP-1, and ATP are all required for the assembly.

Variations of the micrococcal nuclease assay can be used to analyze interactions of DNA-binding transcription factors with the DNA during or after chromatin assembly. The nucleosomal array disruption assay[25] is

---

[25] M. J. Pazin, R. T. Kamakaka, and J. T. Kadonaga, *Science* **266,** 2007 (1994).

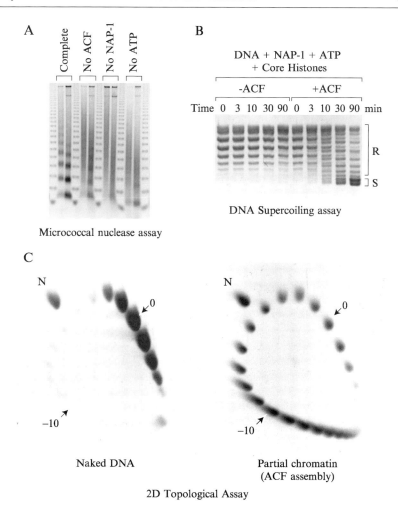

FIG. 2. Analysis of chromatin assembled *in vitro*. A. Partial micrococcal nuclease digests of chromatin assembly reaction products. Complete reaction contained core histones, NAP-1, ACF, DNA, and ATP. The other reactions were lacking individual components as indicated. (Adapted with permission from M. E. Levenstein and J. T. Kadonaga, *J. Biol. Chem.* **277,** 8749 (2002).) B. DNA supercoiling analysis of chromatin assembly reaction products. Standard chromatin assembly reactions were carried out in the presence or in the absence of ACF. The assembly reactions were terminated at the indicated time points. The samples were deproteinized and subjected to agarose gel electrophoresis. The DNA was detected by ethidium bromide staining. The positions of relaxed and highly negatively supercoiled DNA species are indicated. (Reprinted with permission from T. Ito, M. E. Levenstein, D. V. Fyodorov, A. L. Kutach, R. Kobayashi, and J. T. Kadonaga, *Genes Dev.* **13,** 1529 (1999).) C. 2D topological analysis of chromatin assembly reaction products. Completely relaxed plasmid DNA (prior to chromatin assembly) and partially-assembled chromatin (terminated

based on probing Southern transfers of micrococcal nuclease ladders. DNA-binding transcription factors can disrupt the regularity of nucleosome arrays in the vicinity of their binding sites either during chromatin assembly with ACF or after assembly.[11,14,25] Certain transcription factors can pose a physical obstacle for nucleosome deposition or sliding by ACF or other chromatin remodeling factors.[26,27] In this manner, they can help establish translational positioning of nucleosome arrays on non-repetitive DNA.[14,26] The indirect end-labeling assay[25,26] of assembled chromatin is designed to determine whether nucleosomes are positioned relative to a point in the DNA sequence, such as a restriction enzyme site.

Optimization of the Assembly Reaction

The assembly reaction almost invariably requires optimization steps, which can be performed with the help of the micrococcal nuclease assay. Although the standard 70-μl reaction is designed for the assembly of 0.353 μg DNA, it can be performed with up to 10-fold higher DNA concentration, as long as the ratio of protein factors is kept the same. The single most important parameter that has to be optimized is the ratio of core histones to DNA. The reaction is very sensitive to excessive amounts of core histones. Exceeding the optimal concentration by as much as 10% will completely inhibit assembly. On the other hand, low concentration of histones will result in under-assembled chromatin with stretches of naked DNA. The optimal mass ratio of core histones and DNA is usually close to 1. However, standard quantitation methods of histones (Bradford assay or quantitative SDS polyacrylamide gel electrophoresis) and DNA (UV absorption- or fluorescence-based methods) are rarely accurate enough to ensure efficient assembly. Therefore a series of reactions with variable DNA concentrations should be performed for every histone and DNA preparation to establish optimal conditions. The assembled chromatin is then examined by the micrococcal nuclease assay. Typical results of this titration experiment can be found elsewhere.[2]

The assembly reaction is less sensitive to concentrations of NAP-1 or ACF. In a standard reaction that contains 0.353 μg each DNA and

[26] M. J. Pazin, P. Bhargava, E. P. Geiduschek, and J. T. Kadonaga, *Science* **276**, 809 (1997).
[27] S. Lomvardas and D. Thanos, *Cell* **106**, 685 (2001).

after 20 min reaction time) were deproteinized and subjected to the 2D agarose gel electrophoresis. The DNA was detected by ethidium bromide staining. 0 and −10 indicate positions of topoisomers with 0 and 10 negative supercoils, respectively; N specifies the position of nicked DNA.

histones, between 1.42 and 3.53 μg NAP-1 will produce acceptable results (4:1 to 10:1 relative to core histones, w/w). NAP-1 itself can rapidly deposit histone octamers onto supercoiled DNA, but this reconstitution does not result in the assembly of canonical nucleosomes, as evidenced by atomic force microscopy.[28] When relaxed circular DNA is used as a template, purified recombinant NAP-1 possesses an extremely low intrinsic chromatin assembly activity. This activity is ATP-dependent and most likely stems from copurification of ACF or an ACF-like factor. Thus, keeping the concentration of NAP-1 in the reaction low is advisable when it is important to discriminate between ACF-dependent and ACF-independent assembly activities.

It is recommended to keep ACF between 0.05 and 2 ACF protomers per 1 kbp of DNA. ACF appears to assemble chromatin in a processive fashion.[29] Upon initiation of the assembly, it commits to the DNA template and remains associated until it assembles an array of several adjacent nucleosomes. Therefore, committing too many ACF protomers to a single template will result in the assembly of multiple "out-of-phase" arrays. When analyzed by partial micrococcal nuclease digestion, these overlapping arrays will manifest as smeared ladders. On the other hand, lower ACF amounts (much less than one ACF protomer per DNA template) may be insufficient to assemble chromatin completely.

DNA Supercoiling Assay

Following chromatin assembly on circular DNA, histones can be removed by digestion with a protease. After the removal of histones, DNA retains the negative supercoils that were introduced by nucleosome assembly. For every nucleosome that is reconstituted on relaxed circular DNA, the linking number of the template is changed by about -1. While the micrococcal analysis demonstrates periodic spacing of reconstituted nucleosomes, the supercoiling state of the template can be used as a measure of the efficiency of nucleosome deposition (i.e., the average number of reconstituted nucleosomes per template). In this assay, chromatin assembled on a relaxed plasmid in the presence of excess topoisomerase I is deproteinized, and DNA topoisomers are resolved by agarose gel electrophoresis. Completely relaxed DNA and DNA with varying low number of negative supercoils are resolved as discrete bands at the top of the gel. Highly negatively supercoiled DNA (60% to 100% of the highest number of negative supercoils that can be accommodated by a plasmid; 12 to 20 for a 3.2-kbp

[28] T. Nakagawa, M. Bulger, M. Muramatsu, and T. Ito, *J. Biol. Chem.* **276,** 27384 (2001).
[29] D. V. Fyodorov and J. T. Kadonaga, *Nature* **418,** 897 (2002).

plasmid) migrates as a single band of "supercoiled" DNA at the bottom of the ladder. The DNA intercalating agent chloroquine is sometimes added to the electrophoresis gel and buffer to resolve highly negatively super-coiled DNA species. As a variation, DNA supercoiling can be analyzed on a two-dimensional (2D) gel, in which DNA samples are first run in the absence of chloroquine and then in a perpendicular direction after equilibration with chloroquine.

The standard 70-μl chromatin assembly reaction is stopped by adding EDTA to 50 mM, and chromatin is treated with RNase A and prote-inase K, as described above. Precipitated DNA is resuspended in 20 μl loading buffer, and 5 μl (approximately 90 ng DNA) is run on a 0.8% agar-ose gel in 1 $\times$ TBE until the xylene cyanol dye reaches the bottom third of the gel. The gel is stained with ethidium bromide and destained as described above. Fig. 2B presents a DNA supercoiling analysis of the time course of chromatin assembly with and without ACF. Chromatin was assembled with the 4:1 mass ratio of NAP-1, relative to core histones. Note that NAP-1 alone has insignificant histone deposition activity in these conditions.

For a 2D topological analysis, typically only one sample is loaded per gel (up to 0.5 μg DNA). After the electrophoresis in the first direction is complete, the gel is equilibrated to 1$\times$ TBE buffer containing 1 μM chloroquine for 2 to 2.5 hours. The gel is rotated 90 degrees, run in the second direction in TBE buffer containing chloroquine, stained, and de-stained. In Fig. 2C, relaxed plasmid DNA (prior to assembly) and partially assembled chromatin are analyzed by the 2D topological assay. The pos-ition of "0" negative supercoils is defined as the peak of DNA topoisomers in the relaxed DNA sample. Note that DNA in the partially assembled chromatin sample contains two distinct populations of topoisomers: one that corresponds to relaxed (unassembled) template and the other that corresponds to moderately assembled template. This effect is due to the processive character of chromatin assembly by ACF.[29]

Assembly of Chromatin with the Linker Histone H1

An important property of the recombinant chromatin assembly system is its ability to incorporate chromatin-specific proteins other than core his-tones into the assembled arrays. When studied in vitro, chromatin is fre-quently reduced to mononucleosomes or arrays that contain only core histone octamers. However, bulk native chromatin is more complex and in-cludes numerous other structural proteins, most prominently linker his-tones. We have utilized the recombinant system to assemble chromatin that includes *Drosophila* linker histone H1. H1 is efficiently incorporated into nucleosome arrays, and these arrays possess structural and functional features of H1-containing native chromatin.

Assembly of H1-Containing Chromatin

In metazoans, the abundance of H1 in the chromatin is close to 1 molecule H1 per nucleosome particle.[30] Therefore, reconstituted chromatin with 1 molecule of H1 per nucleosome would be an excellent model for native metazoan chromatin. The nucleosomal repeat length of reconstituted chromatin is increased by about 25% on addition of corresponding amounts of H1.[31] Finally, incorporation of H1 inhibits downstream enzymatic reactions that use reconstituted chromatin as a substrate. These considerations dictate several modifications of the standard protocol of chromatin assembly in the presence of H1 and its analysis.

The mass ratio of core histones to DNA in the assembly is reduced by 20% (i.e., to 0.8 from 1.0) to accommodate the longer spacing of the nucleosomes. To facilitate detection of longer spacing, the ionic strength of the reaction buffer is increased (80 to 100 mM from 50 mM monovalent cations). The standard reaction is initiated by pre-binding of NAP-1 with the histones: 1.4 μg bovine serum albumin, 2.83 μg NAP-1 (8:1 mass ratio to core histones), 0.353 μg core histones, 0.088 μg H1 (1:1 molar ratio to core histone octamers), 18.7 μl 300 mM KCl (which will result in the final concentration of 80 mM), and 14 μl PvOH/PEG solution, 5% each, and HEG to 56.1 μl are mixed and incubated on ice. The amount of the DNA template is increased to 0.441 μg (2.5 μl of the standard DNA relaxation reaction). The minimal recommended amount of ACF is 10 units per reaction (i.e., 1 ACF protomer for every 3.2-kbp plasmid molecule). The other parameters of the assembly reaction remain unchanged.

Micrococcal Nuclease Digestion Analysis of H1-containing Chromatin

Like many other enzymes, micrococcal nuclease activity is inhibited by H1. Therefore H1-containing chromatin should be digested with higher concentrations of the nuclease than H1-deficient chromatin to obtain comparable degrees of digestion. It has been empirically determined that chromatin assembled in the presence of one or more H1 molecules per core histone octamer requires 3-fold higher than normal concentration of the micrococcal nuclease. The standard digestion reaction of H1-containing chromatin is performed with 7.5 μl of 1:170 and 1:500 dilutions of the enzyme for 10 min at room temperature, while the H1-free chromatin is

[30] D. L. Bates and J. O. Thomas, *Nucl. Acids Res.* **9,** 5883 (1981).
[31] A. Rodríguez-Campos, A. Shimamura, and A. Worcel, *J. Mol. Biol.* **209,** 135 (1989).

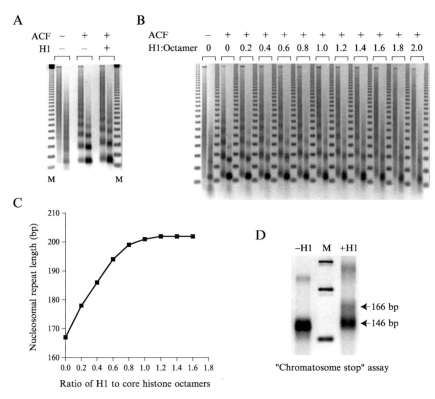

Fig. 3. Assembly of chromatin containing histone H1. A. Micrococcal nuclease analysis of chromatin assembled with and without histone H1. Standard chromatin assembly reactions were carried out in the presence or in the absence of approximately 1 molecule of H1 for every core histone octamer. The reaction products were digested with micrococcal nuclease and analyzed as described. B. Titration of the amount of linker histone H1 in the chromatin assembly reaction. Chromatin assembly was performed in the presence of indicated amounts of histone H1 and analyzed by micrococcal nuclease assay. C. Graphical analysis of nucleosomal repeat length of the chromatin assembled with various concentrations of histone H1. The nucleosomal repeat length was calculated by dividing by 5 the lengths of DNA fragments corresponding to pentanucleosome partial digest products (panel B). The calculated repeat length was plotted against the molecular ratio of histone H1 and core histone octamers in the reactions. D. "Chromatosome stop" assay of chromatin assembled with histone H1. Products of chromatin assembly reactions in the absence or in the presence of histone H1 (1:1, relative to core histone octamers) were digested extensively with micrococcal nuclease (2.5 and 7.5 μl, respectively, of 1:60 dilution of the enzyme stock). After deproteinization, DNA fragments were analyzed by electrophoresis on a 3% agarose gel. The positions of DNA fragments that correspond to the core nucleosome particle (146 bp) and chromatosome particle (166 bp) are indicated.

digested with 2.5 μl of the same dilutions. The rest of the digestion protocol remains unchanged (see above). To digest chromatin assembled in the presence of between 0 and 1 H1 molecules per core histone octamer, intermediate amounts of the micrococcal nuclease can be used.

The results of a typical assembly reaction analyzed by micrococcal nuclease digestion are shown in Fig. 3A. Note the increase in the average nucleosome repeat length from approximately 165 to 200 bp with addition of H1. Fig. 3B shows micrococcal nuclease digestion of chromatin assembled with 0 to 2 H1 molecules per core histone octamer. Fig. 3C presents the graphical analysis of the nucleosome repeat length of reconstituted chromatin shown in Fig. 3B. The repeat length is gradually increasing from 165 to 200 bp with addition of increasing amounts of histone H1. The assembly reaction is inhibited by more than 1.6 H1 per nucleosome.

Extensive digestion of H1-free chromatin with micrococcal nuclease generates DNA fragments of approximately 146 bp, which correspond to the size of a nucleosome core particle. However, the presence of H1 in assembled chromatin results in protection of additional DNA fragments of approximately 166 bp, which correspond to the size of H1-containing chromatosome particle (Fig. 3D). A similar digestion pattern is observed for bulk native chromatin.[32] Finally, incorporated H1 strongly inhibits activated and basal transcription with *Drosophila* nuclear extracts of the reconstituted chromatin templates (data not shown). Thus the addition of histone H1 to the recombinant assembly system recreates many structural and functional properties of native metazoan chromatin.

Acknowledgments

The study of chromatin assembly in the lab of James T. Kadonaga is supported by NIH grant GM58272. We thank Mike Bulger, Takashi Ito, Mark Levenstein, and Wen Jiang for their contributions to the development of the chromatin assembly system. We thank Mark Levenstein, Karl Haushalter, Vassili Alexiadis, and Buyung Santoso for critical reading of the manuscript.

[32] R. T. Simpson, *Biochemistry* **17,** 5524 (1978).

[38] Tying C' Ends of H2A and H2B Using a Molecular Glue, Tissue-Type Transglutaminase

By Hyon E. Choy, Jae-Hong Kim, and Sang Chul Park

Core histones undergo several post-translational modifications such as acetylation, phosphorylation, methylation, ADP ribosylation, ubiquitination, etc.[1] These modifications alter the nucleosomal infrastructure to influence both specific gene activity and the global control of chromosomal activity. These regulatory functions are largely determined by the packaging of the DNA within chromatin. Recently, an additional form of core histone modification has been reported: a covalent cross-linking of histone subunits by tissue-type transglutaminase (tTGase, EC 2.3.2.13).[2–4] The tTGase irreversibly catalyzes covalent cross-linking of proteins by forming isopeptide bonds between peptide-bound glutamine and lysine residues.[5] Although its role has not been fully elucidated, cross-linked H3 and H4 have been identified in starfish sperm.[3,4] In addition, a covalently cross-linked H2A and H2B dimer has been observed in the nuclei prepared from chicken erythrocytes.[2] Subsequently, it was found that C'-terminal tails of H2A and H2B prepared from chicken erythrocytes could be covalently cross-linked by tTGase *in vitro*. Here, we describe the procedure for tTGase purification and the *in vitro* cross-linking reaction by the tTGase using chromatin/nucleosome as a substrate, as well as discuss possible physiological implications of the histone cross-linking.

Procedure

Preparation of tTGase from human erythrocytes

The procedure is essentially described in Signorini *et al.* (1988).[6] Briefly, erythrocytes prepared from 1 pint of blood washed with an equal volume of

[1] A. P. Wolffe, *Academic Press, San Diego*, 2nd ed., pp. 72 (1995).
[2] J. H. Kim, K. H. Nam, O. S. Kwon, I. G. Kim, M. Bustin, H. E. Choy, and S. C. Park, *Biochem. Biophys. Res. Commun.* **293**(5), 1453 (2002).
[3] T. Shimizu, K. Hozumi, S. Horiike, K. Numomura, and S. Ikegami, *Nature* **380**, 32 (1996).
[4] T. Shimizu, T. Takao, K. Hozumi, K. Nunomura, S. Ohta, Y. Shimonishi, and S. Ikegami. *Biochemistry* **36**, 12071 (1997).
[5] J. E. Folk, *Annu. Rev. Biochem.* **49**, 517 (1980).
[6] M. Signorini, F. Bortolotti, L. Polotronieri, and C. M. Bergamini, *Biom. Chem. Hoppe-Seyler.* **369**, 275 (1988).

saline and mixed with equal volume of water (450 ml) were stored at $-20\,^{\circ}$C overnight. The lysate, prepared by thawing the cells at room temperature after addition of an equal volume of cold water, was centrifuged at 20,000 $\times$ g for 20 min. The supernatant was saved for purification of the enzyme.

In the first step of purification, the supernatant was adjusted to pH 7.5 by addition of 1.0 M NaOH. The supernatant was analyzed for protein concentration and mixed with DEAE cellulose ($\sim$1 g of swollen resin per 100 mg of protein) pre-equilibrated in buffer I consisting of 50 mM Tris buffer pH 7.5, 1 mM EDTA, and 5 mM 2-mercaptoethanol. After stirring for 20 min at 4 $^{\circ}$C, the resin was washed with buffer I on a buchner funnel, packed into a wide glass column (less than the inner diameter of 4 cm), and rinsed with buffer I containing 0.15 M NaCl ($\sim$11) until a protein-free elute was obtained. The enzyme was eluted with buffer I containing 0.45 M NaCl.

The fractions showing tTGase activity (see later) were pooled, dialyzed against buffer I containing 0.15 M NaCl overnight, and loaded onto a small heparin-sepharose column, which was washed with buffer I containing 0.15 M NaCl. Proteins were eluted with buffer I containing 1M NaCl. At this stage, we obtained fractions with virtually a single protein ($\sim$75 kD) with tTGase activity as analyzed by SDS-PAGE (Fig. 1A and B).

TTGase Activity Assay

The tTGase activity was determined by measurement of the incorporation of radiolabeled [1,4-^{14}C] putrescine (NEN-Du Pont) into dimethyl casein.[7] The incubation mixture contained 12 mg/ml dimethyl casein, 0.8 mM putrescine (1500 cpm/nmol), 5 mM CaCl$_2$, and 100 mM beta-mercaptoethanol in 0.1 M Tris buffer pH 7.5, in a total volume of 110 μl.[8] The reaction was started by addition of tTGase, appropriately diluted to maintain a linear rate of incorporation of putrescine into dimethyl casein for 60 min at 30°C. Blanks were run in presence of bovine serum albumin instead of the enzyme.

The radioactive dimethyl casein was freed of unbound putrescine by a filter paper procedure[9] and quantified by liquid scintillation counting. One unit of enzyme was defined as the amount that catalyzes formation of 1 μmol of the putrescine per minute. Specific activity is given as the units per mg of protein.

[7] L. Lorand, L. K. Campbell-Wilkes, and L. Cooperstein, *Anal Biochem.* **50**(2), 623 (1972).
[8] Y. Lin, G. E. Means, and R. E. Feeney, *J. Biol. Chem.* **244,** 798 (1969).
[9] J. D. Corbin and E. M. Reimann, *Methods Enzymol.* **38,** 287 (1974).

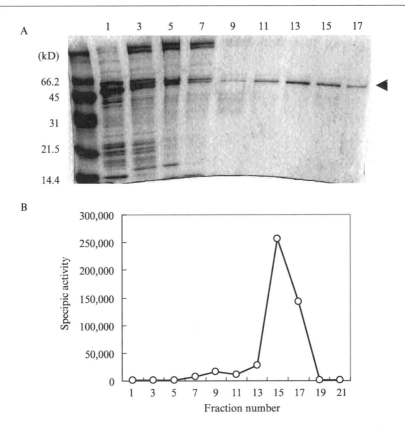

Fig. 1. Protein profile of the fractions eluted from heparin-sepharose column with NaCl gradient. Eluted total proteins in each fraction were analyzed on 8% SDS-PAGE (A) and tTGase activity determined by incorporation of [^{14}C] putrescine into dimethyl casein (B). Numbers represent fraction number of the elutes.

Preparation of nucleosome

Nucleosomes and histone subunits from chicken erythrocyte nuclei were obtained as described by Weintraub et al.[10]

In vitro cross-linking reaction

The standard incubation mixture contained substrate (~5 μg), human tTGase (1 × 10^3 units), 40 mM Tris-Cl (pH 7.5), 4.5 mM CaCl$_2$, 2 mM dithiothreitol, and 0.6 M NaCl in 20 μl volume. The incubation was carried out at 37 °C for 2 h for the result shown in Fig. 2.

[10] H. Weintraboub, K. Palter, and F. Van Lente, Cell 6, 83 (1975).

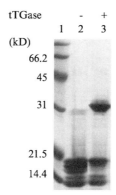

Fig. 2. A specific cross-linking of histone subunits by tTGase. Cross-linking reaction by tTGase was carried out using chromatin as a substrate in the presence of 0.6 M NaCl, and the results were displayed on 15% SDS-PAGE gel. The first lane is protein markers, 2nd lane reaction without tTGase, and 3rd lane with tTGase. Molecular weight of the protein markers are 14.4, 21.6, 31, 45, and 66.2 kd from the bottom.

Purification of tTGase

tTGase was most easily prepared from human erythrocyte following the procedure described previously. The fractions eluted from heparin-sepharose column with NaCl gradient were analyzed for protein profiles on 8% denaturing SDS-PAGE and tTGase activity (Fig. 1A and B). Fractions between 11–17, eluted with ~0.5 M NaCl, showed a single peptide band at about 75 kD (*arrow* in Fig. 1A) with the highest tTGase activity at fraction 15. Fractions between 13–17 were pooled, dialyzed against a large volume of 20 m*M* Tris buffer (pH 7.5) containing 1 m*M* EDTA, and used successfully in the subsequent *in vitro* cross-linking reaction. We estimated that the pooled fractions contain tTGase with more than 95% purity. The dialyzed pooled fractions were stored in the presence of 50% glycerol at −20 °C without significant loss of activity over 6 months.

Cross-Linked Histone Subunits

An *in vitro* experiment was carried out to cross-link histone subunits by the purified tTGase. The chromatin prepared from chicken erythrocyte was reacted with the tTGase in the presence of 0.6 *M* NaCl for 2 hrs at 37 °C, and the reaction mixture was electrophoresed on 15% SDS-PAGE (Fig. 2). Cross-linking of a histone subunit was detected at the position roughly where a dimeric histone subunit should move to, around 25 kD. Formation of the cross-linked products was accompanied by a concomitant decrease

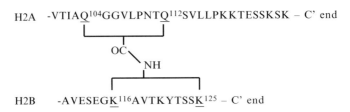

FIG. 3. Schematic diagram of amino residues in H2A and H2B involved in cross-linking by tTGase.

of substrate histone subunits, notably H2A and H2B, two middle bands. Thus, it shows clearly that histones could be crosslinked by tTGase *in vitro* through an acyl transfer reaction between the histone subunits. It was also observed that tTGase-mediated crosslinking of the histone subunits, H2A and H2B, required >0.6 M NaCl when using chromatin as a substrate of a reaction.[2] Subsequent experiments revealed that it was not because the tTGase activity was salt-dependent, but because dissociated H2A and H2B from chromain was the true substrate for tTGase: H2A and H2B dissociate as a heterodimer in the presence of high NaCl at a neutral pH condition.[2,11] Consistently, we observed that the cross-linking reaction using fractionated H2A–H2B by hydroxyapatite column chromatography as substrate was rapid and salt-independent (data not shown).[2]

Further analysis of the cross-linked product by matrix-assisted laser desorption/ionization time-of-flight mass spectrometry (MALDI-TOF-MS) via mass finger printing of in-gel digested sample revealed that the cross-linking was between either one or both glutamines at C-terminal end of H2A ($-$VTIAQ^{104}GGVLPNTQ^{112}SVLLPKKTESSKSK – C'end) and the first and/or third lysine from C-terminal end of H2B ($-$AVESEGK^{116} AVTKYTSSK^{125} – C' end) (Fig. 3).

Perpectives

tTGase is most ubiquitously found in every type of cell and tissue in animals, but its natural substrate has yet to be identified. Several cytosolic proteins have been implicated in the tTGase-mediated cross-linking. Recently, accumulating evidence suggests that tTGase could function in nuclei.[4,5,12–14] Using human neuroblastoma cells, tTGase has been shown

[11] H. J. Greyling, S. Schwager, B. T. Sewell, and C. von Holt, *Eur. J. Biochem.* **137,** 221 (1983).
[12] L. Piredda, M. G. Farrace, M. Lo Bello, W. Malorni, G. Melino, R. Petruzzelli, and M. Piacentini, *FASEB J.* **13,** 355 (1999).
[13] J. Zhang, M. Lesort, R. P. Guttman, and G. V. Johnson, *J. Biol. Chem.* **273,** 2288 (1998).

to translocate from the cytosol into the nucleus in response to the elevation of intracellular calcium concentration.[13] The tTGase in the nucleus is active enough to mediate cross-linking of various nuclear proteins, although the physiological relevance is not known.[14] Subsequently, Piredda et al. reported that the histone H2B was one of several polypeptides immunoprecipitated together with tTGase.[12] Consistent with this finding, a covalently cross-linked form of core histones, most abundant in proteins associated with eukaryotic DNA, has been detected in starfish sperm, specifically between H2B and H4.[4,5] Authors suggested that most likely the transglutaminase was implicated in the cross-linking reaction and therefore, it consequently functions in chromatin compaction during spermatogenesis.

Using chromatin prepared from chicken erythrocyte, we observed that C-terminal end of two core histone subunits, H2A and H2B, could be subject to cross-linking by tTGase.[2] Both the amine donor lysine and acceptor glutamine residues were found to reside in the flexible region of these core histone proteins.[15]

[14] J. A. Han and S. C. Park, *Mol. Cells.* **10,** 612 (2000).
[15] K. Lugaer, A. W. Macler, R. K. Richmond, E. F. Sargent, and S. C. Richmond, *Nature* **389,** 251 (1997).

[39] Chromatin Decompaction Method by HMGN Proteins

By Katherine L. West, Yuri V. Postnikov, Yehudit Birger, and Michael Bustin

In eukaryotic cells, DNA is wrapped around histone octamers to form arrays of nucleosomes, which are then folded and compacted into higher-order chromatin structures. The extent of compaction of the chromatin fiber affects the access of regulatory factors to their targets and thus impacts processes that utilize DNA such as transcription and replication. Therefore, structural proteins that influence the extent of chromatin folding and modulate access to the DNA can be considered regulators of transcription and other DNA-based activities. The linker histone H1 increases chromatin compaction, resulting in reduced DNA accessibility and transcriptional repression. In contrast, HMGN proteins (formerly known as HMG-14/-17) decrease chromatin folding, leading to an increase in the rates of transcription and replication. Here, we describe experimental methods that have provided evidence that HMGN proteins interact with and unfold chromatin both *in vitro* and *in vivo*.

Interaction of HMGN Proteins with Nucleosomes *In Vitro*

The classical method for detecting interactions between HMGN proteins and nucleosomes is the gel mobility shift assay.[1–3] HMGN bind to nucleosomes through a 32 amino acid peptide known as the nucleosome binding domain.[4,5] Each core particle binds two HMGN molecules, and buffer conditions can be chosen to observe either cooperative or non-cooperative binding.[2,6,7]

HMGN purification

HMGN proteins can be purified from vertebrate cells,[8] or recombinant forms can be overexpressed in bacteria.[9] The canonical HMGN1, HMGN2, and highly related family members are acid soluble and can be purified away from most other cellular proteins using perchloric acid. They can then be purified to homogeneity by HPLC.[10] However, this approach is not applicable to HMGN-related proteins, such as Nsbp1, that contain additional large domains that render the protein insoluble in 5% perchloric acid. Recombinant forms of these proteins can be linked to a tag such as FLAG and purified using standard techniques.[11]

Preparation of Nucleosome Core Particles and Other Chromatin Substrates

The substrate for the gel mobility shift assay is usually the nucleosome core particle, which contains 147 bp of DNA. These are most easily purified

[1] J. K. Mardian, A. E. Paton, G. J. Bunick, and D. E. Olins, *Science* **209**, 1534–1536 (1980).
[2] G. Sandeen, W. I. Wood, and G. Felsenfeld, *Nucleic Acids Res.* **8**, 3757–3778 (1980).
[3] S. C. Albright, J. M. Wiseman, R. A. Lange, and W. T. Garrard, *J. Biol. Chem.* **255**, 3673–3684 (1980).
[4] M. P. Crippa, P. J. Alfonso, and M. Bustin, *J. Mol. Biol.* **228**, 442–449. (1992).
[5] L. Trieschmann, Y. Postnikov, A. Rickers, and M. Bustin, *Mol. Cell. Biol.* **15**, 6663–6669 (1995).
[6] H. Schroter and J. Bode, *Europ. J. Biochem.* **127**, 429–436 (1982).
[7] A. E. Paton, S. E. Wilkinson, and D. E. Olins, *J. Biol. Chem.* **258**, 13221–13229 (1983).
[8] E. W. Johns, The HMG Chromosomal Proteins. 1982, London: Academic Press.
[9] M. Bustin, P. S. Becerra, M. P. Crippa, D. A. Lehn, J. M. Pash, and J. Shiloach, *Nucleic Acids Res.* **19**, 3115–3121 (1991).
[10] Y. V. Postnikov, D. A. Lehn, R. C. Robinson, F. K. Friedman, J. Shiloach, and M. Bustin, *Nucleic Acid. Res.* **22**, 4520–4526 (1994).
[11] H. Shirakawa, D. Landsman, Y. V. Postnikov, and M. Bustin, *J. Biol. Chem.* **275**, 6368–74 (2000).

from chicken red blood cells,[12] as it is easy to obtain large numbers of cells. In addition, the extracts are low in nucleases and proteases. Core particles containing acetylated histones can be purified from butyrate-treated cultured cells such as HeLa,[13] which are more transcriptionally active than chicken red blood cells. Nucleosomes and dinucleosomes can also reconstitute onto DNA fragments of defined sequence by mixing either donor nucleosomes or purified histone octamers with the DNA at high salt and gradually reducing the salt concentration of the buffer by dialysis or dilution.[14–16]

Gel Shift Assay

To perform the gel mobility shift assay,[10] core particles are incubated with HMGN protein on ice for 15 minutes. Ficoll is then added, and the sample is loaded on a 4% native acrylamide gel alongside a sample of bromophenol blue dye as a marker. Electrophoresis is carried out at $4\,^\circ$C. Next, the gel is stained with ethidium bromide and the DNA is visualized under UV light. The same buffer is used for the binding reaction and as the electrophoresis buffer. Cooperative binding is observed using a moderate ionic strength buffer, such as 180 mM Tris-borate, while non-cooperative binding is observed at low ionic strength (9 mM Tris-borate). Under cooperative binding conditions, the dissociation constant of HMGN for nucleosomes is about 10^{-7} M, while under non-cooperative conditions, it is 100-fold stronger at around 10^{-9} M.

The gel mobility shift assay has been used extensively to investigate how HMGN proteins interact with nucleosome core particles and to define the nucleosome binding domain.[4,5,17] HMGN proteins bind to nucleosomes to form homodimeric complexes containing two molecules of either HMGN1 or HMGN2. This assay is very sensitive, and it has been used to demonstrate that phosphorylation or mutation of the nucleosome binding domain greatly reduces or even abolishes the interaction with nucleosomes.[10,18]

[12] J. Ausio, F. Dong, and K. E. van Holde, *J. Mol. Biol.* **206,** 451–463 (1989).

[13] J. Ausio and K. E. Van Holde, *Cell Differ.* **23,** 175–189 (1988).

[14] D. J. Steger, A. Eberharter, S. John, P. A. Grant, and J. L. Workman, *Proc. Natl. Acad. Sci. USA* **95,** 12924–12929 (1998).

[15] V. M. Studitsky, D. J. Clark, and G. Felsenfeld, *Methods Enzymol.* **274,** 246–256 (1996).

[16] K. Ura and A. P. Wolffe, *Methods Enzymol.* **274,** 257–271 (1996).

[17] Y. V. Postnikov, L. Trieschmann, A. Rickers, and M. Bustin, *J. Mol. Biol.* **252,** 423–432 (1995).

[18] M. Prymakowska-Bosak, T. Misteli, J. E. Herrera, H. Shirakawa, Y. Birger, S. Garfield, and M. Bustin, *Mol. Cell. Biol.* **21,** 5169–5178 (2001).

Equilibrium Dialysis

Equilibrium dialysis also allows quantitative analysis of the interaction between HMGN proteins and nucleosomes. Using this approach, it was demonstrated that acetylation reduces the affinity of HMGN proteins for nucleosome core particles.[19,20] ^{14}C-acetylated HMGN is mixed with an excess of unmodified protein and added to one side of an equilibrium dialysis chamber. Nucleosome core particles are added to the other side, and the HMGN is allowed to equilibrate between the two chambers. The specific activity of the HMGN in the two chambers is then measured. A reduction in the specific activity of the HMGN in the chamber containing the nucleosome cores indicates that acetylated HMGN has a lower affinity for the nucleosomes than the unmodified form.[19,20] This very sensitive assay allows small differences in binding affinity to be detected within a heterogeneously-labeled population.

Additional Approaches to Study the Interaction of HMGN Proteins with Nucleosomes

DNase I and hydroxyl radical footprinting have been used to determine the binding sites of HMGN proteins on the nucleosome. DNase I footprinting revealed strong protection by HMGN proteins at 25 bp and 125 bp from the end of the 147 bp DNA.[1,2,4] Hydroxyl radical footprinting provided a more detailed analysis and detected additional interactions at either side of the dyad axis.[21] Protein-DNA crosslinking revealed the same interactions as the DNase I footprinting assay, although this type of experiment is harder to perform and interpret.[22] Protein-protein crosslinking supports the interaction of HMGN near the 25 bp and 125 bp positions, as the N-terminal region of HMGN1 can be crosslinked to a part of the H2B histone fold domain that is located in this region of the nucleosome.[23] The interaction of HMGN near the dyad axis is also supported by studies showing that the central domain of HMGN can be crosslinked to part of the globular domain of H3,[24] and that the C-terminal domain of HMGN can be crosslinked to the N-terminal tail of H3.[23] Both of these H3 regions are in the vicinity of the dyad axis.

[19] J. E. Herrera, K. Sakaguchi, M. Bergel, L. Trieschmann, Y. Nakatani, and M. Bustin, *Mol. Cell. Biol.* **19**, 3466–3473 (1999).

[20] M. Bergel, J. E. Herrera, B. J. Thatcher, M. Prymakowska-Bosak, A. Vassilev, Y. Nakatani, B. Martin and M. Bustin, *J. Biol. Chem.* **275**, 11514–11520 (2000).

[21] P. J. Alfonso, M. P. Crippa, J. J. Hayes, and M. Bustin, *J. Mol. Biol.* **236**, 189–198 (1994).

[22] V. V. Shick, A. V. Belyavsky, and A. D. Mirzabekov, *J. Mol. Biol.* **185**, 329–339 (1985).

[23] L. Trieschmann, B. Martin, and M. Bustin, *Proc. Nat. Acad. Sci.* **95**, 5468–5473 (1998).

[24] J. V. Brawley and H. G. Martinson, *Biochemistry* **31**, 364–370 (1992).

Unfolding of Minichromosomes by HMGN Proteins *In Vitro*

Reconstituting HMGN into Chromatin

Insights into the interaction of HMGN with the chromatin fiber have been obtained using nucleosome arrays and minichromosomes, generated by several methods. Most of the studies used minichromosomes assembled using extracts from *Xenopus* eggs.[25–27] Briefly, single-stranded DNA is replicated and assembled into chromatin using *X. laevis* egg extract as a source of assembly factors, core histones, and additional TFIIIA and TFIIC, in either the presence or the absence of HPLC-purified HMGN. Electrophoresis on a chloroquine-containing agarose gel verifies that the assembly reaction is long enough to ensure equal numbers of nucleosomes per template. Extracts from *Drosophila* embryos have also been used to assemble minichromosomes for HMGN studies.[28] In both cases, it is important that the HMGN is added prior to chromatin assembly, as the HMGN has no effect if added after assembly is complete. The interaction of HMGN with chromatin has also been studied using SV40 minichromosomes isolated from cultured cells. In this instance, the same effects were observed whether HMGN was added before or after minichromosome assembly.[29,30] The major difference between SV40 minichromosomes and the chromatin assembled using egg extracts is the presence of H1 in the former system. The effect of HMGN on the chromatin structure and transcriptional potential of SV40 minichromosomes depends on the presence of this linker histone.[29]

Demonstrating an Interaction of HMGN with Minichromosomes

HMGN binding to minichromosomes can be demonstrated directly by centrifuging the complex through a sucrose gradient, followed by a Western analysis to determine whether the fractions containing the minichromosome also contain HMGN.[26]

[25] M. P. Crippa, L. Trieschmann, P. J. Alfonso, A. P. Wolffe, and M. Bustin, *EMBO J.* **12**, 3855–3864 (1993).

[26] L. Trieschmann, P. J. Alfonso, M. P. Crippa, A. P. Wolffe, and M. Bustin, *EMBO J.* **14**, 1478–1489 (1995).

[27] N. Weigmann, L. Trieschmann, and M. Bustin, *DNA and Cell Biology* **16**, 1207–1216 (1997).

[28] S. M. Paranjape, A. Krumm, and J. T. Kadonaga, *Genes Dev.* **9**, 1978–1991 (1995).

[29] H. F. Ding, M. Bustin, and U. Hansen, *Mol. Cell. Biol.* **17**, 5843–5855 (1997).

[30] H. F. Ding, S. Rimsky, S. C. Batson, M. Bustin, and U. Hansen, *Science* **265**, 796–799 (1994).

Unfolding of Minichromosomes

Electron microscopy has been used to demonstrate that assembly of HMGN into minichromosomes extends their conformation (Fig. 1A).[31] The number of nucleosomes incorporated into the template is unaffected by HMGN, as shown by counting the nucleosomes on the electron micrographs and by electrophoresis on a chloroquine-containing agarose gel (Fig. 1B). However, the geometry of the minichromosomes assembled in the presence or absence of HMGN is clearly distinct. The plasmids chromatinized in the presence of HMGN are free of local compacted regions (LC) and have a larger diameter (general decompaction, GD). It can be concluded that minichromosomes assembled in the presence of HMGN have a more extended chromatin structure than minichromosomes devoid of HMGN.

Sucrose gradient sedimentation also demonstrated that incorporation of HMGN into minichromosomes reduces their compaction. HMGN-containing minichromosomes sediment more slowly, most likely because

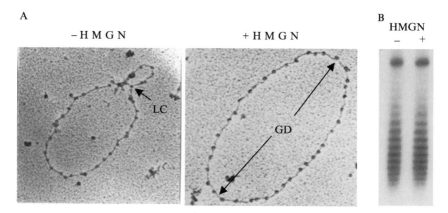

FIG. 1. Visualization of HMGN-mediated chromatin decompaction using electron microscopy. (A) Chromatin assembled in the absence (−) or presence (+) of HMGN2 was purified by sucrose gradients and analyzed by electron microscopy as described.[31] LC: local compacted region. GD: general decompaction. (B) Supercoiling assay. Single-stranded DNA was incubated with *Xenopus* egg extract and labeled α-^{32}P-ATP. Incubation was in the absence (−) or presence (+) of HMGN2. Radiolabeled DNA was extracted and separated on a 0.8% agarose gel in 1 × TPE buffer (40 mM Tris-Cl, pH 8.0, 30 mM NaH$_2$PO$_4$, 1 mM EDTA), containing 20 μM chloroquine. Note that the presence of HMGN increases the diameter of the minichromosomes without affecting the number of nucleosomes on the plasmid.

[31] B. Vestner, M. Bustin, and C. Gruss, *J. Biol. Chem.* **273**, 9409–9414 (1998).

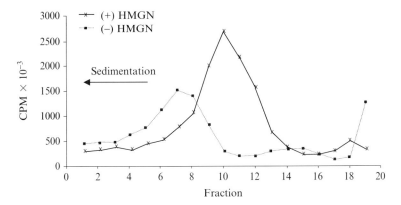

FIG. 2. Sucrose gradient sedimentation analysis of minichromosomes assembled in the presence (crosses) or absence (black squares) of HMGN.[26] Note the HMGN-containing minichromomes sediment slower, suggesting decompaction and an increase in the radius of gyration.

their radius of gyration is increased (Fig. 2).[26,29–32] Decompaction of the minichromosomes is most apparent if all the nucleosomes contain HMGN. The DNA in the minichromosomes is radiolabeled during assembly, and the sample is then loaded onto a 10–30% linear sucrose gradient in HEPES buffer containing 70 mM KCl and 0.25% Triton X-100. The detergent is required to prevent minichromosome aggregation. Centrifugation is carried out using SW40 rotor at 40,000 rpm (200,000 g) for 2.5 h. Fractions are collected from the bottom of the tube and their radioactivity is measured. Under these conditions, minichromosomes assembled in the absence of HMGN produce a band at the level of 24% sucrose. Addition of the HMGN decompacts the chromatin of the minichromosomes and reduces the sedimentation rate, generating a band at 20% sucrose.

Chromatin folding can also be assayed by nucleases, as the accessibility of the linker DNA to nucleases increases with decreasing compaction of the chromatin fiber. The rate of digestion of minichromosomes by restriction enyzmes or micrococcal nuclease is thus increased when the minichromosomes are reconstituted in the presence of HMGN.[25,26] A protocol for investigating chromatin compaction *in vivo* using micrococcal nuclease is presented in detail later in this chapter.

The decrease in chromatin folding has a concomitant effect on processes that require access to the DNA, and the potential for both transcription and replication is increased in the presence of HMGN. HMGN

[32] M. Bustin, *Mol. Cell. Biol.* **19**, 5237–5246 (1999).

potentiates transcription or replication only in the context of chromatin and no effect was observed on naked DNA.[26,31] Interestingly, SV40 minichromosomes isolated from cells demonstrate decompaction and increased transcription whether the HMGN is added before or after purification,[29] which seems to contrast with the observations from minichromosomes assembled using egg or embryo extracts. However, it is important to note that the effect of HMGN on SV40 minichromosomes is dependent on the presence of H1 and is detected only with H1-compacted minichromosomes.[29,30] In both types of system, it has been shown that deletion of the C-terminal region of HMGN abolishes its ability to unfold chromatin and to increase transcription.[5,29] These findings demonstrate a direct link between the ability of HMGN proteins to decompact chromatin and their ability to enhance transcription from chromatin templates.

Use of Purified Nucleosome Arrays

The effect of HMGN on chromatin folding has also been examined by neutron scattering using nucleosome arrays purified from cells.[33] An HMGN-dependent decrease in chromatin compaction was observed, but only when the chromatin had been reconstituted with linker histones.[33] However, similar studies with arrays assembled from purified components onto repeats of the 5S rDNA positioning sequence did not demonstrate an HMGN effect (K. L. West, unpublished data). As described earlier, results from minichromosomes assembled using egg extracts strongly indicate that HMGN must be added prior to chromatin assembly. It is thus unclear whether adding HMGN to purified nucleosome arrays would result in its correct incorporation into the chromatin, and results from such experiments must be interpreted with caution.

Role of HMGN proteins in chromatin unfolding *in vivo*

Association of HMGN Proteins with Chromatin In Vivo

Many studies have demonstrated that HMGN proteins are associated with chromatin when mononucleosomes, oligonucleosomes or SV40 minichromosomes are isolated from cells.[30,34–37] There are also indications that

[33] V. Graziano and V. Ramakrishnan, *J. Mol. Biol.* **214,** 897–910 (1990).

[34] G. H. Goodwin, C. G. Mathew, C. A. Wright, C. D. Venkov, and E. W. Johns, *Nucleic Acids Res.* **7,** 1815–1835 (1979).

[35] S. Weisbrod, M. Groudine, and H. Weintraub, *Cell.* **19,** 289–301. (1980).

[36] C. S. M. Tahourdin, N. K. Neihart, I. Isenberg, and M. Bustin, *Biochemistry* **20,** 910–915 (1981).

HMGN may be preferentially associated with acetylated nucleosomes[38] and/or transcriptionally active genes.[34,35,39,40] However, it is difficult to interpret some of these early studies as many procedures eliminated particular subfractions of chromatin, including those containing linker histones, and steps were not taken to eliminate HMGN redistribution. In a more recent study with nuclei isolated from chicken erythrocytes, it was shown that HMGN2-containing nucleosomes are clustered into domains that, on average, consist of six adjacent HMGN2-containing nucleosomes and no HMGN1-containing nucleosomes.[41]

A powerful method to investigate the DNA sequences to which chromatin-associated proteins are bound *in vivo* is the chromatin immunoprecipitation assay. This assay involves crosslinking the chromatin prior to cell disruption to prevent rearrangement. The crosslinked chromatin is sonicated into short lengths, and then protein-DNA complexes containing HMGN are immunoprecipitated. After the protein-DNA crosslinks are reversed, the DNA is purified and assayed to determine whether a genomic region of interest is enriched in the immunoprecipitated DNA. This technique has shown that HMGN proteins are present at a higher density on transcribed sequences than on non-transcribed genes.[42] The resolution of this method is limited to the size of the chromatin fragments after sonication, and average lengths corresponding to 1–3 nucleosomes are usually desirable.

Chromatin Unfolding by HMGN Proteins In Vivo

Some of the initial papers on HMGN proteins suggested that these proteins induce or stabilize a decompacted chromatin configuration which is more susceptible to DNase I digestion.[35,43,44] Although subsequent reports disagreed with these results,[45,46] the studies with *in vitro* chromatin assembly

[37] L. Levinger, J. Barsoum, and A. Varshavsky, *J. Mol. Biol.* **146,** 287–304 (1981).

[38] N. Malik, M. Smulson, and M. Bustin, *J. Biol. Chem.* **259,** 699–702 (1984).

[39] T. Dorbic and B. Wittig, *Nucleic Acids Res.* **14,** 3363–3376 (1986).

[40] T. Dorbic and B. Wittig, *EMBO J.* **6,** 2393–2399 (1987).

[41] Y. V. Postnikov, J. E. Herrera, R. Hock, U. Scheer, and M. Bustin, *J. Mol. Biol.* **274,** 454–465 (1997).

[42] Y. V. Postnikov, V. V. Shick, A. V. Belyavsky, K. R. Khrapko, K. L. Brodolin, T. A. Nikolskaya, and A. D. Mirzabekov, *Nucleic Acids Res.* **19,** 717–725 (1991).

[43] S. Weisbrod and H. Weintraub, *Proc. Nat. Acad. Sci.* **76,** 630–635 (1979).

[44] S. Weisbrod and H. Weintraub, *Cell* **23,** 391–400 (1981).

[45] R. H. Nicolas, C. A. Wright, P. N. Cockerill, J. A. Wyke, and G. H. Goodwin, *Nucleic Acids Res.* **11,** 753–772 (1983).

[46] G. H. Goodwin, P. N. Cockerill, S. Kellam, and C. A. Wright, *Eur. J. Biochem.* **149,** 47–51 (1985).

systems support the general notion that HMGN proteins reduce the compaction of the chromatin fiber. The advent of *Hmgn1* knockout mice has now allowed this question to be directly addressed, by comparing the nuclease sensitivity of active genes in cells with and without HMGN1.

Micrococcal Nuclease Digestion of Chromatin in $Hmgn1^{-/-}$ and
  $Hmgn1^{+/+}$ Liver Nuclei

Micrococcal nuclease cleaves the linker DNA between nucleosomes, generating a ladder of nucleosomal bands on a DNA gel. As the extent of digestion is increased, there is a progressive decrease in the overall size of the digested fragments. A more compact chromatin fiber is less accessible to the nucleases, and is digested at a slower rate. HMGN1 and HMGN2 have been shown to unfold minichromosomes *in vitro*, making them more accessible to nucleases.[32,47] The possible effects of HMGN proteins on chromatin structure in live cells are likely to be more subtle and difficult to reveal due to the relatively low abundance of the proteins and the extremely dynamic nature of their interaction with chromatin.[48,49] Chromatin immunoprecipitation assays have suggested that HMGN is preferentially associated with active genes.[42,50] Furthermore, immunofluorescence studies have shown the HMGN proteins are associated with sites of transcription and that their localization is dependent on the transcriptional state of the cell.[51] We predict, therefore, that chromatin unfolding by HMGN will occur preferentially on active genes rather than on inactive regions. Here we describe an experimental protocol to compare the nuclease accessibility of an active gene with that of total chromatin in cells with and without HMGN1. Nuclei from the livers of $Hmgn1^{-/-}$ and $Hmgn1^{+/+}$ mice were extracted and digested with increasing concentrations of micrococcal nuclease. The resulting DNA fragments were resolved on an agarose gel, transferred to a nylon membrane, and hybridized with probes of interest.

Nuclear Isolation

  Equipment and Reagents

    Tissue grinder motor and tissue homogenizers
    Refrigerated high-speed centrifuge with fixed-angle rotor

[47] M. Bustin, *Trends Biochem. Sci.* **26,** 431–437 (2001).
[48] F. Catez, D. T. Brown, T. Misteli, and M. Bustin, *EMBO Rep.* **3,** 760–766. (2002).
[49] R. D. Phair and T. Misteli, *Nature* **404,** 604–609 (2000).
[50] S. Druckmann, E. Mendelson, D. Landsman, and M. Bustin, *Exp. Cell Res.* **166,** 486–496 (1986).
[51] R. Hock, F. Wilde, U. Scheer, and M. Bustin, *Embo J.* **17,** 6992–7001 (1998).

Ultracentrifuge with SW50.1 rotor and $2 \times 1/2$ inch Ultraclear tubes (Beckman)

UV-Vis spectrophotometer

Nuclear buffer A: 15 mM HEPES, pH 7.5, 60 mM KCl, 15 mM NaCl, 2 mM EDTA, 0.5 mM EGTA, 0.34 M sucrose, 0.15 mM 2-mercaptoethanol, 0.15 mM spermine, 0.5 mM spermidine, proteinase inhibitor cocktail (Roche)

Nuclear buffer B: 15 mM HEPES, pH 7.5, 60 mM KCl, 15 mM NaCl, 0.1 mM EDTA, 0.1 mM EDTA, 2.1 M sucrose, 0.15 mM 2-mercaptoethanol, 0.15 mM spermine, 0.5 mM spermidine, proteinase inhibitor cocktail (Roche)

Micrococcal nuclease digestion buffer: 10 mM Tris-HCl, 25 mM NaCl, 1 mM $CaCl_2$

Protein digestion buffer: 85 μl water, 100 μl 10% SDS, 15 μl Proteinase K, 50 mg/ml

Organic extraction solution: phenol: chloroform: isoamyl alcohol mixture 50:48:2%

Micrococcal nuclease (MNase, Roche) stock solutions: 100, 10, 1 units/μl

Method

The following volumes are for one adult mouse liver. Rinse the isolated liver in 25 ml ice-cold nuclear buffer A in a 100-ml glass beaker on ice, mince the tissue using a razor blade and submerge it in 8 ml nuclear buffer A. Decant the preparation into a homogenizer. Homogenize tissues with five to ten strokes of a motor-driven tissue grinder, followed by an additional five sharp strokes by hand. Filter each homogenate through several layers of folded cheesecloth that has been prewetted with nuclear buffer A into a 30-ml Corex tube on ice. Layer the homogenate onto cushions of 1.4 ml of 1:1 nuclear buffer A and nuclear buffer B mixture in a 15 ml Corex tube. Centrifuge at 12,000 g for 15 min at 4°C in a fixed-angle rotor (e.g., 10,000 rpm in SS-34 rotor). Discard supernatants and resuspend pellets in 11 ml of nuclear buffer B. Layer 3.6 ml of each suspension over 1.2 ml cushions of nuclear buffer B in three $2 \times 1/2$ inch Ultraclear SW50.1 centrifuge tubes. Spin 90 min at 120,000 g (e.g., 37,000 rpm in SW50.1 rotor) at 4°C. Decant by inverting the tubes. Gently resuspend each pellet into 0.5 ml of nuclear buffer A. Measure OD_{260} by spectrophotometer. The usual yield is approximately 0.5 mg DNA per mouse liver.

Micrococcal Nuclease Digestion of Isolated Nuclei

Each digest should contain at least 15 μg of DNA. Add 0.1 M $CaCl_2$ to a final concentration of 3 mM and incubate for 1.5 min at 37°C to warm the nuclei. Add the desired amount of MNase to each tube. We suggest using a range from 0.05 U to 200 U per one A_{260} unit of nuclei, with the highest MNase concentration sufficient for fairly exhaustive digestion. Small-scale pilot experiments are needed to determine the exact conditions required for each enzyme and nuclei batch. Mix, and incubate tube 5 min at 25°C. Stop the reaction by adding EDTA to 5 mM, and isolate the DNA by SDS/proteinase K digestion for 30 min at 37°C followed by organic solvent extraction, and ethanol precipitation.

Electrophoresis and Southern Blotting

Electrophorese the DNA samples on a 1.5% agarose gel in 1×TBE buffer alongside appropriate DNA molecular weight markers. Stain with ethidium bromide and visualize using a UV light source. Save the picture in a digital bitmap format suitable for quantitation. Blot the gel onto nylon or PVDF membrane using standard Southern blotting techniques. Hybridize the membrane with a radiolabeled probe prepared by random oligonucleotide primed synthesis from the plasmid of interest, and wash extensively to remove background signal. The blot should be visualized by phosphorimager analysis.

Analysis

Figure 3 exemplifies the micrococcal nuclease digestion pattern of liver chromatin from $Hmgn1^{-/-}$ and $Hmgn1^{+/+}$ mice. The average nucleosome length in the digest was calculated as: $La = \Sigma\, N(P_N)$. N is the oligonucleosome size (i.e., mono, di) and P_N is the fraction of a particular oligonucleosomes size out of the total scan. The range N = 1 to N = 6 was used.

The bulk chromatin was digested at a similar rate in both $Hmgn1^{-/-}$ and $Hmgn1^{+/+}$ mice, as indicated by the ethidium bromide-stained DNA in Fig. 3A. However, clear differences were observed after a Southern blot of the same gel was probed for the actively transcribed gene $Hmgn2$ (Fig. 3B). At the last point of digestion, for example, the ratio of dinucleosome to mononucleosome is higher in the $Hmgn1^{-/-}$ chromatin than in the $Hmgn1^{+/+}$ sample. Calculation of the average nucleosome length of the three digestion points reveals that HMGN1-containing chromatin is digested faster in each case (Fig. 3C).

Another important feature visible on a micrococcal nuclease digestion pattern is the length of the nucleosomal repeat. The exonucleolytic activity of micrococcal nuclease tends to induce some nucleosome sliding during digestion, such that the length of the mononucleosome decreases with time. HMGN is thought to minimize this sliding, as it binds near the entry/exit

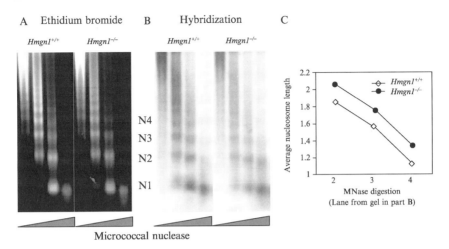

FIG. 3. Micrococcal nuclease digestion analysis of *in vivo* chromatin from *Hmgn1*$^{+/+}$ and *Hmgn1*$^{-/-}$ mouse livers. (A) Nuclei were isolated and digested with increasing concentrations of micrococcal nuclease (lanes shown are from left to right: 0.75, 5, 27.5, and 162.5 units of enzyme per 1 OD$_{260}$ nuclei). Purified DNA was run on a 1.5% agarose gel and stained with ethidium bromide. (B) A Southern blot of the gel in A hybridized with a probe corresponding to the transcribed portion of *Hmgn2* gene. (C) The image from panel B was scanned using a Molecular Dynamics phosphorimager and quantified using ImageQuant software. Average nucleosome length for each lane is plotted for three last lanes. Note that HMGN does not affect the rate of digestion of bulk chromatin, but increases the rate of digestion of an active gene.

point of the nucleosomal DNA. This is manifest as a slight increase in the length of each nucleosome, as has been shown previously for digestion of minichromosomes.[26]

Visualizing HMGN In Vivo

Fluorescence microscopy has proved to be a powerful tool in studying the association of HMGN proteins with chromatin *in vivo*. Indirect immunofluorescence using antibodies specific to each isoform has shown that HMGN1 and N2 are localized in discrete, separate foci within the nucleus.[41] These foci colocalize with sites of transcription and are dependent on the transcriptional status of the cell.[51] Notably, disruption of HMGN function either with anti-HMGN antibodies[52] or with a competing peptide corresponding to the nucleosome binding domain inhibits transcription.[51] Fluorescence recovery after photobleaching (FRAP) and fluorescence loss in photobleaching (FLIP) have been used to study the mobility of

[52] L. Einck and M. Bustin, *Proc. Nat. Acad. Sci.* **80**, 6735–6739 (1983).

GFP-tagged HMGN in living cells.[18,48,49] These techniques showed that HMGN proteins are highly mobile within the nucleus, with a mean residence time on chromatin of less than 30 sec. Importantly, mutations that disrupt the interaction of HMGN with the nucleosome, as indicated by gel shift assays, decrease the residence time as measured by FLIP and FRAP.[18,48] It can therefore be deduced that nucleosome binding is an important component of HMGN localization *in vivo*.

Mechanisms of Chromatin Unfolding by HMGN

Interaction with Core Histone Tails

It has been established that *C*-terminal domain of HMGN proteins is required for chromatin unfolding and transcriptional enhancement,[26,29] although the mechanism of unfolding is still unclear. There is evidence to suggest that several mechanisms may be employed, and one of these is based on the observation that the chromatin unfolding domain of HMGN1 can be crosslinked to the *N*-terminal tail of histone H3. In this approach, a photo-activatable cross-linking reagent was used to define which region of the nucleosome interacts with the *C*-terminal unfolding region of HMGN1.[23] A serine in the *C*-terminal domain of HMGN1 was mutated to cysteine, so that the radiolabeled crosslinking reagent could be covalently attached to that specific residue using a disulphide linkage. The modified protein was bound to nucleosome core particles, and the crosslinker activated with UV light. The core histone that became crosslinked to HMGN was identified by the radiolabel, and the specific peptide involved was identified by proteolytic digests and HPLC. This site-specific cross-linking approach can give much more useful information than cross-linking reagents that simply target all lysine residues, particularly for proteins that are very lysine-rich such as HMGN family members.

The positively charged *N*-terminal tails of the core histones are critical for chromatin folding, and one way they are thought to accomplish this is by neutralizing the negative charge of the linker DNA.[53] They may also promote chromatin compaction by interacting with adjacent nucleosomes and/or adjacent chromatin fibers.[54–56] Removal of the core histone tails, or neutralizing their charge by acetylation prevents the salt-induced

---

[53] D. Angelov, J. M. Vitolo, V. Mutskov, S. Dimitrov, and J. J. Hayes, *Proc. Natl. Acad. Sci. USA* **98**, 6599–6604. (2001).

[54] J. Allan, N. Harborne, D. C. Rau, and H. Gould, *J. Cell. Biol.* **93**, 285–297 (1982).

[55] M. Garcia-Ramirez, F. Dong, and J. Ausio, *J. Biol. Chem.* **267**, 19587–19595 (1992).

[56] T. M. Fletcher and J. C. Hansen, *J. Biol. Chem.* **270**, 25359–25362 (1995).

compaction of the chromatin fiber.[56–58] It is possible that the chromatin unfolding domain of HMGN competes for the core histone tails and induces their rearrangement, leading to decompaction of the nucleosome arrays. HMGN proteins could also influence chromatin folding by affecting the post-translational modification of the core histone tails. For example, full-length HMGN inhibits H3 acetylation, whereas a mutant missing the C-terminal domain does not.[59]

Counteracting Linker Histone Compaction

Several lines of evidence support a second model for HMGN action whereby it counteracts chromatin compaction by linker histones. First, footprinting of HMGN bound to nucleosome core particles indicates that the HMGN binding sites overlap with the proposed binding sites for the linker histones.[21] Second, the ability of HMGN to unfold and activate transcription from SV40 minichromosomes is dependent on the presence of H1.[29] Third, FRAP experiments indicate that the mobility of GFP-labeled linker histone H1 in living cells is increased when full-length HMGN is microinjected into cells as a competitor.[48] No change in the mobility of H1 is observed when C-terminally truncated HMGN is injected instead. Taken together, these results suggest that HMGN may alter the interaction of linker histone with chromatin, so that it can no longer efficiently compact the chromatin fiber.

The two models for chromatin unfolding by HMGN, counteracting H1 and interacting with core histone amino terminal tails, are not mutually exclusive. Several observations point to an interplay between linker histones and the core histone tails,[60–62] and it is conceivable that HMGN could modulate the activity both of the core histone tails and of linker histones within the same nucleosome. It is also possible that one or other of the mechanisms would predominate in different chromatin environments.

[57] J. C. Hansen, C. Tse, and A. P. Wolffe, *Biochemistry* **37**, 17637–17641 (1998).

[58] C. Tse, T. Sera, A. P. Wolffe, and J. C. Hansen, *Mol. Cell. Biol.* **18**, 4629–4638 (1998).

[59] J. Herrera, K. Sakaguchi, M. Bergel, L. Trieschmann, Y. Nakatani, and M. Bustin, *Mol. Cell. Biol.* **19**, 3466–3473 (1999).

[60] J. E. Herrera, K. L. West, R. L. Schiltz, Y. Nakatani, and M. Bustin, *Mol. Cell. Biol.* **20**, 523–529 (2000).

[61] L. J. Juan, R. T. Utley, C. C. Adams, M. Vettese-Dadey, and J. L. Workman, *Embo J.* **13**, 6031–6040 (1994).

[62] K. M. Lee and J. J. Hayes, *Biochemistry* **37**, 8622–8628 (1998).

Concluding Remarks

Three main lines of experimental evidence support the suggestion that HMGN proteins are architectural elements that reduce the compaction of the chromatin fiber. First, HMGN has a higher affinity for nucleosomes than for free DNA. Gel mobility shift, crosslinking and footprinting analyses have provided valuable information as to how HMGN proteins interact with this basic building block of the chromatin fiber. Second, extensive analyses of chromatin reconstituted *in vitro* have demonstrated that HMGN proteins induce physical changes consistent with a reduction in chromatin compaction, with a concomitant increase in the potential for transcription and replication. Third, HMGN proteins associate with sites of transcription *in vivo*, and incubation of permeabilized cells with a competing peptide corresponding to the nucleosome binding domain results in inhibition of transcription. The generation of *Hmgn1* knockout mice has greatly facilitated the study of HMGN1 *in vivo*, and we have presented a protocol for using micrococcal nuclease accessibility to compare chromatin compaction in cells from *Hmgn1*$^{-/-}$ and *Hmgn1*$^{+/+}$ mice. The data demonstrate that HMGN affects the dynamics of higher-order chromatin structure and optimizes the progression of DNA-related activities in the context of chromatin.

[40] Assay of Activator Recruitment of Chromatin-Modifying Complexes

By MICHAEL J. CARROZZA, AHMED H. HASSAN, and JERRY L. WORKMAN

This chapter describes an *in vitro* assay for analyzing activator recruitment of chromatin-modifying complexes to a nucleosomal promoter.[1] The template used in this assay contains five Gal4 binding sites upstream of the Adenovirus E4 core promoter and five sea urchin 5S nucleosome positioning sequences flanking each at end of the 5XGal4-E4 promoter. These positioning sequences result in twelve nucleosomes with two over the 5XGal-E4 promoter. Central to this assay is the use of a nucleosomal template immobilized on paramagnetic beads. The immobilization of the template allows removal and addition of soluble components in the

---

[1] A. H. Hassan, K. E. Neely, and J. L. Workman. Histone acetyltransferase complexes stabilize swi/snf binding to promoter nucleosomes. *Cell* **104,** 817 (2001).

recruitment reactions by collecting templates on a magnetic concentrator and washing the beads. For example, chromatin-modifying complexes, such as the SAGA histone acetyltransferase and SWI/SNF ATP-dependent chromatin-remodeling complexes, can be targeted to a nucleosomal promoter by a transcription activator protein. Afterwards, the activator protein is competed from the template using sequence-specific competitor DNA and the template is washed. The ability of the complexes to remain bound to the promoter in the absence of an activator can then be assessed by western analysis for different components of the reaction, such as acetylated histones, the activator protein, and chromatin-modifying complex components.

Material and Reagents

1. pIC-2085S/G5E4R plasmid DNA[2]
2. *S. cerevisiae* chromatin-modifying complexes, (purification of SAGA and NuA4 complexes,[3-4] and for purification of SWI/SNF[5])
3. HeLa cell core histones and long oligonucleosomes[6]
4. Gal4-activation domain fusion protein[7]
5. Restriction enzymes: Asp 718 (Roche, Indianapolis, IN), Cla I (NEB, Beverly, MA), and BstU I (NEB, Beverly, MA)
6. Klenow DNA polymerase (NEB, Beverly, MA)
7. Deoxynucleotides: dTTP, dCTP, and dGTP (Amersham, Piscataway, NJ), and biotin-14-dATP (Invitrogen, Carlsbad, CA)
8. Qiaquick Gel Extraction Kit (Qiagen, Valencia, CA)
9. Dynabeads M-280 Streptavidin (Dynal, Lake Success, NY)
10. Magnetic particle concentrator (Dynal)
11. Trichostatin A and sodium butyrate (Sigma, St. Louis, MO)

---

[2] K. Ikeda, D. J. Steger, A. Eberharter, and J. L. Workman. Activation domain-specific and general transcription stimulation by native histone acetyltransferase complexes. *Mol. Cell. Biol.* **19,** 855 (1999).

[3] A. Eberharter, S. John, P. A. Grant, R. T. Utley, and J. L. Workman. Identification and analysis of yeast nucleosomal histone acetyltransferase complexes. *Methods* **15,** 315 (1998).

[4] M. J. Carrozza, S. John, A. K. Sil, J. E. Hopper, and J. L. Workman. Gal80 confers specificity on HAT complex interactions with activators. *J. Biol. Chem.* **277,** 24648 (2002).

[5] K. E. Neely, A. H. Hassan, C. E. Brown, L. Howe, and J. L. Workman. Transcription activator interactions with multiple SWI/SNF subunits. *Mol. Cell. Biol.* **22,** 1615 (2002).

[6] T. Owen-Hughes, R. T. Utley, D. J. Steger, J. M. West, S. John, J. Cote, K. M. Havas, and J. L. Workman. Analysis of nucleosome disruption by ATP-driven chromatin remodeling complexes. *Methods. Mol. Biol.* **119,** 319 (1999).

[7] R. J. Reece, R. J. Rickles, and M. Ptashne. Overproduction and single-step purification of GAL4 fusion proteins from *Escherichia coli. Gene* **126,** 105 (1993).

12. Reconstitution buffer A: 50 mM HEPES-NaOH, pH 7.5, 1 mM EDTA, 5 mM DTT, and 0.5 mM PMSF

13. Reconstitution buffer B: 20% glycerol, 10 mM Tris-HCl, pH 7.5, 1 mM EDTA, 100 μg/ml bovine serum albumin, 0.1% NP-40, 5 mM DTT, and 0.5 mM PMSF

14. 1X Binding Buffer: 10 mM HEPES-KOH, pH 7.8, 5% glycerol, 50 mM KCl, 0.25 mg/ml bovine serum albumin, 2mM MgCl$_2$. Add 0.05% NP-40, 5 mM DTT, and 0.5 mM PMSF just prior to use

15. 5X Binding Buffer: 50 mM HEPES-KOH, pH 7.8, 25% glycerol, 250 mM KCl, 1.25 mg/ml bovine serum albumin 10mM MgCl$_2$

Biotinylation and Purification of G5E4-5S Template

In the first part of this method, the 5S array 5XGal4 E4 promoter template (G5E4-5S) is biotinylated at the 3′ end. This is achieved by digesting the plasmid pIC-2085S/G5E4R[2] DNA with a restriction enzyme that linearizes the plasmid just 3′ of the 5S array promoter construct. Both ends of the linear plasmid are biotinylated by filling in both restriction enzyme generated overhangs with Klenow DNA polymerase and a biotinylated nucleotide. Digestion of the DNA with additional restriction enzymes leaves the template with a biotin at only its 3′ end. The G5E4-5S fragment is then gel purified. It is best to start with at least 40 μg of plasmid. The 5S array constitutes only half of the plasmid which should give a predictable yield of 20 μg. However, with losses in the final gel purification step, the actual yield is usually between 10 and 15 μg. It is important to have at least 10 μg to have a DNA concentration suitable for the nucleosome reconstitution step.

Procedure

1. To begin preparation of the G5E4-5S template, the plasmid is digested with a restriction enzyme that linearizes the plasmid just 3′ of the 5S array. The pIC208/G5E4 plasmid DNA is linearized by restriction digestion with the Asp718 restriction enzyme (Roche, Indianapolis, IN). Typically, 40 μg of plasmid is digested with 80 units of Asp718 for 3 h at 37 °C. Complete digestion can be verified by running 0.1 μg of the digested DNA next to uncut plasmid on a 1% agarose TAE gel and visualizing by ethidium bromide staining on a UV illuminator.

2. Both ends of the linear plasmid are biotinylated by filling in both Asp718 overhangs with Klenow DNA polymerase and a biotinylated dATP. This reaction is set up by mixing directly into the completed Asp718 restriction digestion dTTP, dCTP, and dGTP (Amersham, Piscataway, NJ) to final concentration of 100 μM, biotin-14-dATP

(Invitrogen, Carlsbad, CA) to final concentration of 40 μM, and 1 unit of Klenow DNA polymerase (NEB, Beverly, MA) per μg of DNA. The volume of the reaction is adjusted with *E. coli* DNA polymerase buffer in order to maintain a glycerol concentration of $\leq$5%. The fill-in reaction is carried for 15 min at room temperature. The reaction is stopped by adding 0.5 M EDTA, pH8.0 to a final concentration of 10 mM and heating for 10 min at 75 $^{\circ}$C. After heating, a one-tenth volume of 3 M sodium acetate, pH 5.2 is added, the DNA is precipitated with three volumes of ethanol and microcentrifuged at full speed for 20 min. The pellet is washed in 1 ml of 70% ethanol, allowed to dry, and resuspended in water.

3. In order to isolate the G5E4-5S fragment from the plasmid backbone and establish a biotin label on only one end of the template, the DNA is digested with the Cla I restriction enzyme. This enzyme cleaves at the 5$'$ end of the 5S array, thereby removing the backbone DNA and the biotin on the backbone 5$'$ end. The Cla I digestion is carried out by mixing into the DNA suspension from step two the appropriate restriction enzyme buffer and 1 unit Cla I per μg DNA. It is best to carry out the digestion overnight at 37 $^{\circ}$C to achieve complete digestion. Again, complete digestion can be verified by running a small portion of the digestion and linearized plasmid on an agarose gel.

4. The Cla I digestion results in generation of two 2.5 kb fragments. In order to facilitate gel purification of the G5E4-5S, the fragment containing the plasmid backbone is further digested with BstU I. This enzyme cleaves the plasmid backbone into several 600 bp or less fragments. BstU I cuts under the same buffer conditions as Cla I but needs to be incubated at 60 $^{\circ}$C. The BstU I digestion is carried out by increasing the volume of the completed Cla I digestion with water and buffer to maintain a $\leq$5% glycerol concentration and adding 4 units of BstU I per μg DNA. The reaction is incubated for 3 h at 60 $^{\circ}$C.

5. At the completion to BstU I digestion, the sample is prepared for agarose gel purification by adding agarose gel loading dye. It is advisable to use a loading dye that only contains bromophenol blue since the xylene cyanol dye comigrates with the 2.5 kb 5S array fragment. The sample alongside a DNA size ladder in the range of 0.5–10 kb is loaded onto a 0.6% agarose/1X TAE gel. Using a low agarose concentration reduces the amount of agarose that needs to be purified away from the 5S array fragment. Do not overload the gel as this will result in smearing of the vector bands into the 5S array fragment. Resolution of the 2.5 kb G5E4-5S fragment from the 600 bp or less vector fragments may be monitored by including ethidium bromide in the gel and visualizing bands with UV illuminator. Once the DNA fragments are sufficiently resolved,

the 2.5 kb band is excised from the gel using a clean razor blade. The DNA fragment can be purified from the agarose through a wide variety of methods. We find the most efficient and least time-consuming method is to use gel extraction columns from Qiagen according to the supplied instructions. These columns have a binding capacity of 10 μg, so that with a starting plasmid amount of 40 μg the predicted amount in the excised band would be 20 μg and require two columns.

6. Following purification of the fragment from the agarose, the sample is ethanol precipitated by adding a one-tenth volume of 3M sodium acetate, pH 5.2, and three volumes of ethanol, and microcentrifuged at full speed for 20 min. The pellet is washed in 1 ml 70% ethanol, allowed to dry, and resuspended in 1 μl 1X TE per 0.5 μg of starting plasmid. Since this method of preparing template does not give a 100% yield, the sample is quantified by adding 1 μl sample to 100 μl TE in 100 μl quartz cuvette and measuring the OD_{260}. The OD_{280} should also be measured to ensure that DNA is relatively pure. Typical concentration of the G5E4-5S fragment is between 1 and 1.5 μg/μl with an $OD_{260/280}$ ratio of 1.7.

Reconstitution of Nucleosomal 5S Array

Reconstitution of the G5E4-5S array is carried out by mixing a equimolar amount of the DNA template with core histones in 2 M NaCl and performing a step dilution of the sample down to 0.1 M NaCl. We typically use HeLa cells as a core histone source. Methods detailing the purification of HeLa core histones have been described by Owen-Hughes et al.[6]

Procedure

1. The nucleosome reconstitution is set up by mixing in this order: water to bring the final reaction volume to 10 μl, 1 μg of bovine serum albumin, 5M NaCl to adjust the final salt concentration of the reaction to 2 M NaCl, 2.4 μg (1.5 pmol) of G5E4-5S DNA, and 2 μg (18.5 pmol) of HeLa core histones. At twelve nucleosomes per array, 1.5 pmol of template would require 18.5 pmol of core histones for an equimolar reconstitution. If HeLa core histones are purified according to the method of Owen-Hughes et al.[6], the contribution of 2.5 M NaCl in which they are stored should be kept in mind when adjusting the salt concentration of the reconstitution to 2 M NaCl. The reaction is incubated for 15 min at 37°C.

2. The reconstitution is step diluted by adding 3.3, 6.7, 5, 3.6, 4.7, 6.7, 10, 30, and 20 μl of reconstitution buffer A (see section on Materials and Reagents). This results in the stepwise dilution of the NaCl concentration

to 1.5, 1, 0.8, 0.7, 0.6, 0.5, 0.4, 0.25, and 0.2 M. Perform each dilution by slowly adding buffer while gently tapping the tube. Spin the reaction briefly in a microcentrifuge. Following each dilution step, the reconstitution is incubated for 15 min at 30 °C.

3. A final step dilution is performed with 100 μl of reconstitution buffer B (see section on Materials and Reagents), which brings the final NaCl concentration to 0.1 M. Again, slowly add buffer B with gentle mixing, briefly spin in a microcentrifuge, and incubate for 15 min at 30 °C.

4. The reconstitution is verified by briefly running 10 μl of the reconstitution next to an equal amount of naked DNA on a 1.2% agarose/1X TAE gel in absence of ethidium bromide. The nucleosome array is disrupted by ethidium bromide. To prepare the naked DNA, add 0.6 μg of the purified G5E4-5S DNA to 50μl of a 1:1 mixture of reconstitution buffer A and B with 0.1 M NaCl. This buffer mix is prepared by mixing 0.5 ml reconstitution buffer A, 0.5 ml reconstitution buffer B, and 20 μl 5M NaCl. The gel is run at 60–70 V until the bromophenol blue dye front has migrated approximately 2–3 cm, stained with ethidium bromide, and visualized on an UV illuminator. The reconstituted nucleosomal array should appear to as a diffuse band migrating slightly faster than the naked DNA fragment.

5. The reconstitution is stored at 4 °C.

Immobilization of Nucleosomal G5E4-5S Template on Paramagnetic Beads

The G5E4-5S nucleosomal array is immobilized on paramagnetic beads through the biotin incorporated on the 3' end of the array and streptavidin coupled to the beads. Templates are bound under moderate salt condition to promote efficient binding of biotin to streptavidin. Following the binding of the template to the beads, the templates are washed in the recruitment assay binding buffer to lower the salt concentration. Once washed, the nucleosomal template beads are ready for use in the *in vitro* recruitment assay.

The binding conditions described here are suboptimal for binding large biotin-labeled DNA molecules to the streptavidin beads. Optimal conditions require the use of 1 M NaCl for binding naked DNA. However, due to concerns over the stability of the nucleosome array in salt concentrations over 0.5 M, this procedure adds KCl to 0.3 M that along with the 0.1 M NaCl already in the reconstitution brings the total salt concentration to 0.4 M. This results in an approximate 50% binding of the nucleosome array to the paramagnetic beads. Although not very efficient, this binding is adequate for the assay described here.

Procedure

1. Prepare a 1-ml mixture containing 0.5 ml of reconstitution buffer A and 0.5 ml of reconstitution buffer B. To this mixture add 20 μl of 5M NaCl to bring NaCl concentration to 0.1 M and 112 μl of 3 M KCl to bring the KCl concentration to 0.3 M.

2. 60 μl (0.6 mg) of paramagnetic beads (Dynal, Lake Success, NY) is collected on a magnetic particle concentrator (MPC) (Dynal, Lake Success, NY) for 1 min and the storage solution is removed with micropipette.

3. The beads are washed by resuspending in 0.5 ml of the reconstitution buffer A/B mixture from step one and collecting on the MPC for 1 min. The buffer is removed, and the wash is repeated once more.

4. The 200 μl nucleosome reconstitution is brought to 0.3 M KCl by adding 22.4 μl of 3 M KCl dropwise while gently tapping the tube to mix. Give the reconstitution a brief spin in a microcentrifuge.

5. The nucleosome reconstitution now in 0.3 M KCl/0.1 M NaCl is mixed with the washed paramagnetic beads.

6. The nucleosome array is bound to the beads by incubating the mixture for 1 h at 42 °C. The binding reaction is mixed by gently tapping the tube every 10 min to keep the beads in suspension.

7. The binding reaction is given a brief spin in a microcentrifuge, and the beads are collected on the MPC for 1 min.

8. The beads are washed by resuspending in 0.5 ml of 1X binding buffer (see section on Materials and Reagents) and collecting on the MPC for 1 min. The buffer is removed, and the wash is repeated twice.

9. The beads are resuspended 200 μl of 1X binding buffer and stored at 4 °C.

In Vitro Recruitment Assay

The recruitment assay is an order of addition type of experiment where a Gal4 DNA-binding domain-activator domain fusion protein is first bound to the template beads. Methods detailing the purification of Gal4 DNA-binding domain fusion proteins have been described by Reece *et al.*[7] This is followed by additional steps where the remaining components of the experiment are added and allowed to incubate with the activator-bound templates. Such components can include chromatin-modifying complexes (i.e., the SAGA and NuA HAT complexes, or the SWI/SNF chromatin-remodeling complex), their cofactors (i.e., acetyl-CoA for HATs and ATP for chromatin-remodeling and kinase complexes), competitor chromatin, and/or Gal4-specific competitor DNA.

Most chromatin-modifying complexes tested thus far in these types of experiments exhibit a high nonspecific affinity for the nucleosome template. The addition of nonspecific long oligonucleosomes obtained from HeLa cells reduces nonspecific binding of complexes to barely detectable levels. When an activator that exhibits an affinity for the complex being studied is added into the experiment, binding of the complex to the template should be restored and is interpreted as a recruitment or targeting event. Methods for isolation of HeLa oligonucleosomes have been described by Owen-Hughes *et al.*[6]

Procedure

1. Before setting up the recruitment reactions, a protocol should be written. This is usually done by laying out the experiment in a sort of spreadsheet format with the different experimental conditions on the vertical and the various components of the experiment on the horizontal. When developing a protocol, consider the volume of all the factors going into the experiment. This includes a 10 μl template bead suspension, the Gal4-activator fusion protein, competitor chromatin, cofactors, and the chromatin-modifying complexes. The final volume of the recruitment reaction is usually 100 μl. Although a smaller volume may be used in this assay to increase factor concentration, keep in mind the salt contribution from the various factors going into the reaction. The salt concentration should not exceed 70 mM at any point in the reaction.

The reaction is brought to its final volume by adding a master mix containing common components of the reaction. The master mix is made by mixing into water 5X binding buffer (see section on Materials and Reagents) to make the master mix a 1X concentration, NP-40 to bring the entire recruitment reaction concentration to 0.05%, if HAT activity is a component of the experiment add the histone deacetylase inhibitors trichostatin A or sodium butyrate to a final reaction concentration of 1 μM or 10 mM, respectively, DTT to bring the master mix concentration to 5 mM and PMSF to bring the master mix concentration to 0.5 mM.

2. Recruitment reactions are initially set up by binding a Gal4 DNA-binding domain fusion protein to the G5E4-5S nucleosome template beads. The beads are stored in 1X binding buffer so they are ready to use for the activator binding step. Add 10 μl of the template bead suspension to a tube for each experimental condition. Assuming that 50% of the template bound the beads, this amount of bead suspension is equal to 38 fmol of template DNA or 0.19 pmol in terms of the five Gal4 binding sites. For samples that receive activator, add 0.5 pmol of the Gal4-activator fusion protein to the 10 μl bead suspension. Generally, Gal4-activator fusion

proteins are expressed and purified in a highly concentrated form. If this is case, dilute the Gal4-activator fusion in 1X binding buffer and add as 1–2 μl. Mix components into reactions by gently pipetting up and down. Incubate the binding reaction for 20 min at 30°C. Gently tap the tubes at 10-min intervals to keep the beads in suspension. At the completion of the binding reaction, the samples are spun briefly in a microcentrifuge.

3. At this stage, the master mix, competitor oligonucleosomes, cofactors, and chromatin-modifying complexes are mixed into the reaction in this order. The volume of oligonucleosomes should be minimized since they are stored in 0.6 M NaCl. Typically 1 μg of oligonucleosomes in terms of DNA in 1–2 μl is added. Incubate the reactions for 1 h at 30°C. Mix samples every 10 min if tubes are stationary or place a magnetic bead mixer in a 30°C incubator and incubate with constant mixing. The magnetic beads tend not to stay in suspension using most conventional tube mixers. However, the mixer sold by Dynal (Lake Success, NY) works well. Once the reaction is complete, briefly spin the samples in a microcentrifuge. Input controls for the protein(s) being studied should also be made at this point by taking the same amount as added to the reaction and mixing with SDS-gel sample buffer.

4. If the experiment involves removing the Gal4-activator protein from the template, an excess of Gal4 specific competitor DNA, usually 50–100 pmol in 1–2 μl, is added. The samples are incubated for an additional 30 min at 30°C with mixing.

5. The samples are placed in the MPC, the beads are collected for 1 min, and the supernatant is discarded.

6. The samples are washed twice with 0.5 ml of 1X binding buffer by resuspending the beads with a micropipette, collecting the beads on the MPC for 1 min, and removing the buffer with a micropipette.

7. At this point the samples are solubilized in SDS-gel sample buffer, run on SDS-PAGE, and subjected to western blot analysis. Include input controls for the proteins being analyzed.

[41]   Insights into Structure and Function of GCN5/PCAF and yEsa 1 Histone Acetyltransferase Domains:

By ADRIENNE CLEMENTS and RONEN MARMORSTEIN

Eukaryotic genomic DNA is packaged into chromatin, which is a dynamic macromolecule that is assembled with repeating nucleosome subunits.[1] Each nucleosome contains a histone H3/H4 tetramer that interacts with two histone H2A/H2B heterodimers and wraps approximately 146 base pairs of DNA around its core.[2] The histone proteins contain highly helical regions located at the core of the nucleosome particle and highly conserved, but flexible, N-terminal tails that protrude from the core.[2] Originally defined simply as DNA packaging molecules, nucleosomes are now appreciated to play a much broader role in the dynamic regulation of transcription.[3] The N-terminal histone tails are subject to an array of post-translational modifications such as acetylation, phosphorylation, and methylation.[4] These modifications are linked with a variety of transcriptional and cellular events and can function synergistically or antagonistically to modulate transcription.[3,5–11]

Although a highly complex relationship exists between histone modifications and their transcriptional effects, histone acetylation is generally correlated with transcriptional activation.[12,13] The first characterized enzyme with histone acetyltransferase activity was GCN5 from *Tetrahymena* (tGCN5),[14] the paralog of yeast and human GCN5,[15,16] and a homolog

[1] R. D. Kornberg and Y. Lorch, *Cell* **98**, 285–294 (1999).

[2] K. Luger, A. Maeder, R. Richmond, D. Sargent, and T. Richmond, *Nature* **389**, 251–259 (1997).

[3] T. Jenuwein and C. D. Allis, *Science* **293**, 1074–1080 (2001).

[4] E. Badbury, *BioEssays* **14**, 9–16 (1992).

[5] J. Nakayama, J. C. Rice, B. D. Strahl, C. D. Allis, and S. I. Grewal, *Science* **292**, 110–113 (2001).

[6] S. Rea, *et al.*, *Nature* **406**, 593–599 (2000).

[7] W. S. Lo, *et al.*, *Mol. Cell.* **5**, 917–926 (2000).

[8] W. S. Lo, *et al.*, *Science* **293**, 1142–1146 (2001).

[9] P. Cheung, *et al.*, *Mol. Cell.* **5**, 905–915 (2000).

[10] A. L. Clayton, S. Rose, M. J. Barratt, and L. C. Mahadevan, *EMBO. J.* **19**, 3714–3726 (2000).

[11] M. Melcher, *et al.*, *Mol. Cell. Biol.* **20**, 3728–3741 (2000).

[12] T. Hebbes, A. Thorne, and C. Crane-Robinson, *EMBO J.* **7**, 1395–1403 (1988).

[13] B. M. Turner, *Cell* **75**, 5–8 (1993).

[14] J. E. Brownell and C. D. Allis, *Proceedings of the National Academy of Sciences of the United States of America* **92**, 6364–6368 (1995).

of human PCAF (p300/CPB-associated factor).[17] Neuwald and Landsman have identified a number of other proteins that fall into a GCN5-related N-acetyltransferase (GNAT) superfamily based on primary sequence homology, which includes a number of prokaryotic and eukaryotic proteins that acetylate a vast array of substrates.[18] In addition to the GCN5/PCAF family of histone acetyltransferases, other nuclear HAT families include the MYST family (MOZ, YBF2/SAS3, SAS2, Tip60), steroid coactivators (SRC-1, ACTR/AIB-1/pCIP/RAC3/TRAM-1, and TIF2/GRIP1), human TAF$_{II}$250, CBP/p300, TFIIIC, and ATF-2.[19,20] Many of these proteins are also transcriptional coactivators that require HAT activity for their function. *In vivo* nuclear HATs exist in large macromolecular complexes that are associated with diverse biological processes and show specificity for a variety of histone, as well as sometimes non-histone substrates.[20] This chapter discusses recent X-ray crystallographic and biochemical studies that have been carried out to understand the catalytic mechanism and substrate specificity of members of the GCN5/PCAF and MYST HAT families.

The GCN5/PCAF HAT family

The Gcn5 (general control nonderepressible-5) histone acetyltransferase from *Saccharomyces cerevisiae* is the best-characterized member of the GCN5/PCAF HAT family. Originally identified through yeast genetics as an adapter protein that facilitates transcriptional activation,[21–23] yGcn5 has been characterized as the HAT subunit of at least two large macromolecular complexes SAGA (Spt-Ada-Gcn5 acetyltransferase) and ADA.[24] The specific biological role of ADA is not clear. However, SAGA has been shown to stimulate transcription in a subset of yeast genes and requires the HAT activity of the yGcn5 subunit.[24] Both SAGA and ADA acetylate histone H3 and H2B on nucleosome targets.[24] On nucleosome templates, SAGA appears to acetylate lysines 9, 14, 18, and 23 of histone H3, while

[15] T. Georgakopoulos, N. Gounalaki, and G. Thireos, *Mol. Gen. Genet.* **246,** 723–728 (1995).

[16] R. Candau, *et al.*, *Mol. Cell. Biol.* **16,** 593–602 (1996).

[17] X. J. Yang, V. V. Ogryzko, J. Nishikawa, B. H. Howard, and Y. Nakatani, *Nature* **382,** 319–324 (1996).

[18] A. F. Neuwald and D. Landsman, *Trends Biochem. Sci.* **22,** 154–155 (1997).

[19] H. Kawasaki *et al.*, *Nature* **405,** 195–200 (2000).

[20] D. E. Sterner and S. L. Berger, *Microbiol. Mol. Biol. Rev.* **64,** 435–459 (2000).

[21] S. L. Berger, *et al.*, *Cell* **70,** 251–265 (1992).

[22] G. A. Marcus, N. Silverman, S. L. Berger, J. Horiuchi, and L. Guarente, *EMBO J.* **13,** 4807–4815 (1994).

[23] N. Silverman, J. Agapite, and L. Guarente, *Proc. Natl. Acad. Sci. USA* **91,** 11665–11668 (1994).

[24] P. A. Grant, *et al.*, *Genes Dev.* **11,** 1640–1650 (1997).

ADA acetylates only lysines 14 and 18.[25] Purified recombinant yGcn5 acetylates only free histones preferentially at lysine 14 of histone H3,[26] suggesting that the other components of the complexes contribute to its substrate specificity *in vivo*.

In humans, two proteins containing HAT domains that are homologous to yGcn5 are PCAF and hGCN5. Although these proteins have approximately 64% similarity to yGcn5 in a region that extends from the catalytic HAT domain through the C-terminus, PCAF and hGCN5 have approximately 400 additional N-terminal residues.[20] This N-terminal domain enables purified recombinant PCAF to acetylate nucleosomal substrates in addition to free histones, preferentially on lysine 14 of histone H3 and on histone H4 lysine 8 to a lesser extent.[27] *In vivo*, these human HATs are subunits of at least three macromolecular complexes, the PCAF complex,[28] TFTC (TBP-free TAF_{II}-containing complex),[29] and STAGA (SPT3-TA-F_{II}31-ADA-GCN5L acetylase).[30] These complexes bear resemblance to the yeast SAGA complex with subunits that are human homologs to yeast ADA proteins, SPT proteins, and TAF_{II}–related proteins.[28–31] Similar to SAGA, these complexes also preferentially acetylate histones H3 and H2B; this acetylation correlated with transcriptional activation.[28,29,31] Additionally, PCAF has been shown to interact with p300/CBP (hence its name) to promote transcriptional activation.[17]

The MYST HAT Family and Yeast Esa1

Members of the MYST HAT family, named after its founding members, hMOZ,[32] $yYbF^2$/ySas3,[33] ySas2,[33] and hTIP60,[34] have been identified in yeast and higher eukaryotes. All partially purified or recombinant MYST proteins that have been tested can acetylate lysines on histone H2A, H3, and H4, although these proteins generally have a preference for acetylation of histone H4.[20] *In vivo*, these HATs have been associated with a variety of biological activities. Human MOZ (monocytic leukemia zinc finger

---

[25] P. A. Grant, *et al.*, *J. Biol. Chem.* **274,** 5895–5900 (1999).

[26] M. H. Kuo, *et al.*, *Nature* **383,** 269–272 (1996).

[27] R. L. Schiltz, *et al.*, *J. Biol. Chem.* **274,** 1189–1192 (1999).

[28] V. V. Ogryzko, *et al.*, *Cell* **94,** 35–44 (1998).

[29] M. Brand, K. Yamamoto, A. Staub, and L. Tora, *J. Biol. Chem.* **274,** 18285–18289 (1999).

[30] E. Martinez, T. K. Kundu, J. Fu, and R. G. Roeder, *J. Biol. Chem.* **273,** 23781–23785 (1998).

[31] E. Martinez *et al.*, *Mol. Cell. Biol.* **21,** 6782–6795 (2001).

[32] J. Borrow, *et al.*, *Nat. Genet.* **14,** 33–41 (1996).

[33] C. Reifsnyder, J. Lowell, A. Clarke, and L. Pillus, *Nat. Genet.* **14,** 42–49 (1996).

[34] J. Kamine, B. Elangovan, T. Subramanian, D. Coleman, and G. Chinnadurai, *Virology* **216,** 357–366 (1996).

protein) was first identified as a fusion to the CBP HAT protein, which occurs in acute myeloid leukemia (AML).[32] Fusions of MOZ with either p300 or TIF2, and a fusion of MORF (MOZ-related protein) with CBP also have been found in AML patients.[35–37] Although the specific relationship between these fusions and AML is presently unclear, it is possible that the fusion proteins result in misdirected histone acetylation and subsequent aberrant gene expression related to AML. Human TIP60 (Tat-interacting protein) was identified through its interaction with (and inhibition by) the activation domain of the HIV-1 Tat protein,[34] but more recently has been found in a large complex that has been implicated in the process of DNA repair and apoptosis.[38] Other members of the MYST family have been identified more recently. HBO1 (histone acetyltransferase-bound to ORC 1) was identified through its interaction with the origin recognition complex.[39] MOF (males absent on the first) is involved in dosage compensation in *Drosophila melanogaster* and is part of the MSL complex.[40,41]

In yeast, the identified MYST family members are Sas2 (something about silencing 2),[33] Sas3,[33] and Esa1 (essential Sas2-related acetyltransferase).[42] Similar to the other family members, recombinant forms of these proteins can acetylate lysines on histones H4, H3, and H2A to varying degrees.[42] However, significant differences in substrate specificity appear *in vivo* in the context of their respective HAT complexes. Sas2 interacts with the Cac1 subunit of CAF1 (chromatin assembly factor 1) and Asf1, and is a component of the SAS-I complex.[43,44] SAS-I is recruited to newly replicated DNA by chromatin assembly complexes to acetylate K16 on histone H4.[44] Sas3 is an essential component for the histone H3 HAT activity of the NuA3 complex.[45] NuA3 can stimulate transcription and may promote transcription elongation by enabling the transcriptional machinery to overcome nucleosome-mediated blocks on DNA.[45,46] Esa1 is necessary for G2/M cell cycle progression in yeast and is the HAT subunit for the NuA4 complex.[42,47] Similar to SAGA and STAGA, NuA4 contains the

[35] M. Chaffanet, *et al.*, *Genes Chromosomes Cancer* **28**, 138–144 (2000).

[36] M. Carapeti, R. C. Aguiar, J. M. Goldman, and N. C. Cross, *Blood* **91**, 3127–3331 (1998).

[37] I. Panagopoulos, *et al.*, *Hum. Mol. Genet.* **10**, 395–404 (2001).

[38] T. Ikura, *et al.*, *Cell* **102**, 463–473 (2000).

[39] M. Iizuka and B. Stillman, *J. Biol. Chem.* **274**, 23027–23034 (1999).

[40] A. Hilfiker, D. Hilfiker-Kleiner, A. Pannuti, and J. C. Lucchesi, *EMBO J* **16**, 2054–2060 (1997).

[41] E. R. Smith, *et al.*, *Mol. Cell. Biol.* **20**, 312–318 (2000).

[42] S. Allard, *et al.*, *EMBO J.* **18**, 5108–5119 (1999).

[43] S. Osada, *et al.*, *Genes Dev.* **15**, 3155–3168 (2001).

[44] S. H. Meijsing and A. E. Ehrenhofer-Murray, *Genes Dev.* **15**, 3169–3182 (2001).

[45] S. John, *et al.*, *Genes Dev.* **14**, 1196–1208 (2000).

[46] D. J. Steger, A. Eberharter, S. John, P. A. Grant, and J. L. Workman, *Proc. Natl. Acad. Sci. USA* **95**, 12924–12929 (1998).

yeast homolog and human TRRAP protein, Tra1p.[48] However, unlike the GCN5-containing complexes, the NuA4 complex can acetylate all four conserved lysines on the histone H4 amino-terminal tail to stimulate transcription[24] and may be involved in DNA double-stranded break repair.[49] The recombinant HAT domain of Esa1 acetylates histone H4 approximately five-fold more efficiently than histone H3.[50]

While both MYST and GCN5/PCAF HAT family members clearly have related enzymatic activities and substrates, the lysine residue specificities and biological roles for these HAT families differ. In addition, the HAT domains of these two families have no sequence homology, except for a small approximate 20 amino acid region in motif A of the GNAT proteins.[18] Herein, we describe structural and biochemical studies that have been used to study the catalytic mechanism and substrate specificities of the HAT domains of these two families of HAT proteins.

Structure Determination of GCN5/PCAF and Esa1 HAT Domains

Structure of the GCN5/PCAF HAT domain

The minimal HAT domains of the GCN5/PCAF family members were identified through deletion analysis and sequence conservation among family members from different species. All nuclear HATs appear to be acetyl-coenzyme A (acetyl-CoA)-dependent, using it to transfer the acetyl moiety from acetyl-CoA ultimately to the histone substrate.[20] For the GCN5-related HATs studied in our laboratory, we utilized this property to include a purification step consisting of CoA-agarose (Sigma). This step is advantageous for enriching the sample with protein that is competent for CoA-binding. Subsequent elution of the apo-protein from the column is with NaCl. Other purification steps consist of standard chromatographic techniques such as cation-exchange chromatography (SP-sepharose from Pharmacia) and gel filtration (Superdex 75 from Pharmacia).

The crystal structures of the HAT domains from yGcn5,[51] PCAF,[52] tGCN5,[53] and the NMR structure of the tGCN5 HAT domain[54] were

[47] A. S. Clarke, J. E. L., S. J. Jacobson, and L. Pillus, *Mol. Cell. Biol.* **19**, 2515–2526 (1999).

[48] C. E. Brown, *et al.*, *Science* **292**, 2333–2337 (2001).

[49] A. W. Bird, *et al.*, *Nature* **419**, 411–415 (2002).

[50] Y. Yan, N. A. Barlev, R. H. Haley, S. L. Berger, and R. Marmorstein, *Mol. Cell.* **6**, 1195–1205 (2000).

[51] R. Trievel, *et al.*, *Proc. Natl. Acad. Sci. USA* **96**, 8931–8936 (1999).

[52] A. Clements, *et al.*, *EMBO J.* **18**, 3521–3532 (1999).

[53] J. R. Rojas *et al.*, *Nature* **401**, 93–98 (1999).

[54] Y. Lin, C. M. Fletcher, J. Zhou, C. D. Allis, and G. Wagner, *Nature* **400**, 86–89 (1999).

solved as various enzyme/substrate complexes. Each structure reveals a globular protein containing a mixture of 5 α-helices and 6 β-strands (Fig. 1A). The N- and C-terminal protein segments form the walls of an L-shaped cleft in the center of the HAT domain. The ternary tGCN5 complex with CoA and a histone H3 peptide shows that CoA is bound in the short segment of this L-shaped cleft while the interaction surface for the H3 peptide is within the longer segment[53] (Fig. 1A). The general positions of the CoA pantetheine arm and pyrophosphate group is similar in the ternary tGCN5 complex and the binary tGCN5/Acetyl-CoA and PCAF/CoA complexes, while the adenine group is highly flexible and modeled in various positions.[52,53] Motif A defines a β4-loop-α3 structure that is critical to forming the base of the cleft and provides the interaction surface for CoA (Fig. 1A). The pyrophosphate group of CoA appears to stabilize the conformation of the loop in this motif (in aqua), since another conformation in the nascent yGCN5 structure suggests that it is flexible in the absence of CoA.[51] It is interesting to note that nascent HAT enzymes only have been crystallized in the presence of molecule/protein or protein/protein contacts in the substrate-binding cleft. Although the yGcn5 and tGCN5 HAT crystal structures were solved in the absence of cofactor/substrates, the yeast HAT domain contained the C-terminus from a symmetry-related yGen5 molecule in the histone-binding cleft[51] and tGCN5 contained a HEPES buffer molecule bound to the CoA-binding region.[53] This, coupled with the fact that the PCAF HAT domain could only be crystallized in the presence of CoA or Acetyl-CoA in our laboratory suggests that the HAT domains of these proteins become more rigid in the presence of a ligand, making them more amenable to crystallization. Nonetheless, comparisons of the nascent tGCN5 structure, the binary tGCN5/Acetyl-CoA complex, and the ternary tGCN5/CoA/H3 peptide complex reveal that the N- and C-terminal segments of the HAT domain open wider as it accommodates each additional substrate.[53]

Structure of yEsa1

A clue to the location of the yEsa1 HAT domain was provided by the presence of a 20-residue sequence that is similar to the CoA-binding region (within motif A: R/Q-X-X-G-X-G/A) in the GNAT superfamily.[18] However, there was no sequence homology to GCN5/PCAF family members outside of this region, so a radioactive HAT assay (described in the next section) of various Esa1 deletion mutants was used to define a minimal HAT domain within residues 147–445.[50] Recombinant Esa1 (147–445) was clipped at both termini when expressed in bacteria.[50] This result, together with a reinvestigation of the sequence conservation within the

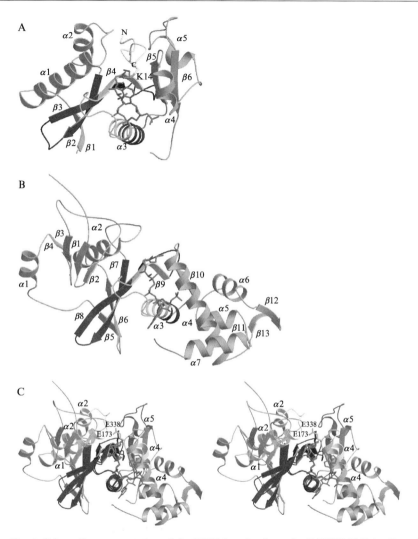

FIG. 1. Schematic representation of the HAT domains from the GCN5/PCAF family and from the Esa1 member of the MYST family. (A) Crystal structure of the tGCN5/CoA/H3 peptide complex. CoA is represented as a red stick figure. The H3 peptide is shown in yellow. The central core domain is shown in blue. Within this domain, the region with primary sequence homology with Esa1 is in aqua. The N- and C-terminal domains are in grey. (B) Crystal structure of the yEsa1/CoA complex. Color coding is similar to A. (C) tGCN5/ Esa1 superposition. The central core domain is shown in blue. The aqua region represents the Esa1 N- and C-terminal regions that are structurally similar to GCN5/PCAF. The yellow region represents the N- and C-terminal regions of GCN5/PCAF that are structurally similar to Esa1. The α5-helix loop region of tGCN5 is shown in magenta. Residues in tGCN5 are numbered according to the yGcn5 sequence. (See color insert.)

MYST family members led to the production of an Esa1 construct encoding residues 160–435 (referred to hereafter as the Esa1 HAT domain), which was stably expressed, harbored HAT activity, and was used for all further structural and biochemical studies.[50] The purification of the Esa1 HAT domain consisted of two of the three chromatographic steps used for GCN5/PCAF purification, cation-exchange chromatography and gel filtration.[50] Interestingly, the CoA agarose affinity chromatography step did not work well for this enzyme, presumably because this domain was too large to efficiently interact with the cofactor that was linked to the agarose matrix with only a one-carbon spacer.

The structure of Esa1 with CoA reveals that the HAT domain has an elongated shape, consisting of 7 α-helices and 13 β-strands[50] (Fig. 1B). Similar to the GCN5/PCAF HAT domain, the CoA binding site is in the center of the HAT domain, flanked by N- and C-terminal protein segments (Fig. 1). Motif A (β9-loop-α3) makes essentially identical interactions with CoA, as observed for the GCN5/PCAF HAT domain (Fig. 1A-B, in aqua). Interestingly, β-strands 7–8 also superimpose well with β-strands 2–3 of the PCAF HAT domain (Fig. 1C), even though there is no obvious sequence homology in this region. In addition, although the N- and C-terminal domains of Esa1 do not superimpose well with corresponding domains of GCN5/PCAF, close inspection of these regions reveal structural homology between the Esa1 α2-loop and the α1-loop of the GCN5/PCAF HAT domain, as well as between the N-terminus of α4 from both HAT families (Fig. 1C, in magenta). It is interesting to note that in the GCN5/PCAF HAT domain, this region plays an important role in histone substrate binding, suggesting that it may play a similar role in Esa1-histone interaction.

X-ray Crystallographic Studies to Probe the Catalytic Mechanism of GCN5/PCAF

Crystal structures of the binary tGCN5/Acetyl-CoA[53] and PCAF/ CoA[52] complexes revealed that the pantetheine arm of the cofactor interacts extensively with the base of the histone-binding cleft, while the adenine moiety is flexible and appears to have no major role in HAT binding. In the tGCN5/Acetyl-CoA complex, the carbonyl group of the acetyl moiety forms a hydrogen bond with a main chain amine from β4 (Fig. 2A). This presumably serves two purposes: 1) to stabilize the position of the acetyl group, marking the active site of the complex and 2) to increase the dipole moment of the carbonyl in order to facilitate a nucleophilic attack and subsequent transfer of the acetyl moiety.

The ternary tGCN5/CoA/H3 peptide crystal structure[53] (Fig. 1A) provides direct evidence that both cofactor and substrate can interact with the

GCN5 simultaneously, consistent with a sequential ordered (ternary complex) mechanism of catalysis. In addition, CoA provides part of the interaction surface for the H3-peptide, whereby acetyl-CoA binds first, followed by H3. Further inspection of all GCN5/PCAF HAT structures revealed a strictly conserved glutamate (E173 in yGcn5, Fig. 2B) in an electronegative patch at the base of the histone-binding cleft on $\beta4$, which provides an attractive surface for interaction with a positively charged histone tail (all GCN5/PCAF residues discussed in this chapter will correspond to yGcn5 numbering). Seven highly conserved hydrophobic residues that surround the glutamate carboxylate create a hydrophobic environment.[51–53] This environment potentially raises the pK_a of the glutamate so that it can function as a general base for catalysis at physiological pH. This residue may act through a water molecule that sits in between the glutamate side chain and the substrate lysine in the tGCN5/CoA/H3 peptide crystal structure (Fig. 2A-B). Studies in yeast demonstrated that an E173Q mutation reduces Gcn5-mediated transcription to similar levels observed when Gcn5 is deleted, demonstrating that this residue is necessary for Gcn5 function *in vivo*.[51] Together, the GCN5/PCAF structures point to a sequential ordered catalytic mechanism (Fig. 2E) in which E173 functions as a general base for catalysis.

Enzymatic Assays of GCN5/PCAF HAT Domains Support an
  Ordered-Sequential Reaction Mechanism and the Role of
  yGcn5 Glu-173 as a Catalytic Base

A radioactive HAT assay was employed to probe the biochemical properties of yGCN5 and to determine the role of E173.[55] This endpoint assay utilizes [³H]acetyl-CoA as the cofactor. The reaction is stopped when the mixture is blotted onto phosphocellulose (Whatman P81) cation-exchange paper. The paper is washed with a pH 9 buffer to displace all molecules except for the highly basic histone or [³H]acetyl-histone peptides. The amount of [³H]acetyl-histone peptide absorbed on the membrane is then quantified by scintillation counting. In agreement with the structural information, steady-state kinetic analysis using this assay demonstrates that yGcn5 participates in a ternary complex mechanism.[55] In order to determine the critical functional groups of residues that are necessary for catalysis, three chemical modifying reagents, DEPC, iodoacetate, or EDAC were used to derivatize yGcn5 either at histidines, cysteines, or carboxylates (respectively) prior to assaying HAT function.[55] This analysis revealed that only EDAC inhibited HAT activity, pointing to the role of a glutamate or

[55] K. G. Tanner, *et al.*, *J. Biol. Chem.* **274,** 18157–18160 (1999).

aspartate as a general base for catalysis, and also providing evidence that an acetyl-cysteine intermediate is unlikely. A pH-dependent rate analysis with either wt-yGcn5 or the yGcn5 E173Q mutant revealed that the enzymatic rate of the mutant was 320–600 fold lower than the wt-yGcn5 HAT domain at pH 7.5, while it was only 6-fold lower than the wt HAT domain at pH 9.75 when the substrate lysine is non-enzymatically deprotonated due to the high pH of the buffer.[55] These results also support the catalytic mechanism derived from the GCN5/PCAF crystal structures, which identifies E173 as the general base for catalysis.

Additional HAT assays have been utilized to further characterize the enzymatic properties and substrate specificity of yGcn5 and PCAF. A radioactive ^{14}C-based assay with labeled acetyl-CoA has been used to characterize the mechanism of PCAF. The resulting ^{14}C acetylated products are isolated using SDS Tris-Tricine polyacrylamide gels and quantified by

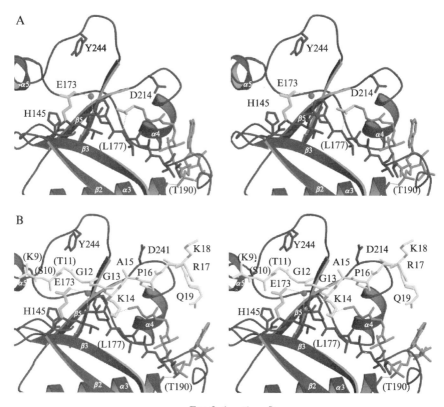

FIG. 2. (continued)

phosphorimage analysis.[56] A sensitive, fluorescence-based assay using the sulfhydryl-sensitive dye, CPM (7-diethylamino-3-(4'-malcimidylphenyl)-4-(methylcoumarin), that reacts with CoA has been developed to measure the enzyme-catalyzed production of CoA.[57] A continuous spectrophotometric enzyme-linked HAT assay also has been developed using either a-ketoglutarate or pyruvate dehydrogenase.[58] These enzymes utilize free CoA(produced from the HAT reaction) and NAD^+ to generate NADH that is quantified by absorbance at 340 nm. Further mutational studies, pH-dependent rate analysis, product inhibition assays, pre-steady state kinetics, and equilibrium dialysis have provided some information about the catalytic mechanism of GCN5/PCAF family members.[59,60] The results of these studies can be summarized as follows (Fig. 2E): 1) catalysis by the GCN5/PCAF

[56] O. D. Lau, *et al.*, *J. Biol. Chem.* **275,** 21953–21959 (2000).
[57] R. C. Trievel, F. Y. Li, and R. Marmorstein, *Anal. Biochem.* **287,** 319–328 (2000).
[58] Y. Kim, K. G. Tanner, and J. M. Denu, *Anal. Biochem.* **280,** 308–314 (2000).
[59] K. G. Tanner, M. R. Langer, and J. M. Denu, *Biochemistry* **39,** 11961–11969 (2000).
[60] K. G. Tanner, M. R. Langer, Y. Kim, and J. M. Denu, *J. Biol. Chem.* **275,** 22048–22055 (2000).

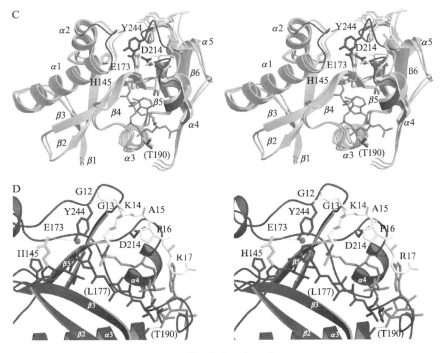

FIG. 2. (*continued*)

E

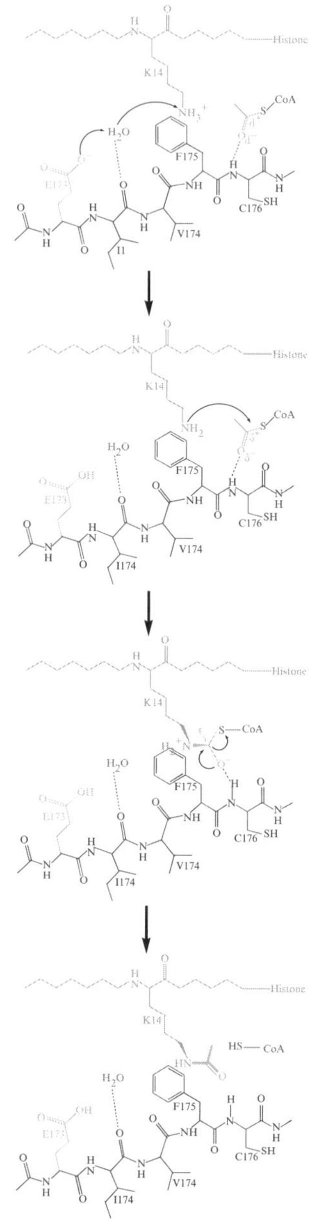

Fig. 2. Detailed view of the GCN5/PCAF active site. (A) A detailed view of the tGCN5 active site from the binary tGCN5/Acetyl-CoA complex. tGCN5 is blue. Acetyl-CoA is shown in red. A highly ordered water that may participate in the catalytic mechanism is in magenta. The tGCN5 residues are numbered according to yGCN5. HAT residues in parenthesis are tGCN5

HAT proteins proceed through a sequential ordered mechanism, 2) acetyl-CoA binds first, followed by the histone substrate, 3) a conserved catalytic glutamate residue abstracts a proton from the substrate lysine, 4) the rate-limiting step is the catalytic step, where the acetyl moiety is transferred directly from acetyl-CoA to the histone substrate, 5) the acetylated histone product is released first, followed by CoA.

In addition to these general points, specific roles are more clearly defined for other GCN5/PCAF residues based on enzymatic studies with H3-derived peptides. First, in enzymes with a Thr at position 190 (which actually is an alanine in yGcn5), acetyl-CoA has a ~10-fold lower K_m.[61] This corresponds to the fact that this residue is making a hydrogen bond to CoA in the PCAF and to tGCN5 crystal structures (Fig. 2B). Second, D214 appears to play a major role in substrate binding by enabling the movement of the β5-loop-α4 segment out of the histone-binding cleft[62] (Fig. 2C). Histidine-145 is deprotonated and is required for optimal histone H3-binding[62] (Fig. 2B). In addition, Y244 plays a role in the catalytic turnover of PCAF and may promote the exclusion of the acetylated histone product by moving the α5-loop region toward the center of the cleft[63] (Fig. 2C-D; discussed in the following sections).

Bi-substrate Analog Inhibitor Studies with the GCN5/PCAF HAT Domain

Inhibition studies of p300 and PCAF with bi-substrate analogs revealed that Lys-CoA (with the lysine zeta nitrogen attached through an acetyl linker to the sulfur of CoA) is a relatively good inhibitor of p300, but not

[61] M. R. Langer, C. J. Fry, C. L. Peterson, and J. M. Denu, *J. Biol. Chem.* **277**, 27337–27344 (2002).
[62] M. R. Langer, K. G. Tanner, and J. M. Denu, *J. Biol. Chem.* **276**, 31321–31331 (2001).
[63] A. N. Poux, M. Cebrat, C. M. Kim, P. A. Cole, and R. Marmorstein, *Proc. Natl. Acad. Sci. USA* **99**, 14065–14070 (2002).

residues that are not conserved in tGCN5: (L177) corresponds to a Cys in yGen5 and all other GCN5/PCAF family members. (T190) is an Ala in yGen5, but is a Thr in other GCN5/PCAF family members. Dotted lines represent hydrogen bonds. (B) A detailed view of the tGCN5 active site from the ternary tGCN5/CoA/H3 peptide complex. Color coding is similar to A. The histone H3 substrate peptide is in yellow. Histone residues that are labeled with parentheses were modeled without their side chains. (C) Residues that play a role in substrate binding and catalysis. Dynamic regions of the protein are color coded as follows: the binary tGCN5/Acetyl-CoA complex is yellow, the tGCN5/bi-substrate inhibitor complex is in blue, and the ternary tGCN5/CoA/H3 peptide complex is in aqua. (D) A detailed view of the tGCN5 active site from the tGCN5/bi-substrate analog inhibitor complex. The peptide portion of the inhibitor is in yellow. The CoA portion of the inhibitor is in red. The linker region is in brown. The acetyl moiety is in orange. (E) Proposed catalytic mechanism for GCN5/PCAF. (See color insert.)

PCAF.[64] However, if lysine-14 within a 20-residue histone H3-derived peptide is similarly linked to CoA, the resulting inhibitor (referred to as H3-CoA-20) is potent for PCAF but not for p300.[64] The potency of this inhibitor had ~10-fold enhanced IC_{50} when compared to the potency of H3-CoA-7 for PCAF. The IC_{50} for PCAF was enhanced further when a methyl group (isopropionyl bridge) was introduced in the linker region between CoA and the histone peptide[63] (H3-(Me)CoA-20; Fig. 2D). The X-ray crystal structure of tGCN5 with H3-(Me)CoA-20 reveals that the methyl group orients and enhances the interactions between the target lysine and the HAT in a hydrophobic pocket. Interestingly, only five of the twenty H3 peptide residues were ordered in the structure[63] (Fig. 2C). These peptide residues are sitting on top of the α5-loop of the HAT apparently because this loop is shifted inward toward the base of the cleft (Fig. 2D). A strictly conserved tyrosine residue from this loop interacts extensively with several hydrophobic residues in the core of the protein, as well as with the aliphatic portion of the lysine-14 side chain of the inhibitor. Since this inhibitor requires the twenty amino acid portion of H3 for its full potency, it is proposed that inhibitor residues terminal to the five that are ordered in the crystal structure are important for initial binding. However, this interaction is short-lived and is consequently not trapped in the crystal structure of the complex. The complex that is trapped in the structure is proposed to represent a late catalytic intermediate in which acetyl transfer promotes the displacement of the histone peptide from its initial binding site. Based on this structure, it also appears that the acetyl-transfer step requires the movement of the α5-loop into the cleft where it may help exclude the acetylated H3 product. Due to the extensive interactions observed between Y244 and the HAT domain in this structure, it is proposed that this residue plays a particularly important role in nucleating the movement of the α5-loop (Fig. 2C-D). In correlation, a PCAF Y244A mutation reduces the k_{cat} of the enzyme by ~15-fold and increases the IC_{50} of the inhibitor by nearly 5-fold.[63] Since the inhibitor was bound to tGCN5 in an unexpected way (i.e., it did not look like the ternary tGCN5/CoA/H3 peptide complex, Fig. 2), this study emphasizes the powerful role that X-ray crystallography can play in rational inhibitor design.

Substrate Specificity of GCN5/PCAF

In addition to histones, several non-histone targets have been identified for the GCN5/PCAF family members.[20] These non-histone proteins are generally classified as non-histone chromatin-associated proteins or

[64] O. D. Lau, et al., Mol. Cell. 5, 589–595 (2000).

transcriptional activators. Some targets, such as p53 and E2F-1, are critically involved in cell-cycle regulation. PCAF has been shown to promote the sequence-specific DNA binding of p53 and E2F through acetylation of these transcription factors.[20] In addition, recent studies have shown that serine-10 phosphorylation on histone H3 enhances yGen5-mediated histone acetylation to activate transcription at some promoters.[7,9,20] In order to understand how the PCAF HAT domain can accommodate these substrates, the fluorescent histone acetyltransferase assay was employed to compare the substrate specificities of PCAF for histone H3, histone H4, and p53-derived peptides.[57] This study demonstrated that the substrate specificity of PCAF for a 19-residue histone H3-derived peptide is 100-fold better than the 11-residue H3-derived peptide shown previously. When this peptide was extended to 26 residues, the substrate specificity was not significantly enhanced ($\sim$2-fold) from the 19-residue peptide. Additionally, histone H4 and p53-derived peptides were $\sim$100 and $\sim$1000 poorer substrates for PCAF, respectively, despite the demonstration of PCAF-mediated acetylation of these proteins in cells.[20] In a separate study, kinetic analysis of the yGcn5 HAT domain shows that the K_m of a phosphorylated-histone H3 peptide is enhanced 6-10 fold over the unphosphorylated version.[9] These studies point to three basic conclusions: 1) histone H3 residues 5 through 23 likely provide the core recognition sequence for the GCN5/PCAF HAT domain, 2) serine-10 phosphorylation enhances the substrate specificity of the GCN5/PCAF HAT domain for the H3 tail, and 3) regions outside of the HAT domain, either on full-length PCAF or in PCAF-associated proteins, likely contribute to the substrate specificity of PCAF for other histone and non-histone substrates (such as histone H4 and p53) *in vivo*, since the HAT domain has much weaker activity with these targets. The structural analysis of other HAT/substrate complexes will be required to understand the mechanism of substrate specificity/selectivity.

X-ray Crystallographic Studies to Probe the Catalytic Mechanism of yEsa1

Initial inspection of the Esa1/CoA structure and a structural comparison with the GCN5/PCAF structures revealed a strictly conserved glutamate residue (E338) of Esa1 that could potentially serve as a general base for catalysis.[50] Despite the close proximity of these corresponding glutamate residues of GCN/PCAF and Esa1 upon superposition, the residues fall in different positions in the primary sequence. GCN5/PCAF E173 is located on β4 and points up from the base of the cleft, whereas Esa1 E338 is on a C-terminal loop and points down into the core of the HAT (Fig. 1C). An Esa1 E338Q mutant transfected into yeast had a severe dominant-negative phenotype, exhibited G2/M arrest, and the purified recombinant mutant showed only background levels of HAT activity.[50] This suggested that Esa1 and GCN5/PCAF proteins employ a similar mechanism for histone lysine deprotonation prior to acetylation. However, further inspection of the Esa1/CoA structure revealed a cysteine residue (C304 in yeast Esa1) that was strictly conserved among the MYST HATs and in close proximity of the sulfhydryl group of CoA (Fig. 3A). Mutation of this cysteine to a serine or alanine reduces the HAT activity of the enzyme to the same background level as the E338Q mutant, demonstrating that this cysteine residue is required for HAT function and that it may play a direct role in catalysis as an acetylated intermediate.[65]

In order to explore the catalytic role of cysteine 304 of Esa1, several crystal structures of Esa1 and various mutants were solved with either CoA or acetyl-CoA.[65] The structure of the wt-Esa1 HAT domain, determined from crystals grown with acetyl-CoA, revealed that Cys304 was acetylated with CoA remaining bound to the HAT domain (Fig. 3B). This was the first direct evidence for a self-acetylated HAT intermediate. Surprisingly, the structure of the E338Q mutant crystallized with acetyl-CoA and revealed acetyl-CoA intact (Fig. 3A), suggesting that E338 not only functions as a catalytic base to abstract a proton from the histone lysine (as in GCN5/PCAF), but also to deprotonate C304 so that it can act as a strong nucleophile in the acetyl transfer from acetyl-CoA to C304.

Biochemical Assays Reveal a Ping-Pong Catalytic Mechanism for Esa1

The tritium-based radioactive HAT assay was modified to probe self-acetylation of Esa1.[65] For this experiment, Esa1 was incubated with [³H]acetyl-CoA and no histone substrate. The reaction was stopped with the addition of excess CoA (to dissociate any unreacted [³H]acetyl-CoA from the HAT). The mixture was blotted onto the p81 phospho-cellulose paper and washed under conditions that retained the protein on the paper. From these studies, wt-Esa1, but no mutants or the PCAF control, exhibited self-acetylation. Additionally, mass spectrometry experiments demonstrated that an isolated acetylated Esa1 catalytic intermediate could produce an acetylated histone product. Other experiments were carried out to show that the acetyl group could be quantitatively transferred from acetyl-CoA to Esa1 and subsequently to a histone substrate. In support of these experiments, bi-substrate kinetic analysis using the fluorescence-based HAT assay revealed that catalysis by Esa1 involves a ping-pong

[65] Y. Yan, S. Harper, D. W. Speicher, and R. Marmorstein, *Nat. Struct. Biol.* **9**, 862–869 (2002).

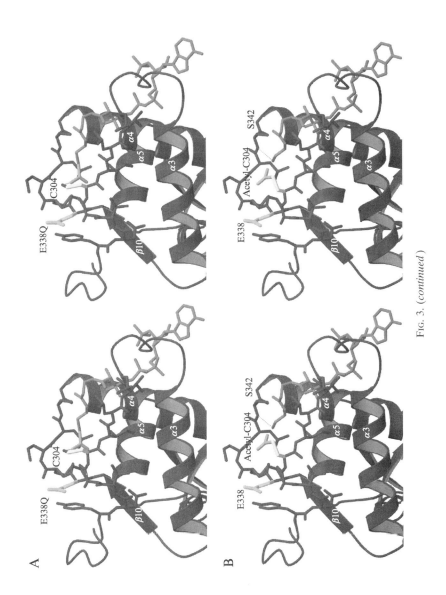

FIG. 3. (continued)

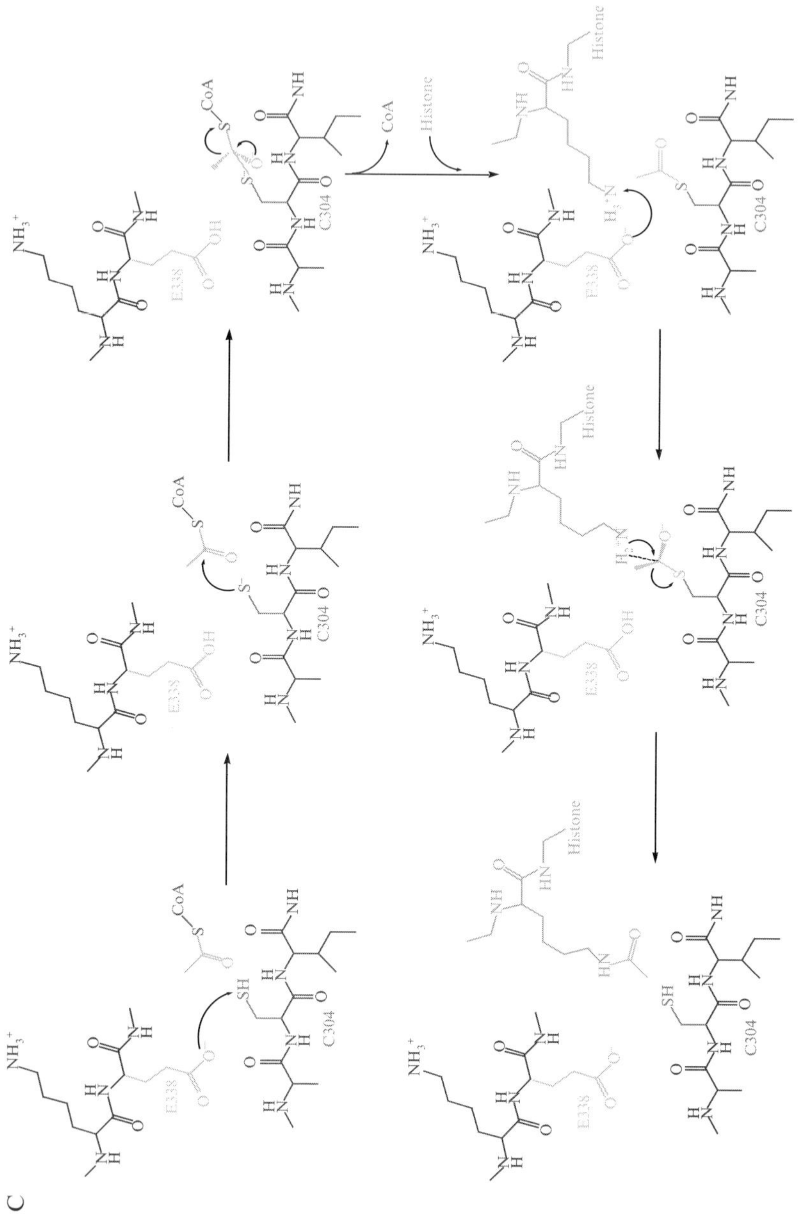

Fig. 3. (continued)

reaction mechanism[65] (Fig. 3C). Taken together, the crystallographic and biochemical data lead to the following proposed catalytic mechanism for yEsa1 and presumably other members of the MYST HAT family: 1) acetyl-CoA binds to Esa1, 2) the acetyl-moiety is transferred directly from acetyl-CoA to C304, 3) the CoA product is released, 4) the histone substrate binds to the acetylated Esa1 intermediate, and 5) the acetyl moiety is then transferred directly from acetyl-C304 to the substrate lysine. These studies also point to the role of E338 as a catalytic base that deprotonates C304. Additionally, E338 may donate the proton to CoA to facilitate the displacement of CoA, while regenerating the basic character of E338 for subsequent lysine deprotonation prior to lysine acetylation (as described for the GCN5/PCAF family).

Conclusions

Despite the fact that the GCN5/PCAF and Esa1 HAT families have similar core acetyl-CoA binding regions, it is clear that these enzymes have different modes of catalysis. GCN5/PCAF participates in an ordered sequential (ternary complex) catalytic mechanism (Fig. 2E), whereas Esa1 uses a ping-pong mechanism (Fig. 3C). It also is interesting to note that, when compared to the GCN5/PCAF structures, Esa1 has a less pronounced putative substrate binding cleft that may not be able to accommodate the co-factor and histone substrate simultaneously (Fig. 1). The fact that different MYST HAT complexes exhibit diverse substrate specificities also suggests that other proteins in MYST complexes contribute greatly to the substrate specificity of this class of enzymes. Conversely, the GCN5/PCAF HAT domain has a more pronounced histone-binding cleft defined partly by the α5-loop region (Fig. 1A). It is important to note that this region has no structural homology with Esa1 (Fig. 1B) or with other GNAT crystal structures.[66] In correlation, the bi-substrate analog inhibitor (H3-(Me)CoA-20) requires the α5-loop region of GCN5/PCAF for full potency. By contrast, Esa1 is not inhibited by this compound and cannot form a ternary complex. Additionally, the GCN5/PCAF cleft appears to influence its substrate specificity *in vivo*, since the GCN5 HAT domain preferentially acetylates histone H3 K14, and a K14A mutation in yeast abolishes GCN5-dependent transcription.

Fig. 3. Detailed view of the Esa1 active site. (A) A detailed view of the E338Q yEsa1 mutant with acetyl-CoA. Esa1 is in blue. Acetyl-CoA is in red. E338Q is in aqua. C304 is shown in yellow. This residue is oxidized with the S-O bond in brown. Dotted lines represent hydrogen bonds. (B) A detailed view of the wt-yEsa1-acetylated intermediate. Color coding is the same as A. E338 is in aqua. (C) Proposed catalytic mechanism for yEsa1. (See color insert.)

It should be useful to apply the structural and biochemical studies described here to other HAT families. Additionally, it is likely that some of the themes uncovered by the GCN5/PCAF and Esa1 studies may apply to other nuclear HAT families. The structurally conserved CoA-binding domain for both of these HAT families suggests that other nuclear HAT families may interact with the cofactor comparably. Similar to Esa1, steady-state kinetic analysis suggests that p300 participates in a ping-pong mechanism, which raises the possibility that the Esa1 structure has some similarity with the p300/CBP HAT domain. However, the complete lack of sequence homology between Esa1 and p300/CBP suggests that there may be important differences. Further mechanistic insights into the chemistry of this fascinating HAT superfamily will require additional structural and biochemical studies on members of other nuclear HAT families, as well as biologically relevant multi-protein *in vivo* HAT complexes.

---

[66] F. Dyda, D. C. Klein, and A. B. Hickman, *Annu. Rev. Biophys. Biomol. Struct.* **29**, 81–103 (2000).

[42] Assay of the Fate of the Nucleosome During Transcription by RNA Polymerase II

By W. Walter, M. L. Kireeva, V. Tchernajenko, M. Kashlev, and
V. M. Studitsky

Eukaryotic genes transcribed by RNA polymerase II (Pol II) retain nucleosomal structure[1]. Thus, Pol II encounters nucleosomes every ~200 bp during transcript elongation *in vivo*. Until now, only two types of Pol II-based experimental systems *in vitro* were available for analyzing the outcome of a Pol II-nucleosome encounter.[2–4] One type includes systems supporting promoter-dependent transcription initiation, either in crude extracts[3] or with highly purified proteins.[4] The main disadvantage of this approach is that only ~1 % of the templates are transcribed.[5] This low efficiency of template utilization makes analysis of the fate of nucleosomes after transcription nearly impossible. In contrast, DNA templates containing a single-stranded, 3′-extending oligo dC "tail" support efficient

[1] G. Orphanides and D. Reinberg, *Nature* **407**, 471 (2000).
[2] R. L. Dedrick and M. J. Chamberlin, *Biochemistry* **24**, 2245 (1985).
[3] M. G. Izban and D. S. Luse, *Genes Dev.* **5**, 683 (1991).
[4] G. Orphanides, G. LeRoy, C. H. Chang, D. S. Luse, and D. Reinberg, *Cell* **92**, 105 (1998).

end-initiation by Pol II *in vitro*.[2] However, in this system, determination of the fate of the nucleosome during transcription is complicated by the formation of extremely stable DNA-Pol II complexes at the end of DNA.[6] Moreover, it is likely that end-initiated and promoter-initiated RNA polymerases differ in the way they progress through the nucleosome.[6]

A recent novel approach for analysis of Pol II elongation complexes (ECs) employs assembly of "authentic" ECs using histidine-tagged yeast Pol II and synthetic RNA and DNA oligonucleotides.[7,8] The structure and functional properties of the assembled and promoter-initiated ECs are very similar.[8,9] In this system, the fate of nucleosomes during transcription can be analyzed after ligation of the ECs to positioned mononucleosomes that are assembled separately.[10] Nucleosomes form an absolute block to Pol II, but this barrier can be reduced by increasing the ionic strength of the reaction, thus allowing more templates to be transcribed. Transcription through nucleosomes by Pol II induces the loss of an H2A/H2B dimer without changing the position of the histones on the DNA.[10]

Use of this experimental system for analysis of the fate of the nucleosome during transcription is described below. This "minimal" model system can be easily adopted for analysis of polynucleosomal templates and for analysis of the role of different elongation factors during transcription through chromatin. Similar approaches can be applied for analysis of a variety of biological processes (such as chromatin remodeling, DNA replication, recombination, and repair) that may involve changes in nucleosome positioning or the histone content of nucleosomes.

2.0. Materials and Methods

Template Preparation

Purification of core histones and donor chromatin
The detailed protocols for purification of core histones on hydroxyapatite[11,12] and purification of donor chromatin[13,14] are described elsewhere.

---

[5] J. A. Knezetic, G. A. Jacob, and D. S. Luse, *Mol. Cell. Biol.* **8**, 3114 (1988).

[6] Y. V. Liu, D. J. Clark, V. Tchernajenko, M. E. Dahmus, and V. M. Studitsky, *Biopolymers*, **68**, 528 (2003).

[7] I. Sidorenkov, N. Komissarova, and M. Kashlev, *Mol. Cell* **2**, 55 (1998).

[8] M. L. Kireeva, N. Komissarova, D. S. Waugh, and M. Kashlev, *J. Biol. Chem.* **275**, 6530 (2000).

[9] M. L. Kireeva, N. Komissarova, and M. Kashlev, *J. Mol. Biol.* **299**, 325 (2000).

[10] M. L. Kireeva, W. Walter, V. Tchernajenko, V. Bondarenko, M. Kashlev, and V. M. Studitsky, *Mol. Cell* **9**, 541 (2002).

Template Design and Purification of DNA

Careful consideration should go into the design of the template to be used for analyzing transcription through the nucleosome by Pol II. First, the ECs and nucleosomes are assembled separately and then ligated together. Thus for efficient ligation, a long, asymmetric sticky end, such as that generated by TspRI cleavage, is preferable. Second, a defined template with a positioned nucleosome(s) is desirable for mechanistic studies. For this, strong and well-characterized nucleosome positioning sequences, such as that of the *Xenopus* 5S RNA gene, can be employed.[10,15] Finally, to map nucleosome positioning before and after the reaction, there should be several restriction enzyme sites along the entire length of the template for restriction endonuclease protection assays.

The pVT1 Template[10]

The template DNA is PCR amplified from pVT1 plasmid by using the primers (Invitrogen Corporation, Carlsbad, CA) pVT1-431-454-up (5' GAC ACT ATA GAA TTA ATG GGG ATC 3') and pVT1-737-716-low (5' CCT TCC AAG TAC TAA CCA GGC C 3'). Some of the lower primer is radiolabeled with $\gamma[^{32}P]ATP$ (7000 Ci/mmol ICN Biomedicals, Inc., Irvine, CA) using T4 polynucleotide kinase (New England Biolabs, Beverly, MA) prior to the PCR. The resulting 306 bp product is digested with TspRI, and the sample is loaded onto an 8% (19:1) polyacrylamide gel containing 1X TAE and 4 M urea (to prevent re-association of the 9 nt, GC-rich sticky ends of the TspRI-digested fragments). The 204 bp TspRI-StuI fragment is cut out of the gel, the gel slice is crushed, and the DNA is extracted overnight at 4°C in 3 to 5 volumes of TE buffer. The ethanol is then precipitated and resuspended in dH$_2$O. The template DNA is further "cleaned up" using QIA quick gel extraction kit columns (Qiagen, Chatsworth, CA).

Reconstitution, characterization, and purification of mononucleosomes and hexasomes

It is desirable to find efficient reconstitution conditions where the amount of free DNA is minimal (less than 10 to 15%) so the nucleosomes do not have to be further purified. The methods described below are for

---

[11] R. H. Simon and G. Felsenfeld, *Nucleic Acids Res.* **6,** 689 (1979).

[12] C. von Holt, W. F. Brandt, H. J. Greyling, G. G. Lindsey, J. D. Retief J. D. Rodrigues, S. Schwager, and B. T. Sewell, *Methods Enzymol.* **170,** 431 (1989).

[13] J. J. Hayes and K. M. Lee, *Methods* **12,** 2 (1997).

[14] R. T. Utley, T. A. Owen-Hughes, L. J. Juan, J. Cote, C. C. Adams, and J. L. Workman, *Methods Enzymol.* **274,** 276 (1996).

[15] J. J. Hayes, D. J. Clark, and A. P. Wolffe, *Proc. Natl. Acad. Sci. USA* **88,** 6829 (1991).

templates that are about 150 to 250 bp in size and, thus, allow for only one nucleosome per molecule of DNA.

Reconstitution of mononucleosomes via octamer exchange from donor chromatin

One method of making nucleosomal templates is to use donor chromatin as a source of histone octamers for exchange[14] onto a template of interest. This method was chosen to reconstitute nucleosomes for transcription because it is generally very efficient (the amount of free DNA is less than 15%), and nucleosomes do not have to be purified. Nucleosomes prepared this way contain excess donor chromatin, but it can be removed by washing after the template is ligated to immobilized ECs.

DNA (1 to 5 μg) is mixed with long $-$H1 donor chromatin at a ratio of 1:60 (wt:wt), respectively, (sample volume is determined by donor chromatin concentration) in buffer containing 1 M NaCl and 0.1% Igepal CA-630 (Sigma, St. Louis, MO). The sample is dialyzed overnight at 4°C against a gradient ($\sim$1 L) starting at 1 M NaCl and ending with no NaCl in buffer containing 10 mM Tris-HCl, pH, 7.5, 0.2 mM EDTA, and 0.1% NP-40.

Reconstitution of mononucleosomes and subnucleosomal particles from purified histones

Nucleosomes can also be reconstituted from purified histones. The protocol described below is a slightly modified version of the method used by the Bradbury laboratory.[16,17] This procedure can also be employed to create subnucleosomal particles by varying the amount of H2A/H2B used.

To make nucleosomes, 5 μg of DNA is mixed with 1.23 μg of H3/H4 and 2.70 μg of H2A/H2B (ratio of H3/H4:H2A/H2B = 0.455) at a volume of 100 μl in buffer containing 2 M NaCl, 0.2 mM EDTA, 10 mM Tris-HCl, pH 7.4, and 0.1% Igepal CA-630 (Sigma, St. Louis, MO). To make hexasome, the amount of H3/H4 is kept constant while the amount of H2A/H2B is reduced to 1.35 and 0.67 μg (with ratios of H3/H4:H2A/H2B = 0.91 and 1.82, respectively). Tetrasome is reconstituted by using the same amount of H3/H4 but eliminating the H2A/H2B. Dialysis is performed at 4°C against the same buffer but with decreasing NaCl concentration (2 M, 1.5 M, 1 M, 0.75 M, 0.5 M, and 10 mM NaCl) for 1 hour at each step.

Characterization and purification of nucleosomes and chromatin subparticles

Nucleosome and hexasome preparations must be analyzed for the amount of free DNA, hexasome, and nucleosome present in each preparation, as well as the location of histones on the template. The position of a

[16] G. Meersseman, S. Pennings, and E. M. Bradbury, *EMBO J.* **11,** 2951 (1992).
[17] S. Pennings, G. Meersseman, and E. M. Bradbury, *Proc. Natl. Acad. Sci. USA* **91,** 10275 (1994).

nucleosome can be determined with about 10 bp resolution based on its mobility during native PAGE.[17] However, this method cannot discriminate between two symmetrically positioned nucleosomes or differently positioned nucleosomes formed on DNA ∼200 bp or less.[17] The exact position can be further narrowed down by restriction enzyme digestion or micrococcal nuclease mapping (described elsewhere[18]).

For analysis, the reconstitutes (10 ng aliquots) are supplemented with buffer providing 20 mM Na-HEPES, pH 7.8, 5 mM MgCl$_2$, 2 mM spermidine (Sigma, St. Louis, MO), and 0.5 mg/ml BSA. One sample is not digested, while appropriate restriction enzymes (10 U) are added to the others. Digestion is performed at room temperature (RT) for 0.5 to 1 hour. Buffer is added, providing a final concentration of 10 mM EDTA, 10% sucrose, and 250 μg/ml sheared herring testes DNA (Intergen, Purchase, NY). The templates are then resolved by native gel electrophoresis (4.5% acrylamide (39:1), 5% glycerol, 20 mM Na-HEPES (pH 8), 0.1 mM EDTA) at 100 V for 2.5 to 4 hrs (depending on the size of the DNA fragment and the degree of resolution desired) as described.[18] Quantitation is performed using a Cyclone Storage Phosphor System (Packard, Meriden, CT). If gel purification is required,[18] the samples are loaded without carrier DNA. The appropriate band is cut out of the gel, the gel is crushed, and the DNA, nucleosome, or hexasome are extracted overnight at 4 °C in 1 to 2 volumes of 10 mM Na-HEPES, pH 8.0, 0.1 mM EDTA, and 0.5 mg/ml BSA. The supernatant is collected, and the concentration of the sample is determined by the specific activity of the DNA.

Analysis of the pVT1 Reconstitutes[10]

Nucleosomes formed via octamer exchange

Nucleosomes are reconstituted very efficiently on the 204 bp (TspRI-StuI) pVT1 template by octamer exchange (Fig. 1, lane 1). There is one nucleosomal band, but the templates are too short to be resolved based on nucleosome position. Digestion with EcoRI (lane 2), EcoRV (lane 3), and EcoRI+EcoRV (not shown) reveals that about half of the nucleosomes are resistant to EcoRI and sensitive to EcoRV (N1), while half are resistant to EcoRV and sensitive to EcoRI (N2). The pVT1 DNA contains the strong 5S nucleosomal positioning sequence, so one nucleosome position (N2) was expected. However, the TspRI end of the DNA is an equally good site for nucleosome formation (N1).

Nucleosomes and hexasomes reconstituted from purified histones

Nucleosomes are not reconstituted as efficiently from purified histones as they are with exchange (compare Fig. 2A, lane 1 and Fig. 1, lane 1).

[18] V. M. Studitsky, *Methods Mol. Biol.* **119,** 17 (1999).

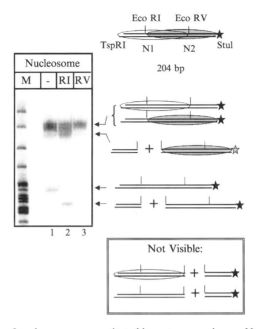

FIG. 1. Analysis of nucleosomes reconstituted by octamer exchange. Nucleosomes formed on 204 bp DNA (labeled at the StuI end, indicated by a star) were loaded onto a native gel before and after digestion with EcoRI and EcoRV. Positions of the EcoRI and EcoRV sites on the template are labeled. Mobilities of the nucleosomes (N1 and N2), free DNA, and the products of their digestion are indicated at the right. Molecular weight markers are an MspI digest of pBR322 DNA.

Restriction enzyme digestion with EcoRI and EcoRV reveal that nucleosomes formed by this method are indistinguishable from those formed by exchange. As the amount of H2A/H2B is decreased, less nucleosome is formed, and more hexasome is present in the sample (Fig. 2A).

Due to the large amount of free DNA in the samples, the nucleosomes and hexasomes have to be gel purified (Fig. 2B). The yield for this procedure is extremely low. Therefore, this was not the protocol of choice for preparing nucleosomes used for transcription.

Transcription of Defined Nucleosomal Templates

Transcription Buffers

TB0 contains 20 mM Tris-HCl, pH, 7.9, 5 mM MgCl$_2$, and 1 to 2 mM β-mercaptoethanol (2-ME). TB40, TB150, TB300, TB700, and TB1000 contain 40 mM, 150 mM, 300 mM, 700 mM, and 1 M KCl, respectively. Acetylated BSA was purchased from Sigma (St. Louis, MO).

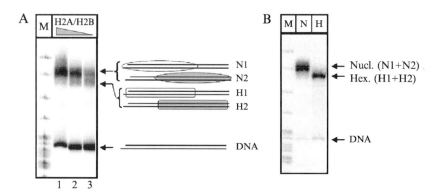

FIG. 2. Analysis of nucleosomes and hexasomes reconstituted from purified histones. (A) Nucleosomes and hexasomes formed on 204 bp pVT1 DNA. Mobilities of the nucleosomes (N1 and N2) and hexasomes (H1 and H2), as well as the free DNA are indicated at the right. (B) Gel-purified nucleosomes (N) and hexasomes (H).

EC assembly and ligation to DNA or nucleosomal templates

Yeast Pol II ECs are assembled from purified components as described in Komissarova et al.[19] using the following oligonucleotides (*Oligos, Etc. Inc.*, Wilsonville, OR): TDS50 (5′ GGTGTCGCTTGGGTTGGCTTTTC GGGCTGTCCC- TCTCGATGGCTGTAAGT 3′), RNA9 (5′ AUCGA-GAGG 3′), and NDS59 (5′ ACTTACAGCCATCGAGAGGGACACG GCG AAAAGCCAACCCAAGCGACACCGGCACTGGG 3′). The TDS50 oligo is labeled with $\gamma[^{32}P]ATP$ (7000Ci/mmol, ICN Biomedicals, Inc., Irvine, CA) using T4 polynucleotide kinase (New England Biolabs, Beverly, MA) as described.[19] The Ni^{2+}/NTA agarose (Qiagen, Chatsworth, CA) is pre-treated with 0.2 mg/ml acetylated BSA for 10 min and washed 2X with 1 ml of TB40 prior to immobilization of Pol II.

ECs assembled with ~2 to 3 pmol of Pol II (immobilized on 50 μl of 50% resin suspension, enough for ~10 to 15 reactions) are incubated in the presence of 100 to 200 ng of template, 100 μM ATP, 1% PEG-8000, and 50 units of T4 DNA ligase (New England Biolabs, Beverly, MA) in a volume of ~100 μl at 12°C for 1 to 2 hours. The ECs are washed with TB40, incubated for 10 min with TB700, and washed twice with TB40. The ligation efficiency achieved by using the 9 nt 3′ overhang generated by TspRI cleavage (5′ CCCAGTGCC 3′) is at least 50%.

[19] N. Komissarova, M. L. Kireeva, I. Sidorenkov, and M. Kashlev, *Methods Enzymol.* in press. (2002).

Analysis of ECs and nucleosome mapping after ligation

ECs (assembled with ~1 to 1.5 pmol of Pol II, ligated to nucleosomes and washed) are eluted in a volume of ~50 μl in TB40 containing 0.5 mg/ml acetylated BSA and 100 mM imidazole (Sigma, St. Louis, MO). The samples are incubated at RT for 10 minutes, diluted 2-fold, mixed, and microcentrifuged for 10 sec. 50 μl of the supernatant is withdrawn and diluted 5-fold with buffer providing 20 mM Na-HEPES, pH 7.6, 5 mM MgCl$_2$, 2 mM spermidine (Sigma, St. Louis, MO), 0.5 mg/ml acetylated BSA, and 1 mM β-mercaptoethanol. The sample is aliquoted (10 μl each), restriction enzyme digestions (with 10 U of enzymes) are performed for 30 to 60 minutes at RT, and the products are analyzed in a native gel as described previously.

Analysis of ECs after ligation to the pVT1 nucleosomal template and mapping of the ligated nucleosomes

Nucleosome positioning on the pVT1 template was analyzed after ligation to labeled EC9 (EC9 denotes an elongation complex with RNA that is 9 nt long) and elution with imidazole. Note the faster mobility of the unligated EC9 as compared to the EC9 ligated to the nucleosomal template (EC9N, Fig. 3B, lane 1). Some of the ECs (5 to 10%) spontaneously dissociate during elution and electrophoresis. The free nucleosomes (N1 and N2) are resolved based on their positions along the DNA. As expected, digestion with EcoRI and EcoRV (lanes 2 and 3) reveals that that about half of the templates (Pol II-bound or not) are sensitive to EcoRI, and half are sensitive to EcoRV. A detailed map of the templates used for transcription is shown in Fig. 3A.

Transcript elongation

Immobilized Pol II can be walked to any point along the DNA by simply adding a subset of NTPs to the reaction, washing the resin, adding a different subset of NTPs, and so on. The concentration of NTPs added and the length of incubation vary based on the template, the location of the EC on the DNA, and the NTPs used.[20] The conditions described below are specific for the pVT1 template.

The nucleosome is an extremely strong barrier for Pol II under physiological conditions. This makes analysis of the transcribed templates difficult as most of the polymerase is stuck in the nucleosome, and the templates are not released. However, at 300 mM KCl, ~30% of the polymerase can

---

[20] M. Kashlev, E. Nudler, K. Severinov, S. Borukhov, N. Komissarova, and A. Goldfarb, *Methods Enzymol.* **274,** 326 (1996).

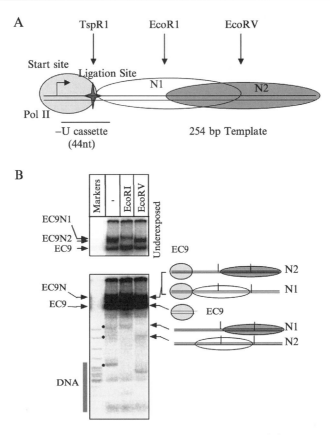

FIG. 3. Analysis of Pol II ECs and nucleosomes after ligation. (A) A map of the pVT1 template. ECs are labeled at the TspRI site (star) and ligated to nucleosomes. The Pol II start site is indicated with an arrow. The first 45 nt to be transcribed by Pol II lack UTP. Positions of nucleosomes (N1 and N2) and restriction enzyme sites used for mapping are indicated. (B) Imidazole-eluted ECs were analyzed in a native gel before and after EcoRI or EcoRV digestion. The upper panel is a shorter exposure of the gel. Mobilities of unligated EC (EC9) and EC ligated to the nucleosomes (EC9N1 and EC9N2) are indicated. The lower panel is a longer exposure of the gel and reveals the free nucleosomes (N1 and N2) and free DNA fragments (marked by dots).

overcome the barrier and finish transcription through the nucleosome.[10] Therefore, to analyze the fate of the nucleosome after transcription, the reaction is carried out at 300 mM KCl so that enough templates are transcribed and released for analysis.

It is important to analyze only those templates that are fully transcribed and released by the polymerase. However, it is common for ECs to become arrested during transcription of the first 15 to 50 nt. Many of these ECs dissociate, and this results in contamination of the supernatant with non-transcribed templates. For this reason, Pol II is walked to the +45 position on the pVT1 template, and the ECs are washed extensively to remove any non-transcribed templates.

Walking Pol II on the pVT1 template to form EC45

The first 45 nt of the pVT1 template consist of a −UTP cassette (see Fig. 3A). Thus, in the absence of UTP, Pol II can be walked to the +45 position (EC45). The volume of the EC9N (assembled with ∼2 to 3 pmol of RNAP) is adjusted to 100 μl (enough for 10 to 10 μl reactions) in TB40 containing 0.2 mg/ml acetylated BSA. The ECs are incubated in the presence of 200 μM ATP, CTP, and GTP for 10 minutes at RT. The sample is mixed, microcentrifuged for about 10 seconds and 50 μl of the supernatant (S1) is withdrawn and divided into 5 μl aliquots. The S1 supernatant contains free templates as a result of early EC dissociation and is representative of non-transcribed templates. The pellet is washed 2 times with 1 ml of TB300, incubated in 1 ml of TB300 for 15 minutes and washed 2 more times with 1 ml of TB300. This procedure is repeated to remove any remaining non-transcribed templates.

Immobilized transcription of the pVT1 template

Many controls are done to make sure that the templates analyzed are actually released as a result of transcription and not present due to further EC dissociation. The volume of the washed pellet (EC45) is adjusted to 100 μl in TB300 containing 0.5 mg/ml acetylated BSA. The sample is aliquoted (10 μl per experimental point) and incubated for 5 minutes at RT with (I) no NTPs (S2), (II) 100 μM ATP, CTP, and GTP (S3), or (III) 100 μM NTPs (S4). The samples are mixed, briefly centrifuged, and the supernatants (S2-4, 5 μl per experimental point) are withdrawn for analysis. Supernatant S2 is a control for EC dissociation without NTP addition, S3 is a mock-transcription control for EC dissociation in the presence of NTPs, and S4 contains the transcribed templates.

Transcription of the pVT1 template in solution

For transcription in solution, the volume of the washed EC45 is adjusted to 100 μl in TB300 containing 0.5 mg/ml acetylated BSA and 100 mM imidazole (Sigma, St. Louis, MO). The ECs are eluted for 10 minutes at RT. The sample is mixed, centrifuged, and 50 μl of the supernatant is withdrawn. The supernatant is diluted 5-fold with TB300 containing 0.5 mg/ml acetylated BSA. The EC45 is aliquoted (10 to 20 μl per experimental point) and incubated in the presence of (I) no NTPs (E2), (II) 100 μM ATP, CTP, and GTP (E3), or (III) 100 μM NTPs (E4) for 5 minutes at RT.

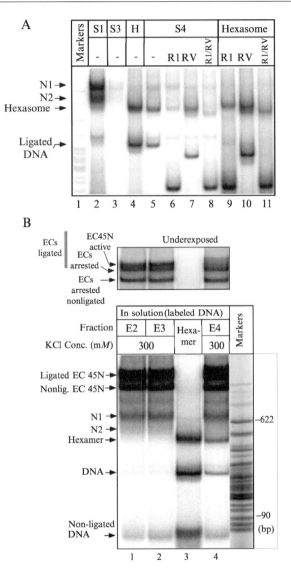

FIG. 4. Analysis of transcribed pVT1 nucleosomal templates. Templates were labeled as in Fig. 3A. (A) Templates transcribed by immobilized polymerase.[21] Samples were analyzed before and after digestion with EcoRI, EcoRV, or EcoRI + EcoRV. The 254 bp reconstituted hexasome control (H) is indicated. Mobilities of the original nucleosomes (N1 and N2),

[21] Reprinted from: M. L. Kireeva, W. Walter, V. Tchernajenko, V. Bondarenko, M. Kashlev, and V. M. Studitsky, *Mol. Cell* **9**, 541 (2002); Copyright 2002, with permission from Elsevier Science.

Analysis of the transcribed templates

For analysis, the samples (S1-4 or E1-4) are supplemented with buffer providing 20 mM Na-HEPES, pH 7.8, 5 mM MgCl$_2$, 2 mM spermidine (Sigma, St. Louis, MO), and 0.5 mg/ml BSA and aliquoted into several tubes. One sample of each fraction is not digested, while appropriate restriction enzymes (10 U) are added to the others, and digestion is performed at RT for 0.5 to 1 hour. Buffer is added, providing a final concentration of 10 mM EDTA, 10% sucrose, 250 μg/ml sheared herring testes DNA (Intergen, Purchase, NY), and 10 ng of carrier nucleosomes (to block non-specific interactions with the wells of the gel, donor chromatin works well). The templates are resolved by native gel electrophoresis as described previously. Quantitation is performed using a Cyclone Storage Phosphor System (Packard, Meriden, CT).

Reconstituted and gel-purified 204 bp hexasome is ligated to annealed NDS59 and ^{32}P-labeled TDS50 to make a mobility control for the 254 bp hexasome. Ligation is performed by incubating equimolar amounts of the annealed TDS50 and NDS59, as well as the 204 bp hexasome in the presence of 100 μM ATP, 1% PEG-8000, and 50 units of T4 DNA ligase (New England Biolabs, Beverly, MA) at 12°C for 1 to 2 hours. The sample is aliquoted, digested, and prepared for electrophoresis as described previously.

Results from analysis of transcription of the pVT1 nucleosome, both with immobilized Pol II and in solution, are shown in Figures 4A and 4B, respectively. The transcribed template has the same mobility in a native gel as the hexasome control. Furthermore, the restriction enzyme digestion profile of the transcribed template and the reconstituted hexasome are very similar. Moreover, the digestion patterns (half sensitive to EcoRI and half sensitive to EcoRV) are similar between the templates before and after transcription. Thus, transcription by Pol II results in the loss of an H2A/H2B dimer, but the remaining histones stay in their original positions on the DNA.

Remarks

This section briefly describes some additional controls that are used for this procedure.

the hexasome, free DNA, and non-ligated promoter fragment are indicated. S1 is non-transcribed templates, S3 is the mock-transcribed control, and S4 is transcribed templates. (B) Templates transcribed in solution. Samples were analyzed by their mobility in a native gel beside the 254 bp reconstituted hexasome control. Mobilities of the ligated EC45, non-ligated EC45, free nucleosomes (N1 and N2), hexasome, DNA, and non-ligated promoter DNA are indicated. An underexposure of the top of the same gel is shown in the upper panel. Active and arrested ECs are indicated.

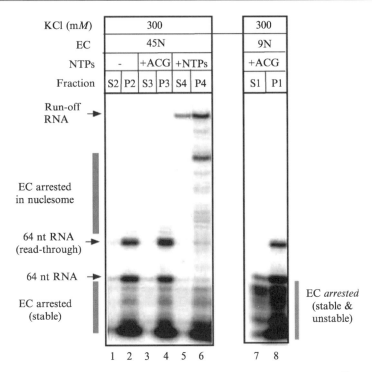

FIG. 5. Analysis of EC stability and NTP-dependent release of RNA/DNA.[21] ECs with labeled RNA9 were ligated to nucleosomes, and EC45N was formed. Supernatant (S1-4) and pellet (P1-4) fractions were collected and loaded onto a 6% denaturing gel. The mobilities of the 45 nt RNA and the run-off product are indicated by arrows. Transcripts from arrested ECs are marked by shaded bars.

Analysis of RNA released after transcription

It can be demonstrated that only the fully transcribed templates are released into the S4 supernatant by analyzing the transcripts that are released into the supernatant. A small amount of smaller transcripts may be present (from EC dissociation), so the controls (S2 and especially S3) are important for determining the background amount of EC dissociation.

EC9 is assembled with γ-[^{32}P]ATP-labeled RNA9 and phosphorylated TDS50 and immobilized on acetylated BSA-treated Ni^{2+}/NTA agarose as described.[19] EC9 (from ~1.5 pmol Pol II) is ligated to the nucleosomal template as described above, and the volume is adjusted to 50 μl, 10 μl of resin is aliquoted into 1 tube, while 40 μl of resin is aliquoted into another tube. EC45 is formed in each tube by the addition of 200 μM ATP, CTP, and GTP for 10 minutes at RT in TB40 and 0.2 mg/ml acetylated BSA.

For tube #1, 5 μl of the supernatant (S1) is withdrawn, and the pellet fraction (P1, also 5 μl) is kept for analysis. For tubes #2 to 4, the resin is washed with 1 ml of TB300 twice, incubated with 1 ml of TB300 for 15 minutes, and washed with 1 ml of TB300 twice again. This procedure is repeated, and the resin is adjusted to a volume of 40 μl in TB300 containing 0.5 mg/ml acetylated BSA. 10 μl of the resin is aliquoted into each of the 3 tubes. 5 μl of the supernatants are withdrawn after a 5-minute incubation with: (I) no additions (S2), (II) 100 μM each ATP, CTP, and GTP (S3), and (III) 100 μM NTPs (S4). The corresponding pellet fractions (P2-4, also 5 μl each) are also kept for analysis. An equal volume (5 μL) of gel-loading buffer containing 50 mM EDTA and 8 M urea is added, the samples are boiled for 5 to 10 minutes at 95°C, and the RNA is analyzed in a 6% (19:1) denaturing polyacrylamide gel. Quantitative analysis is performed with a Cyclone Storage Phosphor System (Packard, Meriden, CT). Figure 5 is an example of the results obtained with this procedure on the pVT1 nucleosomal template.

Analysis of templates released after transcription of free DNA

It is also important to transcribe the free DNA template exactly like the nucleosomal templates are transcribed. The procedures for ligation, transcription, and analysis of the DNA template are the same as those described above for the nucleosomal template.

Acknowledgment

The work was supported by the NIH grant GM58650 to V.M.S.

[43] Probing Chromatin Immunoprecipitates with CpG-Island Microarrays to Identify Genomic Sites Occupied by DNA-Binding Proteins

By Matthew J. Oberley *and* Peggy J. Farnham

Many methods have been developed to identify genomic targets of DNA-binding factors. We outline and discuss our adaptation of the chromatin immunoprecipitation (ChIP) assay to a high-throughput microarray–based method for discovering genomic regions occupied by human DNA-binding proteins. Others, primarily using yeast model systems, have also explored the use of coupled-chromatin immunoprecipitation and microarrays to identify large sets of binding sites of DNA-binding proteins.

For example, binding sites for yeast transcriptional regulators such as SBF, MBF, Gal4, Ste12, and Rap1 have been identified.[1–3] Other studies have focused on chromatin remodeling factors, such as components of the RSC complex,[4,5] and components of the DNA replication machinery, such as ORC and MCMs.[6] In certain cases, the investigators have monitored all intergenic regions and/or known promoters in the yeast genome.[1,3] In other cases, the authors have taken advantage of the relatively small size of the yeast genome to spot probes that span the entire genome (both intergenic regions and open reading frames) onto a microarray.[2,6] A recent Herculean effort by Lee *et al.*[7] has used ChIPs and microarrays to identify the global binding locations of the majority of known yeast transcription factors and used the resulting data along with data from mRNA expression array analysis to define the regulatory networks that exist in yeast.

Because the human genome (3.2 Gb) is three orders of magnitude larger than the yeast genome (12 Mb), a chip containing all human intergenic regions and open reading frames has not yet been created. The largest effort to date consists of a set of oligonucleotide arrays that span the smaller chromosomes 21 and 22 at 35 bp intervals.[8] However, this array has not to date been used to identify sites for DNA-binding factors; rather, it is been used to assess the sum transcriptional activity of these chromosomes. A microarray, more limited in scope but allowing a more detailed study of specific regions, which includes the sequence spanning from −700 to +200 of 1444 human genes, has been used to find novel promoters bound by E2F family members.[9] Although a step in the right direction, such an array by necessity is biased in that the particular 1444 promoters analyzed were chosen by a set of criteria. To eliminate such bias, our

[1] V. R. Iyer, C. E. Horak, C. S. Scafe, D. Botstein, M. Snyder, and P. O. Brown, *Nature* **409**, 533 (2001).

[2] J. D. Lieb, X. Liu, D. Botstein, and P. O. Brown, *Nat. Genet.* **28**, 327 (2001).

[3] B. Ren, F. Robert, J. J. Wyrick, O. Aparicio, E. G. Jennings, I. Simon, J. Zeitlinger, J. Schreiber, N. Hannett, E. Kanin, T. L. Volkert, C. J. Wilson, S. P. Bell, and R. A. Young, *Science* **290**, 2306 (2000).

[4] H. H. Ng, F. Robert, R. A. Young, and K. Struhl, *Genes Dev.* **16**, 806 (2002).

[5] M. Damelin, I. Simon, T. I. Moy, B. Wilson, S. Komili, P. Tempst, F. P. Roth, R. A. Young, B. R. Cairns, and P. A. Silver, *Mol. Cell.* **9**, 563 (2002).

[6] J. J. Wyrick, J. G. Aparicio, T. Chen, J. D. Barnett, E. G. Jennings, R. A. Young, S. P. Bell, and O. M. Aparicio, *Science* **294**, 2357 (2001).

[7] T. I. Lee, N. J. Rinaldi, F. Robert, D. T. Odom, Z. Bar-Joseph, G. K. Gerber, N. M. Hannett, C. T. Harbison, C. M. Thompson, I. Simon, J. Zeitlinger, E. G. Jennings, H. L. Murray, D. B. Gordon, B. Ren, J. J. Wyrick, J. Tagne, T. L. Volkert, E. Fraenkel, D. K. Gifford, and R. A. Young, *Science* **298**, 799 (2002).

[8] P. Kapranov, S. E. Cawley, J. Drenkow, S. Bekiranov, R. L. Strausberg, S. P. Fodor, and T. R. Gingeras, *Science* **296**, 916 (2002).

[9] B. Ren, H. Cam, Y. Takahashi, T. Volkert, J. Terragni, R. A. Young, and B. D. Dynlacht, *Genes Dev.* **16**, 245 (2002).

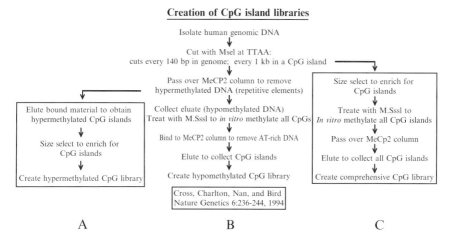

FIG. 1. Creation of CpG-island libraries. Column B outlines the original protocol used by Cross *et al.*[13] to prepare the CpG-island library used to spot the slides discussed in the text. Columns A and C represent alternative experimental strategies that could be used to isolate a hypermethylated and a comprehensive library, respectively. MeCP2, Methyl CpG Binding Protein 2; A, adenine; T, thymine.

laboratory has used a microarray that contains 7776 CpG island clones to identify genomic sites to which E2F and pRb family members are recruited.[10,11] Of particular interest to the study of transcription factors is that CpG islands tend to be found in intergenic regions and at the 5′ ends of genes. Thus, use of CpG arrays now allows an unbiased analysis of many thousands of human intergenic regions.

CpG islands, found in or near approximately 50% of human promoters, are identified by three primary characteristics: (1) they are more than 200 bp long, (2) they have over 50% GC composition, and (3) they retain an observed/expected ratio of CpG dinucleotides greater than 0.6.[12] The CpG library used on the microarrays was originally prepared in the laboratory of Cross *et al.*[13] and consists of approximately 60,000 independent clones that represent CpG islands which were not hypermethylated in human male genomic DNA. Clearly one limitation of this array is that any CpG island, which was hypermethylated in the starting genomic DNA, would not be represented in the library. However, the technique

[10] A. S. Weinmann, P. S. Yan, M. J. Oberley, T. H. Huang, and P. J. Farnham, *Genes Dev.* **16,** 235 (2002).
[11] J. Wells, P. S. Yan, M. Cechvala, T. H. Huang, and P. J. Farnham, *Oncogene.* **22,** 1445–1460 (2003).
[12] I. P. Ioshikhes and M. Q. Zhang, *Nat. Genet.* **26,** 61 (2000).
[13] S. H. Cross, J. A. Charlton, X. Nan, and A. P. Bird, *Nat. Genet.* **6,** 236 (1994).

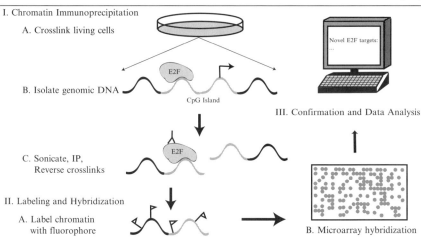

FIG. 2. The chromatin immunoprecipitation and microarray analysis (ChIPs-chip) protocol can be subdivided into three major areas: I. chromatin immunoprecipitation, II. labeling and hybridization, and III. confirmation and data analysis.

could be modified to allow the collection of all CpG islands (Fig. 1). Alternatively, new libraries could be prepared from different starting tissues, allowing the creation of arrays containing CpG islands that are hypomethylated vs. hypermethylated in specific tissues, in tumors, or in different developmental stages. Approximately 8000 clones representing hypomethylated CpG islands were arrayed by the Yan et al.[14] onto slides for high-throughput study of DNA methylation changes in human cancer. As described later, we have now developed protocols that allow these arrays to be used to perform a large-scale identification of genomic regions occupied by DNA-binding proteins under physiologically relevant *in vivo* conditions. As illustrated in Fig. 2, the assay subsumes three major areas: (1) chromatin immunoprecipitation, (2) DNA labeling and hybridization, and (3) data analysis. Therefore this chapter has been divided into these three major areas. Each section begins with the protocol and ends with Critical Points and Reagents; illustrative figures are also provided.

Chromatin Immunoprecipitation

We have used the following protocol to successfully immunoprecipitate sequences specifically associated with E2F family members, pocket proteins (Rb, p107, p130), β-catenin and TCF family members, RNA polymerase, Myc family members, histones, and so forth. Therefore we believe that this

[14] P. S. Yan, C. M. Chen, H. Shi, F. Rahmatpanah, S. H. Wei, and T. H. Huang, *J. Nutr.* **132**, 2430S (2002).

protocol is applicable to a wide variety of different types of DNA-binding factors. We have utilized many different cell types (e.g., epithelial, fibroblast, blood cells) and have also adapted it for use with mouse and human tissues. Although minor modifications may be required to develop the optimal set of conditions for a particular antibody or cell type, we anticipate that the following protocol will provide a good starting point for most transcription factors. For further information on the ChIPs assay, see Weinmann and Farnham.[15]

Day 0

Block Staph A Cells

The preparation of Staph A cells is described later in the reagents section. Thaw 100 μl of prepared Staph A for approximately every 5×10^7 cells used. Add 10 μl of herring sperm DNA (10 mg/ml) and 10 μl of BSA (10 mg/ml) for every 100 μl of Staph A. Incubate the mixture on the rotating platform at 4°C overnight; on the following day, microfuge the mixture for 3 min. Remove the supernatant and wash the Staph A pellet twice with 1× dialysis buffer. Finally, resuspend the Staph A cells in a volume of 1× dialysis buffer equal to the original starting volume.

Day 1[16]

!   1. Add formaldehyde directly to tissue culture media to a final concentration of 1%. Incubate adherent cells on a shaking platform and suspension cells on a stir plate for 10 min at room temperature. We use between 2×10^6 and 1×10^7 cells per antibody per time point for standard ChIP assays. However, for the CpG arrays, we generally start with 2×10^8 cells, that are divided among 20 immunoprecipitation reactions and pooled at a later step.

2. Stop the cross-linking reaction by adding glycine to a final concentration of 0.125 M. Continue to rock or spin the cells at room temperature for 5 min.

3. For adherent cells, pour off the media and rinse the plates twice with cold 1 × PBS. For suspension cells, centrifuge at 1500 rpm and wash the cell pellet twice with cold 1 × PBS (for suspension cells then proceed to step 6).

4. For adherent cells, add an appropriate volume (try 5 ml per 500 cm^2 dish) of 1 × PBS or a 0.0025% trypsin solution (a 1:5 dilution of a 1 × Trypsin-EDTA tissue culture–grade solution diluted in 1 × PBS).

---

[15] A. S. Weinmann and P. J. Farnham, *Methods.* **26**, 37 (2002).

[16] If a step in the protocol is marked with an exclamation point, that indicates that the reader is referred to a Critical Point, located immediately after that section of the protocol.

Incubate the cells at 37°C for 10 min if trypsin is used. The trypsin step is useful for cells that are difficult to swell; for cell types that are easily swelled, this step may not be necessary.

!   5. After the addition of trypsin or 1 × PBS, scrape adherent cells from dishes. If trypsin is used to help swell the cells, inactivate the trypsin by adding a small amount of serum. Centrifuge the scraped adherent cells at 1500 rpm and wash the cell pellet once with 1× PBS plus PMSF (10 μl of the stock solution per ml).

6. Resuspend the cell pellet in cell lysis buffer plus the protease inhibitors PMSF (10 μl of the stock solution per milliliter), aprotinin (1 μl of the stock solution per milliliter), and leupeptin (1 μl of the stock solution per milliliter). The final volume of cell lysis buffer should be sufficient so that no clumps of cells are present. Incubate the cells on ice for 10 min. Cells can also be homogenized on ice with a dounce homogenizer (B pestle) several times to aid in nuclei release. The cell lysis conditions, buffers, and homogenizing may need to be optimized for nuclei isolation from individual cell types.

7. Centrifuge the homogenate at 5000 rpm for 5 min at 4°C to pellet the nuclei.

8. Discard the supernatant and resuspend the nuclei pellet in 1 ml of nuclei lysis buffer plus protease inhibitors. Incubate on ice for 10 min. If the nuclei are too dense, resuspend them in a larger volume of nuclei lysis buffer.

9. Sonicate the chromatin to an average length of approximately 0.5–1 kb, ensuring that the samples are kept cold during the sonication procedure. The sonication time and number of pulses vary depending on the type of sonicator used, cell type, and extent of cross-linking. As a starting point, we have performed the sonication step using 4 pulses of 15 seconds each at setting 7 on a Fisher model 60 sonic dismembrator. Samples should be allowed to cool for 30 seconds on ice between pulses. To aid in sonication, it may also be helpful to add 0.1 g of glass beads (212–300 μm; Sigma, G-1277) prior to sonication. After sonication, microcentrifuge samples at 14,000 rpm for 10 min at 4°C. Carefully remove the supernatant and transfer to a new tube. At this point, the chromatin can be snap-frozen in liquid nitrogen and stored at −80°C.

10. Preclear the chromatin by adding 10 μl of preblocked Staph A cells for every 10^7 cross-linked cells.

11. Incubate the chromatin—Staph A mixture on a rotating platform at 4°C for 15 min, then microcentrifuge the mixture at 14,000 rpm for 4 min.

!   12. Transfer the supernatant to a new tube and divide equally among IP samples. For example, if starting with $2 × 10^8$ cells, then divide into 20 samples—10 to be precipitated with the antibody specific to the DNA-binding protein of interest and 10 for a no-antibody or preimmune serum control. Also include a "mock" sample that contains IP dilution buffer without chromatin. The no-antibody and mock samples are critical to monitor for nonspecific interactions and DNA contamination of IP and wash solutions, respectively. Adjust the final volume of each sample with two times the chromatin volume of IP dilution buffer (plus protease inhibitors). The volume of each sample should be between 200 μl and 600 μl. We generally add 1 μg of the specific antibody to the appropriate samples. However, the optimal antibody concentration may vary for individual antibodies; initial tests can be performed with 0.5, 1.2, and 5 μg of antibody per sample.

13. Incubate the chromatin samples with the antibodies on a rotating platform at 4 °C overnight (or for at least 3 hours).

Day 2

14. If rabbit polyclonal antibodies are not used (i.e., you are using a mouse monoclonal antibody or a goat polyclonal antibody), add 1 μg of an appropriate secondary antibody and incubate for an additional 1 hour. We have found that rabbit polyclonal antibodies associate well with Staph A cells; therefore secondary antibodies should be from a rabbit.

15. Add 10 μl of blocked Staph A cells to each sample and incubate them on a rotating platform at room temperature for 15 min.

16. Microcentrifuge samples at 14,000 rpm for 4 min. Save the supernatant from the no-antibody sample as "total input chromatin." For the other samples, discard the supernatant.

17. Wash pellets twice with 1.4 ml of 1× dialysis buffer (if a monoclonal antibody is used, omit the sarkosyl from the buffer) and four times with 1.4 ml of IP wash buffer (pH 8.0 for monoclonal antibodies). For each wash, dissolve the pellet in 700 μl of buffer and then add an additional 700 μl of buffer and incubate samples on a rotating platform for 3 min. Next, microcentrifuge samples at 14,000 rpm for 4 min. Attempt to remove as much buffer as possible after each wash without aspirating the Staph A cells.

18. After the final wash, remove as much wash buffer as possible and dry spin the tubes for 4 min at 4 °C. Then remove all traces of wash buffer.

19. Add 150 μl of IP elution buffer, and vortex on setting 3 for 15 min at room temp. Spin the tube for 4 min and transfer the elution buffer to a fresh tube.

20. Add an additional 150 μl of IP elution buffer to the Staph A pellet and vortex again on setting 3 for 15 min at room temperature. Spin the tube for 4 min, remove the supernatant, and combine the two elution volumes.

21. After the second elution, microcentrifuge the samples at 14,000 rpm for 4 min to remove any traces of Staph A cells. Transfer the

supernatants to new tubes. Add 1 μl of high-concentration RNase A (Roche, 1579681; 10 mg/ml) and 12 μl of 5 M NaCl to a final concentration of 0.3 M. Remember to include the "total" sample at this point to reverse cross-links. For the total sample, use only 20% of starting volume obtained from the supernatant of a no-antibody reaction (step 16) and increase the volume of the sample to 600 μl with elution buffer. Add 24 μl of 5 M NaCl and 2 μl of RNase A to the total sample and then incubate the samples at 67 °C for 4–5 hours to reverse the formaldehyde crosslinks. After the 4- to 5-hour incubation, add 2.5 volumes of ethanol to each sample and precipitate them at −20 °C overnight.

Day 3

22. Microcentrifuge the samples at 14,000 rpm for 15–20 min at 4 °C. Discard the supernatant and respin the samples to remove any residual ethanol. Allow the pellets to air-dry completely.

23. Dissolve each pellet in 100 μl of TE. Add 25 μl of 5× PK buffer and 1.5 μl of proteinase K (25 mg/ml) to each sample. If more than 20% of the supernatant from a no-antibody reaction is used for the "total" sample (step 21), the sample may be very viscous, requiring dissolution in a larger volume. Incubate the samples at 45 °C for 1–2 hours.

! 24. If the samples are to be used for a standard ChIP assay, then they are phenol extracted, ethanol precipitated, and redissolved in water.[15] However, if the samples are to be used for hybridization to an array, they are treated differently. To remove all protein and contaminants from the immunoprecipitated chromatin, we purify the chromatin with a Qiaquick PCR cleanup kit (Qiagen, 28106). We follow the Qiagen protocol, but we elute the solution by adding 50 μl molecular biologic-grade water to the column and allow it to stand 1 min at room temperature. The samples are then centrifuged at maximum speed for 1 min in a microcentrifuge. A second elution is performed and the two eluates are combined.

! 25. At this point, the success of the ChIP portion of the experiment can be monitored by a standard PCR reaction using primers specific for a region of the genome thought to be bound by the factor of interest.

Critical Points

Step 1

For attached cells, we have noticed that it is critical to formaldehyde cross-link cells on the dish before trypsinizing, as we have observed that DNA-binding factors can relocalize after trypsinization (Fig. 3). This

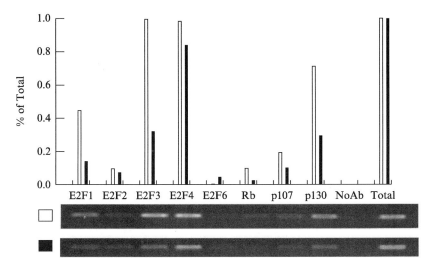

FIG. 3. Decreased specificity of binding of E2F family members is caused by trypsinization. A standard ChIP was performed with gene-specific primers with the cells cross-linked before (black) and after (white) trypsinization. Disruption of the normal cell-surface contact allows certain sites to become more accessible to some transcription factors (e.g., E2F3).

precludes an experimental design that incorporates live cell sorting after trypsinization to create relatively pure populations of cells. However, we have found that nuclei prepared from cross-linked cells can be easily sorted for DNA content and subsequently immunoprecipitated with antibodies specific to chromatin-binding proteins.

Step 5

At this point, the cell pellets can be snap-frozen and stored at $-80\,^{\circ}$C. This is helpful if transcription factor binding in different cell populations (which are not necessarily ready at the same time) is to be compared in the ChIP assay.

Step 12

The number of individual immunoprecipitates required to pool for hybridization to the microarray should be empirically determined. The minimum number we have successfully used to date is 14 IPs sourced from $\sim$15 million cells. A sample of 5 μg of chromatin allows for ample DNA to hybridize to a CpG-island microarray.

Step 24

We have found that the use of buffers containing Tris to elute the chromatin sample from the Qiagen columns will severely degrade the Cyanine dyes that are subsequently coupled to the chromatin. Therefore it is critical that water, and not Tris-containing buffers, be used at this step.

Step 25

Do not proceed with the hybridization portion of the experiment unless the signal obtained in a test PCR experiment shows higher signal in the antibody samples than in the no-antibody (or preimmune) control. Care must be taken to avoid cross-contamination of samples.

REAGENTS

Cell Lysis Buffer

5 mM PIPES pH 8.0
85 mM KCl
0.5% NP40
Add PMSF just before use

Nuclei Lysis Buffer

50 mM Tris-Cl pH 8.1
10 mM EDTA
1% SDS
Add PMSF, aprotinin, and leupeptin just before use

IP Dilution Buffer

0.01% SDS
1.1% Triton X 100
1.2 mM EDTA
16.7 mM Tris-Cl pH 8.1
167 mM NaCl
Add PMSF, aprotinin, and leupeptin just before use

1X Dialysis Buffer

2 mM EDTA
50 mM Tris-Cl pH 8.0
0.2% Sarkosyl (omit for monoclonal antibodies)
Add PMSF just before use

IP Wash Buffer

100 m*M* Tris-Cl pH 9.0 (8.0 for monoclonal antibodies)
500 m*M* LiCl
1% NP40
1% deoxycholic acid
Add PMSF just before use

Elution Buffer

50 m*M* NaHCO$_3$
1% SDS

5X PK Buffer

50 m*M* Tris-Cl pH 7.5
25 m*M* EDTA
1.25% SDS

Protease Inhibitors

100 m*M* phenylmethylsulfonyl fluoride (PMSF; Sigma, P-7626) in
   isopropanol, use at 1:100 dilution
10 mg/ml aprotinin (Sigma, A-1153) in 0.01 M HEPES pH 8.0, use at
1:1,000 dilution.
10 mg/ml leupeptin (Sigma, L-2884) in water, use at 1:1,000 dilution

Staph A Cells

Resuspend 1 g of lyophilized protein A–positive *Staphylococcus aureus*
   cells, whole cells (Cowan 1 strain), heat-killed, fixed in formalin,
   (Calbiochem, 507862) in 10 ml of 1× dialysis buffer. Centrifuge at
   10,000 rpm for 5 min at 4°C. Discard the supernatant and repeat
   wash. Resuspend in 3 ml of 1× PBS plus 3% SDS and 10% BME.
   Boil for 30 min. Centrifuge at 10,000 rpm for 5 min. Wash in
   1× dialysis buffer and centrifuge at 10,000 rpm for 5 min. Repeat
   wash. Resuspend in 4 ml of 1× dialysis buffer. Divide into 100–*μ*l
   aliquots, snap-freeze, and store in liquid nitrogen.

Chromatin Labeling and Hybridization

Day 1

!   1.   The chromatin samples obtained in the last step of the chromatin
immunoprecipitation protocol (see previous text) are first vacuum
desiccated to complete dryness with heat and then resuspended in 33 *μ*l
of molecular biologic-grade H$_2$O. Often the chromatin is difficult to

completely redissolve and, if so, the tube is vortexed vigorously followed by a burst spin to resuspend all the chromatin. Then 30 μl of 2.5 $\times$ random primer buffer (Invitrogen, 18094-011) is added.

! 2. The chromatin is denatured by holding at 95 °C for 5 min on a PCR block, and then it is immediately placed on ice for 3 min. The labeling reaction is initiated by adding 7.5 μl of the 10 $\times$ dNTP mix, 1.8 μl of 10 mM aminoallyl-dUTP (Sigma, A0410), and 2.5 μl of high-concentration Klenow (40 U/μl, Invitrogen, 18094-011), and holding at 37 °C for 2 hours.

! 3. The excess nucleotides are then removed using a Microcon YM30 (Millipore, 42410) concentrator. It is very important at this stage to remove as much of the unincorporated aminoallyl-dUTP as possible, because it will bind with the NHS-ester cyanine dyes and reduce the labeling efficiency of the chromatin. To clean up the chromatin, add 450 μl of H$_2$O to the Microcon YM30 concentrator column, followed by the 74.8 μl of aminoallyl-labeled chromatin sample, and triturate carefully, taking care not to contact the membrane with the pipet tip. The columns are then spun in a microfuge at 12,000 rpm for 5 min, which allows the water, salts, and unincorporated nucleotides to pass through the membrane while the labeled chromatin sample is retained in the column (the retentate). Check the remaining volume of liquid in the column and continue to spin the columns at 1-minute intervals until approximately 10 μl of liquid remains in the retentate. Take care not to spin the columns to dryness, because the chromatin will be irretrievably lost, but if more than 10 μl of retentate is left, the unincorporated nucleotides will not be removed efficiently. When the retentate volume has been reduced to about 10 μl, add an additional 450 μl of H$_2$O to the column and repeat the 12,000-rpm spin process five times. After the last spin, add H$_2$O to around 30 μl and remove the chromatin by inverting the columns and spinning them at maximum speed for 1 min. Dry the chromatin samples in vacuo and store them at −20 °C overnight. Alternatively, the process can be continued with the conjugation step listed in the following text and hybridization to the array in 1 day.

Day 2

! 4. To couple the NHS-ester cyanine dye to the chromatin, resuspend the aminoallyl-labeled chromatin in 4.5 μl of H$_2$O, and resuspend the cyanine dye in 4.5 μl of 0.1 M NaHCO$_3$ (pH 9). Vortex the tubes and spin them several times to ensure that all samples are redissolved. Next, combine the chromatin and cyanine dye and incubate them at room temperature in the dark for 1.5 hours, agitating and microfuging the samples every 15 min of the coupling.

5. Add 4.5 μl of 4 M hydroxylamine to quench the coupling reaction and incubate for 15 min at room temperature. Add 35 μl of 100 mM Na-acetate (pH 5.2) to lower the pH of the solution to allow the chromatin to bind to the Qiaquick columns, and then add 35 μl of H_2O.

6. We have found that removal of the unincorporated dye is quite efficient with Qiaquick PCR columns. We follow the Qiagen protocol with one exception—we elute with 50 μl of water instead of the EB buffer provided with the kit. Repeat and combine the eluates so the total volume is 100 μl of H_2O.

! 7. The total dye incorporation in determined by measuring the absorbance of the entire sample at 650 nm for Cy5 and 550 nm for Cy3. At this stage we also measure the amount of DNA. Vacuum desiccate the labeled chromatin with heat and store it dry at $-20\,^{\circ}$C until ready for hybridization.

! 8. While the labeled chromatin is drying, the CpG island arrays are prehybridized to reduce the amount of nonspecific binding. This is accomplished by heating 52.45 μl of prehybridization solution to 95°C for 5 min and then cooling briefly to 50°C before applying to the microarray. We typically drop the prehybridization solution onto a 24 $\times$ 60 coverslip and place the microarray probe side–down onto the coverslip, and then carefully turn the array upright. Prehybridize for 1 h at 37°C in a humidified chamber, and then remove the coverslip by gently dipping the slide into H_2O. The coverslip should never be forced off, as this will cause the probes to smear. The microarray is then washed twice for 20 min each, and then the slide is dried by spinning it at 600 rpm in a 50-ml conical tube.

9. The labeled chromatin is then resuspended in 10 μl of 1.0 μg/μl Cot-1 DNA (Invitrogen, 15279-011) and 5 μl of H_2O. Cot-1 DNA is included to bind to repeat elements in the chromatin, which helps to prevent nonspecific binding to the array. Vortex the tube several times and spin to completely redissolve the chromatin. Carefully add 35 μl of the hybridization solution without vortexing to avoid bubble formation. Denature the mixture at 95°C for 2 min, cool to 60°C, and then apply the mixture to the microarray using the same technique outlined for the prehybridization.

10. We use a dual hybridization chamber (Genemachines) and a water bath for the hybridization because it maintains a more constant temperature. It is important that the hybridization chamber not directly contact the bottom of the water bath because it will cause inconsistent heat transfer. We typically use a small piece of foam to separate the chamber from the metal bottom of the water bath. Hybridize overnight at 60°C for up to 18 hours.

Day 3

11. To remove the coverslip, the microarray(s) are inverted in a glass dish filled with 1 × SSC and 0.1% SDS preheated to 50 °C. The coverslip should be gently removed by agitation, but it is important to remove it rapidly and place the array in the 50 °C wash so that the chromatin will not start to nonspecifically stick to the array.

12. The arrays are then agitated in 1 × SSC, 0.1% SDS at 50 °C for 5 min, followed by agitation in 1 × SSC, 0.1% SDS at room temperature for 5 min. Finally, we wash the array with 0.2 × SSC for 5 min at room temperature and then dry the slide by centrifugation at 600 rpm for 5 min in a 50-ml conical tube.

13. We then scan immediately using a Packard Biochip Scan Array 5000 scanner (Fig. 4).

Critical Points

Step 1

We have found that starting the labeling procedure with 5 μg of chromatin is the least amount of chromatin that can be used to provide a good signal/noise ratio on the array. The number of IPs required to generate this amount must be determined empirically and is a function of several factors, including cross-linking efficiency, DNA-binding factor abundance, and epitope availability. If the amount of chromatin is limited, the amount of DNA can be amplified by using degenerate oligonucleotide primers (see Lieb *et al.*[2] for a detailed protocol) or a ligation-mediated PCR technique

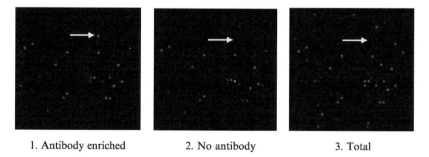

1. Antibody enriched   2. No antibody   3. Total

Fig. 4. Results from a typical microarray experiment. The arrow points to a signal that is significantly enriched in a pool of chromatin immunoprecipitated with an antibody (1), compared with an equivalent amount of a no-antibody immunoprecipitate (2) or genomic DNA (3).

(see Ren *et al.*[3]). In both cases the cyanine dyes can be directly incorporated into the PCR reaction, or the indirect means of labeling discussed previously can be used. It is critical to monitor for amplification bias by examination to determine whether the same amount of enrichment is retained after amplification on a known target before utilizing any particular amplification technique for an array experiment.

Step 2

Two primary methods can be used to label any given chromatin sample: direct vs. indirect. Direct labeling involves denaturing the chromatin, random priming, and polymerizing new duplex DNA with a nucleotide analogue that has the cyanine fluorescent dye conjugated to it. Because these fluorescent dyes are bulky, the polymerase incorporates the conjugates at different rates than it would a natural nucleotide. This leads to labeling bias; for example, Cy5 tends to incorporate more readily than Cy3. This factor must be considered when co-hybridizing both fluorophores onto the same array, and the reciprocal experiment must be done to control for these differences where the test and control samples are each labeled in separate experiments with each dye. For these reasons, our laboratory primarily uses an indirect means of labeling the immunoprecipitated chromatin. Indirect means of labeling incorporate nonfluorescent nucleotide analogues such as aminoallyl dUTP or biotinylated dUTP; these small conjugates help to eliminate the incorporation biases that occur in direct labeling with Cy5-dUTP and Cy3-dUTP. The cyanine dyes can be purchased as an NHS-ester conjugate (Amersham, RPN5661) that covalently binds to the aminoallyl nucleotide analogues, or as streptavidin conjugates that form a tight complex with the biotinylated chromatin. We routinely use aminoallyl labeling for chromatin samples, followed by cyanine dye coupling via the protocol outlined previously.

Step 3

If the aminoallyl incorporation was successful, the amount of the DNA should be amplified 2–5-fold over the starting material.

Step 4

The cyanine dyes are easily photobleached. Thus it is best to perform all steps under minimal light exposure conditions, both during and after the cyanine dyes have been incorporated. The cyanine dyes are also extremely susceptible to degradation by air humidity, so they should be stored dry with a desiccant at all times.

Step 7

We have found that at least 150 pmol of Cy5 and Cy3 needs to be incorporated for a good signal/noise ratio on scanning the arrays. This is measured by using the A650 reading for Cy5 and A550 reading for Cy3 on a spectrophotometer. Determining the amount of incorporated dye (in pmol) is accomplished with the following formula: (A650) (μl of solution) (dilution factor)/(0.25 for Cy5, or 0.15 for Cy3) = pmol of Cy dye.

Step 8

The arrays are spotted using a four-pin Affymetrix arrayer onto poly-L-lysine–coated microscope slides. The slides are processed prior to prehybridization by following the protocol found at http://derisilab. ucsf.edu/microarray/protocols.html. After processing, the slides can be stored dry at room temperature for several months. Because the microarrays are highly susceptible to dust, the arrays are always covered or inverted whenever they are manipulated. After prehybridization, the arrays can be stored dry at room temperature for up to a day before they are used for hybridization.

Reagents

   10 × dNTP Mixture

   2 mM dATP
   2 mM dCTP
   2 mM dGTP
   0.35 mM dTTP

   Prehybridization Solution

   7.5 μl SSS DNA (10 μg/μl)
   7.6 μl 20 × SSC
   1.35 μl 10% SDS
   36 μl H$_2$O

   Hybridization Solution (Genisphere Buffer 6, catalog 100V600)

   0.25 M NaPO$_4$
   4.5% SDS
   1 mM EDTA
   1 × SSC

Wash 1

1 × SSC (7.5 ml 20 × SSC)
0.1% SDS (1.5 ml 10% SDS)
H_2O to 150 ml

Use the 2.5 × Random Primer Buffer and Klenow from the Invitrogen Bio-Prime kit

Wash 2

0.2 × SSC (500 μl)
H_2O to 50ml

Aminoallyl-dUTP: Make a 0.1 M stock in H_2O and store at −20 °C. Hydroxylamine: Make a 4 M stock in H_2O and store at room temperature; this stock can be used for 2–3 months.

Analysis

The identification of positive clones is a multistep procedure that first require normalization of the two arrays to be compared (i.e., the one hybridized with chromatin immunoprecipitated with an antibody to a DNA-binding protein and the one hybridized with the no-antibody or preimmune control), followed by clone selection, confirmation, and ultimately, functional analysis.

1. The hybridized microarrays are analyzed using the Genepix Pro 3.0 (Axon Laboratory) software package. This provides a set of raw values for each clone on each array. To identify clones corresponding to selectively enriched chromatin, the signals on the two arrays must be normalized. To do this, we use the values derived for 10 internal repeat sequence controls that invariably are precipitated in the presence or absence of antibody during the immunoprecipitation procedure. The resultant normalized signals can then be directly compared on two different arrays.

2. In our published studies, we have chosen to pursue putative targets that are enriched more than twofold using chromatin precipitated with a specific antibody over the no-antibody signal in at least two independent hybridizations. Alternatively, genes that are enriched in the experimental sample could be identified as compared to signals seen on an array hybridized with an identical amount of unenriched DNA (see Fig. 4). In this case, several independent total arrays must be compared to determine the experimental derived error for each spot. Then an array is hybridized with chromatin enriched using the specific antibody and spots that provide a significantly higher signal on the antibody-enriched array vs. the "total" array are identified as positive.[3]

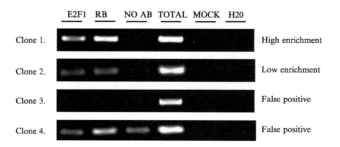

FIG. 5. Preliminary confirmation of microarray results using PCR analysis with gene-specific primers. Primers were made to four CpG islands identified by immunoprecipitating with an Rb antibody followed by hybridization to a CpG array. In these examples, clones 1 and 2 represent true positives; the first is highly enriched and the second is less so. Clones 3 and 4 illustrate two kinds of false positives that can result and which subsequently should be discarded from the positive list.

3. Preliminary confirmation is demonstrated using an independent ChIP assay where 2 μl of an individual IP is examined via PCR with gene-specific primers and compared to a no-antibody IP. This allows false-positive clones to be identified, and discarded from a putative target list. Fig. 5 illustrates two types of false-positive findings and examples of high and low specific-enrichment after a ChIP analysis with gene-specific primers. Ultimately, an ideal experimental validation can be derived from cells with specific gene deletions and their wild-type counterparts to show the PCR signals are the result of the desired specific antibody-epitope interaction. For example, Wells et al.[17] identified several E2F-1–specific genes and validated them using ChIPs in wild-type and E2F-1 knockout mouse embryo fibroblasts.

4. A wealth of functional assays are available to experimentally determine whether an observed target is also regulated by the DNA-binding factor of interest. For example, RNAi represents an ideal experimental system where the expression of any particular gene can be knocked down in a transient way.[18] This system allows perturbation of a normal biological system without some of the concerns with traditional gene-ablation technology in which the knock-out of a specific gene may be compensated for during development of the animal by upregulation of other family members. RNAi could be used to specifically knock down the expression of the DNA-binding factor of interest followed by use of RNase

[17] J. Wells, C. R. Graveel, S. M. Bartley, S. J. Madore, and P. J. Farnham, *Proc. Natl. Acad. Sci. USA* **99,** 3890 (2002).
[18] M. T. McManus and P. A. Sharp, *Nature Rev. Gen.* **3,** 737 (2002).

protection, Northern blot, or traditional microarray gene expression analysis to determine whether the lack of binding to a putative target has a functional consequence. Shi et al.[19] have developed protocols that allow the CpG-island microarray to be probed with mRNA that uses RNA ligase–mediated full-length cDNA synthesis. This protocol allows examination of the effect of DNA-binding factor knock-down on the resultant mRNA expression of the identified target genes using the CpG-island microarray.

Conclusions

Ultimately, an understanding of the sets of genes regulated by the entire spectrum of human transcription factors will greatly facilitate understanding of complex mammalian cellular processes, such as the controlled changes that occur during development and the uncontrolled growth of neoplasia. The use of the ChIP-chip assay to study interactions of DNA-binding factors in living cells allows new insights into cell biology that have not been possible in the past. For example, knowledge of the large set of relatively small genomic stretches identified by ChIP-chip allows the development of consensus binding sites through the use of computer programs such as TransFac or MEME (for an example of how these programs were used to analyze yeast transcription factor binding sites, see http://staffa.wi.mit.edu/cgi-bin/young_public/navframe.cgi?s = 17&f = sequence). Consensus binding sites obtained using such in vivo data allow insights into the function of factors that could not be obtained using typical, in vitro binding site selection protocols. For example, many factors use interaction with other DNA-binding proteins to allow regulation of distinct sets of target genes. The link between such factors and their target genes would be missed if only in vitro binding of purified proteins to isolated sites was examined. Great strides have been made in this area in yeast, where regulatory networks have been identified for the majority of identified yeast transcription factors.[7] To achieve comparable results in mammalian systems, in which many types of cells all have different global transcriptional programs, will require enormous effort. However, the recent advances in ChIP-chip technologies and bioinformatics methodologies now provide confidence that such studies will soon be possible.[20]

[19] H. Shi, P. S. Yan, C. M. Chen, F. Rahmatpanah, C. Lofton-Day, C. W. Caldwell, and T. H. Huang, Cancer Res. 62, 3214 (2002).
[20] For the most recent version of our protocols, see http://mcardle.oncology.wisc.edu/farnham/

Acknowledgments

We thank David Inman, Antonis Kirmizis, and Amy Weinmann for assistance with the development of the hybridization protocol and critical reading of this manuscript. We also thank Pearlly Yan and Tim Huang for their generous gift of the CpG arrays and with assistance in the hybridizations. This work was supported in part by Public Health Service grant CA22484 and CA45240 (to P.J.F.) and by the Molecular and Environmental Toxicology Center training grant NIEHS144KH84 (to M.J.O) (Contribution No. 344 from the Molecular and Environmental Toxicology Center).

[44] Isolation of RNA Polymerase Suppressors of a (p)ppGpp Deficiency

By Helen Murphy and Michael Cashel

In bacteria, a regulatory role in nutritional stress responses has been established for the ppGpp nucleotide analogs of GTP and GDP bearing pyrophosphate residues on the ribose $3'$ hydroxyl. We use ppGpp as an abbreviation for both ppGpp and pppGpp. Regulation by these analogs is generally believed to involve transcription, but mechanistic details have remained elusive.[1–3] A direct effect of ppGpp on transcription is suggested from reports of covalent cross-linking of ppGpp to RNA polymerase sites near the C-terminal end of the β (RpoB) subunit and in the structurally neighboring N-terminal region of the β' (RpoC) subunit.[4,5] A complete deficiency of ppGpp (ppGpp^0) is found when both the *relA* and *spoT* genes are deleted from an *Escherichia coli* K12 strain, MG1655.[6] Various features of the pleiotropic ppGpp^0 regulatory phenotype can be exploited to select extragenic suppressors with somewhat different results. One attribute is the inability of strain CF1693, a ppGpp^0 derivative of MG1655, to grow on glucose minimal media when arg, gly, his, leu, phe, ser, thr, or val, as well as probably ile, are omitted from the otherwise full complement of amino acids.[6] A second selection for suppressor alleles is the poor viability of ppGpp^0 strains during prolonged stationary phase exposure.[7] Alleles

[1] R. Wagner, *J. Mol. Microbiol. Biotechnol.* **4**, 331 (2002).

[2] D. Chatterj and A. K. Ojha, *Curr. Opin. Microbiol.* **4**, 160 (2001).

[3] M. M. Barker, T. Gaal, C. A. Josaitis, and R. L. Gourse, *J. Mol. Biol.* **305**, 673 (2001).

[4] I. I. Toulokhonov, I. Shulgina, and V. J. Hernandez, *J. Biol. Chem.* **276**, 1220 (2001).

[5] D. Chatterji, N. Fujita, and I. Ishihama, *Genes Cells* **3**, 270 (1998).

[6] H. Xiao, M. Kalman, K. Ikehara, S. Zemel, G. Glaser, and M. Cashel, *J. Biol. Chem.* **266**, 5980 (1991).

[7] D. R. Gentry, V. J. Hernandez, L. H. Nguyen, D. B. Jensen, and M. Cashel, *J. Bacteriol.* **175**, 7982 (1993).

suppressing a ppGpp^0 phenotype also are found among mutants[8] selected as rifampicin-resistant (rif-r).

Principles

An old saying in microbial genetics states that mutants of RNA polymerase in general, and rifampicin-resistant mutants in particular, can answer almost any selection. If actually true, this might reflect the unique role played by a single RNA polymerase in all *E. coli* transcription. A rich variety of spontaneous mutations of RNA polymerase subunits are found as the predominant answers for selections calling for the reversal of the apparent multiple amino acid auxotrophy accompanying a deficiency of (p)ppGpp due to deletion of the two genes responsible for the synthesis of ppGpp in *E. coli*.

Method

Standard microbial methods[9] are modified only to accommodate growth peculiarities of ppGpp^0 *E. coli* K12 strains. These include (1) lowering the NaCl concentration in LB to 5 g/L, rather than 10, because ppGpp^0 strains are somewhat salt sensitive; (2) subculturing working isolates every few days to minimize accumulation of unwanted revertants; (3) preservation of stocks by freezing at $-20\,^\circ$C in LB containing 10% glycerol rather than using solid media slants; and (4) using a standard growth temperature of 32$\,^\circ$C because growth of ppGpp^0 strains begins to deviate from wild-type strains above 37$\,^\circ$C.

Selection I: Direct Selection for Mutants Restoring Growth on Minimal Medium

Single colony isolates of ppGpp^0 strain CF1693, previously verified as unable to grow on M9 glucose minimal medium (phenotypically termed M$^-$), are streaked heavily for single colonies on several LB plates and allowed to grow for 24 hours. A sterile cotton swab is swirled in a few separate colonies, which are then rubbed on a glucose minimal plate as a $\sim$2-cm patch, applying about 10^5 to 10^6 cells per patch. This process is repeated with many plates, each containing 10–12 patches. After 48- to 72-hour incubations, about one-third of the patches typically show growth

[8] Y. N. Zhou and D. J. Jin, *Proc. Natl. Acad. Sci. USA* **95**, 2908 (1998).
[9] J. H. Miller, *in* "Experiments in Molecular Genetics," Cold Spring Harbor Laboratory Press, Cold Spring Harbor, New York, 1972.

of prototrophic (M$^+$) revertants with occasional jackpots of many revertants per patch. This pattern comprises a fluctuation test and provides an
estimate of the mutation rate, as well as assurance that isolates for different
patches are truly independent.[10] A single M$^+$ colony is picked from each
patch, streaked again for single colonies on minimal plates, and grown
for 36–48 hours. The ensuing single colonies are picked again, but streaked
both on LB and on minimal plates and similarly grown. Colonies from LB
plates, which appear morphologically homogeneous on both minimal and
LB plates, are suspended in 1 ml of liquid LB + 10% glycerol, frozen in
dry ice, and stored at −70 °C. Of 50 prototrophic suppressor alleles isolated
in this manner, 27 were found to map in *rpoB* (9 were rifampicin-resistant,
7 of which were T563P), 17 mapped in *rpoC*, and 3 mapped in *rpoD*. Three
were discarded because they contained mutations in both *rpoBC* or trans-
duced poorly (see later text). Attempts to isolate differential suppressors of
individual amino acid requirements generally yield the same mutants that
simultaneously suppress all requirements.

Selection II: Isolation of Suppressors Screened as M$^+$ Among Stationary Phase Survivors

Cultures are inoculated as for Selection I but on LB medium and incubated for 10 days–2 weeks before visualizing opaque microcolonies that
emerge at random from within almost transparent, lysed colonies. These
microcolonies are picked (one per parental colony), streaked on glucose
minimal medium, incubated for 36 hours, yielding growth in about half of
the cases. These M$^+$ isolates are purified and stored as for Selection I.
These revertants are lysis resistant when restreaked. The pattern of alleles
found by Selection II is different than for Selection I in that the majority of
(135) isolates show lesions in *rpoC* (77) rather than *rpoB* (58), *rpoB* T563P
is not the most frequently isolated mutation, and no mutations are found in
rpoD. The appearance of mutants by Selection II highlights the need to
minimize chances of accumulating secondary mutations.

Selection III: Rifampicin-Resistant Mutants Screened as M$^+$

Selection of spontaneous rif-r mutants of CF1693 on LB plates containing 50 μg/ml of rifampicin followed by screening for growth on minimal
medium revealed half (10/20) to be M$^+$. Almost all of previously known
spontaneous rif-r alleles[11] occur in *rpoB* codon region 500-575, but these

---

[10] S. E. Luria and M. Delbruck, Genetics **28**, 491 (1943).
[11] D. J. Jin and C. A. Gross, *J. Mol. Biol.* **202**, 45 (1988).

were not the most frequent rif-r mutations screened as prototrophs in the ppGpp^0 parental strain. Instead, this population is enriched for mutations in the *rpoB* codon region 123-181 where only a single spontaneous change of *rpoB* val146 was previously found[12] to be rif-r. It is surprising that this new spectrum of mutant alleles is encountered because 42 spontaneous rif-r alleles in strain MG1655 were reported to yield only 17 unique alleles leading to the conclusion that near-saturation was reached.[11] This hints that a ppGpp deficiency influences the spectrum of rif-r alleles obtained in an otherwise identical strain background. In only two instances have we isolated the same allele by selections I, II, and III. One is *rpoB* T563P, (alias *rpoB3770*); the other is *rpoB* G534C, a new allele. To explore the M$^+$ suppressor properties of known rif-r alleles, a set of 17 "classical" alleles[10] was transferred into strain CF1693 by phage P1 transduction and screened (see later text) for rifampicin resistance and for growth on minimal medium. We found the M$^+$ phenotype was associated with 6 of the 17 known rif-r alleles: *rpoB3595* (S522F); *rpoB114* (S531F); *rpoB3449* (A532Δ); *rpoB3443* (L533P); *rpoB3370* (T563P), and *rpoB111* (P564L). Some of these alleles have been characterized as "stringent mutants" of RNA polymerase.[8]

Other selections in a ppGpp^0 host have yielded additional mutants. Isolation of suppressors restoring complete prototrophy of a Δ*dksA* mutant of MG1655 (wild type for both *relA* and *spoT*) led to finding several instances of the *rpoB* T563P.[13] A selection for mutants in a ppGpp^0 host that restore activity of a *rpoN*-dependent reporter during entry into stationary phase yields a subset of M$^+$ suppressor alleles identical to those found here in *rpoB,C,D*, as well as new M$^+$ mutations and a separate subset of new *rpoBC* M$^-$ and *rpoN* alleles.[14] SpoI mutant suppressors of slow growth due to *hns stpA* double mutants in a *relA1* host also appear to be ppGpp^0 with a M$^-$ phenotype; not surprisingly, selection for rif-r yields M$^+$ suppressors.[15]

Genetic Linkage by P1 Transduction

Stocks of phage P1 vir are grown on MG1655 *btuB*::Tn10 (~50% linked to *rpoBC*), or CAG12152 MG1655 *air*::Tn10 (~90% linked to *rpoD*). These are used to transduce each mutant to tetracycline resistance (LB +

[12] K. Severinov, M. Soushko, A. Goldfarb, and V. Nikiforov, *Mol. Gen. Genet.* **244**, 120 (1994).
[13] L. Brown, D. Gentry, T. Elliott, and M. Cashel, *J. Bacteriol.* **184**, 4455 (2002).
[14] A. D. Laurie, L. M. D. Bernardo, C. C. Sze, E. Skarfstad, A. Szalewska-Palasz, T. Nystrom, and V. Shingler, *J. Biol. Chem.* **278**, 1494 (2003).
[15] J. Johansson, C. Balsalobre, S.-Y. Wang, J. Urbonaviciene, D. J. Jin, B. Sonden, and B. E. Uhlin, *Cell* **102**, 475 (2000).

40 μg/ml TC) to test for genetic linkage. Approximately 50 transductants are picked to grids on LB Tc plates. These grids were replica plated to LB Tc and to glucose minimal plates with Tc, scoring the frequency of recombinants that restore the M$^-$ phenotype as evidence of linkage. The ensuing tetracycline-resistant M$^+$ recombinants can be purified for use as P1 donors for backcrosses to the CF1693 parent to be sure the M$^+$ phenotype does not arise from sites in the mutant parent other than those linked to the transposon.

Complementation with rpoB, rpoC, rpoBC, and rpoD Multicopy Plasmid

Complementation tests of mutants transformed with individual multicopy plasmids, capable of constitutive expression of *rpoB*, *rpoC*, *rpoD*, or *rpoBC*, can be useful to verify localization of suppressor mutations. The plasmids used were *rpoB* (pT7*rpoB*)[16] or (*pDJJ11*),[17] *rpoC* (*pT7rpoC*)[16] *rpoD* (pMRG7),[18] or *rpoBC* (pDJJ12).[17] Complementation can occur from these plasmids without adding inducer. Reversal of the M$^+$ phenotype by multicopy plasmids is not achieved if the chromosomal suppressor mutation is a strong dominant negative, actually observed in very few instances. Many classical rif-r *rpoB* alleles do display dominance (known as rif^d) by failing to become rifampicin sensitive in the presence of an F$'$ 105 merodiploid in a *recA* background.[19] We find that dominance of the M$^+$ phenotype parallels that of rif-r under the same conditions. Nevertheless, a few M$^+$ mutants, including those found among the classical rif^d alleles, retain the M$^+$ phenotype even in the presence of the appropriate multicopy wild-type plasmid. Complementation tests are made by replica plating 5–10 transformants for each plasmid from grids on LB + Ap (100 μg/ml) plates to M9 minimal glucose + Ap plates to score for the M$^+$ phenotype.

Sequence Localization of Mutations

Mutations in *rpoB* and *rpoC* initially were localized by cleavage of mismatches within wild-type–mutant RNA heteroduplexes and later by gene sequencing. Mutations in *rpoD* were identified solely by gene sequencing.[20] The MisMatch Detect II nonisotopic RNase cleavage assay kit from Ambion was used according to the manufacturer's instructions.

---

[16] K. Zalenskaya, J. Lee, C. N. Gujuluva, Y. K. Shin, M. Slutsky, and A. Goldfarb. *Gene.* **89,** 7 (1990).

[17] D. J. Jin and C. A. Gross, *Mol. Gen. Genet.* **216,** 269 (1989).

[18] Kindly provided by Richard R. Burgess.

[19] D. J. Jin and C. A. Gross, *J. Bacteriol.* **171,** 5229 (1989).

[20] V. J. Hernandez and M. Cashel, *J. Mol. Biol.* **252,** 536 (1995).

Chromosomal DNA templates for the production of full-length *rpoB*, *rpoC*, or *rpoD* PCR products were used as templates for both sequencing reactions and for second-stage PCR reactions to generate RNA for RNA-RNA heteroduplexes. Primers for the RNA synthesis reactions were fused to phage T7 or SP6 promoter sequences and were chosen such that each pair of complementary nested RNA products (~1 Kb) overlapped neighbors by approximately 150 base pairs. No RNA-RNA mismatches were found for about 10% of the mutant strains analyzed. Complete gene sequencing is therefore necessary.

Mutant Saturation

It appears that these procedures continue to show promise for generating new mutants. The selections described have led to analyzing nearly 200 M^+ isolates. These have led to the identification of 35 unique alleles of *rpoB*, 24 of *rpoC*, and 2 of *rpoD*. At the time of isolation, 55 of these 61 mutants were previously unknown. The frequency of encountering new alleles remained high throughout the search and hopes of achieving mutant saturation were abandoned. However, the *rpoB* T563P mutation was isolated frequently, especially in Selection I. Deletions between direct sequence repeats were also found frequently, such as *rpoC* Δ(201–204), *rpoC* Δ(212–217), *rpoC* Δ(215–220), *rpoC* Δ(312–214), and *rpoC* Δ(1150–1174). The *rpoC* Δ(212–217) and *rpoC* Δ(215–220) mutations were isolated a total of 21 times, have different junctions within the same 10-base pair direct repeat, and were encountered about equally in Selections I and II. In contrast, we isolated only two *rpoD* alleles (P504L and S506F) repeatedly by Selection I, but never by Selection II, regardless of whether *rpoB* and *rpoC* mutants were minimized by the presence of a multicopy *rpoBC* expressing plasmid during mutant selections.[20]

Comments

Mutants of RNA polymerase subunits that suppress various features of the ppGpp[0] phenotype have been mentioned in the literature as providing support for a variety of views of ppGpp-regulatory effects.[8,14,15,20–23]

[21] C. C. Sze and V. Shingler, *Mol. Microbiol.* **31,** 1217 (1999).
[22] M. S. Bartlet, T. Gaal, W. Ross, and R. L. Gourse, *J. Bacteriol.* **182,** 1969 (2000).
[23] K. Kvint, A. Farewell, and T. Nystrom, *J. Biol. Chem.* **275,** 14795 (2000).

[45] Analysis of Transcriptional Repression by Mig1 in *Saccharomyces cerevisiae* Using a Reporter Assay

By SERGEI KUCHIN and MARIAN CARLSON

Reporter assays for transcriptional repression in the yeast *Saccharomyces cerevisiae* provide a genetic method for analysis of the function and regulation of a repressor *in vivo*. Briefly, the repressor is expressed as a fusion to a heterologous DNA-binding domain (DBD) and tested for its ability to repress transcription when recruited to the promoter of a reporter gene.[1] The promoter carries a binding site for the DBD but otherwise drives reporter transcription independent of the repressor and independent of the physiological or genetic conditions being studied. A major advantage of this approach is that the experimental procedures are relatively easy and do not require knowledge of the DNA-binding site. The application of this approach to such repressors as Mig1 has provided valuable insights into repressor function, regulation, and relationships to corepressors and to RNA polymerase II.

Strategy and Practical Considerations

The method compares the effects of a DBD-fused repressor on the function of two promoters, which differ only in their recognition by the DBD (Fig. 1). Two reporter plasmids are used. The first, parent plasmid contains a reporter in which a relatively strong promoter drives the transcription of a gene whose expression can be easily monitored (Fig. 1A), for example, the *CYC1-lacZ* fusion in which the yeast *CYC1* promoter directs the expression of the *Escherichia coli lacZ* gene.[2] The second plasmid is derived from the parent plasmid by inserting DBD binding sites 5' to the upstream activation sequence (UAS) (Fig. 1B). These plasmids are introduced into yeast by transformation. In the absence of a DBD-fused repressor, the two promoters should support nearly equal *lacZ* expression. When a DBD-repressor fusion is expressed, the reporter with DBD-binding sites is expected to display a significant decrease in β-galactosidase expression, while the reporter lacking the sites should be minimally affected. The repressive effect is numerically expressed as the ratio of the β-galactosidase activity assayed for the reporter lacking DBD-binding sites

---

[1] C. A. Keleher, M. J. Redd, J. Schultz, M. Carlson, and A. D. Johnson, *Cell* **68**, 709 (1992).
[2] L. Guarente and T. Mason, *Cell* **32**, 1279 (1983).

METHODS IN ENZYMOLOGY, VOL. 371

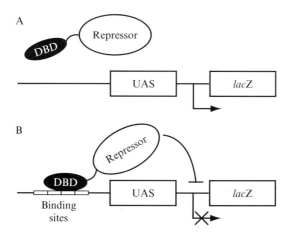

FIG. 1. Schematic representation of the repression assay. (A) A reporter consisting of a gene fusion in which a strong promoter drives the transcription of *lacZ*. The DBD-repressor does not bind. (B) A reporter with DBD binding sites inserted 5′ to the UAS of the same gene fusion. The DBD-repressor binds to the DBD sites to confer transcriptional repression. Representations are not to scale.

to that assayed for the reporter with the sites. The cells being compared both contain DBD-repressor, so that both reporters are equally subjected to any unanticipated effects on the state of the cell caused by DBD-repressor. As a control, the effect of the unfused DBD (DBD alone) on the expression of both reporters should be tested. The candidate repressor is concluded to possess repressor activity if the repression ratio determined for DBD-repressor considerably exceeds that determined for DBD alone.

The choice of the heterologous DBD is dictated by several factors. It should bind to its recognition sites in yeast, and no other protein should bind. It should not behave as a strong transcriptional repressor (or activator) when recruited to a promoter. It should not affect the native function of the fused protein; conversely, its ability to bind DNA should not be impaired by the fusion.

The bacterial LexA protein has been extensively used as a heterologous DBD in yeast. Although integration of a single LexA-binding site into a yeast promoter is often sufficient for binding, a tandem array of sites is usually integrated to improve DNA binding. LexA alone is not a transcriptional activator in yeast[3] and does not confer considerable negative effects, unless it is recruited to a site between the UAS and the TATA

[3] R. Brent and M. Ptashne, *Cell* **43,** 729 (1985).

box.[1,4] Most LexA-fused proteins retain their normal function, as judged by their ability to complement mutations in the respective genes. Although proteins fused to the C terminus of LexA have been reported to affect its DNA-binding efficiency, especially when using the isolated DNA-binding domain of LexA (residues 1-87),[5] this does not seem to pose a serious problem. Further advantages to the use of LexA as the DBD include the availability of reporters, expression vectors, and commercial antibodies against LexA.

Another general consideration concerns the choice of the promoter used for the expression of DBD-repressor fusions. When studying the effects of physiological and genetic conditions on repression, the function of such a promoter should not be largely affected. Western blot analyses must be performed to rule out the possibility that the apparent effects on repression simply reflect altered DBD-repressor protein levels.

Applications

Repressor Activity In Vivo

The use of a reporter assay has proved useful in confirming that a protein identified genetically as a negative regulator actually functions directly to repress transcription. If mutation of a gene encoding protein R results in increased expression of the target gene, it is not an automatic indicator that protein R is a transcriptional repressor. For example, R could inhibit the function of a positive regulator, or R could activate the transcription or the function of another repressor. The reporter assay described here can be used to answer the question: Can R function as a repressor when recruited to a promoter? If so, then further studies can examine whether repression by R is regulated, whether repression depends on other factors such as corepressors or components of RNA polymerase II holoenzyme, and so on. In addition, this assay can be used to map regions within the repressor that are required for repression, independent of DNA binding, by constructing derivatives of the DBD-R fusion.

Mig1 as a Specific Example. In yeast many genes are regulated at the transcriptional level by glucose repression, including the *GAL* and *SUC2* genes, which are required for growth on galactose and sucrose. It was not known which factors bind to the promoters of these genes to confer glucose repression until the *MIG1* gene was identified as a multicopy inhibitor of *GAL1* expression.[6] Overexpression and deletion of *MIG1* conferred

[4] R. Brent and M. Ptashne, *Nature* **314,** 198. (1985).
[5] E. A. Golemis and R. Brent, *Mol. Cell. Biol.* **12,** 3006 (1992).
[6] J. O. Nehlin and H. Ronne, *EMBO J.* **9,** 2891 (1990).

phenotypes suggesting that Mig1 is a negative regulator of the *GAL* and *SUC2* genes, and *MIG1* was found to encode a zinc-finger protein that binds to sites within a negative regulatory region of the *SUC2* promoter.[6] Thus it was proposed that Mig1 directly represses transcription. The use of a reporter assay to demonstrate that LexA-Mig1 represses transcription of a *lexA-CYC1-lacZ* reporter not only provided convincing evidence that Mig 1 is a transcriptional repressor, but also offered a system for genetic and physiological analysis of its function.[7,8]

Regulation of Repression in Response to Specific Signals

Once it has been established that protein R functions as a repressor, another important question is whether R is a downstream effector of a signal transduction pathway that represses transcription in response to a signal. Even if mutation of the repressor relieves repression, R need not be a bona fide member of the signaling pathway; for example, it could be a corepressor serving many different pathways, or it could be a negative factor that simply helps to fully shut down transcription. Use of this reporter assay may provide compelling evidence that the function of R as a repressor is regulated by the signaling pathway. First, the assay can be used to test whether the ability of DBD-R to repress reporter transcription is regulated in response to the signal. Second, the repressor function of the DBD-R fusion can be tested in mutants with defects in the signaling pathway. Controls must be performed to confirm that a particular signal or mutation does not affect the expression of the DBD-R protein.

One special case occurs when R is regulated at the level of binding to its native recognition site; the reporter assay will leave such regulation undetected because DBD-R is constitutively recruited to the reporter via DBD. If the native binding site is known, it can be integrated into the reporter much like the DBD sites, although one limitation of this approach is that other native proteins with similar recognition sequences may also bind.

Mig1 as a Specific Example. Because Mig1 negatively regulates a number of glucose-repressible genes, it was proposed that Mig1 itself is regulated by the glucose signal. Once LexA-Mig1 was shown to repress *lexA-CYC1-lacZ* reporter expression, experiments were performed to demonstrate that repression by LexA-Mig1 is maximal when cells are grown in high glucose.[7] Reporter repression by LexA-Mig1 in high glucose was 20-fold (1.8-fold for LexA alone). On shift to low glucose, repression was only 3.1-fold (2.6-fold for LexA alone). Western blot analysis showed

[7] M. A. Treitel and M. Carlson, *Proc. Natl. Acad. Sci. USA* **92,** 3132 (1995).
[8] M. A. Treitel, S. Kuchin, and M. Carlson, *Mol. Cell. Biol.* **18,** 6273 (1998).

that this difference could not be attributed to a change in LexA-Mig1 protein levels; moreover, this difference correlated with differential phosphorylation of Mig1. Thus these results strongly suggested that Mig1 is regulated by a glucose signal.

Functional Relationships to Corepressors and Components of the Transcriptional Apparatus

This reporter system can be used to analyze the functional hierarchy of components in a repression pathway. If R1 and R2 are functionally related and have been established to repress reporter transcription, then repressor activity of R1 can be examined in a strain lacking R2, and vice versa. For example, this approach was used in experiments leading to the model that the Ssn6-Tup1 complex is a corepressor that is recruited to diverse promoters by specific DNA-binding proteins.[1,7,9,10] The primary functional role appears to be played by Tup1, since reporter repression by LexA-Tup1 occurs in an *ssn6Δ* mutant, while repression by LexA-Ssn6 is abolished in a *tup1Δ* mutant.[1,9] The function of particular DNA-binding repressors was shown to require Ssn6-Tup1 by carrying out the repression assay in a mutant lacking the corepressor.

Mig1 as a Specific Example. Evidence that *mig1*, *ssn6*, and *tup1* mutations relieve glucose repression of many genes suggested that repression by Mig1 depends on the Ssn6-Tup1 corepressor. To test this model, LexA-Mig1 was assayed for ability to repress reporter transcription in *ssn6Δ* and *tup1Δ* mutants; LexA-Mig1 did not repress transcription, indicating dependence on Ssn6-Tup.[7] Two-hybrid assays confirmed that Mig1 and Ssn6 interact. Interestingly, in the absence of Ssn6-Tup1, LexA-Mig1 behaved as a transcriptional activator, suggesting an unexpected possibility that Mig1 is a dual-function transcriptional regulator: a repressor under some conditions and an activator under others.

Evidence implicates proteins associated with the RNA polymerase II holoenzyme in transcriptional repression.[11,12] For a particular repressor, its functional dependence on such components can be assessed by examining repression of a reporter in appropriate mutants. In the case of mutations that affect the general transcriptional apparatus, monitoring expression levels of the DBD-repressor is critical. Moreover, the possibility of indirect effects, such as altered expression of corepressors or other proteins that modify the function of the repressor, must also be considered.

[9] D. Tzamarias and K. Struhl, *Nature* **369**, 758 (1994).

[10] D. Tzamarias and K. Struhl, *Genes Dev.* **9**, 821 (1995).

[11] M. Boube, L. Joulia, D. L. Cribbs, and H.-M. Bourbon, *Cell* **110**, 143 (2002).

[12] A. J. Courey and J. Songtao, *Genes Dev.* **15**, 2786 (2001).

Mig1 as a Specific Example. Genetic evidence suggested that proteins associated with the RNA polymerase II holoenzyme are directly involved in glucose repression, including the Srb10-Srb11 kinase.[13–15] To test the idea that this kinase has roles in repression by Mig1, reporter repression by LexA-Mig1 was examined in *srb10* and *srb11* mutants. Repression was decreased, as was repression by LexA-Ssn6 and LexA-Tup1.[16]

Materials and Methods

Strains

All strains used for a single set of experiments should be isogenic. Strains must carry at least two auxotrophic markers to select for the reporter and expression plasmids. If repressor activity will be examined in mutant strains, these should also be isogenic. The genetic background of the strain may affect the repression ratios.

LexA fusion plasmids

Vectors expressing the LexA DNA-binding domain (residues 1-87) and full-size LexA (residues 1-202) have provided good results. Full-size LexA contains a dimerization domain, which has been reported to improve DNA binding and may reduce the possible adverse effects of the fused protein on the DNA-binding properties. Also, $LexA_{1-87}$ fusions often display somewhat higher than predicted mobilities on SDS-PAGE, suggesting the possibility of conformational aberrations caused by $LexA_{1-87}$. The following expression vectors can be used: pSH2-1 ($LexA_{1-87}$, 2μ, *HIS3*),[17] Lex(1-202) + PL ($LexA_{1-202}$, 2μ, *HIS3*),[18] pEG202 ($LexA_{1-202}$, 2μ, *HIS3*),[19] YCp91 ($LexA_{1-202}$, *CEN*, *TRP1*).[9] Expression is driven by one of two versions of the *ADH1* promoter: the longer promoter contains a strong UAS and generally supports higher levels of expression (in pSH2-1, pEG202, and $LexA_{1-202}$ + PL) than the shorter version of the promoter (in YCp91). Although both versions of the *ADH1* promoter function nearly

[13] L. G. Vallier and M. Carlson, *Genetics* **137,** 49 (1994).

[14] S. Kuchin, P. Yeghiayan, and M. Carlson, *Proc. Natl. Acad. Sci. USA* **92,** 4006 (1995).

[15] D. Balciunas and H. Ronne, *Nucleic Acids Res.* **23,** 4421 (1995).

[16] S. Kuchin and M. Carlson, *Mol. Cell. Biol.* **18,** 1163 (1998).

[17] S. D. Hanes and R. Brent, *Cell* **57,** 1275 (1989).

[18] D. M. Ruden, J. Ma, Y. Li, K. Wood, and M. Ptashne, *Nature* **350,** 250 (1991).

[19] E. A. Golemis, I. Serebriiskii, J. Gjuris, and R. Brent, *in* "Current Protocols in Molecular Biology" (F. M. Ausubel, R. Brent, R. E. Kingston, D. D. Moore, J. G. Seidman, J. A. Smith, and K. Struhl, K., eds.) Vol. 3, p. 20.1. Wiley, New York, 1997.

constitutively, they show differential responses to some genetic and physiological conditions.

Confirmation of expression of the desired LexA-repressor fusion by western blot analysis is advised. Testing whether the LexA-repressor fusion retains the function of the native repressor is also important. This is accomplished by examining the plasmid for ability to complement the corresponding yeast mutation. The previously mentioned vectors are designed for fusing LexA to the N terminus of the repressor. If the fusion protein is not detectable or functional, LexA could be fused at the C terminus of the repressor.

Reporter Plasmids

Plasmid pLGΔ312S, which contains a *CYC1-lacZ* gene,[2] and its derivative pJK1621 (4 lexA-binding sites 5′ to the *CYC1* UAS)[1] are multicopy 2μ-based, *URA3*-selectable reporters that have been widely used.[1,7,9,20–24] Plasmid pBM2762 (2 μ, *URA3*) contains a *HIS3-LEU2-lacZ* fusion with the *HIS3* UAS and *LEU2* TATA box,[25] and its derivative pMT27 has a LexA site 5′ to the *HIS3* UAS.[8] One disadvantage of these reporters is the possibility of copy number variation depending on the physiological or genetic conditions. It may be beneficial to use or construct an integrating reporter; the number of integrated copies should be verified.

Transformation of Yeast to Introduce Plasmids

Reporter plasmids and plasmids expressing the repressor and control constructs are introduced into yeast by transformation with selection for plasmid markers.[26,27] In the case of an integrated reporter, no continuous marker selection is required, except for occasional confirmatory tests.

Stock Solutions

1. $10 \times$ TE (100 m*M* Tris-HCl, pH7.5; 10 m*M* EDTA). Filter-sterilize.
2. $10 \times$ LiAc (1 M lithium acetate, pH7.5; adjust pH with acetic acid). Filter-sterilize.

[20] H. Wang and D. J. Stillman, *Mol. Cell. Biol.* **13,** 1805 (1993).
[21] J. Recht, B. Dunn, A. Raff, and M. A. Osley, *Mol. Cell. Biol.* **16,** 2545 (1996).
[22] W. Song and M. Carlson, *EMBO J.* **17,** 5757 (1998).
[23] S. H. Park, S. S. Koh, J. H. Chun, H. J. Hwang, and H. S. Kang. *Mol. Cell Biol.* **19,** 2044 (1999).
[24] V. K. Vyas, S. Kuchin, and M. Carlson, *Genetics* **158,** 563 (2001).
[25] S. Ozcan, L. G. Vallier, J. S. Flick, M. Carlson, and M. Johnston, *Yeast* **13,** 127 (1997).
[26] H. Ito, Y. Fukuda, K. Murata, and A. Kimura, *J. Bacteriol.* **153,** 163 (1983).
[27] M. D. Rose, F. Winston, and P. Hieter, "Methods in Yeast Genetics: A Laboratory Course Manual," Cold Spring Harbor Laboratory Press, Plainview, NY, 1990.

3. 50% polyethylene glycol, average molecular weight 3,350 (PEG 3,350). Warm in a microwave oven for better solubilization, but avoid overheating. Filter-sterilize. Store this solution tightly closed at room temperature.

Working solutions (for best results, prepare with sterile distilled water immediately before use):

    Solution A: 1× TE
    Solution B: 1× TE, 1× LiAc
    Solution C: 40% PEG, 1× TE, 1× LiAc

Procedure

1. Grow a 50-ml yeast culture in YPD medium (1% yeast extract, 2% bacto-peptone, 2% glucose) at 30°C to mid-log phase ($OD_{600} = 0.5$–1.0). Such a culture provides cells for 20 transformations.

2. Collect the cells by centrifugation at 2,000 g for 5 min at room temperature.

3. Wash the cells by resuspending the cell pellet in 10 ml of Solution A and collecting the cells by centrifugation as previously described.

4. Resuspend the cells in 1.0 ml of Solution B. Incubate at 30°C for 1 h. The cells are then competent for transformation.

5. Competent cells are best if used immediately, but they can be stored several days on ice or at 4°C. Some strains show very little reduction in transformation efficiency after being stored on ice for extensive periods of time; however, transformants obtained from such competent cell stocks may display a growth delay on selective plates and take a day longer to form colonies. After that time, such transformants appear to behave indistinguishably from transformants obtained from freshly prepared competent cells.

6. For transformation of these competent cells with a plasmid DNA, add the following components to a sterile 1.5-ml microcentrifuge tube:

    a. 2 μl of plasmid DNA (about 0.1–0.2 μg/μl). If two or three different plasmids are being introduced, use this amount of DNA for each plasmid. DNA prepared by standard alkaline lysis miniprep procedures works adequately.

    b. 2 μl of carrier DNA (10 mg/ml). Commercially available sheared and phenol-extracted salmon sperm or calf thymus DNA solutions sold for use in hybridization analyses may be used.

    c. 50 μl of competent yeast cell suspension.

    d. 250 μl of Solution C.

Mix gently and incubate at 30°C for 1–2 h or longer. Longer incubation times may improve the recovery of transformants with multiple plasmids.

7. (Optional). Heat-shock the transformation mix at 42°C for up to 15 min. Note that some mutants or wild-type strains of some strain backgrounds may not tolerate the heat shock. It appears that the heat shock step offers no substantial improvement in transformation efficiency if the transformation mix has been incubated for an extended period of time (3–4 h).

8. Collect the cells by centrifugation in a microfuge at 10,000 g for 20 sec. Remove the viscous supernatant using a sterile pipetting tip or by aspirating with a sterile Pasteur pipette. Add 100–200 μl of Solution A to the cell pellets and let stand for several minutes until the cells can be pipetted easily. Plate the cells onto medium[27] that is selective for the markers carried by the plasmids.

Alternate Procedure for Introduction of Multiple Plasmids. If attempts to transform yeast simultaneously with more than one plasmid fail, the plasmids can be introduced sequentially. Because selection is required for plasmid maintenance, transformation of a strain carrying a plasmid requires a modification of the basic protocol.

1. Transform with the first plasmid as previously described.

2. Grow an overnight culture of the transformant in the appropriate selective liquid medium.

3. Transfer the cells to YPD at a density of OD_{600} = 0.15–0.25. Grow the cells for one to two generations. Growing cells in YPD prior to transformation is believed to greatly increase transformation efficiency. At the same time, plasmid loss due to lack of selection for 1–2 generations is practically insignificant.

4. Perform steps 2–8 of the regular transformation procedure.

Assays of β-Galactosidase Activity

β-Galactosidase activity may be assayed in permeabilized cells or in protein extracts.[27–29] Fold repression by a DBD-repressor fusion is calculated as the ratio of the β-galactosidase activity level measured for the reporter containing the DBD-binding sites to the β-galactosidase activity level measured for the control reporter that lacks DBD sites. A control experiment should be performed in parallel to determine the fold repression

[28] J. H. Miller, "Experiments in Molecular Genetics," Cold Spring Harbor Laboratory, Plainview, NY, 1972.
[29] L. Guarente, *Methods Enzymol.* **101**, 181 (1983).

for cells expressing the DBD alone. The DBD-repressor fusion is concluded to have repressor activity if the fold repression it causes exceeds that of the DBD alone.

β-Galactosidase is a stable enzyme. For example, if a substance causes repression, adding the substance to a culture that has already accumulated β-galactosidase will not result in the disappearance of β-galactosidase.

Because of variation among transformants, particularly with respect to plasmid copy number, assay of five or more transformants representing each strain/plasmid combination is advisable. The fold repression values are obtained by dividing mean activities that have been determined with a certain standard error. If the respective relative standard errors for activities A1 and A2 are E1(%) and E2(%), then the relative standard error of the value A1/A2 will be approximately [E1 + E2] (%).

Reagents

1. Z buffer (for 1 L): 16.1 g $Na_2HPO_4 \times 7H_2O$, 5.5 g $NaH_2PO_4 \times H_2O$, 0.75 g KCl, 0.25 g $MgSO_4 \times 7H_2O$, adjust to pH 7.0 if necessary. Filter-sterilize; do not autoclave. Immediately before use, add 27 μl of β-mercaptoethanol per 10 ml of Z buffer.

2. ONPG (*o*-nitrophenyl-β-D-galactopyranoside; 4 mg/ml in H_2O). Prepare fresh solution or store in the dark at $4\,^\circ$C for several days.

3. Chloroform.

4. 0.1% SDS.

5. 1 M Na_2CO_3.

6. (For Protocol 2): Bradford reagent for determining protein concentrations (e.g., made from BioRad dye reagent concentrate [catalog No. 500-0006.])

Protocol 1.   Assays of Permeabilized Cells

1. Grow yeast at $30\,^\circ$C in selective liquid medium to mid-log phase ($OD_{600} = 0.3$–0.7). In the case of flocculent (clumpy) strains, take culture aliquots, add EDTA to 5–20 mM, and disperse the clumps by vortexing the cells, then take the OD_{600} readings. IMPORTANT: Do not add EDTA to the entire culture or to culture aliquots directly intended for β-galactosidase assays, as EDTA inhibits the enzymatic activity of β-galactosidase.

2. Spin down an equivalent of 1 ml at $OD_{600} = 0.5$ in small glass tubes in a clinical centrifuge at 2,000 g for 5 min at room temperature. Aspirate the medium.

3. Add 1 ml of Z buffer, 3 drops of chloroform, and 2 drops of 0.1% SDS. Include a cell-free control sample to be used as a blank.

4. Vortex at room temperature for 10 sec at maximum speed to permeabilize the cells. Equilibrate the tubes at 28°C in a water bath for 5 min.

5. Start timer. Add 0.2 ml of ONPG to the samples, including the control sample. To process multiple samples, add ONPG at equal time intervals (e.g., every 10 sec). Mix by gentle shaking or vortexing.

6. When the sample turns yellow, stop the reaction by adding 0.5 ml of 1 M Na_2CO_3 and mixing gently. Also add 0.5 ml of Na_2CO_3 to the control sample. Record the reaction time.

7. Spin down cell debris at 2,000 g for 5 min. Read the OD_{420} of the supernatant relative to the blank sample. Calculate the β-galactosidase activity in Miller units[28] using the formula:

$$\text{Activity} = (OD_{420} \times 1000)/(OD_{600} \times \text{vol assayed} \times \text{time in min})$$

NOTE: When activity levels are low, it may be necessary to allow the reactions to proceed 2 h or longer. This often results in detectable background OD_{420} readings in the cell-free control samples (relative to Z buffer alone) due to nonenzymatic ONPG decay. For this reason, it is important to use a cell-free control sample processed in parallel, not Z buffer alone, as a blank for establishing the correct OD_{420} readings in the experimental samples. Since two or more sets of samples are usually assayed in one experiment that require different reaction times (e.g., 2 h for set 1, and 4 h for set 2), an individual control sample should be included for each set.

Protocol 2. Assays of Protein Extracts

1. Grow yeast cultures at 30°C in selective liquid medium to $OD_{600} = 0.5$–1.0 (for satisfactory protein recovery, each culture should be at least 5 ml). In the case of flocculent strains, take culture aliquots, add EDTA to 5–20 mM, vortex the cells, and take the OD_{600} readings. IMPORTANT: Do not add EDTA to culture aliquots intended for β-galactosidase assays. Although it is important that the cultures are in exponential phase, the exact OD_{600} readings are not important.

2. Spin down the cultures in small glass tubes in a clinical centrifuge at 2,000 g for 5 min at room temperature. Aspirate the medium. (The cell pellets can be frozen at −20°C or at −70°C and assayed at a later date.)

3. Put the samples on ice. If the cell pellets were stored frozen, allow them to thaw. Add 200 μl of cold Z buffer. Add 0.5–0.6 g of glass beads (0.45 mm in diameter). Add 3 μl of octanol to prevent foam formation during subsequent vortexing.

4. Vortex at maximum speed for 10 sec, followed by 10 sec on ice; repeat 10 times. Keep samples on ice.

5. Transfer the liquid into a 1.5-ml microfuge tube using a 200-μl Pipetman with a tip cut off approximately 3 mm from the nose, to avoid flow blockage by the glass beads.

6. Spin down cell debris in a microfuge at 10,000 g for 5 min. Transfer the cleared supernatants into 1.5-ml microfuge tubes. Keep samples on ice.

7. Assay protein concentrations. The amounts of recovered protein may vary with strain and culture size. For a pilot estimate, place a small (1–5 μl) protein extract aliquot into a 1.5-ml tube, add 1 ml of 5-fold diluted BioRad protein assay reagent (precalibrated using BSA as standard), and allow to stand at room temperature for 10 min. Use Z buffer as a control. Read OD_{595} of the reactions and calculate the protein concentrations according to your reagent calibration curve. Repeat this step if necessary using a different amount of protein extract; if necessary, dilute the extracts in Z buffer.

8. Place 0.9 ml of Z buffer in a small glass tube; add 100 μl of protein extract. Include one or more control samples containing 1 ml of Z buffer. Equilibrate at 28 °C for 5 min.

9. Start timer. Add 0.2 ml of ONPG to the samples, including the control samples. To process multiple samples, add ONPG at equal time intervals (e.g., 10 sec). Mix by gentle shaking or vortexing.

10. When the reaction mixture turns yellow, stop the reaction by adding 0.5 ml of 1M Na_2CO_3. Record the reaction time. Also add 0.5 ml of Na_2CO_3 to the control sample. Gently mix.

11. Check OD_{420} of the supernatant relative to the appropriate control sample. Calculate the β-galactosidase activity using the formula:

$$\text{Activity} = (OD_{420} \times 1000)/(\text{reaction time in min} \times \text{mg of protein assayed})$$

NOTE: Appropriate controls are discussed in Protocol 1.

Western Blot Analysis

It is imperative that western blot analyses be performed to monitor repressor protein levels. The amount of repressor protein in the cell can easily affect the efficacy of repressor function in the assay. Expression levels can vary for many reasons: variation in copy number of the expression plasmid, alterations in expression or stability of the protein under different physiological conditions, or alterations in one of these parameters in certain mutant backgrounds. For key results, at least two transformants of each type should be examined by western blot; use the two transformants with β-galactosidase values closest to the group average.

Reagents

1. 2× protein loading buffer (125 mM Tris-HCl, pH 6.8, 4% SDS, 20% glycerol, 0.05% bromophenol blue, add β-mercaptoethanol to 4% just before use).

2. Antibodies that recognize the repressor fusion protein. Monoclonal (Clontech, catalog No. 5397-1) and polyclonal (Invitrogen, catalog No. R990-25) LexA antibodies are commercially available. IMPORTANT: For fusions to LexA$_{1-87}$, use the polyclonal antibodies.

Protocol

1. It is preferable to save samples of the cultures that are assayed for β-galactosidase activity. These samples can be stored as frozen cell pellets. Otherwise, regrow the cultures under the same conditions.

2. Collect the equivalent of 1.0 ml of cells at OD$_{600}$ = 0.5 by centrifugation in a 1.5-ml microfuge tube at 10,000 g for 1 min. Aspirate off the supernatant. Add 0.1 g of glass beads and 50 μl of 2 × protein loading buffer.

3. Place the tubes in a boiling water bath for 1 min.

4. Vortex the tubes at top speed for 1 min.

5. Repeat steps 3 and 4 three times.

6. (Optional). In most cases, the samples can be stored at $-70\,^{\circ}$C at this step and analyzed later. If stored, the samples should be reboiled for 1 min just before use.

7. Spin down the cell debris at 10,000 g for 1 min.

8. Load 20 μl of supernatant onto an SDS-PAGE minigel.

9. Transfer the proteins onto a membrane and perform western blot analysis using a standard protocol. At a field strength of 7 V/cm, proteins of different size (30–100 kDa) and phosphorylation status appear to be uniformly retained on a PVDF membrane (Immobilon-P from Millipore, pore size 0.45 μm) after a more than 2-h transfer.

Concluding Remarks

This reporter assay provides a useful approach for analysis of the function and regulation of a repressor. The advantages include (1) assessment of function *in vivo*, (2) variation of physiological conditions and genetic background of the host, and (3) no requirement of knowledge of the DNA sequence recognized by the native repressor. The major limitations are inherent in the fact that the fusion protein and reporter may not display all of the regulatory properties exhibited by the native protein at its natural target promoter. Findings require confirmation by other methods that assess the function of the native repressor at its natural target sites.

[46] Mutational Analysis of *Drosophila* RNA Polymerase II

By MARK A. MORTIN

This article recounts the history of the mutational analysis of RNA polymerase II in *Drosophila melanogaster*. It highlights the different methods used to generate these mutations, summarizes the benefits and limitations of RNA polymerase II second-site selections, and discusses the prospects for the future of this type of mutational analysis in *Drosophila*.

Mutational Analysis

An excellent and detailed review of the parameters that must be considered before deciding on a mutagenesis strategy for recovering random mutations in *Drosophila* are presented elsewhere.[1] Targeted gene knockout and the ability to introduce specific mutations are also now available thanks to the development of strategies to achieve homologous recombination.[2] A brief discussion of the pros and cons of these two approaches follows.

Random

A number of parameters should be considered in designing a random screen for mutations. For example, what mutant phenotype will be identified in a screen? For RNA polymerase II, the answer comes from yeast where 12 subunits have been identified and mutated with 9 being essential for viability.[3] The genomic sequence of *Drosophila* is now complete, and homologs of all 12 yeast subunits have been identified.[4,5] Mutations exist in four of these subunits: RpII215, RpII140, RpII33, and RpII15,[6-9]

[1] M. Ashburner, "*Drosphila*: A Laboratory Handbook," Cold Spring Harbor Laboratory Press, Cold Spring Harbor, New York, 1989.
[2] Y. S. Rong and K. G. Golic, *Science* **288**, 2013 (2000).
[3] R. A. Young, *Annu. Rev. Biochem.* **60**, 689 (1991).
[4] M. D. Adams, S. E. Celniker, R. A. Holt, *et al.*, *Science* **287**, 2185 (2000).
[5] N. Aoyagi and D. A. Wassarman, *J. Cell. Biol.* **150**, F45 (2000).
[6] A. L. Greenleaf, *J. Biol. Chem.* **258**, 13403 (1983).
[7] M. A. Mortin, R. Zuerner, S. Berger, and B. J. Hamilton, *Genetics* **131**, 895 (1992).
[8] D. A. Harrison, M. A. Mortin, and V. G. Corces, *Mol. Cell. Biol.* **12**, 928 (1992).
[9] M. Ashburner, S. Misra, J. Roote, S. E. Lewis, R. Blazej, T. Davis, C. Doyle, R. Galle, R. George, N. Harris, G. Hartzell, D. Harvey, L. Hong, K. Houston, R. Hoskins, G. Johnson, C. Martin, A. Moshrefi, M. Palazzolo, M. G. Reese, A. Spradling, G. Tsang, K. Wan, K. Whitelaw, B. Kimmel, S. Celniker, and G. M. Rubin, *Genetics* **153**, 179 (1999).

representing homologs of the yeast subunits Rpb1, Rpb2, Rpb3, and Rpb9, respectively. A reasonable assumption is that the subunits that are essential for viability in yeast are also required in *Drosophila*. In addition, with the added complexity of the development of a multicellular organism, it is possible that mutations that cause reduced viability in yeast will result in recessive lethality in flies. This has already been demonstrated to be the case for the ninth subunit, RpII15, which is essential in flies but not in yeast.[8]

Other questions to consider include what types of mutations are desired and how large a screen is the investigator willing to perform? Performing a random mutagenesis allows the fly to dictate which domains in the subunits are essential for normal activity and what mutant phenotypes they will engender. The size of the screen depends on a number of factors discussed in the following text. A standard screen for recessive-lethal mutations is the simplest method for obtaining a random collection of mutations in a gene of interest. A schematic diagram of such a screen is presented in Fig. 1

1. The first decision diagrammed is the choice of mutagen. The most commonly used mutagens are the chemical ethylmethane sulfonate (EMS), ionizing irradiation including X-rays and gamma rays, and the mobilization of transposable elements via hybrid dysgenesis. The choice of mutagen largely determines the type of mutation recovered and the size of the screen that must be performed.

Hybrid Dysgenesis: The advantage of using the mobilization of transposable elements is that it provides a molecular marker for the site of insertion. The disadvantages of this approach are that some genes appear to be refractory to insertion, the rate of insertion at different sites is highly variable and locus dependent, and mutations often only partially inactivate gene expression; these "leaky" mutations might have sufficient residual activity to permit survival over the tester chromosome and therefore be missed in this screen.

Irradiation: The advantage of using ionizing irradiation is that it is more random than the mobilization of transposable elements. The resulting mutations are often chromosome aberrations, including deletions, inversions, and translocations, which provide molecular markers for the locus of interest. The disadvantages of this approach are that chromosome aberrations often affect multiple genes, making identification of the desired mutation difficult. Furthermore, the rate of mutation, although higher than with the mobilization of transposable elements, is still only approximately 1 in 10,000 irradiated chromosomes mutated at a given locus. Thus very large screens must be performed to recover recessive-lethal mutations at a given locus.

EMS: Ethylmethane sulfonate preferentially alkylates guanine residues, changing its pairing specificity to thymine and resulting in mutations

1. Choice of Mutagen: EMS/Irradiation/Hybrid Dysgenesis

2. Males with Isogenic Chromosome × 3. Female with Balancer Chromsome(s)

Genotype	red e/red e	TM3, Sb e/TM6B, e
Phenotype	Red, Ebony	Stubble, Ebony

4. Mutagenized Isogenic Chromosome/Balancer Chromosome × 5. Tester Chromosome/Balancer Chromosome

Genotype	*red e/TM3, Sb e	Df(3R)red^{P52}/TM3, Sb e
Phenotype	Ebony, Stubble	Stubble

6. Mutagenized Chromosome/Tester Chromosome   7. Mutagenized Chromosome/Balancer Chromosome

Genotype	*red e/Df(3R)red^{P52}	TM3, Sb e/TM3, Sb e	Df(3R)red^{P52}/TM3, Sb e	*red e/TM3, Sb e
Phenotype	Red	Dead	Stubble	Ebony, Stubble

FIG. 1. Diagram of a screen for recessive-lethal mutations in a subunit of RNA polymerase II.

TABLE I
Screens to Recover Recessive-Lethal Mutations in Genes Encoding RNA-Polymerase II Subunits

Mutagen	RpII215		RpII140		RpII33	RpII15
	Irradiation[a]	EMS[b,c,d]	Irradiation[e]	EMS[e]	EMS[f]	EMS[e]
No. of Chromosomes	23,829	25,067	3,416	14,189	NA	14,189
No. of RpII alleles	2	25	1	30	2	1
Frequency	0.0001	0.0010	0.0003	0.0021	NA	0.00007

[a] G. Lefevre, *Genetics* **99**, 461 (1981).
[b] M. A. Mortin and G. Lefevre, Jr., *Chromosome* **82**, 237 (1981).
[c] M. A. Mortin and T. C. Kaufman, *Mol. Gen. Genet.* **187**, 120 (1982).
[d] R. A. Voelker, G. B. Wisely, S.-M. Huang, and H. Gyurkovics, *Mol. Gen. Genet.* **201**, 437 (1985).
[e] M. A. Mortin, R. Zuerner, S. Berger, and B. J. Hamilton, *Genetics* **131**, 895 (1992).
[f] M. Ashburner, S. Misra, J. Roote, S. E. Lewis, R. Blazej, T. Davis, C. Doyle, R. Galle, R. George, N. Harris, G. Harzell, D. Harvey, L. Hong, K. Houston, R. Hoskins, G. Johnson, C. Martin, A. Moshrefi, M. Palazzolo, M. G. Reese, A. Spradling, G. Tsang, K. Wan, K. Whitelaw, B. Kimmel, S. Celniker, and G. M. Rubin, *Genetics* **155**, 179 (1999).
NA, Not applicable.

of G:C into A:T (reviewed in reference 1). In fact, 12 of 13 sequenced mutations recovered in screens (as opposed to selections, see following text) for mutations in RNA polymerase II are G:C to A:T transitions.[8,10,11] EMS can also (1) modify thymidine such that it pairs with guanine, resulting in mutations of A:T into G:C, and (2) modify DNA such that it becomes depurinated, resulting in chromosome aberrations via inaccurate excision repair (reviewed in ref 1). The rate of mutation is higher than with irradiation, with roughly 1 in 1,000 mutagenized chromosomes mutated at an average size locus. The disadvantage of this approach is that there usually is not an obvious molecular lesion responsible for the mutation, which often must be identified by extensive sequence analysis.

On average the relative rate of mutation following irradiation is proportional to the size of the gene, including essential regulatory sequences, whereas the rate of EMS-induced mutation is roughly proportional to the number of essential amino acids. In general, larger genes are bigger targets for mutagenesis than smaller ones, although this may not be true for the genes encoding the two largest subunits of RNA polymerase II. Table I

[10] Y. Chen, J. Weeks, M. A. Mortin, and A. L. Greenleaf, *Mol. Cell. Biol.* **13**, 4214 (1993).
[11] I. Krasnoselskaya, J. Huang, T. Jones, C. Dezan, and M. A. Mortin, *Mol. Gen. Genet.* **258**, 457 (1998).

summarizes the results from six mutagenesis experiments that produced mutations in four RNA polymerase II subunits. RpII215 is approximately 1.5 times large than RpII140, yet the latter appears to be more readily mutable by both irradiation (although the sample size is small) and EMS. With irradiation, 2 RpII215 mutations were recovered in a screen of 23,829 mutagenized chromosomes compared to 1 RpII140 mutation in a screen of 3,416 mutagenized chromosomes (Table I). With EMS, 25 RpII215 mutations were recovered in screens of 25,067 mutagenized chromosomes compared to 30 RpII140 mutations in a screen of 14,189 mutagenized chromosomes. In other words, the rates of EMS-induced mutations are 1 in 1,000 for RpII215 but 1 in 500 for RpII140 (Table I). Consistent with their sizes, both RpII33 and RpII15 are much smaller targets for mutation than the two largest subunits. Mutation rates for RpII140 and RpII15 can be directly compared because they were recovered in the same screen. Although RpII140, is 10 times larger than the RpII15, it is 30 times more mutable (Table I). Because only one subunit is smaller than RpII15, it is likely that mutations in all subunits will prove to be attainable by this approach.

2. Although random mutagenesis mutates all chromosomes, the simplest approach is to isolate and test only one at a time. The 12 known subunits have been mapped to 3 different chromosomes, with RpII215 mapping to the first or X chromosome; RpII33, Rpb5, Rpb11 and Rpb12 to the second, and RpII140, Rpb4, RpII18, Rpb7, Rpb8, RpII15, and Rpb10 to the third. The example in Fig. 1 shows the design of the mutagenesis experiment that generated mutations in RpII140 and RpII15.[7] Flies with an isogenic chromosome carrying recessive-visible mutations, which permit the unambiguous identification of all genotypic classes of progeny, were mutagenized. The isogenic chromosome eliminates polymorphisms that would make identification of new point mutations more difficult. The chromosome used in Fig. 1 has two mutations that cause recessive-visible mutant phenotypes—red, which darkens the eye color, and ebony (e), which darkens the body pigmentation. Young males from this stock were isolated and aged 3–6 days. An EMS solution was made in a fume hood starting with a 1% sucrose solution and 2.4 μl/ml of EMS,[12] which needs to be aspirated several times to disperse well. A folded cloth absorbent wipe (kim wipe) was placed in an 8-dram shell vial, which is then impregnated with approximately 1 ml of the mutagen solution. Up to 50 males are transferred to the vial, without anesthetizing, and left in the hood overnight.

---

[12] E. B. Lewis and F. Bacher, *Dros. Inform. Serv.* **43**, 193 (1968).

3. The mutagenized males are then mated to virgin females carrying a multiply rearranged "balancer" chromosome (the example shown uses TM3, Sb e), which prevents recombination between it and the mutagenized homolog. The balancer shown in this figure can be observed by virtue of its dominant-visible mutation, Stubble (Sb), which shorten the bristles, and the e mutation.

4. When the first-generation progeny enclose, they possess a single mutagenized isogenic chromosome, which is balanced. In the example given, individual progeny carrying a single mutagenized red e chromosome balanced over TM3 are isolated. These flies appear ebony and Stubble.

5. Single flies with balanced mutagenized chromosomes are mated to 2–4 flies of the opposite sex carrying a tester chromosome, in this case a deficiency, Df(3R)red^{P52}, which removes RpII140, RpII15, and several other genes. This deficiency is balanced over TM3. In this way, individual mutagenized chromosomes are tested.

6. Four classes of progeny are produced, each with a different genotype and each of which can be distinguished by their phenotypes. One is the mutagenized red e chromosome over the tester chromosome, Df(3R)red^{52}. If these flies survive to adulthood, they do not have a recessive-lethal mutation in a gene removed by the deficiency. This can be seen as the survival of mutant red-eyed flies. These flies can also be examined for unexpected visible-mutant phenotypes. The entire cross can be discarded if these flies survive and are otherwise wild-type. Another class of progeny carries two copies of the balancer chromosomes; these do not survive to adulthood. A third class carries the tester and balancer chromosomes and produce Sb flies, which are discarded.

7. The fourth class has both the mutagenized red e and balancer chromosomes. If the first class is absent in the aforementioned cross, males and females of this genotype are saved and mated to each other to establish a balanced stock of the putative mutation.

Targeted

Techniques have been developed to target particular *Drosophila* genes, either for disruption or introduction of specific mutations, using homologous recombination.[2] The technique is comparable to a random mutagenesis in that it is neither quick nor easy; however, it could be used to mutate all RNA polymerase II subunits. Its major advantage over random mutagenesis is that it does not require prescient knowledge of the mutant phenotype. In addition, it does not depend on the existence of a mutation or chromosomal deficiency that could be used as a tester chromosome. The only requirement to begin a mutagenesis is knowledge of the sequence, which is now available for all 12 subunits.

Targeted gene mutagenesis is a tripartite system consisting of a genomic clone of the gene of interest, and a site-specific recombinase and endonuclease. The object is to generate a linear DNA molecule, which is highly recombinagenic *in vivo*, with a fly selectable marker and ends that reside within the target gene. This stimulates recombination between the linear extrachromosomal DNA and its homologous chromosomal sequence. Transgenic flies with the inducible genes encoding the enzymes that perform the latter two steps already exist.

The first step in using this system is to clone genomic DNA including the gene of interest. The efficiency of homologous recombination is likely to be proportional to the amount of homology between donor and target, as it was reported that an 8.9-kb donor was fivefold more efficient at targeting than a 2.5-kb donor.[13] The genomic DNA is cloned into a specialized P-element transposable element designed for germline transformation, with P-element ends, a fly selectable marker such as eye color, and containing FLP recombinase target FRT sites on both sides of the inserted genomic DNA. An I-SceI restriction site is inserted into the middle of the gene and mutations are introduced on both sides of this site. Because homologous recombination and therefore insertion occurs between the DNA ends and within the homologous sequence, the further the point mutations are introduced from the I-SceI site, the more likely they are to be recombined into the genomic DNA.

The modified genomic DNA construct is transformed into the *Drosophila* genome and then crossed to flies carrying the inducible enzymes. The recombinase and restriction endonuclease are induced, with the former catalyzing a recombination between the FRT sites, resulting in the looping out of a circular DNA molecule containing the gene of interest and the latter linearizing this molecule. Endogenous DNA repair enzymes are hypothesized to use the linear DNA molecule to recombine with the homologous chromosomal site, creating a duplicated copy of the target gene, each containing a different mutation, with the fly selectable marker in between. These duplicate genes are unstable and often resolved by looping out, resulting in the loss of the fly selectable marker and possibly retaining one of the two intended mutations.

Six targeted genes were all successfully mutated.[2,13] However, the type of mutation recovered is variable. It was possible to create specific allele substitutions with this method, but many other types of events were also recovered. These include the creation of duplicated genes and genes with internal deletions. It is too soon to predict the rates at which these events

[13] Y. S. Rong, S. W. Titen, H. B. Xie, M. M. Golic, M. Bastiani, P. Bandyopadhyay, B. M. Olivera, M. Brodsky, G. M. Rubin, and K. G. Golic, *Genes Dev.* **16,** 1568 (2002).

might be recovered for any given gene. Nevertheless, it appears that targeted gene disruption could be a powerful technique to disrupt specific genes. With regard to genes encoding the 12 subunits of RNA polymerase II, it should be straightforward to target disruption of all genes; however, mutation of the smallest subunits may be problematic. Furthermore, the ability to introduce specific mutations is still in question.

Selections

RNA polymerase II is more sensitive to the mushroom toxin α-amanitin than other polymerases. It had been demonstrated that tissue culture lines could be selected for resistance to α-amanitin.[14] Greenleaf et al.[15] used the death of flies reared on α-amanitin as a selection for mutagenized flies that were resistant to α-amanitin.[15] Males were mutagenized as previously described for screens for recessive-lethal mutations[12] and mass mated to virgin females. Embryos were collected and reared on media with α-amanitin. One mutation, which confers resistance to high concentrations of α-amanitin, was recovered following selection of 3×10^6 progeny from mutagenized fathers. This mutation has the properties that (1) it confers dominant resistance to α-amanitin, that is flies, which have one wild-type and one resistant allele survive on the drug,[16] (2) the resistant mutation still performs sufficient RNA polymerase II functions to support viability, both in the presence and absence of α-amanitin; and (3) flies carrying it are 250-fold less sensitive to α-amanitin than are wild-type flies.[15] This is because the mutant RNA polymerase II has a reduced binding for α-amanitin, as the dissociation constants are 3.8×10^{-7} M vs. 10^{-4}–10^{-5} M for wild-type and mutant polymerase, respectively.[17] This mutagenesis/selection experiment demonstrates the feasibility of selecting mutations that alter protein structure but maintain enzymatic activity.

The theory behind this experiment is diagrammed in Fig. 2 and is modeled after the genetic dissection of bacteriophage assembly.[18] A simple multimeric protein, consisting of three wild-type subunits, A^+, B^+, C^+, is shown. A single amino acid substitution in one subunit, B^m, will disrupt interactions with subunit C, but permit interaction with subunit A. If this is an essential enzyme and either C^+ is an essential subunit or the B^m

[14] V. L. Chan, G. F. Whitmore, and L. Siminovitch, *Proc. Natl. Acad. Sci. USA* **69**, 3119 (1972).

[15] A. L. Greenleaf, L. M. Borsett, P. F. Jiamachello, and D. E. Coulter, *Cell* **18**, 613 (1979).

[16] A. L. Greenleaf, J. R. Weeks, R. A. Voelker, S. Ohnishi, and B. Dickson, *Cell* **21**, 785 (1980).

[17] D. E. Coulter and A. L. Greenleaf, *J. Biol. Chem.* **257**, 1945 (1982).

[18] J. Jarvik and D. Botstein, *Proc. Natl. Acad. Sci. USA* **72**, 2738 (1975).

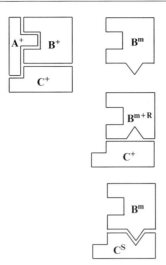

FIG. 2. Second-site suppression of a trimeric enzyme. Wild-type subunits A^+, B^+, and C^+ are shown. The mutant phenotype caused by the B^m mutation can be suppressed (restored to that of wild-type) by a precise intragenic revertant to B^+, an intragenic suppressor with a second amino acid change to B^{m+R}, or an extragenic suppressor mutation C^S. It is not possible to suppress B^m with an extragenic mutation in subunit A.

mutation disrupts a domain necessary for catalytic activity, B^m will cause recessive lethality. It is then possible to select for suppressor mutations that ameliorate the defect caused by B^m as they will restore viability. The positions of these mutations will reveal structural interactions among subunits and within structural domains of a multimeric enzyme. The methodology for such a selection is described in the following text.

Three types of B^m suppressor mutations can be envisioned. The first two are intragenic suppressors, in that they alter the B subunit, and the third is an extragenic suppressor, in that it alters a gene that encodes a protein other than subunit B. First there is a precise revertant of B^m to B^+ whereby the altered amino acid is restored to that of the wild-type protein. Second is an additional mutation in the B^m subunit resulting in B^{m+R}, such that the conformation of B permits interaction with C^+ and this combination has sufficient enzymatic activity for the organism to survive. Third is a compensatory mutation in the C subunit, which allows a mutant C^S subunit to bind to the mutant B^m subunit, as is diagrammed in Fig. 2. I have selected through more than 1.1×10^7 flies for suppressor mutations. Of the RNA polymerase II suppressor mutations that my research group has analyzed to date, 4 are B^m to B^+, 10 are B^m to B^{m+R}, and 8 are $B^m C^S$ (T. Jones and M. A. Mortin, in preparation).[11,19]

The choice of starting mutation (i.e., B^m) is critical in determining the success of the selection and the informational quality of the resulting suppressor mutation. For example, if the mutation used in the selection, B^m, results from a stop codon, most or all of the suppressor mutations will be precise revertants to B^+. Mutations that are loss of function, because they encode rapidly degraded proteins, are likely to be suppressed by intragenic suppressor mutations, B^m to either B^+ or B^{m+R}. Mutations with reduced or altered enzymatic activity are more likely to be suppressed by second-site mutations, which reveal important structural relationships among the mutant proteins. Nearly all mutations in Drosophila RNA polymerase II are of this class.[20] From a technical standpoint, the mutations that are the simplest to use in second-site selections are conditional recessive-lethal mutations, which are viable and fertile homozygotes at permissive condition but lethal at nonpermissive condition. It is trivial to select through millions of dead embryos, larvae, and pupae for the rare suppressor-mutation–carrying adult to emerge from a bottle of conditional lethal mutations. It is possible (although not easy) to visually screen millions of adults for the rare homozygous mutant flies carrying the suppressor mutations of interest.

Methods

Multiple individual lines of each starting mutation are maintained in half-pint bottles on standard *Drosophila* food medium. Because my standard metal rack holds 30 bottles, I maintained 30 individual lines of a given stock, numbered 1–30. Stocks were transferred to fresh food with bottle 1 transferred to a new bottle 1, and so on. The purpose of this was twofold. First, it permitted the rapid discovery that a putative suppressor mutation was instead a contamination event in a particular bottle, as it could be traced to the nonmutagenized parental bottle. Second, if putative suppressor mutations were recovered in more than one numbered bottle in a given experiment, they could be considered separate mutational events.

Where possible, the chromosome carrying each mutation used in a selection experiment was "marked" by also carrying a unique (for a particular laboratory) combination of recessive-visible mutations. Thus the X-linked RpII215^{K1} mutation was combined with the X-linked recessive-visible mutations, ras v, which together caused flies to have light orange eyes. A putative suppressor mutation from such a selection with a different

[19] W. J. Kim, L. P. Burke, and M. A. Mortin, *J. Mol. Biol.* **244,** 13 (1994).
[20] M. A. Mortin, *Proc. Natl. Acad. Sci. USA* **87,** 4864 (1990).

eye color would immediately be suspected of being a contaminant and could almost always be traced to the parental nonmutagenized bottle stock.

The procedure for mutagenizing flies was modified from Lewis and Bacher[12] to facilitate the recovery of larger numbers of progeny from mutagenized parental flies. Thirty numbered half-pint bottles containing standard *Drosophila* media were used for each experiment. Two Whatman No. 1 filters were cut to fit on top of the media. These were then impregnated with 1.5 ml of a mixture of 1% sucrose and 10 μl/ml EMS. This concentration was empirically determined to produce the same rate of recessive-lethal mutations as the protocol described for the generation of random recessive-lethal mutations, although the dose of EMS is approximately fourfold higher. The efficiency of mutagenesis might be improved even further by starving the adults for 12 hours prior to placing them on EMS (reviewed in reference 1). Entire bottles containing males and females (the P or parental generation) from 1–7 days old were transferred to the EMS containing bottles and left overnight. These were then transferred to fresh bottles and embryos were collected in new bottles for each of 6 successive days before being discarded. The progeny (the F_1 or filial$_1$ generation) were permitted to grow at a permissive condition. Initial attempts to shift these embryos to a restrictive condition produced very few suppressor mutations. I believe this occurred for two reasons. First, the selections were performed on single suppressor mutations. Some percentage of these undoubtably died as a result of the rigorous selection conditions. Second, EMS produces a large number of mosaic mutations (reviewed in reference 1). These flies would have some cells with a suppressor mutation and others lacking the suppressor mutation. Those lacking the suppressor mutation would die because the polymerase mutations are cell lethal.[21] Thus mosaic flies with patches of cells lacking a suppressor mutation are likely to die.

Adult F_1 progeny eclosing over a 6-day period were then transferred to fresh bottles, where they were allowed to lay eggs for 24 hours and then discarded. The resulting F_2 (second filial$_2$ generation) embryos were then shifted to restrictive conditions for development. Performing the selection in the F_2 generation provided two important advantages over the F_1 generation. First, a single suppressor mutation in the F_1 generation could produce many progeny in the F_2 generation. A rigorous selection that killed most of the suppressor-bearing flies might still be recovered. Second, the F_2 generation flies are not mosaic for the suppressor mutations. Any surviving progeny were saved as putative mutations. These were examined to determine whether they had the correct recessive-visible markers.

[21] M. A. Mortin, N. Perrimon, and J. J. Bonner, *Mol. Gen. Genet.* **199,** 421 (1985).

The parental stock was also examined for contaminating flies. Putative suppressor mutations were retested by collecting embryos at a permissive condition and shifting the progeny to a restrictive condition for selection; those surviving were saved as bona fide suppressor mutations. These were maintained by continuous selection at what had been the restrictive condition. The suppressor mutations were then mapped to a particular chromosomal location by standard methods.

Concluding Remarks and Recommendations

Fourteen million progeny have been tested in "selections" for resistance to α-amanitin[15] and suppression of recessive lethality,[11,20] resulting in the recovery of 42 mutations. This sample size is large enough to allow some generalizations about selections as opposed to the screens described previously. Although discussion is limited to RNA polymerase II, these general rules are applicable to the study of any multimeric enzyme.

Six selections have been attempted; five were successful, including selections for α-amanitin resistance and suppression of recessive lethality for four of five RNA polymerase II alleles tested (Table II). Most selections were successful, I believe in large part, because the beginning mutations were chosen because they alter the confirmation of the subunit, as evidenced by the mutant flies having residual activity (details presented previously).

Of the successful selections, mutations were recovered at rates ranging from 3.3×10^{-7} for α-amanitin resistance to 2×10^{-3} for RpII140^{z36} suppressors (Table II). The former is artificially low because this selection was carried out as an F_1 selection, whereas most of the other mutations resulted from F_2 selections. This difference can be directly compared in the selection using RpII215^{ts}, in which three suppressors were recovered in F_1 selections of 4×10^6 progeny, whereas two suppressors were recovered in F_2 selections of 1×10^5 progeny.[11] There is no difference in the types of mutations recovered in F_1 and F_2 selections. This can also be directly demonstrated with the suppressors selected using RpII215^{ts}, in which two different suppressors recovered in F_1 selections result from the same nucleotide changes as two suppressors recovered in F_2 selections.[11] I cannot explain why RpII140^{z36} suppressors were selected at such a high rate compared to other mutant alleles. The suppressors may be less specific in their action and could affect numerous target proteins and/or domains within a protein. This does not appear to be the case as the three intragenic suppressor mutations, which have been sequenced, all identify one domain in the second-largest subunit of RNA polymerase II and are therefore of the B^m to B^{m+R} class (Fig. 2).

TABLE II

SELECTIONS TO RECOVER SECOND-SITE SUPPRESSOR MUTATIONS IN GENES ENCODING RNA POLYMERASE II SUBUNITS AND ACCESSORY FACTORS

Starting mutation	[a] RpII215[+]	[b,c] RpII215[12]	[b,c] RpII215[ts]	[b,d] RpII215[K1]	[b,c] RpII215[Ubl]	[b,c,e] RpII140[Z36]
No. of progeny selectec	3×10^6	3.1×10^6	4.1×10^6	3×10^6	1×10^6	0.1×10^6
No. of suppressor mutations	1	0	5	13	3	20
Rate	3.3×10^{-7}	0	1.2×10^{-6}	4.3×10^{-5}	3.0×10^{-5}	2.0×10^{-3}
Intragenic/Extragenic	NA	0/0	5/0	3/10	3/0	$\leq14/\geq6$
G:C → A:T/Other	1/0	NA	3/2	$\geq4/\geq6$	0/3	$\geq0/\geq3$

[a] A. L. Greenleaf, H. M. Borsett, P. F. Jiamachello, and D. E. Coulter, Cell 18, 613 (1979). Selection was based on α-amanitin–caused lethality.

[b] M. A. Mortin, Proc. Natl. Acad. Sci. USA 87, 4864 (1990).

[c] I. Krasnoselskaya. I. Huang, T. Jones, C. Dezan, and M. A. Mortin, Mol. Gen. Genet. 258, 457 (1998).

[d] W. J. Kim, L. P. Burke, and M. A. Mortin, J. Mol. Biol. 244, 13 (1984).

[e] Jones and Mortin, In preparation.

NA, Not applicable.

Extragenic suppressors are allele specific; that is, suppressors select-ed to rescue the recessive lethality caused by the mutation, RpII215^{k1}, preferentially suppress that mutation. When a second mutation is identified that can be rescued by a suppressor, that mutation maps to the same domain in the protein as the original mutation.[11] These data strongly infer that the model presented in Fig. 2, showing a mutation B^m and a compen-satory mutation, C^s, is correct. This model predicts that the mutations B^m and C^s have protein alterations in domains that directly interact with each other.

On a theoretical basis, the suppressor selection described randomly tests every possible amino acid change in every fly protein for its ability to correct the original defect in RNA polymerase II. How large must selections be to achieve this level of saturation, and is this possible in flies?

On a practical basis, this level of saturation is not possible because the selections are not truly random, as outlined in the following text.

1. The most common mutagenic result of EMS is to produce G:C to A:T mutations. As stated, 12 of 13 EMS-induced recessive-lethal RNA polymerase II mutations, which have been sequenced, have this change. If this type of mutation were the only possible event, precise revertants of B^m to B$^+$ would never be recovered, yet four such events were identified.[11] The second most common EMS-induced event results in A:T to G:C mutations, which allows precise revertants. In fact, 8 of 22 selected suppressor mutations are the preferred G:C to A:T change (Table II), and 5 of 22 are the second most common A:T to G:C; four of these are precise revertants.[11] Nevertheless, the number of recoverable suppressor muta-tions must be greatly reduced because of this bias. The identification of a mutagen with less specificity would greatly increase the efficacy of selections.

2. Another major reason for non-randomness is that selections are performed at the level of protein; however, nucleotides are being mutated. It is not possible to mutate a single nucleotide in every codon so that it might encode all the other 19 amino acids. The rate of mutation is too low to expect the simultaneous recovery of any two specific nucleotide alterations that might be required to suppress a recessive-lethal mutation.

3. The existence of codon bias imposes additional non-randomness on the selections.

4. It is possible that differences in chromatin condensation make some genomic regions more or less accessible to mutagens.

Although the methodology presented describes a selection for viability, it requires the restoration of sufficient RNA polymerase II enzymatic activity to support survival of an intact organism. The end result is the identification of single-protein alterations that improve or correct RNA polymerase II function compared with that directed by the beginning polymerase mutation. The selection procedures are simple. A single person working an hour a day and using one incubator could easily select through three million progeny in less than 1 month. This selection size is sufficient to approach saturation for the types of recoverable suppressor mutations, starting from any given suppressible mutation. Evidence for this comes from the observation that of the 22 suppressor mutations recovered in selections and characterized as to the responsible nucleotide change, 9 are unique events, 4 different nucleotide changes were recovered two independent times and 1 change was recovered five times (Jones and Mortin, in preparation).[11,19] Large-scale suppressor selections should produce a wealth of structural and functional data on RNA polymerase II and other multimeric proteins.

Acknowledgment

I thank Jim Kennison for comments on this article.

Author Index

Numbers in parentheses are footnote reference numbers and indicate that an author's work is referred to although the name is not cited in the text.

A

Subject Index

A

ACF, *see* ATP-utilizing chromatin assembly and remodeling factor
Active center, RNA polymerase
cross-linking studies
autocatalysis
principles, 192–193
reaction conditions, 193
mapping of cross-linked sites in protein
advantages, 206
N-bromosuccinimide cleavage, 200
N-chlorosuccinimide cleavage, 200
cyanogen bromide cleavage, 199–200
hydroxylamine cleavage, 200
interpretation, 200–203
limited proteolysis, 202
2-nitro-5-thiocyanobenzoic acid cleavage, 200
overview, 196–197, 199
protein excision from gels, 199
refinement, 203–206
RNA–protein cross-linking in elongation complex active center, 195
RNA–protein cross-linking in initiating complex active center with rifampicin–GTP, 193, 195
structural information elucidation, 191
DNA conformation studies with iodine-125 radioprobing, 118–119
AFM, *see* Atomic force microscopy
AP-endonuclease 1
acetylated protein studies
acetylated lysine residue identification, 296–297
electrophoretic mobility shift assay with nCaRe-B oligodeoxynucleotide, 297–298
prospects, 300
repression of reporter gene expression, 298–299

acetylation conditions and purification
cell line transfectants, 293–294
p300 acetylation *in vitro*, 294–296
base excision repair, 292
p300 purification for acetylation studies, 296
transcriptional regulation, 292–293
APE1, *see* AP-endonuclease 1
Atomic force microscopy, single DNA molecule analysis of transcription complexes
DNA bend angle determination, 41–44
DNA contour length measurements
computation of contour length, 40–41
identification of DNA molecule, 39–40
overview, 38–39
DNA-streptavidin tagging of biotin-labeled RNA polymerase subunits
biotinylated DNA tag preparation and streptavidin fusion, 49
open promoter complex labeling with DNA tags, 49
principles, 47
recombinant subunit preparation with biotinylation signal, 47–48
DNA wrapping in transcription complexes, 44–46
overview, 34–35, 49–50
sample preparation, 37–38
surface equilibrium of DNA molecules, 35, 37
ATP-utilizing chromatin assembly and remodeling factor
chromatin assembly *in vitro*, 509–510
functions, 499–500
purification of *Drosophilia* recombinant protein, 503–504
subunits, 500
6-Azauracil, yeast drug-sensitive growth assay for mutant RNA polymerase II analysis, 286

I

L

Q

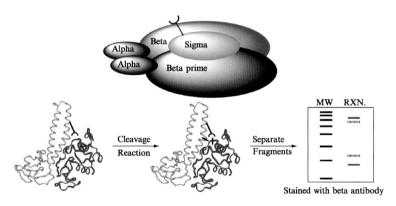

HOLMES *ET AL.*, CHAPTER 5, FIG. 4. Plots of the percentage of complexes that incorporated CMP at position +25. Data for reactions at six CTP concentrations (1, 5, 10, 20, 100, and 500 μM) are shown. The curves were generated using the program KinSim and plotted on the same graph as the data points to evaluate the quality of the fits.

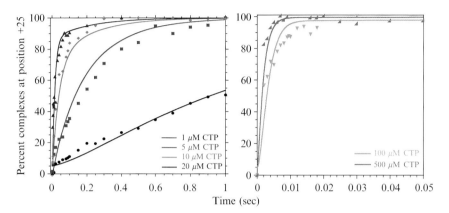

MEARES *ET AL.*, CHAPTER 6, FIG. 2. Determining protein-protein interactions via protein cleavage reactions. The ⊃— symbol represents a protein cleavage reagent that is conjugated to the sigma subunit. For simplicity, the beta subunit is shown being cleaved; beta-prime is also cleaved prominently by conjugated sigma. The cleavage reaction products are subsequently separated via SDS-PAGE and transferred to a membrane, where they are stained with a specific antibody (directed against the *N* terminus of the beta chain in this example).

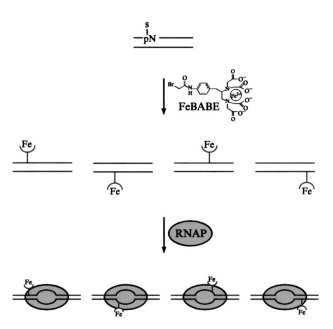

MEARES *ET AL.*, CHAPTER 6, FIG. 7. DNA residue N (representing either A, G, C, or T) is lightly conjugated to produce a random labeling of FeBABE on the DNA backbone. A small percentage of the total number of residues are labeled. After formation of the RNAP-DNA complex, cleavage of the protein occurs where it is proximal to a FeBABE group. [Reprinted in part with permission from B. D. Schmidt and C. F. Meares, *Biochemistry 41*, 4186 (2002). Copyright 2002 American Chemical Society.]

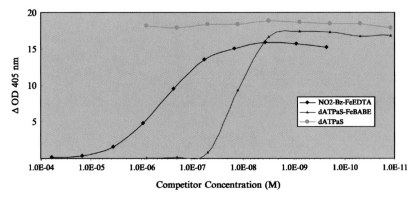

MEARES *ET AL.*, CHAPTER 6, FIG. 8. Competitive ELISA comparing conjugated phosphorothioate DNA (dATPαS-FeBABE) to non-conjugated phosphorothioate DNA (dATPαS). The non-conjugated phosphorothioate DNA does not compete for the anti-chelate antibody. The number of chelates per DNA was derived by dividing the concentration of the Fe-(S)-*p*-nitrobenzyl-EDTA standard at 50% saturation by the concentration of DNA in the DNA-FeBABE conjugate at 50% saturation. Adapted in part with permission from B. D. Schmidt and C. F. Meares, *Biochem.* **41**, 4186 (2002). Copyright 2002 American Chemical Society.

MEARES *ET AL.*, CHAPTER 6, FIG. 11. *Taq* RNAP structure showing cuts inferred using lysine-conjugated *E. coli* 2IT-FeBABE. The panels show the α-carbon trace with cuts rendered as 20aa runs of α-carbon spheres on β (green) and β' (red), first without σ (*left*) and then with σ (lavender) (*right*). The two α subunits at the top of the structure are colored yellow and blue, and the omega subunit is gray.

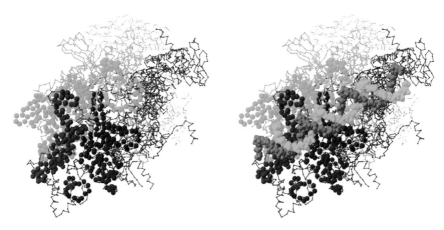

MEARES *ET AL.*, CHAPTER 6, FIG. 12. The cut sites observed by proteolytic *lac* UV5 DNA on *E. coli* RNAP, mapped onto the homologous regions of the structure of the *Taq* RNAP-DNA fork-junction complex, which contains the upstream promoter region but not the downstream template. The panels show the α-carbon trace with cuts rendered as 20aa runs of α-carbon spheres on β (green), β' (red), and σ (lavender), first without DNA (*left*) and then with DNA (*right*). DNA strands brown (*upper*) and light blue (*lower*) run from upper right (*upstream*) to lower left (*fork*). The σ_4 region lies under the upstream end of the DNA (see text). Other colors match those in Figure 11.

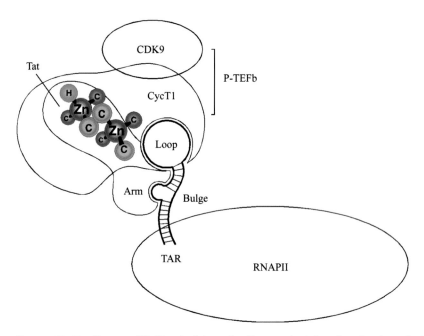

Gomes *et al.*, Chapter 24, Fig. 1. Schematic diagram depicting the zinc-dependent interaction between Tat and Cyclin T1 (*CycT1*) and the loop– and bulge sequence–specific interactions that recruit the Tat:P-TEFb complex to nascent TAR RNA. In addition to recruiting CDK9 to the transcript early in transcription elongation, Tat strongly activates the CTD kinase activity of P-TEFb in cells. TAR RNA-binding studies with recombinant P-TEFb subunits suggest that the complex must undergo autophosphorylation in the presence of Tat in order to bind avidly to TAR RNA (see Fig. 4), indicating that the proteins bound to the RNA may be highly phosphorylated. It has been proposed that additional post-translational modifications, including acetylation or ubiquitination, may play a role in the later steps of elongation or in the release of the P-TEFb complex from TAR RNA, and the biochemical techniques described here can be readily adapted to address these mechanistic questions.

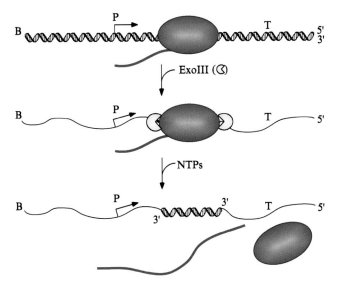

Uptain, Chapter 25, Fig. 1. Diagram of experimental design for dsDNA and ssDNA templates. Transcription is initiated at the promoter (P, bent rightward arrow), and the elongation complexes are halted proximal to the termination site (T) on dsDNA, PCR-generated templates that are biotinylated (B) at one of the 5′ ends. Next, the halted elongation complexes (filled ellipse) are digested with ExoIII (filled, cut-out circles), and then ATP, GTP, CTP, and UTP are added, allowing the elongation complex to resume transcript elongation. The 5′ and 3′ ends of the DNA (black lines) are as indicated. Upon recognizing the intrinsic termination site, the RNA polymerase releases the completed transcript (red line) and dissociates from the DNA template.

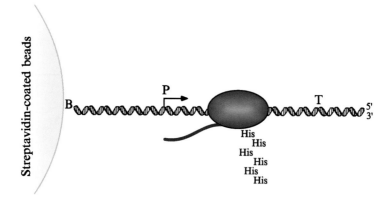

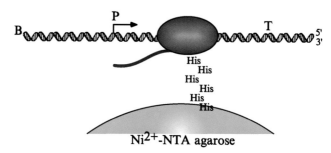

Uptain, Chapter 25, Fig. 2. Diagram of two solid-state transcription methods. There are two methods to perform transcription in the solid phase. RNA polymerase elongation complexes (filled ellipse) can be bound to Ni^{2+}-NTA agarose (blue half-circle) via its hexahistidine tag (His), or the 5′ biotinylated (B) DNA template can be bound to streptavidin-coated paramagnetic beads (yellow half-circle).

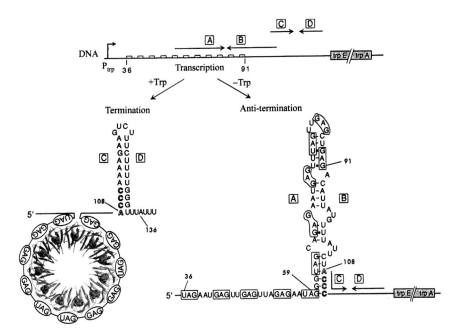

GOLLNICK, CHAPTER 31, FIG. 1. Model of transcription attenuation of the *B. subtilis trp* operon. Numbers indicate the residue positions relative to the start of transcription. The large boxed letters designate the complementary strands of the terminator (C and D) and antiterminator (A and B) RNA structures. Nucleotides 108–111 overlap between the antiterminator and terminator structures and are shown as outlined letters. The TRAP protein is shown as a ribbon diagram with the 11 subunits in different colors, and the bound RNA is shown encircling TRAP, as seen in the crystal structure. The GAG and UAG repeats involved in TRAP binding are shown in ovals surrounding the protein and are also outlined in the sequence of the antiterminator structure.

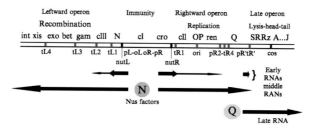

DAS *ET AL.*, CHAPTER 33, FIG. 1. Transcription antitermination in bacteriophage lambda. The diagram shows the genetic map of lambda (not drawn in scale), outlining the three operons that are regulated by N and Q. Relevant genes are shown on top, and the location of the *nut* sites and terminators are indicated at the bottom of the map. The origin of replication (ori) and the cleavage site for DNA packaging (cos) are shown. Transcripts relevant to the lytic program of lambda development are drawn below as arrows, reflecting direction of transcription, the start and stop sites, as well as the temporal order with which they are made. Some basal transcription of cIII and cII-OP-ren occurs without the aid of N, as indicated, owing to inefficient termination at tL1 and tR1, respectively, both of which are Rho-dependent terminators. The terminators tL2 and tR2 are intrinsic terminators that are stimulated by NusA, while termination at tL4 is strongly enhanced by NusA.

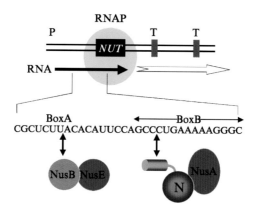

DAS *ET AL.*, CHAPTER 33, FIG. 2. Function of the *nut* site. Sequence of the lambda *nutR* RNA is shown along with the RNA-protein interactions involving BoxA and BoxB. The arrows above BoxB identify the hairpin stem, whose 5′ arm binds the ARM helix of N.

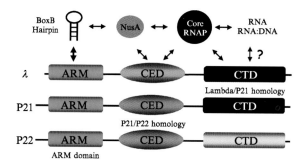

Das *et al.*, Chapter 33, Fig. 3. Architecture and targets of N. A schematic representation illustrates the demonstrated function of the three domains of lambda N protein (top) and the similarities and differences among N homologues from P21 and P22. Note that P21 and P22 ARM domain is extended at the amino terminus. The central domain (CED) of P21 and P22 N proteins are very similar; however, lambda N CED is quite different. Lambda and P21 N proteins share a structurally similar and functionally equivalent carboxy terminal domain (CTD). Although P22 N differs from lambda and P21 N proteins in the CTD, it does show strong similarity in this domain with several other N homologues, including that from the Shiga-toxin phage H19B. The possibility that the CTD of lambda N may bind a nucleic acid to influence antitermination is indicated with a question mark.

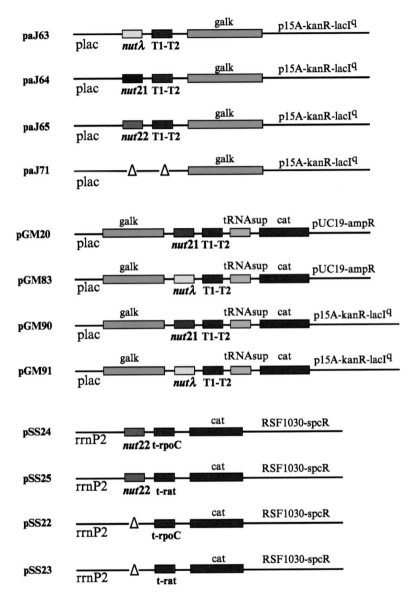

DAS *ET AL.*, CHAPTER 33, FIG. 7. A collection of antitermination reporter plasmids. The pAJ plasmids were derived from constructs made previously. The rest of the plasmids shown represent a new generation of the reporters constructed in this laboratory. The trpoC and t-rat (rho gene attenuator) terminators have been cloned from plasmids provided by C. Squires.

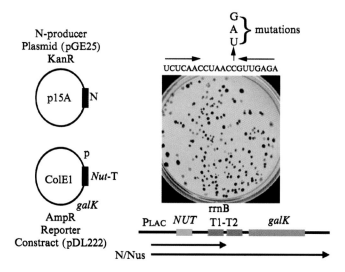

DAS *ET AL.*, CHAPTER 33, FIG. 8. A two-plasmid system for monitoring processive antitermination *in vivo*: screening of *nut* site mutants. One of several available two-plasmid systems is shown here, depicting the basic features of the plasmids, the *galK* reporter construct and the phenotype displayed on MacConkey-galactose agar plates containing 0.1 mM IPTG. The plate displays the colonies of a *galK* mutant strain (N99 *recA::Tn10*) with the P21 N plasmid, pGE25, that was transformed with the mutagenized clones of the reporter plasmid pDL222. Indicated at the top of the P21 BoxB RNA sequence are three substitutions that were introduced by oligonucleotide directed cassette mutagenesis. That certain substitution(s) of a specific loop nucleotide has abolished the *nut* site function is revealed by the colorless nature of some of the colonies. The red colonies in the background, on the other hand, indicate that one or more substitutions of the same nucleotide have little effect on *nut* site function.

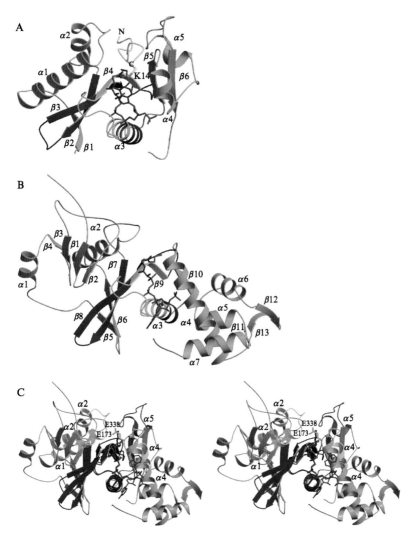

CLEMENTS AND MARMORSTEIN, CHAPTER 41, FIG. 1. Schematic representation of the HAT domains from the GCN5/PCAF family and from the Esa1 member of the MYST family. (A) Crystal structure of the tGCN5/CoA/H3 peptide complex. CoA is represented as a red stick figure. The H3 peptide is shown in yellow. The central core domain is shown in blue. Within this domain, the region with primary sequence homology with Esa1 is in aqua. The N- and C-terminal domains are in grey. (B) Crystal structure of the yEsa1/CoA complex. Color coding is similar to A. (C) tGCN5/Esa1 superposition. The central core domain is shown in blue. The aqua region represents the Esa1 N- and C-terminal regions that are structurally similar to GCN5/PCAF. The yellow region represents the N- and C-terminal regions of GCN5/PCAF that are structurally similar to Esa1. The α5-helix loop region of tGCN5 is shown in magenta. Residues in tGCN5 are numbered according to the yGcn5 sequence.

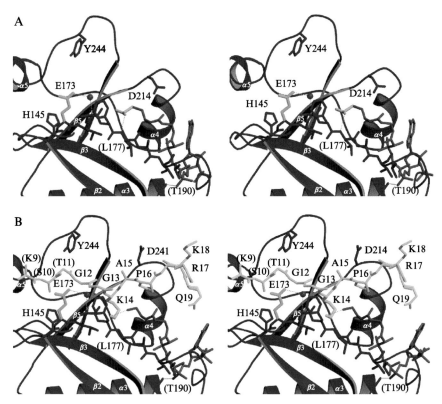

Clements and Marmorstein, Chapter 41, Fig. 2. Detailed view of the GCN5/PCAF active site. (A) A detailed view of the tGCN5 active site from the binary tGCN5/Acetyl-CoA complex. tGCN5 is blue. Acetyl-CoA is shown in red. A highly ordered water that may participate in the catalytic mechanism is in magenta. The tGCN5 residues are numbered according to yGCN5. HAT residues in parenthesis are tGCN5 residues that are not conserved in tGCN5: (L177) corresponds to a Cys in yGen5 and all other GCN5/PCAF family members. (T190) is an Ala in yGen5, but is a Thr in other GCN5/PCAF family members. Dotted lines represent hydrogen bonds. (B) A detailed view of the tGCN5 active site from the ternary tGCN5/CoA/II3 peptide complex. Color coding is similar to A. The histone H3 substrate peptide is in yellow. Histone residues that are labeled with parenthesis were modeled without their side chains.

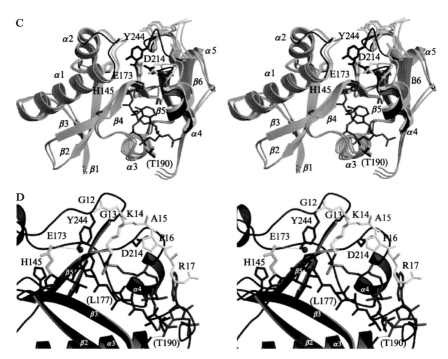

CLEMENTS AND MARMORSTEIN, CHAPTER 41, FIG. 2. Detailed view of the GCN5/PCAF active site. (C) Residues that play a role in substrate binding and catalysis. Dynamic regions of the protein are color coded as follows: the binary tGCN5/Acetyl-CoA complex is yellow, the tGCN5/bi-substrate inhibitor complex is in blue, and the ternary tGCN5/CoA/H3 peptide complex is in aqua. (D) A detailed view of the tGCN5 active site from the tGCN5/bi-substrate analog inhibitor complex. The peptide portion of the inhibitor is in yellow. The CoA portion of the inhibitor is in red. The linker region is in brown. The acetyl moiety is in orange.

CLEMENTS AND MARMORSTEIN, CHAPTER 41, FIG. 2. Detailed view of the GCN5/PCAF active site. (E) Proposed catalytic mechanism for GCN5/PCAF.

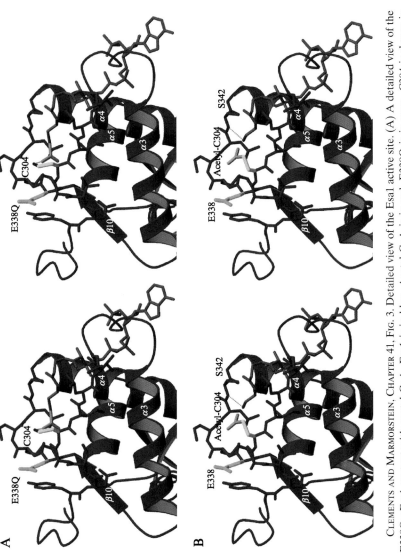

CLEMENTS AND MARMORSTEIN, CHAPTER 41, FIG. 3. Detailed view of the Esa1 active site. (A) A detailed view of the E338Q yEsa1 mutant with acetyl-CoA. Esa1 is in blue. Acetyl-CoA is in red. E338Q is in aqua. C304 is shown in yellow. This residue is oxidized with the S-O bond in brown. Dotted lines represent hydrogen bonds. (B) A detailed view of the wt-yEsa1-acetylated intermediate. Color coding is the same as A. E338 is in aqua.

CLEMENTS AND MARMORSTEIN, CHAPTER 41, FIG. 3. Detailed view of the Esa1 active site. (C) Proposed catalytic mechanism for yEsa1.

3 5282 00549 9689

ISBN: 0-12-039861-3

90000

9 780120 398614